Numerical Analysis 2000, Volume 6

Ordinary Differential Equations and Integral Equations

Numerical Analysis 2000, Volume 6

Ordinary Differential Equations and Integral Equations

Edited by

C. Baker
University of Manchester
Dept. of Mathematics
Oxford Road
Manchester M13 9PL
UK

J. Pryce
Cranfield University
Computer Info. Systems Eng. Dept.
RMCS Shrivenham
Swindon SN6 8LA
UK

G. Monegato
Politecnico di Torino
Dept. of Mathematics
Corso Duca degli Abruzzi 24
Torino 10129
Italy

G. Vanden Berghe
University of Ghent
Dept. of Applied Maths & Comp. Science
Krijgslaan 281-S9
Ghent B-9000
Belgium

2001
ELSEVIER
Amsterdam - London - New York - Oxford - Paris - Shannon - Tokyo

ELSEVIER SCIENCE B.V.
Sara Burgerhartstraat 25
P.O. Box 211, 1000 AE Amsterdam, The Netherlands

First edition 2001

Library of Congress Cataloging in Publication Data
A catalog record from the Library of Congress has been applied for.

ISBN: 0 444 50600 4

⊗ The paper used in this publication meets the requirements of ANSI/NISO Z39.48-1992 (Permanence of Paper).
Transferred to digital printing 2005

JOURNAL OF COMPUTATIONAL AND APPLIED MATHEMATICS

Special Issue

Numerical Analysis 2000
Vol. VI: Ordinary Differential Equations and Integral Equations

Guest editors:

Christopher Baker
Department of Mathematics, The University of Manchester
Oxford Road, Manchester M13 9PL, UK

Giovanni Monegato
Dipartimento di Matematica, Politecnico di Torino
Corso Duca degli Abruzzi 24, 10129 Torino, Italy

John Pryce
Computer Information Systems Engineering Department
Cranfield University, RMCS Shrivenham, Swindon SN6 8LA, UK

Guido Vanden Berghe
Vakgroep Toegepaste Wiskunde en Informatica
Universiteit Gent, Krijgslaan 281 (S9), Gent 9000, Belgium

ELSEVIER Amsterdam – London – New York – Oxford – Paris – Shannon – Tokyo

JOURNAL OF COMPUTATIONAL AND APPLIED MATHEMATICS

Volume 125, Numbers 1–2, 15 December 2000

Contents

ELSEVIER Journal of Computational and Applied Mathematics 125 (2000) xi–xviii

JOURNAL OF COMPUTATIONAL AND APPLIED MATHEMATICS

www.elsevier.nl/locate/cam

Preface

Numerical Analysis 2000
Vol. VI: Ordinary Differential Equations and Integral Equations

This volume contains contributions in the area of differential equations and integral equations. The editors wish to thank the numerous authors, referees, and fellow editors Claude Brezinski and Luc Wuytack, who have made this volume a possibility; it has been a major but personally rewarding effort to compile it. Due to the limited number of pages we were obliged to make a selection when composing this volume. At an early stage it was agreed that, despite the connections between the subject areas, it would be beneficial to allocate the area of partial differential equations to a volume for that area alone.

Many numerical methods have arisen in response to the need to solve "real-life" problems in applied mathematics, in particular problems that do not have a closed-form solution. It is particularly relevant to this comment that our Journal title involves the expression "Computational and Applied Mathematics". Applied mathematicians from differing academic cultures display differing approaches to computational techniques, but one might hazard the prophecy, based on current observations, that future generations of applied mathematicians will (without necessarily sacrificing mathematical rigour) *almost universally* regard the use of, and possibly the analysis and design of, robust numerical algorithms as an essential part of their research activity.

The differences seen in applied mathematics are reflected in differing approaches to the promotion of numerical analysis by those who today think of themselves as "numerical analysts": some, feeling that it is the mathematical modelling that supports the development of the subject, work closely in the area of mathematical modelling. Others, aware of the effect of advances in computers, are more concerned with accurate and efficient computational algorithms. Yet others prefer to construct more abstract mathematical theories that offer insight into, and justification (at least under well-formulated conditions) of, the numerical mathematics. Major contributions to topics represented here have been published in the past; many of the diverse approaches are represented in this volume. At the same time, there is a mixture of original and survey material represented here, often in the same paper.

Contributions on both initial-value problems and boundary-value problems in *ordinary differential equations* appear in this volume. Numerical methods for *initial-value problems* in ordinary differential equations fall naturally into two classes: those which use *one* starting value at each step (one-step methods) and those which are based on *several* values of the solution (multistep methods). Both methods were developed at the end of the 19th century.

- John Butcher has supplied an expert's perspective of the development of numerical methods for ordinary differential equations in the 20th century.

PII: S0377-0427(00)00454-4

For the one-step methods we can refer to the original work of Runge of 1895 (an extract is reproduced in John Butcher's paper), while for the multistep methods a first paper was published by Adams in 1883 (a copy of the title page appears in John's contribution). Advances in analysis and the advent of electronic computers has of course altered the scene drastically.

- Rob Corless and Lawrence Shampine talk about established technology, namely software for initial-value problems using Runge–Kutta and Rosenbrock methods, with interpolants to fill in the solution between mesh-points, but the 'slant' is new — based on the question, "How should such software integrate into the current generation of *Problem Solving Environments*?" They discuss specifically a new code they have written for the Maple environment and the similarities and differences with corresponding solvers for MATLAB and for traditional numerical libraries. The interplay between interface, environment and algorithm raises many subtle issues and provides an excellent example of mathematical software engineering.
- Natalia Borovykh and Marc Spijker study the problem of establishing upper bounds for the norm of the nth power of square matrices. This problem is of central importance in the stability analysis of numerical methods for solving (linear) initial-value problems for ordinary, partial or delay differential equations. In particular, they apply the new bounds in a stability analysis of the trapezoidal rule for delay differential equations (*vide infra*).

The dynamical system viewpoint has been of great benefit to ODE theory and numerical methods. The characterization and computation of *bifurcations* of parametrized ODEs is an active subfield created some 20 years ago. Related is the study of *chaotic behaviour*. To reproduce long-term behaviour realistically, special numerical methods are needed which *preserve invariants* of the exact system: symplectic methods are an example.

- In the first of three articles in the general area of *dynamical systems*, Willy Govaerts discusses the numerical methods for the computation and continuation of equilibria and bifurcation points of equilibria of dynamical systems. The computation of cycles as a boundary value problem, their continuation and bifurcations are considered. The basic numerical methods and the connections between various computational objects are discussed. References to the literature and software implementations are provided.
- In the second of this trio, Arieh Iserles and Antonella Zanna survey the construction of Runge–Kutta methods which preserve algebraic invariant functions. There is a well-known identity on the RK coefficients that is necessary and sufficient for the retention of a quadratic invariant. They extend this result and present a brief introduction to the Lie-group approach to methods that preserve more general invariants, referring to survey articles for fuller discussion.
- Symplectic methods for Hamiltonian systems are an important example of invariant-preserving methods. Valeria Antohe and Ian Gladwell present numerical experiments on solving a Hamiltonian system of Hénon and Heiles with a symplectic and a nonsymplectic method with a variety of precisions and initial conditions. The long-term behaviour of the Hamiltonian error, and the features of Poincaré sections, show interesting and unexpected phenomena.

Stiff differential equations first became recognized as special during the 1950s. In 1963 two seminal publications laid to the foundations for later development: Dahlquist's paper on A-stable multistep methods and Butcher's first paper on implicit Runge–Kutta methods. Later, order stars became a fundamental tool for the theoretical understanding of order and stability properties of stiff differential equations. Variable-order variable-step methods were the next for study.

- Ernst Hairer and Gerhard Wanner deliver a survey which retraces the discovery of the order stars as well as the principal achievements obtained by that theory, which has become a fundamental role for the understanding of order and stability properties of numerical methods for stiff differential equations. Some later extensions as well as recent developments and open questions are discussed.
- Guido Vanden Berghe, Hans De Meyer, Marnix Van Daele and Tanja Van Hecke construct exponentially fitted Runge–Kutta methods with s stages, which exactly integrate differential initial-value problems whose solutions are linear combinations of functions of the form $\{x^j \exp(\omega x), x^j \exp(-\omega x)\}$, ($\omega \in \mathbb{R}$ or $i\mathbb{R}$, $j = 0, 1, \dots, j\max$), where $0 \leqslant j\max \leqslant \lfloor s/2 - 1 \rfloor$, the lower bound being related to explicit methods, the upper bound applicable for collocation methods. Explicit methods with $s \in \{2, 3, 4\}$ belonging to that class are constructed. For these methods a study of the local truncation error is made, out of which follows a simple heuristic to estimate the ω-value. Error and step length control is introduced based on Richardson extrapolation ideas.

Differential-algebraic equations arise in control, in modelling of mechanical systems and in many other fields. They differ fundamentally from ODEs, and an *index* can be defined that measures (in some sense) how far a given DAE is from being an ODE. DAEs can in some cases be regarded as infinitely stiff ODEs, and methods for stiff problems work well for problems of index 1 or 2 at least. Examples are the classical backward differentiation formulas (BDF). Recent modifications of BDF due to Cash, using the so-called super-future points, have also proved very effective on DAEs as well as on highly oscillatory problems. Other variants on the classical Runge–Kutta and multistep approaches have been studied in recent decades, such as special-purpose “exponentially fitted” methods when the solution of the problem exhibits a pronounced oscillatory or exponential character. A good theoretical foundation of this technique was given by Gautschi in 1961 and Lyche in 1972. Automatic error control when solving DAEs is harder than for ODEs because changing dominance of different solution components may make both index and order of accuracy seem to vary over the range, especially for boundary-value problems. New meshing algorithms have been developed to cope with this.

- Jeff Cash describes a fairly recent class of formulae for the numerical solution of initial-value problems for *stiff and differential-algebraic systems*. These are the modified extended backward differentiation formulae (MEBDF), which offer some important advantages over the classical BDF methods, for general stiff nonoscillatory problems, for damped highly oscillatory problems and for linearly implicit DAEs of index up to 3. Numerical results are given for a simple DAE example with references to other performance tests. Pointers to down-loadable software are given.
- In the same area, Shengtai Li and Linda Petzold describe methods and software for *sensitivity analysis* of solutions of DAE initial-value problems: that is, for computing derivatives of a solution of $F(t, y, y', p) = 0$ with respect to one or more parameters p. As with similar methods for ODEs, they integrate auxiliary systems, incorporating local Jacobians which may be computed by *automatic differentiation* using a package such as ADIFOR. Consistent initialization is shown to need special care and suitable algorithms described. Several numerical experiments illustrate the performance; pointers to down-loadable software are given.
- Again in the area of differential-algebraic systems, Neil Biehn, John Betts, Stephen Campbell and William Huffman present current work on mesh adaptation for DAE two-point boundary-value problems. The context is an optimal control problem, which discretizes to a nonlinear

programming problem, but the problem of "order variation" is more general. Not only are different components of a DAE computed to different order of accuracy as a function of step-size, but the "real order" can vary at different points of the interval and from iteration to iteration. They discuss the principles and details of a new meshing algorithm and show its effectiveness on a computational example.

Contrasting approaches to the question of how good an approximation is as a solution of a given equation involve (i) attempting to estimate the actual *error* (i.e., the difference between the true and the approximate solutions) and (ii) attempting to estimate the *defect* — the amount by which the approximation fails to satisfy the given equation and any side-conditions. (In collocation and Galerkin techniques, the defect is required to satisfy certain constraints. Generally speaking, the relationship between defect and error can be analyzed using results on the stability or conditioning of the solution of the original problem.)

- The paper by Wayne Enright on defect control relates to carefully analyzed techniques that have been proposed both for ordinary differential equations and for delay differential equations in which an attempt is made to control an estimate of the size of the defect.

Many phenomena incorporate noise, and the numerical solution of *stochastic differential equations* has developed as a relatively new item of study in the area.

- Kevin Burrage, Pamela Burrage and Taketomo Mitsui review the way numerical methods for solving stochastic differential equations (SDEs) are constructed. SDEs arise from physical systems where the parameters describing the system can only be estimated or are subject to noise. The authors focus on two important topics: the numerical stability and the implementation of the method. Different types of stability are discussed and illustrated by examples. The introduction of variable step-size implementation techniques is stressed under the proviso that different numerical simulations must follow the same Brownian path.

One of the more recent areas to attract scrutiny has been the area of *differential equations with after-effect* (retarded, delay, or neutral delay differential equations) and in this volume we include a number of papers on evolutionary problems in this area. The problems considered are in general initial-function problems rather than initial-value problems. The analytical study of this area was already well-advanced in the 1960s, and has continued to develop (some of the names that spring to mind are: in the fSU, Myskhis, Krasovskii, Kolmanovskii; in the USA, Bellman and Cooke, and later Driver, Hale, etc.; in Europe, Diekmann, Halanay, Verduyn Lunel and Stépán). There has been an increasing interest in the use of such equations in mathematical modelling. For numerical analysts, one significant issue is the problem of the possible lack of smoothness of the solution, for which various strategies have been advanced, whilst another is the richer dynamics encountered in problems with delay.

- The paper of Genna Bocharov and Fathalla Rihan conveys the importance in mathematical biology of models using retarded differential equations. Although mathematical analysis allows deductions about the qualitative behaviour of the solutions, the majority of models can only be solved approximately using robust numerical methods. There are now a number of papers on the use of delay and neutral delay equations in parameter fitting to biological data; the most recent work relates to a sensitivity analysis of this data-fitting process.

- Delay differential equations also arise in mechanical systems, and the paper by John Norbury and Eddie Wilson relates to a form of constrained problem that has application to the control of a motor.
- The contribution by Christopher Baker, whose group has for some years been working in this area, is intended to convey much of the background necessary for the application of numerical methods and includes some original results on stability and on the solution of approximating equations.
- Alfredo Bellen, Nicola Guglielmi and Marino Zennaro contribute to the analysis of stability of numerical solutions of nonlinear neutral differential equations; they look at problems that display a form of contractivity. This paper extends earlier work on nonlinear delay equations by Torelli, Bellen, and Zennaro. We note that Alfredo Bellen and Marino Zennaro are preparing a book on the numerical solution of delay equations.
- In the papers by Koen Engelborghs, Tatyana Luzyanina and Dirk Roose and by Neville Ford and Volker Wulf, the authors consider the numerics of bifurcation in delay differential equations. Oscillations in biological phenomena have been modelled using delay equations. For some time, it has been realized that the onset of periodicity in such equations can be associated with a Hopf bifurcation and that chaotic behaviour can arise in scalar delay differential equations.
- Christopher Paul, who is the author of a code for the numerical solution of retarded equations (named `Archi`) addresses various issues in the design of efficient software and proposes methods for determining automatically information about the delay equation.
- The preceding papers relate to deterministic problems. Evelyn Buckwar contributes a paper indicating the construction and analysis of a numerical strategy for stochastic delay differential equations (SDDEs). The theory of SDDEs has been developed in books written by Mohammed and by Mao, but the numerics have been neglected. Unlike the corresponding results for stochastic differential equations without time lag (represented herein by the paper of Burrage et al.) some of the basic elements required in the numerical analysis have previously been lacking.

One could perhaps argue that stochastic differential equations (since they are really Itô or Stratonovitch integral equations) should be classified under the heading of integral equations. In any event, this volume contains contributions on both *Volterra and Fredholm-type integral equations.*

- Christopher Baker responded to a late challenge to craft a review of the theory of the basic numerics of Volterra integral and integro-differential equations; it is intended to serve as an introduction to the research literature. The very comprehensive book by Hermann Brunner and Pieter van der Houwen on the numerical treatment of Volterra integral equations remains a standard reference work.
- Simon Shaw and John Whiteman discuss Galerkin methods for a type of Volterra integral equation that arises in modelling viscoelasticity, an area in which they have already made a number of useful contributions.

Volterra integral and integro-differential equations are examples of causal or nonanticipative problems, as are retarded differential equations. It seems likely that such causal (abstract Volterra) problems will, increasingly, be treated together, since mathematical models frequently involve both discretely distributed and continuously distributed delays. The basic discretization theory is concerned with replacing the continuous problem with another causal problem (if one considers a vectorized formulation, this statement is true of block-by-block methods as well as step-by-step methods), and the study of *discrete Volterra equations* is a feature of increasing importance in the analysis.

We turn now to a subclass of *boundary-value problems* for ordinary differential equation, that comprises *eigenvalue problems* such as Sturm–Liouville problems (SLP) and Schrödinger equations. They are important for their role in physics and engineering and in spurring the development of spectral theory. For the classical (second-order) SLP there is reliable software for eigenvalues (less reliable for eigenfunctions) for commonly occurring types of singular behaviour. The underlying self-adjointness is important for these methods. Among current developments are new methods for *higher-order problems*, both self-adjoint and nonself-adjoint. Also of interest are the Constant Potential (CP) methods, which have recently been embodied in good software for the classical regular SLP, and whose efficiency makes them likely to supplant existing methods.

- In the first of a number of articles on *ODE eigenvalue problems*, Liviu Ixaru describes the advances made over the last three decades in the field of piecewise perturbation methods for the numerical solution of Sturm–Liouville problems in general and systems of Schrödinger equations in particular. He shows that the most powerful feature of the introduced constant potential (CP) methods is the uniformity of accuracy with respect to the eigen-parameter. He presents basic formulae and characteristics of a class of CP methods, based on piecewise approximation by a constant and a polynomial perturbation term. He illustrates by means of the Coffey–Evans equation — a standard example with pathologically clustered eigenvalues — the superiority of his code over some other standard codes, such as SLEDGE, SLEIGN and SL02F.
- Alan Andrew surveys the asymptotic correction method for regular Sturm–Liouville problems. Simple but unexpectedly powerful, it applies to finite difference and finite element methods which reduce the problem to a matrix eigenproblem. It greatly improves the accuracy of higher eigenvalues and generally improves lower ones as well, which makes it especially useful where long runs of eigenvalues need to be computed economically. The Coffey–Evans equation is used to show the good performance of the method even on tightly clustered eigenvalues.
- Leon Greenberg and Marco Marletta survey methods for higher-order Sturm–Liouville problems. For the self-adjoint case, the elegant basic theory generalizes the Prüfer phase-angle, used in theory and numerical methods for the classical SLP, to a complex unitary matrix. This was introduced by Atkinson in the 1960s, but the obstacles to using it numerically are considerable. The authors' considerable achievement over the last few years has been to identify suitable subclasses of the problem, and a general strategy (coefficient approximation) that leads to a usably efficient numerical method with error control. They also discuss theory and methods for the very different nonself-adjoint case. Numerical examples of both types are given.
- R. Moore in the 1960s first showed the feasibility of *validated solution* of differential equations, that is, of computing guaranteed enclosures of solutions. Validated methods use outward-rounded interval floating point arithmetic as the computing tool, and fixed-point theorems as the mathematical foundation. An important development in this area was the appearance in 1988 of Lohner's code, AWA, for ODE initial-value problems, which he also applied to the validated solution of boundary-value problems. Recently, these techniques have been extended to eigenvalue problems, e.g. to prove the existence of, and enclose, certain kinds of eigenvalue of a singular Sturm–Liouville problem, and the paper of Malcolm (B.M.) Brown, Daniel McCormack and Anton Zettl describes validated SLP eigenvalue calculation using Lohner's code.

We turn to papers on *boundary integral equations*. In the last 20 years, the numerical solution of integral equations associated with boundary-value problems has experienced continuing interest. This is because of the many intriguing theoretical questions that still needed to be answered, and the many complications that arise in applications, particularly in the engineering fields. Coming from the reformulation of PDE boundary value problems in terms of boundary integral equations, such problems often have complex geometries, bad aspect ratios, corners and other difficulties, all of which challenge existing numerical techniques. In particular, much effort has been concentrated on the following research themes:

— equations with singular and hyper-singular kernels,
— equations of Mellin type and equations defined on nonsmooth regions,
— fast solution numerical methods, especially for three-dimensional problems,
— domain decomposition, and
— the coupling of finite element and boundary element methods.

Many numerical analysts have made major contributions to the above themes, and to other important topics. With a limited number of pages at our disposal, we have included the seven papers below:

- Peter Junghanns and Bernd Silbermann present a selection of modern results concerning the numerical analysis of one-dimensional Cauchy singular integral equations, in particular the stability of operator sequences associated with different projection methods. They describe the main ideas and approaches. Computational aspects, in particular the construction of fast algorithms, are also discussed.
- Johannes Elschner and Ivan Graham summarize the most important results achieved in the last years about the numerical solution of one-dimensional integral equations of Mellin type by means of projection methods and, in particular, by collocation methods. They also consider some examples arising in boundary integral methods for the two-dimensional Laplace equation on bounded polygonal domains.
- A survey of results on quadrature methods for solving boundary integral equations is presented by Andreas Rathsfeld. The author gives, in particular, an overview on well-known stability and convergence results for simple quadrature methods based on low-order composite quadrature rules and applied to the numerical solution of integral equations over smooth manifolds. Useful "negative results" are also presented.
- Qualocation was introduced in the late 1980s as a compromise between Galerkin and collocation methods. It aimed, in the context of spline approximation methods for boundary integral equations on smooth curves, to achieve the benefits of the Galerkin method at a cost comparable to the collocation method. Ian Sloan reviews this method with an emphasis on recent developments.
- Wolfgang Hackbusch and Boris Khoromski present a novel approach for a very efficient treatment of integral operators. They propose and analyze a quite general formulation of the well-known panel clustering method for boundary integral equations, introduced by Hackbusch and Z.P. Nowak in the late 1980s. Their approach may be applied for the fast solution of the linear integral equations which arise in the boundary element methods for elliptic problems.
- Ernst Stephan examines multilevel methods for the h-, p- and hp- versions of the boundary element method, including pre-conditioning techniques. In his paper he reviews the additive Schwarz methods for the above versions of the Galerkin boundary element method applied to first kind (weakly singular and hyper-singular) integral equations on surfaces.

- Domain decomposition methods are well suited for the coupling of different discretization schemes such as finite and boundary element methods. George Hsiao, Olaf Steinbach and Wolfgang Wendland analyze various boundary element methods employed in local discretization schemes. They also describe appropriate iterative strategies, using both local and global pre-conditioning techniques, for the solution of the resulting linear systems.

The latter papers not only present overviews of some of the authors' recent research activities, but, in some cases, also contain original results and new remarks. We think that they will constitute fundamental references for any further research work on the numerical resolution of boundary integral equations.

Christopher Baker[a], Manchester
Giovanni Monegato, Turin
John Pryce, Shrivenham
Guido Vanden Berghe, Gent

Guest Editors

[a]*Department of Mathematics*
The Victoria University of Manchester
Oxford Road, Manchester M13 9PL, UK
E-mail address: cthbaker@ma.man.ac.uk (C.T.H. Baker).

ELSEVIER Journal of Computational and Applied Mathematics 125 (2000) 1–29

JOURNAL OF COMPUTATIONAL AND APPLIED MATHEMATICS

www.elsevier.nl/locate/cam

Numerical methods for ordinary differential equations in the 20th century

J.C. Butcher *

The University of Auckland, Department of Mathematics, Private Bag 92019, Auckland, New Zealand

Received 7 July 1999; received in revised form 18 November 1999

Abstract

Numerical methods for the solution of initial value problems in ordinary differential equations made enormous progress during the 20th century for several reasons. The first reasons lie in the impetus that was given to the subject in the concluding years of the previous century by the seminal papers of Bashforth and Adams for linear multistep methods and Runge for Runge–Kutta methods. Other reasons, which of course apply to numerical analysis in general, are in the invention of electronic computers half way through the century and the needs in mathematical modelling of efficient numerical algorithms as an alternative to classical methods of applied mathematics. This survey paper follows many of the main strands in the developments of these methods, both for general problems, stiff systems, and for many of the special problem types that have been gaining in significance as the century draws to an end. © 2000 Elsevier Science B.V. All rights reserved.

Keywords: Initial value problems; Adams–Bashforth method; Adams–Moulton method; Runge–Kutta method; Consistency; Stability and convergence; Order of methods; Stiff problems; Differential equation software

1. Introduction

It is not possible to assess the history of this subject in the 20th century without first recognizing the legacy of the previous century on which it has been built. Notable are the 1883 paper of Bashforth and Adams [5] and the 1895 paper of Runge [57]. Not only did the former present the famous Adams–Bashforth method, which plays an essential part in much modern software, but it also looked ahead to the Adams–Moulton method and to the practical use of Taylor series methods. The paper by Runge is now recognized as the starting point for modern one-step methods. These early contributions, together with a brief introduction to the fundamental work of Euler, will form the subject matter of Section 2.

* Tel.: +64-9-373-7599; fax: +64-9-373-7457.

E-mail address: butcher@scitec.auckland.ac.nz (J.C. Butcher).

PII: S0377-0427(00)00455-6

These early papers each formulates the general initial value problem in much the same form. That is, given a function $f(x,y)$ and an "initial value" y_0, corresponding to a solution value at x_0, we seek to evaluate numerically the function y satisfying

$$y'(x) = f(x, y(x)), \quad y(x_0) = y_0. \tag{1}$$

The basic approach is to extend the set of x values for which an approximation to $y(x)$ is known, in a step-by-step fashion.

In the early writing on this problem, y is regarded as a scalar value function but the generalization to more general problems is suggested by a consideration of a pair of simultaneous equations

$$y'(x) = f(x, y(x), z(x)), \quad y(x_0) = y_0,$$

$$z'(x) = g(x, y(x), z(x)), \quad z(x_0) = z_0.$$

Today it is more natural to use formulation (1) but to interpret y as a vector-valued function. In this case, it is even possible to consider an autonomous system of differential equations

$$y'(x) = f(y(x)), \tag{2}$$

because, if necessary, x can be appended to $y(x)$ as an additional component satisfying the trivial differential equation $\mathrm{d}x/\mathrm{d}x = 1$.

After the section dealing with 19th century contributions, this review paper is divided into a number of further sections dealing either with specific periods of time or with contributions with a unifying theme. The development of algorithms based on linear multistep methods continued with the paper of Moulton [49] and to the predictor–corrector formulation together with local error estimation using Milne's device. This will be discussed in Section 2.6.

Sections follow on Runge–Kutta methods and on Taylor series methods. Special methods are needed for stiff problems, and we review some of the stability and other issues involved with the phenomenon of stiffness in Section 6. The development of software to solve initial value problems is discussed in Section 7. Finally, we discuss in Section 8 a number of identifiable problem classes that call for special techniques and special methods.

2. Early work on numerical ordinary differential equations

2.1. The Adams–Bashforth paper

The famous booklet by Bashforth and Adams [5] has a very long title but, when this is broken into two halves, as it appears on the title page, Fig. 1, the authorship of the two distinct aspects of the work is clearly ascribed to the separate authors. Thus, we may assume that the numerical component of this work is due to Mr Adams.

The numerical discussion begins by pointing out that given, for example, a second-order differential equation

$$\frac{\mathrm{d}^2 y}{\mathrm{d}t^2} = f\left(\frac{\mathrm{d}y}{\mathrm{d}t}, y, t\right),$$

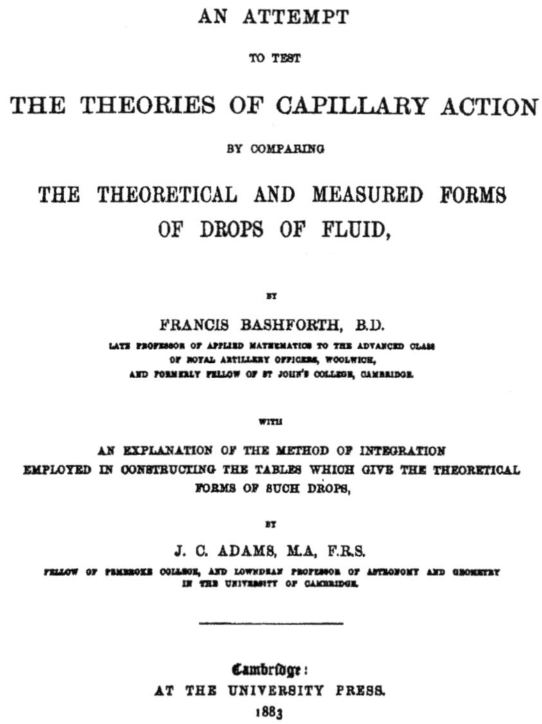

AN ATTEMPT

TO TEST

THE THEORIES OF CAPILLARY ACTION

BY COMPARING

THE THEORETICAL AND MEASURED FORMS

OF DROPS OF FLUID,

BY

FRANCIS BASHFORTH, B.D.

LATE PROFESSOR OF APPLIED MATHEMATICS TO THE ADVANCED CLASS
OF ROYAL ARTILLERY OFFICERS, WOOLWICH,
AND FORMERLY FELLOW OF ST JOHN'S COLLEGE, CAMBRIDGE.

WITH

AN EXPLANATION OF THE METHOD OF INTEGRATION
EMPLOYED IN CONSTRUCTING THE TABLES WHICH GIVE THE THEORETICAL
FORMS OF SUCH DROPS,

BY

J. C. ADAMS, M.A., F.R.S.

FELLOW OF PEMBROKE COLLEGE, AND LOWNDEAN PROFESSOR OF ASTRONOMY AND GEOMETRY
IN THE UNIVERSITY OF CAMBRIDGE.

Cambridge:
AT THE UNIVERSITY PRESS.
1883

Fig. 1. The title page of the Adams–Bashforth paper.

it is possible to find, by repeated differentiation and substitution of $\mathrm{d}^2y/\mathrm{d}t^2$ into the result, formulas for

$$\frac{\mathrm{d}^3y}{\mathrm{d}t^3}, \frac{\mathrm{d}^4y}{\mathrm{d}t^4}, \ldots .$$

From these data evaluated at the initial value, the solution may then be advanced using the Taylor series. Hence, after a small time-step, values of y and of $\mathrm{d}y/\mathrm{d}t$ can be found. Further steps can then be taken in the same manner until a desired value of t is reached.

After these remarks, Adams goes on to derive the Adams–Bashforth method, as we know it today, in the form

$$y_1 - y_0 = \omega\left(q_0 + \tfrac{1}{2}\Delta q_0 + \tfrac{5}{12}\Delta^2 q_0 + \cdots\right), \tag{3}$$

where ω is the stepsize and $q_0, q_{-1}, \ldots$ denote the derivatives computed at the points $t_0, t_{-1}, \ldots$ where the solution values are $y_0, y_{-1}, \ldots$. In the Adams notation, Δ denotes the backward difference $\Delta q_0 = q_0 - q_{-1}$, in contrast to the modern terminology of reserving Δ for the forward difference and using ∇ for the backward difference.

Adams goes on to discuss the relative merits of using, instead of (3), the formula

$$y_0 - y_{-1} = \omega\left(q_0 - \tfrac{1}{2}\Delta q_0 - \tfrac{1}{12}\Delta^2 q_0 + \cdots\right). \tag{4}$$

He correctly observes the advantages of (4) in terms of magnitudes of the error constants. The use of this implicit form of the Adams method was revisited and developed many years later by Moulton [49].

Eine ähnliche Ueberlegung führt nun auch für die Differentialgleichungen zu einer wesentlichen Verbesserung des Euler'schen Verfahrens. Ich will mich zunächst auf Differentialgleichungen erster Ordnung beschränken.

Statt

(1) $$\Delta y = f(x_0 y_0)\,\Delta x \text{ u. s. w.}$$

ist es schon viel besser wenn man

(2) $$\Delta y = f\left(x_0 + \tfrac{1}{2}\Delta x,\ y_0 + \tfrac{1}{2} f(x_0 y_0)\,\Delta x\right)\Delta x$$

u. s. w.

setzt. Diese Art der Berechnung entspricht dem aus der Summe der Tangententrapeze gebildeten Näherungswerthe eines Integrals und deckt sich völlig damit, wenn $f(xy)$ von y unabhängig vorausgesetzt wird.

Oder man kann der Summe der Sehnentrapeze entsprechend setzen:

(3) $$\Delta y = \frac{f(x_0 y_0) + f\left(x_0 + \Delta x,\ y_0 + f(x_0 y_0)\,\Delta x\right)}{2}\,\Delta x$$

u. s. w.

Fig. 2. An extract from the Runge paper.

2.2. The Runge paper

The second great legacy of the 19th century to numerical methods for ordinary differential equations was the work of Runge [57]. Whereas the Adams method was based on the approximation of the solution value for given x, in terms of a number of previously computed points, the approach of Runge was to restrict the algorithm to being "one step", in the sense that each approximation was based only on the most recent point already computed in a previous step. To achieve the required accuracy, approximations are found at a number of internal points within each step and the final result is computed in terms of these various stage values. The short extract from Runge's paper given in Fig. 2, includes the formulations of methods with two derivative calculations per step, based on the mid-point and trapezoidal quadrature rules, respectively.

2.3. The contributions of Heun and Kutta

Following the important and prophetic work of Adams and of Runge, the new century began with further contributions to what is now known as the Runge–Kutta method, by Heun [40] and Kutta [45]. In particular, the famous method in Kutta's paper is often known as *the* Runge–Kutta method. Heun's contribution was to raise the order of the method from two and three, as in Runge's paper, to four. This is an especially significant contribution because, for the first time, numerical methods for differential equations went beyond the use of what are essentially quadrature formulas. Even though second-order Runge methods can be looked at in this light, because the derivatives of the solution are computed from accurate enough approximations so as not to disturb the second-order behaviour, this is no longer true for orders greater than this. Write a three stage method in the form

$$Y_1 = y_0, \quad F_1 = f(x_0, Y_1),$$

$$Y_2 = y_0 + ha_{21}F_1, \quad F_2 = f(x_0 + hc_2, Y_2),$$

$$Y_3 = y_0 + h(a_{31}F_1 + a_{32}F_2), \quad F_3 = f(x_0 + hc_3, Y_3),$$

$$y_1 = y_0 + h(b_1F_1 + b_2F_2 + b_3F_3),$$

where a_{21}, a_{31}, a_{32}, b_1, b_2, b_3, c_2, c_3 are constants that characterize a particular method in this family. We can view computation of the stage values Y_1, identical to the initial value for the step, Y_2, which approximates the solution at x_0+hc_2 and Y_3, which approximates the solution at x_0+hc_3 as temporary steps, whose only purpose is to permit the evaluation of F_1, F_2 and F_3 as approximations to $y'(x_0)$, $y'(x_0+hc_2)$ and $y'(x_0+hc_3)$, respectively. From these derivative approximations, the result at the end of the step is found from the quadrature approximation

$$y(x_0+h) \approx y(x_0) + h(b_1 y'(x_0) + b_2 y'(x_0+hc_2) + b_3 y'(x_0+hc_3)).$$

It is essential that this quadrature formula be sufficiently accurate to integrate polynomials of degree up to 2 exactly. This gives the conditions

$$b_1 + b_2 + b_3 = 1,$$

$$b_2c_2 + b_3c_3 = \tfrac{1}{2},$$

$$b_2c_2^2 + b_3c_3^3 = \tfrac{1}{3}.$$

However, because of possible inaccuracies in the computation of Y_2 and Y_3 as approximations to $y(x_0+hc_2)$ and $y(x_0+hc_3)$, respectively, the quadrature conditions are not enough and it is also necessary that

$$b_3a_{32}c_2 = \tfrac{1}{6},$$

to obtain third-order behaviour.

An example of a method due to Heun which satisfies the four conditions for this order uses the coefficients

$$c_2 = \tfrac{1}{3}, \quad c_3 = 1, \quad a_{21} = \tfrac{1}{3}, \quad a_{31} = 0, \quad a_{32} = \tfrac{2}{3}, \quad b_1 = \tfrac{1}{4}, \quad b_2 = 0, \quad b_3 = \tfrac{3}{4}.$$

Kutta took this investigation further and found a complete classification of the solutions to the eight conditions for four-stage methods with order 4. He also derived the 16 conditions for order 5.

The extract of Kutta's paper given in Fig. 3, includes the formulation of the method, together with the order conditions and the first line of the solution in the case that 0, c_2, c_3 and c_4 are all distinct numbers. In his notation we see that $\kappa = c_2$, $\lambda = c_3$ and $\mu = c_4$. It is an interesting consequence of these order conditions, that μ is necessarily equal to 1.

Of the various four stages, fourth-order methods derived by Kutta, the most famous, and also the most widely used, is

$$Y_1 = y_0, \quad F_1 = f(x_0, Y_1),$$

$$Y_2 = y_0 + \tfrac{1}{2}hF_1, \quad F_2 = f(x_0 + \tfrac{h}{2}, Y_2),$$

$$Y_3 = y_0 + \tfrac{1}{2}hF_2, \quad F_3 = f(x_0 + \tfrac{h}{2}, Y_3),$$

$$Y_4 = y_0 + hF_3, \quad F_4 = f(x_0 + h, Y_4),$$

$$y_1 = y_0 + h(\tfrac{1}{6}F_1 + \tfrac{1}{3}F_2 + \tfrac{1}{3}F_3 + \tfrac{1}{6}F_4).$$

The set of conditions for fifth-order methods is actually a little more complicated than Kutta realised, because there are actually 17 conditions. The reason for the discrepancy is that he was dealing with scalar differential equations, rather than vector-valued differential equations, and for orders five

Zu den Näherungen vierter Ordnuug übergehend finden wir die Berechnung von vier Funktionswerten nötig, und die Vergleichung des Taylorschen Satzes ergiebt die folgenden acht Bedingungsgleichungen für die Koeffizienten:

$$
\begin{aligned}
a + b + c + d &= 1,\\
b\varkappa + c\lambda + d\mu &= \tfrac{1}{2},\\
b\varkappa^2 + c\lambda^2 + d\mu^2 &= \tfrac{1}{3},\\
c\varrho\varkappa + d(\sigma\lambda + \tau\varkappa) &= \tfrac{1}{6},\\
b\varkappa^3 + c\lambda^3 + d\mu^3 &= \tfrac{1}{4},\\
c\varrho\varkappa\lambda + d(\sigma\lambda + \tau\varkappa)\mu &= \tfrac{1}{8},\\
c\varrho\varkappa^2 + d(\sigma\lambda^2 + \tau\varkappa^2) &= \tfrac{1}{12},\\
d\varrho\sigma\varkappa &= \tfrac{1}{24},
\end{aligned}
$$

wenn als gewünschte Näherung angesetzt ist:

$$
\begin{aligned}
\Delta y &= a\Delta' + b\Delta'' + c\Delta''' + d\Delta'''',\\
\Delta' &= f(x, y)\Delta x,\\
\Delta'' &= f(x + \varkappa\Delta x,\ y + \varkappa\Delta')\Delta x,\\
\Delta''' &= f(x + \lambda\Delta x,\ y + \varrho\Delta'' + (\lambda - \varrho)\Delta')\Delta x,\\
\Delta'''' &= f(x + \mu\Delta x,\ y + \sigma\Delta''' + \tau\Delta'' + (\mu - \sigma - \tau)\Delta')\Delta x.
\end{aligned}
$$

Hier läſst sich das Lösungssystem auch noch niederschreiben. Wenn man $\varkappa$ und λ willkürlich läſst, erhält man nach einiger Rechnung:

$$
c = \frac{1 - 2\varkappa}{12\lambda(\lambda - \varkappa)(1 - \lambda)},\quad b = \frac{1 - 2\lambda}{12\varkappa(\varkappa - \lambda)(1 - \varkappa)},\quad d = \frac{6\varkappa\lambda - 4(\varkappa + \lambda) + 3}{12(1 - \lambda)(1 - \varkappa)};
$$

Fig. 3. An extract from the Kutta paper.

or greater the conditions become different. Another difficulty is in actually finding solutions to the algebraic conditions and Kutta presented methods that are slightly incorrect. It is interesting that once the correction is made, the additional condition, to make the method applicable to high-dimensional problems, happens to be satisfied.

2.4. The contributions of E.J. Nyström

The early history of Runge–Kutta methods culminated in the work of Nyström [53] in 1925. He was able to correct some of the fifth-order methods of Kutta and he also showed how to apply the Runge–Kutta method to second-order differential equation systems.

At first sight this is quite straightforward, because every second-order system can be re-formulated as a first-order system with additional dependent variables. However, solving such a problem directly may be much more efficient and the great prevalence of second-order problems in physical modelling makes this sort of gain in efficiency of considerable practical significance.

2.5. Moulton's paper and predictor–corrector methods

Implicit versions of Adams methods were first suggested in the Adams–Bashforth paper, but not studied in their own right until the paper of Moulton [49]. These so-called Adams–Moulton methods have two great advantages over the original explicit methods. The first is that they do not need to use so many past values to obtain the same order and they have smaller error constants. To use them in practice, however, one first has to overcome the difficulty associated with their implicit nature. This difficulty hinges on the fact that y_n is not given in terms of rational operations on known data, but as the solution to an algebraic equation. For example, consider the third-order Adams–Bashforth and Adams–Moulton methods given by

$$y_n = y_{n-1} + h(\tfrac{23}{12}f(x_{n-1}, y_{n-1}) - \tfrac{4}{3}f(x_{n-2}, y_{n-2}) + \tfrac{5}{12}f(x_{n-3}, y_{n-3})), \tag{5}$$

$$y_n = y_{n-1} + h(\tfrac{5}{12}f(x_n, y_n) + \tfrac{2}{3}f(x_{n-1}, y_{n-1}) - \tfrac{1}{12}f(x_{n-2}, y_{n-2})). \tag{6}$$

It is known that the error introduced into the result in a single step is $-\frac{3}{8}y^{(4)}h^4+\mathrm{O}(h^5)$ for the Adams–Bashforth method and $\frac{1}{24}y^{(4)}h^4 + \mathrm{O}(h^5)$ for the Adams–Moulton method. The way that advantage is gained from the desirable properties of each of the methods is to use them in "predictor–corrector mode". This means that a predicted value of y_n is first found using the explicit form of the method. The implicit or Moulton form of the method is then used with the term $f(x_n, y_n)$ replaced by the value calculated using the *predicted* value of y_n. There are many variants of this method in common use, but the most popular is the so-called PECE mode. In this mode, $f(x_n, y_n)$ is re-evaluated for use in later steps using y_n found from the Adams–Moulton method. Thus each step requires two evaluations of the function f and is thus twice as expensive as the simple use of the Adams–Bashforth formula alone. However, the advantages in terms of stability and accuracy resulting from the use of this PECE predictor–corrector mode are usually regarded as well worth the additional computing cost.

2.6. The Milne device

Although Milne preferred methods based on Newton–Cotes quadrature formulas, methods which are largely abandoned today in favour of Adams methods, a proposal he made [47] has been adapted to other situations and widely used. In the context of the predictor pair (5) and (6), implemented, for example in PECE mode, there are two approximations to $y(x_n)$ computed in each step. Since the local truncation errors of the two approximations are in the ratio -9 to 1, it is proposed that the *difference* of the two approximations divided by 10 should be used as an estimate of the error in the corrected formula.

Milne, of course, intended this device to be used to check the accuracy of hand-computed results, but today it is used in automatic solvers, not just to verify the accuracy of any completed step, but also to adjust the size of a subsequent step in the interests both of efficiency and robustness.

Many modern computer codes implement predictor–corrector methods in a different manner than we have described. Specifically, the step number k is chosen to be the same for both the predictor and corrector formulas. This means that the order of the predictor will be k and the order of the corrector, which becomes the overall order of the combined method, will be $p=k+1$. Even though the difference between the predicted and corrected solutions is no longer asymptotically equal to a multiple of the local truncation error, this difference is still used as the basis for stepsize control.

3. The modern theory of linear multistep methods

The modern analysis of linear multistep methods is intimately bound up with the work of Dahlquist [21,22]. This large body of work is in several parts, of which the first deals with the concepts of consistency, stability and convergence, expressed in terms of generating functions for the coefficients of the method. The key result in this phase of the work, is that consistency and stability are together equivalent to convergence. The second principle phase relates order of accuracy to stability and culminates in the famous "Dahlquist barrier" result, which limits the order of a convergent linear k-step method to $k+1$ (if k is odd) and to $k+2$ (if k is even). The remaining phase of Dahlquist's work is more appropriately discussed in Section 6.

3.1. Generating functions

Consider a linear multistep method of the form

$$\begin{aligned}&\alpha_k y_n + \alpha_{k-1} y_{n-1} + \alpha_{k-2} y_{n-2} + \cdots + \alpha_0 y_{n-k}\\ &\quad = h(\beta_k f(x_n, y_n) + \beta_{k-1} f(x_{n-1}, y_{n-1}) + \beta_{k-2} f(x_{n-2}, y_{n-2}) + \cdots + \beta_0 f(x_{n-k}, y_{n-k})),\end{aligned}$$

assuming that $\alpha_k \neq 0$ and that α_0 and β_0 are not both zero (otherwise the value of k could be reduced). Such a method is known as a "linear k-step method" because the solution at step number n depends on exactly k previous step values. Dahlquist introduced polynomials ρ and σ to characterize the method as follows:

$$\begin{aligned}\rho(z) &= \alpha_k z^k + \alpha_{k-1} z^{k-1} + \alpha_{k-2} z^{k-2} + \cdots + \alpha_0,\\ \sigma(z) &= \beta_k z^k + \beta_{k-1} z^{k-1} + \beta_{k-2} z^{k-2} + \cdots + \beta_0.\end{aligned}$$

Although Dahlquist allowed for the generality of allowing the coefficient of z^k to take on any non-zero value, in an actual computation with the method, the value of α_k has to be cancelled out from both polynomials.

It is clear that given any linear multistep method, the corresponding pair of polynomials (ρ, σ) can be written down automatically and, given the polynomials, the method is completely specified. Hence, it has become customary to identify the methods with the pair of polynomials and we can speak of "the method (ρ, σ)". It is convenient to assume that ρ and σ have no common polynomial factor, since it would be possible to describe most aspects of the computational behaviour of the method in terms of simpler polynomials. Following Dahlquist, we will make this assumption.

3.2. Consistency, stability and convergence

There are some natural assumptions that can be made about linear multistep methods to guarantee that they can at least solve certain specific problems. We will consider these one, by one.

The first problem is $y'(x) = 0$, with initial value $y(0) = 0$. Since we are given only this single initial value we will need an algorithm to generate y_0, y_1, $y_2, \ldots, y_{k-1}$ which is, in the limit as $h \to 0$, consistent with the given initial data. Choose some $x > 0$, for example $x = 1$, as the point where the numerical result approximating the solution is supposed to be found. We would like our method to be able to compute $y(1)$ exactly in the limiting case as $x \to 0$.

This requirement is equivalent to the "stability condition": A linear multistep method (ρ, σ) is stable if all zeros of ρ lie in the unit disc and all zeros on the boundary are simple.

The second initial value problem is also based on the equation $y'(x) = 0$ but with $y(0) = 1$. To compute the correct result $y(1) = 1$, in the limit, it is necessary that $\rho(1) = 0$. We will refer to this as the "pre-consistency condition".

Finally, consider the initial value problem $y'(x)=1$, $y(0)=0$. If a method is stable and pre-consistent, then its ability to solve this problem in the limit hinges on the requirement that $\rho'(1) = \sigma(1)$. This condition, when combined with the pre-consistency condition, is known as the "consistency condition".

The definition of convergence is rather technical but deals with the ability of the linear multistep method to solve *any* differential equation system on condition only that f is continuous in its first variable and satisfies a Lipschitz condition in its second variable. The k initial approximations required to start the numerical process must converge to the given initial value as the stepsize tends to zero. This class of problems might seem restrictive but it is easy to extend it to many situations where the Lipschitz condition is replaced by a *local* Lipschitz condition.

The basic theorem connecting these concepts is that a method is convergent if and only if it is both stable and consistent. Of course convergence is not enough to ensure that the method is computationally efficient. In the next section we look at the criteria for the method to have some specific order of accuracy and we review a famous result of Dahlquist which imposes a barrier on what order is really achievable.

3.3. The order of linear multistep methods

Given a linear multistep method characterized by the polynomials ρ and σ, define the operator L on the continuously differentiable functions $I \to \mathbb{R}^N$ by the formula

$$L(y)(x) = \sum_{i=0}^{k} \alpha_{k-i} y(x - ih) - h \sum_{i=0}^{k} \beta_{k-i} y'(x - ih). \tag{7}$$

A method is said to be of order p if $L(P) = 0$ for P any polynomial of degree not exceeding p.

To understand the significance of this definition, assume that y is continuously differentiable at least $p + 1$ times and expand the right-hand side of (7) in a Taylor series about x_n. We have

$$L(y)(x_n) = \sum_{i=0}^{p+1} C_i h^i y^{(i)}(x_n) + \mathrm{O}(h^{p+2}),$$

where

$$C_0 = \sum_{i=0}^{k} \alpha_i,$$

$$C_1 = -\sum_{i=1}^{k} i\alpha_{k-i} - \sum_{i=0}^{k} \beta_{k-i},$$

$$C_j = \frac{(-1)^j}{j!} \left(\sum_{i=1}^{k} i^j \alpha_{k-i} + j \sum_{i=1}^{k} i^{j-1} \beta_{k-i} \right), \quad j = 2, 3, \ldots, p + 1.$$

If y is replaced by a polynomial of degree p, then

$$L(P)(x_n) = \sum_{i=0}^{p} C_i h^i P^{(i)}(x_n)$$

and, for this to vanish for all such polynomials, it is necessary and sufficient that

$$C_0 = C_1 = C_2 = \cdots = C_p = 0.$$

We derive the two methods (5) and (6) using these expressions. The Adams–Bashforth method of order 3 requires $k = 3$ and assumes that $\alpha_3 = 1$, $\alpha_2 = -1$, $\alpha_1 = \alpha_0 = \beta_3 = 0$. We have

$$C_0 = \alpha_3 + \alpha_2 + \alpha_1 + \alpha_0 \quad = 0,$$

$$C_1 = -\alpha_2 - 2\alpha_1 - 3\alpha_0 - \beta_2 - \beta_1 - \beta_0 \quad = 1 - \beta_2 - \beta_1 - \beta_0,$$

$$C_2 = \tfrac{1}{2}(\alpha_2 + 4\alpha_1 + 9\alpha_0 + 2(\beta_2 + 2\beta_1 + 3\beta_0)) \quad = \beta_2 + 2\beta_1 + 3\beta_0 - \tfrac{1}{2},$$

$$C_3 = -\tfrac{1}{6}(\alpha_2 + 8\alpha_1 + 27\alpha_0 - 3(\beta_2 + 4\beta_1 + 9\beta_0)) \quad = \tfrac{1}{6} - \tfrac{1}{2}(\beta_2 + 4\beta_1 + 9\beta_0).$$

The solution of $C_1 = C_2 = C_3 = 0$ is $\beta_2 = \frac{23}{12}$, $\beta_1 = -\frac{4}{3}$, $\beta_0 = \frac{5}{12}$, with the first nonzero coefficient in the Taylor expansion of $L(y)(x_n)$ given by

$$C_4 = -\tfrac{1}{24}(1 - 4(\beta_2 + 8\beta_1 + 27\beta_0)) = \tfrac{3}{8}.$$

The value of this quantity is closely related to the "error constant" for the method which is actually given by $C_4/\rho'(1)$. Note that, in this case, and also for the Adams–Moulton method which we will discuss next, $\rho'(1) = 1$.

For the Adams–Moulton method of order 3, an additional nonzero parameter β_k is available and $k = 2$ is sufficient for this order. We find $C_0 = 0$, $C_1 = 1 - \beta_2 - \beta_1 - \beta_0$, $C_2 = \beta_1 + 2\beta_0 - \frac{1}{2}$ and $C_3 = \frac{1}{6} - \frac{1}{2}(\beta_1 + 4\beta_0)$ and $C_1 = C_2 = C_3 = 0$ implies $\beta_2 = \frac{5}{12}$, $\beta_1 = \frac{2}{3}$, $\beta_0 = -\frac{1}{12}$, with

$$C_4 = -\tfrac{1}{24}(1 - 4(\beta_1 + 8\beta_0)) = -\tfrac{1}{24}.$$

To investigate the order conditions further, it is convenient to consider the expression $(\rho(\exp(z)) - z\sigma(\exp(z)))\exp(-kz)$ which can be expanded by Taylor series

$$\begin{aligned}(\rho(\mathrm{e}^z) - z\sigma(\mathrm{e}^z))\,\mathrm{e}^{-kz} &= \alpha_k + \alpha_{k-1}\mathrm{e}^{-z} + \alpha_{k-2}\mathrm{e}^{-2z} + \cdots \\ &\quad + z(\beta_k + \beta_{k-1}\mathrm{e}^{-z} + \beta_{k-2}\mathrm{e}^{-2z} - \cdots) \\ &= C_0 + C_1 z + C_2 z^2 + \cdots \\ &= C_{p+1} z^{p+1} + \mathrm{O}(z^{p+2}),\end{aligned}$$

if the order is p. The number C_{p+1} does not vanish unless the order is actually higher than p. Hence

$$\rho(\mathrm{e}^z) - z\sigma(\mathrm{e}^z) = \mathrm{O}(z^{p+1}).$$

Because $\rho(1) = 0$ for a consistent method, we can divide by z and we find

$$\frac{\rho(\mathrm{e}^z)}{z} - \sigma(\mathrm{e}^z) = \mathrm{O}(z^p)$$

and substituting $\exp(z)$ by $1 + z$

$$\frac{\rho(1+z)/z}{\ln(1+z)/z} - \sigma(1+z) = \mathrm{O}(z^p), \tag{8}$$

where $\ln(1+z)/z$ is defined in a neighbourhood of 0 by the series

$$\frac{\ln(1+z)}{z} = 1 - \frac{z}{2} + \frac{z^2}{3} - \cdots$$

so that

$$\left(\frac{\ln(1+z)}{z}\right)^{-1} = 1 + \frac{z}{2} - \frac{z^2}{12} + \frac{z^3}{24} - \frac{19z^4}{720} + \frac{3z^5}{160} + \mathrm{O}(z^6).$$

Using this expression, (8) can be used to derive methods with specific choices of ρ. Rewriting in the form

$$\rho(1+z) = \sigma(1+z)\left(z - \frac{z^2}{2} + \frac{z^3}{3} - \cdots\right)$$

enables coefficients to be found for the backward difference and similar methods in which the form of σ is prescribed.

3.4. The Dahlquist barrier

Even though it is possible, in principle, for linear multistep methods to have order as high as $2k$, this does not yield stable methods if $k > 2$. This is a consequence of the so-called "Dahlquist barrier" [21], which states that

Theorem 1. *The order of a stable linear k-step method is bounded by*

$$p \leqslant \begin{cases} k+2, & k \text{ even}, \\ k+1, & k \text{ odd}. \end{cases}$$

Proof. We will give here a vastly abbreviated proof, along the same lines as originally given by Dahlquist. Let

$$r(z) = \rho\left(\frac{1+z}{1-z}\right)\left(\frac{1-z}{2}\right)^k,$$

$$s(z) = \sigma\left(\frac{1+z}{1-z}\right)\left(\frac{1-z}{2}\right)^k,$$

where we note that the order conditions can be rewritten in the form

$$\frac{r(z)/z}{\ln((1+z)/(1-z))/z} - s(z) = \mathrm{O}(z^p). \tag{9}$$

Let $r(z) = a_0 + a_1 z + a_2 z^2 + \cdots + a_k z^k$ and $s(z) = b_0 + b_1 z + b_2 z^2 + \cdots + b_k z^k$, where $a_0 = 0$ by the consistency condition. By the stability condition, $a_1 \neq 0$ and no two of the coefficients in r can have opposite signs. If

$$\frac{1}{\ln((1+z)/(1-z))/z} = c_0 + c_2 z^2 + c_4 z^4 + \cdots,$$

it can be shown that $c_2, c_4, \ldots$ are all negative [21,39]. If (9) is to hold for $p > k+1$, then the coefficient of z^{p+1} in

$$(c_0 + c_2z^2 + c_4z^4 + \cdots)(a_1 + a_2z + \cdots + a_kz^{k-1}), \tag{10}$$

must vanish. If the order is $p > k+2$, then the coefficient of z^{k+2} in (10) must also vanish. The two coefficients are respectively

$$a_kc_2 + a_{k-2}c_4 + \cdots, \tag{11}$$

$$a_{k-1}c_4 + a_{k-3}c_6 + \cdots. \tag{12}$$

If k is odd, (11) cannot vanish because this would imply that

$$a_k = a_{k-2} = \cdots = a_1 = 0.$$

On the other hand, if k is even, then (12) cannot vanish because we would then have

$$a_{k-1} = a_{k-3} = \cdots = a_1 = 0.$$

4. The modern theory of Runge–Kutta methods

The meaning of order looks quite different and is relatively complicated for one-step methods, for the very good reason that the result computed in a step is built up from the derivatives evaluated sequentially from the stages values and, at least for the early stages, these have low accuracy. In contrast, the result computed in linear multistep methods makes use of derivatives evaluated from a number of step values, which themselves have been evaluated in previous steps and all share the same order.

The basic approach to the analysis of Runge–Kutta methods is to obtain the Taylor expansions for the exact and computed solutions at the end of a single step and to compare these series term by term. This idea dates back to Runge, Heun, Kutta and Nyström and we will give as an example the derivation of the conditions for order 3.

For the scalar differential equation

$$y'(x) = f(x, y(x)), \tag{13}$$

we calculate in turn

$$y'' = \frac{\partial f}{\partial x} + \frac{\partial f}{\partial y} f, \tag{14}$$

$$y''' = \frac{\partial^2 f}{\partial x^2} + 2\frac{\partial^2 f}{\partial x \partial y} f + \frac{\partial^2 f}{\partial y^2} f^2 + \frac{\partial f}{\partial x}\frac{\partial f}{\partial y} + \left(\frac{\partial f}{\partial y}\right)^2 f, \tag{15}$$

where we have substituted $y' = f$ in the formula $y'' = (\partial f/\partial x) + (\partial f/\partial y)y'$ to obtain (14) and made similar substitutions in the derivation of (15). From these expressions we can write down the first few terms of the Taylor expansion $y(x_0 + h) = y(x_0) + hy'(x_0) + \frac{1}{2}h^2y''(x_0) + \frac{1}{6}h^3y'''(x_0) + O(h^4)$.

Complicated though these expressions are, they are simple in comparison with the corresponding formulas for the fourth and higher derivatives. To obtain conditions for order 3 we also need the

Table 1
Details of Taylor expansions up to order 3

	Y_1	hF_1	Y_2	hF_2	Y_3	hF_3	y_1	$y(x_0+h)$
y	1	0	1	0	1	1	1	1
hf	0	1	a_{21}	1	$a_{31}+a_{32}$	1	$b_1+b_2+b_3$	1
$h^2\dfrac{\partial f}{\partial x}$	0	0	0	c_2	$a_{32}c_2$	c_3	$b_2c_2+b_3c_3$	$\frac{1}{2}$
$h^2\dfrac{\partial f}{\partial y}f$	0	0	0	a_{21}	$a_{32}a_{21}$	$a_{31}+a_{32}$	$b_2a_{21}+b_3(a_{31}+a_{32})$	$\frac{1}{2}$
$h^3\dfrac{\partial^2 f}{\partial x^2}$	0	0	0	$\frac{1}{2}c_2^2$	$\frac{1}{2}a_{32}c_2^2$	$\frac{1}{2}c_3^2$	$\frac{1}{2}(b_2c_2^2+b_3c_3^2)$	$\frac{1}{6}$
$h^3\dfrac{\partial^2 f}{\partial x\partial y}f$	0	0	0	c_2a_{21}	$a_{32}c_2a_{21}$	$c_3(a_{31}+a_{32})$	$b_2c_2a_{21}+b_3c_3(a_{31}+a_{32})$	$\frac{1}{3}$
$h^3\dfrac{\partial^2 f}{\partial y^2}f^2$	0	0	0	$\frac{1}{2}a_{21}^2$	$\frac{1}{2}a_{32}a_{21}^2$	$\frac{1}{2}(a_{31}+a_{32})^2$	$\frac{1}{2}(b_2a_{21}^2+b_3(a_{31}+a_{32})^2)$	$\frac{1}{6}$
$h^3\dfrac{\partial f}{\partial x}\dfrac{\partial f}{\partial y}$	0	0	0	0	0	$a_{32}c_2$	$b_3a_{32}c_2$	$\frac{1}{6}$
$h^3\left(\dfrac{\partial f}{\partial y}\right)^2 f$	0	0	0	0	0	$a_{32}a_{21}$	$b_3a_{32}a_{21}$	$\frac{1}{6}$

formulas for the first, second and third derivatives of the approximation computed by a Runge–Kutta method, which we will assume is explicit and has exactly 3 stages.

To simplify notation we will denote x, y, f and the various partial derivatives, as being evaluated at the initial point (x_0, y_0) in a step and we will then find Taylor expansions in turn for $Y_1, hF_1, \dots, Y_2$, hF_2, Y_3, hF_3 and finally y_1. We will express the sequence of calculations in tabular form in Table 1, where the coefficients of y, hf, etc. are shown. In addition to the coefficients in the expansion of y_1, we append the corresponding coefficients for the exact solution at x_0+h.

By equating the last two columns of this table, we obtain conditions for order 3. These imply that

$$a_{21}=c_2, \tag{16}$$

$$a_{31}+a_{32}=c_3 \tag{17}$$

and that

$$b_1+b_2+b_3=1,$$

$$b_2c_2+b_3c_3=\tfrac{1}{2},$$

$$b_2c_2^2+b_3c_3^3=\tfrac{1}{3},$$

$$b_3a_{32}c_2=\tfrac{1}{6}.$$

If $s=p$, which turns out to be possible for orders up to 4, conditions such as (16) and (17) always hold. Even for higher orders, where the argument is a little more complicated, there is never any reason for not assuming that

$$\sum_{j=1}^{s} a_{ij} = c_i, \quad i = 1, 2, \ldots, s, \tag{18}$$

where we adopt a convention that $a_{ij}=0$ for $j \geqslant i$ in explicit methods. For the more general implicit methods, we will continue to assume (18).

There are three reasons for abandoning (1) as the standard problem and replacing it instead by (2), where the values of $y(x)$ are now in a finite-dimensional vector space rather than scalars. The first reason for the change to a high-dimensional *autonomous* problem is that there is no need to retain x as an argument of f in the vector case, because nonautonomous problems can always be transformed into equivalent autonomous problems by adding an additional component which always has a value exactly the same as x. A consideration of this formal re-formulation can be used to justify the assumption (18). The second reason is that the analysis is actually more straightforward in the autonomous vector case. Finally, it is found that the conditions for order as derived using the scalar first-order problem (13) are inadequate for specifying the order requirements for the general vector case. The two theories do not diverge until the fifth-order case is reached but after that the families of order conditions for the scalar and vector cases become increasingly different.

4.1. The order of Runge–Kutta methods

The analysis of order for the vector case that we present here is due to the present author [9] and is related to earlier work by Gill [33] and Merson [46]. Since it relates the various terms in the Taylor series expansion of both the exact solution and the approximation computed by a Runge–Kutta method, to the graphs known as "rooted trees" or arborescences, we briefly review rooted trees.

A rooted tree is simply a connected directed graph for which each vertex, except the root, has a single predecessor (or parent). The root has no predecessor. The order of a rooted tree t is the number of vertices. Denote this by $r(t)$. Clearly, the number of arcs in the rooted tree is $r(t)-1$. Let a_n denote the number of distinct rooted trees with order n. Table 2 gives the first few values of a_n together with the sums $\sum_{i=1}^{n} a_i$.

The eight rooted trees for which $r(t) \leqslant 4$ are shown in Table 3, together with the values of $\sigma(t)$, the "symmetry" of t and $\gamma(t)$ the "density" of t. The quantity $\sigma(t)$ is the order of the group of permutations of the vertices which leave the structure unchanged, while $\gamma(t)$ is the product over all vertices of t of the total number of descendents (including the vertex itself) of this vertex. Also

Table 2
Numbers of trees and accumulated sums up to order 8

n	1	2	3	4	5	6	7	8
a_n	1	1	2	4	9	20	48	115
$\sum_{i=1}^{n} a_i$	1	2	4	8	17	37	85	200

Table 3
Various functions on trees

t	•							
$r(t)$	1	2	3	3	4	4	4	4
$\sigma(t)$	1	1	2	1	6	1	2	1
$\gamma(t)$	1	2	3	6	4	8	12	24
$\alpha(t)$	1	1	1	1	1	3	1	1
$\beta(t)$	1	2	3	6	4	24	12	24

shown are the values of $\alpha(t)=r(t)!/\gamma(t)\sigma(t)$ and $\beta(t)=r(t)!/\sigma(t)$. The values of α and β have simple interpretations in terms of possible labellings of the vertices of a tree under various restrictions.

It can be shown that the Taylor expansion of the exact solution has the form

$$y(x_0+h)=y(x_0)+\sum_{t\in T}\frac{\alpha(t)h^{r(t)}}{r(t)!}F(t)(y_0) \tag{19}$$

and that the corresponding expansion for the solution computed using a Runge–Kutta method is

$$y(x_0)+\sum_{t\in T}\frac{\beta(t)\Phi(t)h^{r(t)}}{r(t)!}F(t)(y_0). \tag{20}$$

In each of these formulas, $F(t)$ is the "elementary differential" which we will define below and $\Phi(t)$ is the "elementary weight". The formula for $F(t)(y)$ is defined in terms of the differential equation and $\Phi(t)$ in terms of the Runge–Kutta method being used. Each of these quantities can be defined recursively but, for our present purposes, it will be enough to present one example, using a tree of order 7 and to list these quantities for all trees of order up to 4. In the special example, the tree t is shown with labels i, j, k, l, m, n, o attached to the vertices. The formula for $F(t)$, is given in terms of an expression for component number i, written as a superscript. The summation convention is assumed and f^i_{jkl} denotes the third partial derivative, $\partial^3 f^i/\partial y^j\partial y^k\partial y^l$, of f^i, evaluated at y, with similar meanings for f^j_m, f^k_{no}. The summations in the formula for $\Phi(t)$ are over all subscripts running from 1 to s. Note that the formula is simplified from $\Phi(t)=\sum b_i a_{il}a_{ij}a_{jm}a_{ik}a_{kn}a_{ko}$ by summing over l, m, n and o.

$t =$ 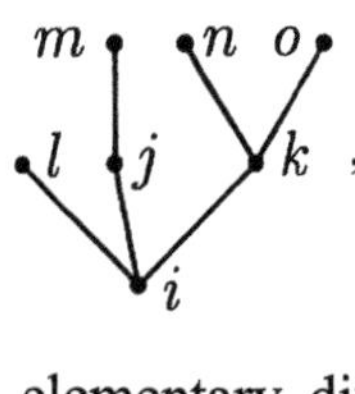

,

$$F^i(t)=f^i_{jkl}f^l f^j_m f^m f^k_{no} f^n f^o,$$
$$\Phi(t)=\sum b_i c_i a_{ij} c_j a_{ik} c_k^2.$$

Because the elementary differentials are independent, in the sense that, given any set of n rooted trees, $t_1, t_2, \ldots, t_n$ and any sequence of real numbers $q_1, q_2, \ldots, q_n$, it is possible to find a function f such that for some specific value of y and some specific coordinate direction, say e_1^{T}, all the equations

$$e_1^{\mathrm{T}}F(t_i)(y)=q_i, \quad i=1,2,\ldots,n,$$

Table 4
Elementary differential and weights up to order 4

t								
$F(t)$	f^i	$f^i_j f^j$	$f^i_{jk} f^j f^k$	$f^i_j f^j_k f^k$	f^i_{jkl}	$f^i_{jk} f^j f^k_l f^l$	$f^i_j f^j_{kl} f^k f^l$	$f^i_j f^j_k f^k_l f^l$
$\Phi(t)$	$\sum b_i$	$\sum b_i c_i$	$b_i c_i^2$	$\sum b_i a_{ij} c_k$	$\sum b_i c_i^3$	$\sum b_i c_i a_{ij} c_j$	$\sum b_i a_{ij} c_j^2$	$\sum b_i a_{ij} a_{jk} c_k$

can be satisfied simultaneously, it is only possible that (19) and (20) agree to within $O(h^{p+1})$ if

$$\alpha(t) = \beta(t)\Phi(t)$$

for every tree t with no more than p vertices.

Inserting the formulas for α and β, we find that

$$\Phi(t) = \frac{1}{\gamma(t)} \tag{21}$$

as the condition corresponding to this tree (Table 4).

It is interesting that, for the single first-order scalar differential equation (13), the independence of the elementary differentials breaks down and it turns out to be possible to obtain methods that have some specified order in this case, but a lower order for the more general system of equations given by (2). This effect occurs for order 5 and higher orders.

Other interpretations of order are of course possible. An alternative derivation of the order conditions, due to Albrecht [1], is based on expressions arising from the Taylor series for

$$y(x_0) + h \sum_{j=1}^{s} a_{ij} y'(x_0 + hc_j) - y(x_0 + hc_i) = \varepsilon^{(2)} h^2 + \varepsilon^{(3)} h^3 + \cdots,$$

where

$$\varepsilon_i^{(2)} = \sum_j a_{ij} c_j - \frac{1}{2} c_j^2,$$

$$\varepsilon_i^{(3)} = \sum_j a_{ij} c_j^2 - \frac{1}{3} c_j^3,$$

$$\vdots \quad \vdots$$

For order 4 for example, it is found to be necessary that

$$\sum_i b_i \varepsilon_i^{(2)} = 0, \quad \sum_i b_i c_i \varepsilon_i^{(2)} = 0, \quad \sum_i b_i a_{ij} \varepsilon_j^{(2)} = 0, \quad \sum_i b_i \varepsilon_i^{(3)} = 0,$$

which, together with the quadrature conditions

$$\sum_i b_i c_i^{k-1} = \frac{1}{k}, \quad k = 1, 2, 3, 4,$$

are equivalent to (21), up to order four. A third approach, due to Hairer and Wanner [36], is based on the use of B-series. This theory, used to study compositions of Runge–Kutta methods, is related to [13], and has applications also to more general problems and methods.

Table 5
Minimum s to obtain order p

p	1	2	3	4	5	6	7	8
s	1	2	3	4	6	7	9	11

4.2. Attainable order of Runge–Kutta methods

For explicit Runge–Kutta methods with s stages, there are $s(s+1)/2$ free parameters to choose. It is easy to show that an order p is possible only if $s \geqslant p$. Up to order 4, $s = p$ is actually possible. However, for $p > 4$, the relationship between the minimum s to obtain order p is very complicated but is partly given in Table 5. The results given for $p > 4$ were proved in [11,15].

For implicit Runge–Kutta methods, which we will discuss below, the relationship is much simpler. In fact, order p can be obtained with s stages if and only if $p \leqslant 2s$.

4.3. Implicit Runge–Kutta methods

One of the earliest references to implicitness, as applied to Runge–Kutta methods, was in the book by Kunz [44] where the method of Clippinger and Dimsdale was quoted. This method with tableau

$$\begin{array}{c|ccc} 0 & 0 & 0 & 0 \\ \frac{1}{2} & \frac{5}{24} & \frac{1}{3} & -\frac{1}{24} \\ 1 & \frac{1}{6} & \frac{2}{3} & \frac{1}{6} \\ \hline & \frac{1}{6} & \frac{2}{3} & \frac{1}{6} \end{array}$$

is the forerunner both of Lobatto methods and of block methods [59].

The method of Hammer and Hollingsworth [38] will be explored in some detail. It is the forerunner of Gauss and other important classes of methods. The coefficients for the method are

$$\begin{array}{c|cc} \frac{1}{2} - \frac{\sqrt{3}}{6} & \frac{1}{4} & \frac{1}{4} - \frac{\sqrt{3}}{6} \\ \frac{1}{2} + \frac{\sqrt{3}}{6} & \frac{1}{4} + \frac{\sqrt{3}}{6} & \frac{1}{4} \\ \hline & \frac{1}{2} & \frac{1}{2} \end{array}$$

This method has order 4. This is a little surprising because the eight conditions for this order have been seemingly satisfied using only the six free parameters in A and b^{T}. Although the order conditions are trivial to check, we will verify them below using an argument that illustrates what happens much more generally.

If the coefficient matrix A is allowed to be fully implicit, that is any element on or above the diagonal may have a non-zero value, then there are clearly more free parameters available to satisfy the order conditions. The advantages, in terms of order, are even greater than might be expected from a mere comparison of the number of conditions with the number of free parameters, because various simplifying assumptions are easier to satisfy. These simplifying assumptions bring about a drastic lowering of the number of further conditions required for order; furthermore they interact and reinforce each other.

The simplifying assumptions we will use, are denoted by $C(\eta)$, $D(\xi)$ and $E(\xi,\eta)$, where we have used the notation of [10]. In each case ξ and η are positive integers and the assumptions refer to some equalities involving the coefficients of a specific method. The specific meanings are

$$C(\eta)\colon\ \sum_{j=1}^{s} a_{ij}c_j^{l-1} = \frac{1}{k}c_i^l, \quad i=1,2,\ldots,s,\ l=1,2,\ldots,\eta,$$

$$D(\xi)\colon\ \sum_{i=1}^{s} b_i c_i^{k-1} a_{ij} = \frac{1}{k}b_j(1-c_j^k), \quad j=1,2,\ldots,s,\ k=1,2,\ldots,\xi,$$

$$E(\xi,\eta)\colon\ \sum_{i=1}^{s} b_i c_i^{k-1} a_{ij} c_j^{l-1} = \frac{1}{l(k+1)}, \quad k=1,2,\ldots,\xi,\ l=1,2,\ldots,\eta.$$

Let us consider the relationship between these assumptions in the case that $\eta=\xi=s$ and the further assumption that c and b^{T} are chosen, as in the two-stage method we are considering, so that the c_i are the zeros of the degree s Legendre polynomial, shifted to the interval $[0,1]$, and b_i are the corresponding Gaussian weights. These assumptions on c and b^{T} will guarantee that $\sum_i b_i c_i^{k-1} = 1/k$ for $k=1,2,\ldots,2s$. Under this condition, $E(s,s)$ follows from $D(s)$ and because the linear combinations used to verify this have coefficients in a nonsingular (Vandermonde) matrix, the argument can be reversed. Similarly, $C(s)$ is also equivalent to $E(s,s)$.

In what has come to be referred to as a Gauss method, the b^{T} and c vectors are chosen to satisfy the requirements of Gaussian quadrature and the elements in each row of A are chosen so that $C(s)$ is satisfied. It then follows that $D(s)$ also holds. The method formed in this way always has order $2s$ and we will verify this for $s=2$ in Table 6. Where no reason is given, the result is because of Gaussian quadrature. In other cases the manipulations are based on $C(2)$ or $D(2)$ and make use of order conditions already verified earlier in Table 6. Gauss methods for arbitrary $s\geqslant 1$ were introduced in [10,17].

Methods also exist with order $2s-1$ based on Radau quadrature of type I ($c_1=0$) or type II ($c_s=1$). The most important of these are the Radau IIA methods. Some variants of Lobatto methods ($c_1=0$ and $c_s=1$) with order $2s-2$, were once considered attractive for practical computation but have been superseded by other implicit methods.

It is now generally believed that the proper role of implicit Runge–Kutta methods is in the solution of stiff problems (see Section 6). There is a conflict between the three aims of high accuracy, good stability, and low implementation cost. Gauss methods seem to be ideal from the stability and accuracy points of views but they are very expensive, because of the fully implicit structure of the coefficient matrix. The accuracy is not as good as might be expected from order considerations alone because of an "order reduction" phenomenon [55,29], but the cost alone is enough to make alternative methods more attractive.

4.4. DIRK and SIRK methods

An obvious alternative to fully implicit methods, is to insist that the coefficient matrix have a lower triangular structure, because in this case the stages can be evaluated sequentially and the cost of each is relatively low. It turns out to be an advantage to have the diagonal elements equal and this

Table 6
Verification of order conditions for 2 stage Gauss method

	$b_1 + b_2 = 1$
	$b_1c_1 + b_2c_2 = \frac{1}{2}$
	$b_1c_1^2 + b_2c_2^2 = \frac{1}{3}$
	$b_1(a_{11}c_1 + a_{12}c_2) + b_2c_2(a_{11}c_1 + a_{12}c_2)$ $= \frac{1}{2}(b_1c_1^2 + b_2c_2^2) = \frac{1}{6}$
	$b_1c_1^3 + b_2c_2^3 = \frac{1}{4}$
	$b_1c_1(a_{11}c_1 + a_{12}c_2) + b_2c_2(a_{11}c_1 + a_{12}c_2)$ $= \frac{1}{2}(b_1c_1^3 + b_2c_2^3) = \frac{1}{8}$
	$(b_1a_{11} + b_2a_{21})c_1^2 + (b_1a_{12} + b_2a_{22})c_2^2$ $= b_1(1 - c_1)c_1^2 + b_2(1 - c_2)c_2^2 = \frac{1}{12}$
	$\sum b_ia_{i1}(a_{11}c_1 + a_{12}c_2) + \sum b_ia_{i2}(a_{21}c_1 + a_{22}c_2)$ $= \frac{1}{2}(\sum b_ia_{i1}c_1^2 + \sum b_ia_{i2}c_2^2) = \frac{1}{24}$

additional requirement has little impact on the availability of methods of a required order with good stability. Methods of this type have been variously named "semi-implicit" [10], "semi-explicit" [51] and "diagonally implicit" or "DIRK" [2]. Although equal diagonals were originally built into the DIRK formulation, common usage today favours using this name more widely and using "SDIRK" (or "singly diagonally implicit") in the more restricted sense. Other key references concerning these methods are [3,18].

Singly implicit methods, *without* necessarily possessing the DIRK structure are those for which A has only a single eigenvalue $\sigma(A) = \{\gamma\}$ [7]. If the stage order is s, it turns out that the abscissae for the method satisfy $c_i = \gamma\xi_i$, where $\xi_1, \xi_2, \ldots, \xi_s$ are the zeros of the Laguerre polynomial L_s. The advantage of these methods is that for many large problems, the component of the computer cost devoted to linear algebra is little more than for a DIRK method. Various improvements to the design of SIRK methods have been proposed.

5. Nontraditional methods

While the traditional methods, linear multistep and Runge–Kutta, are widely used and are generally regarded as satisfactory for solving a wide variety of problems, many attempts have been made to extend the range of available methods. Some of these will be discussed in this section.

5.1. Taylor series methods

Because the Euler method is based on approximations to the Taylor expansion

$$y(x_0+h) \approx y(x_0) + hy'(x_0),$$

it is natural to ask if it is possible to take this expansion further by evaluating $y''(x_0)$, and possibly higher derivatives, by algebraic means. Algorithms for carrying out this process can be constructed using a recursive evaluation scheme. We mention two important early papers which exploit this idea, using a classical view of computer arithmetic but combined with this non-classical method of solution [4,32].

A second interesting and important contribution to Taylor series that has a further distinctive feature is the work of Moore [48]. The distinctive feature is that the work is carried out in the context of interval arithmetic. This means that it becomes possible, not only to advance the solution step-by-step in a relatively efficient manner, but it also becomes possible, owing to the standard bounds on the truncation error of a Taylor expansion, to obtain *rigorous* error bounds. Thus, in principle, it became possible to obtain intervals in which each component of the solution is *certain* to lie for any particular value of x. The difficulty is, of course, that the lengths of these intervals can grow rapidly as x increases.

5.2. Hybrid methods

These methods are similar to linear multistep methods in predictor–corrector mode, but with one essential modification: an additional predictor is introduced at an *offstep* point. This means that the final (corrector) stage has an additional derivative approximation to work from. This greater generality allows the consequences of the Dahlquist barrier to be avoided and it is actually possible to obtain convergent k-step methods with order $2k+1$ up to $k=7$. Even higher orders are available if two or more offstep points are used. The three independent discoveries of this approach were reported in [34,30,12]. Although a flurry of activity by other authors followed, these methods have never been developed to the extent that they have been implemented in general purpose software.

5.3. Cyclic composite methods

It is remarkable that even though a number of individual linear multistep methods may be unstable, it is possible to use them cyclically to obtain a method which, overall, is stable. An example of a fifth-order method given in the key paper on this subject [26] is as follows:

$$\begin{aligned} y_n = &-\tfrac{8}{11}y_{n-1} + \tfrac{19}{11}y_{n-2} \\ &+h(\tfrac{10}{33}f(x_n,y_n) + \tfrac{19}{11}f(x_{n-1},y_{n-1}) + \tfrac{8}{11}f(x_{n-2},y_{n-2}) - \tfrac{1}{33}f(x_{n-3},y_{n-3})), \end{aligned} \tag{22}$$

$$\begin{aligned} y_n = &\tfrac{449}{240}y_{n-1} + \tfrac{19}{30}y_{n-2} - \tfrac{361}{240}y_{n-3} + h(\tfrac{251}{720}f(x_n,y_n) + \tfrac{19}{30}f(x_{n-1},y_{n-1}) - \tfrac{449}{240}f(x_{n-2},y_{n-2}) \\ &- \tfrac{35}{72}f(x_{n-3},y_{n-3})), \end{aligned} \tag{23}$$

$$\begin{aligned} y_n = &-\tfrac{8}{11}y_{n-1} + \tfrac{19}{11}y_{n-2} \\ &+h(\tfrac{10}{33}f(x_n,y_n) + \tfrac{19}{11}f(x_{n-1},y_{n-1}) + \tfrac{8}{11}f(x_{n-2},y_{n-2}) - \tfrac{1}{33}f(x_{n-3},y_{n-3})). \end{aligned} \tag{24}$$

The method is used cyclically in the sense that in each of three steps, the first uses (22), the second uses (23) and the third uses (24) (which in this particular example happens to be the same as the first member of the cycle). Methods of this family have been studied by a number of authors and have also been discovered for use with stiff problems.

5.4. Rosenbrock methods

Rosenbrock in his 1963 paper [56], discusses the problem of evaluating the stages of a diagonally implicit Runge–Kutta methods. Normally this is carried out by an iteration process based on Newton's method. For each stage and each iteration, an evaluation of f is carried out together with a solution of a linear equation system with matrix of coefficients of the form $I - h\lambda J$, where J denotes an approximation to the Jacobian matrix. The question was then asked if improved performance can be obtained by an alternative procedure in which exactly the same amount of work is performed but *only once* per stage, with the proviso that J is exactly the Jacobian evaluated at y_{n-1}. Amongst the examples of this type of "Rosenbrock method" given in the original paper, the following is identified as having order 2 and possessing L-stability:

$$F_1 = (I - h(1 - \tfrac{\sqrt{2}}{2})J)^{-1} f(y_{n-1}),$$
$$F_2 = (I - h(1 - \tfrac{\sqrt{2}}{2})J)^{-1} f(y_{n-1} + h(\tfrac{1}{2}\sqrt{2} - \tfrac{1}{2})F_1),$$
$$y_n = y_{n-1} + hF_2.$$

Amongst the many further contributions to the study of Rosenbrock methods, and their generalizations, we refer to [16,52,43].

6. Methods for stiff problems

The paper by Curtiss and Hirschfelder [20] is usually acknowledged as introducing numerical analysts to the phenomenon of stiffness. Much has been written about what "stiffness" really means but the property is generally understood in terms of what goes wrong when numerical methods not designed for such problems are used to try to solve them. For example, classical explicit Runge–Kutta methods were not intended to solve stiff problems but, when one attempts to use them, there is a severe restriction on stepsize that must be imposed, apparently because of stability rather than accuracy requirements. It is easy to see how this can come about for linear problems of the form

$$y'(x) = My(x),$$

if the matrix M happens to have all its eigenvalues close to zero or else in the left half complex plane and with a large magnitude. Assuming for simplicity that M can be diagonalized and that the problem is solved in its transformed form, the accuracy is determined by the ability of the numerical method to solve problems of the form

$$y'(x) = \mu y(x),$$

where $|\mu|$ is small. However, the stability of the numerical approximation is limited by the fact that we are simultaneously trying to solve a problem of the form

$$y'(x) = \lambda y(x),$$

where $|\lambda|$ is large. In the exact solution, terms of the second type correspond to rapidly decaying transients, whereas in the computed solution they represent unstable parasitic solutions, unless h is so small that $h\lambda$ lies in what is known as the "stability region".

To find the stability region for a numerical method it is necessary to consider the behaviour of the numerical method with a problem of just this type. For classical methods the behaviour depends on the product of h and λ which we will write as z. For the classical fourth-order Runge–Kutta method, the numerical solution for this problem satisfies

$$y_n = R(z)y_{n-1}, \tag{25}$$

where

$$R(z) = 1 + z + \frac{z^2}{2} + \frac{z^3}{6} + \frac{z^4}{24}$$

and the stability region is the set of points in the complex plane for which $|R(z)| \leqslant 1$.

If $z = h\lambda$ is outside this set, as it might well be, then selecting h to make $h\mu$ of a reasonable size will not be satisfactory because the unstable behaviour of the component of the solution associated with λ will swamp the solution and destroy its accuracy.

To analyse this type of possible instability, Dahlquist [23] introduced the concept of A-stability. A numerical method is said to be "A-stable" if its stability region includes all of the left half-plane. Even though the definition was first framed in the context of linear multistep methods, it was soon applied to Runge–Kutta methods, for which it takes a particularly simple form. Write Y for the vector of stage values, then this vector and the output approximation are related by

$$Y = y_{n-1}\mathrm{e} + zAY, \quad y_n = y_{n-1} + zb^{\mathrm{T}}Y,$$

leading to (25) with the stability function given by

$$R(z) = 1 + zb^{\mathrm{T}}(I - zA)^{-1}\mathrm{e}.$$

For explicit methods the stability region is always a bounded set and these methods cannot be A-stable. On the other hand for an implicit method, $R(z)$ has the form $N(z)/D(z)$ where the polynomials N and D can have degrees as high as s. Methods of arbitrarily high orders can be A-stable.

For a general linear multistep method, defined by polynomials ρ and σ, the stability region is the set of points z for which the polynomial $\rho(w) - z\sigma(w)$, of degree k in w satisfies the root condition. It was shown by Dahlquist [23] that for these methods, A-stability cannot exist for orders greater than 2.

6.1. Order stars

Runge–Kutta methods of Gauss type have stability functions of the form

$$R(z) = \frac{N(z)}{N(-z)},$$

where the polynomial N has degree s. Furthermore, $R(z)=\exp(z)+O(z^{2s+1})$. This is an example of a "Padé approximation" to the exponential function, in the sense that the order of approximation is exactly the sum of the numerator and denominator degrees. Not only are the diagonal members of the Padé table significant, but the same can be said of the first subdiagonal (with degree $s-1$ for the numerator and s for the denominator, because these correspond to the stability functions for the Radau IA methods, and for the practically important Radau IIA methods. The second subdiagonals are also important because they are the stability functions for the Lobatto IIIC methods. It is known that the Padé approximations to the exponential function, in each of these three diagonals, correspond to A-stable methods. It is also clear that the approximations above the main diagonal cannot share this property but what can be said about approximations *below* the second subdiagonal? Considerable evidence existed for the "Ehle conjecture" [27] which claimed that *none* of these stability functions can correspond to an A-stable method or, in Ehle's terminology, that they are not A-acceptable.

In 1978 a new method was discovered for settling this, and many related questions. This approach introduced "order stars" [60], based on relative stability regions. Rather than study the regions of the complex plane for which $|R(z)|<1$, the regions are studied for which $|\exp(-z)R(z)|<1$. Since A-stable methods are those for which the stability function has no poles in the left half-plane and has its magnitude bounded by 1 on the imaginary axis, changing from the stability function $R(z)$ to the relative stability function $\exp(-z)R(z)$ leaves this criterion unchanged, but introduces much more structure, because $\exp(-z)R(z)=1+Cz^{p+1}+O(z^{p+2})$, when z is small.

Consider Fig. 4 taken from [60]. Shown in this figure are the order stars of four Padé approximation, with degrees j (denominator) and k (numerator). The shaded parts of the figures, known as the fingers and characterized by $|\exp(-z)R(z)|>1$ and the unshaded parts, the dual fingers characterized by $|\exp(-z)R(z)|<1$ meet at 0 in a pattern determined by the sign of the real part of Cz^{p+1}, for $|z|$ small. This means that there will be exactly $p+1$ fingers and the same number of dual fingers meeting at zero. Furthermore, the angles subtended by each finger is the same $\pi/(p+1)$. It can be shown that all the bounded fingers contain poles and the bounded dual fingers contain zeros. The two upper figures are for A-stable methods, in which all the poles are in the right half-plane and no finger crosses the imaginary axis. The two lower figures, for which $j-k>2$, cannot meet these requirements, because there are too many bounded fingers for it to be possible for them all to leave zero in the right-hand direction. Some of these fingers must leave zero in the left-hand direction and either reach poles in the left half-plane or cross the imaginary axis to reach poles in the right-hand half-plane. A rigorous form of this argument is used to prove the Ehle conjecture and to prove a number of other results concerning both one step and multistep methods.

A recent study of order stars, which reviews most of the work up to the present time, is available in the book by Iserles and Nørsett [42].

6.2. Nonlinear stability

To obtain a deeper understanding of the behaviour of stiff problems, and of the numerical methods used to solve them, Dahlquist in 1975 [24], studied nonlinear problems of the form

$$y'(x)=f(x,y(x)), \tag{26}$$

where f satisfies the dissipativity condition

$$\langle f(x,u)-f(x,v),u-v\rangle \leqslant 0 \tag{27}$$

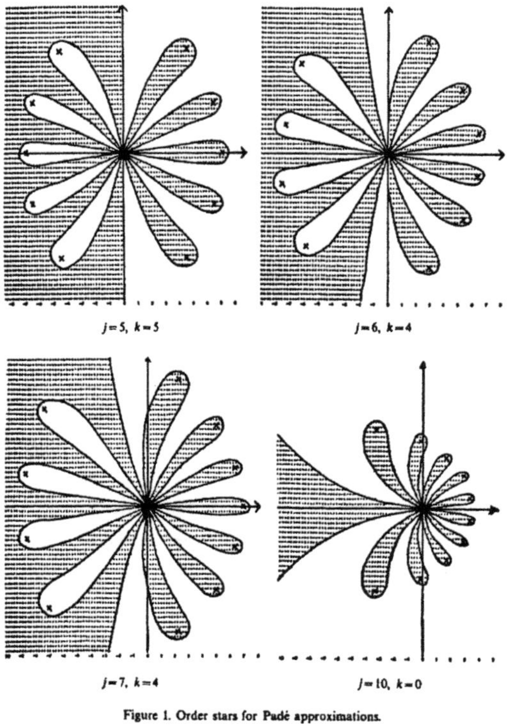

Fig. 4. An extract from the order star paper.

and $\langle\cdot\rangle$ denotes an inner product, with $\|\cdot\|$ the corresponding norm. It is easy to see that two exact solutions to (26) possess the property that

$$\|y(x)-z(x)\| \leqslant \|y(x_0)-z(x_0)\|, \quad \text{for } x \geqslant x_0. \tag{28}$$

The aim is now to find conditions on a method such that a discrete analogue of (28) holds. It turns out to be more convenient to consider instead of a linear multistep method

$$\begin{aligned}&\alpha_k y_n + \alpha_{k-1}y_{n-1} + \alpha_{k-2}y_{n-2} + \cdots + \alpha_0 y_{n-k}\\ &\quad = h(\beta_k f(x_n, y_n) + \beta_{k-1} f(x_{n-1}, y_{n-1}) + \cdots + \beta_0 f(x_{n-k}, y_{n-k})),\end{aligned}$$

the corresponding "one-leg method"

$$\begin{aligned}&\alpha_k y_n + \alpha_{k-1}y_{n-1} + \alpha_{k-2}y_{n-2} + \cdots + \alpha_0 y_{n-k}\\ &\quad = h\left(\sum_{i=0}^{k}\beta_i\right) f\left(\frac{\beta_k}{\sum_{i=0}^k \beta_i} y_n + \frac{\beta_{k-1}}{\sum_{i=0}^k \beta_i} y_{n-1} + \cdots + \frac{\beta_0}{\sum_{i=0}^k \beta_i} y_{n-k}\right).\end{aligned}$$

For this type of method, Dahlquist considered contractivity in the sense that

$$\|Y_n\| \leqslant \|Y_{n-1}\|,$$

where

$$Y_n = \begin{bmatrix} y_n \\ y_{n-1} \\ \vdots \\ y_{n-k+1} \end{bmatrix}$$

and

$$\left\| \begin{bmatrix} \eta_1 \\ \eta_2 \\ \vdots \\ \eta_k \end{bmatrix} \right\| = \sum_{i,j=1}^{k} g_{ij} \langle \eta_i, \eta_j \rangle.$$

It is assumed that

$$G = \begin{bmatrix} g_{11} & g_{12} & \cdots & g_{1k} \\ g_{21} & g_{22} & \cdots & g_{2k} \\ \vdots & \vdots & & \vdots \\ g_{k1} & g_{k2} & \cdots & g_{kk} \end{bmatrix}$$

is a positive-definite matrix.

It is explained in Dahlquist's paper how results for one-leg methods can be interpreted as having a significance also for the corresponding linear multistep methods. He also found necessary and sufficient conditions for this property to hold. In a later paper [25], he showed that for one-leg methods, A-stability and G-stability are essentially equivalent properties.

The corresponding theory for Runge–Kutta methods [14,19,8], leads to a consideration of a matrix M with (i,j) element equal to $b_i a_{ij} + b_j a_{ji} - b_i b_j$. Assuming that this matrix is positive semi-definite, and the same is true for $\operatorname{diag}(b_1, b_2, \ldots, b_s)$, then a Runge–Kutta method applied to two distinct solutions of (26), satisfying (27), satisfies the contractivity property

$$\|y_n - z_n\| \leqslant \|y_{n-1} - z_{n-1}\|.$$

It is interesting that M has a more modern role in connection with symplectic methods for Hamiltonian problems.

A further development, initiated in the paper [29], is connected with the behaviour of truncation error for Runge–Kutta methods applied to stiff problems.

7. The beginnings of differential equation software

Programs to solve differential equations are as old as modern computers themselves. Today, a central aim in the design of differential equation software is the building of general purpose codes, specific only as regards stiffness versus nonstiffness, which adapt their behaviour to that of the computed solution dynamically. Variable stepsize is a characteristic feature of this software and usually variable order is used as well.

The most famous of the early codes in this tradition is the FORTRAN subroutine named by its designer, Gear, as "DIFSUB" [31]. Actually, this name was used generally by Gear for a range of possible subroutines using a range of different methods. We will concentrate for the moment on the linear multistep version of DIFSUB.

As with all linear multistep implementations, the characteristic problems of starting values, local error estimation, change of stepsize and output interpolation have to be solved. A basic principle used in DIFSUB are the use of the Nordsieck representation of the data passed between steps, and this plays a crucial role in the solution of all these implementation questions, as well as the further problem of variable order.

The single paper of Nordsieck [50] explains how it is possible to rewrite a k-step Adams methods so that information on the values of $hf(x_{n-i}, y_{n-i})$ for $i=1,2,\dots,k$ is organized as approximations to $hy'(x_{n-1}), \frac{1}{2!}h^2y''(x_{n-1}), \dots, \frac{1}{k!}h^ky^{(k)}(x_{n-1})$. The rules for integration to the next step, are particularly simple in the case of the Adams–Bashforth method. The solution is first extrapolated using the approximation

$$\begin{bmatrix} y(x_n) \\ hy'(x_n) \\ \frac{1}{2!}h^2y''(x_n) \\ \vdots \\ \frac{1}{k!}h^ky^{(k)}(x_n) \end{bmatrix} \approx \begin{bmatrix} 1 & 1 & 1 & 1 & \cdots & 1 \\ 0 & 1 & 2 & 3 & \cdots & k \\ 0 & 0 & 1 & 3 & \cdots & \binom{k}{2} \\ \vdots & \vdots & \vdots & \vdots & & \vdots \\ 0 & 0 & 0 & 0 & \cdots & 1 \end{bmatrix} \begin{bmatrix} y(x_{n-1}) \\ hy'(x_{n-1}) \\ \frac{1}{2!}h^2y''(x_{n-1}) \\ \vdots \\ \frac{1}{k!}h^ky^{(k)}(x_{n-1}) \end{bmatrix}$$

and a correction is then made to each component using a multiple of $hf(x_n, y_n) - hy'(x_n)$, so as to ensure that the method is equivalent to the Adams–Bashforth method. Adding an Adams–Moulton corrector to the scheme, is equivalent to adding further corrections.

Using the Nordsieck representation, it is possible to change stepsize cheaply, by simply rescaling the vector of derivative approximations. It is possible to estimate local truncation error using the appropriately transformed variant of the Milne device. It is also possible to measure the accuracy of lower and one higher-order alternative methods so that the appropriateness of order-changing can be assessed. Thus the ingredients are present to build a completely adaptive nonstiff solver. By adapting the backward difference methods to a similar form, it is possible to allow also for stiffness.

The DIFSUB program of Gear uses these techniques to obtain an efficient solver and many later programs are based on similar ideas. The first general purpose solver for differential–algebraic equations, the DASSL subroutine of Petzold [54], is also closely related to DIFSUB.

Early success was also achieved in the algorithm of Bulirsch and Stoer [6]. This used extrapolation in a similar manner to the quadrature algorithm of Romberg. The main difference between differential equations and quadrature is that significant efficiency gains are made by reusing some of the abscissae in a quadrature formula; this happens in the traditional version of the Romberg method because the mesh size is halved in each iteration. For differential equations there is no advantage in this, because reuse is not possible. Hence, in later developments of extrapolation methods, for both nonstiff and stiff problems, various sequences of stepsizes have been considered, where the aim is to balance computational cost against the quality of the convergence.

As many programs became available, using a variety of methods and variety of implementations of the same basic method, it became appropriate to consider what is really expected of these automatic solvers. Each person who develops this software needs to apply quality tests and to compare any new implementation against existing codes. In the interests of providing objective standards, a number of test sets have been developed. The earliest of these that has become widely adopted, and which in fact serves as a de facto standard, is the DETEST problem set [41]. While this is appropriate for testing and comparing nonstiff solvers, a stiff counterpart, known as STIFF DETEST, [28] became available a little later.

8. Special problems

While the general aim of providing accurate and efficient general purpose numerical methods and algorithms has been a central activity in the 20th century, there has always been a realization that some problem types have such distinctive features that they will need their own special theory and techniques. Stiff problems were recognized approximately half way through the century as such a problem type and these have received considerable attention, especially in the last 30 years.

Another of the special problem types that has a separate claim for its own special methods, has been second-order differential equations and systems. These have a natural importance as arising in classical mechanical modelling and they were treated as a particular case by Nyström and others. While any problem of this type can be rewritten as a first-order system, it is found that treating them directly can lead to substantial gains in efficiency, especially if the second-order system takes the special form

$$y''(x) = f(x, y(x)),$$

where we note that $y'(x)$ does not occur as an argument of f. The Runge–Kutta approach to this type of problem was studied by Nyström [53] and has been of interest ever since. A modern theory of these methods is given in [37]. Linear multistep methods for this problem were studied as part of an investigation of a more general situation

$$y^{(n)}(x) = f(x, y(x)),$$

by Dahlquist [22].

It is interesting that one of the most active areas of modern research is closely related to this long-standing problem. Mechanical problems that can be expressed in a Hamiltonian formulation, rather than as a second-order system, can be studied in terms of the preservation of qualitative properties. It is found that the symplectic property can be preserved by the use of specially designed Runge–Kutta methods. The burgeoning subject of geometric integration, started from the study of Hamiltonian systems by Feng Kang, J.M. Sanz-Serna and others, and is now a central activity as the century closes. Although it is too early to view geometric integration from a historical perspective, it is at least possible to refer to a recent review of this subject [58].

There are several other evolutionary problems that can be solved by methods closely related to ordinary differential equation methods. Delay differential equations, and other types of functional differential equations can be solved using a combination of a differential equation solver, an interpolator and a means of handling discontinuities.

We have already noted that algebraic differential equations, especially those of low index, can be effectively solved using linear multistep methods. Implicit Runge–Kutta methods also have an important role in the numerical treatment of differential–algebraic equations. The theory of order of these methods can be extended to allow for the inclusion of algebraic constraints in the formulation, using generalizations of rooted trees [35].

References

[1] P. Albrecht, A new theoretical approach to Runge–Kutta methods, SIAM J. Numer. Anal. 24 (1987) 391–406.

[2] R. Alexander, Diagonally implicit Runge–Kutta methods for stiff ODEs, SIAM J. Numer. Anal. 14 (1977) 1006–1021.

[3] R. Alt, Deux théorèms sur la A-stabilité des schémas de Runge–Kutta simplement implicites, Rev. Francais d'Automat. Recherche Opérationelle Sér. R-3 6 (1972) 99–104.

[4] D. Barton, I.M. Willers, R.V.M. Zahar, The automatic solution of ordinary differential equations by the method of Taylor series, Comput. J. 14 (1971) 243–248.

[5] F. Bashforth, J.C. Adams, An attempt to test the theories of capillary action by comparing the theoretical and measured forms of drops of fluid, with an explanation of the method of integration employed in constructing the tables which give the theoretical forms of such drops, Cambridge University Press, Cambridge, 1883.

[6] R. Bulirsch, J. Stoer, Numerical treatment of ordinary differential equations by extrapolation methods, Numer. Math. 8 (1966) 1–13.

[7] K. Burrage, A special family of Runge–Kutta methods for solving stiff differential equations, BIT 18 (1978) 22–41.

[8] K. Burrage, J.C. Butcher, Stability criteria for implicit Runge–Kutta methods, SIAM J. Numer. Anal. 16 (1979) 46–57.

[9] J.C. Butcher, Coefficients for the study of Runge–Kutta integration processes, J. Austral. Math. Soc. 3 (1963) 185–201.

[10] J.C. Butcher, Implicit Runge–Kutta processes, Math. Comp. 18 (1964) 50–64.

[11] J.C. Butcher, On the attainable order of Runge–Kutta methods, Math. Comp. 19 (1965) 408–417.

[12] J.C. Butcher, A modified multistep method for the numerical integration of ordinary differential equations, J. Assoc. Comput. Mach. 12 (1965) 124–135.

[13] J.C. Butcher, An algebraic theory of integration methods, Math. Comp. 26 (1972) 79–106.

[14] J.C. Butcher, The non-existence of ten stage eighth order explicit Runge–Kutta methods, BIT 25 (1985) 521–540.

[15] J.C. Butcher, A stability property of implicit Runge–Kutta methods, BIT 15 (1975) 358–361.

[16] D.A. Calahan, A stable, accurate method of numerical integration for nonlinear systems, Proc. IEEE 56 (1968) 744.

[17] F. Ceschino, J. Kuntzmann, Problèmes Différentiels de Conditions Initiales, Dunod, Paris, 1963.

[18] M. Crouzeix, Sur les méthodes de Runge–Kutta pour l'approximation des problèmes d'évolution, Lecture Notes in Economics and Mathematical Systems, Vol. 134, Springer, Berlin, 1974, pp. 206–223.

[19] M. Crouzeix, Sur la B-stabilité des méthodes de Runge–Kutta, Numer. Math. 32 (1979) 75–82.

[20] C.F. Curtiss, J.O. Hirschfelder, Integration of stiff equations, Proc. Nat. Acad. Sci. 38 (1952) 235–243.

[21] G. Dahlquist, Convergence and stability in the numerical integration of ordinary differential equations, Math. Scand. 4 (1956) 33–53.

[22] G. Dahlquist, Stability and error bounds in the numerical integration of ordinary differential equations, Kungl. Tekn. Högsk. Handl. Stockholm 130 (1959) 1–87.

[23] G. Dahlquist, A special stability property for linear multistep methods, BIT 3 (1963) 27–43.

[24] G. Dahlquist, Error Analysis for a Class of Methods for Stiff Non-Linear Initial Value Problems, Lecture Notes in Mathematics, Vol. 506, Springer, Berlin, 1976, pp. 60–72.

[25] G. Dahlquist, G-stability is equivalent to A-stability, BIT 18 (1978) 384–401.

[26] J. Donelson, E. Hansen, Cyclic composite multistep predictor-corrector methods, SIAM J. Numer. Anal. 8 (1971) 137–157.

[27] B.L. Ehle, A-stable methods and Padé approximations to the exponential, SIAM J. Numer. Anal. 4 (1973) 671–680.

[28] W.H. Enright, T.E. Hull, B. Lindberg, Comparing numerical methods for stiff systems of ODEs, BIT 15 (1975) 10–48.

[29] R. Frank, J. Schneid, C.W. Ueberhuber, The concept of B-convergence, SIAM J. Numer. Anal. 18 (1981) 753–780.

[30] C.W. Gear, Hybrid methods for initial value problems in ordinary differential equations, SIAM J. Numer. Anal. 2 (1965) 69–86.

[31] C.W. Gear, Algorithm 407, DIFSUB for solution of ordinary differential equations, Comm. ACM 14 (1971) 447–451.

[32] A. Gibbons, A program for the automatic integration of differential equations using the method of Taylor series, Comput. J. 3 (1960) 108–111.

[33] S. Gill, A process for the step-by-step integration of differential equations in an automatic digital computing machine, Proc. Cambridge Philos. Soc. 47 (1951) 96–108.
[34] W.W. Gragg, H.J. Stetter, Generalized multistep predictor-corrector methods, J. Assoc. Comput. Mach. 11 (1964) 188–209.
[35] E. Hairer, C. Lubich, M. Roche, in: The Numerical Solution of Differential–Algebraic Systems by Runge–Kutta Methods, Lecture Notes in Mathematics, Vol. 1409, Springer, Berlin, 1989.
[36] E. Hairer, G. Wanner, On the Butcher group and general multi-value methods, Computing 13 (1974) 1–15.
[37] E. Hairer, G. Wanner, A theory for Nyström methods, Numer. Math. 25 (1976) 383–400.
[38] P.C. Hammer, J.W. Hollingsworth, Trapezoidal methods of approximating solutions of differential equations, MTAC 9 (1955) 269–272.
[39] P. Henrici, Discrete Variable Methods in Ordinary Differential Equations, Wiley, New York, 1962.
[40] K. Heun, Neue Methoden zur approximativen Integration der Differentialgleichungen einer unabhängigen Veränderlichen, Z. Math. Phys. 45 (1900) 23–38.
[41] T.E. Hull, W.H. Enright, B.M. Fellen, A.E. Sedgwick, Comparing numerical methods for ordinary differential equations, SIAM J. Numer. Anal. 9 (1972) 603–637.
[42] A. Iserles, S.P. Nørsett, Order Stars, Chapman & Hall, London, 1991.
[43] P. Kaps, G. Wanner, A study of Rosenbrock type methods of high order, Numer. Math. 38 (1981) 279–298.
[44] K.S. Kunz, Numerical solution of ordinary differential equations: methods of starting the solution, in: Numerical Analysis, McGraw-Hill, New York, 1957.
[45] W. Kutta, Beitrag zur näherungsweisen Integration totaler Differentialgleichungen, Z. Math. Phys. 46 (1901) 435–453.
[46] R.H. Merson, An operational method for the study of integration processes, Proceedings of the Symposium on Data Processing, Weapons Research Establishment, Salisbury, South Australia, 1957.
[47] W.E. Milne, A note on the numerical integration of differential equations, J. Res. Nat. Bur. Standards 43 (1949) 537–542.
[48] R.E. Moore, The automatic anlyasis and control of error in digital computation based on the use of interval numbers, Error in Digital Computation, Vol. 1, Wiley, New York, 1964, pp. 61–130.
[49] F.R. Moulton, New Methods in Exterior Balistics, University of Chicago, Chicago, 1926.
[50] A. Nordsieck, On numerical integration of ordinary differential equations, Math. Comp. 16 (1962) 22–49.
[51] S.P. Nørsett, One-step methods of Hermite type for numerical integration of stiff systems, BIT 14 (1974) 63–77.
[52] S.P. Nørsett, A. Wolfbrandt, Order conditions for Rosenbrock type methods, Numer. Math. 32 (1979) 1–15.
[53] E.J. Nyström, Über die numerische Integration von Differentialgleichungen, Acta Soc. Sci. Fennicae 50 (13) (1925).
[54] L.R. Petzold, A description of DASSL: a differential/algebraic system solver, in: Scientific Computing, Montreal, Quebec, North-Holland, Amsterdam, 1983, 65–68.
[55] A. Prothero, A. Robinson, On the stability and accuracy of one-step methods for solving stiff systems of ordinary differential equations, Math Comp. 28 (1974) 145–162.
[56] H.H. Rosenbrock, Some general implicit processes for the numerical solution of differential equations, Comput. J. 5 (1963) 329–330.
[57] C. Runge, Über die numerische Auflösung von Differentialgleichungen, Math. Ann. 46 (1895) 167–178.
[58] J.M. Sanz-Serna, Geometric integration, The State of the Art in Numerical Analysis, Oxford University Press, Oxford, 1997, pp. 121–143.
[59] L.F. Shampine, H.A. Watts, Block implicit one-step methods, Math. Comp. 23 (1969) 731–740.
[60] G. Wanner, E. Hairer, S.P. Nørsett, Order stars and stability theorems, BIT 18 (1978) 475–489.

ELSEVIER

Journal of Computational and Applied Mathematics 125 (2000) 31–40

JOURNAL OF COMPUTATIONAL AND APPLIED MATHEMATICS

www.elsevier.nl/locate/cam

Initial value problems for ODEs in problem solving environments

L.F. Shampine[a,*], Robert M. Corless[b]

[a] *Mathematics Department, Southern Methodist University, Dallas, TX 75275, USA*

[b] *Department of Applied Mathematics, University of Western Ontario, London, Ontario, N6A 5B7, Canada*

Received 21 June 1999; received in revised form 28 February 2000

Abstract

A program is presented for solving initial value problems for ODEs numerically in Maple. We draw upon our experience with a number of closely related solvers to illustrate the differences between solving such problems in general scientific computation and in the problem solving environments Maple and MATLAB. © 2000 Elsevier Science B.V. All rights reserved.

Keywords: Initial value problems; ODEs; Problem solving environment; PSE

1. Introduction

The problem solving environments (PSEs) Maple [8] and MATLAB [7] are in very wide use. Although they have much in common, they are clearly distinguished by the emphasis in Maple on algebraic computation and in MATLAB on numerical computation. We discuss here a program, `IVPsolve`, for solving numerically initial value problems (IVPs) for systems of first-order ordinary differential equations (ODEs), $y' = f(x, y)$, in Maple. We draw upon our experience with a number of closely related solvers to illustrate the differences between solving IVPs in general scientific computation (GSC) and in these PSEs. The RKF45 code of Shampine and Watts [10,11] is based on the explicit Runge–Kutta formulas F(4, 5) of Fehlberg. It has been widely used in GSC. Translations of this code have been the default solvers in both Maple and MATLAB. Neither takes much advantage of the PSE. In developing the MATLAB ODE Suite of solvers for IVPs, Shampine and Reichelt [9]

* Corresponding author.

E-mail address: lshampin@mail.smu.edu (L.F. Shampine).

PII: S0377-0427(00)00456-8

exploit fully the PSE, as well as algorithmic advances. `IVPsolve` is the result of a similar investigation for Maple, though on a much smaller scale. It also uses the F(4,5) formulas for non-stiff problems.

In addition to general issues of solving IVPs in PSEs, we discuss specifics for the Maple PSE. Because the user is allowed to specify the precision, the floating point arithmetic of Maple is implemented in software. As the PSE has evolved, it has added facilities that allow users to work directly with the underlying hardware floating point arithmetic. `IVPsolve` exploits these facilities to solve IVPs faster. Using a continuous extension of the F(4,5) pair and a new design, `IVPsolve` handles output more efficiently and avoids numerical difficulties of the kind pointed out in [2]. The solvers of Maple look different to users and solve different computational problems. In contrast, it is possible to use all the solvers of the MATLAB ODE Suite in *exactly* the same way. `IVPsolve` achieves this in Maple. Methods for the solution of stiff IVPs require (approximations to) Jacobians. To make it possible to use all the solvers of the ODE Suite in the same way, Shampine and Reichelt approximate Jacobians numerically in the codes for stiff IVPs. This is accomplished in `IVPsolve` by using the tools of Maple to evaluate partial derivatives analytically. `IVPsolve` uses a Rosenbrock method for stiff IVPs, an excellent method for the PSE that is not widely used in GSC because it requires analytical partial derivatives.

Conventions for this paper. Because this paper discusses implementations of similarly named solvers in different contexts (GSC, the PSE MATLAB and the PSE Maple), the following notational conventions are used to distinguish the solvers. For GSC, upper-case names such as RKF45, LSODE, and ODE/STEP, INTRP are used. For the MATLAB ODE Suite, we use the exact names `ode45`, `ode15s`, and `ode113`. For the built-in Maple routines of `dsolve[numeric]`, we use their exact names, `rkf45` and `lsode`, which are lower-case versions of the names of the corresponding GSC codes. The new routines that we have written are packaged together under the name `NODES`, but are referred to in this paper by their names `IVPsolve`, `IVPval`, and so on.

2. IVPsolve

A number of issues distinguish solving IVPs in PSEs from solving them in GSC. Obvious ones are interactive computation and graphical output. Less obvious but no less important are advanced language features and interpreted computation. The emphasis on graphical output in PSEs means that IVPs are solved to accuracies modest by the standards of GSC. This makes the convenience of default tolerances quite practical, a distinguishing characteristic of solvers in PSEs. It also means that fewer steps are taken. In GSC the $f(x,y)$ are often so complicated that the cost of evaluating this function dominates all the remaining cost of the computation, the overhead. This is much less true of the problems solved in PSEs. Because of interactive computation, fewer steps, and relatively simple f, overhead is much more important in PSEs. These factors influence greatly the choice of method, a matter we now examine in more detail.

2.1. Choice of non-stiff method

An explicit Runge–Kutta pair of orders 4 and 5 is very attractive for solving non-stiff IVPs in a PSE because the order is appropriate for typical accuracies and the method has a low overhead.

The original ode45 of MATLAB is a translation of RKF45 with its F(4, 5) pair, but the solver of the same name in the ODE Suite [9] that replaced it at version 5 is different in important ways. In particular, it is based on a different explicit Runge–Kutta (4,5) pair due to Dormand. The default solver of Maple, rkf45, is a translation of RKF45. All these codes do local extrapolation so that the integration is advanced at order 5.

Selecting a formula is a complicated matter that involves both theoretical considerations and experimentation. There have been many investigations that attempt to identify the "best" Runge–Kutta pair of moderate order, but most have been concerned with GSC. Some aspects of the formulas are much more important in the present context than in GSC. We have already mentioned the importance of overhead in a PSE. To emphasize the point, we note that the ODE Suite includes a translation, ode113, of the well-known Adams solver ODE/STEP,INTRP. It varies both order and step size so as to minimize the number of evaluations of $f(x, y)$. This is very efficient in GSC, but the overhead is quite high, making it much less attractive in a PSE. That is why the MATLAB documentation recommends that ode45 be tried before ode113. Furthermore, in a code with the convenient user interface and powerful capabilities of ode45, much of the overhead and a sometimes considerable portion of the computations are not directly associated with evaluation of the Runge–Kutta pair. Generally the number of stages is not important in GSC, so it is scaled out in comparisons. It is important in PSEs because generally there is a relatively small number of steps and the tolerances are relatively large so that failed steps are relatively common. Shortly, we take up output and discuss the use of a continuous extension of the Runge–Kutta pair for output. We shall see that both ode45 and IVPsolve evaluate their continuous extensions several times in the course of each step. Obviously, the cost of the continuous extension, both in terms of evaluations of $f(x, y)$ and overhead, is quite important in this context. We implemented several (4, 5) pairs and even a (3, 4) pair with attractive features and compared them experimentally. The decision was not easy, but we chose to use the F(4, 5) pair with local extrapolation and the continuous extension of [3] in IVPsolve.

2.2. Choice of stiff method

By far, the most widely used method for solving stiff IVPs is a variable order implementation of the backward differentiation formulas (BDFs). Included in dsolve[numeric] is a translation, lsode, of the well-known BDF solver LSODE. The MATLAB documentation recommends that a code of this kind, ode15s, be tried first for stiff IVPs, this despite the high overhead due to variation of order and step size. The sparsity structure of the Jacobian is crucial in solving large systems of stiff equations, so an important feature of ode15s is its capabilities for dealing conveniently with this. Analytical partial derivatives improve the robustness of codes for solving stiff IVPs, but they have been avoided in GSC because it may be inconvenient for users to supply them and $f(x, y)$ may be only piecewise smooth. IVPsolve is limited to the stiff problems for which Maple can form analytical partial derivatives. Because Maple does not yet have functions for sparse linear algebra in hardware floating point, IVPsolve is also limited to relatively small systems for which Jacobians are treated as full matrices. We use the Rosenbrock (3, 4) pair implemented as METH = 1 in the RODAS code [5]. This pair was constructed for local extrapolation, so the integration is advanced at order 4. Although the order is lower than that of the method for non-stiff problems, it is adequate for the PSE. The formulas are stiffly accurate, A-stable, and have small error constants. The overhead is low compared to a BDF code because the order is fixed and the formulas are linearly implicit.

2.3. Use of HF arithmetic

Speed is not one of the primary goals in a PSE. Notwithstanding this, the computation is interactive, so the faster it can be done, the better. In developing `IVPsolve` we aimed to accelerate the solution of IVPs in Maple by exploiting hardware floating point arithmetic (HF). This is an issue particular to Maple with its emphasis on algebraic computation because its floating point arithmetic is implemented in software (SF). The user can control the precision of computations by means of the environment variable `Digits`. Naturally SF is slower, but how much slower depends strongly on how the precision specified relates to HF. Invoking functions inside a call to `evalhf` causes them to be carried out in HF. We make heavy use of this function and in particular, use it to speed up the evaluation of the ODEs in the procedure supplied by the user. Unfortunately, some of the special functions cannot yet be evaluated in HF. This causes a considerable complication to the solver: it must begin by testing whether the ODEs can be evaluated in HF and if they cannot, it must use SF for all the computations. For this case, we set `Digits := trunc(evalhf(Digits))` so that these computations are carried out in approximately the same precision as is available in HF.

The `hfarray` data structure was added to Maple to hold an array of numbers in HF. Converting between SF and HF representations of numbers is costly, so it is important to avoid this. At present some of the fast builtin functions cannot be applied to `hfarray`s, forcing either a conversion or a slower way of processing the data. Because we do not know in advance the number of steps required for an integration, we might have to adjust the sizes of the arrays that hold the output. `hfarrays` have fixed dimension and cannot be extended inside a call to `evalhf`, so we have to return from the function that advances the integration to the main program in order to create a larger array for the output, copy the current output array into the new one, and then return to the function through `evalhf` to continue the integration with this larger array. Because Maple does not yet provide for solving linear systems stored as `hfarray`s, we translated some FORTRAN programs of C.B. Moler for this that we use inside a call to `evalhf` to obtain the speed of HF.

2.4. Output considerations

RKF45 provides its output in the course of the integration, either at specific points or at each internal step, by returning control to the user along with the computed solution. Exploiting the dynamic storage allocation of MATLAB, the translation of RKF45 into this PSE returns control to the user only when the integration is complete (or fails). It returns two arrays, one containing the mesh points and the other, corresponding solution values. This form of the output is perfectly suited to the plot routines of MATLAB. Because of the emphasis on graphical output, the translation did not follow RKF45 in allowing output at specific points, a deficiency remedied in the ODE Suite.

RKF45 obtains output at a specific point by stepping to the point. Reducing the step size for this reason reduces the efficiency of the integration. The natural step size of the $F(4,5)$ pair is so large that solution values at each step may not result in a smooth graph. The development of continuous extensions made it possible to deal with these matters. The solvers of the MATLAB ODE Suite have two output modes. The user either specifies all the points where output is desired or accepts output at the points selected by the solver. In the first mode, the solvers use the largest step size possible and obtain output at the specified points by evaluating a continuous extension. In the second mode, the explicit Runge–Kutta solver `ode45` supplements the solution values at mesh points by evaluating its continuous extension at four points in the span of each step. It was found by experimentation that four additional solution values generally suffice for a smooth graph. If they do not, the user has to increase the number of additional solution values by means of an option and solve the problem again.

Output is handled very differently in Maple. `dsolve` makes available a number of methods for solving ODEs analytically. Because the numerical solution of IVPs is treated as an option, it is natural that the design of `rkf45` resembles as much as possible the analytical solution of an IVP. In particular, the solver returns a procedure for computing the solution at any specified value of the independent variable. Because the computational problem is not the same as the analytical problem, this design leads to anomalies. A fundamental difference is that the stability of the IVP is crucial to numerical integration. This stability depends on both the direction of the integration and the length of the interval. For the sake of efficiency, `rkf45` remembers the last approximate solution and advances the integration from it to the value of the independent variable specified in the current call. This is faithful to RKF45 except for allowing the direction to change. Changing direction can have serious consequences noted by Coombes et al. [2] that are revealed by considering what happens if we start with initial value at $x=0$, compute a solution at $x=1$, then at $x=2$, and again at $x=1$. It is obviously inefficient to recompute the solution at $x=1$. However, the more important point is that if the IVP is very stable in one direction, it is very unstable in the other, so integrating back to $x=1$ is expensive and inaccurate. The two values computed at $x=1$ are different and can be dramatically different. Some of the methods of `dsolve[numeric]` have an error control that depends on the history of the integration and so get different answers for an entirely different reason. Getting different answers at the same point is rather unsatisfactory. There is a more subtle effect of the design. If the output points chosen by the plot routines do not result in a smooth graph, the user can increase the number of points plotted. This causes `rkf45` to solve the problem again with shorter steps, resulting in a more accurate solution. It is rather unsatisfactory that simply asking for more plot points changes the computed solution. Generally this change is not visible in the plots, but we give an example in Section 3 for which the plotted solutions are qualitatively different.

`IVPsolve` requires the user to specify that the integration is to go from a to b so as to avoid dangers inherent in the design of `dsolve[numeric]`. `IVPsolve` integrates the IVP over the whole interval and returns the mesh, the solution at the mesh points, and some other information in a structure. We exploit the more complex data structures available in the PSEs to simplify the user interface. In our design all information about the solution is encapsulated as a structure. The user has no reason to examine the contents of this structure because auxiliary procedures are used to evaluate and plot the solution.

Some details about the form of the solution are required. Enright et al. [3] developed a continuous extension for the Fehlberg (4,5) pair that does not require any additional evaluations of $f(x,y)$. It furnishes a numerical solution $S(x) \in C^1[a,b]$ that is a quartic polynomial in the span of the step from x_n to x_n+h. In `IVPsolve` this polynomial is evaluated at $x_n+jh/4$ for $j=1,2,3$. The numerical solution is uniformly accurate, so these approximations at intermediate points are given equal treatment in the output array. A great deal of experience with the F(4,5) pair in `ode45` and a limited amount with `IVPsolve` shows that this many equally spaced solution values generally yields a smooth graph. There is another reason for choosing five values that we take up shortly. We proceed similarly with the method for stiff IVPs. The Rosenbrock (3,4) pair was derived with a continuous extension that furnishes a numerical solution $S(x) \in C^0[a,b]$ that is a cubic polynomial in $[x_n, x_n + h]$. The continuous extension is evaluated at $x_n + jh/3$ for $j = 1,2$.

2.5. Auxiliary procedures

`IVPval` is an auxiliary procedure for evaluating the solution at any point in the interval of integration. The user interface is simple: the user supplies the solution structure computed by `IVPsolve` and the x at which the solution is desired. The solution structure contains the information needed to evaluate $S(x)$ anywhere in $[a,b]$. In more detail, the number of output points in the span of each step depends on whether the F(4,5) or R(3,4) pair was used, so this number is included in the solution structure. The number of solution values we chose to output for the step is not just enough to get a smooth graph, but also enough that polynomial interpolation of the appropriate degree on the span of the step reproduces the continuous extension there. `IVPsolve` first locates the step containing the x input and then evaluates $S(x)$ by interpolation. In contrast to `dsolve[numeric]`, the answer at a given point is always exactly the same. Furthermore, the IVP is integrated only once, no matter the number and location of points where an answer is desired. An attempt to evaluate the solution outside the interval where it has been computed results in a message to this effect and a reminder of the interval $[a,b]$ corresponding to the solution structure input. The MATLAB IVP solvers return the solution on a mesh and there is no way to obtain solutions at other points except by solving the IVP again. Kierzenka and Shampine [6] have written a MATLAB code for boundary value problems that deals with output much like `IVPsolve`. A version of this BVP solver that exploits new capabilities of the PSE will appear in MATLAB 6. The situation with the IVP solvers is different and illustrates an important point: The user interface of the Suite is uniform, but how capabilities are realized in the various solvers necessarily depends on the underlying method and these methods are quite different in nature. Adding a new capability can be challenging because it must be provided for all the solvers. In the present instance, it would be easy enough to follow `IVPsolve` in adding a new mode of output to `ode45`, but not to the variable order solvers, especially `ode113` with its high orders.

The traditional description of ODEs and initial conditions of `dsolve[numeric]` is convenient for small systems of equations, but not for large. For this reason, in `IVPsolve` we follow the standard in GSC of expecting the ODEs to be provided as a procedure for evaluating a system of first order equations and the initial conditions as a vector. For convenience we provide a procedure, `ODE2proc`, for converting the conventional description of an IVP in Maple to the procedure and vector expected by `IVPsolve`. Unlike MATLAB, Maple distinguishes clearly a scalar and a vector of one component. This complicates the user interface for a scalar ODE and the coding of the solver: We allow vectors of one component for the sake of consistency, but `IVPsolve` and its auxiliary procedures also accept scalar IVPs because we expect that users will find them more natural. For example, it is more natural to specify a scalar ODE like $y' = y^2 - x$ with an operator like $f{:=}(x,y) \to y^2 - x$ than a procedure.

Output from `dsolve[numeric]` in the form of procedures for evaluating the solution is convenient for the plot routines of Maple. Unfortunately, returning the solution in arrays as in MATLAB is not as convenient because it is necessary to form a list of pairs of points for plotting. In the PSEs it is best to use the builtin functions as much as possible because they are compiled and execute much faster. Plotting proved to be unacceptably slow in Maple because the builtin functions for forming such lists do not yet accept `hfarrays`. Eventually we learned that we could convey `hfarrays` to the plotting procedures by means of the CURVE data structure. This increased the speed of all plots by as much as two orders of magnitude. We also found that a low-level plot routine handled logarithmic scales inefficiently. Avoiding it increased the speed of such plots by more than another order of magnitude.

By default, `IVPplot` plots all the solution values returned, just as the MATLAB programs do. Generally this results in a smooth graph, but when the user zooms in on an area using the `view` option or when plotting in the phase plane, it is not unusual that straight-line segments are visible in the graph. The `refine` option is used to deal with this. Invoking `IVPplot` with, say, `refine` = 2 instructs it to double the number of plot points. The additional solution values are equally spaced in the span of each step. They are formed by interpolation using `IVPval`. This way of getting a smooth graph is much more efficient than solving the IVP again as would be done in `dsolve[numeric]` or one of the MATLAB solvers.

A number of the solvers of `dsolve[numeric]` are translations of FORTRAN codes written for GSC. Each of the FORTRAN codes is a good one, but their authors treated important matters in very different ways and the methods themselves require certain differences in the user interface. These differences were not resolved in `dsolve[numeric]`, so it is fairly described as a collection of solvers. In developing the MATLAB ODE Suite, it was considered essential that it be possible to use all the solvers in *exactly* the same way. In the context of this PSE, it was natural that the different methods be implemented in codes with different names. Methods for the solution of stiff IVPs use the Jacobian matrix and perhaps $\partial f/\partial x$. If it is to be possible to solve stiff problems in the same way that non-stiff problems are solved, these partial derivatives must be formed internally. In MATLAB this is done numerically, so the only difference visible to the user between solving a non-stiff IVP with `ode45` and a stiff problem with, say, `ode15s` is that the name of the solver is different. We have implemented an explicit Runge–Kutta method for the solution of non-stiff IVPs and a Rosenbrock method for the solution of stiff problems. In the context of Maple, it is more natural to implement both methods in the same solver with a keyword used to indicate whether the problem is stiff. Because we form partial derivatives analytically, the only difference visible to the user between solving a stiff and a non-stiff problem with `IVPsolve` is that the keyword `stiff` is set to `true`.

3. Numerical Examples

In writing `IVPsolve` for Maple, we aimed to make it easy and efficient to solve an IVP, evaluate the solution, and plot the solution. We were confident that the new solver would perform much better than those of `dsolve[numeric]` in several ways. We hoped that it might be somewhat competitive with the `ode45` and `ode15s` solvers of MATLAB, at least when solving problems that do not require the additional capabilities of those solvers. In this section we compare the solvers using standard test problems. The NODES package and a tutorial will be made available from http://www.apmaths.uwo.ca/~rcorless.

Comparing solvers is always difficult and the present situation is no exception. Other solvers we consider have different aims and this has affected choices made in the codes. For example, the error control of `rkf45` and `IVPsolve` is the same, but the default error tolerances in `rkf45` are much more stringent than those of `IVPsolve`. They are too stringent if the aim is to plot the

Table 1
Non-stiff examples. Times to integrate and plot averaged over 5 runs

	ode45	IVPsolve	rkf45	lsode(Adams)
3 body	0.59	0.24	2.20	4.58
Lorenz	2.10	2.81	6.95	

solution and arguably too stringent in general for an explicit Runge–Kutta formula of only moderate order. Nevertheless, we use default tolerances for our examples because we believe that is what the typical user does. As discussed earlier, the fundamentally different designs of `IVPsolve` and `dsolve[numeric]` make it impossible simply to compare the costs of integrating the IVPs; we must include the cost of a plot or evaluation of the solution. Furthermore, the designs imply that the cost of experimentation with plot options differs greatly because `dsolve[numeric]` causes the IVP to be integrated for every plot and `IVPsolve` merely evaluates a solution already computed. Although some of our examples do call for such experimentation, we report only the cost of an acceptable plot. To obtain representative run times, we computed the average over five independent runs. Only the relative run times matter, and only in a gross way at that, but the times reported are in seconds of computing with a PC running at 450 MHz.

A standard test problem for codes that solve non-stiff IVPs is a restricted three body problem – spaceship, earth, and moon. Initial conditions leading to a periodic orbit and its period were found numerically. Because the orbit is sensitive to perturbations as the spaceship passes close to the earth, it is necessary to use tolerances more stringent than the default values to reproduce the qualitative behavior of the orbit. This IVP is provided as `orbitode.m` with MATLAB 5 and we follow it in using relative error tolerance 1e–5 and absolute error tolerance 1e–4. When solving the IVP with `rkf45` we had to make a number of runs with different values of `numpoints` to find a value that results in a smooth graph. A value of 800 was used for the computations of Table 1. The solver based on Adams–Moulton methods, `lsode` with `adamsfunc`, was also used. It measures error in an RMS norm, so the tolerances must be divided by the square root of the number of equations to make them comparable to the maximum norm used by the Runge–Kutta solvers. Again it was necessary to experiment with `numpoints` and again 800 seemed adequate. The default output of both `IVPsolve` and `ode45` provides a satisfactory graph.

As in the computations resulting in Fig. 9.8.6 of Boyce and DiPrima [1], we integrated the Lorenz equations with $\sigma=10$, $r=28$, $b=8/3$; initial conditions $y_1(0)=5$, $y_2(0)=5$, $y_3(0)=5$; and plotted $y_3(t)$ against $y_1(t)$. `IVPsolve` monitors the work expended and when a maximum number of evaluations of $f(x,y)$ is exceeded, it terminates the run. An option allows this maximum to be increased as needed. We integrated over a longer interval than [1], namely $[0,50]$, to exercise this option. Also, the large number of steps in this integration exercises the portion of `IVPsolve` that extends storage arrays. The cost of solving with `rkf45` depended strongly on `numpoints`. The default `numpoints` produces a graph that is completely unacceptable and increasing it to 1000 results in a graph for which straight-line segments are still obvious. The entry in Table 1 corresponds to a `numpoints` of 5000, which gives an acceptable graph.

We turn now to some stiff IVPs. In solving them with `IVPsolve` we used the Rosenbrock method specified by `stiff = true` and in solving them with `lsode` we used the BDF method specified

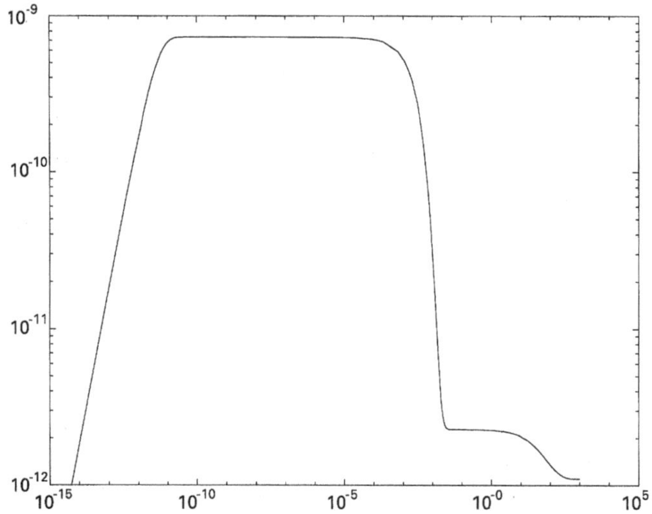

Fig. 1. A log–log plot of the second component of the solution of CHM6.

Table 2
Stiff examples. Times to integrate and plot averaged over 5 runs

	ode15s	IVPsolve	lsode(BDF)
van der Pol	1.92	0.81	2.69
CHM6	0.59	0.37	

by `backfull`. A standard test problem is the van der Pol equation, $y'' + 1000(y^2 - 1)y' + y = 0$, with initial conditions $y(0) = 2$, $y'(0) = 0$ on the interval $[0, 3000]$. The solution $y(x)$ converges quickly to a relaxation oscillation with very sharp changes. When we solved this IVP with `lsode`, we found that the graph did not have sufficiently sharp corners, so we increased `numpoints` to 100. This IVP is provided as `vdpode.m` with MATLAB 5 where it is solved with analytical Jacobian. How Jacobians are handled is important when solving stiff IVPs: `IVPsolve` creates internally procedures for analytical partial derivatives; `ode15s` of MATLAB approximates the Jacobian numerically, but we have supplied a function for the analytical Jacobian in this instance; and `lsode` approximates the Jacobian numerically. (The code LSODE that underlies `lsode` does accept analytical Jacobians, but that capability was not implemented in the Maple version.)

Shampine and Reichelt [9] use the CHM6 problem of [4] to illustrate the stiff IVP solvers of MATLAB. The log–log plot of the second solution component seen in Fig. 1 shows regions of exceedingly sharp change. Correspondingly, the step sizes used by the solvers range over many orders of magnitude. The default relative error tolerance is acceptable, but some solution components are so small that an absolute error tolerance of 1e–20 is appropriate. The solution components have such different behavior that it is appropriate to plot them in different ways. In particular, it is appropriate to use linear scales for the first component and a log–log plot for the second component. Table 2 gives the times taken to integrate the IVP and display a log–log plot of the second component. `ode15s` is a BDF solver that solved the IVP without difficulty, but that was not the case with the BDF solver `lsode`. Indeed, we were not able to make a meaningful comparison. The differential equations of CHM6 have three constant solutions, steady states. The solution of the initial value problem tends to one of these steady states as the independent variable tends to infinity. If the integration is not sufficiently accurate, the numerical solution might tend to the wrong steady state. We found this much more likely to happen with `lsode` than with the other two solvers. Particularly disconcerting is the fact that merely increasing the number of plot points sometimes caused `lsode` to compute a solution that tended to a different steady state, a consequence of the Maple design discussed in Section 2.

4. Conclusions

We have discussed briefly a new code, `IVPsolve`, for solving numerically IVPs for ODEs in Maple and its auxiliary codes `IVPval` for evaluating the solution, `IVPplot` for plotting it, and `ODE2proc` for converting a conventional description of ODEs in Maple to a first order system. The solver provided with Maple, `dsolve[numeric]`, is a collection of programs with different user interfaces. `IVPsolve` implements two methods, one for non-stiff problems and one for stiff, that are used in *exactly* the same way. `IVPsolve` is significantly faster than `dsolve[numeric]` because its algorithms are tailored to the PSE and it exploits hardware floating point arithmetic. The design of `IVPsolve` avoids certain numerical difficulties inherent in the design of `dsolve[numeric]`.

Our discussion of `IVPsolve` has served as a framework for discussing the solution of ODEs in general scientific computation and problem solving environments. In this a common thread has been the solution of non-stiff problems with the F(4,5) pair of Runge–Kutta formulas. We have seen that there are important and interesting differences between solving ODEs in GSC and in PSEs and also between solving ODEs in a PSE oriented towards algebraic computation like Maple and one oriented towards numerical computation like MATLAB.

References

[1] W.E. Boyce, R.C. DiPrima, Elementary Differential Equations and Boundary Value Problems, Wiley, New York, 1997.
[2] K.R. Coombes, B.R. Hunt, R.L. Lipsman, J.E. Osborn, G.J. Stuck, Differential Equations with Maple, Wiley, New York, 1996.
[3] W.H. Enright, K.R. Jackson, S.P. Nørsett, P.G. Thomsen, Interpolants for Runge–Kutta formulas, ACM Trans. Math. Software 12 (1986) 193–218.
[4] W.H. Enright, T.E. Hull, Comparing numerical methods for the solution of ODE's arising in chemistry, in: L. Lapidus, W. Schiesser (Eds.), Numerical Methods for Differential Systems, Academic Press, New York, 1976.
[5] E. Hairer, G. Wanner, Solving Ordinary Differential Equations II, 2nd Edition, Springer, Berlin, 1996.
[6] J. Kierzenka, L.F. Shampine, A BVP solver based on residual control and the MATLAB PSE, Rept. 99-1, Math. Dept., Southern Methodist University, Dallas, 1999.
[7] MATLAB 5, The MathWorks, Inc., 3 Apple Hill Dr., Natick, MA 01760, 1998.
[8] Maple V Release 5, Waterloo Maple Inc., 450 Phillip St., Waterloo, ON N2L 5J2, Canada, 1998.
[9] L.F. Shampine, M.W. Reichelt, The MATLAB ODE Suite, SIAM J. Sci. Comput. 18 (1997) 1–22.
[10] L.F. Shampine, H.A. Watts, in: J.R. Rice (Ed.), The art of writing a Runge–Kutta code I, Mathematical Software III, Academic Press, New York, 1977.
[11] L.F. Shampine, H.A. Watts, The art of writing a Runge–Kutta code II, Appl. Math. Comput. 5 (1979) 93–121.

ELSEVIER

Journal of Computational and Applied Mathematics 125 (2000) 41–56

JOURNAL OF COMPUTATIONAL AND APPLIED MATHEMATICS

www.elsevier.nl/locate/cam

Resolvent conditions and bounds on the powers of matrices, with relevance to numerical stability of initial value problems

N. Borovykh, M.N. Spijker *

Mathematical Institute, University of Leiden, Niels Bohrweg 1, 2333 CA Leiden, The Netherlands

Received 1 April 1999; received in revised form 11 September 1999

Abstract

We deal with the problem of establishing upper bounds for the norm of the nth power of square matrices. This problem is of central importance in the stability analysis of numerical methods for solving (linear) initial value problems for ordinary, partial or delay differential equations. A review is presented of upper bounds which were obtained in the literature under the resolvent condition occurring in the Kreiss matrix theorem, as well as under variants of that condition. Moreover, we prove new bounds, under resolvent conditions which generalize some of the reviewed ones. The paper concludes by applying one of the new upper bounds in a stability analysis of the trapezoidal rule for delay differential equations.

Keywords: Resolvent conditions; Stability analysis; Error growth; Numerical method; Discretization; Initial value problem; Delay differential equation; Trapezoidal rule

1. Introduction

1.1. The purpose of the paper

This paper is concerned with the analysis of numerical methods for the solution of (linear) initial value problems. Most methods in current use are applied in a step-by-step fashion so as to obtain numerical approximations corresponding to consecutive discrete values t_n of the time variable t. A crucial question about these methods is whether they behave *stably* or not. Here we use the term stable to designate the situation where any (numerical) errors, introduced at some stage of the calculations, are propagated mildly — i.e., do not blow up unduly in the subsequent applications of the numerical method.

* Corresponding author. Tel.: +071-527-2727; fax: 071-527-6985.
E-mail address: spijker@rulwinw.leidenuniv.nl (M.N. Spijker).

PII: S0377-0427(00)00457-X

Fourier transformations, and the corresponding famous Von Neumann condition for stability, are classical tools for assessing a priori the stability of methods for solving (partial) differential equations. However, in many practical cases these tools fail to be relevant for analysing stability: e.g., for pseudo-spectral methods applied to initial-boundary value problems, and for finite volume or finite element methods based on unstructured grids.

Recently, progress was made in analysing stability without using Fourier transformation techniques. Conditions for stability were studied which are related to the so-called *resolvent condition of Kreiss*. These conditions apply in some cases where Fourier techniques fail. Moreover, due to the framework in which the conditions are formulated, applications are possible in the solution of ordinary and partial differential equations as well as of delay differential equations. The purpose of the present paper is threefold: we shall review various (recent) results related to the Kreiss resolvent condition; furthermore, we shall present a substantial generalization of some of the reviewed material; finally, we apply our generalization in deriving a new stability estimate in the numerical solution of delay differential equations.

1.2. Organization of the paper

Section 2 is still introductory in nature. In Section 2.1 we relate the stability analysis of numerical processes specified by square matrices B to the problem of deriving upper bounds on the norm $\|B^n\|$ (for $n=1,2,3,\dots$). Further, in Section 2.2 we recall that the eigenvalues of B can be an unreliable guide to stability.

Section 3 gives a review of various upper bounds for $\|B^n\|$ obtained in the literature. In Section 3.1 we review two bounds for $\|B^n\|$ which are valid under the resolvent condition of Kreiss. The sharpness of these bounds is discussed in Section 3.2. In Section 3.3 we review some stronger versions as well as weaker versions of the Kreiss condition and corresponding bounds for $\|B^n\|$.

Section 4 deals with a quite general resolvent condition, which generalizes some of the conditions reviewed in Section 3. In Section 4.1 we formulate this resolvent condition, and we give a lemma on the arc length of the image, under a rational function, of a subarc of a circle in the complex plane. In Section 4.2 we prove Theorem 4.2 making use of this lemma. Theorem 4.2 gives upper bounds for $\|B^n\|$ under the general resolvent condition. Most of these bounds are new. Section 4.3 shortly discusses how the estimates for $\|B^n\|$, given in Theorem 4.2, depend on certain parameters. Moreover, a short discussion is given of the sharpness of these estimates.

In Section 5 we use one of the new estimates of $\|B^n\|$, given by Theorem 4.2, in a stability analysis of the trapezoidal rule applied to delay differential equations.

2. Stability analysis of linear numerical processes

2.1. Relating stability to bounds on $\|B^n\|$

We deal with an abstract numerical process of the form

$$u_n = Bu_{n-1} + b_n \quad (n=1,2,3,\dots). \tag{2.1}$$

Here b_n denote given vectors in the s-dimensional complex space $\mathbb{C}^s$, and B denotes a given complex $s \times s$ matrix. Further, the vectors $u_n \in \mathbb{C}^s$ (for $n \geqslant 1$) are computed by applying (2.1), starting from a given $u_0 \in \mathbb{C}^s$.

Recurrence relations of the form (2.1) arise in the numerical solution of initial value problems for linear (ordinary, partial or delay) differential equations. The vectors u_n then provide numerical approximations to the solution of the problem under consideration. For instance, finite difference schemes for solving initial-boundary value problems in linear partial differential equations can be written in the form (2.1), as soon as the time step is constant and the space steps as well as the coefficients in the differential equation only depend on the space variables. In this situation, the dimension s is related to the space steps, and will tend to infinity if the steps approach zero. For actual numerical processes, written in the form (2.1), see e.g. [5] and the Sections 2.2, 5.1 of the present paper.

Suppose the numerical computations based on (2.1) were performed using a slightly perturbed starting vector $\tilde{u}_0$ instead of u_0. For $n \geqslant 1$, we then would obtain approximations $\tilde{u}_n$, instead of u_n, satisfying the recurrence relation $\tilde{u}_n = B\tilde{u}_{n-1} + b_n$ $(n = 1, 2, 3, \ldots)$. In the stability analysis of (2.1) the crucial question is whether, for $n \geqslant 1$, the *propagated errors* $v_n = \tilde{u}_n - u_n$ can be bounded suitably in terms of the *initial error* $v_0 = \tilde{u}_0 - u_0$. One may thus be looking for bounds of the form

$$|v_n| \leqslant M \cdot |v_0| \quad (n \geqslant 1). \tag{2.2}$$

Here M denotes a constant of moderate size. Further, $|\cdot|$ stands for a norm on $\mathbb{C}^s$ which is considered suitable for measuring error vectors; e.g. the familiar l_p-norm for vectors $x \in \mathbb{C}^s$, with components ξ_i, defined by

$$|x|_p = \left(\sum_{i=1}^{s} |\xi_i|^p \right)^{1/p} \quad (\text{if } 1 \leqslant p < \infty), \qquad |x|_p = \max_{1 \leqslant i \leqslant s} |\xi_i| \quad (\text{if } p = \infty).$$

By subtracting the recurrence relations satisfied by $\tilde{u}_n$ and by u_n from each other, we find $v_n = Bv_{n-1} = B^n v_0$. By defining, for $s \times s$ matrices A,

$$\|A\| = \max\{|Ax|/|x|: 0 \neq x \in \mathbb{C}^s\}, \tag{2.3}$$

we thus see that the stability analysis of process (2.1) amounts to deriving bounds on $\|B^n\|$. The following bound (2.4) would match (2.2):

$$\|B^n\| \leqslant M \quad (n \geqslant 1). \tag{2.4}$$

In this paper we shall deal with the general problem of deriving suitable upper bounds on $\|B^n\|$.

2.2. Eigenvalue conditions

In this subsection we review some simple conditions for (2.4) formulated in terms of the *eigenvalues* λ of the matrix B. We denote the *spectral radius* of B by

$$r(B) = \max\{|\lambda|: \lambda \text{ is an eigenvalue of } B\}.$$

It follows from the Jordan canonical form of B (see, e.g., [8]) that an M with property (2.4) exists if and only if

$$r(B) \leqslant 1, \text{ and any Jordan block corresponding to an eigenvalue } \lambda \text{ of } B, \text{ with } |\lambda| = 1, \text{ has order } 1. \tag{2.5}$$

However, it was noticed already long ago (see, e.g., [16]) that the eigenvalue condition (2.5) can be a very misleading guide to stability. The fact is, that under condition (2.5) the smallest M satisfying (2.4) can be prohibitively large. This phenomenon occurs in practice, even under the subsequent condition (2.6), which is stronger than (2.5).

$$r(B) < 1. \tag{2.6}$$

An instructive example, illustrating that (2.5), (2.6) are unreliable, is provided by the $s \times s$ bidiagonal matrix

$$B = \begin{pmatrix} \lambda_1 & & & \\ 2 & \lambda_2 & & \\ & \ddots & \ddots & \\ & & 2 & \lambda_s \end{pmatrix}. \tag{2.7}$$

We consider the situation where s is large and all $|\lambda_i| < 1$, so that (2.6) holds.

For any $s \times s$ matrix A and $1 \leqslant p \leqslant \infty$, we use the notation

$$\|A\|_p = \max\{|Ax|_p/|x|_p\colon 0 \neq x \in \mathbb{C}^s\}. \tag{2.8}$$

It is easy to see that, for $1 \leqslant p \leqslant \infty$, the matrix B defined by (2.7) satisfies

$$\|B^n\|_p \geqslant 2^n \quad (n = 1, 2, \ldots, s-1). \tag{2.9}$$

For moderately large values of s, say $s \approx 100$, we have $\|B^{s-1}\|_p \gtrsim 10^{30}$, so that actually instability manifests itself although (2.6) is fulfilled.

We note that matrices of the form (2.7) exist which may be thought of as arising in the numerical solution of initial-boundary value problems, e.g.,

$$u_t(x,t) + u_x(x,t) = u(x,t), \quad u(0,t) = 0, \quad u(x,0) = f(x),$$

where $0 \leqslant x \leqslant 1$, $t \geqslant 0$ and f is a given function. Consider the difference scheme

$$\frac{1}{\Delta t}(u_{m,n} - u_{m,n-1}) + \frac{1}{\Delta x}(u_{m,n-1} - u_{m-1,n-1}) = u_{m,n-1},$$

where $\Delta t > 0$, $\Delta x = 1/s < 1$, $m = 1, 2, \ldots, s$ and $n = 1, 2, 3, \ldots$. We define $u_{0,n-1} = 0$ and $u_{m,0} = f(m\Delta x)$, so that $u_{m,n}$ approximates $u(m\Delta x, n\Delta t)$. Clearly, when $\Delta t/\Delta x = 2$, the vectors u_n with components $u_{m,n}$ $(1 \leqslant m \leqslant s)$ satisfy $u_n = Bu_{n-1}$ where B is of the form (2.7) with $\lambda_i = -1 + \Delta t \in (-1, 1)$. Further, since $\Delta x = 1/s$ it is natural to focus on large values of s.

The above example (2.7) shows that under the general conditions (2.5), (2.6) the size of M in (2.4) is not under control and errors can grow exponentially — see (2.9). In the rest of this paper we focus on reliable conditions on arbitrary $s \times s$ matrices B under which such disastrous error growth cannot take place.

3. Stability estimates and resolvent conditions from the literature

3.1. The resolvent condition of Kreiss

Throughout this Subsection 3.1 we assume, unless stated otherwise, that $\|\cdot\|$ is a matrix norm induced by an arbitrary vector norm in $\mathbb{C}^s$, according to (2.3).

We shall relate property (2.4) (with moderate M) to the condition that

$$r(B) \leqslant 1 \quad \text{and} \quad \|(\zeta I - B)^{-1}\| \leqslant \frac{L}{|\zeta| - 1} \quad \text{for all } \zeta \in \mathbb{C} \text{ with } |\zeta| > 1. \tag{3.1}$$

Here I denotes the $s \times s$ identity matrix, and L is a real constant. One usually calls $(\zeta I - B)^{-1}$ the *resolvent* of B at ζ, and we shall refer to (3.1) as the *Kreiss resolvent condition.* We use the latter terminology because (3.1) was used, with $\|\cdot\| = \|\cdot\|_2$, by Kreiss [10] in formulating what nowadays is called the Kreiss matrix theorem. In many cases of practical interest it is easier to verify (3.1) than (2.4).

If (2.4) holds, then $r(B) \leqslant 1$. Moreover, a power series expansion of the resolvent, for $|\zeta| > 1$, then yields

$$\|(\zeta I - B)^{-1}\| = |\zeta|^{-1} \left\| \sum_{n=0}^{\infty} (\zeta^{-1} B)^n \right\| \leqslant |\zeta|^{-1} (1 - |\zeta^{-1}|)^{-1} \max\{1, M\}.$$

It follows that (2.4) implies (3.1), with $L = \max\{1, M\}$. For the case where $\|\cdot\| = \|\cdot\|_2$, Kreiss [10] succeeded in proving that conversely (3.1) implies (2.4) with $M = M_{L,s}$ only depending on L and s.

In the following we shall be interested in the case where s is large. Therefore, it is important to understand how $M_{L,s}$ depends on s. The original proof of Kreiss does not provide a sharp value for $M_{L,s}$, and many subsequent authors studied the size of this quantity; see [31] for a historical survey. Eventually, for arbitrary matrix norms (2.3), the following theorem was obtained — for its proof see, e.g., [5, pp. 208, 209].

Theorem 3.1. *For any real constant L and any $s \times s$ matrix B satisfying* (3.1), *we have*

$$\|B^n\| \leqslant \mathrm{e} L s \quad (n \geqslant 1,\ s \geqslant 1), \tag{3.2a}$$

$$\|B^n\| \leqslant \mathrm{e} L (n+1) \quad (n \geqslant 1,\ s \geqslant 1). \tag{3.2b}$$

According to this theorem, under the Kreiss resolvent condition, the size of $\|B^n\|$ is rather well under control. Exponential error growth cannot occur — at the worst there may be weak instability in that the propagated errors increase linearly with n or s.

For applications of the above theorem (and its predecessors), one may consult [5,7,10,15,17,18,23, 25]; for diverse theoretical issues related to the theorem, we refer to [5,13,14,26,29].

3.2. *The sharpness of the stability estimates (3.2)*

In this subsection we discuss the sharpness of the estimates (3.2) for the interesting case where $\|\cdot\| = \|\cdot\|_\infty$. We focus on this norm because of the following three reasons: there exist rather complete results about the sharpness of (3.2) for the norm $\|\cdot\|_\infty$; moreover, important practical situations exist where $\|(\zeta I - B)^{-1}\|_\infty$ can rather easily be estimated; finally, error estimates in terms of the l_∞-norm allow of a useful and easy interpretation. For sharpness results pertinent to other norms, we refer to [5,12,24].

It is known that $s \times s$ matrices B_s exist, satisfying (3.1) with $\|\cdot\| = \|\cdot\|_\infty$ and with some finite $L = L_s$ (for $s = 1, 2, 3, \ldots$), such that the quotient $\|(B_s)^{s-1}\|_\infty/(sL_s)$ tends to e when $s \to \infty$ (see [5, Corollary 2.3]). It follows that the estimates given in Theorem 3.1 are sharp in that the constant e, occurring in the right-hand members of (3.2a) and (3.2b), cannot be replaced by any smaller constant.

Unfortunately, the values L_s, in the above counterexample, tend to ∞ when $s \to \infty$. Therefore, the nice sharpness result just mentioned is related to the fact that the estimates in Theorem 3.1 are required to follow from (3.1) simultaneously for *all possible values* of L. The above counterexample fails to be relevant to the important question in how far the stability estimates (3.2a) and (3.2b) are also best possible, when L is an arbitrary but *fixed* constant. In fact, for $\|\cdot\| = \|\cdot\|_\infty$ and $L = 1$, these estimates can substantially be improved: in this situation the resolvent condition (3.1) is known to imply $\|B^n\|_\infty \leqslant 1$ $(n \geqslant 1,\ s \geqslant 1)$ — see, e.g., [5, Theorem 2.6].

The important problem arises as to whether the upper bounds (3.2a) and (3.2b) can be improved, for all fixed values L, to bounds on $\|B^n\|_\infty$ which do not grow, or which grow (much) slower than linearly with s or n.

This problem was solved by Kraaijevanger [9]. He succeeded in constructing $s \times s$ matrices B_s satisfying (3.1), with $\|\cdot\| = \|\cdot\|_\infty$ and $L = \pi + 1$, such that

$$\|(B_s)^n\|_\infty = 2s - 1 = 2n - 1 \quad (\text{whenever } n = s \geqslant 1). \tag{3.3}$$

In view of (3.3), we conclude that the upper bounds (3.2a) and (3.2b) cannot be improved, for all fixed values L, into bounds which grow slower than linearly with s or n.

3.3. Variants to the Kreiss resolvent condition

Throughout this subsection we assume again, unless specified otherwise, that $\|\cdot\|$ is a matrix norm induced by an arbitrary vector norm in $\mathbb{C}^s$, according to (2.3). We shall deal with two *stronger* versions of condition (3.1) as well as two *weaker* versions.

In view of the conclusion at the end of Section 3.2, the question poses itself of whether bounds on $\|B^n\|$ which grow slower than linearly with s or n can still be established under conditions that are slightly *stronger* than (3.1) (and fulfilled in cases of practical interest). Below we review shortly two conclusions, obtained in the literature, pertinent to this question. For additional results, see [3,13,22].

Consider for arbitrary $s \times s$ matrices B the condition that

$$r(B) \leqslant 1 \quad \text{and} \quad \|(\zeta I - B)^{-m}\| \leqslant \frac{L}{(|\zeta| - 1)^m} \quad \text{for } |\zeta| > 1 \quad \text{and} \quad m = 1, 2, 3, \ldots. \tag{3.4a}$$

Clearly, this so-called *Hille-Yosida* or *iterated resolvent condition* implies (3.1). Unlike (3.1), condition (3.4a) implies the stability estimate

$$\|B^n\| \leqslant Ln!(\mathrm{e}/n)^n \leqslant \mathrm{e}L\sqrt{n} \quad (n \geqslant 1,\ s \geqslant 1). \tag{3.4b}$$

This estimate was obtained by various authors. It follows for instance easily from the material in [2, p. 41] in combination with [11], or directly from [13].

A still better stability estimate can be established under the following condition (3.5a), which was introduced in [28].

$$r(B)\leqslant 1 \quad \text{and} \quad \|(\zeta I - B)^{-1}\| \leqslant \frac{L}{|\zeta - 1|} \quad \text{for } |\zeta| > 1. \tag{3.5a}$$

Since $|\zeta - 1|^{-1} \leqslant (|\zeta| - 1)^{-1}$, also this condition implies (3.1). Moreover, as was shown in [3], condition (3.5a) implies the inequality

$$\|B^n\| \leqslant \mathrm{e}L^2/2 \quad (n \geqslant 1,\ s \geqslant 1), \tag{3.5b}$$

the right-hand member of which does not grow with n or s. We refer to the paper just mentioned for an application of (3.5b) in proving numerical stability for a class of Runge–Kutta methods in the numerical solution of initial-boundary value problems for parabolic partial differential equations.

Clearly, in order to apply the general stability estimates (3.2), (3.4b) or (3.5b) in any practical situation, one has to check whether the corresponding resolvent conditions are actually fulfilled. Sometimes this may be difficult, and there are cases where one cannot even prove (3.1) (see, e.g., Section 5.2). Therefore, it is an important issue of whether estimates similar to (3.2a) and (3.2b) still hold under resolvent conditions which are *weaker* than (3.1). Below we mention two results pertinent to this issue.

Consider, for arbitrary $s \times s$ matrices B and a given constant $\delta > 0$, the condition that

$$r(B)\leqslant 1 \quad \text{and} \quad \|(\zeta I - B)^{-1}\| \leqslant L\frac{|\zeta|^{\delta \cdot s}}{|\zeta| - 1} \quad \text{for } |\zeta| > 1. \tag{3.6a}$$

This weaker version of (3.1) is known to imply that

$$\|B^n\| \leqslant \mathrm{e}L[\delta s + \min\{s, n+1\}] \quad (n \geqslant 1,\ s \geqslant 1); \tag{3.6b}$$

see [21] for a proof and an application of (3.6b). We conclude that under condition (3.6a), similarly as under the stronger condition (3.1), the norm $\|B^n\|$ cannot grow faster than linearly with s.

A further weaker version of (3.1), considered in the literature, requires that, for a given fixed value $\alpha > 0$, the $s \times s$ matrix B satisfies

$$r(B)\leqslant 1 \quad \text{and} \quad \|(\zeta I - B)^{-1}\| \leqslant L\frac{|\zeta|^{\alpha}}{(|\zeta| - 1)^{1+\alpha}} \quad \text{for } |\zeta| > 1 \tag{3.7a}$$

(cf. [6,17,27]). Under this condition the norm of the resolvent is allowed to grow (when $|\zeta| \to 1+$) like $(|\zeta| - 1)^{-1-\alpha}$, which is faster than in the situation (3.1). In [6] it was shown that, under condition (3.7a),

$$\|B^n\| \leqslant \mathrm{e}L(n+1)^{1+\alpha} \quad (n \geqslant 1,\ s \geqslant 1). \tag{3.7b}$$

Further, by the arguments in [27], condition (3.7a) is seen to imply, for the case where $\|\cdot\| = \|\cdot\|_2$, that

$$\|B^n\| \leqslant cLs(n+1)^{\alpha} \quad (n \geqslant 1,\ s \geqslant 1), \tag{3.7c}$$

where $c = 32\,\mathrm{e}^{1+\alpha}/\pi$.

We conclude this section by noting that slightly modified versions of (3.1), (3.5a), (3.6a), (3.7a) were considered in the literature as well: most of the papers mentioned above also deal with the situation where the inequality for the norm of the resolvent is required to hold only for $1 < |\zeta| < \rho$ — where ρ is a finite constant — rather than for all ζ with $|\zeta| > 1$. In this situation upper bounds for $\|B^n\|$ were proved which equal the original bounds, specified in (3.2), (3.5b), (3.6b), (3.7b) and (3.7c), respectively, multiplied by a factor γ only depending on ρ (and α). For the special case (3.5b), the corresponding factor γ is exceptionally simple in that $\gamma = 1$ for any ρ with $1 < \rho < \infty$: it can be proved that $\|B^n\| \leqslant \mathrm{e}L^2/2 \quad (n \geqslant 1,\ s \geqslant 1)$ whenever $r(B) \leqslant 1$ and $\|(\zeta I - B)^{-1}\| \leqslant L|\zeta - 1|^{-1}$ $(1 < |\zeta| < \rho)$.

4. Stability estimates under a general resolvent condition

4.1. Preliminaries

Throughout this Section 4 we assume, unless specified otherwise, that $\|\cdot\|$ is an arbitrary norm on the vector space of all complex $s \times s$ matrices (i.e., $\|A\| > 0$ for $A \neq 0$, and $\|A + B\| \leqslant \|A\| + \|B\|$, $\|\lambda \cdot A\| = |\lambda| \cdot \|A\|$ for all $\lambda \in \mathbb{C}$ and all $s \times s$ matrices A, B).

We shall present upper bounds for $\|B^n\|$, under the following general resolvent condition:

$$r(B) \leqslant 1 \quad \text{and} \quad \|(\zeta I - B)^{-1}\| \leqslant \frac{L}{(|\zeta| - 1)^k |\zeta - 1|^l} \quad (1 < |\zeta| < \rho). \tag{4.1}$$

Here L is a positive constant, k and l are nonnegative fixed integers with $k + l \geqslant 1$, and $1 < \rho \leqslant \infty$. Clearly, condition (4.1) generalizes some of the resolvent conditions reviewed in Section 3.

In deriving our upper bounds for $\|B^n\|$, we shall make use of

Lemma 4.1. *Let $\alpha \leqslant \beta \leqslant \alpha + 2\pi$, $r > 0$, and let Γ denote the subarc of a circle given by $\zeta = r\mathrm{e}^{\mathrm{i}t}$ $(\alpha \leqslant t \leqslant \beta)$. Assume $R(\zeta) = P(\zeta)/Q(\zeta)$, where $P(\zeta)$, $Q(\zeta)$ are polynomials of a degree not exceeding s, with $Q(\zeta) \neq 0$ on Γ. Then*

$$\int_\Gamma |R'(\zeta)|\,|\mathrm{d}\zeta| \leqslant \pi s \operatorname{diam} R(\Gamma) \leqslant 2\pi s \max_\Gamma |R(\zeta)|. \tag{4.2}$$

In (4.2) we denote by $\operatorname{diam} R(\Gamma)$ the diameter of the set $\{R(r\mathrm{e}^{\mathrm{i}t})\colon \alpha \leqslant t \leqslant \beta\}$. We note that this lemma allows of a simple geometrical interpretation since the integral in (4.2) equals the arc length of the image, under (the mapping) R, of Γ. A version of this lemma with $\beta = \alpha + 2\pi$ was already proved in [20] and in [31]. The more general Lemma 4.1 is no consequence of that version. But, property (4.2), for the general case $\alpha \leqslant \beta \leqslant \alpha + 2\pi$, can easily be proved by a straightforward adaptation of the arguments used in [20]. We omit the details.

4.2. Formulation and proof of general stability estimates

The following theorem summarizes our upper bounds for $\|B^n\|$, under the resolvent condition (4.1).

Theorem 4.2. *There is a constant γ depending only on k, l, ρ such that, for all $n \geqslant 1$, $s \geqslant 1$ and for each $s \times s$ matrix B satisfying* (4.1),

$$\|B^n\| \leqslant \gamma L n^{k-1} \min\{s, n\} \quad (\textit{if } k \geqslant 1,\ l = 0), \tag{4.3a}$$

$$\|B^n\| \leqslant \gamma L n^{k} \min\{\log(s+1), \log(n+1)\} \quad (\textit{if } k \geqslant 0,\ l = 1), \tag{4.3b}$$

$$\|B^n\| \leqslant \gamma L n^{k+l-1} \quad (\textit{if } k \geqslant 0,\ l \geqslant 2). \tag{4.3c}$$

Clearly, the bound (4.3a) is closely related to the estimates (3.2) and (3.7b), (3.7c) (with $\alpha = k-1$). The proof below of (4.3a) will consist in a straightforward application of arguments used earlier in the literature. It will rely among other things on Lemma 4.1 with $\beta = \alpha + 2\pi$.

To the best of our knowledge, the estimates (4.3b) and (4.3c) are new. Our proof of (4.3b) will require an application of Lemma 4.1 with $\beta < \alpha + 2\pi$.

Proof of Theorem 4.2. (1) Let $n \geqslant 1$, $s \geqslant 1$ and let the $s \times s$ matrix B satisfy (4.1). We shall use the Dunford–Taylor representation (see, e.g., [19, Chapter 10])

$$B^n = \frac{1}{2\pi \mathrm{i}} \int_\Gamma \zeta^n (\zeta I - B)^{-1}\, \mathrm{d}\zeta,$$

where Γ is the positively oriented circle $|\zeta| = r$ with $r = \min\{\rho, 1 + 1/(n+1)\}$.

By a well known corollary to the Hahn–Banach theorem (see, e.g., [19, Chapter 3]), there is a linear mapping F from the vector space of all complex $s \times s$ matrices to $\mathbb{C}$, with $F(B^n) = \|B^n\|$ and $|F(A)| \leqslant \|A\|$ for all $s \times s$ matrices A. Consequently,

$$\|B^n\| = \frac{1}{2\pi \mathrm{i}} \int_\Gamma \zeta^n R(\zeta)\, \mathrm{d}\zeta, \tag{4.4a}$$

where $R(\zeta) = F((\zeta I - B)^{-1})$ and

$$|R(\zeta)| \leqslant L(|\zeta| - 1)^{-k} |\zeta - 1|^{-l} \quad \text{for } 1 < |\zeta| \leqslant \rho.$$

(2) Let $l = 0$. Similarly as in [12], [5, pp. 208, 209] we perform a partial integration so as to obtain from (4.4a)

$$\|B^n\| = \frac{-1}{2\pi \mathrm{i}(n+1)} \int_\Gamma \zeta^{n+1} R'(\zeta)\, \mathrm{d}\zeta \leqslant \frac{r^{n+1}}{2\pi(n+1)} \int_\Gamma |R'(\zeta)|\, |\mathrm{d}\zeta|. \tag{4.4b}$$

By still using arguments similar to those in the above references, one can see that $R(\zeta) = P(\zeta)/Q(\zeta)$, where $P(\zeta)$, $Q(\zeta)$ are polynomials of a degree not exceeding s. Furthermore, for $|\zeta| = r$, we have $Q(\zeta) \neq 0$. Consequently, we can conclude from (4.4a), (4.4b) and Lemma 4.1 (with $\beta = \alpha + 2\pi$) that

$$\|B^n\| \leqslant \frac{r^{n+1}}{(n+1)(r-1)^k} L \min\{s, n+1\}.$$

This inequality implies the relations (4.5a) and (4.5b), which in their turn prove (4.3a):

$$\|B^n\| \leqslant \mathrm{e} L (n+1)^{k-1} \min\{s, n+1\} \quad \text{if } n+1 \geqslant (\rho - 1)^{-1}, \tag{4.5a}$$

$$\|B^n\| \leqslant \frac{L\rho^{n+1}}{(n+1)(\rho-1)^k} \min\{s, n+1\} \quad \text{if } n+1 < (\rho - 1)^{-1}. \tag{4.5b}$$

(3) Let $l=1$. We decompose the circle Γ into two subarcs Γ_0 and Γ_1, where Γ_0 is given by $\zeta=re^{it}$ $(-\delta\leqslant t\leqslant\delta)$, and Γ_1 by $\zeta=re^{it}$ $(\delta\leqslant t\leqslant 2\pi-\delta)$. Here δ is a value with $0\leqslant\delta\leqslant\pi$ to be specified below. Putting $\zeta_0=re^{i\delta}$, we obtain from (4.4a), by partial integration, the representation

$$\|B^n\|=\frac{1}{2\pi i}\int_{\Gamma_0}\zeta^n R(\zeta)\,d\zeta+\frac{-1}{2\pi i(n+1)}\int_{\Gamma_1}\zeta^{n+1}R'(\zeta)\,d\zeta$$
$$+\frac{1}{2\pi i(n+1)}(\bar{\zeta}_0^{\,n+1}R(\bar{\zeta}_0)-\zeta_0^{n+1}R(\zeta_0)).$$

We denote the three successive terms in the right-hand member of the last equality by I_0, I_1, I_2, respectively.

We define $y=2\delta[\pi(r-1)]^{-1}$ and assume that $y\geqslant 4$. We have

$$|I_0|\leqslant\frac{Lr^{n+1}}{\pi(r-1)^k}\int_0^\delta\frac{dt}{\sqrt{(r-1)^2+(2t/\pi)^2}}=\frac{Lr^{n+1}\log(y+\sqrt{1+y^2})}{2(r-1)^k}$$
$$\leqslant\frac{Lr^{n+1}}{(r-1)^k}[1/2+\log(y/2)].$$

By applying (among other things) Lemma 4.1, with $\alpha=\delta$, $\beta=2\pi-\delta$, we also have

$$|I_1|\leqslant Ks,\ |I_2|\leqslant K/\pi\quad\text{where } K=\frac{Lr^{n+1}}{(r-1)^{k+1}(n+1)\sqrt{1+y^2}}.$$

We first choose $\delta=\pi$. We have $\|B^n\|=|I_0|$ and $y=2(r-1)^{-1}\geqslant 4$ so that $\|B^n\|\leqslant Lr^{n+1}(r-1)^{-k}\{1+\log[(r-1)^{-1}]\}$.

Next, we assume that $s<n$ and we choose $\delta=\pi(s+1)/(n+1)$. We now have $y=2(s+1)[(r-1)(n+1)]^{-1}\geqslant 2(s+1)\geqslant 4$. Combining the inequality $\|B^n\|\leqslant|I_0|+|I_1|+|I_2|$ and our upper bounds for $|I_0|$, $|I_1|$, $|I_2|$ we arrive at the estimate $\|B^n\|\leqslant Lr^{n+1}(r-1)^{-k}\{1+\log[(r-1)^{-1}(s+1)/(n+1)]\}$.

The two bounds for $\|B^n\|$ just obtained are equivalent to

$$\|B^n\|\leqslant\frac{r^{n+1}}{(r-1)^k}L\left\{1+\log\left[\frac{\min\{s+1,n+1\}}{(n+1)(r-1)}\right]\right\}.$$

This inequality implies the relations (4.6a) and (4.6b), which in their turn prove (4.3b):

$$\|B^n\|\leqslant eL(n+1)^k[1+\log(1+\min\{s,n\})]\quad\text{if } n+1\geqslant(\rho-1)^{-1},\tag{4.6a}$$

$$\|B^n\|\leqslant\frac{L\rho^{n+1}}{(\rho-1)^k}\left\{1+\log\left[\frac{1+\min\{s,n\}}{(n+1)(\rho-1)}\right]\right\}\quad\text{if } n+1<(\rho-1)^{-1}.\tag{4.6b}$$

(4) Let $l\geqslant 2$. In order to prove (4.3c), we use (4.4a) so as to obtain

$$\|B^n\|\leqslant\frac{Lr^{n+1}}{\pi(r-1)^{k+l}}J\quad\text{with } J=\int_0^\pi\frac{dt}{(1+(\mu t)^2)^{l/2}}\quad\text{and}\quad\mu=\frac{2}{\pi(r-1)}.$$

Introducing the variable x by the relation $\mu t = (e^x - e^{-x})/2$, we have

$$J \leqslant 2^{l-1}\mu^{-1}\int_0^{\infty}(e^x + e^{-x})^{1-l}\,dx \leqslant 2^{l-1}\mu^{-1}(l-1)^{-1}\left[\frac{2}{e^x + e^{-x}}\right].$$

Combining this estimate of J and the above bound for $\|B^n\|$, one obtains the relations (4.7a) and (4.7b), which in their turn prove (4.3c):

$$\|B^n\| \leqslant \frac{2^{l-2}e}{l-1}L(n+1)^{k+l-1} \quad \text{if } n+1 \geqslant (\rho-1)^{-1}, \tag{4.7a}$$

$$\|B^n\| \leqslant \frac{2^{l-2}}{l-1}\frac{L\rho^{n+1}}{(\rho-1)^{k+l-1}} \quad \text{if } n+1 < (\rho-1)^{-1}. \quad \square \tag{4.7b}$$

4.3. Remarks in connection with Theorem 4.2

The estimates (4.3) in Theorem 4.2 have deliberately been formulated concisely without indicating how γ may depend on the parameters k, l, ρ. Bounds for $\|B^n\|$ in which the dependence on these parameters is explicit can be obtained from (4.5)–(4.7). As an illustration we mention that (4.5) can be used in proving, for $k \geqslant 1$, $l = 0$ and any $s \times s$ matrix B satisfying (4.1), that

$$\|B^n\| \leqslant cL(n+1)^{k-1}\min\{s, n+1\} \quad (n \geqslant 1, s \geqslant 1),$$

where $c = e$ (for $\rho \geqslant 3/2$), $c = \max\{e, \rho^2 2^{-k}(\rho-1)^{-k}\}$ (for $1 < \rho < 3/2$). We note that this bound can be applied in the situation (3.7a) (with $\alpha = k-1$), so as to yield (3.7c) with a smaller value for c than the one given in Section 3.3.

We conclude this section by a short discussion of the sharpness of the stability estimates, given in Theorem 4.2, for the important case $\|\cdot\| = \|\cdot\|_\infty$. We focus on the question of whether these estimates can be improved, for all fixed L and ρ, to bounds on $\|B^n\|_\infty$ which grow slower with n or s than the bounds in (4.3).

Kraaijevanger's result (3.3) makes clear that, when $k = 1$, $l = 0$, the estimate in (4.3a) cannot be improved, for all L, ρ, into a bound of the form $\|B^n\|_\infty \leqslant c\min\{\phi(s), \psi(n)\}$, where $c = c(\rho, L)$ only depends on ρ, L and either $\phi(s)$ or $\psi(n)$ grows slower than linearly with s or with n, respectively. On the other hand, an essential improvement over the estimate in (4.3b) is possible when $k=0$, $l=1$: in Section 3.3 we have seen that in this case $\|B^n\| \leqslant c$ with $c = eL^2/2$.

The authors found that, somewhat surprisingly, a conclusion, similar to the one just mentioned for $k = 1$, $l = 0$, can be reached whenever $k \neq 0$ or $l \neq 1$. In fact, for each $k \geqslant 1$, the estimate in (4.3a) cannot be improved into a bound of the form $\|B^n\|_\infty \leqslant c(\rho, L, k)n^{k-1}\min\{\phi(s), \psi(n)\}$ with any functions $\phi(s)$, $\psi(n)$ as considered above. Further, for each $k \geqslant 1$, the estimate in (4.3b) cannot be improved into $\|B^n\|_\infty \leqslant c(\rho, L, k)n^k\min\{\phi(s), \psi(n)\}$ with either $\phi(s)$ or $\psi(n)$ growing slower than $\log(s+1)$ or $\log(n+1)$, respectively. Finally, for each $k \geqslant 0$ and $l \geqslant 2$, the estimate in (4.3c) cannot be improved into $\|B^n\|_\infty \leqslant c(\rho, L, k, l)\psi(n)$ with $\lim_{n\to\infty}\psi(n)/n^{k+l-1} = 0$. More details are given in [4].

5. Stability analysis in the numerical solution of delay differential equations

5.1. Applying the trapezoidal rule to a linear test problem

The above general considerations will be illustrated in the numerical solution of the initial value problem

$$Z'(t) = f(Z(t), Z(t-\tau)) \quad (t \geqslant 0), \qquad Z(t) = g(t) \quad (t \leqslant 0).$$

Here f, g are given functions, $\tau > 0$ is a fixed delay, and $Z(t)$ is unknown (for $t > 0$).

We focus on the following well-known version of the *trapezoidal rule*:

$$z_n = z_{n-1} + \frac{h}{2}[f(z_n, z_{n-s+1}) + f(z_{n-1}, z_{n-s})] \quad (n \geqslant 1). \tag{5.1}$$

Here s denotes an integer with $s \geqslant 2$, and $h = \tau/(s-1)$ is the so-called stepsize. Further, z_n are approximations to $Z(t)$ at the gridpoints $t = t_n = nh$. Putting $z_n = g(t_n)$ $(n \leqslant 0)$, one may compute successively approximations z_n (for $n = 1, 2, 3, \ldots$) from (5.1).

Many authors (see, e.g., [1,30,32]) studied the stability of numerical methods, for the above initial value problem, by analysing the behaviour of the methods in the solution of the following linear test problem:

$$Z'(t) = \lambda Z(t) + \mu Z(t-\tau) \quad (t \geqslant 0), \qquad Z(t) = g(t) \quad (t \leqslant 0).$$

Here λ, μ denote fixed complex coefficients, and $g(t)$, $Z(t) \in \mathbb{C}$.

Method (5.1), when applied to the test equation, reduces to the recurrence relation

$$z_n = az_{n-1} + bz_{n-s+1} + bz_{n-s} \quad (n \geqslant 1),$$

where $a = (2+x)(2-x)^{-1}$, $b = y(2-x)^{-1}$ and $x = h\lambda$, $y = h\mu$. This recurrence relation can be written in the form

$$u_n = Bu_{n-1} \quad (n \geqslant 1) \quad \text{where } u_n = (z_n, z_{n-1}, \ldots, z_{n-s+1})^{\mathrm{T}}.$$

Here the $s \times s$ companion matrix $B = (\beta_{ij})$ is defined, for $s \geqslant 3$, by $\beta_{ij} = a$ (if $i = j = 1$), $\beta_{ij} = b$ (if $i = 1$ and $j = s-1, s$), $\beta_{ij} = 1$ (if $1 \leqslant j = i-1 \leqslant s-1$), and $\beta_{ij} = 0$ otherwise. For $s = 2$, we have $\beta_{11} = a + b$, $\beta_{12} = b$, $\beta_{21} = 1$, $\beta_{22} = 0$. Clearly, B depends (only) on x, y and s. Accordingly, we shall write $B = B_s(x, y)$.

Following standard practice in dealing with the above test problem, we consider the so-called *stability region*

$$S = \{(x, y)\colon r(B_s(x, y)) < 1 \text{ for all } s \geqslant 2\}.$$

It is known that all pairs (x, y) with $\mathrm{Re}\, x < -|y|$ belong to S, and that all $(x, y) \in S$ satisfy $\mathrm{Re}\, x \leqslant -|y|$ (see, e.g., [30]).

But, as highlighted in Section 2, with regard to error propagation the crucial question is not of whether the spectral radius condition $r(B_s(x, y)) < 1$ is fulfilled, but of whether $\|B^n\|$ is of moderate size, where $B = B_s(x, y)$ and $\|\cdot\|$ is related to a suitable vector norm according to (2.3).

In the following we focus on estimating $\|B^n\|_\infty$ for $B = B_s(x, y)$, $n \geqslant 1$, $s \geqslant 2$, uniformly for all $(x, y) \in S$.

5.2. Obtaining stability results by using resolvents

In [21] it was proved that, corresponding to any given fixed $s \geqslant 2$, there exists no finite L such that $B = B_s(x, y)$ satisfies the Kreiss condition (3.1) (with $\|\cdot\| = \|\cdot\|_\infty$) uniformly for all $(x, y) \in S$. Since (2.4) implies (3.1) (with $L = \max\{1, M\}$, see Section 3.1), it follows that the quantity M_s, defined by

$$M_s = \sup\{\|B^n\|_\infty : n \geqslant 1,\ B = B_s(x, y),\ (x, y) \in S\},$$

satisfies

$$M_s = \infty \quad (\text{for } s = 2, 3, 4, \ldots).$$

In spite of this negative stability result, it is still possible to establish an upper bound for $\|B^n\|_\infty$ (uniformly for $B = B_s(x, y)$ with $(x, y) \in S$) which is only slightly weaker than (3.2). This bound can be obtained by a combination of Theorem 4.2 and the following lemma.

Lemma 5.1. *Let* $\mathrm{Re}\, x \leqslant -|y|$. *Then the matrix* $B = B_s(x, y)$ *satisfies*

$$r(B) \leqslant 1 \quad \text{and} \quad \|(\zeta I - B)^{-1}\|_\infty \leqslant \frac{11}{(|\zeta| - 1)|\zeta + 1|} \quad \left(\text{for } 1 < |\zeta| < \tfrac{3}{2}\right). \tag{5.2}$$

Proof. Let $x, y \in \mathbb{C}$, with $\mathrm{Re}\, x \leqslant -|y|$, and let $s \geqslant 2$, $B = B_s(x, y)$. The polynomial $P(\zeta) = \det(\zeta I - B)$ can be written in the form

$$P(\zeta) = (\zeta - a)\zeta^{s-1} - (\zeta + 1)b.$$

Let $\zeta \in \mathbb{C}$, with $|\zeta| > 1$. In [21, pp. 243, 244] it was shown that the spectral radius $r(B) \geqslant 1$, and that

$$\|(\zeta I - B)^{-1}\|_\infty \leqslant 2\left\{\frac{1}{|\zeta| - 1} + \frac{|\zeta|^{s-1}}{|P(\zeta)|}\right\}.$$

We write ζ in the form

$$\zeta = \frac{2 + z}{2 - z} \quad \text{with} \quad \mathrm{Re}\, z > 0,\ z \neq 2.$$

By straightforward calculations it can be seen that

$$\zeta - a = \frac{4(z - x)}{(2 - z)(2 - x)}, \qquad (\zeta + 1)b = \frac{4y}{(2 - z)(2 - x)},$$

$$|\zeta|^2 - 1 = \frac{8\mathrm{Re}\, z}{|2 - z|^2}, \qquad |\zeta + 1| = \frac{4}{|2 - z|}.$$

These equalities imply that

$$\frac{|P(\zeta)|}{|\zeta|^{s-1}} \geqslant |\zeta|^{-1}\{|(\zeta - a)\zeta| - |(\zeta + 1)b|\} \geqslant \frac{4(|(z - x)\zeta| + \mathrm{Re}\, x)}{|\zeta(2 - z)(2 - x)|}.$$

Since $\mathrm{Re}\, z - \mathrm{Re}\, x \leqslant |z - x|$, we have

$$|(z - x)\zeta| + \mathrm{Re}\, x \geqslant |z - x|(|\zeta| - 1) + \mathrm{Re}\, z$$

$$\geqslant (|\zeta| - 1)\big||2 - x| - |2 - z|\big| + \tfrac{1}{8}(|\zeta|^2 - 1)|2 - z|^2.$$

Combining this bound for $|(z-x)\zeta| + \operatorname{Re} x$ with the above lower bound for $|P(\zeta)|/|\zeta|^{s-1}$, we obtain

$$\begin{aligned}\frac{|P(\zeta)|}{|\zeta|^{s-1}} &\geqslant \frac{|\zeta|-1}{|\zeta|}\left[\left|1-\left|\frac{2-z}{2-x}\right|\right|\cdot|\zeta+1|+\frac{1}{2}(|\zeta|+1)\left|\frac{2-z}{2-x}\right|\right]\\ &\geqslant \frac{|\zeta|-1}{|\zeta|}\left[\left||\zeta+1|-\frac{4}{|2-x|}\right|+\frac{2}{|2-x|}\right]\\ &\geqslant \frac{|\zeta|-1}{|\zeta|}\max\left\{\frac{2}{|2-x|},\frac{|\zeta+1|}{2}\right\}.\end{aligned}$$

In view of the above upper bound for $\|(\zeta I-B)^{-1}\|_\infty$, we arrive at

$$\|(\zeta I-B)^{-1}\|_\infty \leqslant 2\left[1+|\zeta|\min\left\{\frac{|2-x|}{2},\frac{2}{|\zeta+1|}\right\}\right](|\zeta|-1)^{-1}.$$

We conclude that, for $|\zeta|>1$,

$$\|(\zeta I-B)^{-1}\|_\infty \leqslant (2|\zeta+1|+4|\zeta|)|\zeta+1|^{-1}(|\zeta|-1)^{-1}.$$

This implies (5.2). □

The following neat stability result for the trapezoidal rule can now easily be proved.

Theorem 5.2. *There is a constant c such that $B=B_s(x,y)$ satisfies*

$$\|B^n\|_\infty \leqslant cn\min\{\log(s+1),\log(n+1)\} \quad (n\geqslant 1, s\geqslant 2),$$

uniformly for all $(x,y)\in S$.

Proof. Lemma 5.1 shows that the matrix $-B$ satisfies (4.1) with $\|\cdot\| = \|\cdot\|_\infty$ and $k=l=1$, $L=11$, $\rho=3/2$. In view of Theorem 4.2, we have $\|B^n\|_\infty = \|(-B)^n\|_\infty \leqslant cn\min\{\log(s+1),\log(n+1)\}$ with $c=11\gamma$. Here γ is constant. □

We note that the upper bound for $\|B^n\|_\infty$ given by the above theorem can be interpreted as a stability result of the form

$$|\tilde{z}_n - z_n| \leqslant \phi(s,n)\max\{|\tilde{z}_0-z_0|,|\tilde{z}_{-1}-z_{-1}|,\ldots,|\tilde{z}_{-s+1}-z_{-s+1}|\},$$

valid for any two sequences z_n, $\tilde{z}_n$, computed from (5.1) with $f(\xi,\eta)=\lambda\xi+\mu\eta$ and $\operatorname{Re}\lambda\leqslant-|\mu|$. Here $\phi(s,n)=cn\min\{\log(s+1),\log(n+1)\}$ is growing only slightly faster than linearly with n, so that a mild error propagation is present.

Finally, we note that, in line with the first paragraph of Section 4.3, a fully explicit upper bound for $\|B^n\|_\infty$ can be obtained as well. From (4.6a) and Lemma 5.1, we easily obtain, for $B=B_s(x,y)$ and all $(x,y)\in S$,

$$\|B^n\|_\infty \leqslant 11\mathrm{e}(n+1)[1+\min\{\log(s+1),\log(n+1)\} \quad (n\geqslant 1, s\geqslant 2).$$

Acknowledgements

The authors profited from comments, about the presentation of the material, made by an anonymous referee.

References

[1] V.K. Barwell, Special stability problems for functional differential equations, BIT 15 (1975) 130–135.
[2] F.F. Bonsall, J. Duncan, Numerical ranges, in: R.G. Bartle (Ed.), Studies in Functional Analysis, The Mathematical Association of America, 1980, pp. 1–49.
[3] N. Borovykh, D. Drissi, M.N. Spijker, A note about Ritt's condition, related resolvent conditions and power bounded operators, Numer. Funct. Anal. Optim., in press.
[4] N. Borovykh, M.N. Spijker, The sharpness of stability estimates corresponding to a general resolvent condition, Linear Algebra Appl. 311 (2000) 161–175.
[5] J.L.M. van Dorsselaer, J.F.B.M. Kraaijevanger, M.N. Spijker, Linear stability analysis in the numerical solution of initial value problems, Acta Numer. 1993 (1993) 199–237.
[6] S. Friedland, A generalization of the Kreiss matrix theorem, SIAM J. Math. Anal. 12 (1981) 826–832.
[7] E. Hairer, G. Wanner, Solving Ordinary Differential Equations, Springer, Berlin, 1991.
[8] R.A. Horn, C.R. Johnson, Matrix Analysis, Cambridge University Press, Cambridge, 1990.
[9] J.F.B.M. Kraaijevanger, Two counterexamples related to the Kreiss matrix theorem, BIT 34 (1994) 113–119.
[10] H.-O. Kreiss, Über die Stabilitätsdefinition für Differenzengleichungen die partielle Differentialgleichungen approximieren, BIT 2 (1962) 153–181.
[11] H.W.J. Lenferink, M.N. Spijker, A generalization of the numerical range of a matrix, Linear Algebra Appl. 140 (1990) 251–266.
[12] R.J. LeVeque, L.N. Trefethen, On the resolvent condition in the Kreiss matrix theorem, BIT 24 (1984) 584–591.
[13] Ch. Lubich, O. Nevanlinna, On resolvent conditions and stability estimates, BIT 31 (1991) 293–313.
[14] O. Nevanlinna, On the growth of the resolvent operators for power bounded operators, in: J. Janas, F.H. Szafraniec, J. Zemánek (Eds.), Linear Operators, Banach Center Publications, vol. 38, Inst. Math. Pol. Acad. Sciences, Warszawa, 1997, pp. 247–264.
[15] C. Palencia, Stability of rational multistep approximations of holomorphic semigroups, Math. Comp. 64 (1995) 591–599.
[16] S.V. Parter, Stability, convergence, and pseudo-stability of finite-difference equations for an over-determined problem, Numer. Math. 4 (1962) 277–292.
[17] S.C. Reddy, L.N. Trefethen, Stability of the method of lines, Numer. Math. 62 (1992) 235–267.
[18] R.D. Richtmyer, K.W. Morton, Difference Methods for Initial-Value Problems, 2nd Edition, Wiley, New York, 1967.
[19] W. Rudin, Functional Analysis, McGraw-Hill, New York, 1973.
[20] M.N. Spijker, On a conjecture by LeVeque and Trefethen related to the Kreiss matrix theorem, BIT 31 (1991) 551–555.
[21] M.N. Spijker, Numerical stability, resolvent conditions and delay differential equations, Appl. Numer. Math. 24 (1997) 233–246.
[22] M.N. Spijker, F.A.J. Straetemans, Stability estimates for families of matrices of nonuniformly bounded order, Linear Algebra Appl. 239 (1996) 77–102.
[23] M.N. Spijker, F.A.J. Straetemans, Error growth analysis, via stability regions, for discretizations of initial value problems, BIT 37 (1997) 442–464.
[24] M.N. Spijker, S. Tracogna, B.D. Welfert, About the sharpness of the stability estimates in the Kreiss matrix theorem, Math. Comp., in press.
[25] J.C. Strikwerda, Finite Difference Schemes and Partial Differential Equations, Wadsworth, Belmont, 1989.
[26] J.C. Strikwerda, B.A. Wade, A survey of the Kreiss matrix theorem for power bounded families of matrices and its extensions, in: J. Janas, F.H. Szafraniec, J. Zemánek (Eds.), Linear Operators, Banach Center Publications, vol. 38, Inst. Math. Pol. Acad. Sciences, Warszawa, 1997, pp. 329–360.

[27] E. Tadmor, The equivalence of L_2-stability, the resolvent condition, and strict H-stability, Linear Algebra Appl. 41 (1981) 151–159.
[28] E. Tadmor, The resolvent condition and uniform power boundedness, Linear Algebra Appl. 80 (1986) 250–252.
[29] K.C. Toh, L.N. Trefethen, The Kreiss matrix theorem on a general complex domain, Report no. 97/13, Oxford Univ. Comp. Lab, 1997.
[30] D.S. Watanabe, M.G. Roth, The stability of difference formulas for delay differential equations, SIAM J. Numer. Anal. 22 (1985) 132–145.
[31] E. Wegert, L.N. Trefethen, From the Buffon needle problem to the Kreiss matrix theorem, Amer. Math. Monthly 101 (1994) 132–139.
[32] M. Zennaro, P-stability properties of Runge–Kutta methods for delay differential equations, Numer. Math. 49 (1986) 305–318.

ELSEVIER

Journal of Computational and Applied Mathematics 125 (2000) 57–68

JOURNAL OF COMPUTATIONAL AND APPLIED MATHEMATICS

www.elsevier.nl/locate/cam

Numerical bifurcation analysis for ODEs

W. Govaerts [1]

Department of Applied Mathematics and Computer Science, University of Gent, Krijgslaan 281-S9, B-9000 Gent, Belgium

Received 4 April 1999; received in revised form 4 October 1999

Abstract

We discuss numerical methods for the computation and continuation of equilibria and bifurcation points of equilibria of dynamical systems. We further consider the computation of cycles as a boundary value problem, their continuation and bifurcations. Homoclinic orbits can also be computed as (truncated) boundary value problems and numerically continued. On curves of homoclinic orbits further bifurcations can be detected and computed. We discuss the basic numerical methods, the connections between various computational objects, and provide references to the literature and software implementations. © 2000 Elsevier Science B.V. All rights reserved.

Keywords: Equilibrium; Continuation; Cycle

1. Introduction

Bifurcation of ODEs is a subfield of dynamical systems theory. We consider parameterized ordinary differential equations of the form

$$\frac{dx}{dt} \equiv x' = G(x, \alpha), \tag{1}$$

where $x \in \mathbb{R}^N$ is called the state variable, $\alpha \in \mathbb{R}^m$ is called the parameter and $G(x,\alpha) \in \mathbb{R}^N$ is a nonlinear function of x, α. The space in which x lives is called the state space. Examples of systems of the form (1) are ubiquitous in mathematical modelling. Classical application fields are many branches of physics, engineering, chemistry, economy and finance. In our opinion the most interesting new examples are in biology and medicine. A large collection of biological examples can be found in [20], e.g., insect outbreak models (the spruce budworm), harvesting of fish, predator–prey models, pest control, and biochemical reactions. For a recent work on epidemiology see [8]. A complicated neurobiological model (a nerve cell of the crab, Cancer Borealis) is considered in [13].

E-mail address: willy.govaerts@rug.ac.be (W. Govaerts).

[1] Research Director of the Fund for Scientific Research FWO — Vlaanderen.

PII: S0377-0427(00)00458-1

For a review of modelling the dynamics of HIV infection see [21]. Many of these models compete with or are combined with models of another type, e.g. discrete dynamical systems, differential-algebraic equations, delay differential equations and PDEs. For the study of these other approaches the methods and results from bifurcation in ODEs often provide the basic ideas and inspiration.

If an initial condition $x(0)=x_0$ is given then for a fixed parameter α the system (1) has a unique solution $x(t)$ under very general conditions on $G(x,\alpha)$. The domain of definition may be small but always includes a neighborhood of $t=0$. Such a solution is called an orbit of (1). A relevant region of the state space typically contains a family of orbits which do not intersect each other (because of the uniqueness for a given initial condition). A picture of such a region is called a phase portrait.

The focus of dynamical systems theory is the behaviour of phase portraits under variations of the parameter α. Suppose that for (1) we have a particular value α_0 of α in mind. Then for values of α near α_0 the system (1) is called an unfolding of

$$\frac{dx}{dt} \equiv x' = G(x,\alpha_0). \tag{2}$$

In a region in the state space, (2) is structurally stable if the phase portraits of (1) are qualitatively the same as those of (2) for $\|\alpha-\alpha_0\|$ sufficiently small, i.e., they can be obtained by smooth nonlinear transformations of coordinates. Values α_0 for which (2) is not structurally stable are called bifurcation values. To make this mathematically precise a sophisticated machinery is required. We refer to [14,15] for details. For most practical applications and numerical work it is enough to understand intuively that phase portraits are qualitatively the same if there exists a homeomorphism that preserves the orbits and the direction of the flow.

One important aim of analytical bifurcation theory is the classification of bifurcations. To this end an equivalence relation is defined and in each equivalence class a bifurcation with a mimimal number of state variables and of simplest polynomial form is chosen. This is called a *normal form*. A normal form usually contains one or more coefficients which determine the dynamic behaviour of the normal form itself and of its possible unfoldings.

The simplest solutions to (1) are the equilibria, i.e., solutions to the equation

$$G(x,\alpha)=0. \tag{3}$$

If (x,α) is a solution to (3) then the question arises whether α is a bifurcation value. Since in many models parameters may be introduced in various ways, the essential problem is to give conditions on $G(x,\alpha)$ for a fixed value of α such that in all unfoldings of the problem the phase portraits of the slightly perturbed systems are qualitatively the same as in the unperturbed system. If no bifurcation can occur in an unfolding, we say that the equilibrium is structurally stable. It turns out that the necessary and sufficient condition for structural stability of an equilibrium is that the Jacobian matrix G_x has no eigenvalues on the imaginary axis of the complex plane. This is not to be confused with the dynamic stability of the equilibrium itself. The equilibrium is asymptotically stable if all eigenvalues have a strictly negative real part, it is unstable if there is at least one eigenvalue with a stricly positive real part.

Now eigenvalues on the imaginary axis can generically appear in one-parameter problems ($m=1$) in two ways, either as a simple zero eigenvalue or as a conjugate pair $\pm i\omega$ ($\omega>0$) of pure imaginary eigenvalues. The first case is called a Fold point, the second case a Hopf point. The Fold bifurcation typically leads to a change in the stability properties of the solution under parameter perturbations, the Hopf bifurcation to the emergence of another type of solution, namely *periodic orbits*, i.e.,

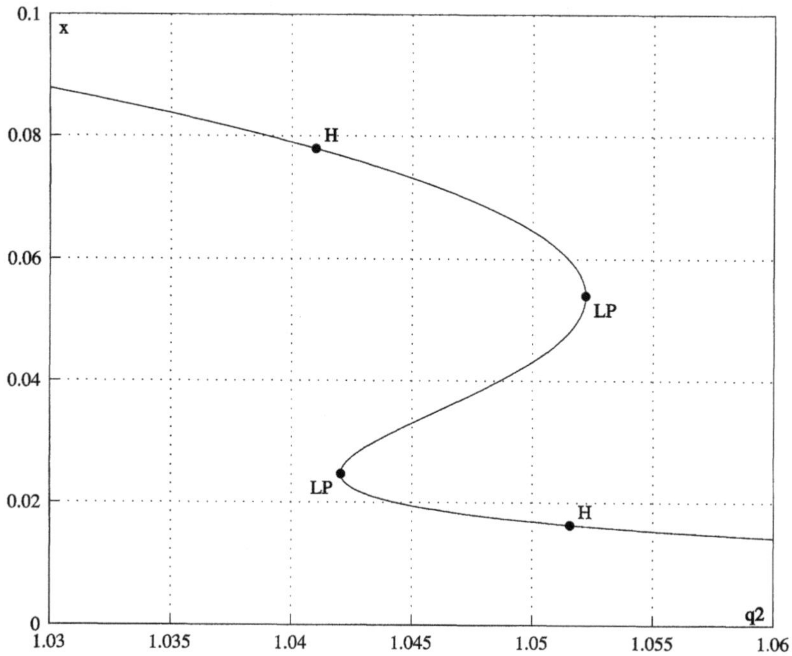

Fig. 1. Fold and Hopf points in the catalytic oscillator model.

solutions for which $x(T) = x(0)$ for a number $T > 0$ called the period of the cycle. If a periodic orbit is isolated (which is generically the case) they it is also called a *(limit) cycle*.

As an example, we consider the catalytic oscillator model in [5] for the catalytic reaction of CO-oxidation. It has three state variables x, y, s and seven parameters $q_1, q_2, q_3, q_4, q_5, q_6, k$. Explicitly, we have

$$x' = 2q_1(1 - x - y - s)^2 - 2q_5x^2 - q_3xy,$$

$$y' = q_2(1 - x - y - s) - q_6y - q_3xy,$$

$$s' = q_4(1 - x - y - s) - kq_4s.$$

We used CONTENT [16] to study this system and produce some relevant pictures. The system has a stable equilibrium at $x = 0.0014673$, $y = 0.826167$, $s = 0.123119$ for the parameter values $q_1 = 2.5$, $q_2 = 1.92373$, $q_3 = 10$, $q_4 = 0.0675$, $q_5 = 1$, $q_6 = 0.1$, $k = 0.4$. Freeing the parameter q_2 we can compute a curve of equilibria. In Fig. 1 we present a part of this curve in (q_2, x)-space. It has two Fold points (LP in Fig. 1) and two Hopf points (H in Fig. 1). The equilibria in the upper left and bottom right corners are stable, the equilibria between the two Hopf points are unstable. As can be expected generically, the Fold points are turning points with respect to the parameter q_2.

The Hopf point in the bottom right of Fig. 1 is found for the parameter value $q_2 = 1.051556$; the state values are $x = 0.01635814$, $y = 0.5239649$, $s = 0.3283407$. From this point we can start a family of periodic orbits, again with free parameter q_2. We present it in Fig. 2 in (x, y)-space. We note that close to the Hopf point the periodic orbit looks like an ellipse (the boundary of the inner white space in Fig. 2) but later becomes irregular (the outer contour of the drawing in Fig. 2).

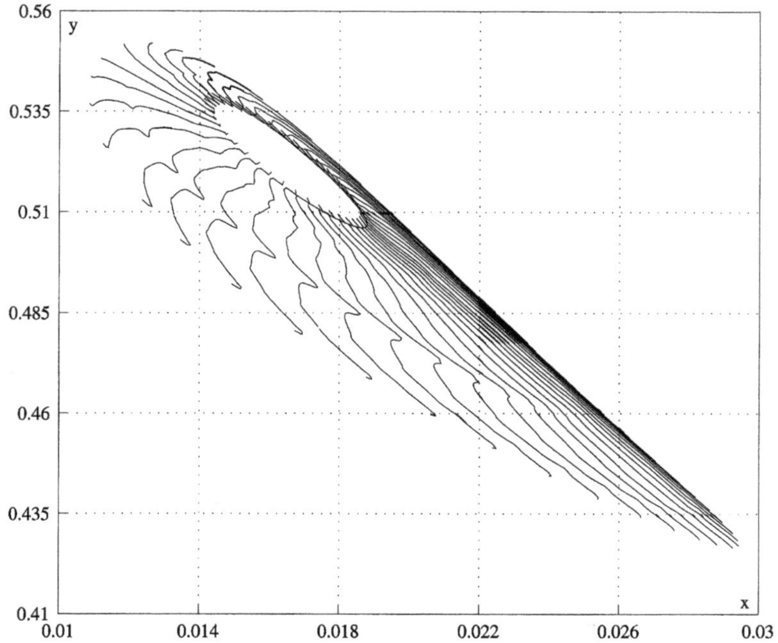

Fig. 2. Growing periodic orbits in the catalytic oscillator model.

In the case of more parameters ($m > 1$) more complicated equilibrium bifurcations can be expected. We deal with these in Section 3.2.

In generic one-parameter problems cycles can bifurcate in several ways. One possibility is that the period of the orbit tends to infinity if α approaches a bifurcation value. Then the orbit becomes a homoclinic orbit at the bifurcation value (see Section 5). Some other possibilities are characterized by the properties of the *monodromy matrix*, i.e., the linearized return map of the cycle. The eigenvalues of this matrix are called the *multipliers* of the cycle. We deal with cycles in Section 4.

A third well-studied solution type of (1) are the connecting orbits, i.e., solutions $x(t)$ which converge to finite limit values x_+ an x_- if t tends to $\pm\infty$, respectively. If the two limit values are the same, then the connecting orbit is called a homoclinic connection, otherwise a heteroclinic connection.

There are other types of solutions as well. We mention invariant tori. Chaotic behaviour is also a solution type.

2. Numerical continuation

Numerical continuation is one basic ingredient of the numerical study of bifurcation problems (for good reasons the interactive software package [16] is called CONTENT = CONTinuation ENvironmenT). It is a technique to compute curves of solutions to an underdetermined system of equations. In bifurcation problems it allows to find bifurcations of more and more complicated types, starting from simple objects like equilibria.

Consider a system of nonlinear equations

$$f(z)=0, \tag{4}$$

where $z \in \mathbb{R}^{N+1}$, $f(z) \in \mathbb{R}^N$. In the context of (1) z might consist of the N components of x and one free parameter. If z_0 is a solution to (4) and the Jacobian matrix $f_z(z_0)$ has full rank N then by the implicit function theorem (4) has a curve of solutions through z_0. Numerical continuation is a technique to "compute" this curve, i.e., to compute sequences of points on the curve. Practically all currently used packages (in particular, AUTO [9], CONTENT [16]) use a tangent predictor step and a Newton-type corrector step. For details we refer to [4, Vol. III]; a full discussion is given in [12, Chapter 2].

The tangent vector $v \in \mathbb{R}^{N+1}$ to the solution curve of (4) is determined by the equation $f_z v = 0$ and some normalization condition that ensures that v is nonzero and points in the desired direction along the curve. Typically one adds a condition of the form $v_0^{\mathrm{T}} v = 1$ where v_0 is an approximation to v, e.g., from a previous continuation step. If in z one distinguishes a state component $x \in \mathbb{R}^N$ and a parameter component $\alpha \in \mathbb{R}$ then this leads to a bordered system of equations

$$\begin{pmatrix} f_y & f_\alpha \\ v_{0y}^{\mathrm{T}} & v_{0\alpha} \end{pmatrix} \begin{pmatrix} v_y \\ v_\alpha \end{pmatrix} = \begin{pmatrix} 0_N \\ 1 \end{pmatrix}. \tag{5}$$

Such bordered systems are ubiquitous in numerical continuation and bifurcation. Then we normalize v by replacing $v \leftarrow v/\|v\|$ and the prediction for the next step along the curve is $z^1 = z_0 + (\Delta s)v$ where Δs is the predicted steplength along the tangent, a quantity which is updated during the continuation. Starting with z^1 as an initial guess, a new point along the curve is computed. A simple way to achieve this is to look for a point in the hyperplane orthogonal to v, i.e., to solve by Newton's method the system

$$f(z)=0, \tag{6}$$

$$v^{\mathrm{T}}(z-z^1)=0. \tag{7}$$

This method is used in AUTO [10]; it is called pseudo-arclength continuation. The linearized equations again form a bordered matrix. Other software packages use slightly different corrector algorithms (and so compute slightly different points on the curve). The convergence properties of the iteration can be used (in one of several ways) to update the choice of the steplength. If the iteration does not converge in a satisfactory way, then a popular strategy is to halve the stepsize and try again. If convergence is obtained, then the number of Newton iteration steps can be used to update the steplength.

3. Bifurcation of equilibria

3.1. Codimension-1 bifurcations

If an equilibrium is found and one parameter is freed ($m=1$) then one can numerically continue a path of equilibria of (1). The equilibrium is structurally stable if G_x has no eigenvalues on the imaginary axis. Bifurcations occur if either a real eigenvalue crosses the imaginary axis (Fold bifurcation) or a conjugate pair of complex eigenvalues crosses it (Hopf bifurcation). In the Fold case the dynamic behaviour at the Fold point depends on a normal form coefficient

$$a = \tfrac{1}{2} p^{\mathrm{T}} G^0_{xx} qq, \tag{8}$$

where p, q are the left and right singular vectors of G_x, respectively, normalized so that $\langle p,q\rangle = \langle q,q\rangle = 1$. Generically, $a \neq 0$ and the system behaves like

$$w' = aw^2. \tag{9}$$

The case $a=0$ is a codimension-2 situation (see Section 3.2). In the Hopf case the dynamic behaviour depends on another coefficient, called the first Lyapunov value

$$\ell_1 = \tfrac{1}{2} Re\langle p, C(q,q,\bar{q}) + B(\bar{q},\zeta) - 2B(q, A^{-1}B(q,\bar{q}))\rangle, \tag{10}$$

where $\zeta = (2\mathrm{i}\omega I_N - A)^{-1}B(q,q)$, $A = G_x$ and B, C denote the tensors of second- and third-order derivatives of G at the Hopf point; p, q are determined by the conditions $Aq = \mathrm{i}\omega q$, $p^{\mathrm{H}}A = \mathrm{i}\omega p^{\mathrm{H}}$, $\langle p,q\rangle = 1$. If stability is lost in the Hopf point and $\ell_1 < 0$ then stable periodic orbits arise at the side where the equilibrium is unstable (soft loss of stability). If stability is lost and $\ell_1 > 0$ then there are unstable cycles at the side where the equilibrium is stable (hard loss of stability).

3.2. Codimension-2 bifurcations

If two parameters are freed ($m = 2$) then bifurcations of a more complex type can be expected. In fact, there are five such bifurcations. Three of them are determined by the Jacobian structure of G_x; they are called the Bogdanov–Takens (BT), Zero-Hopf (ZH) and double-Hopf (DH) bifurcations, respectively. In the BT-case G_x has a double-zero eigenvalue with geometric multiplicity one. In the ZH-case a zero eigenvalue and a conjugate pair of pure imaginary eigenvalues are simultaneously present. In the DH-case there are two (different) conjugate pairs of pure imaginary eigenvalues. The two other codimension-2 bifurcations are the cusp point (CP) and the Generalized Hopf point (GH). A CP point is found on a Fold curve if the value a in (8) vanishes, a GH point is found on a Hopf curve if ℓ_1 vanishes. The dynamic behaviour near codimension-2 points is quite complicated: bifurcations of cycles, homoclinic connections and chaotic behaviour are all generic. Therefore applications of bifurcation methods require a good understanding of the mathematical background, as can be obtained from, e.g., [15]. On the other hand, the results obtained in this way can be quite powerful in explaining the complex behaviour of dynamical systems. We refer to [12, Chapter 9] for examples in the case of GH.

3.3. Detection, computation and continuation of codimension-1 bifurcations

In a computational study of parameterized dynamical systems we typically start by computing equilibria for a given set of parameters. These can be found by solving (3) by any appropriate method. The eigenvalues of the Jacobian G_x tell us the stability properties of the equilibrium, i.e., some information on the behaviour of the dynamical system in a nearby region. Asymptotically stable equilibria can also be found by direct simulation of (1). If an equilibrium is found then we free a parameter of the problem and continue numerically the curve of equilibria. To *detect* bifurcations on this curve (Fold or Hopf) we need *test functions*. A test function is a function that can be computed in each continuation point and changes sign at the bifurcation point. It can also be used to *compute* the bifurcation point, e.g., by the bisection method. If it has a regular zero at the bifurcation point, then it can further be used to continue a curve of bifurcation point if another parameter is freed.

The determinant function $\det(G_x)$ provides a test function for Fold which is often available as a byproduct of an LU-factorization of G_x. A related but somewhat more sophisticated function is the value s obtained by solving

$$\begin{pmatrix} G_x & b \\ c^{\mathrm{T}} & d \end{pmatrix} \begin{pmatrix} v \\ s \end{pmatrix} = \begin{pmatrix} 0_N \\ 1 \end{pmatrix}, \tag{11}$$

where $b, c \in \mathbb{R}^N, d \in \mathbb{R}$ must be such that the square matrix in (11) is nonsingular. So it is necessary to update b, c, d along the equilibrium curve.

A test function for Hopf is the determinant of the bialternate product matrix $2A \odot I_N$ where $A = G_x$. We recall that if A, B are $N \times N$ matrices, then $A \odot B$ is an $m \times m$ matrix where $2m = N(N-1)$. The rows and columns of $A \odot B$ are labeled by multi-indices (p, q) and (r, s), respectively, where $1 \leqslant q < p \leqslant N$, $1 \leqslant s < r \leqslant N$ and the elements of $A \odot B$ are given by

$$(A \odot B)_{(p,q),(r,s)} = \frac{1}{2} \left\{ \begin{vmatrix} a_{pr} & a_{ps} \\ b_{qr} & b_{qs} \end{vmatrix} + \begin{vmatrix} b_{pr} & b_{ps} \\ a_{qr} & q_{ps} \end{vmatrix} \right\}. \tag{12}$$

The eigenvalues of $2A \odot I_N$ are precisely all sums of the form $\lambda_i + \lambda_j$ where $i < j$ and $\lambda_1, \ldots, \lambda_N$ are the eigenvalues of A. So $\det(2A \odot I_N)$ detects not only Hopf pairs where $\lambda_{i,j} = \pm i\omega$ but also *neutral saddles* where G_x has two real eigenvalues with opposite sign. Neutral saddles do not have a dynamic meaning for general equilibria (there is no bifurcation) but are quite useful in the numerical study of dynamical systems because they help to find Hopf points. Furthermore, Bogdanov–Takens points are often connected by curves of neutral saddles.

Also, they get a dynamic meaning in the case when the equilibrium is the hyperbolic fixed point of a homoclinic connection, see Section 5.

If a Fold or Hopf point is computed then we want to know its dynamic properties; this information is contained in the *normal form coefficients*, i.e., a for Fold and ℓ_1 for Hopf.

If a second parameter is freed, then we can compute curves of Fold or Hopf points. To this end we need *defining systems*, i.e., systems of equations whose regular solutions are the Fold, respectively, Hopf points.

A classical defining system for Fold curves [19,24] is

$$\begin{aligned} &G(x, \alpha) = 0, \\ &G_x(x, \alpha)v = 0, \\ &c^{\mathrm{T}} v = 1. \end{aligned} \tag{13}$$

This is a system of $2N + 1$ equations in the $2N + 2$ unknowns x, α, v. It is a typical example of a so-called *large augmented system* in the sense that it contains unknowns additional to the state and parameter variables. The additional unknown is the singular vector v of G_x. The vector c is an *auxiliary variable*, it is fixed and chosen somewhat arbitrarily. Formally, it is only required that it is not orthogonal to v. In practice, it is updated along the Fold curve; an obvious choice for c is the value of v obtained in the previous computed equilibrium point.

It is also possible to compute Fold curves using a minimally extended system, i.e., one whose only unknowns are the state and parameter variables. A typical example is the system

$$\begin{aligned} G(x,\alpha) &= 0, \\ s(x,\alpha) &= 0, \end{aligned} \tag{14}$$

where s is obtained by solving (11). An advantage of this method is that the derivatives of s can be obtained easily from the derivatives of G_x. The determinant function can replace s in this definition; however, it is usually not so well scaled and the computation of derivatives is more difficult.

For Hopf bifurcations, a rich variety of large extended systems and at least one minimally extended system is known; several systems for Fold and Hopf are incorporated into [16].

3.4. Detection, computation and continuation of bifurcations with higher codimension

To detect codimension-2 bifurcations on Fold and Hopf curves we need new test functions. BT, ZH and CP points can be expected on Fold curves. ZH, DH and GH points can be expected on Hopf curves. Large and minimally extended systems are available in all cases, cf. [16].

Many techniques for codimension-2 cases can be extended to codimension-3 and higher cases. However, each new codimension also introduces essentially new situations that require a novel approach. For example, for codimension-3 the case of 1 : 1 resonant Double Hopf (two identical Hopf pairs) was considered (numerically) in [13]. A survey of existing methods is given in [12].

Since the mathematical analysis of codimension-3 points is extremely complicated and can hardly be applied in practical situations, further numerical work in this direction does not seem urgently needed.

4. Bifurcation of cycles

4.1. Computation and continuation of cycles

To compute a cycle of period T of (1) one fixes the interval of periodicity by rescaling time. Then (1) becomes

$$x'(t) = TG(x(t),\alpha), \tag{15}$$

and we want solutions of period 1, i.e.,

$$x(0) = x(1). \tag{16}$$

The period T is one of the unknowns of the problem. In a continuation context we assume that a solution $(x_{k-1}(.), T_{k-1}, \alpha_{k-1})$ is already known and we want to find $(x_k(.), T_k, \alpha_k)$. Eqs. (15) and (16) together do not fix the solution completely since any solution can be translated freely in time, i.e., if $x(t)$ is a solution then so is $x(t+\tau)$ for any τ. So it is necessary to add a "phase condition" to fix the solution. In AUTO [10] and CONTENT [16] this is done as follows. Let $x(t)$ be a solution of (15) and (16). We want the phase-shifted solution that is closest to x_{k-1}, i.e., that minimizes

$$D(\sigma) \equiv \int_0^1 \|\tilde{x}(t+\sigma) - x_{k-1}(t)\|_2^2 \, dt. \tag{17}$$

The optimal solution must satisfy $(\mathrm{d}D(\sigma)/\mathrm{d}\sigma) = 0$, which leads to the condition

$$\int_0^1 x(t)x'_{k-1}(t)\,\mathrm{d}t = 0. \tag{18}$$

Now the cycle is determined by (15), (16) and (18) which together form a boundary value problem with integral condition. In a continuation context x, T, α vary along the curve of cycles in a function space and the continuation condition is

$$\int_0^1 (x(t) - x_{k-1}(t))^{\mathrm{T}}\dot{x}_{k-1}(t)\,\mathrm{d}t + (T - T_{k-1})\dot{T}_{k-1} + (\alpha - \alpha_{k-1})\dot{\alpha}_{k-1} = \Delta s, \tag{19}$$

where the derivatives are taken with respect to arclength in the function space and should not be confused with the time derivatives in (18).

The most widely used method to discretize this problem is the method of orthogonal collocation with piecewise polynomials. It is used in the code COLSYS [1] as well as in AUTO and CONTENT. The method is known for its high accuracy [7] and particularly suitable because of its known mesh adaptation techniques [23]. The numerical continuation of the discretized equations leads to linear systems with a structured sparsity. These are solved in AUTO by a sophisticated elimination strategy that allows to recover the multipliers as a byproduct.

4.2. Starting a cycle from a Hopf point

Asymptotically stable cycles can be found by time integration of (1). A more general way to find stable or unstable cycles is to start from a Hopf point. Let it be (x_0, α_0) with Hopf eigenvalues $\pm \mathrm{i}\omega_0$. The Hopf bifurcation theorem ensures the existence of a bifurcating branch of cycles. This branch can be locally parameterized by a real number ε and the asymptotic estimates hold:

$$x(t,\varepsilon) = x_0 + \varepsilon\phi(t) + \mathrm{O}(\varepsilon^2), \quad T(\varepsilon) = T_0 + \mathrm{O}(\varepsilon^2), \quad \alpha(\varepsilon) = \alpha_0 + \mathrm{O}(\varepsilon^2), \tag{20}$$

where $T(\varepsilon)$ is the period, $T_0 = (2\pi/\omega_0)$. The function $\phi(t)$ is the normalized nonzero periodic solution of the linearized, constant coefficient problem

$$\frac{\mathrm{d}\phi(t)}{\mathrm{d}t} = G_x^0\phi(t). \tag{21}$$

To compute a first cycle we can, in principle, solve (15), (16), (18) and (19) if in the last two we replace $x_{k-1}, T_{k-1}, \alpha_{k-1}$ by known quantities. The obvious choice is $x_0 + \varepsilon\psi(t), T_0, \alpha_0$, respectively, for some small value of ε where $\psi(t)$ is a time-scaled version of $\phi(t)$, i.e., $\psi(t) = \sin(2\pi t)w_s + \cos(2\pi t)w_c$, where $w_s, w_c \in \mathbb{R}^N$ are such that $w_s + \mathrm{i}w_c$ is a right eigenvector of G_x^0 for the eigenvalue $i\omega_0$. By rescaling, if necessary, we may assume $\|\psi\| = 1$. So (18) is replaced by

$$\int_0^1 x(t)\frac{\mathrm{d}\psi(t)}{\mathrm{d}t}\,\mathrm{d}t = 0. \tag{22}$$

Next, in (19) we set $\dot{T}_{k-1} = 0$, $\dot{\alpha}_{k-1} = 0$ and $\dot{x}_{k-1} = \psi(t)$ so that (19) is actually replaced by

$$\int_0^1 (x(t) - x_0)^{\mathrm{T}}\psi(t)\mathrm{d}t = \Delta s. \tag{23}$$

4.3. Codimension-1 bifurcations of cycles

A cycle always has a multiplier equal to 1. If all other multipliers are strictly inside the unit circle in the complex plane, then the cycle is asymptotically stable. If at least one multiplier has modulus greater than one, the cycle is unstable. Three bifurcations are generic on a curve of cycles: Fold, Flip and Neimark–Sacker. In a Fold point 1 is a double multiplier with geometric multiplicity one. Typically, this indicates a turning point in the curve of cycles. In a Flip point there is a multiplier equal to -1. Typically, this indicates a *period doubling* of the cycle, i.e., the cycle is replaced by a cycle with initially double period. In a Neimark–Sacker point there is a conjugate pair of complex eigenvalues with modulus 1. Typically, this indicates a bifurcation to an invariant torus, i.e., the periodic behaviour of the dynamical system is replaced by a much more complicated movement on an invariant torus. The Fold, Flip and Neimark–Sacker bifurcations can be detected by monitoring the multipliers. It is also possible to give test functions for these bifurcations. To numerically compute and continue curves of codimension-1 bifurcations of cycles, defining systems can be obtained in the form of generalized boundary value problems. See [4, Vol. III] for details.

5. Connecting orbits

The numerical methods to compute homoclinic orbits can easily be extended to heteroclinic orbits. We restrict to homoclinic orbits since they are particularly important in global bifurcation theory. In fact, the appearance of a homoclinic orbit is often related to the disappearance of cycle solutions and the onset of chaotic behaviour. Wild dynamic behaviour is the rule near homoclinic orbits and numerical time integration can be expected to have problems (this applies also to heteroclinic connections, for an example see [22]).

Nevertheless, homoclinic orbits are quite common. In fact, they are codimension-1 phenomena, i.e., they appear generically in problems with one free parameter. Let x_0 be the limit of $x(t)$ for $t \to \pm\infty$. Obviously, x_0 has to be an unstable equilibrium. Actually, there are two types of codimension-1 homoclinic orbits. In one type x_0 is a hyperbolic equilibrium of (1), i.e., $G_x(x_0)$ has no eigenvalues with real part zero. In the second type x_0 is a Fold point with one zero eigenvalue and no other eigenvalue with real part zero.

Homoclinic orbits can be computed as a boundary value problem in the following way. If x_0 is hyperbolic with parameter value α_0 then there exists a unique equilibrium $x_e(\alpha)$ for α in a neighborhood of α_0 such that $x_e(\alpha_0)=x_0$. If x_0 is a Fold, then we set $x_e(\alpha) \equiv x_0$. Now consider the following boundary value problem on an infinite interval:

$$x'(t) = G(x(t), \alpha), \tag{24}$$

$$\lim_{t\to\pm\infty} x(t) = x_e(\alpha). \tag{25}$$

As in the case of cycles, any time-shift of a solution to (24) and (25) is again a solution, so we need a phase condition. If an initial guess $\tilde{x}(t)$ is known, then an integral phase condition

$$\int_{-\infty}^{+\infty} (x(t) - \tilde{x}(t))^{\mathrm{T}} \tilde{x}'(t)\,\mathrm{d}t = 0 \tag{26}$$

similar to (18) can be used to fix the phase.

In [2,3] Beyn proposes to truncate the boundary-value problem (24)–(26) to a finite interval $[-T_-, T_+]$ with suitable boundary conditions. Suppose that $G_x(x_e(\alpha), \alpha)$ has n_s eigenvalues with negative real part, n_c eigenvalues with zero real part and n_u eigenvalues with positive real part. So n_c is either 0 or 1 and $n_s + n_c + n_u = N$. Now (25) can be interpreted as

$$x(-T_-) \in B^u(x_e(\alpha)), \quad x(T^+) \in B^s(x_e(\alpha)). \tag{27}$$

Here B^u and B^s denote the unstable and stable sets of $x_e(\alpha)$, respectively. The right eigenvectors corresponding to the eigenvalues with positive (respectively, negative) real part span the tangent space to the unstable (respectively, stable) manifold. So let $L_s(\alpha)$ (respectively, $L_u(\alpha)$) be an $n \times n_s$ matrix (respectively, an $n \times n_u$ matrix) whose columns span the left-singular space that corresponds to the eigenvalues with negative real part (respectively, positive real part). So we can linearize the conditions (25) by

$$L_s^T(x(-T_-) - x_e(\alpha)) = 0, \tag{28}$$

$$L_u^T(x(T_+) - x_e(\alpha)) = 0. \tag{29}$$

These conditions force $x(-T_-)$ to be in the center-unstable and $x(T_+)$ to be in the center-stable eigenspaces of $G_x(x_e(\alpha))$. Finally, the phase condition (26) is simply truncated to $[-T_-, T_+]$.

The above method is implemented in AUTO, using a collection of routines called HomCont, cf. [6]. We note that starting points for the continuation of homoclinic orbits may be hard to find. One approach is to compute cycles to large period, cf. [10]. Also, it is known that some codimension-2 equilibrium bifurcations, in particular BT-points, are starting points for homoclinic orbits.

We note that there is another approach to the computation of connecting orbits which avoids the truncation of the interval by parametrizing the orbits in a different way, say by arclength instead of time. We refer to [17,18] for more details.

Along branches of homoclinic orbits several codimension-2 bifurcations can appear. Among other possibilities, n-homoclinic orbits may emerge which follow the homoclinic loop n times. Another possibility is a change in the stability of the cycles that accompany the homoclinic orbit. For details of the dynamics near codimension-2 homoclinic orbits we refer to [4,11].

As in the case of equilibria and cycles, codimension-2 homoclinic bifurcation points are detected by locating zeroes of certain *test functions*. In the simplest cases, test functions can be obtained from the eigenvalues of the equilibrium. For example, in the case of a hyperbolic equilibrium, a *resonant saddle* bifurcation is detected by the test function $\lambda_1^s + \lambda_1^u$ where λ_1^s (respectively, λ_1^u) is the smallest in absolute value stable (respectively, unstable) eigenvalue. We note that test functions for neutral saddles of equilibria can be used in this situation.

In other cases, the computation of the test functions requires the homoclinic solution or the solution to an adjoint variational equation. We refer to [4, Vol. III] for details.

References

[1] U.M. Ascher, J. Christiansen, R.D. Russell, A collocation solver for mixed order systems of boundary value problems, Math. Comp. 33 (1979) 659–679.

[2] W.-J. Beyn, Global bifurcations and their numerical computation, in: D. Roose, A. Spence, B. De Dier (Eds.), Continuation and Bifurcations: Numerical Techniques and Applications, Kluwer, Dordrecht, 1990, pp. 169–181.

[3] W.-J. Beyn, The numerical computation of connecting orbits in dynamical systems, IMA J. Numer. Anal. 9 (1990) 379–405.
[4] H. Broer, F. Takens (Eds.), Handbook of Dynamical Systems, Vol. I (eds. B. Hasselblatt, A. Katok): Ergodic Theory; Vol. II (eds. H. Broer, F. Takens): Bifurcation Theory; Vol. III (eds. B. Fiedler, G. Iooss, N. Kopell): Towards Applications, Elsevier, Amsterdam, to be published.
[5] V.I. Bykov, G.S. Yablonski, V.F. Kim, On the simple model of kinetic self-oscillations in catalytic reaction of CO-oxidation, Dokl. AN SSSR 242(3) (1978) 637–639 (in Russian).
[6] A. Champneys, Yu.A. Kuznetsov, B. Sandstede, A numerical toolbox for homoclinic bifurcation analysis, Internat. J. Bifur. Chaos. 6 (1996) 867–887.
[7] C. De Boor, B. Swartz, Collocation at Gaussian points, SIAM J. Numer. Anal. 10 (1973) 582–606.
[8] O. Diekmann, H.J. Heesterbeek, Mathematical Epidemiology of Infectious Diseases, Wiley, New York, 1999.
[9] E.J. Doedel, A.R. Champneys, T.F. Fairgrieve, Yu.A. Kuznetsov, B. Sandstede, X.J. Wang, AUTO97: Continuation and Bifurcation Software for ordinary differential Equations (with HomCont), User's Guide, Concordia University, Montreal, Canada, 1997 (ftp.cs.concordia/pub/doedel/auto).
[10] E.J. Doedel, J.-P. Kernévez, AUTO: Software for continuation problems in ordinary differential equations with applications, California Institute of Technology, Applied Mathematics, 1986.
[11] B. Fiedler, Global pathfollowing of homoclinic orbits in two-parameter flows, in: G. Dangelmayr, B. Fiedler, K. Kirchgässner, A. Mielke (Eds.), Dynamics of Nonlinear Waves in Dynamical Systems: Reduction, Bifurcation and Stability, Pitman Research Notes in Mathematics, Vol. 352, Pitman, London, 1996.
[12] W. Govaerts, Numerical Methods for Bifurcations of Dynamical Equilibria, SIAM Publications, Philadelphia, PA, 2000.
[13] W. Govaerts, J. Guckenheimer, A. Khibnik, Defining functions for multiple Hopf bifurcations, SIAM J. Numer. Anal. 34 (1997) 1269–1288.
[14] J. Guckenheimer, P. Holmes, in: Nonlinear Oscillations, Dynamical Systems, and Bifurcations of Vector Fields, Applied Mathematical Sciences, Vol. 42, Springer, Berlin, 1983.
[15] Yu.A. Kuznetsov, in: Elements of Applied Bifurcation Theory, Applied Mathematical Sciences, Vol. 112, Springer, Berlin, 1995, 1998.
[16] Yu.A. Kuznetsov, V.V. Levitin, CONTENT: A multiplatform environment for analyzing dynamical systems, Dynamical Systems Laboratory, CWI, Amsterdam 1995–1997 (ftp.cwi.nl/pub/CONTENT).
[17] L. Lui, G. Moore, R. Russell, Computation and continuation of homoclinic and heteroclinic orbits with arclength parametrizations, SIAM J. Sci. Comput. 18 (1997) 69–93.
[18] Y. Lui, L. Lui, T. Tang, The numerical computation of connecting orbits in dynamical systems: a rational spectral approach, J. Comput. Phys. 111 (1994) 373–380.
[19] G. Moore, A. Spence, The calculation of turning points of nonlinear equations, SIAM J. Numer. Anal. 17 (1980) 567–576.
[20] J.D. Murray, in: Mathematical Biology, Biomathematics Texts, Vol. 19, Springer, Berlin, 1989.
[21] A.S. Perelson, P.W. Nelson, Mathematical analysis of HIV-1 dynamics in vivo, SIAM Rev. 41 (1999) 3–44.
[22] S. Rinaldi, G. Feichtinger, F. Wirl, Corruption dynamics in democratic societies, Complexity 3 (5) (1998) 53–64.
[23] R.D. Russell, J. Christiansen, Adaptive mesh selection strategies for solving boundary value problems, SIAM J. Numer. Anal. 15 (1978) 59–80.
[24] R. Seydel, Numerical computation of branch points in nonlinear equations, Numer. Math. 33 (1979) 339–352.

ELSEVIER

Journal of Computational and Applied Mathematics 125 (2000) 69–81

JOURNAL OF COMPUTATIONAL AND APPLIED MATHEMATICS

www.elsevier.nl/locate/cam

Preserving algebraic invariants with Runge–Kutta methods

Arieh Iserles[a,*], Antonella Zanna[b]

[a]*Department of Applied Mathematics and Theoretical Physics, University of Cambridge, Silver Street, Cambridge CB3 9EW, UK*

[b]*Department of Informatics, University of Bergen, Bergen N-5020, Norway*

Received 22 September 1999; received in revised form 6 March 2000

Abstract

We study Runge–Kutta methods for the integration of ordinary differential equations and the retention of algebraic invariants. As a general rule, we derive two conditions for the retention of such invariants. The first is a condition on the coefficients of the methods, the second is a pair of partial differential equations that otherwise must be obeyed by the invariant. This paper extends previous work on multistep methods in Iserles (Technical Report NA1997/13, DAMTP, University of Cambridge, 1997). The cases related to the retention of quadratic and cubic invariants, perhaps of greatest relevance in applications, are thoroughly discussed. We conclude recommending a generalized class of Runge–Kutta schemes, namely Lie-group-type Runge–Kutta methods. These are schemes for the solution of ODEs on Lie groups but can be employed, together with group actions, to preserve a larger class of algebraic invariants without restrictions on the coefficients.

1. Background and notation

In this paper we study the numerical solution by Runge–Kutta methods of the ordinary differential system

$$\boldsymbol{y}' = \boldsymbol{f}(t,\boldsymbol{y}), \quad \boldsymbol{y}(0) = \boldsymbol{y}_0 \tag{1}$$

for $t \geqslant 0$, where $\boldsymbol{y} \in \mathbb{R}^d$ and $\boldsymbol{f} : \mathbb{R}^+ \times \mathbb{R}^d \to \mathbb{R}^d$ is a Lipschitz function. We assume that the exact solution $\boldsymbol{y}(t)$ of (1) is known to obey the condition that there exists a nontrivial function $\rho : \mathbb{R}^d \times \mathbb{R}^d \to \mathbb{R}$ (or a family of such functions) such that

$$\rho(\boldsymbol{y}(t), \boldsymbol{y}_0) \equiv 0, \quad t \geqslant 0. \tag{2}$$

* Corresponding author.

E-mail address: a.iserles@amtp.cam.ac.uk (A. Iserles).

PII: S0377-0427(00)00459-3

We say, in this case, that the solution $\boldsymbol{y}$ is *ρ-invariant.* [1] Sometimes, we say that ρ is a first integral of (1) or that it is a conservation law or that it defines a manifold $\mathcal{M}$ on which the solution $\boldsymbol{y}$ evolves. All these terms will be used interchangeably in the course of the paper. The degree of smoothness of ρ is related to the degree of smoothness of the function $\boldsymbol{f}$ defining the differential equation (1). Moreover, we say that ρ is a *strong invariant* if there exists a nonempty open set $\mathcal{U}$ in $\mathbb{R}^d$ such that for all $\boldsymbol{y}_0 \in \mathcal{U}$ the solution y with initial value $y(0) = y_0$ satisfies $\rho(\boldsymbol{y}(t), \boldsymbol{y}_0) \equiv 0$ for $t \geqslant 0$. In the present paper, we restrict our attention to the case when ρ is a strong invariant.

There exist numerous problems in applied mathematics that can be paraphrased in the above formalism. Just to mention a few, many physical systems evolve in time and yet their total energy or the phase-space volume or angular momentum stay put. In particular, the Hamiltonian energy of *Hamiltonian systems* is preserved. See [11,26,27,19] for further examples and applications.

Given the differential equation (1) in tandem with the invariance condition (2) and having introduced a subdivision $t_0 = 0 < t_1 < \cdots < t_n < \cdots$ of the integration interval, we say that a one-step numerical method

$$\boldsymbol{y}_{n+1} = \boldsymbol{\phi}_h(\boldsymbol{y}_n), \quad h = t_{n+1} - t_n \tag{3}$$

is *ρ-invariant* (or equivalently *$\mathcal{M}$-invariant*) if

$$\rho(\boldsymbol{y}_n, \boldsymbol{y}_0) = 0, \quad \forall n \geqslant 0 \tag{4}$$

for all $h < \bar{h}$, or, equivalently,

$$\boldsymbol{y}_0 \in \mathcal{M} \Rightarrow \boldsymbol{y}_n \in \mathcal{M} \quad \text{for all } n \geqslant 0, \tag{5}$$

$\mathcal{M}$ being the manifold defined by the function ρ [7].

Conditions that ensure preservation of invariants by Runge–Kutta methods have been already considered in a number of papers. Let us mention first the work of Cooper [4] who proved that there exists a subclass of Runge–Kutta methods that preserve quadratic invariants: all the functions ρ of the form $\rho(\boldsymbol{y}, \boldsymbol{y}_0) = \sum_{i,j=1}^{d} \alpha_{i,j} y^i y^j + \sum_{i=1}^{d} \beta_i y^i + \gamma$ where $\alpha_{i,j}$, β_i and γ are coefficients allowed to depend on y_0. The very same schemes that preserve quadratic invariants preserve also canonical symplectic structure, a result independently discovered in [26]. Later, in their investigation on numerical methods and isospectral flows, Calvo et al. proved that there is no subclass of such schemes that preserves also cubic laws: in other words, given an RK method that preserves quadratic manifolds, it is always possible to construct a differential equation with a cubic invariant ρ for which (4) does not hold for $n = 1, 2, \ldots$ (see [3,27]).

With regard to other classes of methods and their preservation of conservation laws, we mention that results along similar lines have been derived by Iserles for *multistep methods* and for *Taylor-type* methods [14] and independently by Hairer and Leone in the context of symplecticity [10]. We will follow here the approach of [3,27,14] and show that cubic invariance is equivalent to requiring a very strong condition on the invariant, namely that it has to be the solution of a partial differential equation called the Bateman equation. Although the Bateman-equation condition arises also in the case of multistep and Taylor-type methods [14], Runge–Kutta methods are a case apart since their error term is not a constant times a derivative of the function but is a linear combination

[1] Note that in this paper we discuss algebraic invariance. Therefore, ρ should be confused with neither a symmetry nor a differential invariant. We refer the reader to [15] and to the references therein for a treatment of symmetry invariants and to [2] for discussion of differential invariants.

of mixtures of derivatives of various orders (elementary differentials [12]), a feature that makes RK schemes different from many other numerical schemes for integration of ODEs.

The paper is organised as follows. In Section 2 we discuss classical RK methods and their condition for invariance, deriving the Bateman equation (a second-order partial differential equation), and a third-order partial differential equation, for the algebraic invariant ρ. The main result of the paper is presented in Section 3. First, we analyse the third-order counterpart of the Bateman equation. Secondly, we focus on the case of polynomial conservation laws and deduce that Runge–Kutta schemes cannot preserve any polynomial conservation law except for linear and quadratic.

We conclude with Section 4, relating Runge–Kutta methods with a larger class of numerical scheme on Lie groups of which classical RK schemes are but one representative. Numerical methods that stay on Lie groups are nowadays a very active area of research, and constitute an alternative approach to more classical stabilization and projection techniques, and differential-algebraic equations.

Although the material of Section 4 is not original, it furnishes an important example how to bypass the restrictions of this paper and of [14], which limit the applicablity of classical time-stepping methods when the retention of algebraic invariants is at issue. The material of Section 4 is relevant not just because Lie groups represent a major instance of invariants and symmetries, with a wide range of applications, but also for a deeper reason. Traditionally, numerical analysis of differential equations concerned itself mainly with methods that minimise error and cost. Lately, greater attention is being paid to correct modelling of geometric features of differential equations: invariants, asymptotics, symmetries etc. The main thrust of [14] and of this paper is that little can be expected of classical methods insofar as invariants are concerned. The lesson of Section 4 and of much of contemporary effort in geometric integration is that a very powerful approach toward correct rendition of invariants originates in the introduction of ideas from differential geometry and topology to numerical mathematics [2]. We firmly believe that this will increasingly become a major area of computational activity.

2. Necessary condition for invariance: the Bateman equation and its third-order counterpart

Without loss of generality, let us assume that the differential equation (1) is *autonomous*, namely that the function $\boldsymbol{f} \equiv \boldsymbol{f}(\boldsymbol{y})$ does not depend explicitly on time. Throughout the exposition, we also assume that $\boldsymbol{f}$ and ρ are analytic functions.

The exact solution of (1) is approximated numerically by means of a ν-stage Runge–Kutta method,

$$\begin{aligned} \boldsymbol{Y}_i &= \boldsymbol{y}_n + h\sum_{j=1}^{\nu} a_{i,j}\boldsymbol{K}_j, \\ \boldsymbol{K}_i &= \boldsymbol{f}(\boldsymbol{Y}_i), \quad i=1,2,\ldots,\nu, \\ \boldsymbol{y}_{n+1} &= \boldsymbol{y}_n + h\sum_{i=1}^{\nu} b_i\boldsymbol{K}_i, \end{aligned} \tag{6}$$

defined in terms of the *RK matrix* $A=(a_{i,k})$ and the *RK weights* $\boldsymbol{b}=(b_i)$ [12,13].

Recall that any system (1) that is ρ-invariant and autonomous can be written in the *skew-gradient* form

$$y' = S(y)\nabla\rho(y), \tag{7}$$

whereby $S(\cdot)$ is a $d\times d$ skew-symmetric matrix [18,25], and in particular we will restrict our attention to the case of two variables, i.e. $d=2$, whereby (7) yields

$$\begin{aligned} y_1' &= \psi(y)\frac{\partial\rho(y)}{\partial y_2},\\ y_2' &= -\psi(y)\frac{\partial\rho(y)}{\partial y_1} \end{aligned} \tag{8}$$

for some arbitrary smooth function $\psi:\mathbb{R}^2\to\mathbb{R}$. Since we wish to derive necessary conditions for invariance, we may assume without loss of of generality that $\psi\equiv 1$.

Proposition 1. *Assume that the function $\rho\in C^2[\mathbb{R}^2]$ is not a solution of the* Bateman equation

$$\mathscr{B}(u) = \left(\frac{\partial u}{\partial y_2}\right)^2\frac{\partial^2 u}{\partial y_1^2} - 2\frac{\partial u}{\partial y_1}\frac{\partial u}{\partial y_2}\frac{\partial^2 u}{\partial y_1\partial y_2} + \left(\frac{\partial u}{\partial y_1}\right)^2\frac{\partial^2 u}{\partial y_2^2} = 0 \tag{9}$$

[5], *where $u\equiv u(y_1,y_2)$. A necessary condition for the RK method* (6) *to preserve ρ for all $h<\bar h$ is*

$$b_i a_{i,j} + b_j a_{j,i} = b_i b_j,\quad i,j=1,2,\ldots,\nu. \tag{10}$$

Proof. For clarity's sake we suppress the dependence of ρ on the initial condition y_0. Expanding in powers of h and using (6), we have

$$\begin{aligned} \rho(y_{n+1}) &= \rho\left(y_n + h\sum_{i=1}^{\nu} b_i K_i\right)\\ &= \rho(y_n) + \sum_{k=1}^{\infty}\frac{h^k}{k!}\sum_{i_1,\ldots,i_k=1}^{\nu} b_{i_1}\cdots b_{i_k}\\ &\quad\times\sum_{l=1}^{k}\binom{k}{l}\frac{\partial^k\rho(y_n)}{\partial^{k-l}y_1\partial^l y_2} f_1(Y_{i_1})\cdots f_1(Y_{i_{k-l}}) f_2(Y_{i_{k-l+1}})\cdots f_2(Y_{i_k}), \end{aligned} \tag{11}$$

whereby the index of f denotes either its first or its second component, namely,

$$f_1(Y_i) = \frac{\partial\rho(Y_i)}{\partial y_2},\quad f_2(Y_i) = -\frac{\partial\rho(Y_i)}{\partial y_1},\quad i=1,2,\ldots,\nu.$$

Let us assume that $\rho(y_n)=0$ and let us focus on the terms up to order 2 in h. Using the identity

$$y_n = Y_i - h\sum_{j=1}^{\nu} a_{i,j}K_j,$$

and expanding the functions $\partial\rho/\partial y_l$, $l = 1, 2$, we obtain

$$\frac{\partial\rho}{\partial y_1}(\boldsymbol{y}_n)\frac{\partial\rho}{\partial y_2}(\boldsymbol{Y}_i) - \frac{\partial\rho}{\partial y_2}(\boldsymbol{y}_n)\frac{\partial\rho}{\partial y_1}(\boldsymbol{Y}_i)$$

$$= -h\sum_{j=1}^{v} a_{i,j}\left[\frac{\partial^2\rho(\boldsymbol{Y}_i)}{\partial y_1^2}\frac{\partial\rho(\boldsymbol{Y}_i)}{\partial y_2}\frac{\partial\rho(\boldsymbol{Y}_j)}{\partial y_2} - \frac{\partial^2\rho(\boldsymbol{Y}_i)}{\partial y_1\partial y_2}\frac{\partial\rho(\boldsymbol{Y}_i)}{\partial y_2}\frac{\partial\rho(\boldsymbol{Y}_j)}{\partial y_1}\right.$$

$$\left. - \frac{\partial^2\rho(\boldsymbol{Y}_i)}{\partial y_1\partial y_2}\frac{\partial\rho(\boldsymbol{Y}_i)}{\partial y_1}\frac{\partial\rho(\boldsymbol{Y}_j)}{\partial y_2} + \frac{\partial^2\rho(\boldsymbol{Y}_i)}{\partial y_2^2}\frac{\partial\rho(\boldsymbol{Y}_i)}{\partial y_1}\frac{\partial\rho(\boldsymbol{Y}_j)}{\partial y_1}\right].$$

Taking into account that $\boldsymbol{Y}_j = \boldsymbol{Y}_i + \mathcal{O}(h)$, we expand the above expression at $\boldsymbol{Y}_i$ to obtain

$$\frac{\partial\rho}{\partial y_1}(\boldsymbol{y}_n)\frac{\partial\rho}{\partial y_2}(\boldsymbol{Y}_i) - \frac{\partial\rho}{\partial y_2}(\boldsymbol{y}_n)\frac{\partial\rho}{\partial y_1}(\boldsymbol{Y}_i) = -h\sum_{j=1}^{v} a_{i,j}\mathscr{B}(\rho)(\boldsymbol{Y}_i) + \mathcal{O}(h^3).$$

By the same token,

$$\frac{h^2}{2}\sum_{i,j=1}^{v} b_i b_j \sum_{l=1}^{2}\binom{2}{l}\frac{\partial^2\rho(\boldsymbol{y}_n)}{\partial^{2-l}y_1\partial^l y_2} f_1(\boldsymbol{Y}_{i_1})\cdots f_1(\boldsymbol{Y}_{i_{2-l}})f_2(\boldsymbol{Y}_{i_{2-l+1}})\cdots f_2(\boldsymbol{Y}_{i_2})$$

$$= \frac{h^2}{2}\sum_{i,j=1}^{v} b_i b_j \mathscr{B}(\rho)(\boldsymbol{Y}_i) + \mathcal{O}(h^3).$$

Hence, reordering indices, we obtain

$$\rho(\boldsymbol{y}_{n+1}) = \frac{1}{2}h^2\sum_{i,j=1}^{v}(b_i b_j - b_i a_{i,j} - b_j a_{j,i})\mathscr{B}(\rho)(\boldsymbol{Y}_i) + \mathcal{O}(h^3).$$

Thus, unless ρ is a solution of the Bateman equation (9), annihilation of the $\mathcal{O}(h^2)$ term requires relations (10). □

The Bateman equation (9) plays a very important role also in the context of linear multistep methods and retention of conservation laws. As a matter of fact, Iserles proves that a necessary condition for ρ-invariance of a multistep method is that ρ obeys the Bateman equation [14].

The following result characterises the level sets of the solutions of the Bateman equation (9): essentially, they are determined by linear functions!

Proposition 2 (Iserles [14]). *Solutions $\rho(x,y)$ of the Bateman equation* (9) *such that $\rho(x,y)$=const have the form*

$$\rho(x,y) = \omega(\alpha x + \beta y + \gamma),$$

where α, β and γ are arbitrary constants and $\omega \equiv \omega(z)$ is an arbitrary analytic function.

An important consequence of the above result is that linear multistep methods (and in general Taylor-type methods) can be invariant solely in linear manifolds.

We have seen that

$$\rho(y_{n+1}) = \frac{1}{2}h^2 \sum_{i,j=1}^{\nu} (b_i b_j - b_i a_{i,j} - b_j a_{j,i})\mathscr{B}(\rho)(\boldsymbol{Y}_i) + \mathcal{O}(h^3),$$

whereby the $\mathcal{O}(h^3)$ term contains partial derivatives of ρ of order greater than two. Hence, if ρ is a quadratic manifold, all these derivatives are zero and we are left with the condition

$$\rho(y_{n+1}) = \frac{1}{2}h^2 \sum_{i,j=1}^{\nu} (b_i b_j - b_i a_{i,j} - b_j a_{j,i})\mathscr{B}(\rho)(\boldsymbol{Y}_i),$$

which implies that condition (10) is necessary and sufficient for the retention of quadratic conservation laws, a result well known and understood in the literature of Runge–Kutta methods [4,3,27].

Theorem 3. *A necessary condition for preserving a nonquadratic algebraic invariant* ρ *is that* $\rho \in C^3[\mathbb{R}]$ *is a solution of the partial differential equation*

$$\mathscr{L}(u) = \left(\frac{\partial u}{\partial y_2}\right)^3 \frac{\partial^3 u}{\partial y_1^3} - 3\left(\frac{\partial u}{\partial y_2}\right)^2 \frac{\partial u}{\partial y_1}\frac{\partial^3 u}{\partial y_1^2 \partial y_2} + 3\frac{\partial u}{\partial y_2}\left(\frac{\partial u}{\partial y_1}\right)^2 \frac{\partial^3 u}{\partial y_1 \partial y_2^2} - \left(\frac{\partial u}{\partial y_1}\right)^3 \frac{\partial^3 u}{\partial y_2^3} = 0. \tag{12}$$

Proof. Proceeding as in Proposition 1 but carrying the expansions a step further, we obtain, as a first contribution, the term

$$\frac{1}{3}h^3 \sum_{i,j,l=1}^{\nu} b_i b_j b_l \mathscr{L}(\rho)(\boldsymbol{Y}_i) + \mathcal{O}(h^4)$$

when $k=3$ in (11). The second contribution is obtained from the term for $k=2$ in (11): substituting $y_n = \boldsymbol{Y}_i - h\sum_{l=1}^{\nu} a_{i,l}\boldsymbol{K}_l$ and collecting similar terms, we obtain

$$\frac{1}{2}h^2 \sum_{i,j=1}^{\nu} b_i b_j \left[\frac{\partial^2 \rho(y_n)}{\partial y_1^2} f_1^2 + 2\frac{\partial^2 \rho(y_n)}{\partial y_1 \partial y_2} f_1 f_2 + \frac{\partial^2 \rho(y_n)}{\partial y_2^2} f_2^2\right]$$

$$= \frac{1}{2}h^2 \sum_{i,j=1}^{\nu} b_i b_j \mathscr{B}(\rho)(\boldsymbol{Y}_i) - \frac{1}{2}h^3 \sum_{i,j,l=1}^{\nu} b_i b_j a_{i,l} \mathscr{L}(\rho)(\boldsymbol{Y}_i) + \mathcal{O}(h^4),$$

whereby $f_1 \equiv f_1(\boldsymbol{Y}_i)$ and $f_2 \equiv f_2(\boldsymbol{Y}_i)$. Finally, the last contribution arises from the series expansion of the $\mathcal{O}(h)$ term. We have

$$\frac{\partial \rho}{\partial y_1}(y_n)\frac{\partial \rho}{\partial y_2}(\boldsymbol{Y}_i) - \frac{\partial \rho}{\partial y_2}(y_n)\frac{\partial \rho}{\partial y_1}(\boldsymbol{Y}_i)$$

$$= \frac{\partial \rho(\boldsymbol{Y}_i)}{\partial y_2}\left\{\frac{\partial \rho(\boldsymbol{Y}_i)}{\partial y_1} - h\sum_{j=1}^{\nu} a_{i,j}[\rho_{y_1 y_1}(\boldsymbol{Y}_i) f_1(\boldsymbol{Y}_j) + \rho_{y_1 y_2}(\boldsymbol{Y}_i) f_2(\boldsymbol{Y}_j)]\right.$$

$$+\frac{1}{2}h^2\sum_{j,l=1}^{\nu} a_{i,j}a_{i,l}[\rho_{y_1y_1y_1}(\boldsymbol{Y}_i)f_1(\boldsymbol{Y}_j)f_1(\boldsymbol{Y}_l)+2\rho_{y_1y_1y_2}f_1(\boldsymbol{Y}_j)f_2(\boldsymbol{Y}_2)+\rho_{y_1y_2y_2}f_2(\boldsymbol{Y}_j)f_2(\boldsymbol{Y}_l)]\Bigg\}$$

$$+\frac{\partial\rho(\boldsymbol{Y}_i)}{\partial y_1}\Bigg\{\frac{\partial\rho(\boldsymbol{Y}_i)}{\partial y_2}-h\sum_{j=1}^{\nu} a_{i,j}[\rho_{y_1y_2}(\boldsymbol{Y}_i)f_1(\boldsymbol{Y}_j)+\rho_{y_2y_2}(\boldsymbol{Y}_i)f_2(\boldsymbol{Y}_j)]$$

$$+\frac{1}{2}h^2\sum_{j,l=1}^{\nu} a_{i,j}a_{i,l}[\rho_{y_1y_1y_2}(\boldsymbol{Y}_i)f_1^2(\boldsymbol{Y}_j)+2\rho_{y_1y_2y_2}f_1(\boldsymbol{Y}_j)f_2(\boldsymbol{Y}_j)+\rho_{y_2y_2y_2}f_2^2(\boldsymbol{Y}_j)]\Bigg\}+\mathcal{O}(h^3).$$

Let us focus on the term

$$\sum_{i,j=1}^{\nu} b_i a_{i,j}\rho_{y_2}(\boldsymbol{Y}_i)\rho_{y_1y_1}(\boldsymbol{Y}_i)f_1(\boldsymbol{Y}_j)$$

and similar expressions. We write this term in the form

$$\frac{1}{2}\sum_{i,j=1}^{\nu} b_i a_{i,j}\rho_{y_2}(\boldsymbol{Y}_i)\rho_{y_1y_1}(\boldsymbol{Y}_i)f_1(\boldsymbol{Y}_j)+\frac{1}{2}\sum_{i,j=1}^{\nu} b_j a_{j,i}\rho_{y_2}(\boldsymbol{Y}_j)\rho_{y_1y_1}(\boldsymbol{Y}_j)f_1(\boldsymbol{Y}_i)$$

and expand in series whilst exploiting the relation

$$\boldsymbol{Y}_j=\boldsymbol{Y}_i+h\sum_{l=1}^{\nu}(a_{j,l}-a_{i,l})\boldsymbol{K}_l.$$

We have

$$\frac{1}{2}\sum_{i,j=1}^{\nu} b_i a_{i,j}\rho_{y_2}(\boldsymbol{Y}_i)\rho_{y_1y_1}(\boldsymbol{Y}_i)f_1(\boldsymbol{Y}_j)$$

$$=\frac{1}{2}\sum_{i,j=1}^{\nu} b_i a_{i,j}\rho_{y_2}(\boldsymbol{Y}_i)\rho_{y_1y_1}(\boldsymbol{Y}_i)\rho_{y_2}(\boldsymbol{Y}_i)$$

$$-h\sum_{i,j,l=1}^{\nu} b_i a_{i,j}(a_{i,l}-a_{j,l})\rho_{y_2}(\boldsymbol{Y}_i)\rho_{y_1y_1}(\boldsymbol{Y}_i)\,[\rho_{y_1y_2}f_1(\boldsymbol{Y}_l)+\rho_{y_2y_2}f_2(\boldsymbol{Y}_l)]+\mathcal{O}(h^2)$$

and

$$\frac{1}{2}\sum_{i,j=1}^{\nu} b_j a_{j,i}\rho_{y_2}(\boldsymbol{Y}_j)\rho_{y_1y_1}(\boldsymbol{Y}_j)\rho_{y_2}(\boldsymbol{Y}_i)$$

$$=\frac{1}{2}\sum_{i,j=1}^{\nu} b_j a_{j,i}\rho_{y_2}(\boldsymbol{Y}_i)-h\sum_{l=1}^{\nu}(a_{j,l}-a_{i,l})\,(\rho_{y_1y_2}f_1+\rho_{y_2y_2}f_2)\rho_{y_1y_1}(\boldsymbol{Y}_j)\rho_{y_2}(\boldsymbol{Y}_i)$$

$$-\frac{1}{2}h\sum_{i,j,l=1}^{\nu} b_j a_{j,i}(a_{j,l}-a_{i,l})\rho_{y_2}(\boldsymbol{Y}_i)\,[\rho_{y_1y_1y_1}(\boldsymbol{Y}_i)f_1+\rho_{y_1y_1y_2}(\boldsymbol{Y}_i)f_2]\rho_{y_2}(\boldsymbol{Y}_i)+\mathcal{O}(h^2).$$

Expanding in an identical manner all similar terms, we observe that the terms containing two second derivatives of ρ sum up to zero, hence we are left with terms containing just one second derivative and one third derivative of the function ρ. After some tedious algebra along the lines of [3,27], the contribution of the $k=1$ term in (11) reduces to

$$-\frac{1}{2}h^2\sum_{i,j=1}^{v}(b_i a_{i_j}+b_j a_{j,i})\mathscr{B}(\rho)(\boldsymbol{Y}_i)+\frac{1}{2}h^3\sum_{i,j,l=1}^{v}b_j a_{j,i}a_{i,l}\mathscr{L}(\rho)(\boldsymbol{Y}_i)+\mathcal{O}(h^4).$$

Collecting all the relevant terms, we obtain

$$\rho(\boldsymbol{y}_{n+1})=\frac{1}{2}h^2\sum_{i,j=1}^{v}(b_i b_j-b_i a_{i,j}-b_j a_{j,i})\mathscr{B}(\rho)(\boldsymbol{Y}_i)$$

$$+\frac{1}{6}h^3\sum_{i,j,l=1}^{v}(b_i b_j b_l-3b_i b_j a_{i,l}+3b_i a_{i,j}a_{j,l})\mathscr{L}(\rho)(\boldsymbol{Y}_i)+\mathcal{O}(h^4),$$

where the coefficients of $\mathscr{B}$ and $\mathscr{L}$ are exactly those derived in [3,27], in the context of cubic invariants.

Assume now that ρ does not obey the Bateman equation (see above), whose level sets are straight lines. Hence, in order to annihilate the $\mathcal{O}(h^2)$ term, condition (10) must be satisfied by the coefficients of the RK scheme in question.

In order to annihilate the $\mathcal{O}(h^3)$ term, we have two possibilities: either the coefficients of the scheme obey $\Upsilon=O$, where

$$\Upsilon_{i,j,l}=b_i b_j b_l-(b_i b_j a_{i,l}+b_j b_l a_{j,i}+b_l b_i a_{lj})$$

$$+(b_i a_{i,j}a_{j,l}+b_j a_{j,l}a_{l,i}+b_l a_{l,i}a_{i,j})=0,\quad \forall i,j,l=1,\ldots,v \tag{13}$$

(which has been already encountered in [3,27] in a discussion of cubic invariants) or ρ obeys the differential equation (12). However, it is well known that condition (10) and $\Upsilon=O$ are contradictory [3,27], therefore the only possibility is that the differential condition (12) is satisfied. □

3. On the solutions of the equation $\mathscr{L}(u) = 0$

In this section we wish to analyse some properties of the solutions of the partial differential equation (12). Following the same approach as [14], we distinguish two cases. Firstly, we note that when $\partial\rho/\partial y_2$ then ρ does not depend on the second variable, hence the system (8) can be reduced to the univariate case which is trivial to integrate: from $\rho(y_1,y_2)=f(y_1)$, we have

$$y_1'=0,$$
$$y_2'=-f(y_1),$$

hence $y_1=c$ is constant and $y_2=-f(c)t+y_2(0)$.

Let us assume thus that $\partial\rho/\partial y_2\neq 0$ at some point, hence, as a consequence of the analyticity of ρ, the same is true in a proper neighbourhood $\mathscr{U}$. Because of the implicit function theorem, there exists a function η, such that

$$\rho(y_1,y_2)=0 \;\Leftrightarrow\; y_2=\eta(y_1)\quad \forall \boldsymbol{y}\in\mathscr{U},$$

hence $\rho(y_1,\eta(y_1))=0$. To avoid confusion, let us denote such independent variable by x; thus,

$$\rho(x,\eta(x))=0. \tag{14}$$

Differentiating $\rho(x,\eta(x))=0$ with regards to x, we have

$$\frac{\partial\rho(x,\eta(x))}{\partial y_1}+\frac{\partial\rho(x,\eta(x))}{\partial y_2}\eta'(x)=0,$$

from which we deduce that

$$\eta'=-\frac{\partial\rho(x,\eta(x))}{\partial y_1}\left[\frac{\partial\rho(x,\eta(x))}{\partial y_2}\right]^{-1}.$$

Further differentiation of (14) implies that

$$\mathscr{B}(\rho)(x,\eta(x))+\left[\frac{\partial\rho(x,\eta)}{\partial y_2}\right]^3\eta''=0,$$

as in [14]. In particular, we deduce that

$$\eta''=-\mathscr{B}(\rho)(x,\eta(x))\left[\frac{\partial\rho(x,\eta)}{\partial y_2}\right]^{-3}. \tag{15}$$

Differentiating (14) for a third time, we obtain

$$\mathscr{L}(\rho)(x,\eta)-3\mathscr{B}(\rho)(x,\eta)\left[\frac{\partial^2\rho(x,\eta)}{\partial y_1\partial y_2}+\frac{\partial^2\rho(x,\eta)}{\partial y_2^2}\eta'\right]+\left[\frac{\partial\rho(x,\eta)}{\partial y_2}\right]^4\eta'''=0. \tag{16}$$

Assume that ρ obeys the partial differential equation (12). Substituting in (16) the expression for η'' and dividing by $(\partial\rho/\partial y_2)^2$, which we are assuming not equal identically to zero, we deduce

$$\eta'''\frac{\partial\rho(x,\eta)}{\partial y_2}+3\eta''\frac{\mathrm{d}}{\mathrm{d}x}\left(\frac{\partial\rho(x,\eta)}{\partial y_2}\right)=0. \tag{17}$$

Lemma 4. *Assume that $\partial\rho/\partial y_2\neq 0$. Then all solutions of the equation (17) obey the differential equation*

$$\mathscr{B}(\rho)(x,\eta(x)))=\text{const.}$$

Proof. We distinguish two cases. Firstly $\eta''=0$, in which case the assertion is satisfied because of (15), choosing the constant equal to zero. Otherwise, it is true that $\eta''\neq 0$ in a certain neighbourhood of x. Hence, we can write

$$\frac{\eta'''}{\eta''}=-3\frac{(\mathrm{d}/\mathrm{d}x)(\partial\rho/\partial y_2)}{\partial\rho/\partial y_2},$$

and, integrating both sides with respect to x, we obtain

$$\log\eta''=-3\log\frac{\partial\rho}{\partial y_2}+\text{ an integration constant,}$$

from which we deduce that

$$\eta''=K\left(\frac{\partial\rho}{\partial y_2}\right)^{-3},$$

K being an arbitrary constant of integration. The result follows by comparing the above expression for η'' with (15). □

Theorem 5. *Assume that $\rho(x,y)$ is a polynomial in x,y of degree $n>2$, that $\rho_y\neq 0$ and that* (1) *has no other conservation laws except for ρ. Then the RK scheme* (6) *cannot preserve ρ for all sufficiently small $h>0$.*

Proof. As a consequence of the above lemma, the problem reduces to studying solutions that render the Bateman operator $\mathscr{B}$ constant. Note that if $\rho(x,y)$ is a polynomial of degree n in x and y, then

$$\rho_{xx}\rho_y^2-2\rho_{xy}\rho_x\rho_y+\rho_{yy}\rho_x^2$$

is a polynomial of degree $(n-2)+2(n-1)=3n-4$. In particular, it follows that $3n-4=n$ for $n=2$, while $3n-4>n$ for all $n\geqslant 3$. Assume that

$$\rho(x,y)=ax^2+bxy+cy^2+dx+ey+f=0.$$

Direct computation reveals that

$$\tfrac{1}{2}\mathscr{B}(\rho)=(4ac-b^2)[\rho(x,y)-f]+ae^2-bde+cd^2,$$

hence $\mathscr{B}(\rho)$ is constant, provided that so is ρ. If $n>2$ then $\mathscr{B}(\rho)$ is a proper polynomial in x and y of degree strictly greater then n. Therefore, the system must admit a conservation law other than ρ, of order lower then n if ρ is a factor of the polynomial $\mathscr{B}(\rho)$, larger than n otherwise. This, rules out the important case when ρ is the only integral of the system. □

It has been already established in [3] that Runge–Kutta schemes cannot preserve all arbitrary cubic algebraic invariants. The method of proof in [3] is based on the construction of a specific cubic integral, depending on the coefficients of a scheme which cannot be preserved by the method.

In passing, we mention that, as in the case of Proposition 2, the results of Theorem 5 can be extended to analytic functions of polynomials in the following manner. Assume that ρ is not a polynomial but an analytic function ω of $q(x,y)$, namely $\rho(x,y)=\omega(q(x,y))$. Differentiating and substituting into $\mathscr{B}(\rho)=\text{const}$, we obtain $\omega'(q(x,y))\mathscr{B}(q)=K$, hence, if q is such that $\mathscr{B}(q)=\text{const}$, also $\omega'(q)$ is constant and we obtain a new solution. Thus, the solutions of $\mathscr{B}(\rho)=\text{const}$ are defined up to an arbitrary analytic function ω. This reflects the observation that the manifolds $\{\rho(\boldsymbol{x})=c\}$ and $\{\omega(\rho(\boldsymbol{x}))=\omega(c)\}$ are identical for bijective ω.

Theorem 5 does not rule out the existence of 'proper' sufficiently smooth functions that may be automatically preserved by the Runge–Kutta scheme for sufficiently small h. Seeking an example of such function, we employ the technique of separation of variables. Assume that $\rho(x,y)=v(x)w(y)$, whereby v and w are two C^3 functions of x and y respectively such that $v',w'\neq 0$, hence v and w are at least linear functions. Then the condition $\mathscr{L}(\rho)=0$ is equivalent to

$$\frac{v'''v^2}{v'^3}-3\frac{v''v}{v'^2}+3\frac{w''w}{w'^2}-\frac{w'''w^2}{w'^3}=0$$

(the prime denoting differentiation with respect to the independent variable) which results in an identical ordinary differential equation for the functions v and w, namely

$$\frac{z'''z^2}{z'^3}-3\frac{z''z}{z'^2}=K,\quad z\equiv z(t),$$

where K is an arbitrary constant. When $K = 0$, we can reduce the above third-order differential equation into a second-order one by integration,

$$z'' = cz^3,$$

whereby c is an arbitrary integration constant. The solution of the latter is given in implicit form by

$$t = \pm \int_0^{z(t)} \frac{2\mathrm{d}s}{\sqrt{2cs^4 + 4C_1}} + C_2,$$

where c, C_1 and C_2 are arbitrary integration constants. This, however, is unlikely to represent an invariant of practical importance. In general, the determination of all level sets of (12) is incomplete, although we believe that virtually all nonquadratic invariants of interest are excluded and, anyway, it is trivial to check by direct differentiation whether $\mathscr{L}(\rho) = 0$ for any specific function ρ.

4. Runge–Kutta methods in a Lie-group formulation

Although we have already seen that the equation $\mathscr{L}(\rho)$ admits solutions that are not necessarily linear or quadratic in y_1, y_2, the sheer complexity of (17) reveals that such manifolds described by $\rho(y_1, y_2) = 0$ are exceptional. Moreover, recall that $\mathscr{L}(\rho) = 0$ is merely a necessary condition for invariance. We deduce that generic retention of conservation laws by means of classical RK integration cannot be achieved easily, if at all.

A standard way to treat ODEs with invariants that classically are not automatically preserved by RK methods is to reformulate the invariants as constraints and use a differential–algebraic approach [20]. Discussion on numerical preservation of invariants can be traced already to the early 1970s, especially in the fields of constrained mechanics and electronic circuits [6]. There exists a rich literature on Runge–Kutta methods applied to the solution of differential equations with algebraic invariants (DAEs) and these methods, essentially based on projections, have proved themselves to be very effective and successful in many practical applications [12]. It is sometimes argued that numerical schemes that employ projection damage geometric properties of the underlying problem, and this has provided strong motivation to devise numerical schemes that intrinsically retain the underlying invariants. New types of symmetric projections have been recently introduced by Hairer [8] so that not only the invariant, but most of the remaining geometric properties are retained under discretization. Other successful methods for the exact or almost-conservation of invariants are based on splitting of the vector field $\boldsymbol{f}$ into simpler vector fields that are easy to integrate or can be integrated exactly. We refer to the surveys of Hairer [9] and of McLachlan and Quispel [21] for an up-to-date list of techniques for various problems that posses invariants, or, more generally, geometrical structure that one would like to preserve under discretization.

In the last few years there has been a growing interest in devising Lie-group methods that somehow follow the logic of Runge–Kutta schemes in a different manner from the RK schemes for DAEs above. Let us present here the main ideas, referring the reader to [23,17,28] and to the review article [16] for further details.

Lie groups are smooth manifolds endowed with a multiplicative group operation, and without loss of generality, we can identify them with subgroups of $\mathrm{GL}(d, \mathbb{R})$, the set of all $d \times d$ real matrices. (Identical theory can be extended to the complex field.) Familiar examples are $\mathrm{O}(d, \mathbb{R})$, the set of all $d \times d$ orthogonal matrices, and $\mathrm{SL}(d,\mathbb{R})$, the *special linear group* of all $d \times d$ matrices with unit determinant. A (finite-dimensional) Lie algebra is a linear space, closed under commutation. The tangent space at identity of a Lie group is a Lie algebra, hence the importance of the latter construct in any discussion of ODEs evolving on a Lie group. For example, the Lie algebra corresponding to $\mathrm{O}(d,\mathbb{R})$ is $\mathfrak{so}(d,\mathbb{R})$, the linear space of $d \times d$ skew-symmetric matrices, while the Lie algebra of $\mathrm{SL}(d,\mathbb{R})$ is $\mathfrak{sl}(d,\mathbb{R})$, the set of all $d \times d$ matrices with zero trace.

An ordinary differential system on a Lie group G can be always written in the form

$$y' = \gamma(t,y)y, \quad y(0) = y_0,$$

where $y \in G$ and $\gamma : \mathbb{R}^+ \times \mathfrak{g}$, where $\mathfrak{g}$ is the Lie algebra of G, and can be solved so that the numerical approximation resides in G,

$$y_n \in G \quad n = 0,1,2,\ldots,$$

provided that $y_0 \in G$, by using a Lie-group modification of classical Runge–Kutta schemes. The main idea is to translate the original ODE in each step from G to $\mathfrak{g}$ by means of the exponential map, $y(t) = \exp(\sigma(t))y_0$, by means of the so-called *dexpinv equation*,

$$\sigma' = \mathrm{dexp}_\sigma^{-1}\gamma, \quad \sigma(t_n) = 0,$$

which acts in $\mathfrak{g}$ instead of G. The function dexp^{-1} is defined as

$$\mathrm{dexp}_\sigma^{-1}(\gamma) = \sum_{k=0}^{\infty} \frac{\mathrm{B}_k}{k!}\mathrm{ad}_\sigma^k\gamma,$$

where the B_k's are Bernoulli numbers [1] and the *adjoint operators* ad^k are k-times iterated commutators of σ with γ, namely $\mathrm{ad}_\sigma^k\gamma = [\sigma,[\sigma,\cdots[\sigma,\gamma]\cdots]]$ (see [23,17,28,16]).

The redeeming feature of this transformation is that $\mathfrak{g}$ is a linear space, while G is usually described by nonlinear conservation laws. Thus, following a construction of Munthe-Kaas [22], an arbitrary Runge–Kutta method can be employed in $\mathfrak{g}$ to produce a numerical approximation $\sigma_{n+1} \approx \sigma(t_{n+1})$, so that

$$y_{n+1} = \exp(\sigma_{n+1})y_n \in G$$

is a numerical approximation for $y(t_{n+1})$ which has the same order as the original RK scheme while remaining in the Lie group. Thus, for example, if $G = \mathrm{SL}(d,\mathbb{R})$, such Lie-group-based RK schemes allow us to preserve to machine accuracy the algebraic invariant $\det y = 1$, a polynomial equation of degree d, while, as we have seen in Section 2, standard RK schemes are bound to fail. Similarly, when $G = \mathrm{O}(d,\mathbb{R})$, with Lie-groups schemes we can use an explicit Lie-group RK method and obtain an orthogonal approximation, while with standard schemes we would require that the RK method obeys condition (10), hence being an implicit scheme.

Such Lie-group schemes do not apply only to Lie groups, but also to a wider class of problems, evolving on *homogeneous spaces* [24], i.e. manifolds on which the dynamics is described by a Lie-group action. (Examples include a d-sphere, a d-torus, isospectral matrices, symmetric matrices, Stiefel and Grassmann manifolds.) In this setting one can obtain the classical Runge–Kutta schemes as a special case of Lie-group Runge–Kutta methods for which the acting group is $\mathbb{R}^d$ with the group operation '+' and the manifold acted upon is also $\mathbb{R}^d$. Although such schemes are not yet fully competitive in comparison with the more established DAE methods, a pleasing feature of this approach is that one might choose a different group action to preserve different underlying geometrical features of the problem in question. The search for a good action has to take into account qualitative features that need be preserved, as well as the computational cost of the scheme. This is an area currently under active investigation.

References

[1] M. Abramowitz, I.A. Stegun, Handbook of Mathematical Functions, Dover, New York, 1965.
[2] C.J. Budd, A. Iserles, Geometric integration: numerical solution of differential equations on manifolds, Philos. Trans. Roy. Soc. London Ser. A 357 (1999) 945–956.
[3] M.P. Calvo, A. Iserles, A. Zanna, Numerical solution of isospectral flows, Math. Comput. 66 (1997) 1461–1486.
[4] G.J. Cooper, Stability of Runge–Kutta methods for trajectory problems, IMA J. Numer. Anal. 7 (1987) 1–13.
[5] D.B. Fairlie, J.A. Mulvey, Integrable generalizations of the 2-dimensional Born–Infeld equation, J. Phys. A 27 (1994) 1317–1324.
[6] C.W. Gear, The simultaneous numerical solution of differential–algebraic equation, IEEE Trans. Circuit Theory CT-18 (1971) 89–95.
[7] V. Guillemin, A. Pollack, Differential Topology, Prentice-Hall, Englewood Cliffs, NJ, 1974.
[8] E. Hairer, Symmetric projection methods for differential equations on manifolds, BIT, to appear.
[9] E. Hairer, Numerical geometric integration, University of Geneva Technical report 1988.
[10] E. Hairer, P. Leone, Order barriers for symplectic multi-value methods, Technical Report, Université de Genéve, 1997.
[11] E. Hairer, S.P. Nørsett, G. Wanner, Solving Ordering Differential Equations, I. Nonstiff Problems, 2nd Revised Edition, Springer, Berlin, 1993.
[12] E. Hairer, G. Wanner, in: Solving Ordinary Differential Equations II, Springer Series in Computational Mathematics, Vol. 14, Springer, Berlin, 1991.
[13] A. Iserles, A First Course in Numerical Analysis of Differential Equations, Cambridge University Press, Cambridge, 1996.
[14] A. Iserles, Multistep methods on manifolds, Technical Report NA1997/13, DAMTP, University of Cambridge, 1997.
[15] A. Iserles, R. McLachlan, A. Zanna, Approximately preserving symmetries in numerical integration, European J. Appl. Math. 10 (1999) 419–445.
[16] A. Iserles, H. Munthe-Kaas, S.P. Nørsett, A. Zanna, Lie-group methods, Acta Numer. 9 (2000) 215–365.
[17] A. Iserles, S.P. Nørsett, On the solution of differential equations in Lie groups, Philos. Trans. Roy. Soc. London Ser. A 357 (1999) 983–1019.
[18] T. Itoh, K. Abe, Hamiltonian-conserving discrete canonical equations based on variational difference quotients, J. Comput. Phys. 76 (1988) 85–102.
[19] B. Leimkhuler, S. Reich, Manuscript, in preparation.
[20] R. März, Numerical methods for differential–algebraic equations, Acta Numer. 1 (1992) 141–198.
[21] R. McLachlan, G.R. Quispel, Six lectures in geometric integration, in: R. Devore, A. Iserles, E. Sűli (Eds.), Foundations of Computational Mathematics, to appear.
[22] H. Munthe-Kaas, Runge–Kutta methods on Lie groups, BIT 38 (1998) 92–111.
[23] H. Munthe-Kaas, High order Runge–Kutta methods on manifolds, J. Appl. Numer. Math. 29 (1999) 115–127.
[24] H. Munthe-Kaas, A. Zanna, Numerical integration of differential equations on homogeneous manifolds, in: F. Cucker, M. Shub (Eds.), Foundations of Computational Mathematics, Springer, Berlin, 1997, pp. 305–315.
[25] G.R.W. Quispel, H.W. Capel, Solving ODE's numerically while preserving all first integrals, Technical Report, La Trobe University, 1997.
[26] J.M. Sanz-Serna, M.P. Calvo, Numerical Hamiltonian Problems, Vol. ACCM-7, Chapman & Hall, London, 1994.
[27] A. Zanna, On the Numerical Solution of Isospectral Flows, Ph.D. Thesis, Newnham College, University of Cambridge, 1998.
[28] A. Zanna, Collocation and relaxed collocation for the Fer and the Magnus expansion, SIAM J. Numer. Anal. 36 (4) (1999) 1145–1182.

ELSEVIER
Journal of Computational and Applied Mathematics 125 (2000) 83–92
JOURNAL OF COMPUTATIONAL AND APPLIED MATHEMATICS
www.elsevier.nl/locate/cam

Performance of two methods for solving separable Hamiltonian systems

V. Antohe *, I. Gladwell
Department of Mathematics, Southern Methodist University, Dallas, TX 75275, USA

Received 15 December 1999; received in revised form 19 March 2000

Abstract

We make qualitative comparisons of fixed step symplectic and variable step nonsymplectic integrations of the separable Hénon–Heiles Hamiltonian system. Emphasis is given to interesting numerical phenomena. Particularly, we observe the relationship of the error in the computed Hamiltonian to the presence and absence of chaos, when computing with a symplectic (fixed step) method, qualitative phenomena in the Hamiltonian error for a variable step method, and the sensitivity of the chaotic behavior and of the computation of features in Poincaré sections to very small changes in initial conditions, step sizes and error tolerances.

1. Introduction

Assume the autonomous Hamiltonian $H(\boldsymbol{p},\boldsymbol{q})$ is a smooth real function where $\boldsymbol{p}$ represents the generalized momenta and $\boldsymbol{q}$ the generalized coordinates, and $(\boldsymbol{p}^{\mathrm{T}},\boldsymbol{q}^{\mathrm{T}})=(p_1,p_2,\dots,p_d,q_1,q_2,\dots,q_d)$; d is the number of degrees of freedom. The Hamiltonian system corresponding to $H(\boldsymbol{p},\boldsymbol{q})$ is

$$\frac{\mathrm{d}p_i}{\mathrm{d}t}=-\frac{\partial H}{\partial q_i},\quad \frac{\mathrm{d}q_i}{\mathrm{d}t}=\frac{\partial H}{\partial p_i},\quad i=1,2,\dots,d \tag{1}$$

and we need initial conditions $\boldsymbol{p}(t_0)=\boldsymbol{p}_0$, $\boldsymbol{q}(t_0)=\boldsymbol{q}_0$. Note that $H(\boldsymbol{p},\boldsymbol{q})$ is constant with time (i.e., $\mathrm{d}H/\mathrm{d}t=0$).

A separable Hamiltonian has the structure

$$H(\boldsymbol{p},\boldsymbol{q})=T(\boldsymbol{p})+V(\boldsymbol{q}). \tag{2}$$

In mechanics, $T=\frac{1}{2}\boldsymbol{p}^{\mathrm{T}}M^{-1}\boldsymbol{p}$ represents kinetic energy (M is the mass matrix) and V potential energy. The Hamiltonian system has 'partitioned form':

$$\frac{\mathrm{d}\boldsymbol{p}}{\mathrm{d}t}=-\nabla_{\boldsymbol{q}}V,\qquad \frac{\mathrm{d}\boldsymbol{q}}{\mathrm{d}t}=\nabla_{\boldsymbol{p}}T=M^{-1}\boldsymbol{p}. \tag{3}$$

* Corresponding author.

0377-0427/00/$ - see front matter
PII: S0377-0427(00)00460-X

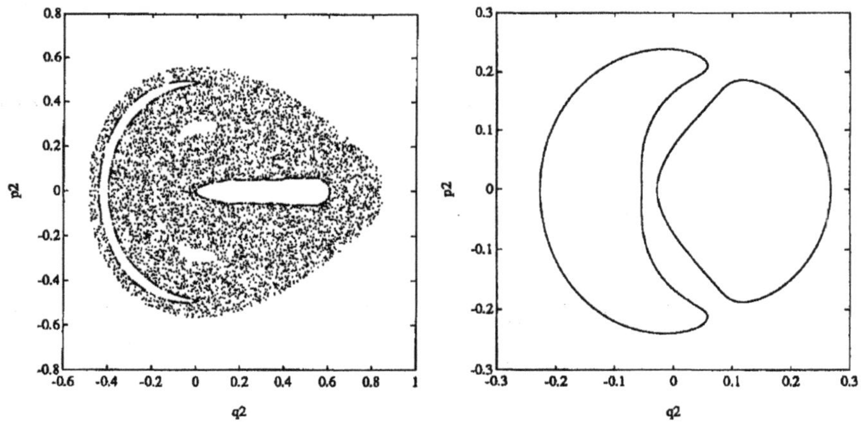

Fig. 1. Poincaré sections [15, pp. 12,13].

This system can be integrated using specially designed methods, for example the partitioned Runge–Kutta and certain Runge–Kutta–Nyström methods.

Hénon and Heiles (see [5]) formulated the Hamiltonian system

$$\frac{\mathrm{d}p_1}{\mathrm{d}t} = -(q_1 + 2q_1q_2), \quad \frac{\mathrm{d}p_2}{\mathrm{d}t} = -(q_2 + q_1^2 - q_2^2), \quad \frac{\mathrm{d}q_i}{\mathrm{d}t} = p_i, \quad i = 1,2 \tag{4}$$

with $d = 2$ but with only one conserved quantity, the Hamiltonian

$$T = \tfrac{1}{2}(p_1^2 + p_2^2), \qquad V = \tfrac{1}{2}(q_1^2 + q_2^2) + q_1^2 q_2 - \tfrac{1}{3}q_2^3. \tag{5}$$

For later comparison, in Fig. 1, we show 'Poincaré cross sections' of Sanz-Serna and Calvo [5, pp. 12,13], intersections of the solution (p_2, q_2) with the plane $q_1 = 0$. The left figure corresponds to the initial condition IC1:

$$q_1 = q_2 = p_2 = 0, \quad p_1 = \sqrt{2H} \tag{6}$$

$(H = 0.15925)$ integrating to $t_{\mathrm{end}} = 3 \times 10^4$. In [5], this solution is described as 'randomly scattered' or 'chaotic' or 'stochastic'. The right figure depicts the corresponding 'quasiperiodic' Poincaré section for initial condition IC2:

$$q_1 = q_2 = p_1 = p_2 = 0.12 \tag{7}$$

$(H = 0.029952)$ generated by integrating to $t_{\mathrm{end}} = 2 \times 10^5$.

In the remainder of the paper, we introduce numerical methods for Hamiltonian problems and discuss the qualitative numerical phenomena observed in a variety of integrations of the Hénon and Heiles problem (4). In Section 2, we discuss a symplectic method used in a fixed-step integration and, in Section 3, a nonsymplectic method in a variable step integration. It is *not* intended that this discussion provides a comparison of the efficiency of the two types of methods. Such a comparison would require a more careful choice of methods.

Table 1
Fixed step RUTH4; initial condition IC1, DP

h	t_{end}	Poincaré	H $error_{max}$
0.200	3×10^4	C	5.27×10^{-4}
0.100	3×10^4	NC	2.31×10^{-5}
0.050	3×10^4	NC	1.42×10^{-6}
0.025	3×10^4	C	1.23×10^{-7}
0.010	3×10^4	C	3.16×10^{-9}
0.100	3×10^5	C	3.19×10^{-5}
0.050	3×10^5	C	1.98×10^{-6}

2. Symplectic methods

A discrete method is symplectic if it is a symplectic map. A symplectic map is one which, operating on any two-dimensional set D in the $2d$-dimensional space in which the system is formulated, preserves the sum of the areas of the projections of D onto the (p_i, q_i) planes. Symplectic discrete methods are often preferred as they have bounded Hamiltonian error.

For partitioned Hamiltonian systems, we can apply (explicit) partitioned RK (PRK) methods. We use a symplectic fourth-order PRK method (RUTH4) of [2,3], as implemented in the Fortran 90 code *new-hamint* [4].[1] (We have checked numerically that this implementation is fourth-order.) RUTH4 was used with initial condition IC1 and step sizes $h = 0.2, 0.1, 0.05, 0.025, 0.01$ in

(1) single precision (SP) with an approximate unit roundoff 1.19×10^{-7},
(2) double precision (DP) with an approximate unit roundoff 2.22×10^{-16},
(3) extended precision (EP) with an approximate unit roundoff 1.93×10^{-34}.

(Precision is adjusted simply using the "module" feature of Fortran 90.) Initial tests used double precision (see Table 1). For all step sizes except $h = 0.1$ and $h = 0.05$, the solution displays chaotic behavior (C) when integrating to $t_{end} = 3 \times 10^4$. Longer integrations (to $t_{end} = 3 \times 10^5$) using these step sizes also show chaotic behavior. The relative error in the Hamiltonian (Herror) is oscillatory; this relative error is computed over all output points. As anticipated, the maximum error decreases as h decreases. Fig. 2 depicts the Poincaré section for a typical chaotic solution ($h = 0.2$) while Fig. 3 shows the nonchaotic solution ($h = 0.1$). (The Poincaré section is computed by calculating the root of cubic Hermite interpolating polynomial fitted to the output values immediately on opposite sides of the section. Of course, the accuracy of the calculated section depends on the precision of the arithmetic and on the integration error.) Figs. 2 and 3 also present the corresponding relative Hamiltonian error showing the oscillatory behavior. Note, in Fig. 2, that the oscillatory band in the error widens around $t = 2.2 \times 10^4$, approximately where the chaotic behavior in the Poincaré section begins. For this reason the anticipated fourth-order behavior of the Hamiltonian error computed using the RUTH4 formula is not apparent in Table 1. In our computations, for all choices of h the quasiperiodic precedes the chaotic behavior. We computed a nonchaotic solution step size range

[1] The software used here, the integrator *new-hamint.f90* and a driver code for the Hénon–Heiles problem, may be accessed at the ftp site ftp.cygnus.math.smu.edu/pub/gladwell/new-hamint.f90.

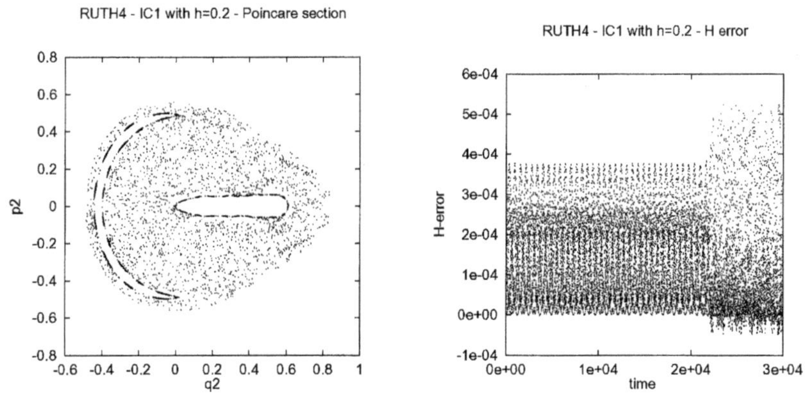

Fig. 2. RUTH4, $h = 0.2$, IC1, DP; Poincaré section and Hamiltonian error.

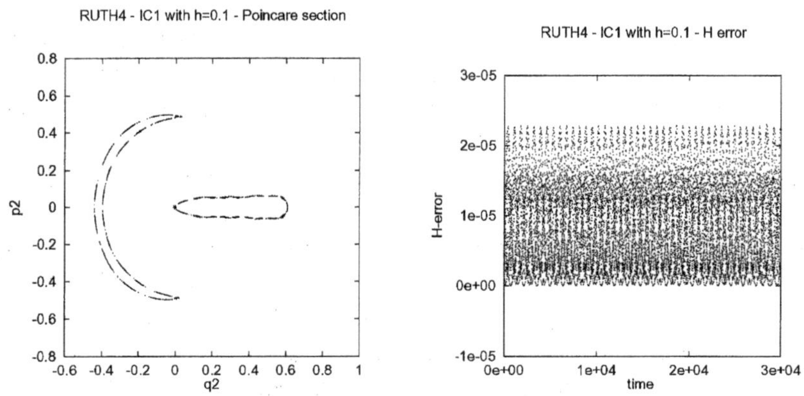

Fig. 3. RUTH4, $h = 0.1$, IC1, DP; Poincaré section and Hamiltonian error.

Table 2
Fixed step RUTH4; initial condition IC1, EP, $t_{\text{end}} = 3 \times 10^4$

h	Poincaré	H $\text{error}_{\text{max}}$
0.10	NC	2.29×10^{-5}
0.05	C	1.98×10^{-6}
0.01	NC	2.27×10^{-9}

around $h = 0.1$, that is $0.099 < h < 0.102$. *That this range exists implies a real possibility of being deceived numerically about long-term solution behavior.*

Extended precision integration to $t_{\text{end}} = 3 \times 10^4$ gives a nonchaotic solution around $h = 0.1$ and $h = 0.01$ but not at $h = 0.05$ (see Table 2), the Hamiltonian error is always oscillatory and essentially the same as for the double-precision calculation. Integrating further to $t_{\text{end}} = 3 \times 10^5$ gives chaotic solutions for all three step sizes. In single precision, the relative Hamiltonian error is characterized by both low- and high-frequency oscillations (see Fig. 4). The corresponding Poincaré section is similar to that on the left of Fig. 2. A different step size, $h = 0.097$, produces an (isolated) nonchaotic solution.

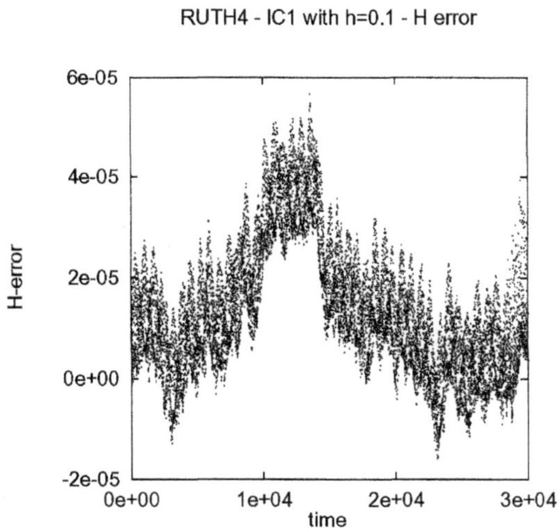

Fig. 4. RUTH4, $h = 0.1$, IC1, SP; Hamiltonian error.

Note that the effect of precision can be sufficient to induce different qualitative behaviors. For example, for $h = 0.1$ double and extended-precision integrations to $t_{\text{end}} = 3 \times 10^4$ are nonchaotic whereas single-precision integration is chaotic. Starting with the single precision initial values in both the double and extended-precision calculations still gives nonchaotic behavior. Looking more closely, if we compare the coordinates of corresponding points in the Poincaré sections then double- and extended-precision integrations agree to 15 digits initially but only to seven at the end. In contrast, double- and single-precision agree to seven digits initially but to no digits at the end of the integration, and the two integrations exhibit different qualitative behaviors.

In an extended-precision integration with $h = 0.05$, additional structures appear in the Poincaré section, particularly symmetrical (about the q_2 axis) clusters of points on the right side of the figure and an "envelope" for the section. Most of these features appear early in the integration; however, some do not, particularly the "envelope", but after a brief chaotic regime following the initial quasiperiodic regime. In Fig. 5, we show the results for an integration to $t_{\text{end}} = 3 \times 10^4$ and, in Fig. 6, for an integration to $t_{\text{end}} = 3 \times 10^5$. Note that, in general, the chaotic regime corresponds to the largest relative Hamiltonian errors and the smaller errors correspond to the initial quasiperiodic state and to later quasiperiodic states where structures internal to the section are computed. In contrast, the part of the integration where the envelope is computed is in the time period $t = (2.23\text{–}4.76) \times 10^4$ approximately, and *the relative Hamiltonian error is about the size seen in the chaotic regime.* Here again the sensitivity of the system may be observed. The qualitative features (the envelope and the special interior structures) may vanish if the initial value for p_1 is perturbed slightly. So, computing $p_1 = \sqrt{2H}$ in extended precision then solving the Hamiltonian system in extended precision leads to a rich set of structures. Just changing the initial value by computing it in double precision then solving the Hamiltonian system in extended precision is sufficient that the qualitative features are lost. The corresponding computations for initial condition IC2 are summarized in Table 3. The solution is quasiperiodic for both double and extended precision. Typically, for $h = 0.05$ (Fig. 7) the Poincaré section resembles that in [5]. For both double and extended precision, the relative error in the Hamiltonian oscillates, see Fig. 7 for $h = 0.05$.

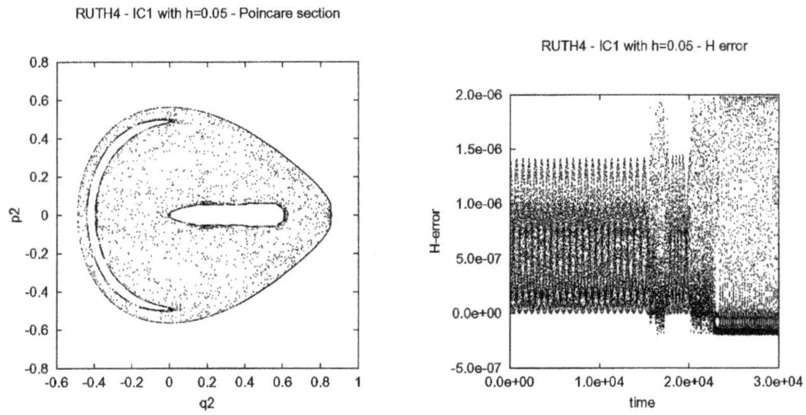

Fig. 5. RUTH4, $h=0.05$, IC1, EP; Poincaré section, Hamiltonian error, $t_{\mathrm{end}}=3\times10^4$.

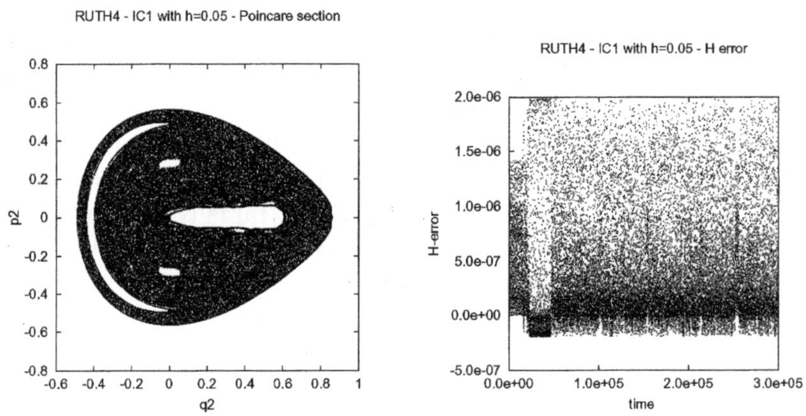

Fig. 6. RUTH4, $h=0.05$, IC1, EP; Poincaré section, Hamiltonian error, $t_{\mathrm{end}}=3\times10^5$.

Table 3
Fixed step RUTH4; initial condition IC2, $t_{\mathrm{end}}=3\times10^5$

h	Precision	H $\mathrm{error}_{\max}$
0.200	DP	1.61×10^{-4}
0.100	DP	9.74×10^{-6}
0.050	DP	6.04×10^{-7}
0.025	DP	3.77×10^{-8}
0.010	DP	9.64×10^{-10}
0.200	EP	1.61×10^{-4}
0.100	EP	9.74×10^{-6}

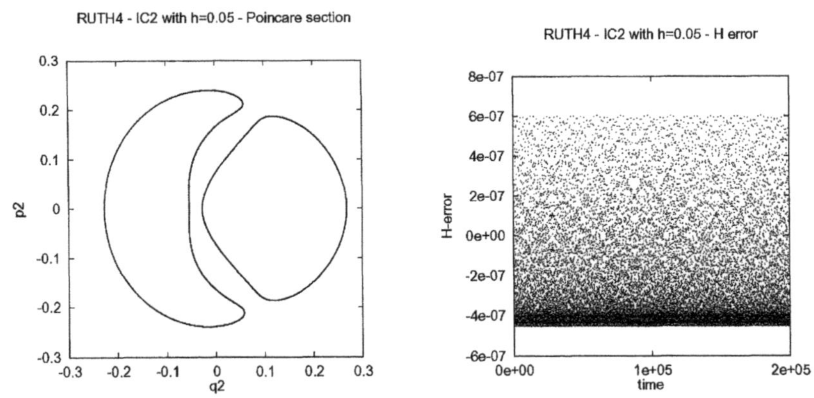

Fig. 7. RUTH4, $h = 0.05$, IC2, DP; Poincaré section and Hamiltonian error.

Table 4
Variable step RKL; initial condition IC1

Tolerance	t_{end}	Precision	Poincaré	H $error_{max}$
10^{-3}	3×10^4	DP	C	1.70×10^{-2}
10^{-4}	3×10^4	DP	C	1.66×10^{-3}
10^{-5}	3×10^4	DP	C	9.92×10^{-5}
10^{-5}	3×10^4	EP	NC	8.20×10^{-6}
10^{-6}	3×10^4	DP	C	4.97×10^{-6}
10^{-7}	3×10^4	DP	C	2.48×10^{-7}
10^{-8}	3×10^4	DP	C	1.12×10^{-8}
10^{-8}	3×10^4	EP	C	1.06×10^{-8}
10^{-9}	3×10^4	DP	C	5.49×10^{-10}
10^{-5}	3×10^5	EP	C	9.92×10^{-4}
10^{-8}	3×10^5	EP	C	1.18×10^{-7}
10^{-9}	3×10^5	EP	C	6.07×10^{-9}

3. Runge–Kutta–Nyström methods

We may integrate systems

$$\frac{d\boldsymbol{p}}{dt} = -\nabla_q V = f(\boldsymbol{q}), \quad \frac{d\boldsymbol{q}}{dt} = \boldsymbol{p} \tag{8}$$

using nonsymplectic Runge–Kutta–Nyström methods as implemented in *new-hamint* [4]. Here we use RKL — a sixth-order explicit method with a fourth-order embedded error estimate, originally developed for integrating the special second-order systems of celestial mechanics [1].

Results for initial condition IC1 are given in Table 4. All Poincaré sections are chaotic except for the isolated instance of the extended-precision integration at tolerance $= 10^{-5}$, for which the section (see Fig. 8) is concentrated around the quasiperiodic structure in Fig. 3 but with more 'spreading' than in Fig. 1. A refined search (near tolerance $= 10^{-5}$) for other nonchaotic solutions failed to reveal any.

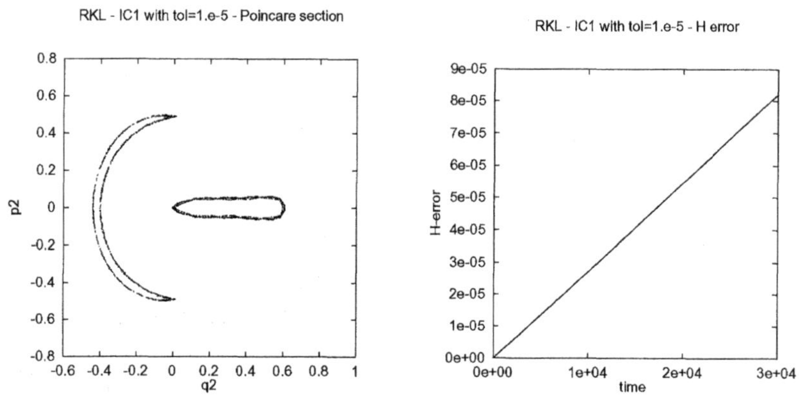

Fig. 8. RKL, tolerance $= 10^{-5}$, IC1, DP; Poincaré section and Hamiltonian error.

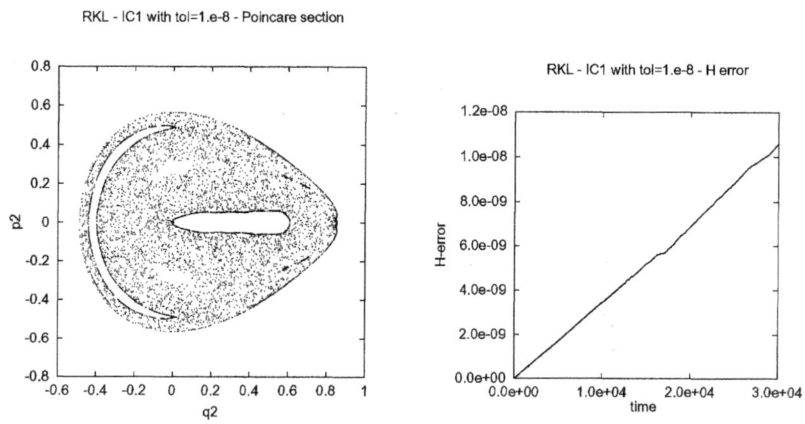

Fig. 9. RKL, tolerance $= 10^{-8}$, IC1, EP; Poincaré section and Hamiltonian error.

In Fig. 8 the Hamiltonian error has an almost linearly increasing trend on which a small oscillation is superposed. This behavior is observed in all accurate integrations in double and extended precision throughout the quasiperiodic regime. For less-accurate integrations this behaviour is less marked and does not appear immediately. Particularly, the oscillations are flatter for the less accurate integrations. Fig. 9 shows a chaotic case in extended precision for tolerance$=10^{-8}$. Linear changes with superposed oscillations in the relative Hamiltonian error in the quasiperiodic regimes are observable, as in other similar integrations. See the left figure in Fig. 10 which presents a detail of the error for a typical double precision case for the range $t = 0\text{–}10^3$. This is a macroscopic view of the error. At the microscopic level, within a single oscillation there are (mainly) small differences in error behavior. These are most marked at relaxed tolerances. In the chaotic regime the relative Hamiltonian error is essentially linearly increasing (without superposed oscillations); the average slope of the error is slightly greater in the chaotic regime than in the quasiperiodic regime. In addition, other features (slope changes and flat regions) are observed; see the right figure in Fig. 10 which presents a detail of the error for a typical extended-precision case for the range $t = (1.2\text{–}1.8) \times 10^4$. These correspond

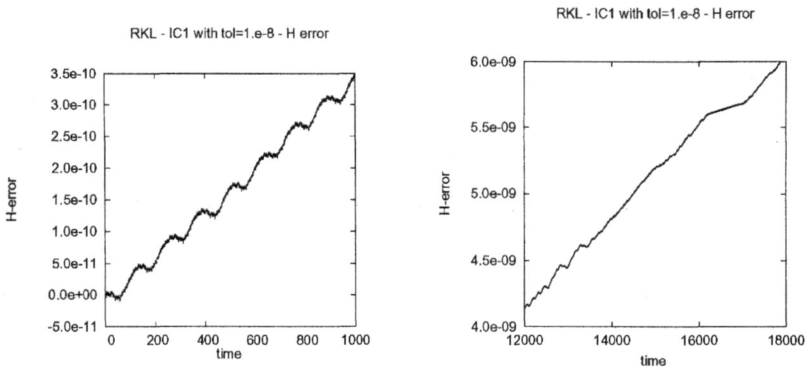

Fig. 10. RKL, tolerance $= 10^{-8}$, IC1; Hamiltonian error detail.

Table 5
Variable step RKL; initial condition IC2, $t_{\text{end}} = 3 \times 10^5$, DP

Tolerance	H error$_{\text{max}}$
10^{-3}	5.51×10^{-2}
10^{-4}	6.04×10^{-3}
10^{-5}	4.25×10^{-4}
10^{-6}	2.59×10^{-5}
10^{-7}	1.21×10^{-6}
10^{-8}	5.49×10^{-8}
10^{-9}	3.02×10^{-9}

to the appearance of additional structures in the Poincaré section, particularly symmetrical (about the q_2 axis) clusters of points on the right side of the figure and an "envelope" for the section partially completed on the right. These features appear early in the integration. However, some features do not appear at the beginning of the integration, particularly the "envelope", but after a brief chaotic regime following the initial quasiperiodic regime.

Table 5 displays the results for initial condition IC2. All tolerances produce a quasiperiodic solution similar to that on the right of Fig. 1 and to that obtained with the fixed step size RUTH4 (Fig. 7). The relative Hamiltonian error has a similar linear with superposed oscillations behavior as that seen in Fig. 10. *We observe that for given initial conditions, the number of oscillations in the error per unit time seems approximately constant.* Of course, the size of the error at a given time depends almost linearly on the tolerance. So, for initial conditions IC1 and any tolerance we observe almost eight oscillations per thousand units of time. For IC2, the corresponding number is close to two.

In either Table 4 or Table 5, plotting the logarithm of the maximum error against the logarithm of required tolerance, a closely fitted straight line has slope slightly greater than one, i.e., close to *tolerance proportionality*. Assuming that when we measure the error we are already in the final integration regime, this enables us to predict a tolerance to control the Hamiltonian error below a given threshold for a given time interval.

References

[1] R.W. Brankin, J.R. Dormand, I. Gladwell, P.J. Prince, W.L. Seward, Algorithm 670: a Runge–Kutta–Nyström code, ACM Trans. Math. Software 15 (1989) 31–40.

[2] J. Candy, W. Rozmus, A symplectic integration algorithm for separable Hamiltonian functions, J. Comput. Phys. 92 (1991) 230–256.

[3] E. Forest, R.D. Ruth, Fourth-order symplectic integration, Physica D 43 (1990) 105–117.

[4] I. Gladwell, K. Bouas-Dockery, R.W. Brankin, A Fortran 90 separable Hamiltonian system solver, Appl. Numer. Math. 25 (1997) 207–217.

[5] J.M. Sanz-Serna, M.P. Calvo, Numerical Hamiltonian Problems, Chapman & Hall, London, 1994.

ELSEVIER

Journal of Computational and Applied Mathematics 125 (2000) 93–105

JOURNAL OF COMPUTATIONAL AND APPLIED MATHEMATICS

www.elsevier.nl/locate/cam

Order stars and stiff integrators

Ernst Hairer*, Gerhard Wanner

Section de Mathématiques, Université de Genève, 2-4, rue du Lievre, CH-1211 Genève 24, Switzerland

Received 31 December 1999

Abstract

Order stars, introduced in G. Wanner, E. Hairer, S.P. Nørsett (Order stars and stability theorems, BIT 18 (1978) 475–489), have become a fundamental tool for the understanding of order and stability properties of numerical methods for stiff differential equations. This survey retraces their discovery and their principal achievements. We also sketch some later extensions and describe some recent developments. © 2000 Elsevier Science B.V. All rights reserved.

Keywords: Stiff differential equations; Order stars; A-stability

1. Ehle's conjecture

Stiff differential equations first became popular mainly during the fifties; for an overview of the early literature see the first section of [13]. In 1963, two seminal publications laid the foundations for later development: Dahlquist's paper on A-stable multistep methods [8], and Butcher's first paper on implicit Runge–Kutta methods [4]. One year later, Butcher developed the general class of Gaussian implicit Runge–Kutta methods of order $2s$ [5], as well as his efforts to find Radau and Lobatto methods with fewer implicit stages [6]. The merger of the two subjects, i.e., the study of *A-stable implicit Runge–Kutta methods* began some 5 years later. Of great influence was the elegant paper of Axelsson [2] on Radau methods as well as the comprehensive Thesis of Ehle [10].

Standard stability analysis. This proceeds as follows (see the scheme in Table 1): The differential equation is linearized and diagonalized, so that it becomes a linear scalar difference equation. The latter is solved by putting $y_n := R^n \cdot y_0$, which leads to a characteristic equation $\Phi(R,z) = 0$, where $z = h\lambda$ is a complex variable, λ an eigenvalue of the Jacobian and h is the step size. Numerical stability requires that $|R(z)| \leqslant 1$ for all roots of $\Phi(R,z) = 0$. The method is called *A-stable* if the *stability domain* $S := \{z; |R(z)| \leqslant 1$ for all roots of $\Phi(R,z) = 0\}$ covers the entire left half plane $\mathbb{C}^-$.

* Corresponding author. Tel.: +41-22-309-1416; fax: +41-22-309-1499.

E-mail addresses: ernst.hairer@math.unige.ch (E. Hairer), gerhard.wanner@math.unige.ch (G. Wanner).

PII: S0377-0427(00)00461-1

Table 1
Scheme of stability analysis

Stiff Differential Eq.	$\xrightarrow{\textit{num. meth.}}$	Numerical Sol.
linear. $\downarrow$ *+diag.*		*linear.* $\downarrow$ *+diag.*
$y' = \lambda y$	$\xrightarrow{\textit{num. meth.}}$	linear scalar Difference Eq.
		$y_n :=$ $\downarrow$ $R^n \cdot y_0$
		Alg. Eq. $\Phi(R, z) = 0$

One-step methods. Here, the equation $\Phi(R,z)=0$ is of degree 1 in R, thus $R(z)=P(z)/Q(z)$, where P and Q are polynomials. Examples are:

$$\begin{aligned} R(z) &= 1+z \quad \text{(explicit Euler method, not } A\text{-stable)}, \\ R(z) &= \frac{1+z/2}{1-z/2} \quad \text{(trap. rule and impl. midpoint, } A\text{-stable)}, \\ R(z) &= \frac{1}{1-z} \quad \text{(implicit Euler method, } A\text{-stable)}. \end{aligned} \tag{1}$$

These are particular entries of the *Padé table of the exponential function* with, in general,

$$P_{kj}(z) = 1 + \frac{k}{j+k}z + \cdots + \frac{k(k-1)\ldots 1}{(j+k)\ldots(j+1)} \cdot \frac{z^k}{k!} \tag{2}$$

and $Q_{kj}(z) = P_{jk}(-z)$. Low-order explicit Runge–Kutta methods have their stability functions in the first row ($j=0$) while Butcher's implicit Gauss methods are on the diagonal ($j=k$).

Major significance of Ehle's thesis. Starting from the known result, that diagonal Padé fractions are A-stable for all degrees [3], Ehle concluded that Butcher's implicit Gauss methods were all A-stable, but found that Butcher's Radau and Lobatto methods were *above* the diagonal and therefore, not A-stable. Ehle then extended the result of Birkhoff–Varga to the first and second subdiagonal $j=k+1$, $j=k+2$ and constructed A-stable modifications of Radau and Lobatto methods. Next he showed that the entries for the third and fourth subdiagonals $j=k+3$, $j=k+4$ were *never* A-stable. The proofs of these results were based on two criteria for A-stability:

Criterion A. All poles ($=$ zeros of Q) are in $\mathbb{C}^+$.

Criterion B. $E(y) = Q(iy)Q(-iy) - P(iy)P(-iy) \geqslant 0$ for all $y \in \mathbb{R}$.

While Criterion B is easy for the first two subdiagonals ($E(y)$ is of the form $C \cdot y^{2j}$), the verification of Criterion A required tedious algebraic developments based on the explicit formulas (2) [10, pp. 37–62]. Ehle then stated his famous conjecture:

Conjecture (Ehle [10, p. 65]). *With the exception of the diagonal and first two subdiagonals, i.e., of* $k \leqslant j \leqslant k+2$, *no entry of the Padé table is A-stable.*

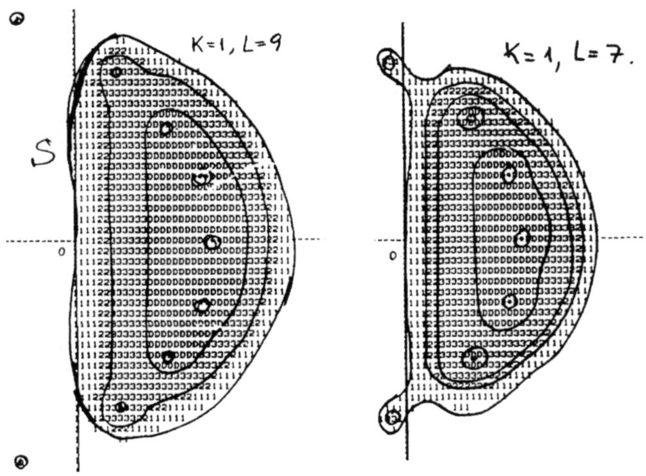

Fig. 1. Early plots of $|R(z)|$ for Padé fractions.

At first sight, a general proof of the conjecture seemed to be a difficult task, because numerical computations revealed that sometimes Criterion A was violated, but not B (as for example with $k=0$ and $j=6$); sometimes Criterion B was violated, but not A (as for example with $j=k+3$, $j=k+4$); and sometimes both were violated.

2. The discovery of order stars

Hamilton discovered the quaternions 1843 while walking over Broughham Bridge in Dublin; in 1964, Buchberger had the crucial idea which lead to Gröbner bases 1964 while riding a bicycle in the Bürgerstrasse in Innsbruck; fresh air seems to be beneficial to good mathematical ideas. The idea for order stars also came at a windy place, the railway station in Lausanne, waiting on track 8 for the train to Geneva, on Friday evening, February 17, 1978.

Many numerical computations of the roots of the polynomial $E(y)$ as well as the poles of R, i.e., the zeros of Q, were performed for various values of k and j. Computer plots of $|R(z)|$ were also made with the graphical tools of the seventies (see reproductions in Fig. 1). One observes in these pictures vertical contour lines, which imitate the contour lines of $|e^z| = e^x$ in the neighbourhood of the origin.

Idea (Wanner, Hairer, Nørsett [25]). Try to look at "contour lines" which follow the slope of $|e^z|$, in hoping for more information:

$$A := \{z; |R(z)| > |e^z|\} = \left\{z; \left|\frac{R(z)}{e^z}\right| > 1\right\} \tag{3}$$

(see reproductions in Fig. 2). Since on the imaginary axis $|e^z| = 1$, the order star is there precisely complementary to S and Criterion B is equivalent to

Criterion B′. $A \cap i\mathbb{R} = \emptyset$.

The "star" with $p+1$ sectors at the origin is produced by the error term

$$e^z - R(z) = C \cdot z^{p+1} + \cdots \quad \text{or} \quad 1 - \frac{R(z)}{e^z} = C \cdot z^{p+1} + \cdots, \tag{4}$$

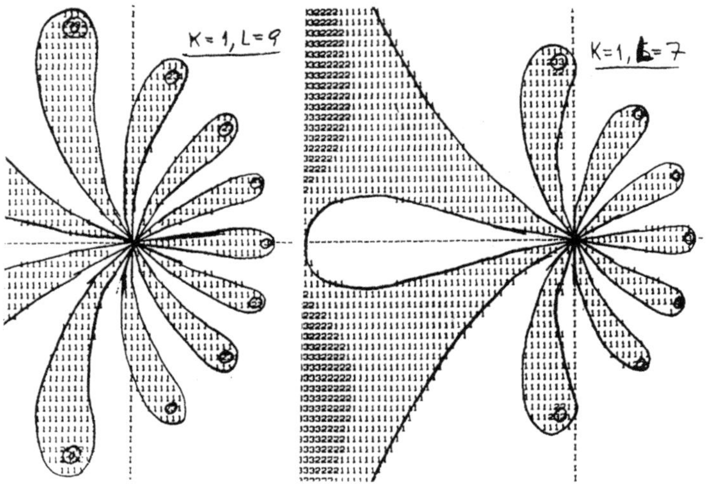

Fig. 2. Early plots of order stars for Padé fractions.

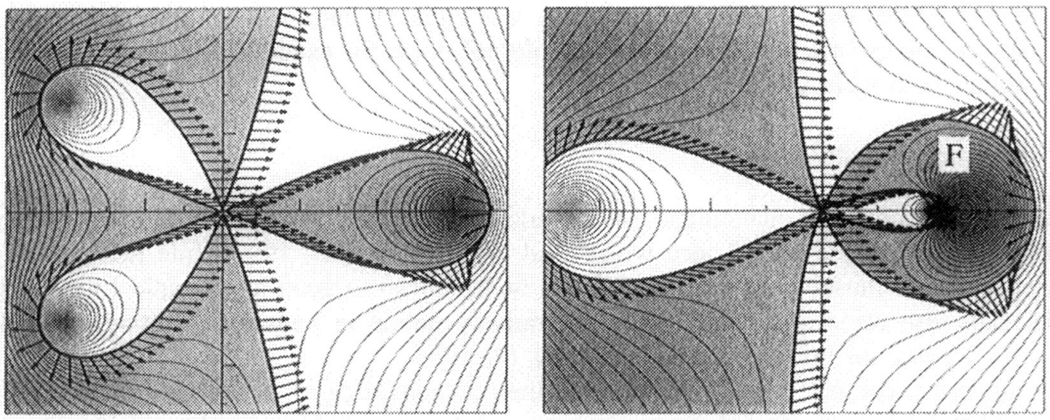

Fig. 3. Two possible order stars for Padé with $k = 2$ and $j = 1$.

which suggested the name "order star". By the maximum principle, each bounded "black finger" of A must contain a pole, and, by the maximum principle applied to $e^z/R(z)$, each bounded "white finger" contains a zero of R.

Proof of Ehle's Conjecture. If there are not enough zeros, there must be black fingers growing out from the origin in $\mathbb{C}^-$, which either cross $i\mathbb{R}$ (hence, violating Criterion B′), or carry a pole in $\mathbb{C}^-$, violating Criterion A (see Fig. 2).

Lemma on the number of poles and zeros. For many further results, the use of the maximum principle, as above, is not sufficient: for example, each of the order stars in Fig. 3 could represent the Padé approximant with $k = 2$ and $j = 1$: The clarification comes from a new idea: along the boundary of A, the argument of $R(z)/e^z$ turns clockwise. This is a consequence of the Cauchy–

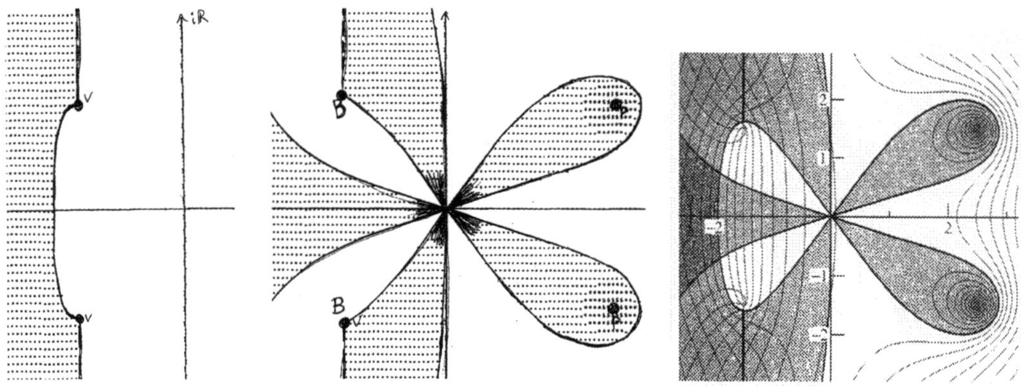

Fig. 4. Early (and modern) plot of the order star for method (5).

Riemann equations written in polar coordinates. At the origin, the argument is 0. Hence, *if the boundary curve of a bounded subset* $F \subset A$ *with* $\partial F \subset \partial A$ *returns m times to the origin, F must contain at least m poles* (Argument Principle). This allows us to exclude the right picture of Fig. 3 (because F requires 2 poles) and leads to the conclusion that all Padé fractions possess order stars with j simple black fingers to the right and k simple white fingers to the left, as in Fig. 2. All results of Ehle are now an easy consequence.

The above result has been made more precise by Iserles [14]: The number of poles in F is *equal* to the number of *exponential interpolation points* of R on ∂F.

Multistep and general linear methods. Here, the difference equation in Table 1 involves y_n-values from *several* steps, consequently, the characteristic equation $\Phi(R,z)=0$ will be of *higher* degree in R. Example:

$$(2z^2-10z+17)R^2-(8z+16)R-1=0$$
$$\Rightarrow R_{1,2}=((4z+8)\pm 3\sqrt{2z^2+6z+9})/(2z^2-10z+17). \quad (5)$$

We can graph *each* of these roots (divided by e^z) separately (see Fig. 4). At some places the roots have discontinuities, but extend continuously on *another* sheet. If we superpose the two sheets, cut and glue them together appropriately, we obtain a *Riemann surface* $\mathcal{M}$, on which the order star is now located (see Fig. 4, right). The order condition produces a star on the *principal sheet*, but not for the "parasitic roots". For A-stability, Criterion B′ must hold on *all* sheets.

3. Main results

The Daniel and Moore Conjecture. This conjecture [9, p. 80] extended the famous theorem of Dahlquist [8], to multistep methods with higher stage or derivative number: the highest order of such an A-stable method is $2j$ (where j is the number of implicit stages, i.e., the number of poles), and "of those A-stable methods", the smallest error constant is that of the diagonal Padé approximation. Fig. 5 demonstrates, in the case of $j=3$, why the first part of this conjecture ($p \leqslant 2j$) follows from the order star. This is because for $p>2j$ the order star either covers $i\mathbb{R}$ or needs additional poles.

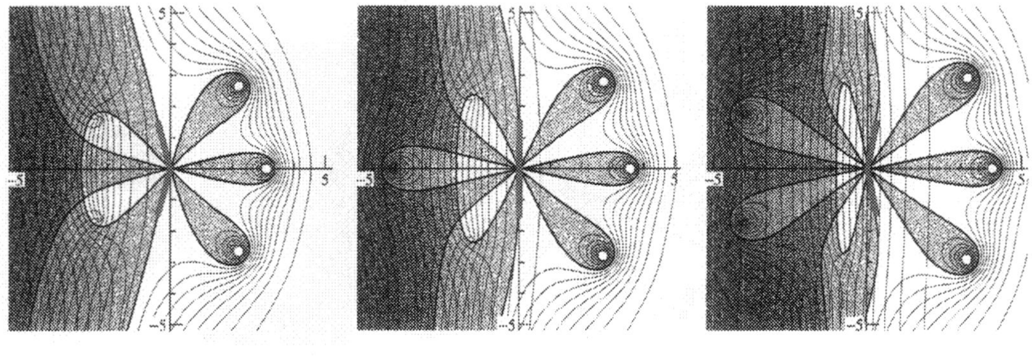

Fig. 5. Method with 3 poles, order 5 (left), order 6 (middle), order 7 (right; not A-stable); $R(z)$ given by (14) and (16).

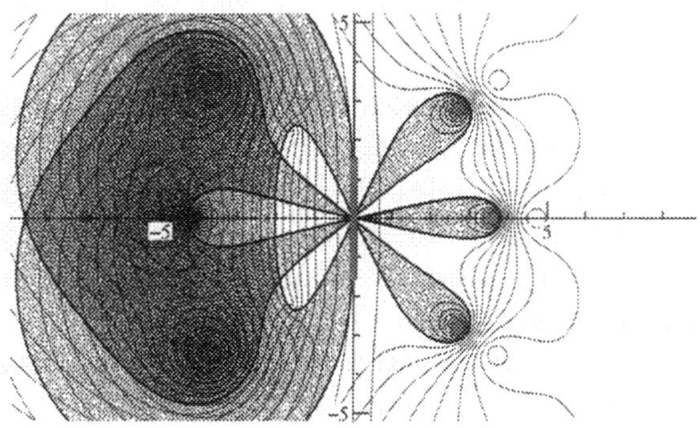

Fig. 6. Order star B for Daniel and Moore Conjecture, Part II (R same as in Fig. 5(b)).

Daniel and Moore Conjecture, Part II. The solution of the "error-constant"-part of this conjecture is based on the observation, that the fingers of the order star represent oscillations of $|R(z)/e^z|$, whose *amplitude* is governed by the error constant (see (4)). Thus, an inequality of the form $|C| > |C_0|$ is expressed by the *colours of the fingers of the relative order star*

$$B := \{z; |R(z)| > |R_0(z)|\} = \left\{z; \left|\frac{R(z)}{R_0(z)}\right| > 1\right\}, \tag{6}$$

where $R_0(z)$ is the stability function of any other method, here the diagonal Padé method (see Fig. 6).

Jeltsch–Nevanlinna Theorem. It came as a surprise that the order star (6) had a much wider field of application than initially intended: *If the two methods, which are compared, are both explicit and have the same number of function evaluations per step, then the two stability domains cannot be included one in the other* [20]. The reason is that explicit methods have all their poles at infinity and, with equal multiplicity, the quotient $R(z)/R_0(z)$ has no pole left at all outside S_0. A condition for this result is that the instability of methods R_0 must be produced on *one* Riemann sheet only. This is called "Property C" by [20]. An illustration is presented in Fig. 7. This result called in question the general belief in the 1950s and 1960s, that multistep methods were superior to Runge–Kutta methods ("simply because they use more information ...") and that the latter were, if at all, just useful for starting the computation. Many extensions for implicit methods are also possible [21].

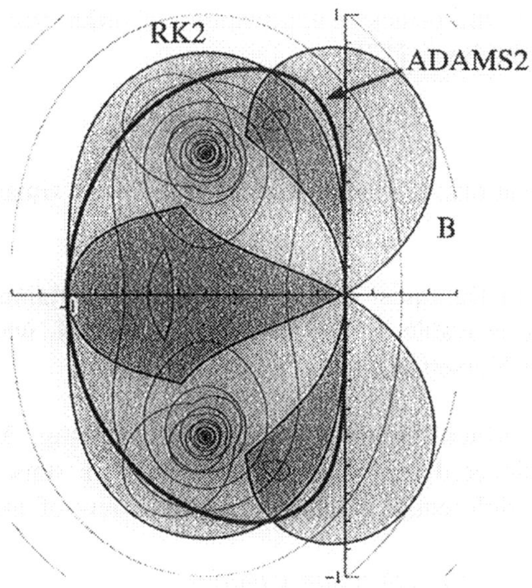

Fig. 7. Order star B for Adams2 versus (scaled) RK2.

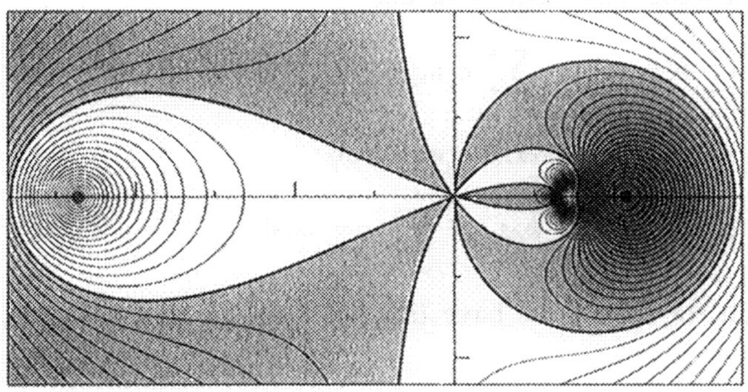

Fig. 8. Approximations with real poles.

Nørsett–Wolfbrandt barrier. Since the computational complexity of fully implicit Runge–Kutta methods, specially when applied to high-dimensional nonlinear problems, seemed enormous, many publications in the late 1960s and early 1970s, especially the Thesis of A. Wolfbrandt [26], were developing so-called “diagonally implicit Runge–Kutta (DIRK) methods”, Rosenbrock methods, and “singly implicit Runge–Kutta (SIRK) methods”. These methods avoided complex linear algebra, but always encountered the order barrier

$$p \leqslant s+1, \tag{7}$$

where s is the number of stages. Once again, order stars delivered a natural explanation for this barrier: since all poles must be on the real axis, only two of the $p+1$ white sectors are “free” and require no zero (see Fig. 8). Thus, $p+1 \leqslant s+2$ and (7) follows.

This concludes the survey of the principal applications of order star theory. For more details see [12, Sections IV.4, V.4] and the book [17].

4. Further applications

Many extensions and applications to generalized or specialized situations have been made in the meantime.

Dahlquist's first barrier. When Dahlquist's *second* barrier was obtained so easily with order stars, his *first* barrier should also be provable by this method. This proof, which indeed was possible, has been published by Iserles and Nørsett [16].

Partial differential equations. Much research, initiated by G. Strang, A. Iserles, R. Jeltsch and J.H. Smit (see [15,18,19,24]) is devoted to the application of oder stars to obtain order bounds for difference schemes for partial differential equations. An overview of many of these results is given in Chapters 6 and 7 of [17].

A prominent model problem is the *advection equation*

$$\frac{\partial u}{\partial t} = c\frac{\partial u}{\partial x} \tag{8}$$

solved by a difference scheme, say, with 3 levels

$$\sum_{\ell=-r_2}^{s_2} a_\ell u_{m+2,\ell} + \sum_{\ell=-r_1}^{s_1} b_\ell u_{m+1,\ell} + \sum_{\ell=-r_0}^{s_0} c_\ell u_{m,\ell} = 0. \tag{9}$$

Stability analysis leads to the characteristic equation

$$\Phi(R,z) := R^2 \sum_{\ell=-r_2}^{s_2} a_\ell z^\ell + R \sum_{\ell=-r_1}^{s_1} b_\ell z^\ell + \sum_{\ell=-r_0}^{s_0} c_\ell z^\ell = 0 \tag{10}$$

with the requirement that for $|z|=1$ we have $|R(z)| \leqslant 1$ for all roots of (10). The *order condition* is here

$$z^\mu - R_1(z) = \mathbf{O}((z-1)^{p+1}), \quad z \to 1 \tag{11}$$

for the principal root, where $\mu = c\Delta t/\Delta x$ is the Courant number. The fact that the exponential function of (4) is now replaced by z^μ with its branching point at the origin, is an additional complication to these studies.

The main result, *proved* in a series of papers for two and three level methods and *conjectured* in general (see [22]), is that

$$p \leqslant 2\min(D, U), \tag{12}$$

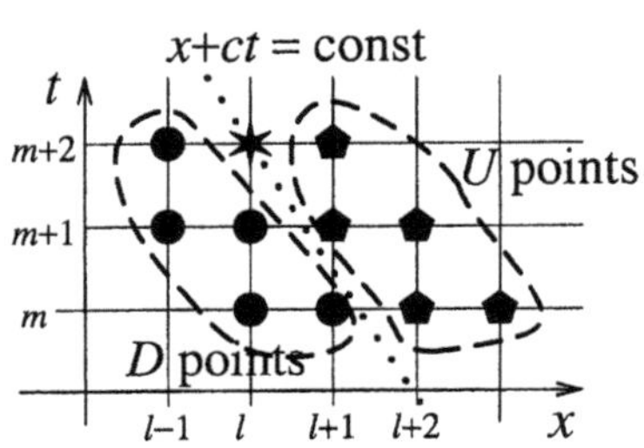

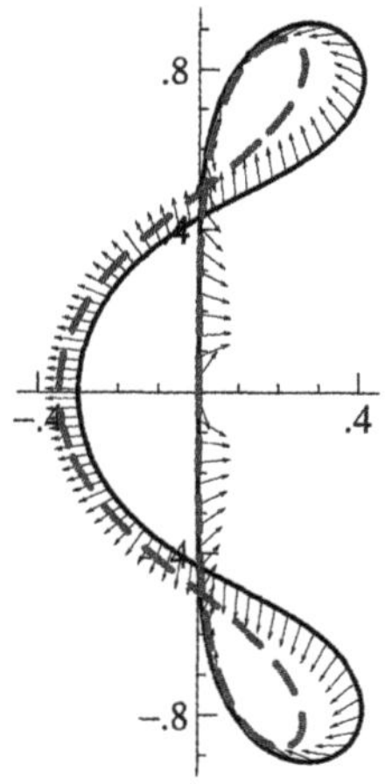

$$\alpha_0 = 0 + \tfrac{1}{2}\varepsilon \qquad \beta_0 = -\tfrac{3}{8} - \tfrac{3\varepsilon}{16}$$
$$\alpha_1 = 0 + \tfrac{1}{2}\varepsilon \qquad \beta_1 = \tfrac{37}{24} - \tfrac{35\varepsilon}{48}$$
$$\alpha_2 = 0 \qquad \beta_2 = -\tfrac{59}{24} - \tfrac{59\varepsilon}{48}$$
$$\alpha_3 = -1 - \varepsilon \qquad \beta_3 = \tfrac{55}{24} - \tfrac{17\varepsilon}{48}$$
$$\alpha_4 = 1 \qquad \beta_4 = 0$$

$\varepsilon = 0.36$

Fig. 9. Increasing stability domain for explicit Adams4.

where D and U are, respectively, the number of downwind and upwind points of the stencil. This bound has the quality of being an extension of the Courant–Friedrichs–Lewy condition, to which it reduces for $p = 1$.

Chebyshev methods for large problems. Important for the cheap integration of large and mildly stiff equations are the so-called Chebyshev methods. Their stability polynomials have a certain order and maximal stability along the negative real axis. The implementation of these methods in the "Lebedev style" (see [12, p. 33]) requires information about the number of complex zeros of $R(z)$. The application of order stars has led to a complete description of this question and also delivered inequalities for the error constants (see [1]).

Delay equations. For the study of unconditional stability with respect to the test equation $y'(t) = a \cdot y(t) + b \cdot y(t-1)$ with real a and b, a sufficient condition, for symmetric stability functions $R(z)$, is that $\operatorname{Arg} R(iy) < y$ for all y. This condition can be elegantly seen from the shape of the order star close to the imaginary axis (see [11]).

5. Three counter examples

These examples show that many hypotheses of the theory in Section 3 cannot be removed.

"Jeltsch–Nevanlinna" *without Property C*. The necessity of Property C for the validity of the Jeltsch–Nevanlinna Theorem is illustrated in [20, p. 69] (see also "Exercise 4" in [12, p. 297] and "Open Problem 8" in [17, p. 218]). The following example has been computed by R. O'Donovan. It perturbs the explicit Adams4 method, whose root locus curve presents "loops", in such a way, that the stability domain increases in *all* directions. For the result see Fig. 9.

Breaking the Nørsett–Wolfbrandt barrier. The above proof of (7) used the fact that two symmetric branches joining a real pole include a white sector, i.e., employs in a certain sense the Jordan Curve

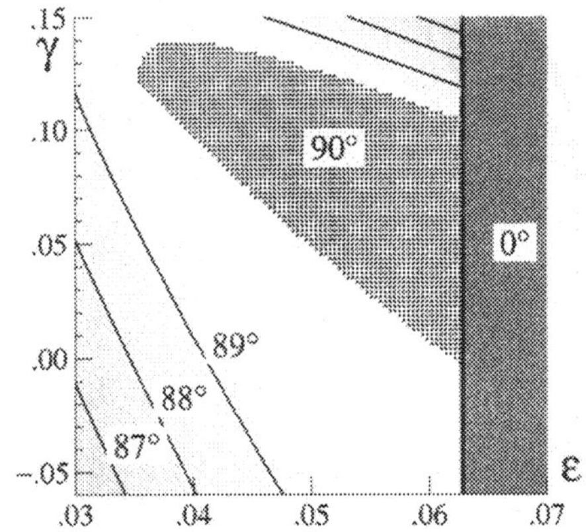

Fig. 10. Region for which the Psihoyios–Cash method is *A*-stable.

Theorem. On a general Riemann surface this is no longer valid and the barrier (7) is, for example, broken by BDF3. Psihoyios and Cash in [23] even found an *A-stable method* with real poles which breaks this barrier. The method is an extension of the MEBDF-method of Cash (see [12, p. 269]): we use *three* consecutive steps of BDF5 to produce three "future points" $\bar{y}_{n+5}$, $\bar{y}_{n+6}$, and $\bar{y}_{n+7}$, and compute y_{n+5} by the corrector

$$\sum_{j=0}^{5} \alpha_j y_{n+j} = h(\beta - \gamma) f_{n+5} + h\gamma \bar{f}_{n+5} + h\delta \bar{f}_{n+6} + h\varepsilon \bar{f}_{n+7} \tag{13}$$

where ε and γ are taken as free parameters. The remaining parameters are determined such that (13) has order 6. This then leads to a 5-step general linear method of order 6 with 4 real poles. If the stability analysis is done by modifying the formulas of [12, p. 269], one obtains a characteristic equation of degree 5. The result, depending on the free parameters ε and γ, is presented in Fig. 10 and exhibits a small region where A-stability actually occurs. The order star of the method for $\varepsilon = 0.05$ and $\gamma = 0.1$ is presented in Fig. 11. On three sheets, lurk boundary curves of A ready to invade the imaginary axis when the point (ε, γ) dares to leave the tiny region for *A*-stability.

The Butcher–Chipman Conjecture. The search for high order general linear methods motivates the following problem: given integers k, l, m, find $K(z)$, $L(z)$, $M(z)$ of degree k, l, m, respectively, such that

$$K(z)R^2 + L(z)R + M(z) = 0 \tag{14}$$

produces an algebraic approximation of highest possible order to e^z,

$$e^z - R(z) = \mathbf{O}(z^{k+l+m+2}). \tag{15}$$

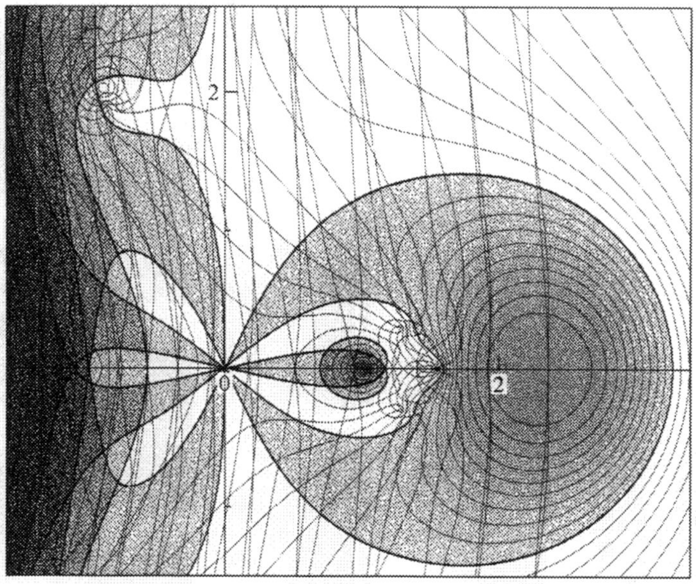

Fig. 11. Order star on 5-sheet Riemann surface for the Psihoyios–Cash method.

Answer [7]:

$$
\begin{aligned}
K(z) &= (D+2)^{-m-1}(D+1)^{-l-1}\frac{z^k}{k!} \qquad k \overset{+1}{\longrightarrow} l, \quad k \underset{+2}{\longrightarrow} m,\\
L(z) &= (D+1)^{-m-1}(D-1)^{-k-1}\frac{z}{l!} \qquad k \overset{-1}{\longleftarrow} l \overset{+1}{\longrightarrow} m,\\
M(z) &= (D-1)^{-l-1}(D-2)^{-k-1}\frac{z^m}{m!} \qquad k \underset{-2}{\longleftarrow} m, \quad l \overset{-1}{\longleftarrow} m.
\end{aligned} \tag{16}
$$

Here $D = \mathrm{d}/\mathrm{d}z$ and $(D+a)^{-n-1} = a^{-n-1}(1+D/a)^{-n-1}$ must be replaced by its binomial series in order to produce generalizations of the formulas (2).

Examples. For $(k,l,m) = (3,1,1)$ and $(3,0,2)$ we obtain

$$
\begin{aligned}
\Phi(R,z) &= \left(-54 + \frac{69}{2}z - 9z^2 + z^3\right)R^2 + (48+24z)R + \left(6 + \frac{3}{2}z\right),\\
\Phi(R,z) &= \left(-\frac{63}{2} + 24z - \frac{15}{2}z^2 + z^3\right)R^2 + 48R + \left(-\frac{33}{2} - 9z - \frac{3}{2}z^2\right),
\end{aligned} \tag{17}
$$

which correct some sign errors in [7, p. 123].

The Butcher–Chipman Conjecture. After extensive numerical testing for A-stable approximations, Butcher and Chipman arrive at the conjecture that

$$0 \leqslant (k+1) - (l+1) - (m+1) \leqslant 2 \tag{18}$$

were necessary and sufficient for A-stability. This would nicely generalize the Ehle Conjecture to the case of quadratic Padé approximations.

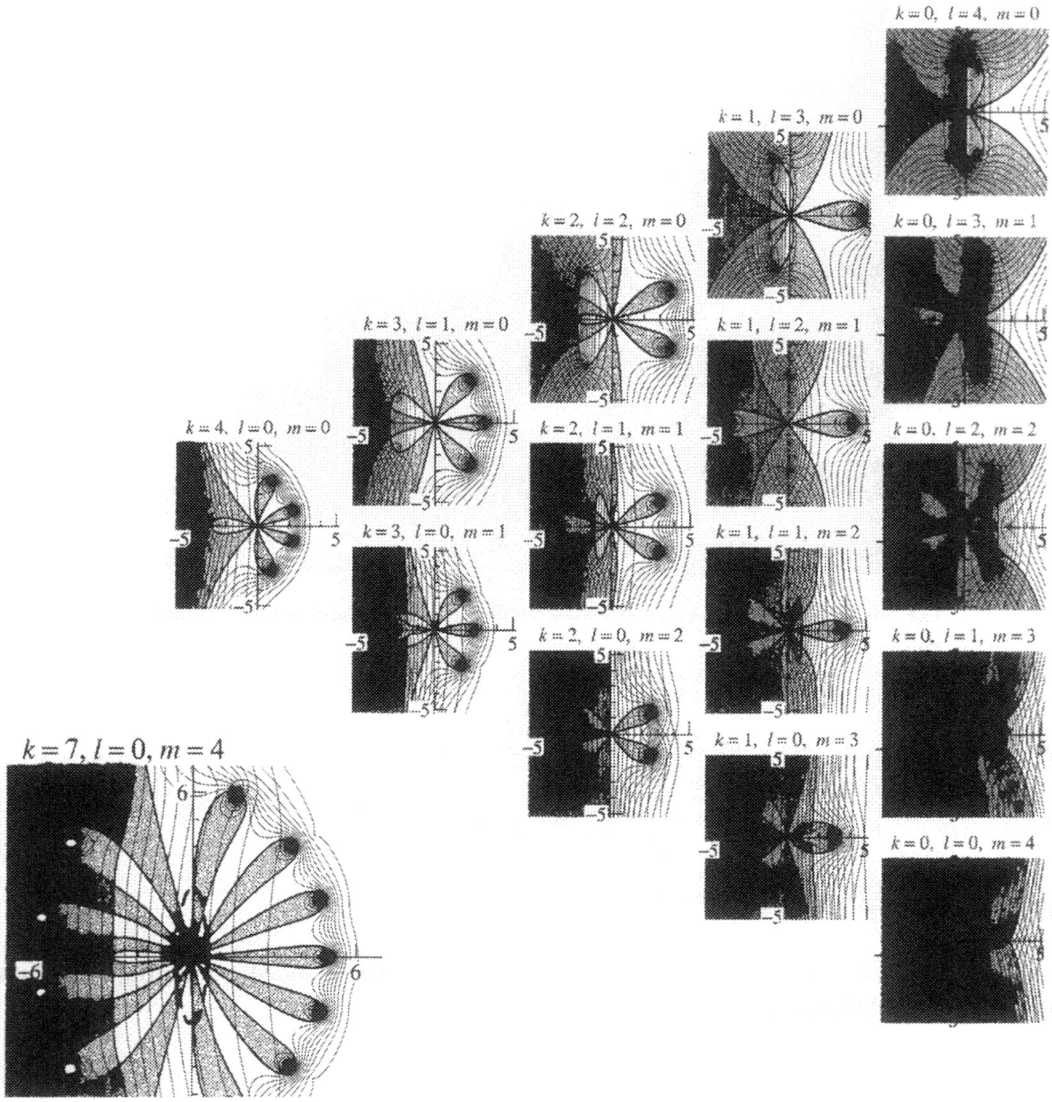

Fig. 12. Order stars for order 5 Butcher–Chipman approximations (right); counter example (left).

First counter example. When $k=m$, Eq. (16), as well as the order star, become symmetric. The order of the particular roots drop, as described in [12, p. 292], and 1 becomes a double eigenvalue of the matrix A. Nevertheless, the corresponding general linear method can remain A-stable. This happens, for example, with the A-stable case $k=m=2, l=0$ in Fig. 12, which violates (18).

Second counter example. The approximations satisfying (18) remain A-stable only for low orders. The first counter-example occurs for $k=7, l=0, m=4$, an approximation of order 12. See, again, Fig. 12. Computations using the Schur criterion reveal that the leftmost black finger crosses the imaginary axis for $|y| \leqslant 1.97$.

Open problem. It appears that Condition (18) remains *necessary* for all non-symmetric examples. Hiervon wäre allerdings ein strenger Beweis zu wünschen.

References

[1] A. Abdulle, On roots and error constants of optimal stability polynomials, BIT 40 (2000) 177–182.
[2] O. Axelsson, A class of A-stable methods, BIT 9 (1969) 185–199.
[3] G. Birkhoff, R.S. Varga, Discretization errors for well-set Cauchy problems, I, J. Math. Phys. 44 (1965) 1–23.
[4] J.C. Butcher, Coefficients for the study of Runge–Kutta integration processes, J. Aust. Math. Soc. 3 (1963) 185–201.
[5] J.C. Butcher, Implicit Runge–Kutta processes, Math. Comp. 18 (1964) 50–64.
[6] J.C. Butcher, Integration processes based on Radau quadrature formulas, Math. Comp. 18 (1964) 233–244.
[7] J.C. Butcher, F.H. Chipman, Generalized Padé approximations to the exponential function, BIT 32 (1992) 118–130.
[8] G. Dahlquist, A special stability problem for linear multistep methods, BIT 3 (1963) 27–43.
[9] J.W. Daniel, R.E. Moore, Computation and Theory in Ordinary Differential Equations, W.H. Freeman, San Francisco, 1970, pp. 172.
[10] B.L. Ehle, On Padé approximations to the exponential function and A-stable methods for the numerical solution of initial value problems, Research Report CSRR 2010, Department of AACS, University of Waterloo, Ontario, Canada, 1969.
[11] N. Guglielmi, E. Hairer, Order stars and stability for delay differential equations, Numer. Math. 83 (1999) 371–383.
[12] E. Hairer, G. Wanner, Solving ordinary differential equations, vol. II (Stiff and differential-algebraic problems), *SCM 14*, Springer, Berlin, 1996, pp. 614.
[13] E. Hairer, G. Wanner, Stiff differential equations solved by Radau methods, J. Comput. Appl. Math. 111 (1999) 93–111.
[14] A. Iserles, Generalized order star theory, in: M.G. de Bruin, H. van Rossum (Eds.), Padé Approximations and its Applications, Amsterdam 1980, Lecture Notes in Math., vol. 888, Springer, Berlin, 1981.
[15] A. Iserles, Order stars and a saturation theorem for first-order hyperbolics, IMA J. Numer. Anal. 2 (1982) 49–61.
[16] A. Iserles, S.P. Nørsett, A proof of the first Dahlquist barrier by order stars, BIT 24 (1984) 529–537.
[17] A. Iserles, S.P. Nørsett, Order Stars, Chapman & Hall, London, 1991, pp. 248.
[18] A. Iserles, G. Strang, The optimal accuracy of difference schemes, Trans. Amer. Math. Soc. 277 (1983) 779–803.
[19] R. Jeltsch, Stability and accuracy of difference schemes for hyperbolic problems, J. Comput. Appl. Math. 12 & 13 (1985) 91–108.
[20] R. Jeltsch, O. Nevanlinna, Stability of explicit time discretizations for solving initial value problems, Numer. Math. 37 (1981) 61–91; Corrigendum: Numer. Math. 39, 155.
[21] R. Jeltsch, O. Nevanlinna, Stability and accuracy of time discretizations for initial value problems, Numer. Math. 40 (1982) 245–296.
[22] R. Jeltsch, R.A. Renaut, J.H. Smit, An accuracy barrier for stable three-time-level difference schemes for hyperbolic equations, IMA J. Numer. Anal. 18 (1998) 445–484.
[23] G.Y. Psihoyios, J.R. Cash, A stability result for general linear methods with characteristic function having real poles only, BIT 38 (1998) 612–617.
[24] J.H. Smit, Order stars and the optimal accuracy of stable, explicit difference schemes, Quaestiones Math. 8 (1985) 167–188.
[25] G. Wanner, E. Hairer, S.P. Nørsett, Orderstars and stability theorems, BIT 18 (1978) 475–489.
[26] A. Wolfbrandt, A study of Rosenbrock processes with respect to order conditions and stiff stability, Thesis, Chalmers University of Technology, Göteborg, Sweden, 1977.

ELSEVIER

Journal of Computational and Applied Mathematics 125 (2000) 107–115

JOURNAL OF COMPUTATIONAL AND APPLIED MATHEMATICS

www.elsevier.nl/locate/cam

Exponentially fitted Runge–Kutta methods ☆

G. Vanden Berghe *, H. De Meyer [1], M. Van Daele, T. Van Hecke [2]

Department of Applied Mathematics and Computational Science, Universiteit Gent, Krijgslaan 281-S9, B-9000 Gent, Belgium

Received 14 June 1999; received in revised form 6 January 2000

Abstract

Exponentially fitted Runge–Kutta methods with s stages are constructed, which exactly integrate differential initial-value problems whose solutions are linear combinations of functions of the form $\{x^j \exp(\omega x), x^j \exp(-\omega x)\}$, ($\omega \in \mathbb{R}$ or $i\mathbb{R}$, $j = 0, 1, \ldots, j\max$), where $0 \leqslant j\max \leqslant \lfloor s/2 - 1 \rfloor$, the lower bound being related to explicit methods, the upper bound applicable for collocation methods. Explicit methods with $s \in \{2, 3, 4\}$ belonging to that class are constructed. For these methods, a study of the local truncation error is made, out of which follows a simple heuristic to estimate the ω-value. Error and step length control is introduced based on Richardson extrapolation ideas. Some numerical experiments show the efficiency of the introduced methods. It is shown that the same techniques can be applied to construct implicit exponentially fitted Runge–Kutta methods. © 2000 Elsevier Science B.V. All rights reserved.

MSC: 65L05; 65L06; 65L20

Keywords: Runge–Kutta method; Exponential fitting; Ordinary differential equations

1. Introduction

In the last decade, a lot of research has been performed in the area of the numerical solution of initial value problems related to systems of first-order ordinary differential equations, i.e.,

$$y' = f(x, y), \qquad y(x_0) = y_0. \tag{1}$$

Particular tuned methods have been proposed when the solution of the above problem exhibits a pronounced oscillatory character. If a good estimate of the frequency is known in advance, one can use linear multistep methods, whereby exact integration of a given set of linearly independent

☆ Supported by the Bilateral Scientific and Technological Cooperation 1998, Flanders-Romania, Grant BIL 98/47.

* Corresponding author.

E-mail address: guido.vandenberghe@rug.ac.be (G.V. Berghe).

[1] Research Director of the Fund for Scientific Research FWO – Vlaanderen.

[2] Postdoctoral fellow of the Fund for Scientific Research FWO – Vlaanderen.

PII: S0377-0427(00)00462-3

functions is achieved. A first good theoretical foundation of this technique was given by Gautschi [2] and Lyche [5]. Since then, a lot of exponentially fitted linear multistep methods have been constructed; most of them were developed for second-order differential equations where the first derivative is absent, and applied to solve equations of the Schrödinger type. Also for first-order equations special tuned algorithms have been constructed. For an exhaustive list of references we refer to [4] and references cited therein. The study of exponentially fitted Runge–Kutta (EFRK) methods is limited and of a very recent date. Paternoster [6] used the linear stage representation of a Runge–Kutta method given in Albrecht's approach and derived some examples of implicit Runge–Kutta–(Nyström) methods of low algebraic order (for the definition of that property see [6]). On the other hand, Simos [7,8] constructed an explicit Runge–Kutta–(Nyström) method of algebraic order 4, which integrates certain particular first-order initial-value problems with periodic or exponential solutions. In the present paper a general technique for the construction of exponentially fitted Runge–Kutta methods is introduced.

An s-stage Runge–Kutta method can integrate exactly at most a set of s linearly independent functions. This maximum is reached if the method is a collocation method. In the latter case, classically the set $\{1, x, x^2, \dots, x^s\}$ is considered; in our approach we choose the set $\{x^m \exp(\omega x), x^m \exp(-\omega x)\}$ or equivalently for $\omega = \mathrm{i}\lambda$ the set $\{x^m \sin(\lambda x), x^m \cos(\lambda x)\}$ with $m = 0, 1, \dots, \lfloor s/2 - 1 \rfloor$. For explicit EFRK methods only two functions are present in the basis, i.e. $m = 0$; this is due to the fact that classical explicit Runge–Kutta methods have a stage order of at most 1.

In order to realize this goal we define an EFRK method as

$$y_{n+1} = y_n + h \sum_{i=1}^{s} b_i f(x_n + c_i h, Y_i) \tag{2}$$

with

$$Y_i = \gamma_i y_n + h \sum_{j=1}^{s} a_{ij} f(x_n + c_j h, Y_j) \tag{3}$$

or in tableau form

$$\begin{array}{c|c|cccc} c_1 & \gamma_1 & a_{11} & a_{12} & \dots & a_{1,s} \\ c_2 & \gamma_2 & a_{21} & a_{22} & \dots & a_{2,s} \\ & & & \dots & & \\ c_s & \gamma_s & a_{s1} & a_{s2} & \dots & a_{ss} \\ \hline & & b_1 & b_2 & \dots & b_s \end{array} \tag{4}$$

This formulation is an extension of the definition of a classical Butcher tableau. A motivation for the introduction of γ_i-values is given in [10]. In this paper we define that a function $g(x)$ is integrated exactly by a (EF)RK method if $y_n = g(x_n)$ for all problems whose solution is $g(x)$. If each stage equation and the final step equation exactly integrate $g(x)$ then the (EF)RK scheme itself also integrates $g(x)$ exactly. Indeed, for small h the uniqueness of the solution of system (3) is guaranteed by a proof similar to that of Theorem 341 of [1], while on the other hand this unique solution is $Y_i = g(x_n + c_i h)$ by construction, given that each equation in (2) and (3) exactly integrates $g(x)$ and that the solution of the problem to be solved is $g(x)$. The above conditions give rise to a system of equations for the components of b and A of the following form (we present the exponential form and $v = \omega h$, $\omega \in \mathbb{R}$). The number of equations to retain depends on the number $2(m+1)$ of elements present in the basic set.

$$\begin{aligned}
&\exp(\pm v) - 1 \mp v \sum_{i=1}^{s} b_i \exp(\pm c_i v) = 0, \\
&\exp(\pm v) - \sum_{i=1}^{s} b_i (1 \pm v c_i) \exp(\pm c_i v) = 0, \\
&\qquad\vdots
\end{aligned} \tag{5}$$

together with

$$\begin{aligned}
&\exp(\pm c_i v) - \gamma_i \mp v \sum_{j=1}^{s} a_{ij} \exp(\pm c_j v) = 0, \\
&c_i \exp(\pm c_i v) - \sum_{j=1}^{s} a_{ij} (1 \pm v c_j) \exp(\pm c_j v) = 0, \\
&\qquad\vdots
\end{aligned} \tag{6}$$

with $i = 1, \ldots, s$.

2. Explicit EFRK methods

In the case of explicit EFRK methods the number of equations in (5) and (6) is restricted to two. Below we give some examples.

For the two-stage method we have chosen $c_0 = 0$ and $c_1 = 1$. Eqs. (5) and (6) fully define $b_1, b_2, \gamma_1, \gamma_2$ and a_{21} as (here we always present the trigonometric solution, i.e. $\omega = \mathrm{i}\lambda, v = h\lambda$):

$$\begin{array}{c|c|cc}
0 & 1 & & \\
1 & \cos(v) & \sin(v)/v & \\
\hline
 & & \dfrac{\sin(v)}{v[\cos(v)+1]} & \dfrac{\sin(v)}{v[\cos(v)+1]}
\end{array} \tag{7}$$

This EFRK method reduces in the limit $v \to 0$ to the classical scheme with $a_{21} = 1$ and $b_1 = b_2 = \frac{1}{2}$, which has algebraic order 2.

For the three-stage explicit method we have chosen $c_1 = 0$, $c_2 = \frac{1}{3}$ and $c_3 = \frac{2}{3}$. Eqs. (5) leave one degree of freedom. In order to fully define all components of b we have added the following supplementary equation:

$$b_1 + b_2 + b_3 = 1, \tag{8}$$

which expresses that (2) reproduces exact results whenever $f(x,y)=1$. Eqs. (6) again leave one degree of freedom; we have chosen $a_{31}=0$. The following exponentially fitted method emerges:

$$\begin{array}{c|c|ccc}
0 & 1 & & & \\
\frac{1}{3} & \cos(v/3) & \frac{\sin(v/3)}{v} & & \\
\frac{2}{3} & 1 & 0 & \frac{\sin(2v/3)}{v\cos(v/3)} & \\
\hline
 & & b_1 & b_2 & b_3
\end{array} \tag{9}$$

with

$$b_1=\frac{-4\cos(v/3)\sin(v/3)+v\cos(v/3)-2\sin(v/3)+v}{2v\sin^2(v/3)},$$

$$b_2=\frac{-2\cos(v/3)\sin(v/3)+v\cos(v/3)-\sin(v/3)}{v(\cos(v/3)-1)},$$

$$b_3=\frac{-4\cos(v/3)^2\sin(v/3)-2\cos(v/3)\sin(v/3)+v\cos(v/3)+v}{v\sin^2(v/3)}.$$

This method reduces for $v\to 0$ to the classical scheme with $a_{21}=\frac{1}{3}$, $a_{32}=\frac{2}{3}$ and $b_1=\frac{1}{4}$, $b_2=0$, $b_3=\frac{3}{4}$ which has algebraic order 3.

For the four-stage method two additional equations can be added to (5) [10]. We require that the first equation (2) reproduces exact results whenever $f(x,y)$ is 1 or x. This delivers the following two equations for the b components:

$$\sum_{i=1}^{4} b_i = 1, \qquad \sum_{i=1}^{4} b_i c_i = \tfrac{1}{2}. \tag{10}$$

Eqs. (5) and (10) fully determine the components of the b-vector. Eqs. (6) leave three degrees of freedom. Inspired by the classical fourth-order scheme we choose $a_{31}=0$ and $a_{42}=0$; additionally, we also set $\gamma_4=1$. The following tableau emerges:

$$\begin{array}{c|c|cccc}
0 & 1 & & & & \\
1/2 & \cos(v/2) & \frac{\sin(v/2)}{v} & & & \\
1/2 & \frac{1}{\cos(v/2)} & 0 & \frac{\tan(v/2)}{v} & & \\
1 & 1 & 0 & 0 & \frac{2\sin(v/2)}{v} & \\
\hline
 & & b_1 & b_2 & b_3 & b_4
\end{array} \tag{11}$$

with

$$b_1 = b_4 = \frac{2\sin(v/2) - v}{2v(\cos(v/2) - 1)}, \quad b_2 = b_3 = \frac{v\cos(v/2) - 2\sin(v/2)}{2v(\cos(v/2) - 1)}.$$

For $v \to 0$ this method reduces to the well-known classical scheme with $a_{21} = \frac{1}{2}$, $a_{32} = \frac{1}{2}$, $a_{43} = 1$ and $b_1 = b_4 = \frac{1}{6}$, $b_2 = b_3 = \frac{1}{3}$, which has algebraic order 4.

In order to have an idea of the form of the local truncation error (LTE) for the explicit EFRK method, we have calculated for the three above-mentioned schemes the difference $y(x_{n+1}) - y_{n+1}$ with $y(x_{n+1})$ a notation for the exact solution in the point x_{n+1}. Note that the well-known theory of Butcher based on rooted trees cannot be applied here, since the row-sum rule is not fulfilled. We present the LTE for a non-autonomous scalar equation. All occurring partial derivatives of $f(x, y(x))$ with respect to x and y are collected in order to express these LTEs in terms of total derivatives of $y(x)$. The following results were obtained.

- For the two-stage method (7)

$$\text{LTE} = -\frac{h^3}{12}[y^{(3)} - 3f_y y^{(2)} + \lambda^2(y' - 3f_y y)] + \mathrm{O}(h^4). \tag{12}$$

- For the three-stage method (9)

$$\text{LTE} = \frac{h^4}{216}[y^{(4)} + 2f_y y^{(3)} + 6f_y^2 y^{(2)} + \lambda^2(y^{(2)} + 2f_y y' + 6f_y^2 y)] + \mathrm{O}(h^5). \tag{13}$$

- For the four-stage method (11)

$$\begin{aligned}\text{LTE} = -\frac{h^5}{2880}&[y^{(5)} - 5f_y y^{(4)} - 10f_{xy} y^{(3)} - 10f_{yy} f y^{(3)} + 10f_y^2 y^{(3)}\\ &+15f_{yy}(y^{(2)})^2 + 30f_{xy} f_y y^{(2)} + 30f_{yy} f f_y y^{(2)} - 30f_y^3 y^{(2)}\\ &+\lambda^2(y^{(3)} - 5f_y y^{(2)} - 10f_{xy} y' - 10f_{yy} f y'\\ &+10f_y^2 y' + 15f_{yy} y y^{(2)} + 30f_{xy} f_y y + 30f_{yy} f f_y y - 30f_y^3 y)\\ &+15\lambda^2((f_y + f_{yy} y)y^{(2)} + \lambda^2(f_y + f_{yy} y)y)] + \mathrm{O}(h^6).\end{aligned} \tag{14}$$

All functions are evaluated at $x = x_n$ and $y = y_n$. Note that the leading order terms in the given LTEs become zero for $y = \sin(\lambda x)$ or $y = \cos(\lambda x)$. Moreover in each of the LTE-expressions the derivative of lowest order occurring for $\lambda = 0$ is $y^{(2)}$, showing that classically only the set $\{1, x\}$ is integrated exactly.

3. Local error estimation and a good choice for λ

There exists no mathematical theory to determine λ in an exact way. The only goal we can put forward is to make the LTE as small as possible by calculating λ by means of a heuristically chosen algorithm. Since for a scalar equation in each of the expressions for the LTEs (12)–(14) the term

$y^{(2)} + \lambda^2 y$ is present we propose to calculate λ in each integration interval $[x_n, x_{n+1}]$ in the following way:

$$\lambda = \sqrt{-\frac{y^{(2)}(x_n)}{y(x_n)}}, \quad n = 0, \dots \quad \text{if } y(x_n) \neq 0 \text{ and } \lambda = 0 \text{ otherwise.}$$

Note that if $y(x)$ is a linear combination of $\sin(\alpha x)$ and $\cos(\alpha x)$, and $y(x_n) \neq 0$, we obtain from this recipe $\lambda = \alpha$ and $y(x)$ will be integrated exactly (if infinite precision arithmetic is assumed). For a system of n first-order equations, we propose to make the Euclidean norm of all $({}^t y^{(2)} + \lambda^2\, {}^t y,\ t = 1, \dots n)$ present as small as possible. This results in

$$\lambda = \sqrt{-\frac{\sum_{t=1}^{n} {}^t y(x_n)\, {}^t y^{(2)}(x_n)}{\sum_{t=1}^{n} {}^t y(x_n)^2}}. \tag{15}$$

The expressions for the occurring second derivatives can be obtained analytically from the given ODEs or calculated numerically using previously derived $y(x_{n-j})$ values. The λ-values used are then in each integration interval taken as the positive square root of the numerically obtained λ^2. If negative λ^2-values are obtained, λ is replaced in the corresponding formulae by $\mathrm{i}\lambda$ ($\mathrm{i}^2 = -1$). In fact, in this case, the exponential functions instead of the trigonometric ones are integrated.

Since the λ-values used are calculated, they are never exact due to finite precision arithmetics. As a consequence, the leading term of the LTE does not vanish. This means that one can try to estimate numerically, for a chosen λ, the LTE. A technique which can be used is Richardson extrapolation. First we fix λ by (15). We consider a Runge–Kutta method of order $p \in \{2, 3, 4\}$ to obtain the solution y_{n+1} at x_{n+1}. Under the usual localizing assumption that $y_n = y(x_n)$ it follows from (12) to (14) that the LTE can be written in the form $T_{n+1} = y(x_{n+1}) - y_{n+1} = C(y, f)h^{p+1} + \mathrm{O}(h^{p+2})$, where $C(y, f)$ is some function of y, its derivatives, $f(x, y)$ and its partial derivatives with respect to x and y, all evaluated at the point (x_n, y_n). Let us now compute a second numerical solution at x_{n+1} by applying the same method twice with steplength $h/2$ and also starting in x_n; denote the solution so obtained by z_{n+1}. By starting in both calculations at the same point x_n one can work during these two processes with the same value of λ. The error is now

$$T_{n+1} = y(x_{n+1}) - z_{n+1} = 2C(y, f)(h/2)^{p+1} + \mathrm{O}(h^{p+2}).$$

From these two estimates for the LTE one can derive that the error in the second calculation is given by: error $= 2C(y, f)(h/2)^{p+1} \approx (z_{n+1} - y_{n+1})/(2^p - 1)$ If the user asks for a given tolerance tol, he can control the steplength and the error in the following way:

if $|\text{error}| <= \text{tol}$ accept the step and progress with the z_{n+1} value
if $|\text{error}| > \text{tol}$ reject the step

The step is then adapted in the following way:

$$h_{\text{new}} = h_{\text{old}} \min(\text{facmax}, \max(\text{facmin}, \text{fac}(\text{tol}/\text{error})^{1/(1+p)})),$$

with facmax and facmin, respectively, the maximum and minimum acceptable increasing or decreasing factors. The symbol *fac* represents a safety factor in order to have an acceptable error in the following step. The method for estimating the LTE and the notation used to define h_{new} is given in Hairer et al. [3]. In the code we have developed the following values for these factors were taken: facmax $= 2$, facmin $= 0.5$, fac $= 0.9$.

4. Numerical experiments

In this section we solve some initial-value problems having as solution a combination of sine or cosine functions. The following cases have been considered.

Example 1.

$$y'' = -30\sin(30x), \quad y(0) = 0, \quad y'(0) = 1 \tag{16}$$

with $y(0) = 0$, $y'(0) = 1$; its exact solution is $y(x) = \sin(30x)/30$. Eq. (16) has been solved in the interval $0 \leqslant x \leqslant 10$ with $\mathrm{tol} = 10^{-10}$.

Example 2.

$$y'' + y + y^3 = B\cos(\Omega x), \quad y(0) = 0.20042672806900, \quad y'(0) = 0 \tag{17}$$

with $B = 0.002$, $\Omega = 1.01$; accuracy is judged by comparison with a Galerkin approximation obtained by Van Dooren [9]

$$y(x) \approx 0.200179477536\cos(\Omega x) + 0.246946143 \cdot 10^{-3}\cos(3\Omega x)$$
$$+0.304014 \cdot 10^{-6}\cos(5\Omega x) + 0.374 \cdot 10^{-9}\cos(7\Omega x).$$

Eq. (17) has been solved in the interval $0 \leqslant x \leqslant 300$ with $\mathrm{tol} = 10^{-8}$.

Example 3.

$$y'' + y = 0.001\cos(x), \quad y(0) = 1, \quad y'(0) = 0, \tag{18}$$

with exact solution $y(x) = \cos(x) + 0.0005x\sin(x)$. Eq. (18) has been solved in the interval $0 \leqslant x \leqslant 1000$ with $\mathrm{tol} = 10^{-8}$.

Example 4.

$$y'' + 0.2y' + y = 0, \quad y(0) = 1, \quad y'(0) = 0, \tag{19}$$

with exact solution $y(x) = \exp(-0.1x)(\sin(\sqrt{0.99}x)/\sqrt{99} + \cos(\sqrt{0.99}x))$. Eq. (19) has been solved in the interval $0 \leqslant x \leqslant 2\pi$ with $\mathrm{tol} = 10^{-8}$.

After having rewritten the above equations as equivalent first-order systems, we have solved them with the new fourth-order EFRK method (11) and the equivalent classical fourth-order RK method; in both cases we applied the above-described Richardson extrapolation technique. We have calculated at the endpoint of the integration interval for each example the Euclidean norm of the error vector with components defined as the difference between the numerical and the exact values of the solution and its first derivative. These data are collected in Table 1 together with the number of accepted and rejected steps.

It is clear from Table 1 that for equations with a purely trigonometric solution (Example 1), the new method is superior to the classical one. In this example, (15) initially gives a quite accurate value for λ, such that there is a small error and a huge increase in the stepsize in each step.

Table 1
Comparison of the Euclidian norms of the endpoint global errors obtained by using the fourth-order EFRK and classical RK-methods with step-length and order control based on Richardson extrapolation techniques

		Accepted steps	Rejected steps	Error
Example 1	Exponentially fitted	33	16	$4.01 \cdot 10^{-9}$
	Classical	3495	5	$1.01 \cdot 10^{-8}$
Example 2	Exponentially fitted	1191	0	$1.01 \cdot 10^{-7}$
	Classical	2167	0	$2.79 \cdot 10^{-5}$
Example 3	Exponentially fitted	1946	1	$1.14 \cdot 10^{-5}$
	Classical	9826	0	$5.75 \cdot 10^{-5}$
Example 4	Exponentially fitted	46	3	$1.96 \cdot 10^{-7}$
	Classical	57	0	$2.70 \cdot 10^{-7}$

However, these errors accumulate and at a certain moment such a large stepsize causes a rejected step. After decreasing the stepsize, this process is repeated. In cases where the solution is of a mixed trigonometric form, although one of the terms dominates the solution (Examples 2 and 3), the new method is still more efficient than the classical one. In Example 4 where the solution is a mixture of exponential and trigonometric functions both methods considered are practically equivalent.

5. Implicit EFRK schemes

For a collocation RK scheme with s stages the set of functions $\{1, x, \ldots, x^s\}$ are exactly integrated. It is obvious that in the case of EFRK schemes the couples $\{x^m \exp(\omega x), x^m \exp(-\omega x)\}$, $m = 0, 1, \ldots, \lfloor s/2 - 1 \rfloor$ have to be taken into account together. Two- and three-stage implicit methods have been studied. For the two-stage case, where we have chosen $\gamma_1 = \gamma_2 = 1$ identical results have been obtained as the ones derived by Paternoster [6]. For general, three-stage methods the length of the resulting expressions is large; therefore, we present a rather simple case, the LobattoIIIA EFRK method; we have fixed beforehand the c_i-values, i.e. $c_1 = 0$, $c_2 = \frac{1}{2}$ and $c_3 = 1$. For these methods as well the $m = 0$ as the $m = 1$ couples are taken into account in the set of basic functions. The following results emerge:

$$a_{11} = 0, \qquad a_{12} = 0, \qquad a_{31} = 0, \qquad \gamma_1 = 0,$$

$$a_{21} = \frac{-v^2 - 4 + (4 + 2v^2)\cos^2(v/2)}{2v^2 D}, \qquad a_{23} = \frac{v^2 - 4 + 4\cos^2(v/2)}{2v^2 D},$$

$$a_{22} = -\frac{(v^2 - 4)\cos(v/2) + 2v\sin(v/2) - 2v\cos^2(v/2)\sin(v/2) + 4\cos^3(v/2)}{v^2 D},$$

$$\gamma_2 = \frac{-v\cos(v/2)\sin(v/2) - 2 + 2\cos^2(v/2)}{D},$$

with

$$D = -v\sin(v/2) - 2\cos(v/2) + 2\cos^3(v/2),$$

$$b_1 = a_{31} = \frac{2\sin(v/2) - v\cos(v/2)}{v^2 \sin(v/2)}, \qquad b_2 = a_{32} = \frac{-2\sin(v) + 2v}{v^2 \sin(v/2)},$$

$$b_3 = a_{33} = \frac{2\sin(v/2) - v\cos(v/2)}{v^2 \sin(v/2)}, \quad \gamma_3 = 1.$$

For $\lambda \to 0$ this method reduces to the classical LobattoIIIA method and has algebraic order 4.

6. Conclusions

We have developed a technique to construct explicit and implicit exponentially fitted Runge–Kutta methods. Some examples of both types are given. A heuristic way of determining λ is given for the case of explicit methods, which is based on the typical expression of the LTEs. In cases where the solutions are purely trigonometric the use of Richardson extrapolation is doubtful. The large percentage of rejected steps in the first example supports this statement. Better methods for error estimation are welcome. It is shown that the introduced construction method can also be straightforwardly applied to obtain exponentially fitted implicit methods.

Acknowledgements

The authors are very grateful to two anonymous referees, whose comments greatly improved this paper.

References

[1] J.C. Butcher, The Numerical Analysis of Ordinary Differential Equations, Wiley, Chichester, UK, 1987.
[2] W. Gautschi, Numerical integration of ordinary differential equations based on trigonometric polynomials, Numer. Math. 3 (1961) 381–397.
[3] E. Hairer, S.P. Nørsett, G. Wanner, Solving Ordinary Differential Equations I, Nonstiff Problems, 2nd Edition, Springer, Berlin, 1993.
[4] L. Ixaru, Operations on oscillatory functions, Comput. Phys. Comm. 105 (1997) 1–19.
[5] T. Lyche, Chebyshevian multistep methods for ordinary differential equations, Numer. Math. 19 (1972) 65–75.
[6] B. Paternoster, Runge–Kutta(–Nyström) methods for ODEs with periodic solutions based in trigonometric polynomials, Appl. Numer. Math. 28 (1998) 401–412.
[7] T.E. Simos, An exponentially fitted Runge–Kutta method for the numerical integration of initial-value problems with periodic or oscillating solutions, Comput. Phys. Comm. 115 (1998) 1–8.
[8] T.E. Simos, E. Dimas, A.B. Sideridis, A Runge–Kutta–Nyström method for the numerical integration of special second-order periodic initial-value problems, J. Comput. Appl. Math. 51 (1994) 317–326.
[9] R. Van Dooren, Stabilization of Cowell's classical finite difference methos for numerical integration, J. Comput. Phys. 16 (1974) 186–192.
[10] G. Vanden Berghe, H. De Meyer, M. Van Daele, T. Van Hecke, Exponentially fitted explicit Runge–Kutta methods, Comput. Phys. Comm. 123 (1999) 7–15.

ELSEVIER
Journal of Computational and Applied Mathematics 125 (2000) 117–130

JOURNAL OF COMPUTATIONAL AND APPLIED MATHEMATICS

www.elsevier.nl/locate/cam

Modified extended backward differentiation formulae for the numerical solution of stiff initial value problems in ODEs and DAEs

J.R. Cash

Department of Mathematics, Imperial College of Science, South Kensington, London SW7 2BZ, UK

Received 20 August 1999; received in revised form 24 November 1999

Abstract

For many years the methods of choice for the numerical solution of stiff initial value problems and certain classes of differential algebraic equations have been the well-known backward differentiation formulae (BDF). More recently, however, new classes of formulae which can offer some important advantages over BDF have emerged. In particular, some recent large-scale independent comparisons have indicated that modified extended backward differentiation formulae (MEBDF) are particularly efficient for general stiff initial value problems and for linearly implicit DAEs with index $\leqslant 3$. In the present paper we survey some of the more important theory associated with these formulae, discuss some of the practical applications where they are particularly effective, e.g., in the solution of damped highly oscillatory problems, and describe some significant recent extensions to the applicability of MEBDF codes. © 2000 Elsevier Science B.V. All rights reserved.

Keywords: Stiff differential equations; Differential algebraic equations; MEBDF

1. Introduction

In the 1950s Curtiss and Hirschfelder [12] published one of the first papers which identified clearly the difficulties of solving stiff initial value problems of the form

$$y' = f(x,y), \quad y(x_0) = y_0, \quad y \in \mathbb{R}^s. \tag{1}$$

Since that time a whole variety of methods have been proposed for the numerical solution of (1). The fact that this class of problems has remained so challenging is not at all surprising given the fact

E-mail address: j.cash@ic.ac.uk (J.R. Cash).

PII: S0377-0427(00)00463-5

that it is still not clear exactly what is meant by the term stiffness. Although numerous attempts have been made to give a rigorous definition of this concept, it is probably fair to say that none of these definitions is entirely satisfactory. Indeed the authoritative book of Hairer and Wanner [18] deliberately avoids trying to define stiffness and relies instead on an entirely pragmatic definition given in [12]. What is clear, however, is that numerical methods for solving stiff initial value problems have to satisfy much more stringent stability requirements than is the case for methods intended for nonstiff problems. One of the first, and still one of the most important, stability requirements particularly for linear multistep methods is that of A-stability which was proposed in [13]. However, the requirement of A-stability puts a severe limitation on the choice of suitable linear multistep methods. This is articulated in the so-called Dahlquist second barrier which says, among other things, that the order of an A-stable linear multistep method must be $\leqslant 2$ and that an A-stable linear multistep method must be implicit.

This pessimistic result has encouraged researchers to seek other classes of numerical methods for solving stiff equations. For Runge–Kutta methods, for example, the situation regarding stability is much more satisfactory. In fact, there exist A-stable Runge–Kutta methods of arbitrarily high order. In particular, the s-stage Gauss Runge–Kutta methods have order $2s$ and are A-stable for all s. However, as is well known, these fully implicit Runge–Kutta methods can be very expensive to implement.

We are therefore faced with the classic dilemma that, generally speaking, linear multistep methods are relatively cheap to implement but suffer severe degradation of stability as their order increases while implicit Runge–Kutta methods can have excellent stability properties but tend to be expensive to implement. Most attempts to 'get around' the Dahlquist barrier have involved either lessening the requirement of A-stability to something which is less restrictive, but which is still appropriate for some classes of stiff equations, or proposing a totally different class of formulae. However, the desirability of having methods which are A-stable is sufficiently well documented that it seems the only way to achieve real efficiency for general stiff problems is to consider formulae other than linear multistep methods.

One very successful proposal in this direction was by Hairer and Wanner [18] who developed a code Radau5 based on Radau Runge–Kutta formulae. We will return to this code at a later stage. A second proposal that we wish to describe is to use what have become known as boundary value methods. These methods are able to achieve excellent stability by using information at advanced step points (also known in the literature as superfuture points). As with Runge–Kutta methods it is the efficiency of implementation of these methods rather than their stability which is the real challenge and we will discuss this in the next section.

2. Boundary value methods

To introduce the general class of boundary value methods that we will be interested in we consider again the linear multistep method

$$y_{n+k} + \sum_{j=0}^{k-1} \bar{a}_j y_{n+j} = h\bar{b}_k f(x_{n+k},\ y_{n+k}). \tag{2}$$

In the limit $h=0$ this formula reduces to the linear recurrence relation

$$y_{n+k}+\sum_{j=0}^{k-1}\bar{a}_j y_{n+j}=0. \tag{3}$$

Eq. (3) can be regarded as a linear multistep method 'integrating forward' with a step $h=0$. It is clear that the required solution of (3) is

$$y_n=y_{n+1}=\cdots=y_{n+s}=c. \tag{4}$$

If, however, we appeal to the theory of linear recurrence relations it is well known that if we solve (3) by direct forward recurrence starting with the initial conditions

$$y_{n+i}=c, \quad 0\leqslant i\leqslant k-1 \tag{5}$$

in an attempt to compute the solution

$$y_{n+m}=c, \quad m>k-1, \tag{6}$$

then this process is stable if and only if

(i) $r=1$ is the root of largest modulus of

$$\sum_{j=0}^{k}\bar{a}_j r^j=0 \tag{7}$$

and

(ii) all roots of (7) of unit modulus are distinct.

If these conditions are not satisfied then forward recurrence is unstable. In the parlance of the theory of linear recurrence relations, requirements (i) and (ii) simply impose the condition that $r=1$ is the dominant zero of (7) so that $y_n=c$, for all n, is the dominant solution of (3). In essence the theory tells us that only the dominant solution of (3) can be generated in a stable manner by forward recurrence. However, conditions (i) and (ii) are precisely the conditions for (2) to be zero-stable. Thus, an alternative way of looking at this is to realize that we have to impose the condition of zero-stability on (2) precisely because we demand that we should solve (2) by forward recurrence; that is we solve for y_{n+k} from given values $y_n, y_{n+1}, y_{n+2}, \ldots, y_{n+k-1}$. If we were to interpret (2) not as a prescription for y_{n+k} but as an equation which, if satisfied for $k=0,1,2,\ldots,$ determines the sequence of approximations we are interested in then the relevant question becomes how we solve the resulting simultaneous equations in a stable way without excessive cost. If we were to solve (2) in a different way then we would no longer need to impose the condition of zero-stability and this in turn offers us the possibility of obtaining high order A-stable formulae.

One possible alternative way of solving (3) is to rewrite it as a boundary value problem. (This is the basis of some celebrated algorithms for finding nondominant solutions of linear recurrence relations [23,22,8,9].) To describe one variant of this approach we consider the third order, linear 2 step method:

$$y_{n+2}+4y_{n+1}-5y_n=h(4f_{n+1}+2f_n). \tag{8}$$

It is well known that this formula does not satisfy the condition of zero-stability and so is unsuitable for the solution of initial value problems. It was shown in [10] that if we apply (8) to the linear scalar equation

$$\frac{dy}{dx} = \lambda y, \quad \lambda \in \mathbb{R} \tag{9}$$

to give

$$y_{n+2} + (4 - 4\lambda h)y_{n+1} - (5 + 2h\lambda)y_n = 0, \tag{10}$$

then the required solution of (10) is subdominant for all $h\lambda$. This fact suggests that, instead of solving (10) by forward recurrence starting from two initial conditions on y which would be an unstable process, we should instead generate the required solution using the boundary value formulation

$$\begin{aligned} &y_0 = y(x_0), \\ &y_{n+2} + 4y_{n+1} - 5y_n = h(4f_{n+1} + 2f_n), \quad n = 0, 1, 2, \ldots, N-2, \\ &y_N = 0 \end{aligned} \tag{11}$$

for some large N. Note, in particular, that this defines a tridiagonal system of linear algebraic equations of size $N+1$ for the $N+1$ unknowns. It was shown in [10] that this formulation produces an A-stable algorithm which has order 3. Theoretically, this approach is a very successful one due to the high-order A-stability that we have achieved. In fact, we have used a linear multistep method and achieved A-stability with order > 2. However computationally this algorithm is not, in general, satisfactory for solving initial value problems since it does not allow easy change of stepsize or order and for large systems the storage requirement can be prohibitive. Even more important is the problem that there may be a lack of convergence in the solution of the simultaneous nonlinear algebraic equations. Ideally, what we need is a special kind of boundary value approach which shares the improved stability obtained by (11) but which does allow variable stepsize and order. One of the easiest ways of achieving this is to develop special classes of formulae to which a boundary value approach can be applied.

An early attempt in this direction was in [9] where such a method was derived from a standard Adams–Moulton formula. This early work has since been extended in several ways. In particular, Brugnano and Trigiante [2] have developed a whole theory of boundary value methods suitable for several important classes of initial value problems, such as stiff problems, Hamiltonian problems and differential algebraic equations. However, a major difference is that we set up a 'local' boundary value problem so that when computing y_{n+k} the boundary condition is imposed at x_{n+k+1}. The approach of Brugnano and Trigiante can be regarded as a more conventional one where the boundary condition is imposed at a large distance from the initial point. Their results are too extensive to quote here and we refer the reader to [2].

In what follows, we will extend the approach suggested by (11) to formulae which are particularly suitable for stiff problems. There are numerous ways in which this could be done and in the next section we will describe one particular approach which is based on modified extended backward differentiation formulae and which has proved to be particularly efficient for general stiff initial value problems and differential algebraic equations.

3. Modified extended backward differentiation formulae

Modified extended backward differentiation formulae (MEBDF) were originally proposed as a class of formulae to which an efficient variable order, variable step boundary value approach could easily be applied. The precise form taken by the general k step MEBDF is

$$y_{n+k} + \sum_{j=0}^{k-1} \hat{a}_j y_{n+j} = h[\hat{b}_{k+1} f_{n+k+1} + \hat{b}_k f_{n+k}], \tag{12}$$

where the coefficients are chosen so that this formula has order $k+1$. This order requirement uniquely specifies the coefficients of (12). Starting from given data $y_n, y_{n+1}, \dots, y_{n+k-1}$, a predictor is first used to predict y_{n+k+1}, the derivative approximation y'_{n+k+1} is then computed and finally y_{n+k} is computed from $y_n, y_{n+1}, \dots, y_{n+k-1}, y'_{n+k+1}$. Of course, the accuracy and stability of this method is critically dependent on the predictor used to compute y_{n+k+1} and in particular this predictor must be of order at least k if the whole process is to be of order $k+1$. A natural kth-order predictor is the k-step BDF and this leads to the so-called EBDF algorithm

Stage 1: Use a standard BDF to compute $\bar{y}_{n+k}$:

$$\bar{y}_{n+k} + \sum_{j=0}^{k-1} \bar{a}_j y_{n+j} = h\bar{b}_k f(x_{n+k}, \bar{y}_{n+k}). \tag{13}$$

Stage 2: Use a standard BDF to compute $\bar{y}_{n+k+1}$:

$$\bar{y}_{n+k+1} + \bar{a}_{k-1}\bar{y}_{n+k} + \sum_{j=0}^{k-2} \bar{a}_j y_{n+j+1} = h\bar{b}_k f(x_{n+k+1}, \bar{y}_{n+k+1}). \tag{14}$$

Stage 3: Compute a corrected solution of order $k+1$ at x_{n+k} using

$$y_{n+k} + \sum_{j=0}^{k-1} \hat{a}_j y_{n+j} = h[\hat{b}_{k+1} \bar{f}_{n+k+1} + \hat{b}_k f_{n+k}]. \tag{15}$$

Note that at each of these three stages a nonlinear set of equations must be solved in order that the desired approximations can be computed. The boundary value nature of this approach can be seen from Stage 3 where y_{n+k} is computed from past values y_{n+i} as well as from the future value $\bar{y}_{n+k+1}$.

One of the main drawbacks of this approach is the need to compute and factorize the two iteration matrices arising in the application of a modified Newton iteration at each stage. To avoid this, the EBDF approach described above can be modified (to give the so-called MEBDF approach [5]) by changing Stage 3 to

Stage 3*:

$$y_{n+k} + \sum_{j=0}^{k-1} \hat{a}_j y_{n+j} = h[\hat{b}_{k+1} \bar{f}_{n+k+1} + \bar{b}_k f_{n+k} + (\hat{b}_k - \bar{b}_k)\bar{f}_{n+k}]. \tag{16}$$

Fortunately, this modification not only improves the computational efficiency of this approach, it also improves its stability. The stability properties of the MEBDF together with the reasons for their computational efficiency are fully described in [6,18] and a code MEBDFDAE based on the MEBDF approach is available from NETLIB and from the author's web page. In particular, the MEBDF are

A-stable for order up to 4 and $A(\alpha)$-stable for order up to and including 9 and this is considerably better than the stability achieved by BDF. In the next section we will consider one particular class of problems for which this enhanced stability is particularly appropriate.

4. Damped stiff highly oscillatory problems

As was explained earlier, attempts to give a precise general mathematical definition of stiffness have been largely unsuccessful. However, some important insights into the concept of stiffness can be gained by considering a suitably restricted class of problems. Normally, the aim of considering these problems has been to define a new stability concept which is appropriate for dealing with stiff problems rather than to define stiffness per se. Often these new definitions have evolved from consideration of problems for which there is a contractivity property for the solutions and, for a survey of this, the interested reader is referred to [3]. The most straightforward problem to analyse for linear methods is the constant coefficient equation

$$\frac{\mathrm{d}y}{\mathrm{d}x} = Ay. \tag{17}$$

A definition of stiffness for this problem has been given, for example, in [21]. This definition is not the whole story, however, because other quantities such as the initial conditions and the interval of integration also need to be considered. However if we assume for the time being that all components of the general solution of (17) are in the solution that we require, and that we are integrating over a sufficiently long interval of x, then we can say something about the likely performance of standard codes on (17). In particular if all the eigenvalues of A are real, or lie close to the real axis, then we can expect codes based on BDF and MEBDF to normally perform well since they have linear stability properties appropriate for dealing with such problems. However, if some of the eigenvalues of A have relatively large imaginary part then we would expect BDF to perform rather poorly since they are A-stable only up to order 2. In the case where there exist eigenvalues lying close to or on the imaginary axis whose imaginary part is large in modulus then there are two distinct classes of problems that we need to distinguish between rather carefully. The first is where the large eigenvalues are purely imaginary so that there is no damping in the components of the solution corresponding to these eigenvalues. In this case it is necessary to follow the oscillations exactly and we would expect nonstiff methods to be more efficient than (implicit) stiff methods for this problem. However if the large eigenvalues lie close to the imaginary axis but not on it, so that the rapid oscillations are damped, then it is only necessary to follow them for a short time and in this case highly stable implicit methods normally perform very well. We can conveniently characterize highly oscillatory problems of the form (17) as ones where the eigenvalues λ_j of A satisfy

$$\lambda_j = \mu_j + \mathrm{i}\nu_j, \tag{18}$$

where $\mu_j < 0$ for all j, $\max_{1\leqslant j\leqslant n}|\mu_j| \gg \min_{1\leqslant j\leqslant n}|\mu_j|$ and $|\mu_j| \ll |\nu_j|$ for at least one pair of eigenvalues of large modulus. In their famous DETEST test set [15], Enright et al. devised a problem denoted by $B5$ for which one component of the solution is of the form

$$\exp(-10x)(A\cos\omega x + B\sin\omega x) \quad \text{where } \omega = 50. \tag{19}$$

This problem was specially designed to trap BDF-based methods (and succeeded in doing so!). However, it is important to realize that the performance of BDF codes on these problems can be very different depending on which code is used. For example, when faced with stiff oscillatory problems, LSODE [20] will often use high-order methods with small stepsizes rather than correctly reducing the order to allow the possibility of large stepsize increases. However, the BDF code DASSL is geared towards selecting lower order and this code will often reduce the order to 2 (so that A-stable formulae are being selected) when faced with such problems [1]. We feel that this strategy of DASSL of being biased towards lower order is an extremely important one and the incorporation of this strategy into the MEBDF code MEBDFDAE has had a strong positive influence on its performance. This strategy is fully described in [11]. For a long time these stiff highly oscillatory problems were regarded as being rather intractable mainly because of the relatively poor performance of BDF. Indeed writing in 1984 Gaffney [16] concluded that none of the currently available codes was satisfactory for dealing with these problems. However, recently some excellent codes have become available and this allows these problems to be solved very routinely. In particular, we mention the codes MEBDFDAE, Radau5 and DESI [4] which all have excellent stability properties and can be recommended for these problems. For a survey of the performance of these codes on the highly oscillatory problem the reader is referred to [11].

5. Extensions to the MEBDF approach

In the particular formulation of MEBDF that was considered in Section 3 we set up the boundary value approach using just one superfuture point. The main theoretical result that is indicated by numerical experiment for this particular algorithm is that it is A-stable for order up to and including 4. It would, of course, be valuable to have a proof of this result. It is possible to develop this approach in various directions. For example, we could use more super future points. In particular, it is of interest to see what can be achieved by using two superfuture points. The natural way to define such an algorithm would be to have three predictor steps based on BDF as described in Section 3 and then to apply a corrector of the general form

$$y_{n+k} + \sum_{j=0}^{k-1} \hat{a}_j y_{n+j} = h[\hat{b}_{k+2}\bar{f}_{n+k+2} + \hat{b}_{k+1}\bar{f}_{n+k+1} + \hat{b}_k f_{n+k}]. \tag{20}$$

Here the coefficients are chosen so that the corrector has order $k+1$ and this defines a one parameter family of coefficients for (20). The stability properties of this approach were investigated in detail in [24]. He found that by using this approach it is possible to find A-stable formulae with order up to and including 6. However, rather disappointingly, it is not possible to achieve this stability and still retain the property that only one iteration matrix needs to be factorized. There is, however, a very important theoretical result that comes out of this investigation. This concerns the conjecture that a well-known result of Norsett and Wolfbrandt [25] for one-step methods carries over to the multistep case. Basically, this conjecture is that the order p of an A-stable general linear method whose characteristic function has s poles, all of which are real, satisfies $p \leqslant s+1$. Remarkably, the order 6 MEBDF, with two advanced step points, has 4 real poles and $p=6$. This serves as a rather surprising counterexample to this conjecture. For more details on this the reader is referred to [26].

If we summarize what can be achieved in the way of stability using superfuture points we note that

(1) For linear multistep methods with no superfuture points we have that A-stability implies the order is $\leqslant 2$.
(2) With one superfuture point we have that A-stability implies the order is $\leqslant 4$.
(3) With two superfuture points we have that A-stability implies the order is $\leqslant 6$

It is tempting to conjecture that with k superfuture points we have that A-stability implies that the order p satisfies $p \leqslant 2k+2$. This would be an interesting and important result but, based on the difficulty of finding A-stable methods of order 6 with $k=2$, we expect this conjecture to be false although a proof of this is elusive.

Due to the fact that when using 2 superfuture points we need to factorize two iteration matrices in order to obtain A-stability with order 6 it seems unlikely that the 'two superfuture points' approach will be competitive with the standard MEBDF for the solution of general stiff problems. It may however have a role to play in the highly oscillatory case where high accuracy is requested. However for parallel implementation the situation is quite different. There are many ways in which the MEBDF approach can be parallelized and, in particular, in a parallel environment the need to factorize two iteration matrices is no longer a problem. One possible way of deriving a parallel MEBDF code was investigated in [24]. He developed an approach whereby all predicted solutions can be computed simultaneously and he showed that there is a significant gain in efficiency using this approach. A different and rather ingenious method of parallelization was proposed in [27]. He modified the EBDF approach with two superfuture points to obtain new classes of formulae which are immediately parallelizable. His results indicate that he is able to achieve significant speed ups using this approach and it seems likely that this will be one of the most effective of all parallel algorithms for the solution of general stiff initial value problems.

The second extension we wish to consider in this section is where extra derivative terms are introduced. We illustrate this by considering the one-step case which is of order 2. Here the standard MEBDF is replaced by the three stages:

Stage 1:

$$\bar{y}_{n+1} = y_n + h[\theta f(x_{n+1}, \bar{y}_{n+1}) + (1-\theta) f(x_n, y_n)]. \tag{21}$$

Stage 2:

$$\bar{y}_{n+2} = \bar{y}_{n+1} + h[\theta f(x_{n+2}, \bar{y}_{n+2}) + (1-\theta) f(x_{n+1}, \bar{y}_{n+1})]. \tag{22}$$

Stage 3:

$$\begin{aligned} y_{n+1} = y_n + h[&(c-\tfrac{1}{2}) f(x_{n+2}, \bar{y}_{n+2}) + (\tfrac{3}{2} - 2c - \theta) f(x_{n+1}, \bar{y}_{n+1}) \\ &+ \theta f(x_{n+1}, y_{n+1}) + c f(x_n, y_n)]. \end{aligned}$$

Note that the standard MEBDF is of this form with $\theta = 1$, $c = 0$. Applying these three stages to the standard scalar test equation $y' = \lambda y$ we obtain an expression of the form

$$\frac{y_{n+1}}{y_n} = R(q), \quad q = \lambda h. \tag{23}$$

In order to get the correct asymptotic behaviour, that is

$$\lim_{q \to -\infty} R(q) = 0, \tag{24}$$

we require

$$(c - \tfrac{1}{2})(1-\theta)^2 - (\tfrac{3}{2} - 2c - \theta)(1-\theta)\theta + c\theta^2 = 0. \tag{25}$$

This defines c in terms of θ and leaves θ as a free parameter to improve the accuracy and/or stability of the method. In general, for a k-step formulation, we will again have two free parameters, one of which will be used to give the correct asymptotic behaviour and the other will be used to improve stability. Research is at present in progress to see by how much this approach improves stability and, in particular, whether it is possible to obtain A-stability with $k=4$. However as k increases the situation becomes very complicated since there are many ways in which the extra derivative terms can be added. What is really needed is some theory linking the order and stability of these methods to the step number k.

6. Differential algebraic equations

One of the important properties of MEBDF is that, in common with BDF, they can be extended to the solution of differential algebraic equations in a straightforward way. The extension to linearly implicit DAEs is the most natural and we consider this first of all. Many important classes of differential algebraic equations can be written in the linearly implicit form

$$M\frac{\mathrm{d}y}{\mathrm{d}x} = f(x,y), \tag{26}$$

where the coefficient matrix M is singular. In particular, the constrained system

$$\frac{\mathrm{d}y}{\mathrm{d}x} = F(x,y,z), \quad 0 = g(y,z) \tag{27}$$

can be rewritten as

$$\begin{pmatrix} 1 & 0 \\ 0 & 0 \end{pmatrix}\begin{pmatrix} y' \\ z' \end{pmatrix} = \begin{pmatrix} F(x,y,z) \\ g(y,z) \end{pmatrix}, \tag{28}$$

which is of the form (26). The MEBDF approach described in the previous sections is very straightforward to extend for (26), from ODEs to linearly implicit DAEs. This can be done simply by using the algorithm of Section 3 and being careful to arrange the computation so that we never call for the inverse of the singular matrix M. The one-step MEBDF, for example, would be expressed in a completely analogous way to the one-step BDF as

$$M(y_{n+1} - y_n) = h(-\tfrac{1}{2}f(x_{n+2}, y_{n+2}) + \tfrac{3}{2}f(x_{n+1}, y_{n+1})). \tag{29}$$

An important concept when dealing with DAEs is that of the index. Perhaps the most widely used definitions are of differentiation index and perturbation index. In particular, the differentiation index is i, if i is the minimum number of analytic differentiations that need to be performed on the system to allow an explicit ODE to be extracted. For more on this important concept the reader is referred to [18, p. 445]. The major change from the ODE case is the way in which the errors (i.e., both the local truncation error and the error in the Newton iteration) are controlled. Following the approach described in [19], the error E that we control when using a steplength h is defined as

$$E = \mathrm{Error}_{\mathrm{Index1}} + h\,\mathrm{Error}_{\mathrm{Index2}} + h^2\,\mathrm{Error}_{\mathrm{Index3}}. \tag{30}$$

Here we use the obvious notation that, for example, $\text{Error}_{\text{Index1}}$ is the error in the index 1 variables [19, p. 124]. As explained in [19] this approach is needed essentially to deal with the near singularity (for small h) of the iteration matrix. Numerical results presented on the author's web page [7] indicate the good performance of MEBDFDAE on a variety of linearly implicit test problems of indices 1–3 and this has recently been confirmed by some extensive independent comparisons [14].

It would perhaps be valuable to extend the MEBDF approach to more general equations such as the fully implicit equation

$$F(x, y, y') = 0. \tag{31}$$

This problem could be solved by rewriting (31) in the linearly implicit form

$$y' = z, \quad F(x, y, z) = 0, \tag{32}$$

but at the cost of increasing the size and more importantly the index of the system since one more analytic differentiation now has to be performed to allow an explicit ODE to be extracted.

Another problem we may wish to deal with directly is

$$C(y)\frac{dy}{dx} = g(y), \tag{33}$$

where $C(y)$ is singular. This can be rewritten in the form (28) where the left-hand side of this equation remains the same and the right-hand side is now

$$f(x, y) = \begin{pmatrix} z \\ C(y)z - g(y) \end{pmatrix},$$

providing that the matrix $C(y)$ satisfies some rather general conditions [18, p. 445]. One way of dealing with (33) is by adding extra variables so making it again of the form (26). If the linear algebra is carried out in a careful way [18, p. 576] then the computational effort is increased by relatively little. However, users may not be prepared to change their problems to fit into a more restrictive framework and a more satisfactory approach may be to develop MEBDFDAE to deal directly with these more general equations. The necessary theory to allow this is relatively straightforward to develop and MEBDF codes for the direct solution of (33) and (31) are now available on the author's web page.

A second major problem concerning MEBDF follows from the well-known phenomenon of order reduction. There are at present no theoretical results concerning the order of MEBDF when applied to DAEs. However, the numerical results that have been obtained are highly suggestive. This leads us to make the following conjecture:

A $(p-1)$th step MEBDF when applied to a DAE of index i has order $p+1-i$. The MEBDF code is implemented on the assumption that this is indeed the correct behaviour and it is a serious gap in the theory that we do not have a proof of this result. Finally, we note that this order reduction is particularly serious when dealing with damped highly oscillatory problems of index 3 in the case where it is not required to follow the oscillations. If our conjecture is correct, and the order is indeed reduced by 2, then we in effect have A-stability only up to order 2. In this case the extensions described in the previous section where either two superfuture points are used or possibly where extra derivatives are used will be potentially very important.

7. Numerical results

In this section we will present some numerical results to illustrate the performance of the code MEBDFDAE. The drivers used to obtain these results are available on the web page of the author [7]. The example we consider is that of a simple constrained mechanical system, namely the pendulum. This is a particularly nice example since it is straightforward to derive systems of indices 1, 2 and 3 which describe the equations of motion. In what follows, we will consider the equations of motion of a point mass, m, at the end of a massless rod of length 1 oscillating under the influence of gravity g in the cartesian coordinates (p,q). The first four equations of motion are

$$\begin{aligned} p' &= u, \\ q' &= v, \\ mu' &= -p\lambda, \\ mv' &= -q\lambda - g. \end{aligned} \tag{34}$$

Here u and v are velocities and λ is the rod tension. The fifth equation which completes index 3 formulation is

$$0 = p^2 + q^2 - l^2. \tag{35}$$

To obtain index 2 formulation we differentiate constraint (35) to obtain

$$0 = pu + qv. \tag{36}$$

Eqs. (34)–(36) give index 2 formulation. If we differentiate (36) again we obtain

$$0 = m(u^2 + v^2) - qg - l^2\lambda. \tag{37}$$

Eqs. (34) together with (37) give index 1 formulation. We note that, starting from the original index 3 formulation, there are several ways of rewriting the constraint to reduce the index. In particular, the process of differentiating the constraint may result in the original constraint not being satisfied. This 'drift off' phenomenon is described for example in [18, p. 468; 19, p. 7]. In an attempt to avoid this problem Gear et al. [17] proposed adding in the original constraint via a Lagrange multiplier which vanishes on the exact solution. Thus, index 2 reformulation of index 2 problem (34)–(36) proposed in [17] is

$$\begin{aligned} p' &= u - p\mu, \\ q' &= v - q\mu, \\ mu' &= -p\lambda, \\ mv' &= -q\lambda - g, \\ 0 &= p^2 + q^2 - l^2, \\ 0 &= pu + qv. \end{aligned} \tag{38}$$

In Table 1 we present the results for (34), (35); (34), (36); and (34), (37). We normalize the equations by taking $m = g = l = 1$. The initial conditions are

$$p(0) = v(0) = \lambda(0) = 1, \quad q(0) = u(0) = 0 \tag{39}$$

Table 1
MEBDF results for index 1 pendulum problem

Tol	Fn	Jac	Steps	Time	Figs	3	2	1
10^{-2}	18	4	13	0.01	2.78	0.3d−3	0.6d−3	0.5d−2
10^{-3}	23	5	16	0.01	3.89	0.1d−3	0.2d−3	0.2d−3
10^{-4}	30	4	21	0.02	5.49	0.8d−5	0.5d−5	0.4d−5
10^{-5}	43	5	27	0.03	6.31	0.1d−6	0.1d−6	0.3d−6
10^{-6}	51	7	34	0.04	7.70	0.1d−7	0.5d−8	0.3d−7
10^{-7}	77	11	53	0.06	8.30	0.5d−8	0.5d−8	0.1d−8
10^{-8}	86	10	61	0.06	8.88	0.1d−10	0.4d−9	0.4d−8
10^{-9}	119	12	77	0.10	10.25	0.1d−10	0.6d−10	0.3d−10

Table 2
MEBDF results for index 2 pendulum problem

Tol	Fn	Jac	Steps	Time	Figs	3	2	1
10^{-2}	18	4	13	0.02	2.42	0.3d−2	0.1d−3	0.8d−2
10^{-3}	23	4	16	0.02	3.29	0.9d−4	0.9d−3	0.1d−2
10^{-4}	31	4	21	0.02	4.54	0.1d−6	0.2d−5	0.3d−4
10^{-5}	52	4	28	0.03	5.10	0.1d−5	0.2d−5	0.3d−4
10^{-6}	60	6	39	0.04	5.93	0.4d−7	0.8d−7	0.3d−5
10^{-7}	69	8	45	0.05	7.50	0.6d−8	0.1d−8	0.8d−7
10^{-8}	117	11	68	0.08	8.40	0.8d−10	0.2d−9	0.9d−8
10^{-9}	138	15	84	0.10	9.41	0.2d−10	0.3d−10	0.1d−8

and the range of integration is [0,1]. The results given in Table 1 should be largely self-explanatory. In particular, Tol is the specified local tolerance, Fn is the number of function evaluations, Jac is the number of Jacobian evaluations, Steps is the number of integration steps, Time is the time in seconds taken on an IBM RS6000 and Figs is the number of correct figures at $x=1$. Under columns 3, 2, 1 we also give the amounts by which the constraints (35), (36), (37), which are index 3, index 2 and index 1, respectively, are not satisfied at $x=1$. We see from Table 1 that the code performs well for all three problems. As the index is increased the code obtains less accuracy, as would be expected, but is still satisfactory (see Tables 2 and 3).

8. Conclusions

These results back up the claims made in this paper regarding the promise of MEBDF. In particular, it is clear that the MEBDF have better theoretical properties than the BDF methods. The MEBDF are also excellently suited to stiff oscillatory ODEs. The results presented in [7,14], particularly on the FEKETE problem, indicate that MEBDF perform well on some difficult DAE systems although there are still some gaps in the theory which have been highlighted in this paper and which need to be filled in. However, the BDF codes and Radau5 are powerful codes in their own right. In

Table 3
MEBDF results for index 3 pendulum problem

Tol	Fn	Jac	Steps	Time	Figs	3	2	1
10^{-2}	15	3	9	0.01	1.85	0.5d−3	0.1d−2	0.4d−1
10^{-3}	25	4	14	0.01	2.23	0.2d−2	0.4d−4	0.2d−1
10^{-4}	52	5	25	0.03	2.73	0.1d−4	0.5d−4	0.5d−2
10^{-5}	99	12	38	0.05	3.67	0.8d−9	0.5d−5	0.7d−3
10^{-6}	73	9	41	0.04	4.57	0.2d−8	0.1d−6	0.5d−4
10^{-7}	100	9	55	0.06	5.01	0.1d−7	0.4d−6	0.3d−4
10^{-8}	130	14	81	0.08	6.42	0.3d−9	0.1d−6	0.1d−4
10^{-9}	154	15	96	0.11	6.78	0.1d−10	0.3d−8	0.4d−6

particular, BDF codes are often well suited to large ODE/DAE systems and it remains to be seen how competitive MEBDFDAE is compared to BDF on such problems.

Acknowledgements

The author is grateful to two referees for their careful reading of this manuscript.

References

[1] K. Brenan, S.L. Campbell, L. Petzold, Numerical Solution of Initial Value Problems in Differential-Algebraic Equations, North-Holland, New York, 1989.
[2] L. Brugnano, D. Trigiante, Solving Differential Problems by Multistep Initial and Boundary Value Methods, Gordon and Breach, London, 1997.
[3] J. Butcher, The Numerical Analysis of Ordinary Differential Equations, Wiley, New York, 1987.
[4] J. Butcher, J.R. Cash, M.T. Diamantakis, DESI methods for stiff initial value problems, ACM Trans. Math. Software 22 (1996) 401–422.
[5] J.R. Cash, The integration of stiff initial value problems in ODEs using modified extended backward differentiation formulae, Comput. Math. Appl. 9 (1983) 645–660.
[6] J.R. Cash, S. Considine, An MEBDF code for stiff initial value problems, ACM Trans. Math. Software 18 (1992) 142–160.
[7] http://www.ma.ic.ac.uk/~jcash/IVP_ Software/readme.html.
[8] J.R. Cash, An extension of Olver's method for the numerical solution of linear recurrence relations, Math. Comp. 32 (1978) 497–510.
[9] J.R. Cash, Stable Recursions with Applications to Stiff Differential Equations, Academic Press, London, 1979.
[10] J.R. Cash, Stability concepts in the numerical solution of difference and differential equations, Comput. Math. Appl. 28 (1994) 45–53.
[11] J.R. Cash, A comparison of some codes for the stiff oscillatory problem, Comp. Math. Appl. 36 (1998) 51–57.
[12] C.F. Curtiss, J.D. Hirschfelder, Integration of stiff equations, Proc. Nat. Acad. Sci. 38 (1952) 235–243.
[13] G. Dahlquist, A special stability problem for linear multistep methods, BIT 3 (1963) 27–43.
[14] W.M. Lioen, J.J.B. de Swart, Test set for initial value problem solvers, Report MAS-R9832, CWI, Amsterdam, 1998 and http://www.cwi.nl/cwi/projects/IVPtestset.
[15] W.H. Enright, T.E. Hull, B. Lindberg, Comparing numerical methods for stiff systems of ODEs, BIT 15 (1975) 10–48.
[16] P.W. Gaffney, A performance evaluation of some FORTRAN subroutines for the solution of stiff oscillatory ordinary differential equations, ACM Trans. Math. Software 10 (1984) 58–72.

[17] C.W. Gear, G.K. Gupta, B. Leimkuhler, Automatic integration of Euler–Lagrange equations with constraints, J. Comput. Appl. Math. 12 (1985) 77–90.
[18] E. Hairer, G. Wanner, Solving Ordinary Differential Equations II Stiff and Differential Algebraic Problems, Springer, Berlin, 1996.
[19] E. Hairer, C. Lubich, M. Roche, The Numerical Solution of Differential-Algebraic Systems by Runge–Kutta Methods, Lecture Notes in Maths, Vol. 1409, Springer, Berlin, 1989.
[20] A.C. Hindmarsh, LSODE and LSODI, two initial value ordinary differential equation solvers ACM SIGNUM Newslett. 15 (1980).
[21] J.D. Lambert, Computational Ordinary Differential Equations, Wiley, New York, 1973.
[22] F.W.J. Olver, Numerical solution of second order linear difference equations, J. Res. Nat. Bur. Standards Math. Math. Phys. 71B (1967) 111–129.
[23] J.C.P. Miller, Bessel Functions, Part II, Math. Tables, Vol. X, British Association for the Advancement of Sciences, CUP, 1952.
[24] G. Psihoyios, Advanced step-point methods for the solution of initial value problems, Ph.D. Thesis, University of London, Imperial College, 1995.
[25] S.P. Norsett, A. Wolfbrandt, Attainable orders of rational approximations to the exponential function with only real poles, BIT 17 (1977) 200–208.
[26] G. Psihoyios, J.R. Cash, A stability result for general linear methods with characteristic function having real poles only, BIT 38 (1998) 612–617.
[27] P. Vander Houwen, private communication, 1999.

ELSEVIER Journal of Computational and Applied Mathematics 125 (2000) 131–145

JOURNAL OF COMPUTATIONAL AND APPLIED MATHEMATICS

www.elsevier.nl/locate/cam

Software and algorithms for sensitivity analysis of large-scale differential algebraic systems ☆

Shengtai Li, Linda Petzold *

Department of Computer Science, University of California, Santa Barbara, CA 93106, USA

Received 22 June 1999; received in revised form 21 March 2000

Abstract

Sensitivity analysis for DAE systems is important in many engineering and scientific applications. The information contained in the sensitivity trajectories is useful for parameter estimation, optimization, model reduction and experimental design. In this paper we present algorithms and software for sensitivity analysis of large-scale DAE systems of index up to two. The new software provides for consistent initialization of the solutions and the sensitivities, interfaces seamlessly with automatic differentiation for the accurate evaluation of the sensitivity equations, and is capable via MPI of exploiting the natural parallelism of sensitivity analysis as well as providing an efficient solution in sequential computations.

1. Introduction

This paper is concerned with the solution and sensitivity analysis of initial value problems for differential–algebraic equation systems (DAEs) in the general form

$$F(t, y, y') = 0, \tag{1}$$

where F, y, and y' are N-dimensional vectors. A number of powerful solvers have been written for the solution of DAEs, including RADAU5 [9], SPRINT [1], PSIDE [17], DASSL [3] and DASPK [4]. Many of these solvers are compared in [13], where it was found that DASSL/DASPK works very well compared with other methods and can solve a broader class of problems than the other codes tested. Several methods and codes have been designed in the last decade to compute sensitivities for DAEs [14,6]. In this paper we outline the algorithms and issues for sensitivity analysis of large-scale DAE systems, describe our new software DASPK3.0 for these problems and present some numerical

☆ This work was partially supported by DOE contract number DE-FG03-98ER25354, NSF grant CCR-9896198, NSF KDI grant ATM-9873133, NSF/DARPA grant DMS-9615858, and LLNL ISCR 99-006.

* Corresponding author.

PII: S0377-0427(00)00464-7

examples illustrating the effectiveness of this approach. The sensitivity algorithms presented here are applicable to most solvers. We have based our software on DASPK so as to make use of the excellent properties, particularly for large-scale systems, of this powerful and widely used code.

We begin by giving some background in Section 2 on the basic algorithms used in the DAE solver DASSL and its extension DASPK for large-scale DAE systems. In Section 3 we describe three algorithms for sensitivity analysis of DAEs using a direct approach. These are the staggered direct method [6], simultaneous corrector method [14] and staggered corrector method [8]. Accurate evaluation of the sensitivity residuals is an important problem; an adaptive increment finite difference method is presented, and methods for computing residuals via automatic differentiation are described. In Section 4 we present methods for determining consistent initial conditions for several classes of index-0, index-1 and index-2 DAEs and the associated sensitivities. The methods, which are new for index-2 DAEs, have the important property that they can be implemented with very little additional information required from the user. In Section 5, several issues which are critical for a robust and efficient implementation of sensitivity analysis are explored, along with their resolution in DASPK3.0. These include the error and convergence tests and formulation of the Krylov method for sensitivity analysis. The method used for parallelization in DASPK3.0 is described in Section 6. Finally, numerical experiments illustrating the effectiveness of this approach on both sequential and parallel computers are given in Section 7.

Further details on the implementation and use of DASPK3.0 are given in [11]. DASPK3.0 is available via "http://www.engineering.ucsb.edu/~cse".

2. Background – Algorithms in DASSL and DASPK

DASSL was developed by Petzold [3] and has become one of the most widely used software packages for DAEs. DASSL uses backward differentiation formula (BDF) methods [3] to solve a system of DAEs or ODEs. The methods are variable step size, variable order. The system of equations is written in implicit ODE or DAE form as in (1). Following discretization by the BDF methods, a nonlinear equation

$$F(t, y, \alpha y + \beta) = 0 \tag{2}$$

must be solved at each time step, where $\alpha = \alpha_0/h_n$ is a constant which changes whenever the step size or order changes, β is a vector which depends on the solution at past times, and t, y, α, β are evaluated at t_n. DASSL solves this equation by a modified version of Newton's method

$$y^{(m+1)} = y^{(m)} - c\left(\alpha\frac{\partial F}{\partial y'} + \frac{\partial F}{\partial y}\right)^{-1} F(t, y^{(m)}, \alpha y^{(m)} + \beta), \tag{3}$$

where the linear system is solved via a dense or banded direct solver. The iteration matrix

$$A = \alpha\frac{\partial F}{\partial y'} + \frac{\partial F}{\partial y}$$

is computed and factored and is then used for as many time steps as possible. The reader can refer to [3] for more implementation details.

DASPK2.0 was developed by Brown et al. [4] for the solution of large-scale systems of DAEs. It is particularly effective in the method-of-lines solution of time-dependent PDE systems in two and three dimensions. In contrast to DASSL, which is limited in its linear algebra to dense and banded systems, DASPK2.0 is able to make use of the preconditioned GMRES iterative method [16] for solving the linear system at each Newton iteration.

When solving DAEs, the integration must be started with a consistent set of initial conditions y_0 and y'_0. This is a set of initial conditions which satisfy the algebraic constraints for the DAEs (and for higher-index DAEs, the hidden constraints). The initialization algorithms in DASPK3.0 are new, and will be discussed further in Section 4.

3. Sensitivity analysis of DAEs

3.1. Methods for sensitivity analysis

Several approaches have been developed to calculate sensitivity coefficients [6,14]. Here we summarize the direct methods for sensitivity analysis of ODEs and DAEs.

To illustrate the basic approach for sensitivity analysis, consider the general DAE system with parameters,

$$F(t, y, y', p) = 0, \quad y(0) = y_0, \tag{4}$$

where $y \in \mathbb{R}^{n_y}, p \in \mathbb{R}^{n_p}$. Here n_y is the number of time-dependent variables y as well as the dimension of the DAE system, and n_p is the number of parameters in the original DAE system. Sensitivity analysis entails finding the derivative of the solution y with respect to each parameter. This produces an additional $n_s = n_p \cdot n_y$ sensitivity equations which, together with the original system, yield

$$\begin{aligned} &F(t, y, y', p) = 0, \\ &\frac{\partial F}{\partial y} s_i + \frac{\partial F}{\partial y'} s'_i + \frac{\partial F}{\partial p} = 0, \quad i = 1, \ldots, n_p, \end{aligned} \tag{5}$$

where $s_i = \mathrm{d}y/\mathrm{d}p_i$. Defining

$$Y = \begin{bmatrix} y \\ s_1 \\ \vdots \\ s_{n_p} \end{bmatrix}, \quad \mathrm{F} = \begin{bmatrix} F(t, y, p) \\ \dfrac{\partial F}{\partial y} s_1 + \dfrac{\partial F}{\partial y'} s'_1 + \dfrac{\partial F}{\partial p_1} \\ \vdots \\ \dfrac{\partial F}{\partial y} s_{n_p} + \dfrac{\partial F}{\partial y'} s'_{n_p} + \dfrac{\partial F}{\partial p_{n_p}} \end{bmatrix},$$

the combined system can be rewritten as

$$\mathrm{F}(t, Y, Y', p) = 0, \quad Y(0) = \begin{bmatrix} y_0 \\ \dfrac{\mathrm{d}y_0}{\mathrm{d}p_1} \\ \vdots \\ \dfrac{\mathrm{d}y_0}{\mathrm{d}p_{n_p}} \end{bmatrix}.$$

This system can be solved by the kth order BDF formula with step size h_{n+1} to yield a nonlinear system

$$G(Y_{n+1}) = \mathrm{F}\left(t_{n+1}, Y_{n+1}, Y_{n+1}^{\prime(0)} - \frac{\alpha_s}{h_{n+1}}(Y_{n+1} - Y_{n+1}^{(0)}), p\right) = 0, \tag{6}$$

where $Y_{n+1}^{(0)}$ and $Y_{n+1}^{\prime(0)}$ are predicted values for Y_{n+1} and Y'_{n+1}, which are obtained via polynomial extrapolation of past values [3]. Also, α_s is the fixed leading coefficient which is defined in [3]. Newton's method for the nonlinear system produces the iteration

$$Y_{n+1}^{(k+1)} = Y_{n+1}^{(k)} - \boldsymbol{J}^{-1} G(Y_{n+1}^{(k)}),$$

where

$$\boldsymbol{J} = \begin{bmatrix} J & & & & \\ J_1 & J & & & \\ J_2 & 0 & J & & \\ \vdots & \vdots & \vdots & \ddots & \\ J_{n_p} & 0 & \cdots & 0 & J \end{bmatrix} \tag{7}$$

and

$$J = \alpha \frac{\partial F}{\partial y'} + \frac{\partial F}{\partial y}, \quad J_i = \frac{\partial J}{\partial y} s_i + \frac{\partial J}{\partial p_i}$$

and $\alpha = \alpha_s / h_{n+1}$.

There are three well-established methods to solve the nonlinear system (6):

- Staggered direct method, described in [6].
- Simultaneous corrector method, described in [14].
- Staggered corrector method, described in [8].

Analysis and comparison of the performance of these three methods have been given in [8,12]. Because the relative efficiencies of the methods depend on the problem and on the number of parameters, all of them were made available as options in DASPK3.0. Here we describe briefly the three methods.

The staggered direct method first solves Eq. (6) for the state variables. After the Newton iteration for the state variables has converged, the sensitivity equations in (6) are updated with the most recent values of the state variables. Because Eq. (8) is linear with a matrix J for the sensitivity equations, it is solved directly without Newton iteration. However, to solve the linear system in this way requires computation and factorization of the Jacobian matrix at each step and also extra storage for the matrix $\partial F/\partial y'$. Since the Jacobian is updated and factorized only when necessary in DASPK, the additional matrix updates and factorizations may make the staggered direct method unattractive compared to the other methods. However, if the cost of a function evaluation is more than the cost of factorization of the Jacobian matrix and the number of sensitivity parameters is very large (see [12]), the staggered direct method is more efficient. We have modified the implementation of [6] to make the staggered direct method more reliable for ill-conditioned problems.

The simultaneous corrector method solves (6) as one whole nonlinear system, where Newton iteration is used. The Jacobian matrix $\boldsymbol{J}$ in (7) is approximated by its block diagonal in the Newton iteration. Thus, this method allows the factored corrector matrix to be reused for multiple steps. It has been shown in [14] that the resulting iteration is two-step quadratically convergent for full Newton, and convergent for modified Newton iteration.

The staggered corrector method is similar to the staggered direct method. However, instead of solving the linear sensitivity system directly, a Newton iteration is used

$$s_i^{(k+1)} = s_i^{(k)} - J^{-1}G_{s_i}(s_i^{(k)}), \tag{8}$$

where G_{s_i} is the residual for the ith sensitivity and J is the factored Jacobian matrix which is used in the Newton iteration for the state variables. Like the simultaneous corrector method, this method does not require the factorization of the Jacobian matrix at each step. One of the advantages of the staggered corrector method is that we do not need to evaluate the sensitivity equations during the iteration of solving for the state variables. This can reduce the computation time if the state variables require more iterations than the sensitivity variables. After solving for the state variables in the corrector iteration, only the diagonal part of $\boldsymbol{J}$ in (7) is left. We can expect the convergence of the Newton iteration will be improved over that of using an approximate iteration matrix in the simultaneous corrector method. This has been observed in our numerical experiments.

3.2. Methods for evaluating sensitivity residuals

Several approaches have been developed to calculate the sensitivity residuals that may be used with either the staggered corrector or the simultaneous corrector methods. Maly and Petzold [14] used a directional derivative finite difference approximation. For example, the ith sensitivity equation may be approximated as

$$\frac{F(t, y+\delta_i s_i, y'+\delta_i s_i', p+\delta_i e_i) - F(t, y, y', p)}{\delta_i} = 0, \tag{9}$$

where δ_i is a small scalar quantity, and e_i is the ith unit vector. Proper selection of the scalar δ_i is crucial to maintaining acceptable round-off and truncation error levels [14]. If $F(t, y, y', p)$ is already available from the state equations, which is the case in the Newton iteration of DASPK, (9) needs only one function evaluation for each sensitivity. The main drawback of this approach is that it may be inaccurate for badly scaled problems.

The selection of the increment δ_i for Eq. (9) in DASPK3.0 is an improvement over the algorithms of [14] which was suggested by Hindmarsh [10]. The increment is given by

$$\delta_i = \Delta \max(|p_i|, 1/\|u_i\|_2), \tag{10}$$

where Δ is a scale factor

$$u_i = \left(WT^{in_y+j}/WT^j\colon j = 1,\dots,n_y\right)$$

and WT is a vector of weights determined by the relative and absolute user error tolerances and the solution y,

$$WT^j = \mathrm{RTOL}_j \cdot |y_j| + \mathrm{ATOL}_j.$$

Alternatively, the sensitivity residuals can be evaluated analytically by an automatic differentiation tool such as ADIFOR [2] or other automatic differentiation (AD) methods. We recommend using AD to evaluate the sensitivity equations. Even for some well-scaled problems, the ADIFOR-generated routine has better performance in terms of efficiency and accuracy than the finite difference approximation.

All of the ADIFOR-generated routines require the support of the ADIFOR library [2]. For further information, see http://www-unix.mcs.anl.gov/autodiff/-ADIFOR/. DASPK3.0 can be used without the ADIFOR library, for problems where automatic differentiation is not needed.

4. Consistent initial condition calculation for solution and sensitivity analysis

4.1. Consistent initial conditions for index-one problems

The basic initialization technique in DASPK3.0 for index-1 problems is an extension of the method proposed in [5]. It is applicable to two classes of index-1 initialization problems. Initialization problem I is posed for systems of the form

$$f(t,u,v,u') = 0,$$

$$g(t,u,v) = 0, \tag{11}$$

where $u, f \in \mathbb{R}^{N_d}$ and $v, g \in \mathbb{R}^{N_a}$, with the matrices $f_{u'} = \partial f/\partial u'$, $g_v = \partial g/\partial v$ square and nonsingular. The problem is to find the initial value v_0 of v when the initial value u_0 for u is specified. Hence it is required for the user to specify which variables are algebraic and which are differential.

In initialization problem II, which is applicable to the general index-1 system (1), the initial derivatives are specified but all of the dependent variables are unknown. That is, we must solve for y_0 given y'_0. For example, beginning the DAE solution at a steady state corresponds to specifying $y'_0 = 0$.

Both of these initial condition problems are solved with the help of mechanisms already in place for the solution of the DAE system itself, rather than requiring the user to perform a special computation. It is also possible in DASPK3.0 to specify some of the solution variables and derivatives at the initial time, and solve for the rest. This will be described in more detail in the next subsection.

The sensitivity problem for (11) is given by

$$f(t,u,v,u',p) = 0,$$

$$g(t,u,v,p) = 0,$$

$$\frac{\partial f}{\partial u}s_u + \frac{\partial f}{\partial v}s_v + \frac{\partial f}{\partial u'}s_{u'} + \frac{\partial f}{\partial p} = 0,$$

$$\frac{\partial g}{\partial u}s_u + \frac{\partial g}{\partial v}s_v + \frac{\partial g}{\partial p} = 0. \tag{12}$$

The algebraic variables in Eq. (11) generate algebraic sensitivity variables in Eq. (12). Eq. (12) also has the same index as (11).

DASPK3.0 uses a staggered approach to compute the consistent initial conditions for the sensitivity variables. First the consistent initial conditions are computed for the state variables, and then for the sensitivity variables. It is easy to see by differentiating the initial conditions that the sensitivity variables fall into the same class of initialization problems as the state variables.

4.2. Consistent initial conditions for index-two problems

With partial error control (excluding the algebraic variables from the error control), DASPK3.0 can solve Hessenberg index-2 problems with given consistent initial conditions. However, consistent initial conditions may not be readily available in many cases. Taking the Hessenberg index-2 problem as an example,

$$\begin{aligned} u' &= f(t,u,v), \\ 0 &= g(u), \end{aligned} \tag{13}$$

the objective for index-2 initialization is to compute a new triple $(\hat{u}'_0, \hat{u}_0, \hat{v}_0)$ that satisfies the constraints and consistent initial conditions. The problem is under-determined. Following the idea of [5] for index-1 problems, we solve the consistent initialization problem with the help of mechanisms already in place for the DAE solution itself. We search for the consistent initial conditions in the direction given by the differential equations. This method has a potential advantage that the hidden constraints derived from the equations may also be satisfied. To do that, we should increment the derivative u' by $(1/h)\delta u$ if the solution u is incremented by δu. Consider a general DAE system

$$F(t,u,u',v) = 0. \tag{14}$$

After introducing two new variables δu and δv and an artificial time step h, we transform Eq. (14) into

$$F\left(t, u_0 + \delta u, u'_0 + \frac{1}{h}\delta u, v_0 + \delta v\right) = 0. \tag{15}$$

Then δu and δv in (15) are computed by Newton iteration with initial values of zero. The iteration matrix is

$$J = \left(\frac{1}{h}F_{u'} + F_u, F_v\right), \tag{16}$$

where h is chosen to be the initial step size that satisfies the error tolerance for a zeroth-order method.

It is easy to fix some of the differential variables in (15). However, fixing a differential variable u does not imply fixing the derivative u' in our algorithm, and vice versa. For example, if we fix the first element u_{01} of the vector u in (15), the equation becomes

$$F\left(t, u_{01}, u_{0r} + \delta u_r, u_0' + \frac{1}{h}\delta u, v_0 + \delta v\right) = 0,$$

where u_{0r} is the rest of u (excluding u_{01}), and δu_r is the rest of δu (excluding δu_1). In the algorithm of [5] for initialization problem I, all of the differential variables are fixed and Eq. (15) becomes

$$F\left(t, u_0, u_0' + \frac{1}{h}\delta u, v_0 + \delta v\right) = 0. \tag{17}$$

The initialization problem II for Eq. (11) can also be cast into (15) by fixing all of the derivatives u', which yields

$$F(t, u_0 + \delta u, u_0', v_0 + \delta v) = 0.$$

As in [5], the implementation is designed so that forming the Jacobian for the initialization and solving the nonlinear system requires no information from the user beyond what is already needed for the time stepping.

If the constraint $g(u) = 0$ in (13) is satisfied, all of the differential variables can be fixed and Eq. (17) becomes

$$\begin{aligned} \delta u' + u_0' &= f(t, u_0, v_0 + \delta v), \\ 0 &= g(u_0), \end{aligned} \tag{18}$$

where $\delta u' = (1/h)\delta u$. Since $\delta u'$ and δv are not related to $g(u_0)$, the iteration matrix for (18) is singular, which means the solution is not unique. However, we can replace the constraint equation in (18) with $g_u u' = 0$, which yields

$$\begin{aligned} \delta u' + u_0' &= f(t, u_0, v_0 + \delta v), \\ 0 &= g_u(u_0)(\delta u' + u_0'). \end{aligned} \tag{19}$$

It is easy to evaluate the first equation in (19), however the second equation is not available to DASPK3.0. To evaluate it requires the user to specify which equations are algebraic. We can avoid evaluating $g_u u' = 0$ if $f(t, u, v)$ is linear in v. Note that if (19) is linear with respect to $u' = \delta u' + u_0'$ and $v = v_0 + \delta v$, it has a unique solution for u' and v. The u' and v can be solved via only one iteration for a linear system, independent of the initial values. If we set $u_0' = 0$ and $\delta u' = 0$ in our first guess, the value of the second equation in (19) is zero, which is also the result of $g(u_0)$. Therefore, the residual evaluations can be used without modification. If $f(t, u, v)$ is nonlinear with respect to v, then it might take more than one iteration to solve for u' and v. Since $g_u u'$ might not be zero during the intermediate iterations, $g_u u'$ must be evaluated in addition to the residual evaluations. If $f(t, u, v)$ is nonlinear with respect to v, the user can either evaluate the $g_u u'$ in the residual routine or specify which equations are algebraic and DASPK3.0 will compute $g_u u'$ automatically, via finite difference approximation or automatic differentiation.

In our implementation, a linesearch backtracking algorithm [5] has been used to improve the robustness of the Newton algorithm for the initial condition calculation.

5. Implementation issues

5.1. Error and convergence tests

In DASPK3.0, the norm is evaluated separately for the state variables and for the sensitivities with respect to each parameter. For the error and convergence tests, we choose the largest one among all the norms. We have found from experience that this leads to the most robust and efficient implementation. It is possible to exclude sensitivity variables from the error test, but not from the Newton iteration convergence test.

5.2. Staggered direct method

In the staggered direct method as described in Caracotsios and Stewart [6], system (6) is transformed into

$$Js_{i(n+1)} = \left(-\frac{\partial F}{\partial y'_{n+1}}\beta - \frac{\partial F}{\partial p_i} \right), \tag{20}$$

where $\beta_i = s'^{(0)}_{i(n+1)} - \alpha s^{(0)}_{i(n+1)}$. To solve a linear system in this way requires extra storage for the matrix $\partial F/\partial y'_{n+1}$. Moreover, this implementation often fails when the matrix J is ill-conditioned. This is because the right-hand side of Eq. (20) can be very large and can introduce large round-off errors when J is ill-conditioned [12].

In DASPK3.0, the following linear system is solved for the sensitivities:

$$J\delta = Js^{(0)}_{i(n+1)} + \frac{\partial F}{\partial y'_{n+1}}\beta\frac{\partial F}{\partial p_i}, \tag{21}$$

where $\delta = s^{(0)}_{i(n+1)} - s_{i(n+1)}$. The right-hand side of (21) is easy to obtain in DASPK3.0 by evaluation of the sensitivity equations. It does not require any extra storage or special handling. What is important is that it works well for ill-conditioned problems. This is because the right-hand side of Eq. (21) is usually much smaller than that of Eq. (20) for a successful step (which means the predicted value $s^{(0)}$ is close enough).

5.3. Krylov method

Since the sensitivity equations are linear with respect to the sensitivity variables, Newton iteration is not necessary for the staggered method. Therefore the staggered corrector method and staggered direct method are the same for the preconditioned Krylov iteration. The matrix–vector product Jv_{s_i} is evaluated directly via directional derivative finite difference approximation

$$\begin{aligned} Jv_{s_i} &= (\alpha F_{y'} + F_y)v_{s_i} \\ &\approx \frac{F(t, y + \sigma v_{s_i}, \alpha(y + \sigma v_{s_i}) + \beta, p) - F(t, y, \alpha y + \beta, p)}{\sigma}, \end{aligned} \tag{22}$$

where $F(t, y', y, p)$ are the state equations. The function evaluations in (22) involve only the state equations. Because there is no coupling between different sensitivity variables, the linear iteration for each sensitivity equation can be done separately, which allows us to split the large linear system into several small ones and reduce the length of each orthonormal basis to the number of state variables. For the simultaneous corrector method, we approximate the Newton–Krylov iteration matrix by its block diagonal as for the direct method. Then (22) can be used to calculate the matrix–vector product.

One might consider replacing all finite differencing with ADIFOR-generated routines. However, this does not turn out to be a good idea for the Krylov iteration. There is a trade-off when we consider the efficiency and accuracy of the computations. The ADIFOR-generated routine not only computes the derivatives but also the original functions. To compute one matrix–vector product in an ADIFOR-generated routine requires at least one evaluation of the original function and possibly more than one evaluation of the derivatives. But the matrix–vector product approximated by first-order finite difference requires only one evaluation of the original functions. Since the finite difference approximation in the matrix–vector product for the Krylov iteration has the scaling incorporated into its implementation, in practice it has been quite robust.

6. Parallel implementation of DASPK for sensitivity analysis

Several parallel implementations for sensitivity analysis of DAEs via message-passing interface (MPI) [7] have been compared in [15]. In this section, we describe the parallelization in DASPK3.0. Although all the tests in [15] are for DASSL with the direct method, the comparative results are similar for DASPK3.0 with both the direct method and the Krylov method with the new implementation described in Section 5.3. We have found that the distributed parameter only (DPO) approach of [15] is also the fastest for DASPK3.0.

Our implementation distributes the sensitivity parameters inside the DASPK code so as to reduce the burden on the user. To balance the workload between processors, we allocate the parameters randomly to each processor: if we have NP processors and NPAR parameters, $N = \mathrm{NPAR}/\mathrm{NP}$, we distribute parameter numbers

$$\begin{aligned} &j,\ldots,j+i*\mathrm{NP},\ldots,j+N*\mathrm{NP} \quad \text{if } j\leqslant \mathrm{mod}(\mathrm{NPAR},\ \mathrm{NP}),\\ &j,\ldots,j+i*\mathrm{NP},\ldots,j+(N-1)*\mathrm{NP} \quad \text{if } j> \mathrm{mod}(\mathrm{NPAR},\ \mathrm{NP}), \end{aligned} \tag{23}$$

to the jth processor. Each processor computes the state variables locally, and the Jacobian matrix is also computed and factorized locally when needed. To minimize the storage and memory requirements in each processor, we assume that each processor has distributed memory, i.e., each processor has a local value of the same variable. Therefore, the work space in each processor can be reduced to approximately 1/NP of the total work space. Since the sensitivities are independent of each other, each processor can work independently without communicating with the others.

We have attempted to develop software for which both parallel and serial computation can run efficiently. We enforce the same step size control for all the processors in the parallel implementation. The communication overhead is very small. In each time step, each processor may be using different orders of the BDF formulae. Since this implementation requires an MPI-related routine and the support of the MPI library, which may not be accessible by users doing serial computation, we provide a dummy routine which can be linked without involving the MPI library, for use with serial computation

7. Numerical experiments

In this section we describe several numerical experiments. All tests were run on an SGI O2 workstation. The following quantities are used to compare different methods:

METH	Integration method
NSTP	Number of time steps used
NRES	Number of calls to residual subroutine
NSE	Number of sensitivity evaluations
NJAC	Number of Jacobian evaluation
NNI	Number of nonlinear iterations
NLI	Number of linear iterations (only for Krylov method)
CPU	The total cpu time taken to solve the problem

The integration methods we use include the direct method (D) and Krylov method (K). The integration methods for the sensitivity equations include the staggered corrector method (ST), the staggered direct method (SD) and simultaneous corrector method (SI). Therefore we use STD to represent the staggered corrector direct method, STK to represent the staggered corrector Krylov method, SID to represent the simultaneous corrector direct method, SIK to represent the simultaneous corrector Krylov method, and SDK to represent the staggered direct Krylov method.

The first example models a multi-species food web [5], in which mutual competition and/or predator–prey relationships in the spatial domain are simulated. Specifically, the model equations for the concentration vector $c=(c^1,c^2)^{\mathrm{T}}$ are

$$c_t^1 = f_1(x,y,t,c) + (c_{xx}^1 + c_{yy}^1),$$

$$0 = f_2(x,y,t,c) + 0.05(c_{xx}^2 + c_{yy}^2)$$

with

$$f_i(x,y,t,c) = c^i\left(b_i + \sum_{j=1}^{2} a_{ij}c^j\right).$$

The coefficients a_{ij}, b_i are

$$a_{11} = a_{22} = -1, \quad a_{12} = -0.5\cdot 10^{-6}, \quad a_{21} = 10^4,$$

$$b_1 = -b_2 = 1 + \alpha xy + \beta\sin(4\pi x)\sin(4\pi y).$$

The domain is the unit square $0\leqslant x,y\leqslant 1$ and $0\leqslant t\leqslant 10$. The boundary conditions are of Neumann type with normal derivative equal to 0. The PDEs are discretized by central differencing on an $M\times M$ mesh, for $M=20$. Therefore the resulting DAE system has size $\mathrm{NEQ}=2M^2=800$. The tolerances used were $\mathrm{RTOL}=\mathrm{ATOL}=10^{-5}$.

Table 1
Results for multi-species food web. The upper part is for the ADIFOR option with error control including the algebraic variables. The middle part is for the ADIFOR option with error control excluding the algebraic variables. The bottom part is for the finite difference option with error control excluding the algebraic variables

METH	NSTP	NRES	NSE	NJAC	NNI	NLI	NLIS	NETF	CPU
STD	312	770	389	45	381	0	0	4	30.84
SID	335	508	508	42	508	0	0	3	36.56
STK	341	2712	353	36	406	732	0	1	22.98
SIK	505	4262	617	47	617	1532	0	9	39.12
STD	128	377	190	42	205	0	0	0	17.90
SID	128	228	228	40	228	0	0	0	18.91
STK	133	1456	147	38	165	329	425	0	11.36
SIK	131	1888	202	38	202	332	697	0	15.47
SDK	133	1442	133	38	165	329	425	0	11.03
STD	128	3589	190	42	187	0	0	0	24.85
SID	128	3240	228	40	228	0	0	0	26.11
STK	133	1442	147	38	165	329	425	0	10.36
SIK	131	1818	201	38	201	332	700	0	14.37
SDK	133	1442	133	38	165	329	425	0	10.12

For sensitivity analysis, α and β were taken as the sensitivity parameters with initial values $\alpha=50$ and $\beta=100$. The initial conditions were taken as

$$c^1 = 10 + (16x(1-x)y(1-y))^2,$$
$$c^2 = 10^5,$$

which does not satisfy the constraint equations. The initial conditions for the sensitivity variables were taken as zero, which are not consistent either. We solved this problem with both the direct and Krylov methods. For the Krylov methods, we used the block-grouping preconditioner (which is included in the package DASPK2.0 [5]). To eliminate the effect of finite differencing when comparing different methods, we used the ADIFOR option in DASPK3.0 to generate the Jacobian matrix (only for the direct method) and sensitivity equations. Without initialization, the integration failed because of too many convergence test failures. The consistent initial conditions were computed quickly with both the direct and Krylov methods. Table 1 shows the results of the staggered corrector method and the simultaneous corrector method. Full error control (including the sensitivity variables) was used. Although there were no convergence test failures for this problem, the staggered corrector method (ST) performed better than the simultaneous corrector method (SI).

The finite differencing options for the sensitivity equations were also tested. We used the central difference and $\Delta = 10^{-3}$ (default value). The results are shown in Table 1.

The next example is the heat equation,

$$\frac{\partial u}{\partial t} = p_1 u_{xx} + p_2 u_{yy},$$

Table 2
Results for heat equation with ADIFOR evaluation. The upper half is for partial error control (excluding the sensitivity variables). The bottom half is for full error control

METH	NSTP	NRES	NSE	NJAC	NNI	NLI	NETF	CPU
STD	64	160	65	22	95	0	3	36.20
SID	64	97	97	22	97	0	3	46.35
STK	71	1527	72	18	100	149	1	25.67
SIK	71	1572	102	18	102	184	1	29.40
STD	92	220	103	23	123	0	2	53.58
SID	93	130	130	25	130	0	3	63.63
STK	106	1823	114	24	141	182	2	35.68
SIK	116	1776	155	24	155	213	2	39.06

Table 3
Results for heat equation with finite difference approximation for sensitivities and full error-control

METH	NSTP	NRES	NSE	NJAC	NNI	NLI	NETF	CPU
STD	92	2118	103	23	117	0	2	64.07
SID	93	2175	130	25	130	0	3	75.76
STK	107	3917	114	24	143	187	2	38.39
SIK	116	3695	157	24	157	207	2	44.24

where $p_1 = p_2 = 1.0$, posed on the 2-D unit square with zero Dirichlet boundary conditions. An $M+2$ by $M+2$ mesh is set on the square, with uniform spacing $1/(M+1)$. The spatial derivatives are represented by standard central finite difference approximations. At each interior point of the mesh, the discretized PDE becomes an ODE for the discrete value of u. At each point on the boundary, we pose the equation $u = 0$. The discrete values of u form a vector U, ordered first by x, then by y. The result is a DAE system $G(t, U, U') = 0$ of size $(M+2) \times (M+2)$. Initial conditions are posed as

$$u(t=0) = 16x(1-x)y(1-y).$$

The problem was solved by DASPK3.0 on the time interval $[0, 10.24]$ with $M = 40$. To compute the sensitivities, we took 10 sensitivity parameters; p_1 and p_2 were two of them. The other eight were chosen from the initial conditions. The error tolerances for DASPK are RTOL = ATOL = 1.0D − 4. For the direct method, we used ADIFOR with SparsLinC to generate the Jacobian. For the Krylov method, we used the incomplete LU (ILU) preconditioner, which is part of the DASPK package. The Jacobian for the ILU preconditioner was evaluated by ADIFOR with SparsLinC. The sensitivity residuals were evaluated by ADIFOR with the seed matrix option. Table 2 gives the results of the staggered corrector and simultaneous corrector methods.

Because this problem is linear and well scaled, finite-differencing in the Jacobian and/or sensitivity equation evaluation gets a good result. Table 3 shows the results when central differencing is used for evaluation of the sensitivity equations. The default perturbation factor (10^{-3}) is used in evaluating

Table 4
Results for heat equation with finite difference approximation and partial error-control. MPI was used in all the parallel computations. The same step size control was enforced on all the processors

METH	NPROC	NSTP	NRES	NJAC	NNI	NLI	CPU
	1	64	1810	19	91	0	35.33
Direct	2						19.64
	4						12.50
	8						8.78
	1	71	3410	18	100	181	24.59
Krylov	2	71	1935	18	100	159	12.94
	4	71	1109	18	100	160	7.43
	8	71	696	18	100	156	4.75

the sensitivity equations. The Jacobian is also evaluated by finite-differencing. Only the data for full error-control are listed.

We tested DASPK3.0 on a cluster of DEC alpha machines at Los Alamos National Laboratory. Each processor is 533 MHz with 64 MB memory. The heat equation with 24 sensitivity parameters was used as the test problem. The staggered corrector method was used. The synchronization to achieve the same step size on each processor does not introduce much overhead to the computation, as shown in Table 4.

The next example models a single pendulum

$$y_1' = y_3,$$

$$y_2' = y_4,$$

$$y_3' = -y_1 y_5,$$

$$y_4' = -y_2 y_5 - g,$$

$$0 = y_1 y_3 + y_2 y_4,$$

where g=1.0. This is an index-two problem. The initial conditions are y_1=0.5, $y_2=-\sqrt{p^2 - y_1^2}$, y_3= 10.0, $y_4 = 10.0$, and $y_5 = 0.0$. The sensitivity parameter is p, which has initial value $p = 1.0$. The initial conditions for the sensitivity variables are $(0.0, -1.1547, 0.0, 0.0, 0.0)$. All of the derivatives were set to 0 initially. The tolerance for DASPK was taken as RTOL = ATOL = 10^{-6}. Because

$$g(y) = y_1 y_3 + y_2 y_4 = -3.660254 \neq 0,$$

the consistent initial conditions were first computed via the initialization algorithm for index-2 problems. During the initial condition computation, we monitored three constraints,

$$g_1 = y_1^2 + y_2^2 - p,$$

$$g_2 = y_1 y_3 + y_2 y_4,$$

$$g_3 = y_3^2 + y_4^2 - (y_1^2 + y_2^2) y_5 - y_2.$$

Table 5
Results for consistent initial conditions for pendulum problem

Fixed	y_1	y_2	y_3	y_4	g_1	g_2	g_3
No	0.512	−0.859	11.777	7.016	−1.88e − 4	0.0	3.77e − 15
y_1, y_2	0.5	−0.866	11.83	6.83	−1.1e − 16	0.0	9.28e − 13

Initially, we have

$$g_1 = 0, \quad g_2 = -3.66, \quad g_3 = 200.866.$$

We also tried to fix y_1, y_2 during the experiments on the initial condition computation. The results are shown in Table 5. Note that if y_1 and y_2 are not fixed, g_1 may be violated.

References

[1] M. Berzins, P.M. Dew, R.M. Furzeland, Developing software for time-dependent problems using the method of lines and differential algebraic integrators, Appl. Numer. Math. 5 (1989) 375–397.
[2] C. Bischof, A. Carle, G. Corliss, A. Griewank, P. Hovland, ADIFOR-Generating derivative codes from Fortran programs, Sci. Programming 1(1) (1992) 11–29.
[3] K.E. Brenan, S.L. Campbell, L.R. Petzold, Numerical Solution of Initial-Value Problems in Differential-Algebraic Equations, Elsevier, New York, 1989, second edition, SIAM, Philadelphia, PA, 1996.
[4] P.N. Brown, A.C. Hindmarsh, L.R. Petzold, Using Krylov methods in the solution of large-scale differential-algebraic systems, SIAM J. Sci. Comput. 15 (1994) 1467–1488.
[5] P.N. Brown, A.C. Hindmarsh, L.R. Petzold, Consistent initial condition calculation for differential-algebraic systems, SIAM J. Sci. Comput. 19 (1998) 1495–1512.
[6] M. Caracotsios, W.E. Stewart, Sensitivity analysis of initial value problems with mixed ODEs and algebraic equations, Comput. Chem. Eng. 9 (4) (1985) 359–365.
[7] N. Doss, W. Gropp, E. Luck, A. Skjellum, A model implementation of MPI, Technical Report, Argonne National Laboratory, 1993.
[8] W.F. Feehery, J.E. Tolsma, P.I. Barton, Efficient sensitivity analysis of large-scale differential-algebraic systems, Appl. Numer. Math. 25 (1997) 41–54.
[9] E. Hairer, G. Wanner, Solving Ordinary Differential Equations II: Stiff and Differential-Algebraic Problems, Springer, New York, 1991.
[10] A.C. Hindmarsh, personal communication.
[11] S. Li, L.R. Petzold, Design of new DASPK for sensitivity analysis, Technical Report, Department of Computer Science, University of California Santa Barbara, 1999.
[12] S. Li, L.R. Petzold, W. Zhu, Sensitivity analysis of differential-algebraic equations: a comparison of methods on a special problem, Appl. Numer. Math. 32 (2000) 161–174.
[13] W.M. Lioen, J.B. De Swart, Test set for initial value problem solvers, CWI Report MAS-R9832, 1998.
[14] T. Maly, L.R. Petzold, Numerical methods and software for sensitivity analysis of differential-algebraic systems, Appl. Numer. Math. 20 (1996) 57–79.
[15] L.R. Petzold, W. Zhu, Parallel sensitivity analysis for DAEs with many parameters, Concurrency: Practice Experience 11 (1999) 571–585.
[16] Y. Saad, M.H. Schulz, GMRES: A general minimal residual algorithm for solving nonsymmetric linear systems, SIAM J. Sci. Statist. Comput. 7 (1986) 856–869.
[17] J.B. De Swart, W.M. Lioen, W.A. van der Veen, Parallel software for implicit differential equations (PSIDE), http://www.cwi.nl/cwi/projects/PSIDE.

ELSEVIER Journal of Computational and Applied Mathematics 125 (2000) 147–158

JOURNAL OF COMPUTATIONAL AND APPLIED MATHEMATICS

www.elsevier.nl/locate/cam

Compensating for order variation in mesh refinement for direct transcription methods

John T. Betts[a], Neil Biehn[b], Stephen L. Campbell[c,*,1], William P. Huffman[d]

[a] *Mathematics and Engineering Analysis, The Boeing Company, P.O. Box 3707, MS 7L-21, Seattle, WA 98124-2207, USA*

[b] *Operations Research Program, North Carolina State University, Raleigh, NC 27695-8205, USA*

[c] *Department of Mathematics, College of Phys. and Math. Sci., North Carolina State University, Raleigh, NC 27695-8205, USA*

[d] *Mathematics and Engineering Analysis, The Boeing Company, P.O. Box 3707, MS 7L-21, Seattle, WA 98124-2207, USA*

Received 15 December 1999; received in revised form 24 March 2000

Abstract

The numerical theory for Implicit Runge–Kutta methods shows that there can be order reduction when these methods are applied to either stiff or differential algebraic equations. This paper discusses how this theory can be utilized in direct transcription trajectory optimization by modifying a currently used mesh refinement strategy. © 2000 Elsevier Science B.V. All rights reserved.

Keywords: Implicit Runge–Kutta; Optimization; Differential algebraic equation

1. Introduction

One approach to solving optimal control problems is to parameterize the dynamic variables using values at mesh points on the interval. The resulting optimal control problem is thus *transcribed* into a finite-dimensional nonlinear programming (NLP) problem. Since the discrete variables directly optimize the approximate problem this approach is referred to as the *direct transcription* method. It is necessary to insure that the solution to the finite-dimensional nonlinear programming problem is a good discrete approximation to the original continuous optimal control problem. A method for refining the mesh such that the discrete problem is an adequate approximation to the continuous one

* Corresponding author.

E-mail address: slc@math.ncsu.edu (S.L. Campbell).

[1] Research supported in part by the National Science Foundation under DMS-9714811 and DMS-9802259.

PII: S0377-0427(00)00465-9

was presented in [8]. This method assumes the order of the discretization is known and constant. However, during the course of the optimization process the actual order may vary with iteration and location either because the activation of constraints means we really have differential algebraic equations (DAEs) [10] on subarcs or there is a local change in stiffness. There is a complex interaction between order and the optimization process. As noted in [9] while the DAE theory correctly predicts order reduction it does not always correctly predict what that order reduction is for optimization problems. Having the wrong value for the order can seriously impact on mesh refinement algorithms [9].

In this paper we consider a major modification of the mesh refinement strategy in [8] in order to compensate for this order reduction. In practical optimization problems it can be extremely difficult to get the order right theoretically due to the large number of constraints and events. Accordingly, we are interested in estimating what the order reduction is at different places in the discretization and then using this estimate. It is not important to us whether the order reduction comes from stiffness, the presence of differential algebraic equations, or other sources. The new mesh refinement strategy not only computes an order estimate that varies with mesh interval and mesh iteration but also has a different logic than [8] for determining the location and number of mesh points.

Our discussion will be in terms of a particular industrial optimization code SOCS developed at Boeing. However, the comments and observations are relevant to any other optimization code with a similar overall design philosophy. Section 2 provides needed background. The new mesh refinement algorithm is described in Section 3. A computational example is in Section 4. This paper describes work in progress. Additional analysis and testing will be described more fully in upcoming reports.

2. Transcription formulation

Typically, the dynamics of the system are defined for $t_{\mathrm{I}} \leqslant t \leqslant t_{\mathrm{F}}$ by a set of equations

$$\text{(State equations)} \quad \boldsymbol{y}' = \boldsymbol{f}(\boldsymbol{y}(t), \boldsymbol{u}(t), t), \tag{1a}$$

$$\text{(Initial conditions at time } t_{\mathrm{I}}) \quad \boldsymbol{\psi}_{\mathrm{IL}} \leqslant \boldsymbol{\psi}(\boldsymbol{y}(t_{\mathrm{I}}), \boldsymbol{u}(t_{\mathrm{I}}), t_{\mathrm{I}}) \leqslant \boldsymbol{\psi}_{\mathrm{IU}}, \tag{1b}$$

$$\text{(Terminal conditions at } t_{\mathrm{F}}) \quad \boldsymbol{\psi}_{\mathrm{FL}} \leqslant \boldsymbol{\psi}(\boldsymbol{y}(t_{\mathrm{F}}), \boldsymbol{u}(t_{\mathrm{F}}), t_{\mathrm{F}}) \leqslant \boldsymbol{\psi}_{\mathrm{FU}}, \tag{1c}$$

$$\text{(Algebraic path constraints)} \quad \boldsymbol{g}_{\mathrm{L}} \leqslant \boldsymbol{g}(\boldsymbol{y}(t), \boldsymbol{u}(t), t) \leqslant \boldsymbol{g}_{\mathrm{U}}, \tag{1d}$$

$$\text{(Simple state bounds)} \quad \boldsymbol{y}_{\mathrm{L}} \leqslant \boldsymbol{y}(t) \leqslant \boldsymbol{y}_{\mathrm{U}}, \tag{1e}$$

$$\text{(Simple control bounds)} \quad \boldsymbol{u}_{\mathrm{L}} \leqslant \boldsymbol{u}(t) \leqslant \boldsymbol{u}_{\mathrm{U}}. \tag{1f}$$

Equality constraints can be imposed if an upper and lower bound are equal.

The optimal control problem is to determine the $\boldsymbol{u}(t)$ that minimizes the performance index

$$J = \phi(\boldsymbol{y}(t_{\mathrm{I}}), t_{\mathrm{I}}, \boldsymbol{y}(t_{\mathrm{F}}), t_{\mathrm{F}}). \tag{1g}$$

This control problem is in Mayer form. SOCS can handle other formulations and problems with multiple phases [7] (that is, where different systems (1) are allowed on different intervals). Transcription has been discussed in detail elsewhere [1,2,5,6,12–14,16,17]. SOCS is described in detail in the references. We focus here on mesh refinement.

There are three primary operations that are performed when solving an optimal control problem using a transcription method; transcribing the optimal control problem into a nonlinear programming (NLP) problem by discretization; solving the sparse NLP (SOCS uses sequential quadratic programming), and assessing the accuracy of the approximation and if necessary refining the discretization by carrying out a mesh refinement, and then repeating the optimization steps.

At each mesh iteration, the time interval is divided into n_s segments or subintervals $t_I = t_1 < t_2 < \cdots < t_M = t_F$, where there are $M = n_s + 1$ mesh points. SOCS allows for t_I, t_F to be variable. Let $\boldsymbol{y}_k$, $\boldsymbol{u}_k$ be the computed estimates of $\boldsymbol{y}(t_k)$, $\boldsymbol{u}(t_k)$. The control variable estimates at the midpoint $\bar{t} = \frac{1}{2}(t_k + t_{k-1})$ of a mesh subinterval are $\bar{\boldsymbol{u}}_k$. The two primary discretization schemes used in SOCS are the trapezoidal (TR) and Hermite–Simpson (HS). Both are equivalent to Implicit Runge–Kutta (IRK) methods. Each scheme produces a distinct set of NLP variables and constraints. The default strategy is to do one or two iterations with the TR discretization and then switch to the HS discretization for the remaining mesh refinement iterations.

For the trapezoidal discretization, the NLP variables are $\{\{\boldsymbol{y}_j, \boldsymbol{u}_j\}_{j=1}^M, t_I, t_F\}$. The state equations (1a) are approximately satisfied by solving the *defect* constraints

$$\boldsymbol{\zeta}_k = \boldsymbol{y}_{k+1} - \boldsymbol{y}_k - \frac{h_k}{2}[\boldsymbol{f}_{k+1} + \boldsymbol{f}_k] = 0, \tag{2}$$

where $h_k \equiv t_{k+1} - t_k$, and $\boldsymbol{f}_k \equiv \boldsymbol{f}(\boldsymbol{y}_k, \boldsymbol{u}_k, t_k)$. For the Hermite–Simpson discretization, the NLP variables are $\{\{\boldsymbol{y}_j, \boldsymbol{u}_j\}_{j=1}^M, \{\bar{\boldsymbol{u}}_j\}_{j=2}^M, t_I, t_F\}$. The defects are given by

$$\boldsymbol{\zeta}_k = \boldsymbol{y}_{k+1} - \boldsymbol{y}_k - \frac{h_k}{6}[\boldsymbol{f}_{k+1} + 4\bar{\boldsymbol{f}}_{k+1} + \boldsymbol{f}_k], \tag{3}$$

where

$$\bar{\boldsymbol{f}}_{k+1} = \boldsymbol{f}(\bar{\boldsymbol{y}}_{k+1}, \bar{\boldsymbol{u}}_{k+1}, \bar{t}), \quad \bar{\boldsymbol{y}}_{k+1} = \frac{1}{2}[\boldsymbol{y}_k + \boldsymbol{y}_{k+1}] + \frac{h_k}{8}[\boldsymbol{f}_k - \boldsymbol{f}_{k+1}]. \tag{4}$$

As a result of the transcription, the optimal control constraints (1a)–(1d) are replaced by the NLP constraints. The boundary conditions are enforced directly by the constraints on $\boldsymbol{\psi}$, and the nonlinear path constraints are imposed at the mesh points. In a similar fashion the state and control variable bounds (1e) and (1f) become simple bounds on the NLP variables. The path constraints and variable bounds are always imposed at the mesh points and for the Hermite–Simpson discretization the path constraints and variable bounds are also imposed at the subinterval midpoints. This large, sparse NLP can be solved efficiently using a sequential quadratic programming (SQP) method as described in [3,4].

3. Mesh refinement

The first step in the mesh refinement process is to construct an approximation to the continuous solution from the information available at the solution of the NLP. For the state variable $\boldsymbol{y}(t)$ we use the C^1 cubic B-splines approximation $\tilde{\boldsymbol{y}}(t)$. We require the spline approximation to match the state estimates at the mesh points, $\tilde{\boldsymbol{y}}(t_k) = \boldsymbol{y}_k$ and the derivative of the spline approximation to match the right-hand side of the differential equations, $(d/dt)\tilde{\boldsymbol{y}}(t_k) = \boldsymbol{f}_k$. We also require the spline approximation $\tilde{\boldsymbol{u}}(t)$ to match the control estimates at the mesh points, $\tilde{\boldsymbol{u}}(t_k) = \boldsymbol{u}_k$. When a Hermite–Simpson solution is available it is possible to utilize a higher order approximation for the control using a basis for C^0 quadratic B-splines. The coefficients can be defined from the values at the mesh points as well as the values of the computed control at the midpoint of the subinterval, $\tilde{\boldsymbol{u}}(\bar{t}_{k+1}) = \bar{\boldsymbol{u}}_{k+1}$.

3.1. Estimating the error on a mesh

The true optimal solution $\boldsymbol{y}(t)$, $\boldsymbol{u}(t)$ must satisfy a set of necessary conditions which leads to a two-point boundary value problem in $\boldsymbol{y}(t)$, $\boldsymbol{u}(t)$ and the adjoint variables $\boldsymbol{\lambda}(t)$. A direct transcription method does not explicitly form the necessary conditions. One reason direct transcription methods are popular is that it is not necessary to derive expressions for the necessary conditions and it is not necessary to estimate values for the adjoint variables $\boldsymbol{\lambda}(t)$. On the other hand, because the necessary conditions are not available, they cannot be used to assess the accuracy of the solution. When estimating the error we *assume* $\tilde{\boldsymbol{u}}(t)$ is correct (and optimal), and estimate the error between $\tilde{\boldsymbol{y}}(t)$ and $\boldsymbol{y}(t)$. This is a subtle but very important distinction, for it implies that optimality of the control history $\tilde{\boldsymbol{u}}(t)$ is not checked when estimating the local error. However, it does mean that $\tilde{\boldsymbol{y}}(t)$ will accurately reflect what $\boldsymbol{y}$ will be if $\tilde{\boldsymbol{u}}(t)$ is used in (1a).

The controls will be represented as C^0 quadratic B-splines. This implies that one can expect to accurately solve an optimal control problem provided the optimal state variable $\boldsymbol{y}(t)$ is C^1 and the optimal control variable $\boldsymbol{u}(t)$ is C^0 within each phase. The solution to (1) may require discontinuities in the control and/or state derivatives. In particular, when the path constraints do not involve the control variable explicitly, the optimal solution may contain corners. Similarly, when the control appears linearly in the differential equations, bang–bang control solutions can be expected. Consequently, if the transcription method described here is applied to problems of this type, some inaccuracy must be expected unless the location of discontinuities are introduced explicitly as phase boundaries. We will be satisfied to accurately solve problems when the control is continuous and the state is differentiable. If this is not true, we will be satisfied if the method "does something reasonable".

When analyzing an integration method for (1a) it is common to ascribe an *order of accuracy* to the algorithm [11]. For a fixed control $\boldsymbol{u}(t)$, the *global error* at t_{k+1} is the difference between the computed solution $\boldsymbol{y}_{k+1}$ and the exact solution $\boldsymbol{y}(t_{k+1})$. The method is order p if this error is $O(h^p)$ where $h = \max_k h_k$. For a fixed control $\boldsymbol{u}(t)$, the *local error* is the difference between the computed solution $\boldsymbol{y}_{k+1}$ and the solution of the differential equation which passes through the computed point $\boldsymbol{y}_k$. The Hermite–Simpson discretization (3) is order 4, while the trapezoidal discretization (2) is order 2. Their local error orders are 5 and 3, respectively. The local error associated with the kth mesh subinterval can be shown to statisfy

$$\varepsilon_k \approx \|\boldsymbol{c}_k h_k^{p+1}\|. \tag{5}$$

On a subinterval where constraints are active or there is stiffness, we assume that (5) is replaced by

$$\varepsilon_k \approx \|\boldsymbol{c}_k h_k^{p-r+1}\|, \tag{6}$$

where r is the order reduction. For our purposes r also measures any change in $\boldsymbol{c}_k$. Unfortunately, the amount of order reduction is not known and often difficult to predict theoretically. There is a further problem in which the theory for IRK methods applied to DAEs usually leads to different amounts of order reduction in different variables [10]. This is true for both the trapezoid and Hermite–Simpson discretizations [9,15]. In addition, the difference between the order of the local and global error can sometimes be greater than one. In a complex optimization problem, the activation and deactivation of constraints can not only change the index but can change what the "index" of a particular variable is. We distinguish only between the control and the state so that order reduction is always taken to be the largest order reduction in all the state variables. To estimate the order reduction we need to first estimate the local error on a given mesh. Consistent with the philosophy of SOCS discussed earlier, in estimating the local error we assume the computed control is correct.

Consider a single subinterval $t_k \leqslant t \leqslant t_k + h_k$. Suppose the NLP has produced a spline solution $\tilde{\boldsymbol{y}}(t)$, $\tilde{\boldsymbol{u}}(t)$. Assume $\boldsymbol{y}_k$ is correct so that $\boldsymbol{y}(t_k) = \tilde{\boldsymbol{y}}(t_k)$. Then

$$\boldsymbol{y}(t_k + h_k) = \boldsymbol{y}(t_k) + \int_{t_k}^{t_k+h_k} \boldsymbol{y}'\,\mathrm{d}t = \boldsymbol{y}(t_k) + \int_{t_k}^{t_k+h_k} \boldsymbol{f}(\boldsymbol{y}, \boldsymbol{u}, t)\,\mathrm{d}t. \tag{7}$$

Both $\boldsymbol{y}$ and $\boldsymbol{u}$ are unknown. Consequently, on $[t_k, t_k + h_k]$ we might consider the following two approximations to $\boldsymbol{y}(t)$:

$$\hat{\boldsymbol{y}}(t) = \boldsymbol{y}(t_k) + \int_{t_k}^{t} \boldsymbol{f}(\tilde{\boldsymbol{y}}(t), \tilde{\boldsymbol{u}}(t), t)\,\mathrm{d}t, \tag{8}$$

$$\hat{\boldsymbol{y}}(t) = \boldsymbol{y}(t_k) + \int_{t_k}^{t} \boldsymbol{f}(\boldsymbol{y}(t), \tilde{\boldsymbol{u}}(t), t)\,\mathrm{d}t. \tag{9}$$

When working on a fixed mesh subinterval $[t_k, t_k + h_k]$, we let $z_i(t)$ denote the ith component of the vector function $\boldsymbol{z}$. With either (8) or (9) we could estimate the local error on the kth subinterval as

$$\eta_k = \max_i a_i |\tilde{y}_i(t_k + h_k) - \hat{y}_i(t_k + h_k)|, \tag{10}$$

where the weights a_i are chosen to appropriately normalize the error.

Our particular use for these estimates imposes certain special restrictions. First, we want them to be part of a code that will be used on a wide variety of problems. Secondly, we want to use the estimates on coarse meshes. While (9) might be the most accurate, its computation would require an explicit integration of $\boldsymbol{y}' = \boldsymbol{f}(\boldsymbol{y}, \tilde{\boldsymbol{u}}, t)$ on a possibly coarse mesh. An estimate based on an explicit integrator may be unstable on coarse meshes. The trapezoidal and Hermite–Simpson discretizations are implicit schemes with very good stability properties. Suppose that we use (8). Then

$$\begin{aligned}\tilde{y}_i(t_k + h_k) - \hat{y}_i(t_k + h_k) &= \tilde{y}_i(t_k + h_k) - \tilde{y}(t_k) - (\hat{y}_i(t_k + h_k) - \tilde{y}(t_k)) \\ &= \int_{t_k}^{t_k+h_k} \frac{\mathrm{d}\tilde{y}_i}{\mathrm{d}t} - f_i(\tilde{\boldsymbol{y}}, \tilde{\boldsymbol{u}}, t)\,\mathrm{d}t.\end{aligned} \tag{11}$$

This would seem to suggest using

$$\eta_{i,k} = \left| \int_{t_k}^{t_k+h_k} \frac{\mathrm{d}\tilde{y}_i}{\mathrm{d}t} - f_i(\tilde{\boldsymbol{y}}, \tilde{\boldsymbol{u}}, t)\,\mathrm{d}t \right| := \left| \int_{t_k}^{t_k+h_k} \varepsilon_i(t)\,\mathrm{d}t \right| \tag{12}$$

as the estimate of the error in the ith variable on the kth mesh interval. However, $\varepsilon_i(t)$ is an oscillatory function on the interval $[t_k, t_k + h_k]$ when using HS. At the solution to an NLP, the collocation conditions are satisfied at the mesh points and the subinterval midpoint, so $\varepsilon_i(t)$ is zero at these three points. If one uses (12) the error tends to cancel. For some problems test results show that (12) indicates zero error with the initial coarse mesh and no mesh refinement is done. This means that not only is (12) not a good error estimate but $\tilde{y}_i(t_k + h_k) - \hat{y}_i(t_k + h_k)$ will also be too small. Thus we turn to estimating $\tilde{y}_i(t) - \hat{y}_i(t)$ on $[t_k, t_k + h_k]$ instead.

We define the *absolute local error estimator* in the ith variable on the kth mesh subinterval by

$$\eta_{i,k} = \int_{t_k}^{t_{k+1}} |\varepsilon_i(t)|\,\mathrm{d}t = \int_{t_k}^{t_{k+1}} |\dot{\tilde{y}}_i(t) - f_i(\tilde{y}(t),\tilde{u}(t),t)|\,\mathrm{d}t. \tag{13}$$

Notice that the integrand utilizes the spline approximations for the state and control evaluated over the interval. We define the *relative local error estimate* by

$$\hat{\varepsilon}_k = \max_i \frac{\eta_{i,k}}{(w_i+1)}. \tag{14}$$

The scale weight w_i is the maximum value for the ith state variable or its derivative over the M mesh points in the phase. ODE integrators typically emphasize the error in prediction while (14) emphasizes the error in solving the equations. This is closer to the SOCS termination criteria. We utilize (14) in the ensuing algorithms.

Now let us consider how to compute an estimate for $\eta_{i,k}$. Since this error is essential for estimating the order reduction we construct an accurate estimate for integral (13). Because the spline approximations for the state and control are used, integral (13) can be evaluated using a standard quadrature method. We use a Romberg quadrature algorithm.

We first look at the relationship between (8), (10) and (14). We are working on one mesh subinterval $[t_k, t_k+h_k]$. Let $\underline{y}$ be the solution of $\underline{y}' = \boldsymbol{f}(\underline{y},\tilde{\boldsymbol{u}},t)$, $\underline{y}(t_k) = \tilde{y}(t_k)$ The next theorem shows that (13) measures the difference between the functions $\tilde{y}(t), \underline{y}(t)$ on $[t_k, t_k+h_k]$. One might try to show this by considering $\boldsymbol{\delta} = \tilde{y} - \underline{y}$ and try to linearize directly but we are especially interested in what happens on possibly coarse meshes for problems which might be stiff.

Theorem 1. *Let $\langle\cdot,\cdot\rangle$ be the Euclidian inner product. Assume that $\boldsymbol{f}(\boldsymbol{y}(t),\tilde{\boldsymbol{u}},t)$ satisfies a one-sided Lipschitz condition. That is, $\langle \boldsymbol{y}-\boldsymbol{z}, \boldsymbol{f}(\boldsymbol{y},\tilde{\boldsymbol{u}},t) - \boldsymbol{f}(\boldsymbol{z},\tilde{\boldsymbol{u}},t)\rangle \leqslant \nu\|\boldsymbol{y}-\boldsymbol{z}\|^2$ for all $\boldsymbol{y},\boldsymbol{z}$. Let $K = \max\{\mathrm{e}^{\nu h}, 1\}$. Then*

$$\sup_{t_k \leqslant t \leqslant t_k+h_k} |\underline{y}_i(t) - \tilde{y}_i(t)| \leqslant K \int_{t_k}^{t_k+h_k} |\varepsilon_i(s)|\,\mathrm{d}s.$$

We omit the proof because of page limitations. If the original problem is stiff, then K is close to 1 for larger h also.

3.2. Estimating the order reduction

In order to utilize (6) it is necessary to know the order reduction r. We compute an estimate for this quantity by comparing the behavior on two successive mesh refinement iterations. Assume that the current mesh was obtained by subdividing the old mesh, that is, the current mesh has more points than the old mesh. Consider a *single* subinterval in the old mesh. Denote the local error on this subinterval by

$$\theta = ch^{p-r+1}, \tag{15}$$

where p is the order of the discretization on the old mesh, and h is the stepsize for the subinterval. If the subinterval on the old mesh is subdivided by adding I points, the resulting local error estimate is

$$\eta = c\left(\frac{h}{1+I}\right)^{q-r+1}, \tag{16}$$

where q is the order of the discretization on the current mesh which can be different if the discretization has changed. If we assume the order reduction r and the constant c are the same on the old and current meshes, then we can solve (15) and (16) for these quantities:

$$\hat{r} = q + 1 - \frac{\log(\theta\eta^{-1}h^{q-p})}{\log(1+I)}. \tag{17}$$

The estimated order reduction is then given by

$$r = \max[0, \min(\text{nint}(\hat{r}), q)], \tag{18}$$

where nint denotes the "nearest integer". Note that θ, η in (18) will be the estimates of the error computed on the two meshes. As a final practical matter we assume that the order reduction is the same for all $I+1$ subdivisions of the old subinterval. Thus the "resolution" of our order reduction estimates is dictated by the old "coarse" mesh. We have found this more robust than estimating different orders on each new subinterval.

Let

$$Q(a,b) = q + 1 - \frac{\log(h^{q-p}a/b)}{\log(1+I)}.$$

Then

$$Q(p_1 a, p_2 b) = Q(a,b) - \frac{\log(p_1/p_2)}{\log(1+I)}.$$

Here p_1/p_2 is the ratio in the weights. Note that $p_1/p_2=1.07$ and $I=1$ gives $r=1$ while $p_1/p_2=1.15$ gives $r=2$. Thus a change of 15% in the weights could easily affect an r estimate. Accordingly for the order reduction calculation we take $\eta_k = \max_i(\bar{w}_i\eta_{i,k})$, where $\bar{w}_i$ are fixed weights computed on the first mesh iteration and then $r = Q(\eta_k, \eta_{k+1})$, However, we still use variable weights w_i in the mesh refinement algorithm and termination criteria.

3.3. Constructing a new mesh

We now describe an approach for constructing a new mesh using the estimates for order reduction and local error on a current mesh. "Old mesh" is the previous refinement iteration, "current mesh" is the current iteration, and "new mesh" is the *next* iteration. Simply adding a large number of points to the current mesh increases the size of the NLP problem which must be solved, thereby causing a significant computational penalty. The goal is to reduce the local error as much as possible using a limited number of new points.

The preceding section described how to compute a local error estimate for each segment or subinterval in the current mesh. Equating the local error ε_k in (6) with the relative local error estimate $\hat{\varepsilon}_k$ from (14) we obtain

$$\|\boldsymbol{c}_k\| h^{p-r_k+1} = \max_i \frac{\eta_{i,k}}{(w_i+1)} \tag{19}$$

so that

$$\|c_k\| = \max_i \frac{\eta_{i,k}}{(w_i+1)h^{p-r_k+1}}. \tag{20}$$

Let I_k be the number of points to add to subinterval k, so that from (6) and (20) we may write

$$\varepsilon_k \approx \|c_k\| \left(\frac{h}{1+I_k}\right)^{p-r_k+1} = \max_i \frac{\eta_{i,k}}{(w_i+1)} \left(\frac{1}{1+I_k}\right)^{p-r_k+1} \tag{21}$$

for integers $I_k \geqslant 0$. This is an approximation for the local error on each of the $1+I_k$ subintervals. The new mesh can be constructed by choosing integers I_k to solve the nonlinear *integer programming problem*:

$$\text{minimize:} \quad \phi(I_k) = \max_k \varepsilon_k \tag{22a}$$

$$\sum_{k=1}^{n_s} I_k \leqslant M - 1, \tag{22b}$$

$$I_k \leqslant M_1 \text{ for all } k. \tag{22c}$$

That is, we minimize the maximum error over all of the subintervals in the current mesh, by adding at most $M-1$ total points. At most M_1 points are added to a single subinterval. Typically we use $M_1 = 5$.

When the local errors on each subinterval of the current mesh are approximately the same, the error is *equidistributed*. If the estimated local error for the current mesh is dominated by the estimated error on a single subinterval α, that is, $\varepsilon_\alpha \gg \varepsilon_k$ with $k \neq \alpha$, the solution to (22) will require adding as many as M_1 points into subinterval α.

If the desired local error tolerance is δ, we would like the new mesh to be constructed such that it has an estimated local error below this tolerance. In fact when making predictions we would like the *predicted* error estimates to be "safely" below, $\hat{\delta} = \kappa\delta$ where $0 < \kappa < 1$. Typically we set $\kappa = 1/10$. The procedure begins with estimates for the discretization error ε_k on all subintervals in the current mesh and we initialize $I_k = 0$. When $I_\alpha \neq 0$ the error ε_α is "predicted" and presumably less reliable. In this case we force it to be "safely" less than the tolerance before stopping.

Mesh Construction Procedure

(1) *Subinterval with Maximum Error*; Determine subinterval α, such that $\varepsilon_\alpha = \max_k \varepsilon_k$.
(2) *Discretization Error*; Terminate if:
  (a) $\varepsilon_\alpha \leqslant \delta$ and $I_\alpha = 0$ or;
  (b) $\varepsilon_\alpha \leqslant \kappa\delta$ and $0 < I_\alpha < M_1$.
(3) *Update Information*:
  (a) Set $I_\alpha \leftarrow I_\alpha + 1$ (subdivide subinterval α),
  (b) Update ε_α from (21).
(4) *Next Iteration*;
  (a) Terminate if: $I_\alpha = M_1$, or if $\sum_{k=1}^{n_s} I_k = M - 1$,
  (b) Otherwise return to step 1.

3.4. The mesh refinement algorithm

We now outline the complete mesh refinement algorithm. Denote the mesh refinement iteration number by j_r. Assume that the sparse NLP has been solved for the current discretization. Assume the current mesh has M points and the desired accuracy is δ. The goal of the mesh refinement procedure is to select the number and location of the mesh points in the new mesh as well as the order of the new discretization. Typically, we begin with the trapezoidal discretization and switch to a Hermite–Simpson discretization at some point in the process.

Mesh Refinement Algorithm

(1) *Construct Continuous Representation*; Compute the cubic spline representation from the discrete NLP solution.
(2) *Estimate Discretization Error*; Compute an estimate for the discretization error ε_k in each segment of the current mesh by evaluating (13) using Romberg quadrature; compute the average error $\bar{\varepsilon} = (1/M)\sum_{k=1}^{M} \varepsilon_k$.
(3) *Select Primary Order for New Mesh*;
  (a) If $p < 4$ and $\varepsilon_\alpha \leqslant 2\bar{\varepsilon}$, then set $p = 4$ and go to step 1.
  (b) otherwise if $(p < 4)$ and $j_r > 2$, then set $p = 4$ and go to step 1.
  (c) otherwise continue.
(4) *Construct New Mesh*; containing the M points in the current mesh, and $M_1 = M' - M \geqslant 0$ new points chosen to minimize the maximum estimated error according to (22) using the *Mesh Construction Procedure*.

This algorithm is somewhat heuristic and is designed to be efficient on most applications. Computational experience demonstrates the value of initiating the process with a coarse mesh and low-order method. In SOCS [7], it is possible to specify the initial discretization, which may be effective when the user can provide a good initial guess for the solution. If the discretization error appears to be equidistributed it is reasonable to switch to a higher order discretization (Hermite–Simpson). When the error is badly distributed at least two different discrete solutions are obtained before the order is increased. The new mesh always results in a subdivision of the current mesh, which has been found desirable in practice. The min/max approach to adding points is designed to emphasize equidistributing the error.

4. Computational example

To illustrate the effect of the new mesh refinement algorithm we consider an example (23) due to A.V. Rao which has fast transients very close to either end of the long interval. In these regions we expect order reduction. Fig. 1 shows the solution x of (23).

The problem is

$$\min_u J = \min_u \int_0^{t_f} x^2 + u^2 \,\mathrm{d}t \tag{23a}$$

$$x' = -x^3 + u, \tag{23b}$$

$$x(0) = 1, \; x(t_f) = 1.5, \; t_f = 10\,000. \tag{23c}$$

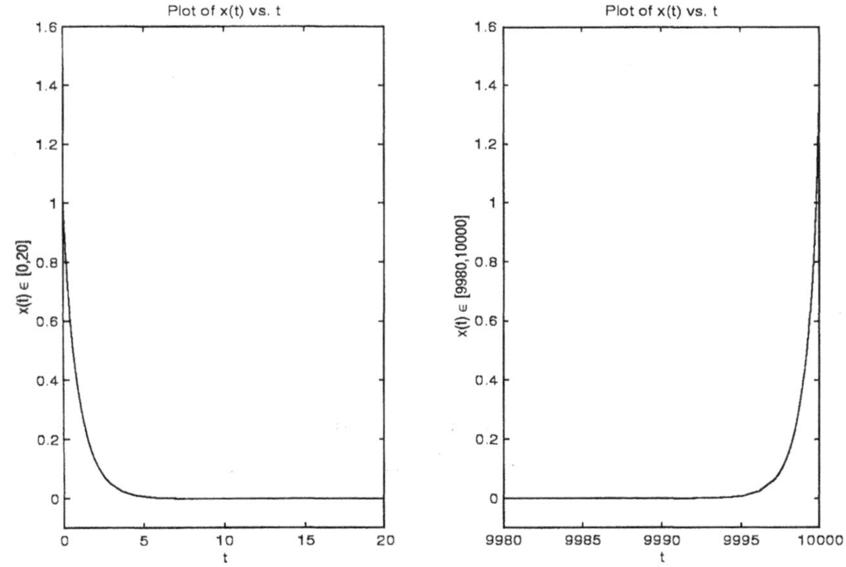

Fig. 1. State solution of optimal control problem (24).

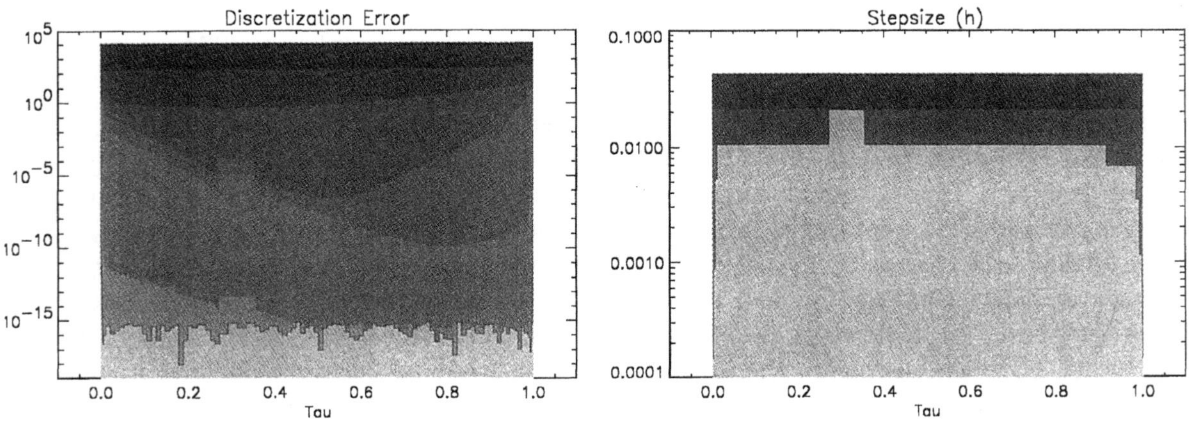

Fig. 2. Discretization error and mesh size.

Starting with an initial mesh of 25 points, the new mesh refinement strategy proposed here used 11 iterations and gave a final mesh with 226 points. The old mesh refinement strategy not using order estimation took 10 iterations but had 374 mesh points. The new algorithm gave a slightly more accurate answer on a substantially smaller mesh.

An examination of the order reduction estimate shows that starting on the third iteration there was a small layer of points at the ends of the interval where $r=4$. It was zero elsewhere. As the iteration progressed, the order reduction region was always several mesh intervals at each end but the actual length of the region shrunk in keeping with the narrow boundary layer.

Fig. 2 shows the mesh selection on each iteration and graphs the estimated discretization error. Here $\tau = t/10000$ and darker corresponds to earlier iterations.

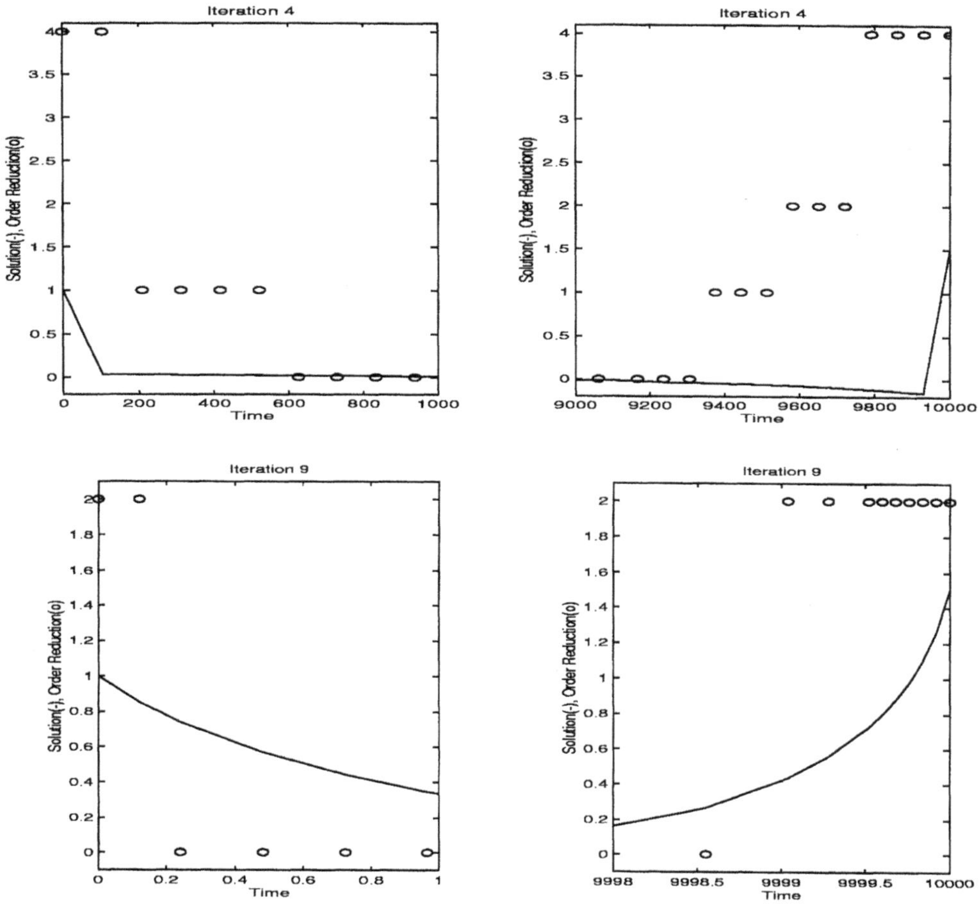

Fig. 3. Order reduction in the boundary layers for iterations 4 and 9.

Fig. 3 shows the order reduction in the boundary layers for iterations 4 and 9. The circles plot the order reduction at mesh points. The dark line is the state solution estimate on that mesh.

5. Summary and conclusions

This paper describes an approach for mesh refinement in the direct transcription method. It differs from many other mesh refinement algorithms in which it dynamically estimates what the real order of the method is at different parts of the interval and on different iterations. A computational example illustrates how the algorithm can be advantageous on stiff problems.

References

[1] J.T. Betts, Trajectory optimization using sparse sequential quadratic programming, in: R. Bulirsch, A. Miele, J. Stoer, K.H. Well (Eds.), Optimal Control, International Series of Numerical Mathematics, Vol. 111, Birkhäuser Verlag, Basel, 1993, pp. 115–128.

[2] J.T. Betts, Issues in the direct transcription of optimal control problems to sparse nonlinear programs, in: R. Bulirsch, D. Kraft (Eds.), Computational Optimal Control, International Series of Numerical Mathematics, Vol. 115, Birkhäuser Verlag, Basel, 1994, pp. 3–18.
[3] J.T. Betts, M.J. Carter, W.P. Huffman, Software for nonlinear optimization, mathematics and Engineering Analysis Library Report MEA-LR-83 R1, Boeing Information and Support Services, The Boeing Company, PO Box 3707, Seattle, WA 98124-2207, June 1997.
[4] J.T. Betts, P.D. Frank, A sparse nonlinear optimization algorithm, J. Optim. Theory Appl. 82 (1994) 519–541.
[5] J.T. Betts, W.P. Huffman, Application of sparse nonlinear programming to trajectory optimization, AIAA J. Guidance Control Dyn. 15 (1992) 198–206.
[6] J.T. Betts, W.P. Huffman, Path constrained trajectory optimization using sparse sequential quadratic programming, AIAA J. Guidance Control Dyn. 16 (1993) 59–68.
[7] J.T. Betts, W.P. Huffman, Sparse optimal control software SOCS, Mathematics and Engineering Analysis Technical Document MEA-LR-085, Boeing Information and Support Services, The Boeing Company, PO Box 3707, Seattle, WA 98124-2207, July 1997.
[8] J.T. Betts, W.P. Huffman, Mesh refinement in direct transcription methods for optimal control, Optim. Control Appl. Methods 19 (1998) 1–21.
[9] N. Biehn, S. Campbell, L. Jay, T. Westbrook, Some comments on dae theory for irk methods and trajectory optimization, J. Comput. Appl. Math. 120 (2000) 109–131.
[10] K.E. Brenan, S.L. Campbell, L.R. Petzold, Numerical Solution of Initial-Value Problems in Differential-Algebraic Equations, SIAM, Philadelphia, 1996.
[11] G. Dahlquist, Å. Björk, Numerical Methods, Prentice-Hall, Inc., Englewood Cliffs, NJ, 1974.
[12] E.D. Dickmanns, Efficient convergence and mesh refinement strategies for solving general ordinary two-point boundary value problems by collocated hermite approximation, 2nd IFAC Workshop on Optimisation, Oberpfaffenhofen, September 1980.
[13] P.J. Enright, B.A. Conway, Optimal finite-thrust spacecraft trajectories using collocation and nonlinear programming, AIAA J. Guidance Control Dyn. 14 (1991) 981–985.
[14] C.R. Hargraves, S.W. Paris, Direct trajectory optimization using nonlinear programming and collocation, AIAA J. Guidance Control Dyn. 10 (1987) 338.
[15] L. Jay, Convergence of a class of Runge–Kutta methods for differential–algebraic systems of index 2, BIT 33 (1993) 137–150.
[16] D. Kraft, On converting optimal control problems into nonlinear programming problems, in: K. Schittkowski (Ed.), Computational Mathematical Programming, NATO ASI Series, Vol. F15, Springer, Berlin, 1985, pp. 261–280.
[17] O. von Stryk, Numerical solution of optimal control problems by direct collocation, in: R. Bulirsch, A. Miele, J. Stoer, K.H. Well (Eds.), Optimal Control, International Series of Numerical Mathematics, Vol. 111, Birkhäuser Verlag, Basel, 1993, pp. 129–143.

JOURNAL OF
COMPUTATIONAL AND
APPLIED MATHEMATICS

ELSEVIER Journal of Computational and Applied Mathematics 125 (2000) 159–170

www.elsevier.nl/locate/cam

Continuous numerical methods for ODEs with defect control ☆

W.H. Enright

Department of Computer Science, University of Toronto, Toronto, Ontario, Canada M5S 3G4

Received 7 July 1999; received in revised form 3 December 1999

Abstract

Over the last decade several general-purpose numerical methods for ordinary differential equations (ODEs) have been developed which generate a continuous piecewise polynomial approximation that is defined for all values of the independent variable in the range of interest. For such methods it is possible to introduce measures of the quality of the approximate solution based on how well the piecewise polynomial satisfies the ODE. This leads naturally to the notion of "defect-control". Numerical methods that adopt error estimation and stepsize selection strategies in order to control the magnitude of the associated defect can be very effective and such methods are now being widely used. In this paper we will review the advantages of this class of numerical methods and present examples of how they can be effectively applied. We will focus on numerical methods for initial value problems (IVPs) and boundary value problems (BVPs) where most of the developments have been introduced but we will also discuss the implications and related developments for other classes of ODEs such as delay differential equations (DDEs) and differential algebraic equations (DAEs).

1. Introduction

Consider the ordinary differential equation

$$y' = f(t, y), \tag{1}$$

with exact solution denoted by $y(t)$. Traditional discrete numerical methods partition the interval of interest, $[t_0, t_F]$, by introducing a mesh $t_0 < t_1 \cdots < t_N = t_F$ and generate a discrete approximation $y_i \approx y(t_i)$ for each associated meshpoint. The number of meshpoints, N, and the distribution of these meshpoints is usually determined adaptively by the method in an attempt to deliver acceptable accuracy at a minimum cost. These methods generally accomplish this objective by keeping N as small as possible subject to a constraint that an indirect measure of $\max_{i=1,\ldots,N} \|y(t_i) - y_i\|$ be kept

E-mail address: enright@cs.toronto.edu (W.H. Enright).

☆ This research was supported by the Natural Science and Engineering Research Council of Canada and by Communication and Information Technology Ontario.

PII: S0377-0427(00)00466-0

small (relative to an accuracy parameter TOL). (We use $\|\cdot\|$ to represent the max–norm for vectors and the induced matrix norm for matrices.) Different methods implement very different strategies in an attempt to achieve this indirect error control and this can make it particularly challenging to interpret the accuracy of a numerical solution.

In recent years, the notion of a continuous extension (of an underlying discrete method) has received considerable attention. With this approach, one associates with each discrete approximation, $\{t_i, y_i\}_{i=1}^N$, a piecewise polynomial, $u(t)$, defined for all t in the interval $[t_0, t_F]$ and satisfying $u(t_i)=y_i$ for $i=1,2,\dots,N$. (For a detailed discussion of the advantages and costs of generating such extensions see [7,13,17,23,26]). For example, consider applying a method based on a standard s-stage, pth-order Runge–Kutta formula to (1). The corresponding discrete approximations will then satisfy

$$y_i = y_{i-1} + h_i \sum_{j=1}^{s} \omega_j k_j, \tag{2}$$

where

$$k_j = f\left(t_{i-1} + h_i c_j, y_{i-1} + h_i \sum_{r=1}^{s} a_{jr} k_r\right),$$

for $j=1,2,\dots,s$; $i=1,2,\dots,N$ and $h_i = t_i - t_{i-1}$. A continuous extension is derived by introducing the additional stages, $k_{s+1}, k_{s+2}, \dots, k_{\bar{s}}$ and polynomials $b_j(\tau)$ of degree $\leqslant p$ for $j=1,2,\dots,\bar{s}$ such that the polynomial $u_i(t)$ defined by

$$u_i(t) = y_{i-1} + h_i \sum_{j=1}^{\bar{s}} b_j\left(\frac{t-t_{i-1}}{h_i}\right) k_j \tag{3}$$

satisfies $u_i(t)=y(t)+\mathrm{O}(h^p)$ for $t \in (t_{i-1}, t_i)$ and $h=\max_{i=1}^N h_i$. Formulas of this type are called continuous Runge–Kutta (CRK) formulas. The polynomials $\{u_i(t)\}_{i=1}^N$ then define a piecewise polynomial, $u(t)$, which will be continuous on $[t_0, t_F]$ and interpolate the underlying discrete approximation if $b_j(1) = w_j$ for $j=1,2,\dots,s$ and $b_{s+1}(1) = b_{s+2}(1) \cdots = b_{\bar{s}}(1) = 0$.

In deriving CRK formulas of order p, several issues arise which can have a significant impact on the implementation and hence on the performance of the resulting continuous method. If the extra stages are restricted to be "explicit", that is if

$$k_{s+j} = f\left(t_{i-1} + c_{s+j} h_i, y_{i-1} + h_i \sum_{r=1}^{s+j-1} a_{s+j,r} k_r\right)$$

for $j=1,2,\dots,(\bar{s}-s)$, then implementation is straightforward and the cost of obtaining the continuous approximation is only the additional $\bar{s}-s$ evaluations of the differential equation on each step. It is, therefore, generally preferable to derive and implement explicit CRK methods although there are classes of problems where one can only use implicit CRK formulas. Differential algebraic problems, DAEs, are one such class and we will consider them in more detail in Section 5.

Another issue that can be important is the magnitude of the defect or residual that is associated with the continuous approximation. This quantity can be interpreted as a measure of the quality of the numerical solution. For a piecewise polynomial approximation $u(t)$ associated with the ODE (1) one defines the defect, $\delta(t)$, for $t \in (t_0, t_F)$ by

$$\delta(t) = u'(t) - f(t, u(t)).$$

Note that, in deriving CRK formulas, our assumption that $u(t)$ be order p implies, for sufficiently smooth problems,

$$u'(t) = y'(t) + O(h^{p-1}). \tag{4}$$

Furthermore, since

$$\begin{aligned}\delta(t) &= u'(t) - f(t,u(t)) - y'(t) + f(t,y(t))\\ &= (u'(t) - y'(t)) + (f(t,y(t)) - f(t,u(t))),\end{aligned} \tag{5}$$

then for differential equations that are Lipschitz continuous, $\|\delta(t)\|$ will be at worst $O(h^{p-1})$.

If we let $z_i(t)$ be the solution of the local initial value problem

$$z_i' = f(t,z_i), \quad z_i(t_{i-1}) = y_{i-1},$$

for $i = 1,2,\ldots,N$; then the local error, $le_i(t)$, associated with the continuous approximation on step i is defined for $t \in (t_{i-1}, t_i)$ to be

$$le_i(t) = z_i(t) - u_i(t).$$

It is well known that the discrete local error of a pth-order Runge–Kutta formula (2) must be $O(h_i^{p+1})$. That is

$$\begin{aligned}le_i(t_i) &= z_i(t_i) - u_i(t_i)\\ &= z_i(t_i) - y_i\\ &= O(h_i^{p+1}).\end{aligned} \tag{6}$$

If we derive order p CRK formulas which satisfy the additional constraint that the associated local error be $O(h_i^{p+1})$ for $t \in (t_{i-1}, t_i)$ then we will have

$$u_i'(t) = z_i'(t) + O(h_i^p) \tag{7}$$

and therefore, for $t \in (t_{i-1}, t_i)$,

$$\begin{aligned}\delta(t) &= (u_i'(t) - z_i'(t)) + (f(t,z_i(t)) - f(t,u_i(t)))\\ &= O(h_i^p).\end{aligned} \tag{8}$$

Furthermore, if we derive order p CRK formulas with the stronger additional constraint that, for $t \in (t_{i-1}, t_i)$, the local error be $O(h_i^{p+1})$ and satisfy

$$le_i(t) = \psi_i(\tau)\, D(t_{i-1}) h_i^{p+1} + O(h_i^{p+2}), \tag{9}$$

where $\tau = (t - t_{i-1})/h_i$, $D(t)$ is a function depending only on the problem, and $\psi_i(\tau)$ is a polynomial in τ whose coefficients are independent of the problem and the stepsize, it can be shown from (8) (see [3] for details) that

$$\delta(t) = \psi_i'(\tau)\, D(t_{i-1})\, h_i^p + O(h_i^{p+1}). \tag{10}$$

In this paper we are considering continuous methods which are designed to directly monitor and control the maximum magnitude of an estimate of the defect of the piecewise polynomial $u(t)$ that is delivered as the approximate solution. We will focus on methods based on an order p CRK formula but most of the discussion and analysis will apply to other classes of continuous methods such as those based on multistep formulas. Note that an order p CRK will always satisfy (5) but, without additional constraints, such formulas may be more difficult to implement in an effective and reliable way (when the objective is to control the magnitude of $\|\delta(t)\|$) since:

- The magnitude of the associated defect will generally only be $O(h_i^{p-1})$,
- Although the defect can be sampled at any point in the interval of interest it may not be easy to justify a rigorous, inexpensive estimate of its maximum magnitude over each step.

The first of these difficulties can be overcome by considering only order p CRK formulas with local error that is $O(h_i^{p+1})$ and both difficulties can be overcome by considering only those order p CRK formulas that satisfy (9). In this latter case as $h_i \to 0$, for any $t \in (t_{i-1}, t_i)$, $\delta(t)$ will satisfy (10) and therefore for $\bar{t} \in (t_{i-1}, t_i)$ (corresponding to a local maximum of $|\psi_i'((t - t_{i-1})/h_i)|$), $\|\delta(\bar{t})\|$ will be an asymptotically correct estimate of the maximum magnitude of the defect on the ith step. Note that since $\psi_i'(\tau)$ is a polynomial which is independent of the problem and the stepsize, the location of its local maximum magnitude, $\bar{\tau}$ (for $\tau \in (0,1)$), is known and the corresponding value for $\bar{t}$ is $\bar{t} = t_{i-1} + \bar{\tau} h_i$.

In Section 2, we will consider continuous methods for IVPs based on defect-control and in subsequent sections we will consider such methods for BVPs and DDEs. In each of these areas there are some general purpose software packages available. Finally, we will consider the development of methods for DAEs based on defect-control. We will discuss some prototype and experimental codes that implement this approach.

2. Initial value methods

The development of software based on defect-control for the numerical solution of IVPs in ODEs has a history that goes back several decades and is closely related to the notion of backward error analysis. Consider the standard IVP

$$y' = f(t, y), \quad y(t_0) = y_0, \quad t \in [t_0, t_F]. \tag{11}$$

Hull [18] and Stetter [24] investigated the reliability (or "effectiveness") of various error control strategies for discrete methods applied to (11) by establishing conditions which would guarantee the existence of a piecewise approximating function, $\hat{u}(t) \in C^0[t_0, t_F]$, which interpolates the discrete approximate solution y_i and satisfies a slightly perturbed IVP

$$\hat{u}' = f(t, \hat{u}) + \hat{\Delta}(t), \quad \hat{u}(t_0) = y_0, \tag{12}$$

where $\hat{\Delta}(t) \in C^0[t_0, t_F]$ and satisfies

$$\|\hat{\Delta}(t)\| \leqslant \hat{k}\,\mathrm{TOL}, \tag{13}$$

for some modest value of $\hat{k}$ (independent of both the problem and the method), and TOL is the specified error tolerance. In these investigation, $\hat{u}(t)$ and $\hat{\Delta}(t)$ were generally not computable but one could use standard results from mathematics, such as the Grobner–Alekseev variation of constants formula (see [22] for details), to obtain appropriate global error bounds. For example, if (11), (12) and (13) are satisfied then it is straightforward to show

$$y(t_i) - y_i = y(t_i) - \hat{u}(t_i) = \int_{t_0}^{t_i} K(t, s)\hat{\Delta}(s)\,\mathrm{d}s, \tag{14}$$

Table 1
Cost per step of relaxed and strict defect control for some CRK formulas

Formula	p	s	$\bar{s}$	$\tilde{s}$
CRK4	4	4	6	7
CRK5	5	6	9	11
CVSS6B	6	7	11	14
CVSS7	7	9	15	20
CVSS8	8	13	21	28

where $K(t,s)$ is a variational matrix that depends only on the problem, and this implies

$$\|y(t_i) - y_i\| \leqslant (t_i - t_0)\hat{k}\; K_{\max}\, \mathrm{TOL}, \tag{15}$$

where $K_{\max}$ is a bound on $\|K(s,t)\|$. Note that this is an example of backward error analysis where the computed solution is guaranteed to exactly satisfy a slightly perturbed IVP and one can interpret $K_{\max}$ as a type of condition number for the problem (which quantifies how sensitive the solution can be to small changes in the problem specification).

Subsequently, several investigators, who were primarily interested in dense output (or off-mesh approximations), analysed and developed computable piecewise polynomial interpolants, $u(t)$, which could be efficiently implemented with new or existing discrete methods (see, for example, [7,17,23,25,27]). It was soon recognized that, once $u(t)$ was generated, the corresponding defect could be sampled and new reliable error and stepsize-control strategies could be developed with the objective of directly satisfying relationships analogous to (12) and (13).

As we have noted earlier, when one considers CRK formulas, the requirement that $\|\delta(t)\|$ be of optimal order and easy to bound by an inexpensive error estimate (for $t \in (t_{i-1}, t_i)$) will generally impose additional constraints on what would comprise a suitable local interpolating polynomial, $u_i(t)$. In a series of investigations [1–3,14–16], several order p CRK formulas have been developed and compared. These investigations have also considered the relative advantages of several alternative defect-control strategies for $3 \leqslant p \leqslant 8$. We will now consider two of the most promising strategies.

The first strategy assumes that the local interpolant $u_i(t)$ defined by (3) satisfies (7) and that the estimate of the maximum magnitude of the defect is obtained by the heuristic

$$\begin{aligned} \mathrm{est}_i &= \|\delta(t_{i-1} + \hat{\tau} h_i)\| \\ &= \|u_i'(t_{i-1} + \hat{\tau} h_i) - f(t_{i-1} + \hat{\tau} h_i, u_i(t_{i-1} + \hat{\tau} h_i))\|, \end{aligned} \tag{16}$$

where $\hat{\tau}$ is chosen in a careful way (see [1] for a detailed discussion of how $\hat{\tau}$ can be chosen). We will refer to this strategy as the "relaxed" defect-control strategy. This heuristic for controlling the maximum magnitude of the defect works well on most problems but the strategy is not asymptotically justified (as $h \to 0$) and it can severely underestimate the size of the defect for some problems. Table 1 reports for $4 \leqslant p \leqslant 8$ the value of s and $\bar{s}$ for some order p CRK which have been found to be particularly effective. The higher-order formulas are members of families of formulas derived in [26] (the particular coefficients are identified in [3]). Note that the cost per step of a method using this error-control strategy for these particular CRKs is $\bar{s}$ derivative evaluations. This follows since although one must perform an additional derivative evaluation to sample the defect (at $t_{i-1}+\hat{\tau}h_i$), each of the formulas is designed to ensure that $u(t) \in C^1[t_0, t_F]$ by requiring that $u_i'(t_{i-1}+h_i) = f(t_{i-1}+h_i, y_i)$. This implies that, on all but the first step, k_1 will be available (as an internal stage of the previous step).

A more rigorous error-control strategy (which we will refer to as the "strict" defect-control strategy) for the same set of underlying discrete formulas can be developed by requiring that the corresponding continuous extension satisfy (9) as well as (7). With this additional constraint we have available (as discussed earlier) an asymptotically correct estimate of the maximum magnitude of the defect (on the ith step) given by $\text{est}_i = \|\delta(t_{i-1}+\bar{\tau}h_i)\|$ where $\bar{\tau}$ is fixed and independent of the problem or stepsize. One way to generate such a CRK (although it may not be optimal) is to begin with an $\bar{s}$-stage order p CRK, $u_i(t)$, satisfying (3) and (7) and replace the "extra" stages, k_{s+j}, $j=1,\dots,(\bar{s}-s)$ (whose corresponding $c_{s+j} \neq 1$) with the more accurate values

$$\tilde{k}_{s+j} = \begin{cases} f(t_{i-1}+c_{s+j}h_i, u_i(t_{i-1}+c_{s+j}h_i)) & \text{if } c_{s+j} \neq 1, \\ k_{s+j} & \text{if } c_{s+j} = 1. \end{cases} \tag{17}$$

A new, more suitable interpolant, $\tilde{u}_i(t)$ can then be defined by

$$\tilde{u}_i(t) = y_{i-1} + h_i \sum_{j=1}^{s} b_j\left(\frac{t-t_{i-1}}{h_i}\right) k_j + \sum_{j=1}^{\bar{s}-s} b_{j+s}\left(\frac{t-t_{i-1}}{h_i}\right)\tilde{k}_{s+j}. \tag{18}$$

It can easily be shown that $\tilde{u}_i(t)$ will satisfy (7), (9) and, when $c_j = 1$ for one value of j in the range $1 \leqslant j \leqslant (\bar{s}-s)$ (as is the case for each of the CRK formulas identified in Table 1), it will be an $\tilde{s}$-stage order p CRK with $\tilde{s} = s + 2(\bar{s}-s) - 1$. Table 1 also reports the value of $\tilde{s}$ for this strict defect-control strategy using the CRK corresponding to (18).

Clearly a trade-off between efficiency and reliability needs to be addressed when choosing which defect-control strategy should be used to solve a particular problem. Fortunately, it is convenient and straightforward to implement a numerical method which can apply either strategy (using either (3) with $\hat{\tau}$ or (18) with $\bar{\tau}$ to define the respective interpolants and defect estimates) and thus the choice can be an option which can be selected by the user. From Table 1 we can observe that the cost per step of using the strict defect-control strategy can be between 20 and 33% more than that for the relaxed strategy but better control of the size of the defect should be realised.

To illustrate and quantify this trade-off we have run both versions of a numerical method based on the CRK formula CVSS6B, on the 25 standard nonstiff problems of the DETEST package [11] at nine error tolerances and assessed how well the defect was controlled. The performance on a typical problem, problem B4, is summarised in Table 2.

These results are typical of that observed on all problems. Both methods are robust in that they are able to deliver, over a wide range of tolerances, a close and consistent relationship between the size of the defect and the specified error tolerance.

Both versions of CVSS6B required over 13 500 steps to solve all of the problems at all tolerances. With the strict defect-control strategy the maximum magnitude of the defect rarely (on fewer than 1.8% of the steps) exceeded TOL and never exceeded 7 TOL. With the relaxed defect-control strategy the maximum magnitude of the defect exceeded TOL on 20% of the steps but it rarely exceeded 5 TOL (on fewer than 1.2% of the steps) and it never exceeded 18 TOL.

An alternative rigorous defect control strategy based on the use of a different norm has been proposed and justified in [19]. On each step one introduces a weighted L_2-norm (which can be

Table 2
Performance of CVSS6B on problem B4 of DETEST

Strategy	TOL[a]	Time[b]	FCN[c]	Steps[d]	GL Err[e]	Max Def[f]	% Succ[g]
Relaxed defect control	10^{-2}	0.010	166	15	4.7	1.9	73
	10^{-3}	0.015	254	22	18.0	1.2	91
	10^{-4}	0.023	375	32	28.4	1.3	84
	10^{-5}	0.031	507	46	34.5	1.7	74
	10^{-6}	0.046	749	68	37.6	1.5	62
	10^{-7}	0.067	1101	100	39.0	1.5	49
	10^{-8}	0.099	1618	147	40.5	1.5	45
	10^{-9}	0.145	2377	216	41.1	1.5	45
Strict defect control	10^{-2}	0.014	253	17	1.2	0.88	100
	10^{-3}	0.021	379	24	9.5	0.94	100
	10^{-4}	0.027	491	35	15.6	0.98	100
	10^{-5}	0.041	743	53	16.6	0.76	100
	10^{-6}	0.061	1093	78	16.6	0.66	100
	10^{-7}	0.090	1625	116	16.5	0.61	100
	10^{-8}	0.131	2381	170	16.5	0.58	100
	10^{-9}	0.193	3501	250	16.5	0.56	100

[a]TOL, specified error tolerance.
[b]Time, computer time required to solve the problem measured in seconds on a SUN Sparc4.
[c]FCN, number of derivative evaluations required to solve the problem.
[d]Steps, number of time steps required to solve the problem.
[e]GL Err, maximum observed global error measured in units of TOL and determined by measuring the global error at 100 equally spaced points per step.
[f]Max Def, maximum magnitude of the defect measured in units of TOL and determined by sampling the defect at 100 equally spaced points per step.
[g]% Succ, percentage of steps where the magnitude of the defect is less than TOL.

interpreted as an average magnitude of the defect),

$$\|\delta_i(t)\|_2 = 1/h_i \left(\int_{t_{i-1}}^{t_i} \|\delta_i(s)\|^2 \, \mathrm{d}s \right)^{1/2}, \tag{19}$$

A method can then be developed with an error control strategy that attempts to ensure that

$$\|\delta_i(t)\|_2 \leqslant \mathrm{TOL}. \tag{20}$$

As is pointed out in [19], $\delta_i(t)$ is usually known at the meshpoints and therefore, for sufficiently smooth problems, one can derive a low cost, asymptotically correct estimate of $\|\delta_i(t)\|_2^2$ using a suitably chosen Lobatto quadrature formula (which would require only a few additional evaluations of the defect). This approach has been implemented and shown to be very effective for a large class of problems.

3. Boundary value methods

Numerical methods for BVPs of the form

$$y' = f(t, y), \quad t \in [t_0, t_F], \quad g(y(t_0), y(t_F)) = 0 \tag{21}$$

generally produce a discrete approximation on a mesh $t_0 < t_1 < \cdots < t_N = t_F$ by solving a large coupled nonlinear system of equations. If the underlying formula that determines the discrete solution is a Runge–Kutta or collocation formula then it is straightforward to introduce a continuous extension $u(t)$ and the associated defect $\delta(t)$ (as we have done for IV methods). From Table 1 we see that the cost per step to compute $u(t)$ and estimate the size of the corresponding defect can be as great as applying the underlying discrete formula. For BV methods the cost per step to determine $u(t)$ and $\delta(t)$, after the discrete solution has been computed, remains the same while the cost of solving for the discrete solution is generally much greater. A consequence of this is that once a converged discrete solution is determined by a BV method (based on the use of a CRK or collocation formula), a continuous extension with an optimum $O(h^p)$ defect can be computed at very little incremental cost (see, for example, [10,12]).

When these formulas are used to determine the discrete solution, defect-based error control and mesh-refinement strategies can be particularly attractive. This approach has been followed in the development of the methods MIRKDC [9] and bvp4c [19] which have been found to be effective for solving a wide class of problems.

In the numerical solution of BVPs, one often encounters difficulties with convergence of the iteration scheme that is used to solve the nonlinear system associated with the discrete mesh. This can be the result of a poor choice of mesh and/or a poor initial guess for the discrete solution. In either case, if the method has available a piecewise polynomial approximation $\bar{u}(t)$ with an associated defect $\bar{\delta}(t)$ (as would be the case, for example, if $\bar{u}(t)$ were associated with a mesh and previously computed discrete solution that was judged not to be sufficiently accurate), then these deficiencies can often be overcome by using the size of the defect to help guide the mesh refinement and using $\bar{u}(t)$ to generate the required initial guess. With this approach one can also use the estimates of the maximum magnitude of the defect to help ensure that the approximate solution that is ultimately delivered by the method, $u(t)$, satisfies

$$\|\delta(t)\| = \|u'(t) - f(t, u(t))\| \leqslant \text{TOL}, \tag{22}$$

and

$$g((u(t_0), u(t_F)) = 0.$$

Note that, with this approach, one could consider using inexpensive interpolants for the mesh refinement strategies, and the more expensive rigorous interpolants for assessing the accuracy of the numerical solution.

When such strategies are adopted by a method, one only has to compute the interpolant and defect estimate on the final iteration after the underlying discrete approximation has converged. Intermediate calculations associated with preliminary coarse meshes or initial iterations (before convergence) either would not require the determination of any interpolant or would only require the less expensive relaxed interpolant.

Numerical experience reported in [9,19] shows that BV methods that implement such defect-based strategies can outperform methods based on more traditional strategies especially when a strongly nonuniform mesh is appropriate. Even on problems where asymptotic analysis is not necessarily relevant, carefully designed defect-based strategies can quickly lead to a suitable mesh and rapid convergence on that mesh.

4. Delay differential equation methods

A class of numerical methods based on CRK formulas with defect control has been analysed [6] for systems of retarded and neutral DDEs of the form

$$y' = f(t, y, y(t - \sigma_1(t, y(t))), \; y'(t - \sigma_2(t, y(t)))), \quad t \in [t_0, t_F]. \tag{23}$$

$$y(t) = \phi(t), \quad t \leqslant t_0, \tag{24}$$

where $\sigma_1(t, y)$ and $\sigma_2(t, y)$ are positive scalar functions. One particular sixth-order formula from this class (the formula CVSS6B discussed in Section 2) has been implemented in a software package, DDVERK [4], and shown to be effective for these DDEs [5].

For this class of problems a discrete method must be able to approximate the solution at off-mesh points in order to evaluate the differential equation at an arbitrary point, $t \in [t_0, t_F]$. Therefore, the requirement that the numerical solution be a piecewise polynomial, $u(t)$, does not impose any extra cost and one can associate an approximation $u(t)$ (with corresponding defect $\delta(t)$) with any numerical method.

To be effective for this class of problems a numerical method must be able to detect and efficiently handle the discontinuities that inevitably arise and are propagated as the integration proceeds. Automatic techniques for detecting and accurately crossing points of discontinuity for standard IVPs based on monitoring changes in the associated defect have been proposed and justified in [8]. This technique has been adapted and refined for DDEs in the solver DDVERK (see [4] for a discussion of the details) where it has proved to be very effective for a wide class of problems. It is certainly competitive with the alternative strategy which explicitly checks for discontinuities at all possible locations where propagation is possible. This is particularly true for systems of equations with multiple delays where the number of potential points of discontinuity can be quite large relative to the number of significant or relevant points of discontinuity. The defect based strategy for coping with discontinuities essentially adjusts the stepsize selection (as well as the error-control mechanism) only on those steps where the presence of a point of discontinuity has severely reduced the stepsize.

5. Differential algebraic equation methods

In recent years, there has been considerable interest and progress made in the development of numerical methods for special classes of DAEs. Nevertheless, very few methods can be applied directly to a system of DAEs in the most general form

$$F(t, y, y') = 0, \quad y(t_0) = y_0, \quad t \in [t_0, t_F], \tag{25}$$

with $(\partial F/\partial y')$ known to be singular. Note that if this matrix is nonsingular for all $t \in [t_0, t_F]$ the problem is said to have index 0. In this case one can solve the nonlinear system associated with

(25) to determine $y'(t)$ for any prescribed value of t and $y(t)$. Any initial value method can be applied and special DAE methods are not necessary.

In general the "index" of a problem of the form (25) refers to the minimum number of times one has to differentiate the equation, $F(t,y,y')=0$, in order to derive an equivalent initial value problem where $y'(t)$ can be determined uniquely in terms of $t, y(t), F$ and various partial derivatives of F. The higher the index of a problem, the more sensitive the solution can be to perturbations in the data and the more difficult it becomes to develop reliable numerical methods. Currently, there are several reliable general-purpose numerical methods for index 1 problems and other reliable methods designed for special classes of index 2 and index 3 problems.

The DAEs that arise in application areas, such as the modelling of constrained mechanical systems or the design of electrical circuits often are of index 2 or index 3 but they possess special structure and numerical methods which exploit this structure have been developed and have received wide acceptance. For example, the algebraic constraints can often be explicitly identified and the system decoupled, $y(t)=[y_1(t), y_2(t)]^{\mathrm{T}}$, and written in the semi-explicit form

$$y_1'(t) = f(t, y_1(t), y_2(t)), \tag{26}$$

$$0 = g(t, y_1(t), y_2(t)). \tag{27}$$

When one considers the development of defect-based error control for DAE methods two key questions must first be answered:

(1) How does one define a sufficiently accurate continuous extension, $u_i(t)$, of the discrete approximation (for $t \in [t_{i-1}, t_i]$)?
(2) What measure of the size of the defect is appropriate to control? That is, can one introduce a measure $\mu_i(\delta)$ such that for $t \in [t_{i-1}, t_i]$ the condition that $\mu_i(\delta(t)) \leqslant \mathrm{TOL}$ will ensure that the global error will be proportional to TOL and $\mu_i(\delta)$ will be inexpensive to estimate on each step?

These questions were considered in [21] where defect-based error-control strategies suitable for important classes of index 2 and index 3, semi-explicit problems were introduced and justified. The approach that was introduced can be applied with any discrete, order p, implicit Runge–Kutta formula to generate, on each step, interpolating polynomials $u_i(t)$ and $v_i(t)$ that approximate $y_1(t)$ and $y_2(t)$, respectively. If one defines the defect of the resulting vector of piecewise polynomials associated with $u(t)$ and $v(t)$ we have (from (26) and (27))

$$\delta_1(t) = u'(t) - f(t, u(t), v(t)), \tag{28}$$

$$\delta_2(t) = g(t, u(t), v(t)). \tag{29}$$

The global errors $\|y_1(t) - u(t)\|$ and $\|y_2(t) - v(t)\|$ were analysed and shown to be bounded by a suitable multiple of TOL provided $\delta_1(t), \delta_2(t)$, and $\delta_2'(t)$ were all suitably bounded in norm. Corresponding measures $\mu_i(\delta)$ were proposed and associated estimates introduced which could be the basis for an effective defect-based numerical method for semi-explicit DAEs.

Another approach has been considered in [20] where no assumptions are made on the structure of the DAE. In order to determine the piecewise polynomial, $u(t)$ which approximates the solution to (25) one begins with an implicit continuous extension of a discrete, order p, implicit Runge–Kutta formula. One then introduces an associated overdetermined system of nonlinear equations on each time step by requiring that the corresponding approximating polynomial, $\tilde{u}_i(t)$ satisfy (in a least squares sense) the defining equations of the underlying continuous extension as well as additional "collocation" equations (which are equivalent to asking that (25) be satisfied at a prescribed set of sample points). The defect, $\tilde{\delta}(t)$, associated with the resulting piecewise polynomial, $\tilde{u}(t)$, is defined by

$$\tilde{\delta}(t) = F(t, \tilde{u}(t), \tilde{u}'(t)). \tag{30}$$

Conditions on the choice of underlying implicit CRK formulas and on the number and choice of collocation points are identified which result in $\|\tilde{\delta}(t)\|$ being $O(h^p)$ for sufficiently differentiable index 1 and index 2 problems. Estimates of $\|\tilde{\delta}(t)\|$ are justified and an experimental code introduced to illustrate the validity of this approach. A general-purpose numerical method based on this approach is under development.

6. Summary and conclusions

As is clear from our discussion so far there are now several general purpose numerical methods for important classes of ODEs that produce piecewise polynomial approximate solutions and attempt to directly control the magnitude of the associated defect. These methods, although more costly than the classical discrete methods, can be efficiently implemented and they produce solutions whose accuracy can be more readily interpreted and compared.

We have also shown that when implementing numerical methods using defect control one must address a trade-off between reliability and efficiency. This trade-off arises from a choice between the use of an inexpensive heuristic or a more expensive (but asymptotically correct) estimate of the maximum magnitude of the defect. This choice can be left to the user but the implications must be understood when interpreting the numerical results.

There are two difficulties that have not been discussed which limit the applicability of this class of methods and which should be addressed in future investigations. If the underlying problem is not sufficiently smooth, then one is restricted to the use of lower-order methods and defect control can be less competitive with the more classical approach at low orders. Also, at limiting precision, where the effect of round-off error may dominate the local error, the currently employed defect estimates are unreliable. More research is required to develop effective strategies for detecting and coping with this situation.

References

[1] W.H. Enright, A new error control for initial value solvers, Appl. Math. Comput. 31 (1989) 288–301.

[2] W.H. Enright, Analysis of error control strategies for continuous Runge–Kutta methods, SIAM J. Numer. Anal. 26 (3) (1989) 588–599.

[3] W.H. Enright, The relative efficiency of alternative defect control schemes for high order Runge–Kutta formulas, SIAM J. Numer. Anal. 30 (5) (1993) 1419–1445.

[4] W.H. Enright, H. Hayashi, A delay differential equation solver based on a continuous Runge–Kutta method with defect control, Numer. Algorithms 16 (1997) 349–364.

[5] W.H. Enright, H. Hayashi, The evaluation of numerical software for delay differential equations, in: R. Boisvert (Ed.), The quality of Numerical Software: Assessment and Enhancement, Chapman & Hall, London, 1997, pp. 179–197.

[6] W.H. Enright, H. Hayashi, Convergence analysis of the solution of retarded and neutral delay differential equations by continuous methods, SIAM J. Numer. Anal. 35 (2) (1998) 572–585.
[7] W.H. Enright, K.R. Jackson, S.P. Nørsett, P.G. Thomsen, Interpolants for Runge–Kutta formulas, ACM Trans. Math. Software 12 (1986) 193–218.
[8] W.H. Enright, K.R. Jackson, S.P. Nørsett, P.G. Thomsen, Effective solution of discontinuous IVPs using a Runge–Kutta formula pair with interpolants, Appl. Math. Comput. 27 (1988) 313–335.
[9] W.H. Enright, P.H. Muir, A Runge–Kutta type boundary value ODE solver with defect control, SIAM J. Sci. Comput. 17 (1996) 479–497.
[10] W.H. Enright, P.H. Muir, Superconvergent interpolants for the collocation solution of BVODEs, SIAM J. Sci. Comput. 21 (2000) 227–254.
[11] W.H. Enright, J.D. Pryce, Two FORTRAN packages for assessing initial value methods, ACM Trans. Math. Software 13 (1) (1987) 1–27.
[12] W.H. Enright, R. Sivasothinathan, Superconvergent interpolants for collocation methods applied to mixed order BVODEs, ACM Trans. Math. Software (2000), to appear.
[13] I. Gladwell, L.F. Shampine, L.S. Baca, R.W. Brankin, Practical aspects of interpolation in Runge–Kutta codes, SIAM J. Sci. Statist. Comput. 8 (2) (1987) 322–341.
[14] D.J. Higham, Defect estimation in Adams PECE codes, SIAM J. Sci. Comput. 10 (1989) 964–976.
[15] D.J. Higham, Robust defect control with Runge–Kutta schemes, SIAM J. Numer. Anal. 26 (1989) 1175–1183.
[16] D.J. Higham, Runge–Kutta defect control using Hermite–Birkhoff interpolation, SIAM J. Sci. Comput. 12 (1991) 991–999.
[17] M.K. Horn, Fourth- and fifth-order scaled Runge–Kutta algorithms for treating dense output, SIAM J. Numer. Anal. 20 (3) (1983) 558–568.
[18] T.E. Hull, The effectiveness of numerical methods for ordinary differential equations, SIAM Stud. Numer. Anal. 2 (1968) 114–121.
[19] J. Kierzenka, L.F. Shampine, A BVP solver based on residual control and the MATLAB PSE SMV Math., Report 99–001.
[20] C. MacDonald, A new approach for DAEs, Ph.D. Thesis, Department of Computer Science, University of Toronto, 1999 (also appeared as DCS Technical Report No. 317/19).
[21] H. Nguyen, Interpolation and error control schemes for algebraic differential equations using continuous implicit Runge–Kutta methods, Ph.D. Thesis, Department of Computer Science, University of Toronto, 1995 (also appeared as DCS Technical Report No. 298/95).
[22] S.P. Nørsett, G. Wanner, Perturbed collocation and Runge–Kutta methods, Numer. Math. 38 (1981) 193–208.
[23] L.F. Shampine, Interpolation for Runge–Kutta methods, SIAM J. Numer. Anal. 22 (5) (1985) 1014–1027.
[24] H.J Stetter, Cosiderations concerning a theory for ODE-solvers, in: R. Bulirsch, R.D. Grigorieff, J. Schroder (Eds.), Lecture notes in Mathematics, Vol. 631, Numerical Treatment of Differential Equations, Springer, Berlin, 1978, pp. 188–200.
[25] H.J. Stetter, Interpolation and error estimation in Adams PC-codes, SIAM J. Numer. Anal. 16 (2) (1979) 311–323.
[26] J.H. Verner, Differentiable interpolants for high-order Runge–Kutta methods, SIAM J. Numer. Anal. 30 (5) (1993) 1446–1466.
[27] M. Zennaro, Natural continuous extensions of Runge–Kutta methods, Math. Comput. 46 (1986) 119–133.

ELSEVIER

Journal of Computational and Applied Mathematics 125 (2000) 171–182

JOURNAL OF COMPUTATIONAL AND APPLIED MATHEMATICS

www.elsevier.nl/locate/cam

Numerical solutions of stochastic differential equations – implementation and stability issues

Kevin Burrage[a,*], Pamela Burrage[a], Taketomo Mitsui[b]

[a]*Department of Mathematics, University of Queensland, St Lucia, Queensland, 4072, Australia*

[b]*Graduate School of Human Informatics, Nagoya University, Furo-cho Chikusa-ku, Nagoya 464-8601, Japan*

Received 25 October 1999

Abstract

Stochastic differential equations (SDEs) arise from physical systems where the parameters describing the system can only be estimated or are subject to noise. There has been much work done recently on developing numerical methods for solving SDEs. This paper will focus on stability issues and variable stepsize implementation techniques for numerically solving SDEs effectively. © 2000 Elsevier Science B.V. All rights reserved.

Keywords: Stability; Stochastic differential equations; Variable stepsize

1. Introduction

This paper presents an overview of stability and implementation issues of numerical methods for solving stochastic differential equations. Due to space constraints it is not possible to give details behind the construction of numerical methods suitable for solving SDEs, instead the paper will focus on the stability and implementation of numerical methods. Thus Section 2 discusses numerical stability both of SDEs and of numerical methods for solving these SDEs, while the implementation of numerical methods using a fixed stepsize is discussed in Section 3; in Section 4 a variable stepsize implementation is presented.

This section continues with some necessary background details covering the form of an SDE together with definitions of order of convergence for numerical methods to solve such SDEs.

Stochastic differential equations describe physical systems where noise is present, with the noise being modelled by a Wiener process that is nowhere differentiable. The general form of an autonomous

* Corresponding author.

E-mail address: kb@maths.uq.edu.au (K. Burrage).

PII: S0377-0427(00)00467-2

SDE is

$$\mathrm{d}y(t) = f(y(t))\,\mathrm{d}t + g(y(t))\,\mathrm{d}W(t), \quad y \in \mathbb{R}^m, \quad y(0) = y_0, \tag{1}$$

where f is the drift coefficient (an m-vector-valued function), g is the diffusion coefficient (an $m \times d$ matrix-valued function), and $W(t)$ is a d-dimensional process having independent scalar Wiener process components ($t \geqslant 0$). A Wiener process W is a Gaussian process with the property that

$$\boldsymbol{E}(W(t)) = 0, \quad \boldsymbol{E}(W(t)W(s)) = \min\{t, s\}.$$

The Wiener increments $W(t) - W(s)$ are independent Gaussian processes with mean 0 and variance $|t-s|$.

Eq. (1) can be written as a stochastic integral equation

$$y(t) = y(t_0) + \int_{t_0}^{t} f(y(s))\,\mathrm{d}s + \int_{t_0}^{t} g(y(s))\,\mathrm{d}W(s)$$

where the first integral is a regular Riemann–Stieltjes integral and the second integral is a stochastic integral, commonly interpreted in either Itô or Stratonovich form. The Stratonovich interpretation follows the usual rules of Riemann–Stieltjes calculus, and for this reason is the form used in this paper (the symbol $\circ$ in front of $\mathrm{d}W(s)$ will serve to confirm a Stratonovich integral). However, an SDE presented in Itô form can be converted to Stratonovich form using a simple formula which relates the two interpretations. Indeed the solution of (1) and its related Stratonovich SDE are exactly the same:

$$\mathrm{d}y(t) = \bar{f}(y(t)) + g(y(t)) \circ \mathrm{d}W(t), \tag{2}$$

$$\bar{f}_i(y(t)) = f_i(y(t)) - \tfrac{1}{2}\sum_{j=1}^{m}\sum_{k=1}^{d} g_{jk}(y(t))\frac{\partial g_{ik}(y(t))}{\partial y_j}, \quad i = 1, \ldots, m. \tag{3}$$

A multiple Stratonovich integral is given by

$$J_{j_1 j_2 \cdots j_l}(t_0, t) = \int_{t_0}^{t}\int_{t_0}^{s_l}\cdots\int_{t_0}^{s_2} \circ\,\mathrm{d}W_{s_1}^{j_1} \circ \cdots \circ\,\mathrm{d}W_{s_l}^{j_l},$$

where $j_l \in \{0, 1, \ldots, d\}$ for d Wiener processes. Note that the integral $J_0(t_0, t) = \int_{t_0}^{t} \circ\,\mathrm{d}W_{s_1}^0 = \int_{t_0}^{t} \mathrm{d}s_1$. For ease of notation, the written dependence on t_0 and t will be dropped when the meaning is clear from the context.

There are two ways of measuring the accuracy of a numerical solution of an SDE – these are strong convergence and weak convergence – only strong convergence will be considered in this paper. Strong convergence is required when each trajectory of the numerical solution must be close to the exact solution:

Definition 1. Let $\bar{y}_N$ be the numerical approximation to $y(t_N)$ after N steps with constant stepsize $h = (t_N - t_0/N)$; then $\bar{y}$ is said to *converge strongly* to y with strong global order p if $\exists C > 0$ (independent of h) and $\delta > 0$ such that

$$\boldsymbol{E}(\|\bar{y}_N - y(t_N)\|) \leqslant Ch^p, \quad h \in (0, \delta).$$

This definition is for global order – the local error can behave as $O(h^{p+1/2})$; fractional orders arise as the root mean square order of the Wiener process is $h^{1/2}$.

Numerical methods for SDEs are derived by comparing the stochastic Taylor series expansion of the numerical solution with that of the exact solution, over one step assuming exact initial values. This comparison results in a set of order conditions to be satisfied – see [2,13] for the development of these order conditions using Rooted Tree theory in the case of Stratonovich problems.

This section has provided an overview of the basic definitions required for studying numerical methods for solving SDEs; in the next section, stability of the SDE and of numerical methods is discussed.

2. Numerical stability analysis

As in other areas of numerical analysis, *numerical stability* is significant in the case of SDEs which usually require a long (numerical) time-integration.

2.1. Stochastic stability

Consider the scalar version of (1). We assume that there exists a unique solution $y(t;t_0,y_0)$ of the equation for $t>t_0$. Moreover, we suppose that the equation allows a steady solution $y(t)\equiv 0$. This means that $f(0)=g(0)=0$ holds. A steady solution is often called an *equilibrium position.* Has'minskii [10] gave the following three definitions of stability.

Definition 2. The equilibrium position of the SDE is said to be stochastically stable, stochastically asymptotically stable and stochastically asymptotically stable in the large, respectively, if the following conditions hold:
(i) For all positive ε and for all t_0 the following equality holds.

$$\lim_{y_0\to 0} P\left(\sup_{t\geqslant t_0}|y(t;t_0,y_0)|\geqslant\varepsilon\right)=0$$

(ii) In addition to the above,

$$\lim_{y_0\to 0} P\left(\lim_{t\to\infty}|y(t;t_0,y_0)|=0\right)=1.$$

(iii) Moreover to the above two,

$$P\left(\lim_{t\to\infty}|y(t;t_0,x_0)|=0\right)=1 \quad \text{for all } y_0.$$

Each item in Definition 2 can be seen as the stochastic counterparts of stability, asymptotic stability and asymptotic stability in the large, respectively, in the ODE case.

Actually we can derive a criterion of the asymptotic stochastic stability for the SDE. Assume that the functions f and g are uniformly asymptotically linear with respect to x; that is, for certain real constants a and b,

$$f(x)=ax+\bar{f}(x),\quad g(x)=bx+\bar{g}(x)$$

with

$$\lim_{|x|\to 0} \frac{|\bar{f}(x)| + |\bar{g}(x)|}{|x|} = 0$$

hold uniformly in t. The solution $y(t)$ of the SDE is stochastically asymptotically stable if $a - b^2/2 < 0$. This criterion found in [8, p. 139] strongly suggests a possibility of analogous *linear stability* analysis for numerical schemes of SDE to those of ODE. We can consider that the linear parts of f and g are dominant in the asymptotic behaviour of solutions around the equilibrium position.

2.2. Numerical asymptotic stability

To cope with linear stability analysis, we introduce a linear test equation (supermartingale equation)

$$\mathrm{d}y(t) = \lambda y(t)\,\mathrm{d}t + \mu y(t)\,\mathrm{d}W \quad (t > 0) \quad \text{with } \lambda, \mu \in \mathbb{C} \tag{4}$$

with the initial condition $y(0) = 1$ to the numerical stability analysis. Since the exact solution of (4) is written as

$$y(t) = \exp\{(\lambda - \tfrac{1}{2}\mu^2)t + \mu W(t)\},$$

it is quite easy to show that the equilibrium position $y(t) \equiv 0$ is stochastically asymptotically stable if

$$\mathrm{Re}\,(\lambda - \tfrac{1}{2}\mu^2) < 0. \tag{5}$$

We can arrive at the following definition which is found in [11].

Definition 3. When a numerical scheme is applied to the stochastically asymptotically stable equation (4) and generating the sequence $\{y_n\}$, it is said to be numerically asymptotically stable if

$$\lim_{n\to\infty} |y_n| = 0 \quad \text{with probability 1.}$$

To analyze the behaviour of real stochastic processes derived from various numerical schemes, the following lemma given in [11] is useful.

Lemma 4. *Given a sequence of real-valued, nonnegative, independent and identically distributed random variable* $\{Z_n\}$, *consider the sequence random variable* $\{Y_n\}$ *defined by*

$$Y_n = \left(\prod_{i=0}^{n-1} Z_i\right) Y_0,$$

where $Y_0 \geqslant 0$ *and* $Y_0 \neq 0$ *with probability* 1. *Suppose that the random variable* $\log(Z_i)$ *are square-integrable. Then* $\lim_{n\to\infty} Y_n = 0$ *with probability* 1 *iff* $\boldsymbol{E}(\log(Z_i)) < 0$ *for all* i.

However, the numerical asymptotic stability criterion does not work well. The reason is that criterion (5) allows the case $\mathrm{Re}\,\lambda > 0$. It implies that some sample paths of the solution surely *decrease* to 0, whereas their distributions possibly *increase*. This can be understood through the fact that when $\mathrm{Re}\,\lambda > 0$ the equation cannot be asymptotically stable even in the ODE sense. Henceforth, it is impossible to carry out a numerical scheme *until* all the sample paths of the exact solution diminish to 0 if two conditions $\mathrm{Re}\,\lambda > 0$ and $\mathrm{Re}\,(\lambda - \frac{1}{2}\mu^2) < 0$ are valid simultaneously. Since the numerical solution would reflect this statistical property, nobody can expect a numerically stable solution. Even in the case of the stochastic θ-method given by

$$y_{n+1} = y_n + (1-\theta)f(y_n)h + \theta f(y_{n+1})h + g(y_n)\Delta W_n \quad (\theta \in [0,1]) \tag{6}$$

there are combinations of the parameters λ and μ in (4) which do not give numerical asymptotic stability with any h (see [11]).

This investigation implies the necessity of another stability concept for SDEs. That is, we try to answer the question what SDE has all sample paths whose distribution tends to 0 as $t \to \infty$.

2.3. MS-stability

Analysis of the previous subsection suggests an introduction of a norm of the SDE solution with respect to the underlying stability concept.

Definition 5. The equilibrium position $y(t) \equiv 0$ is said to be asymptotically stable in pth mean if for all positive ε there exists a positive δ which satisfies

$$\boldsymbol{E}(|y(t)|^p) < \varepsilon \quad \text{for all } t \geqslant 0 \quad \text{and} \quad |y_0| < \delta \tag{7}$$

and, furthermore, if there exists a positive δ_0 satisfying

$$\lim_{t\to\infty} \boldsymbol{E}(|y(t)|^p) = 0 \quad \text{for all } |y_0| < \delta_0. \tag{8}$$

The most frequently used case $p=2$ is called the *mean-square case*. Thus we introduce the norm of the solution by $\|y\| = \{\boldsymbol{E}|y|^2\}^{1/2}$.

The necessary and sufficient condition is rather simple (see [18]).

Lemma 6. *The linear test equation* (*supermartingale equation*) (4) *with the unit initial value is asymptotically stable in the mean-square sense* (*abbreviated as MS-stability*) *iff* $\mathrm{Re}\,\lambda + |\mu|^2/2 < 0$.

Note that since the inequality $\mathrm{Re}(2\lambda - \mu^2) \leqslant 2\mathrm{Re}\,\lambda + |\mu|^2$ is always valid, the asymptotic stability in the mean-square sense implies the stochastic stability.

2.4. Numerical MS-stability

For asymptotically MS-stable problems of SDEs, what conditions are imposed to derive numerically asymptotically MS-stable solutions? That is to say, what conditions should be for the numerical solution $\{y_n\}$ of the linear test equation (4) to achieve $\|y_n\| \to 0$ as $n \to \infty$?

Denote $\mathbf{E}|y_n|^2$ by Y_n. When we apply a numerical scheme to the linear test equation and take the mean-square norm, we obtain a one-step difference equation of the form $Y_{n+1} = R(\bar{h},k)Y_n$ where two scalars $\bar{h}$ and k stand for $h\lambda$ and μ^2/λ, respectively. We can call $R(\bar{h},k)$ the *stability function* of the scheme, and arrive at the following.

Definition 7 (Saito and Mitsui [18]). The scheme is said to be numerically MS-stable for $\bar{h}$ and k if its stability function $R(\bar{h},k)$ is less than unity in magnitude. The set in $\mathbb{C}^2$ given by $\mathscr{R} = \{(\bar{h},k);\ |R(\bar{h},k)| < 1 \text{ holds}\}$ is called the domain of MS-stability of the scheme.

In addition, we can say that a numerical scheme is A-stable if it is MS-stable for arbitrary h which is sufficiently small for the convergence.

We will derive the stability function of some numerical schemes known in the literature. Details with figures can be seen in [18].

First is the Euler–Maruyama scheme (6) (with $\theta = 0$), whose application to (4) implies

$$y_{n+1} = y_n + h\lambda y_n + \mu y_n \Delta W_n.$$

We obtain the stability function as

$$R(\bar{h},k) = |1+\bar{h}|^2 + |k\bar{h}|.$$

Fortunately, the function depends on $\bar{h}$ and $|k|$, not on k. Therefore, we obtain the domain of MS-stability in the three-dimensional space of $(\bar{h},|k|)$.

Next is the stochastic θ-method (6). Note, we assume the implicitness only on the drift term f. A calculation leads to the stability function

$$R(\bar{h},k,\theta) = \frac{|1+(1-\theta)\bar{h}|^2 + |k\bar{h}|}{|1-\theta\bar{h}|^2}.$$

By comparing the regions of MS-stability of the Euler–Maruyama and the semi-implicit Euler schemes under the restriction of real $\bar{h}$ and k we can see that the latter is superior to the former with respect to the stability. Further discussion is carried out in [11].

2.5. T-stability

From the viewpoint of computer implementation, MS-stability may still cause difficulty. To evaluate the quantity of the expectation $Y_n = \mathbf{E}(|y_n|^2)$ where y_n is an approximating sequence of the solution sample path, in a certain probability y_n *happens to overflow* in computer simulations. This actually violates the evaluation of y_n.

The above situation suggests an introduction of another stability notion with respect to the approximate sequence of sample path (trajectory). It must take into account the *driving process,* whose way of realization a numerical scheme for SDE requires for the increment ΔW_n of the Wiener process. For example, in the Euler–Maruyama scheme given in (6) ΔW_n, which stands for $W(t_{n+1}) - W(t_n)$, can be exactly realized with $\xi_n\sqrt{h}$. More sophisticated schemes need more complicated normal random variables. And these random variables are to be realized through an *approximation* with pseudo-random numbers on computer, for the normal random number requires infinitely many trials. Therefore, we arrive at the following.

Definition 8. Assume that the test equation (4) is stochastically asymptotically stable in the large. The numerical scheme equipped with a specified driving process said to be T-stable if $|y_n| \to 0$ $(n \to \infty)$ holds for the driving process.

The above definition gives rise to another problem: a criterion of T-stability depends not only on the scheme but also on the driving process. It causes our analysis more difficulty. To resolve it, we can employ Lemma 4 again. For example, if the Euler–Maruyama scheme is applied to (4) then the quantity $T(h;\lambda,\mu)$ defined through

$$\log T(h;\lambda,\mu)=\int_{-\infty}^{\infty}\log|1+\lambda h+\mu\sqrt{h}x|\,p(x)\,\mathrm{d}x$$

can stand for the T-stability function of the scheme, for $T(h;\lambda,\mu)<1$ implies the T-stability.

For an illustration, we treat the Euler–Maruyama scheme with three-point random variables. The random variable ζ_n is taken as $U_n\sqrt{h}$ whose probability distribution is given by

$$P(U_n=\pm\sqrt{3})=1/6,\quad P(U_n=0)=\tfrac{2}{3}.$$

Since the density function is discrete, the integral is easily calculated to derive

$$\begin{aligned}A^6(h;\lambda,\mu)&=(1+\lambda h+\mu\sqrt{3h})(1+\lambda h)^4(1+\lambda h-\mu\sqrt{3h})\\&=(1+\lambda h)^4\{(1+\lambda h)^2-3\mu^2h\}.\end{aligned}$$

Similar to the Euler–Maruyama case, we may introduce the T-stability function for other schemes (see [16,17]).

In [5], a more practical restriction of T-stability is introduced. To avoid stability violation due to T-stability function close to 1, for a certain positive constant A less than 1 the scheme is said to be $T(A)$-stable if the T-stability function is smaller than A.

Stability analysis for numerical schemes of SDEs is still in a premature stage, although much work has been devoted to it. One of the present difficulties is, contrary to the ODE case, linear stability on the supermartingale equation cannot straightforwardly be extended to the multi-dimensional case, for then we have two matrices for the drift and the diffusion terms, not necessarily commuting with each other. Therefore, much more study is expected.

3. Fixed stepsize implementation

The first method for solving SDEs numerically was the Euler–Maruyama method which is inefficient due to its strong order of convergence $\frac{1}{2}$. Because of this limitation in order, numerical methods of higher order have been developed. Burrage and Burrage [1] have focussed their attention on stochastic Runge–Kutta methods (SRKs) of the form (for $i=1,\ldots,s$)

$$\begin{aligned}Y_i&=y_n+h\sum_{j=1}^{i-1}a_{ij}f(Y_j)+\sum_{j=1}^{i-1}\left(J_1b_{ij}^{(1)}+\frac{J_{10}}{h}b_{ij}^{(2)}\right)g(Y_j),\\y_{n+1}&=y_n+h\sum_{j=1}^{s}\alpha_jf(Y_j)+\sum_{j=1}^{s}\left(J_1\gamma_j^{(1)}+\frac{J_{10}}{h}\gamma_j^{(2)}\right)g(Y_j).\end{aligned}\tag{9}$$

If the method does not include J_{10}, then the maximum strong order is 1.0; the inclusion of this Stratonovich integral allows methods with strong order greater than 1 to be developed (see [1]). Methods formulated from (9) can be extended for use in the d-Wiener process case (as long as the SDE system coefficients are fully commutative — otherwise the order of the method is reduced to 0.5) by sampling additionally from $2,\ldots,d$ random number generators. One way of overcoming this order reduction is to include commutators in the method formulation (see [3]). However, implementation costs are increased for methods with commutators, due to the expense of calculating derivatives, leading to the development of suitable methods without commutators [6].

A fixed stepsize implementation of a SRK involves sampling the random variables in the method (represented by J_1 and J_{10}, for example). The built-in random number generator that produces samples from a $N(0,1)$-distribution can be used; an alternative is to obtain samples from the uniform distribution and to use the Polar–Marsaglia technique to produce pairs of normally distributed random numbers. Thus, given a pair (g_1, g_2) of normally distributed random variables, $J_1 = \sqrt{h}g_1$, $J_{10} = h^{3/2}(g_1 + g_2/\sqrt{3})/2$.

With an initial value for the SDE, and with the means of sampling the necessary random variables, the numerical method can be implemented step by step to obtain a trajectory of the solution. However, fixed stepsize implementations of numerical methods have limitations when, for example, the SDE being solved is stiff in some subdomain of the integration as this forces the stepsize to be very small for the entire range of the integration. Thus it is natural to adapt the implementation technique to use a variable stepsize, and it is this approach that is discussed in the next section.

4. Variable stepsize implementation

In order to use a variable stepsize technique, it is necessary to estimate the error at each step so that a new and appropriate stepsize can be determined. This error estimation must be cheap, and in this paper the errors are estimated via the process of embedding. In this paper, a two-stage SRK of strong order 1 is embedded within a four-stage SRK of strong order 1.5, and the error at each step is determined by comparing the numerical results from each of the two methods; only two extra function evaluations are required to calculate the update value from the two-stage method, and so the error estimate is achieved with minimal overhead.

Let $\hat{y}_{n+1}$ be the numerical result obtained from the implementation of the s-stage method, and let y_{n+1} be that obtained from the higher stage method (where the methods have order $\hat{p}$ and p, respectively). Then y_{n+1} is used to advance the numerical computation on the next step, while both $\hat{y}_{n+1}$ and y_{n+1} are used to estimate the error. Here it is absolute error that is under consideration. For an m-dimensional system, let tol_i be the tolerance permitted for the ith component; then an error estimate of order $q + \frac{1}{2}$ (where $q = \min(\hat{p}, p)$) is given by

$$\text{error} = \sqrt{\frac{1}{m}\sum_{i=1}^{m}\left(\frac{y_{n+1,i} - \hat{y}_{n+1,i}}{\text{tol}_i}\right)^2}.$$

For the (R2,E1)-embedded pair of methods, in which $q = 1$, we extend the variable stepsize strategy in [9], and decrease the optimal stepsize by a safety factor (for example, fac=0.8) to avoid oscillatory behaviour in the stepsize, and place bounds so that the stepsize does not increase or decrease too quickly. Thus

$$h_{\text{new}} = h\min(\text{facmx}, \max(\text{facmn}, \text{fac}(1/\text{error})^{2/3})), \tag{10}$$

where facmx and facmn are the maximal and minimal stepsize scaling factors allowed, respectively, for the problem being solved.

The embedded pair used to produce the numerical results in this paper consists of method R2 defined by

$$\begin{array}{c|c} 0 & 0 \\ \frac{2}{3} & 0 \\ \hline \frac{1}{4} & \frac{3}{4} \end{array} \qquad \begin{array}{c|c} 0 & 0 \\ \frac{2}{3}J_1 & 0 \\ \hline \frac{1}{4}J_1 & \frac{3}{4}J_1 \end{array}$$

and the four-stage method E1 given by (9) with parameters

$$A = \begin{pmatrix} 0 & 0 & 0 & 0 \\ \frac{2}{3} & 0 & 0 & 0 \\ \frac{3}{2} & -\frac{1}{3} & 0 & 0 \\ \frac{7}{6} & 0 & 0 & 0 \end{pmatrix}, \quad B^{(1)} = \begin{pmatrix} 0 & 0 & 0 & 0 \\ \frac{2}{3} & 0 & 0 & 0 \\ \frac{1}{2} & \frac{1}{6} & 0 & 0 \\ -\frac{1}{2} & 0 & \frac{1}{2} & 0 \end{pmatrix}, \quad B^{(2)} = \begin{pmatrix} 0 & 0 & 0 & 0 \\ 0 & 0 & 0 & 0 \\ -\frac{2}{3} & 0 & 0 & 0 \\ \frac{1}{6} & \frac{1}{2} & 0 & 0 \end{pmatrix},$$

$$\alpha^{\mathrm{T}} = (\tfrac{1}{4}, \tfrac{3}{4}, -\tfrac{3}{4}, \tfrac{3}{4}), \quad \gamma^{(1)\mathrm{T}} = (-\tfrac{1}{2}, \tfrac{3}{2}, -\tfrac{3}{4}, \tfrac{3}{4}), \quad \gamma^{(2)\mathrm{T}} = (\tfrac{3}{2}, -\tfrac{3}{2}, 0, 0).$$

Most implementations of numerical methods for solving SDEs use a fixed stepsize, and indeed convergence of the method was only demonstrated for such stepsizes. However, recently [7] have proved that a method must have strong order at least 1 to guarantee convergence to the correct solution if variable stepsizes are used. This result demonstrates that the embedded pair (R2,E1) will converge to the correct solution in a variable stepsize implementation.

It is important when using a variable stepsize implementation to remain on the correct Brownian path. The Brownian path consists of the Wiener increments sampled from the $N(0,1)$ distribution; these increments are scaled according to the stepsize currently being used, so when a stepsize is rejected, the increment must be rescaled in such a way that the integration remains on the true path. This approach ensures that the same Brownian path can be traversed if the numerical calculations are repeated with a different initial value or a different initial stepsize.

The approach in [7] was to use a Brownian tree of increments in the integration. The tree was formed by fixing a top level of increments with a nominal stepsize h_0 and then successively halving the stepsize and calculating the new increments on the subintervals so that the top level path was adhered to. These increments accounted for J_1, while any higher-order Stratonovich integrals (for example, J_{10}) could be calculated using a Lévy area (see [14,7,3]). At any stage of the integration, if the current stepsize h was rejected, the step would be retried with a stepsize of $h/2$, while if the step was successful the next step would proceed with either h or $2h$, depending on the alignment of the current position within the tree. This binary tree structure necessitates only a halving or doubling of stepsize, and in practice this can be too restrictive.

Another approach is in [15] who demonstrates that, given J_1 and J_{10} on a fixed Brownian path, then for $0 < h_1 < h$ and $h_2 = h - h_1$,

$$\begin{pmatrix} J_1(t_0, t_0 + h_1) \\ J_{10}(t_0, t_0 + h_1) \\ J_1(t_0 + h_1, t_0 + h) \\ J_{10}(t_0 + h_1, t_0 + h) \end{pmatrix} = AU \begin{pmatrix} N_1 \\ N_2 \\ A_h^{-1} \begin{pmatrix} j_1 \\ j_{10} \end{pmatrix} \end{pmatrix}, \tag{11}$$

is also on the same Brownian path; here $(j_1, j_{10})^{\mathrm{T}}$ are the sampled values corresponding to $J_1(t_0, t_0 + h) = \int_{t_0}^{t_0+h} \circ \mathrm{d}W_s^{(1)}$ and $J_{10}(t_0, t_0+h) = \int_{t_0}^{t_0+h} \int_{t_0}^{s} \circ \mathrm{d}W_{s_1}^{(1)} \mathrm{d}s$ respectively, $N_1, N_2 \sim N(0,1)$,

$$A_h = \begin{pmatrix} \sqrt{h} & 0 \\ \frac{1}{2}h^{3/2} & 1/2\sqrt{3}h^{3/2} \end{pmatrix}, \quad A = \begin{pmatrix} A_{h_1} & 0 \\ 0 & A_{h_2} \end{pmatrix}, \quad \theta = \frac{h_2}{h_1},$$

$$U = \begin{pmatrix} 0 & -(\theta^2 - \theta + 1)/c_2 & 1/c_3 & \sqrt{3}\theta/c_4 \\ \theta^{3/2}/c_1 & \sqrt{3}/c_2 & 0 & 1/c_4 \\ 0 & 1 - \theta + \theta^2/c_2\sqrt{\theta} & \sqrt{\theta}/c_3 & -\sqrt{3}\theta/c_4 \\ -1/c_1 & \sqrt{3}\theta^{3/2}/c_2 & 0 & \theta^{3/2}/c_4 \end{pmatrix},$$

$$c_1 = \sqrt{\theta^3 + 1}, \quad c_2 = \sqrt{\frac{(1 - \theta + \theta^2)(1 + \theta)^3}{\theta}}, \quad c_3 = \sqrt{\theta + 1}, \quad c_4 = \sqrt{(\theta + 1)^3}.$$

Setting $h_1 = h_2 = h/2$ yields the transformation required when a simple halving of h takes place. Indeed, Mauthner [15] only develops this latter case, due to the ease of storing the simulated values in a binary tree as well as the reduced cost associated with their simulation.

However, in this paper, the case with arbitrary stepsize change is developed as this provides the most flexibility for a variable stepsize implementation. First, the Brownian path is fixed for a nominated stepsize h_{fix} — this can represent a series of output points, for example. If this stepsize is the maximum allowed for the integration, then all subsequent simulations are generated 'downwards'; however, if the integration requires $h > h_{\text{fix}}$, the simulated Stratonovich integrals can just as easily be generated 'upwards' from the fixed path. Given the fixed Brownian path, the integration proceeds, using the desired stepsize h_1; the values of J_1 and J_{10} on these subintervals do not need to be stored — they can be merely generated as required based on the fixed path. At the end of the integration, the sum of the J_1-values along the path actually followed equals the sum of the J_1-values along the fixed path. Similarly, the J_{10}-values adhere to the definition

$$\begin{aligned} J_{10}(t_1, t_3) &= \int_{t_1}^{t_3} \int_{t_1}^{s} \circ \mathrm{d}W_{s_1} \mathrm{d}s = \int_{t_1}^{t_2} \int_{t_1}^{s} \circ \mathrm{d}W_{s_1} \mathrm{d}s + \int_{t_2}^{t_3} \int_{t_1}^{s} \circ \mathrm{d}W_{s_1} \mathrm{d}s, \\ &= J_{10}(t_1, t_2) + \int_{t_2}^{t_3} \left(\int_{t_1}^{t_2} \circ \mathrm{d}W_{s_1} + \int_{t_2}^{s} \circ \mathrm{d}W_{s_1} \right) \mathrm{d}s, \\ &= J_{10}(t_1, t_2) + J_{10}(t_2, t_3) + (t_3 - t_2) J_1(t_1, t_2), \end{aligned}$$

for the subintervals $[t_1, t_3] = [t_1, t_2] \cup [t_2, t_3]$. Further details using this approach, together with other examples, can be found in [4].

This section will conclude with the presentation of an example which demonstrates the efficacy of the variable stepsize approach.

Example 9. This SDE is taken from [12], (Eq. 4.4.46) and has been converted to Stratonovich form

$$\mathrm{d}y = -\alpha(1 - y^2)\mathrm{d}t + \beta(1 - y^2) \circ \mathrm{d}W$$

with $\alpha = 1$ and $\beta =$ each of 0.8, 1.5 and 2.0. The fixed stepsize results (using method E1) are presented in Table 1, with the variable implementation results for a range of tolerances in Table 2

Table 1
Fixed stepsize

$\beta = 0.8$			$\beta = 1.5$			$\beta = 2.0$		
h	Error	Steps	h	Error	Steps	h	Error	Steps
$\frac{1}{3}$	—	30	$\frac{1}{4}$	—	40	$\frac{1}{5}$	—	50
$\frac{1}{5}$	8.88(−5)	50	$\frac{1}{8}$	—	80	$\frac{1}{10}$	—	100
$\frac{1}{10}$	6.67(−5)	100	$\frac{1}{24}$	2.51(−2)	240	$\frac{1}{35}$	3.94(−2)	350
$\frac{1}{36}$	1.01(−5)	360	$\frac{1}{84}$	6.90(−3)	840	$\frac{1}{140}$	8.30(−3)	1400

Table 2
Variable stepsize

	$\beta = 0.8$			$\beta = 1.5$			$\beta = 2.0$		
Tol	Error	Tried	OK	Error	Tried	OK	Error	Tried	OK
0.1	1.63(−2)	29	23	2.17(−2)	44	35	2.52(−2)	105	78
0.01	1.86(−3)	47	37	1.89(−3)	81	61	3.08(−3)	112	82
0.001	1.58(−4)	114	86	1.78(−4)	261	191	2.69(−4)	399	289
0.0001	1.64(−5)	383	279	1.61(−5)	961	694	3.00(−5)	1588	1143

(average steps tried and steps accepted are given too). The initial value is 0, the integration is carried out from 0 to 10, and for the variable implementation an arbitrary initial stepsize of $\frac{1}{32}$ was used. The results were averaged over 100 trajectories.

5. Conclusions

It is clear from the discussion in this paper that stability is a critical aspect in designing useful numerical methods. Just as crucial, and what has been given less attention until recently, is that any effective implementation must consider a number of important issues, one of which is a variable stepsize implementation (under the proviso that different numerical simulations must follow the same Brownian path).

The numerical results in this paper have demonstrated that the variable stepsize implementation is far superior to that of fixed stepsize unless the stochasticity is small enough (e.g., $\beta = 0.8$) for the numerical solution to be smooth (in which case any variable stepsize implementation does not have a chance to perform under conditions suited to it). Also, although there is not exact tolerance proportionality when the tolerance is reduced by a factor of 10, the decrease in error is nearly in proportion. Clearly, our approach and the approach in [4] is very promising and offers great flexibility.

References

[1] K. Burrage, P.M. Burrage, High strong order explicit Runge–Kutta methods for stochastic ordinary differential equations, Appl. Numer. Math. 22 (1996) 81–101.

[2] K. Burrage, P.M. Burrage, General order conditions for stochastic Runge–Kutta methods for both commuting and non-commuting stochastic differential equation systems, Appl. Numer. Math. 28 (1998) 161–177.

[3] K. Burrage, P.M. Burrage, High strong order methods for non-commutative stochastic ordinary differential equation systems and the Magnus formula, in: S. Chen, L. Margolin, D. Sharp (Eds.), Predictability: Quantifying Uncertainty in Models of Complex Phenomena, Elsevier Science B.V., The Netherlands, Physica D 133(1–4) (1999) 34–48.

[4] K. Burrage, P.M. Burrage, A variable stepsize implementation for stochastic differential equations (2000), submitted for publication.

[5] K. Burrage, T.-H. Tian, The composite Euler method for stiff stochastic differential equations (2000), submitted for publication.

[6] K. Burrage, P.M. Burrage, Without-commutator methods for stochastic ordinary differential equations, in preparation.

[7] J. G. Gaines, T.J. Lyons, Variable step size control in the numerical solution of stochastic differential equations, SIAM J. App. Math 57 (5) (1997) 1455–1484.

[8] T. C. Gard, Introduction to Stochastic Differential Equations, Marcel Dekker, New York, 1988.

[9] E. Hairer, S.P. Nørsett, G. Wanner, Solving Ordinary Differential Equations I, Nonstiff Problems, 2nd Edition (revised), Springer, Berlin, 1993.

[10] R.Z. Has'minskii, Stochastic Stability of Differential Equations, Stijthoff & Noordhoff, Alphen a/d Rijn, 1980.

[11] D. J. Higham, Mean-square and asymptotic stability of numerical methods for stochastic ordinary differential equations, University of Strathclyde Mathematics Research Report, Vol. 39, 1998.

[12] P.E. Kloeden, E. Platen, Numerical Solution of Stochastic Differential Equations, Springer, Berlin, 1992.

[13] Y. Komori, T. Mitsui, H. Sugiura, Rooted tree analysis of the order conditions of ROW-type scheme for stochastic differential equations, BIT 37 (1) (1997) 43–66.

[14] P. Levy, Processus Stochastiques et Mouvement Brownien, Monographies des Probabilités, Gauthier-Villars, Paris, 1948.

[15] S. Mauthner, Thesis, Chapter 4, private communication, 1999.

[16] T. Mitsui, Stability analysis of numerical solution of stochastic differential equations, Kokyuroku (Res. Inst. Math. Sci., Kyoto Univ.) 850 (1995) 124–138.

[17] Y. Saito, T. Mitsui, T-stability of numerical scheme for stochastic differential equations, in: R.P. Agarwal (Ed.), Contributions in Numerical Mathematics, World Scientific Series in Applicable Analysis, Vol. 2, World Scientific, Singapore, 1993, pp. 333–344.

[18] Y. Saito, T. Mitsui, Stability analysis of numerical schemes for stochastic differential equations, SIAM J. Numer. Anal. 33 (1996) 2254–2267.

ELSEVIER

Journal of Computational and Applied Mathematics 125 (2000) 183–199

JOURNAL OF
COMPUTATIONAL AND
APPLIED MATHEMATICS

www.elsevier.nl/locate/cam

Numerical modelling in biosciences using delay differential equations

Gennadii A. Bocharov[a,*,1], Fathalla A. Rihan[b,2]

[a] *Institute of Numerical Mathematics, Russian Academy of Sciences, 117951 Moscow, Russia*
[b] *Department of Mathematics, The University of Manchester, Manchester M13 9PL, England, UK*

Received 10 December 1999

Abstract

Our principal purposes here are (i) to consider, from the perspective of applied mathematics, models of phenomena in the biosciences that are based on delay differential equations and for which numerical approaches are a major tool in understanding their dynamics, (ii) to review the application of numerical techniques to investigate these models. We show that there are prima facie reasons for using such models: (i) they have a richer mathematical framework (compared with ordinary differential equations) for the analysis of biosystem dynamics, (ii) they display better consistency with the nature of certain biological processes and predictive results. We analyze both the qualitative and quantitative role that delays play in basic time-lag models proposed in population dynamics, epidemiology, physiology, immunology, neural networks and cell kinetics. We then indicate suitable computational techniques for the numerical treatment of mathematical problems emerging in the biosciences, comparing them with those implemented by the bio-modellers.

Keywords: Delay differential equations; Biological systems; Numerical modelling; Parameter estimation

1. Introduction

Retarded functional differential equations (RFDEs),

$$y'(t) = f\left(t, y(t), y(\alpha(t, y(t))), \int_{-\infty}^{t} \mathcal{K}(t, s, y(t), y(s))\,\mathrm{d}s\right), \quad t \geqslant t_0, \tag{1}$$

wherein $\alpha(t, y(t)) \leqslant t$ and $y(t) = \psi(t)$, $t \leqslant t_0$, form a class of equations which is, in some sense, between ordinary differential equations (ODEs) and time-dependent partial differential equations (PDEs). Such retarded equations generate infinite-dimensional dynamical systems. RFDEs (1) where

* Corresponding author.

[1] Supported by the Alexander von Humboldt-Stiftung, Germany.

[2] Supported by the Egyptian Government.

0377-0427/00/$ - see front matter
PII: S0377-0427(00)00468-4

the integral term is absent are usually called delay differential equations (DDEs) and they assume forms such as $y'(t)=f(t,y(t),y(\alpha(t,y(t))))$ with $\alpha(t,y(t))\leqslant t$. The introduction of the "lagging" or "retarded" argument $\alpha(t,y(t))$ is to reflect an "after-effect".

Two early references for DDEs are the books by Bellman and Cooke [4], and Elsgol'ts and Norkin [8]. These are rich sources for analytical techniques and many interesting examples. Kolmanovskii et al. [14,15] gave a rigorous treatment of a wide class of problems. Starting from the first edition, the monograph of Hale [12] (subsequently Hale and Verduyn Lunel [13]) is a standard source on the theory of delay equations. Another substantial monograph is by Diekmann et al. [6]. Kuang [16] and Banks [3] pay particular attention to problems in population dynamics; the former also looked at "neutral" equations. Marchuk [18] presented various issues of numerical modelling with delay equations in immunology. Gopalsamy [9] and Györi and Ladas [10] addressed the question of oscillations in delay differential equations. Early books by Cushing [5], Driver [7], Halanay [11], MacDonald [1, Ref. 111], [17], May [19], Maynard Smith [20], and Waltman [22] have been very stimulating for the development of the field.

An early use of DDEs was to describe technical devices, e.g., control circuits, where the delay is a measurable physical quantity (for example, the time that the signal takes to travel to the controlled object and return plus the reaction time). In most applications in the life sciences a delay is introduced when there are some hidden variables and processes which are not well understood but are known to cause a time-lag (see [3] Cooke and Grossman [1, Ref. 45], Murray [21]). A well-known example is the phenomenon of Cheyne–Stokes breathing: some people show, under constant conditions, periodic oscillations of breathing frequency [21]. This behaviour is considered to be caused by a delay in the physiological circuit controlling the carbon dioxide level in the blood. However, in some cases, e.g., in simplistic ecological models, it seems that delays have been introduced rather ad hoc, thus putting subsequent researchers on the wrong track (Cushing [1, Ref. 49]). Research on qualitative analysis of RFDEs has profited greatly by considering models from theoretical biology. In order to develop appropriate computational strategies, numerical analysts should classify the various types of mathematical problems with delay (there are point delays, distributed delays, state-dependent delays, integrals within or taken over the delay) and it is helpful to follow recent developments in the life sciences to see what problems require further study. We shall be somewhat selective in the areas of biomathematics that we detail, describing qualitative and quantitative studies based on DDEs.

2. Delay equations and population dynamics

A standard delay model for population dynamics was introduced by Hutchinson [1, Ref. 93], when he modified the classical model of Verhulst to account for hatching and maturation periods

$$y'(t)=ry(t)\left(1-\frac{y(t-\tau)}{K}\right). \tag{2}$$

Here the nonnegative parameters r and K are known, respectively, as the intrinsic growth rate and the environmental carrying capacity. Although Eq. (2) appears at first sight to be simple, the solution can display complicated dynamics and, in particular, an oscillatory behaviour. The basic assumption

[3] As the list of publications referred to in this paper exceeds space limitations, we cite the bibliography in [1]: [1, Ref. XX] refers to citation [XX] given in [1].

underlying Eq. (2) is that the present change in population size depends exactly on the population size of time τ units earlier. Commencing with the early work of Volterra [1, Ref. 172], modellers considered more general equations

$$y'(t) = ry(t)\left(1 - \frac{1}{K}\int_{-\tau}^{0} y(t+s)\,\mathrm{d}\sigma(s)\right). \tag{3}$$

One may formally ask: Why does the delay enter the removal term $-y^2/K$ and not the production term y, or both terms, as suggested, for example, in [3]? This question was examined by applying the theory of structured populations by Hadeler and Bocharov in [1, Ref. 35], where the connection between some models of population dynamics using neutral delay differential equations and the widely accepted Sharpe–Lotka–McKendrick model and its extension due to Gurtin and MacCamy [1, Ref. 131],

$$u_t(t,a) + u_a(t,a) + \mu(a,W)u(t,a) = 0, \quad u(t,0) = \int_0^{\infty} b(a,W)u(t,a)\,\mathrm{d}a \tag{4}$$

was studied. This hyperbolic PDE provides the standard model in the theory of age-structured populations. In Eq. (4) $u(t,a)$ stands for the density of the population with respect to age a, the mortality μ and the fertility b depend on age and on some functional W of the population density, traditionally the total population size $W(t) = \int_0^{\infty} u(t,a)\,\mathrm{d}a$. Under the assumption that (i) there is a maturity age $\tau > 0$, separating juveniles from adults; (ii) μ is a step function $\mu(a) = \mu_0 + (\mu_1 - \mu_0)H_\tau(a)$; and (iii) b is a combination of a step function and a sharp peak $b(a) = b_1 H_\tau(a) + b_2\delta_\tau(a)$, where $H_\tau(\cdot)$, $\delta_\tau(\cdot)$ denote the Heaviside and delta function, respectively, it was shown that for $t \geqslant \tau$, the populations of juveniles $U(t) = \int_0^{\tau} u(t,a)\,\mathrm{d}a$ and adults $V(t) = \int_{\tau}^{\infty} u(t,a)\,\mathrm{d}a$ satisfy a system of neutral DDEs:

$$\begin{aligned} U'(t) &= b_1 V(t) + (b_2 - 1)(b_1 + b_2\mu_1)\mathrm{e}^{-\mu_0\tau}V(t-\tau) \\ &\quad + (b_2 - 1) \times b_2\mathrm{e}^{-\mu_0\tau}V'(t-\tau) - \mu_0 U(t), \end{aligned} \tag{5}$$

$$V'(t) = ((b_1 + b_2\mu_1)V(t-\tau) + b_2 V'(t-\tau))\mathrm{e}^{-\mu_0\tau} - \mu_1 V(t). \tag{6}$$

For $t \in [0,\tau]$, the variables $U(t)$ and $V(t)$ satisfy a nonautonomous system of ODEs and this time interval of length τ is "needed" to "forget" the information contained in the initial data for the PDE (4) (see [1, Ref. 35] for further details). The neutral character of Eq. (6) for the adult population is a consequence of the fertility peak at age τ. If this peak is absent, i.e., $b_2=0$, then one gets the standard DDE: $V'(t) = b_1 V(t-\tau)\mathrm{e}^{-\mu_0\tau} - \mu_1 V(t)$. A similar approach was applied to yield a nonlinear equation with state-dependent b_1, μ_1. For example, if one assumes that the birth and the death coefficients depend on W, defined above, and chooses $W = V$, then instead of the previous equation one has

$$V'(t) = b_1(V(t-\tau))V(t-\tau)\mathrm{e}^{-\mu_0\tau} - \mu_1(V(t))V(t).$$

An equation of this form has been used in modelling an oscillating insect population [1, Refs. 80,141]. A framework for deriving delay models for age-structured populations and further references can be found in [1, Refs. 35,77,162,165].

3. Qualitative features of delay equations

It is generally accepted that the presence of delays in biological models is a potent source of nonstationary phenomena such as periodic oscillations and instabilities. This can manifest itself as the loss of stability of an otherwise stable-steady state if the delay exceeds a certain threshold related to the dominant time-scale of a system. However, there also exists evidence that a time delay can sometimes enhance stability, and short delays were shown to stabilize an otherwise unstable dynamical system discussed in [19], [1, Refs. 141,143]. Recently, it has been suggested that delays can damp out oscillations; this was shown with models for coupled oscillators under the condition that the delays in mutual interactions exceed a threshold value [1, Refs. 152,164].

A simple delay model of cell population growth is given by the linear DDE ([1, Ref. 13]) $y'(t) = \alpha y(t) + \beta y(t - \tau)$. This equation is used as a standard test equation in the analysis of numerical methods for DDEs. Its qualitative behaviour is well understood and can be summarized as follows: the equilibrium solution $y(t) \equiv 0$ becomes unstable when the value of the delay exceeds the threshold given by $\tau^* = \cos^{-1}[-\alpha/\beta]/(\sqrt{\beta^2 - \alpha^2})$ and there is a solution that demonstrates oscillatory behaviour, with a period $T = 2\pi/(\sqrt{\beta^2 - \alpha^2})$.

For a long time, the classical delayed logistic equation (2) was a subject of qualitative and numerical studies in mathematical biology (see [3,17,18,21]). Its solution $y(t)$ converges monotonically to the carrying capacity K for $0 < r\tau < e^{-1}$; it converges to K in an oscillatory fashion for $e^{-1} < r\tau < \pi/2$; it oscillates in a stable limit cycle pattern for $\tau > \pi/2$. Eq. (2) assumes (by a simple re-scaling of the variables) the form

$$y'(t) = -\alpha(1 + y(t))y(t - \tau) \quad (y(t) > 0, \quad t \geqslant -\tau), \tag{7}$$

known as Wright's equation, which has been investigated in number theory. With $x(t) = \ln\{y(t)\}$, and $f(x) = e^x - 1$, Eq. (7) can be further transformed into $x'(t) = -\alpha f(x(t - \tau))$. In the early 1970s equations of this form became the standard subject for qualitative analysis; see [1, Refs. 56,145,174]. Extending the last equation by a feedback term one arrives at the equation

$$w'(t) = -vw(t) + \alpha f(w(t - \tau)), \tag{8}$$

which was used to explain bursting in neurons by delays and also used as a model of blood-cell dynamics (see earlier work by an der Heiden, Glass, Mackey, Milton [1, Refs. 2,113,115]). For this equation, the existence of nontrivial periodic solutions has been shown by Hadeler and Tomiuk [1, Ref. 82]. For a fixed $\tau > 0$ and every $v \geqslant 0$, there is a critical α_v such that the zero solution is stable for $\alpha \in (-v, \alpha_v)$, and unstable for $\alpha > \alpha_v$. For $\tau = 1$ and $v = 0$, the critical value is $\alpha_0 = \pi/2$. Thus, for $\alpha > \alpha_v$, the constant solution becomes unstable, and a stable periodic solution appears. This transition can be treated as a Hopf bifurcation.

Gopalsamy [1, Ref. 74] considered the linear NDDE of the form

$$y'(t) + by'(t - \sigma) + ay(t - \tau) = 0. \tag{9}$$

He proved that if a, b, τ, and σ are nonnegative constants and $[a/(1 + b)](\tau - \sigma)e > 1$, then bounded solutions of (9) are oscillatory. This result was extended to the nonlinear case $y'(t) + f(y'(t - \sigma)) + g(y(t - \tau)) = 0$, where f and g are continuous functions. Under the conditions $yf(y) > 0$ and $yg(y) > 0$ for $y \neq 0$, $0 \leqslant f(x)/x < b \leqslant 1$, $g(x)/x \geqslant a > 0$, and $[a/(1 + b)](\tau - \sigma)e > 1$, all bounded solutions of this equation are oscillatory.

So far, we have discussed delay equations with constant time-lag τ. One can imagine situations (e.g., remote control problems) where the delay is not constant but depends on the state of the system. If such state-dependence is introduced, we get a state-dependent DDE, $y'(t) = f(y(t - \tau(y(t))))$, where $\tau : \mathbb{R} \to [0, \bar{\tau}]$ is a given function, with some upper bound $\bar{\tau}$. It was shown by Mallet–Paret and Nussbaum [1, Ref. 118] that for the main branch of periodic solutions, the equation with state-dependent delay behaves about the same as a constant delay equation. In some biological applications the delay itself can be governed by a differential equation that represents adaption to the system state. Such systems of the form: $y'(t) = f(y(t-\tau(t)))$; $\tau'(t) = g(y(t), \tau(t))$, with $g(y, \tau) = \tilde{g}(y) - \tau$ have been studied recently by Arino and Hadeler in [1, Ref. 4].

Although infinite delays are biologically unrealistic, it is sometimes mathematically convenient to allow for arbitrarily large delays, as in the RFDE $y'(t) = f(\int_{-\infty}^{0} y(t+s)\,\mathrm{d}\sigma(s))$. A particular class of problems consists of systems of equations where the weight function σ is an exponential polynomial $\sigma(s) = \mathrm{e}^{-s}$; these could be reduced to a system of ODEs: $y'(t) = f(z(t))$, $z'(t) = y(t) + z(t)$, where $z(t) = \int_{-\infty}^{0} \mathrm{e}^{-s} y(t+s)\,\mathrm{d}s$; see Fargue [1, Ref. 66] and Wörz-Busekros [1, Ref. 177]. We note in passing that distributed delay models with a gamma-distribution function $F_m(t) = ((t^{m-1}a^m)/(m-1)!)\mathrm{e}^{-at}$ used as the kernel function, with real-valued parameter a and integer-valued parameter m are quite popular in biomodelling. One of the reasons is that the corresponding system of integro–differential equations can be transformed to an equivalent system of ODEs through a linear chain trick technique [1, Ref. 111], [17].

The authors are indebted to Prof. Hadeler (Tübingen) for his input (private communication) to the above.

4. Numerical studies

Numerical studies using mathematical models with delays are undertaken in various branches of biosciences in order to understand the system dynamics, estimate relevant parameters from data, test competing hypotheses, assess the sensitivity to changes in parameters or variations in data and optimize its performance with the least possible cost. These objectives are associated with an increasing complexity of the numerical procedures.

4.1. Ecology

Mathematical studies using delay models of ecological and chemostat systems are built upon various generalizations of Volterra's integro-differential system of predator–prey dynamics:

$$\begin{aligned} y_1'(t) &= b_1 y_1 \left(1 - c_{11} y_1 - c_{12} \int_{-\infty}^{t} y_2(s) k_1(t-s)\,\mathrm{d}s\right), \\ y_2'(t) &= b_2 y_2 \left(-1 + c_{21} \int_{-\infty}^{t} y_1(s) k_2(t-s)\,\mathrm{d}s\right), \end{aligned} \tag{10}$$

where $y_1(t), y_2(t)$ represent the populations of the prey and the predator [4] (see [5]). These equations can be extended naturally to describe the dynamics of multi-species ecological systems. In chemostat models the delay indicates that the growth of a species depends on the past concentration of nutrients. One can, however, face some difficulties in introducing delays in chemostat models as reported in Cunningham and Nisbet [1, Ref. 47]. For early studies of the chemostat see references cited in the book by Smith and Waltman [1, Ref. 163] and the recent exposition of delay models given in Wolkowicz et al. [1, Ref. 180]. Numerical modelling was used to study the behaviour of

[4] There are variations of these equations, including forms with differing limits of integration and forms that incorporate Stieltjes integrals, in the literature.

periodic orbits around the stability–instability boundary. It was reported that the numerical simulations provided evidence that the models with distributed delays are more realistic and accurate in reproducing the observed dynamics [1, Ref. 158].

Various classes of differential equations are used as building blocks for increasingly complex models, with a recent example from parasitology [1, Ref. 36]. The mechanism of oscillations in the populations of hosts and parasitoids was studied with a mixed model that combines the McKendrick–von Foerster equation (PDE) for the juvenile host population, an ODE for the adult hosts and a DDE for the adult parasitoid population. Numerical simulations suggested that it is the delayed density dependence in the parasitoid birth rate that can induce the cycles, in addition to the classic Lotka–Volterra consumer-resource oscillations. Other examples are reaction–diffusion systems with delays in the reaction terms used to model the Lotka–Volterra competition system [1, Ref. 75]. It was found numerically that the stability diagrams in the case of fixed delays have a much more complicated structure than for gamma-distribution delays.

4.2. Epidemiology

In modelling the spread of infections the population is usually considered to be subdivided into disjoint epidemiological classes (or compartments) of individuals in relation to the infectious disease: susceptible individuals, S, exposed individuals, E, infectious individuals, I and removed individuals, R. The development of the infection is represented by transitions between these classes. The assumption that individuals remain for a constant length of time in any of the compartments leads to DDEs. There are examples of delay models using fixed delays to represent the duration of the infectious period (*SIS*-model [1, Ref. 91]), the immune period (*SIRS*-model [1, Ref. 92]), or the periods of latency and temporary immunity (*SEIRS*-model [1, Ref. 44]). A distributed duration of the latent period was considered in a distributed delay *SIR*-model [1, Ref. 23]. In epidemic models that seek to take into account the age structure of the population, the delay represents the maturation period [1, Ref. 79]. It is considered that the major effect of delays in the epidemic models is to make them less stable than the analogous models without delays. Numerical studies are usually carried out to support analytical results and provide some insight into more general situations which are difficult to treat analytically. For references on epidemic models with delays we refer to the recent paper by Hethcote and van der Driessche [1, Ref. 91].

4.3. Immunology

Immunology presents many examples of mathematical models formulated using DDEs (see [1, Refs. 50,51,121,122,130,151] for references). Marchuk and associates [1, Refs. 9,121], developed a hierarchy of immune response models of increasing complexity to account for the various details of the within-host defense responses. The delays are used to represent the time needed for immune cells to divide, mature, or become "destined to die". The numerical approaches to parameter estimation allowed one to quantify relevant parameters of pathogen–host interactions for human infections caused by influenza A [1, Ref. 32], hepatitis B viruses [1, Refs. 124, 125, 161], bacterial infections in lungs [1, Ref. 101], mixed infections [1, Ref. 127] and murine LCMV and influenza infections [1, Refs. 30, 59, 121]. In [1, Refs. 34, 121], two adaptive numerical codes were developed, based (i) on embedded Runge–Kutta–Fehlberg methods of order 4 and 5 supplemented by the Hermite interpolation, and (ii) on Gear's DIFSUB, with the Nordsieck interpolation technique to approximate the variables with delays.

A recent example of fixed time-lag equations in immunology is the nonlinear model of humoral immune response to *H. influenzae* by Rundell et al. [1, Ref. 160]. The parameter identification problem was treated as a sequence of "reduced" parameter estimation problems by splitting the observation interval into a sequence of smaller subintervals. The numerical approach to the optimal control problem with a simpler version of the model [1, Ref. 159] suggested continuous drug administration, as opposed to the standard approach based on periodic dosages.

In studies of lymphocyte migration through various anatomical compartments by Mohler et al. [1, Refs. 68, 135] the delays represent (i) the time that cells reside in a particular compartment, or (ii) the transit times through compartments, or (iii) the duration of inter-compartmental transfer. The problem addressed numerically was the estimation of the lymphocyte intra- and inter-compartment transfer rates (directional permeabilities) using experimental data on the dynamics of labelled cell radioactivity distribution to various organs of the immune system. For other compartmental delay models see Györi and Ladas [10].

4.4. HIV infection

The key problem in mathematical studies of the within-host dynamics of HIV infection undertaken by Nowak with associates and by Perelson with co-workers is to get reliable estimates for the turnover of virus and infected cells [1, Refs. 89, 132]. It was shown that adding more realism to the models by accounting for intracellular delay in virus production, either in the form of a fixed delay or a gamma-distributed delay [1, Ref. 132], could give better accuracy in estimating the viral clearance rate provided detailed patient data are available. A similar problem of reliable estimation of the HIV turnover rate has been addressed recently in the model by Grossman et al. [1, Ref. 76] that takes into account that virus producing cells die after a certain time-lag rather than at an exponential rate.

4.5. Physiology

The great potential for simple DDEs to capture complex dynamics observed in physiological systems was convincingly shown in a series of related works by an der Heiden, Bélair, Glass, Mackey and co-workers [1, Refs. 1–3, 113, 114, 142]. A key element in the models is an assumption that either the production or elimination rates are nonlinear functions of the past state: $f(y(t-\tau)) = y^m(t-\tau)/(\theta + y^n(t-\tau))$ with $m \leqslant n$ and $n \geqslant 1$. The delay models were used to study unstable patterns (periodic solutions and chaotic regimes) of (i) the human respiration system and regulation of blood concentration of CO_2, (ii) the production of blood cells, (iii) hormone regulation in the endocrine system, and (iv) recurrent inhibition in neural networks. In the respiratory system, delays represent the transport time between the lung and the peripheral and central chemoreceptors [1, Refs. 43, 113, 153]. In the study of periodic haematological diseases, DDEs with time and state-dependent lags have been used to formulate physiologically realistic mathematical models [1, Ref. 117], which were solved by a modified fourth-order *RK* scheme with fixed stepsize and a linear interpolation for the delay variable. The same authors advanced a model of granulopoiesis using a non-linear integro–differential equation [1, Ref. 87]: $y'(t) = -\alpha y(t) + M_0(\int_{-\infty}^{t-\tau_m} y(s)g(t-s)\,ds)$, where the kernel is defined by a gamma-distribution function. It was reported that use of a *noninteger* value of the parameter m in the gamma-distribution function provides a good fit to real data. Mathematically, the transition from a normal state to a disease can be associated with a loss of stability of the unique steady state in the model and a supercritical bifurcation of a periodic solution. Numerical examination of the model allowed the authors to test whether the realistic period and amplitude of oscillations in cell numbers can be obtained under a systematic variation of parameters, within their physiological ranges. To treat the model numerically a trapezoidal scheme was used both to advance the time and to evaluate the integral term.

A model for neural reflex mechanisms [1, Ref. 8] is an example of an implicit DDE: $\varepsilon y'(t) = -\alpha(\varepsilon y'(t))y(t) + f(y(t-\tau))$. The need for such models is related to the fact that neuromuscular reflexes with retarded negative feedback have different rates depending on the direction of movement. Both the qualitative and numerical studies of such equations represent a challenge to be addressed.

4.6. Neural networks

The modelling of neural networks (NNs) is an important area of application of delay equations. In the living nervous system the delays can represent the synaptic processing time or the time for action potential to propagate along axons. In artificial NNs the delays arise from the hardware implementation due to finite switching and transmission times of circuit units. Marcus and Westervelt were the first to include the delay in Hopfield's equations [1, Ref. 126] and various generalizations have been suggested [1, Refs. 27,41,57,90,181]. A "standard" delayed Hopfield NN model assumes the form

$$C_i y_i'(t) = -\frac{y_i(t)}{R_i} + \sum_{j=1}^{n} T_{ij} f_j(y_j(t-\tau_j)) + I_i, \quad i = 1,\dots,N. \tag{11}$$

The origin of instabilities in NNs is in the focus of qualitative and numerical studies, which seek to relate the values of the delay, the network structure/connection topology and the properties of the function f_j in Eq. (11) to the emergence of sustained or transient oscillations. For numerical treatment of DDEs modifications of Euler's method and of Gear's code are reported.

4.7. Cell kinetics

Cell growth provides a rich source of various types of delay models. A biochemical model of the cell cycle, describing the dynamics of concentration of two peptides and their complex, which transforms after some time lag into active maturation promoting factor (MPF), was formulated using fixed delay equations [1, Ref. 38], where the cell division was manifested by periodic fluctuations of MPF. In studies of tumor growth the standard avascular model was modified by incorporating a time-delay factor into the net proliferation rate and numerical and asymptotic techniques were used to show how the tumor growth dynamics is affected by including such delay terms [1, Ref. 39].

Cell populations are, in general, structured by their age, size, etc. The generic means for modelling structured cell populations are provided by first-order hyperbolic PDEs (4). Cell populations, which are made synchronous, have the fertility peak (delta function) at some age τ and exhibit a step-like growth. *Neutral* DDEs have been shown numerically to provide a better qualitative and quantitative consistency with the step-like growth patterns [1, Ref. 13] than do ODEs or DDEs with constant time-lag.

Attention has been paid [1, Refs. 46, 112, 154–156] to the analysis of cell-population dynamics using retarded PDEs of hyperbolic type:

$$\frac{\partial u(t,a)}{\partial t} + \frac{\partial u(t,a)}{\partial a} = f(t, u(t,a), u(t-\tau, h(a))) \tag{12}$$

with $\tau > 0$ and $h(a) < a$, for $a > 0$. The model considers proliferation and maturation processes simultaneously, where the kinetic/reaction terms are dependent on the cell population at a previous time represented by a delay τ and at a previous maturity level specified by $h(a)$. Numerical studies proved to be instructive in getting some insight into the possible dynamics of (12), with a maturation delay as a critical parameter. It was observed that many of the time-dependent modes of the retarded PDE are directly associated with a limit-cycle behaviour in the pure birth-and-death cell population balance equation $z' = f(t, z(t), z(t-\tau))$.

4.8. A stochastic approach

The random perturbations which are present in the real world imply that deterministic equations are often an idealization [1, Refs. 116, 134, 141]. For example, neurological control systems operate in a noisy environment, and the effect of noise needs to be considered in the analysis of the experimental traces of the state variables (such as, the electro-encephalogram, pupil area, displacement of the finger position in patients with Parkinson's disease). To model the dynamics of delay systems under random perturbations, stochastic delay differential equations (SDDEs) are used:

$$\mathrm{d}y(t) = f(t, y(t), y(t-\tau))\,\mathrm{d}t + g(t, y(t))\,\mathrm{d}w(t). \tag{13}$$

Such SDDEs can be driven by white noise ($\mathrm{d}w(t) = \xi(t)\,\mathrm{d}t$, where $\xi(t)$ stands for a stationary Gaussian white noise process) or coloured noise ($\mathrm{d}w(t) = \eta(t)\,\mathrm{d}t$, here $\eta(t)$ is so-called Ornstein–Uhlenbeck process) and the choices of driving processes depend on the real-life phenomenon being modelled. There exist two frameworks, namely the Itô and Stratonovich calculus, to deal with (13). If one argues that the SDDEs are serving as approximations to stochastic difference equations with autocorrelated noise, the Itô calculus may provide the more useful approximation. The Stratonovich framework may be more appropriate when the white noise can be considered as the limiting case of a smooth real noise process.

A well-documented example of a biological system where noise is an important component is the pupil light reflex, which displays complicated dynamics [1, Ref. 109]. Noise is introduced into the reflex at the level of the brain-stem nuclei. The noise correlation time, the system response time and the delay in signal transmission are all of the same order of magnitude, and indicative of coloured noise. The spontaneously occurring aperiodic oscillations in the pupil area were explained with the mathematical model: $y'(t) = -\alpha y(t) + c\theta^n/(\theta^n + y^n(t-\tau)) + k$, by assuming the effect of an additive noise ($k = \bar{k} + \eta(t)$) or multiplicative coloured noise ($c = \bar{c} + \eta(t)$). The numerical simulations provided the only possible means to establish a major role of the noise in the dynamic behaviour. The numerical approximations to sample trajectories $y(t)$ were computed using a combination of an integral Euler method for the equation defining the Ornstein–Uhlenbeck process (η), and a fourth-order RK-method with a linear interpolation formula for the delay terms. The effect of additive white noise on the transitions between different limit-cycle attractors of human postural sway was studied in [1, Ref. 65] using a scalar SDDE. Further discussions about modelling with SDDEs can be found in [1, Refs. 119, 133, 169].

5. Numerical methods for delay equations

We shall embark on a brief review of numerical strategies for DDEs. First we remark that some of those undertaking numerical studies of delay equations in biology devise an indirect approach, rather than use purpose-built numerical codes for DDEs; they try to reduce the study to that of a set of ODEs. Thus they eliminate lag-terms from delay differential equations by introducing additional variables on one of the following bases:

(1) the methods of steps [4] allows one to represent a DDE on successive intervals $[0,\tau],[\tau,2\tau],\ldots,[(N-1)\tau,N\tau]$ by successive systems of ODEs with increasing dimension;
(2) a process represented by a delay can be approximated by introducing a number of intermediate stages using an ODE system to mimic the transition through the stages [1, Refs. 70, 121] (for other strategies see [1, Ref. 69]);
(3) the effect of the time-lag can be modelled by using "gearing up" variables [1, Ref. 52].

We note, however, that the long-term dynamics of DDEs and of approximating finite-dimensional ODEs can differ substantially. There are occasions when (2) (given above) may have appeal, but a familiarity with numerical methods for DDEs will often reap dividends.

5.1. Difference approximation

Numerical methods for ODEs provide approximate values $\tilde{y}(t_i)$ to the solution $y(t_i)$ at a sequence of points $(t_0 < t_1 < t_2 < t_3 \cdots < t_N)$ using estimates of the local truncation error or of the defect. Supplementary approximations provide dense output that defines approximate values $\tilde{y}(t)$ (densely defined) for $t \in [t_0, t_N]$. Such ODE methods can be modified, with varying degrees of success, to provide approximate solutions for DDEs. To indicate the principal features, consider the initial function problem for the system of DDEs with parameter $\boldsymbol{p}$:

$$\boldsymbol{y}'(t) = \boldsymbol{f}(t, \boldsymbol{y}(t), \boldsymbol{y}(t-\tau), \boldsymbol{p}), \quad t \geqslant t_0; \qquad \boldsymbol{y}(t) = \boldsymbol{\psi}(t, \boldsymbol{p}), \quad t \in [t_0 - \tau, t_0], \tag{14}$$

in which $\tau > 0$ does not vary with t and the initial function $\boldsymbol{\psi}$ is specified on the interval $t \in [t_0 - \tau, t_0]$. A simplistic approach to solving system (14) numerically consists of replacing (14) by the ODE: $\boldsymbol{y}'(t) = \boldsymbol{f}(t, \boldsymbol{y}(t), \tilde{\boldsymbol{y}}(t-\tau), \boldsymbol{p})$, for $t \geqslant t_n$, where we assume that $\tilde{\boldsymbol{y}}(t)$ for $t \leqslant t_n$ is computed using dense-output techniques. At the risk of over-simplification, numerical methods for DDEs (derived in this manner) amount, in essence, to a combination of two basic elements: a method π_q for approximation of delayed variables with order q in the spirit of a dense-output routine, and an ODE-based pth order method Ψ_p to advance the solution with a step-size h_n (on the assumption $\tau > h_n$). This said, a third feature of an adaptive algorithm concerns the control of step-size and adaptation of the formulae or their implementation. Some features of delay equations can seriously affect the reliability and performance of a naive numerical method based on a pair (Ψ_p, π_q). In general, the solution to (14) is not smooth and has jump discontinuities in its ith derivatives at times $\zeta_i = t_0 + i\tau$, $i \in N^+$. The effect and propagation of the jump discontinuities in the derivatives of the solution have to be addressed when adapting any ODE solver to the problem with delays [1, Ref. 16]. Theoretical analysis of the convergence and asymptotic error expansion issues of the adapted method (Ψ_p, π_q) tells us that we require $q \geqslant p-1$ in order to retain (asymptotically) the global convergence order and $(q \geqslant p)$ the error expansion form characteristic of the ODE method

[1, Refs. 6,28,34]. The scenario outlined above can be modified to provide numerical methods for a wide range of deterministic retarded differential equations. Note that rigorous development of effective numerical techniques for stochastic DDEs is a relatively unexplored area requiring further attention from numerical analysts; see however [2].

5.2. DDE solvers

From a modeller's viewpoint, two historical periods in the production of numerical codes for delay equations can be distinguished. During the first period, a number of experimental codes were developed by modellers or numerical analysts. Fourth-order RK methods and two-point Hermite-type interpolation polynomials were used by Neves [1, Ref. 140], and algorithms based on fourth- and seventh-order Runge–Kutta–Fehlberg methods together with Hermite interpolation polynomials were presented by Oberle and Pesch [1, Ref. 146]. Thompson [1, Ref. 167] developed a code based on a continuously embedded RK method of Sarafyan [1, Ref. 168]. An algorithm based on a predictor–corrector mode of a one-step collocation method at k Gaussian points has been constructed by Bellen and Zennaro [1, Ref. 20].

The second period can be characterized by the availability of more sophisticated DDE solvers. Recently, numerical analysts have developed a number of professional adaptive solvers (based on LMMs, RK or collocation schemes) producing numerical solutions for a wide range of requested tolerances and various classes of problems with delays. The major problems that the designers of such codes try to accommodate are: automatic location or tracking of the discontinuities in the solution or its derivatives, efficient handling of any "stiffness" (if possible), dense output requirements, control strategy for the local and global error underlying the step-size selection, the cost and consistency of interpolation technique for evaluating delayed terms (to name but a few of them). The code `Archi` [1, Ref. 149] is based on the successful Dormand & Prince fifth-order RK method for ODEs due to Shampine and a fifth-order Hermite interpolant [1, Ref. 146]. In addition to `Archi`, which is available from the internet, we mention `DDESTRIDE` (Baker et al. [1, Ref. 15]), `DELSOL` (Willé and Baker [1, Ref. 179]), `DRKLAG6` (Corwin, Sarafyan and Thomson [1, Ref. 42]), `SNDDELM` (by Jackiewicz and Lo [1, Ref. 97]) and the code of Enright and Hayashi [1, Ref. 62]. The Numerical Algorithms Group (Oxford) supported, in part, the construction of the codes written by Paul (`Archi`) and Willé (`DELSOL`).

5.3. Stiffness

Several authors have reported difficulties, which they identified as due to "stiffness", in the numerical modelling of biological processes using delay equations. An example of a variable stiffness problem appearing in modelling the acute immune response is given [1, Ref. 34]. In simulating hepatitis B infection, the "stiffness" emerges at the peak of acute infection, and is associated with the increase in sizes of lymphocytes and antibody populations (by a factor of about 10^5) that accelerates the damping of virus and infected cells by the same scale. The *BDF*-based codes performed nicely, whereas the Adams- and explicit RK based codes failed to produce a numerical solution after the day indicated because of very small step-sizes required. The recent model of immune response by Rundell et al. [1, Ref. 160] also generates apparently stiff computational problems as one can conclude by analyzing the values of parameters being used and they refer to the stiff solver `ode15s` from the SIMULINK collection.

Stiffness is a phenomenon identified in the numerical solution of ODEs, and is variously defined. It is often characterized in terms of the largest and smallest real parts of the zeros of the stability function corresponding to a stable solution. The main symptom of "stiffness" is that one requires a highly stable numerical formula in order to use large step-sizes reliably [1, Ref. 28]. The same symptom could be used to identify "stiffness" in the delay case. Experimental solvers for stiff DDEs based on LMMs were suggested by Kahaner and Sutherland (see discussion of `SDRIV2` in [1, Ref. 139]), Watanabe and Roth [1, Ref. 176], and those using an implicit RK methods were developed by In 't Hout [1, Ref. 94], and Weiner and Strehmel [1, Ref. 178].

The application of delay equations to biomodelling is in many cases associated with studies of dynamical phenomena like oscillations, Hopf bifurcations, chaotic behaviour [1, Ref. 81]. The analysis of the periodic orbits in delay equations and their discretizations based on the RK methods showed that the discretizations possess invariant curves when step-sizes are sufficiently small [1, Ref. 95]. Further studies of spurious numerical solutions of finite-difference approximations to the delay equations, which can be generated at critical (bifurcation) values of model parameters are needed.

6. Fitting models and parameter estimation

6.1. Objective functions and their continuity

Suppose that the general form of a delay model is given by (14). The task of parameter estimation for such mathematical models is one of minimizing a suitable objective function $\Phi(\boldsymbol{p})$ depending on the unknown parameters $\boldsymbol{p} \in \mathbb{R}^L$ and the observed data $\{\mathfrak{y}_j\}_{j=1}^N$ that represent values $\{y(t_j,\boldsymbol{p})\}_{j=1}^N$. This can additionally include estimating τ, the position of the initial time point t_0 and the parameters of the initial function $\psi(\cdot,\boldsymbol{p})$. Possible objective functions are, for example, the least squares (LS) function $\Phi(\boldsymbol{p}) = \sum_{j=1}^N \sum_{i=1}^M [y^{(i)}(t_j,\boldsymbol{p}) - \mathfrak{y}_j^{(i)}]^2$, or the log-least squares function $\Phi(\boldsymbol{p}) = \sum_{j=1}^N \sum_{i=1}^M [\log((\mathfrak{y}_j^{(i)})/(y^{(i)}(t_j,\boldsymbol{p})))]^2$. The second choice provides metrics in $\mathbb{R}_+^M$ and has been used for parameter estimation of immune responses [1, Refs. 33, 137]. The numerical technique for finding the best-fit parameter values for a given mathematical model and objective function involves solving the model equations for the current values of the parameters to compute $\Phi(\boldsymbol{p})$ with high precision. The parameter values are then adjusted (by a minimization routine, for example `E04UPF` in the `NAG` library, `LMDIF` from `NETLIB` or `FMINS` in `MATLAB`) so as to reduce the value of the objective function (see [1, Refs. 13, 14]). One obvious difficulty with such procedures is that solutions of DDEs are not, in general, differentiable with respect to the delay [1, Refs. 11, 86, 108]. Discontinuities in the solution of a DDE and its derivatives, at points $\{\zeta_i\}$, can come from the initial point t_0 and the initial function $\psi(t,\boldsymbol{p})$, and may propagate into $\Phi(\boldsymbol{p})$ via the solution values $\{y(\zeta_i,\boldsymbol{p})\}$ if $\zeta_i \in \{t_j\}$. Consider for simplicity the scalar equation case. From the formula (and a similar one for the second derivative)

$$\left(\frac{\partial \Phi(\zeta_j,\boldsymbol{p})}{\partial p_l}\right)_{\pm} = 2\sum_{j=1}^{N} [y(\zeta_j,\boldsymbol{p}) - \mathfrak{y}_j] \left(\frac{\partial y(\zeta_j,\boldsymbol{p})}{\partial p_l}\right)_{\pm}, \tag{15}$$

Table 1
Best-fit estimates $\hat{p}$, mean of perturbed parameters $\tilde{p}$ and their nonlinear biases to the model (16), NLB $= (\hat{p}/\tilde{p} - 1) \times 100\%$

Best-fit, standard deviation, nonlinear biases					
$\hat{\tau}$	σ	$\hat{\rho}_1$	σ	$\hat{\rho}_2$	σ
5.45	0.038	0.443	0.014	0.864	0.019
$\tilde{\tau}$	NLB(τ)	$\tilde{\rho}_1$	NLB(ρ_1)	$\tilde{\rho}_2$	NLB(ρ_2)
5.446	0.0066%	0.4426	0.0284%	0.8645	0.0772%

it follows that, unless $\eta_j = y(\zeta_j, \boldsymbol{p})$, jumps can arise in the first (second) partial derivatives of $\Phi(\boldsymbol{p})$ with respect to p_l, if the first (second) partial derivatives of $y(t,\boldsymbol{p})$, with respect to p_l, has a jump at $t = \zeta_j$ (one of the data-points). These jumps can also propagate into the second derivative of $\Phi(\boldsymbol{p})$ if the first derivative of $y(t,\boldsymbol{p})$ with respect to p_l has a jump at one of the data-points $t = \zeta_j$, even when $\eta_j = y(\zeta_j, \boldsymbol{p})$. Therefore, for correct numerical parameter estimation in DDEs attention has to be given to the differentiability of the solution $\boldsymbol{y}(t,\boldsymbol{p})$ with respect to the parameters $\boldsymbol{p}$, and the existence and position of the jump discontinuities.

6.2. Analysis of the best fit: uniqueness, nonlinear bias

A fundamental difference between DDE and ODEs is that solutions corresponding to different initial function data can intersect. Of course, solutions that are computed with different parameters can intersect in both the ODE and DDE case. In the context of the parameter estimation problem, this implies that for a given set of $\{t_j\}_{j=1}^{N}$ and an arbitrary function $\boldsymbol{f}$ in (14), there is no reason to suppose that there exists a *unique* minimizer $\hat{\boldsymbol{p}}$ of $\Phi(\boldsymbol{p})$. A straightforward example is provided by the Hutchinson equation (2): one can observe that for the same initial data a range of different values of the carrying capacity parameter K gives solutions that intersect. If the data correspond to the points of intersection, K is not uniquely determined.

The parameter estimation problem for DDE models is an example of nonlinear regression. The nonlinear regression differs, in general, from linear regression in that the LS parameter estimates can be biased, nonnormally distributed, and have a variance exceeding the minimum variance bound. These characteristics depends on the model, the data and the best fit estimates. It is important to assess the effect of nonlinearity, i.e., the *biases* of parameter estimates. There is a convention that if the bias is $< 1\%$ then the effect of nonlinearity is not significant and estimates of both parameters and their deviations are confident. We give an example of such analysis of estimated parameters for a simple DDE growth model for fission yeast [1, Ref. 13]

$$\begin{aligned} y'(t) &= \rho_1 y(t-\tau), \quad t \geqslant 0, \\ y(t) &= (2.25 y_0 \rho_2/\rho_1) E(t+1.5), \quad t \in [-\tau, 0), \; y(0) = y_0, \end{aligned} \tag{16}$$

where y_0 stands for the initial number of cells, $E(\cdot)$ is a bell-shaped initial distribution function. Estimated are the components of $\boldsymbol{p} = [\rho_1, \rho_2, \tau]$. Fig. 1 shows the best-fit solution and the shape of the LS function in the vicinity of the minima and Table 1 provides an insight in how biased are the best-fit estimates of parameters.

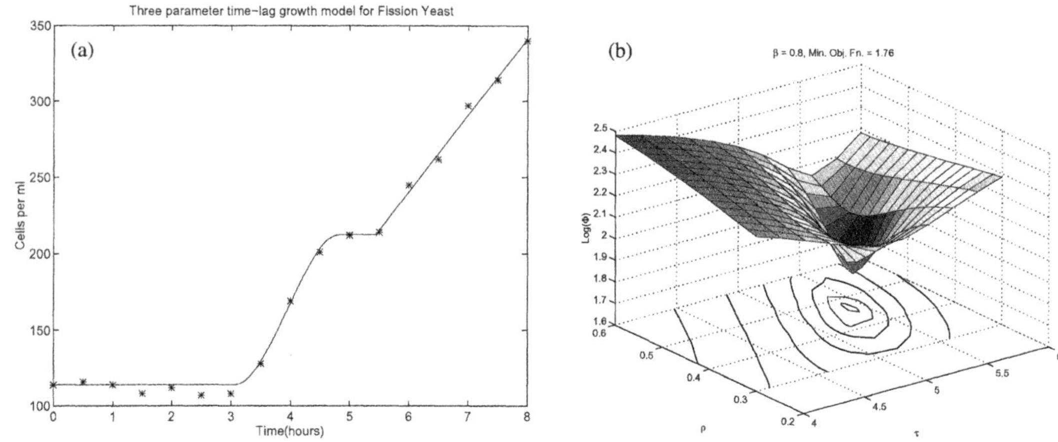

Fig. 1. (a) shows the best fit solution of time-lag model (16) with three parameters, fitted to the observed data. (b) indicates local uniqueness of the best fit and the dependence of Φ on parameters τ, ρ_1 for given $\rho_2 \equiv \beta$.

7. Sensitivity analysis: direct and adjoint methods

Sensitivity analysis (SA) of mathematical models is an important tool for assessing their properties. The following types of sensitivity can be investigated: sensitivity of the solution $\boldsymbol{y}(t,\hat{\boldsymbol{p}})$ to changes in the parameter values $\hat{\boldsymbol{p}}$; sensitivity of the parameter estimates $\hat{\boldsymbol{p}}$ to variations in the observation data $\{t_j;\mathfrak{y}_j\}_{j=1}^N$; sensitivity of biologically meaningful functionals $J(\boldsymbol{y})$ to variations in parameters (see [1, Ref. 12]). The first two types of SA are examined by direct methods and rely upon the computation of the sensitivity coefficients $s_i(t,\boldsymbol{p}) = \partial \boldsymbol{y}(t,\boldsymbol{p})/\partial p_i$ using the variational system

$$\mathscr{A}(\boldsymbol{y}(t,\hat{\boldsymbol{p}}),\hat{\boldsymbol{p}})s_i(t,\hat{\boldsymbol{p}}) = \frac{\partial \boldsymbol{f}}{\partial p_i}, \quad t \geqslant 0, \qquad s_i(t,\hat{\boldsymbol{p}}) = \frac{\partial \boldsymbol{\psi}}{\partial p_i}, \quad t \in [-\tau, 0]. \tag{17}$$

The operator $\mathscr{A} \equiv \mathrm{d}/\mathrm{d}t - [\partial \boldsymbol{f}/\partial \boldsymbol{y}]_t - [\partial \boldsymbol{f}/\partial \boldsymbol{y}_\tau]_t D_\tau$, where $[\cdot]_t$ denotes a matrix-function evaluated at time t, D_τ is a backward shift operator. The overall sensitivity of the solution $\boldsymbol{y}(\boldsymbol{t},\hat{\boldsymbol{p}})$ is given by the matrix-function $\boldsymbol{S}(t,\boldsymbol{p}) = \partial \boldsymbol{y}(t,\boldsymbol{p})/\partial \boldsymbol{p}$ evaluated at $\boldsymbol{p} = \hat{\boldsymbol{p}}$, which characterizes the effect of small variations in the ith datum $\mathfrak{y}_j$ on parameter estimates via the formula

$$\frac{\partial \hat{\boldsymbol{p}}}{\partial \mathfrak{y}_j} = \left[\sum_{i=1}^N \boldsymbol{S}^{\mathrm{T}}(t_i,\hat{\boldsymbol{p}})\boldsymbol{S}(t_i,\hat{\boldsymbol{p}})\right]^{-1} \boldsymbol{S}(t_j,\hat{\boldsymbol{p}}).$$

Numerical sensitivity analysis by the direct method requires solution of the main system (14) and the variational system (17) of $M \times L$ equations taken jointly. This implies that for large-scale multiparameter models, numerical methods that take into account the structure of whole set of DDEs at the linear algebra level are needed.

7.1. Adjoint equations

The sensitivity of nonlinear functionals $J(\boldsymbol{y})$ depending on the solution to the delay models can also be examined using an approach based on adjoint equations; see Marchuk [1, Ref. 120]. Consider, as an example, the quadratic functional and its first-order variation caused by perturbations of the basic parameter set $\hat{\boldsymbol{p}}$ (where $\hat{\boldsymbol{y}} \equiv \boldsymbol{y}(t,\hat{\boldsymbol{p}})$) $J(\hat{\boldsymbol{y}}) = \int_0^{\mathrm{T}} \langle \hat{\boldsymbol{y}}, \hat{\boldsymbol{y}} \rangle \, \mathrm{d}t$, $\delta J(\hat{\boldsymbol{y}}) = 2\sum_{i=1}^{L} \int_0^{\mathrm{T}} \langle \hat{\boldsymbol{y}}, \boldsymbol{s}_i(t,\hat{\boldsymbol{p}})\delta p_i \rangle \, \mathrm{d}t$, where $\boldsymbol{s}_i(t,\hat{\boldsymbol{p}})$ is a solution to (17) on $[0,T]$. The linear operator $\mathscr{A}$ in (17) acts on some Hilbert space H with domain $\mathscr{D}(\mathscr{A})$. Given $\mathscr{A}$, the adjoint operator $\mathscr{A}^*$ can be introduced satisfying the Lagrange identity $\langle \mathscr{A}(\hat{\boldsymbol{y}},\hat{\boldsymbol{p}})\boldsymbol{s}, \boldsymbol{w} \rangle = \langle \boldsymbol{s}, \mathscr{A}^*(\hat{\boldsymbol{y}},\hat{\boldsymbol{p}})\boldsymbol{w} \rangle$, where $\langle \cdot, \cdot \rangle$ is an inner product in H, $\boldsymbol{s} \in \mathscr{D}(\mathscr{A})$, $\boldsymbol{w} \in \mathscr{D}(\mathscr{A}^*)$. Using the solution $\boldsymbol{w}(t)$ of the adjoint problem

$$\mathscr{A}^*(\hat{\boldsymbol{y}},\hat{\boldsymbol{p}})\boldsymbol{w}(t) \equiv -\frac{\mathrm{d}\boldsymbol{w}(t)}{\mathrm{d}t} - \left[\frac{\partial \boldsymbol{f}}{\partial \boldsymbol{y}}\right]_t^{\mathrm{T}} \boldsymbol{w}(t) - \left[\frac{\partial \boldsymbol{f}}{\partial \boldsymbol{y}_\tau}\right]_{t+\tau}^{\mathrm{T}} \boldsymbol{w}(t+\tau) = \boldsymbol{y}(t,\hat{\boldsymbol{p}}),$$
$$0 \leqslant t \leqslant T, \ \boldsymbol{w}(t) = 0, \ t \in [T, T+\tau], \tag{18}$$

the variation of $J(\boldsymbol{y})$ can be estimated via formula $\delta J = \sum_{i=1}^{L} 2 \int_0^{\mathrm{T}} \langle \boldsymbol{w}, (\partial \boldsymbol{f}/\partial p_i)\delta p_i \rangle \, \mathrm{d}t$.

Thus, instead of solving an $M \times L$-dimensional system of equations within a direct approach, one needs to solve, only once, the main and adjoint systems, each being of dimension M. The adjoint technique was used [1, Ref. 121] to select the parameters mostly affecting the severity of influenza and hepatitis infections from the set of 50 ($=L$) parameters. The experience with DIFSUB adapted for constant DDEs indicates that numerical sensitivity analysis using adjoint equations requires particular attention to the following issues: (i) the adjoint problem inherits the jump discontinuities of the forward problem, so the smoothness of the matrix-function $\mathscr{A}^*$ decreases as t approaches 0; (ii) the stiffness properties of the main and adjoint problems are opposite and in general, both display variable stiffness behaviour; (iii) adaptive codes generate different step-size sequences for the main and adjoint problems and $\boldsymbol{y}(t)$ has to be re-evaluated on every integration step of the adjoint problem; therefore, numerical schemes with dense output would give an advantage.

8. Optimal control problems using delay models

Although there are many problems in the biosciences that can be addressed within an optimal control (OC) framework for systems of DDEs (harvesting, chemostat, treatment of diseases, physiological control), the amount of real-life experience is quite small. The general formulation of an OC problem for delay system is as follows: *For a system with the state vector* $\boldsymbol{y}(t,\boldsymbol{u})$ *governed by a DDE, find a control function* $\boldsymbol{u}(t)$, *defined on* $[-\tau_u, T]$, *that gives a minimum to the objective functional* $J_0(\boldsymbol{u})$, *where*

$$\boldsymbol{y}'(t) = \boldsymbol{f}(t, \boldsymbol{y}(t), \boldsymbol{y}(t-\tau_y), \boldsymbol{u}(t), \boldsymbol{u}(t-\tau_u)), \quad 0 \leqslant t \leqslant T, \tag{19}$$

$$J_0(\boldsymbol{u}) = \Phi_0(\boldsymbol{y}(T)) + \int_0^{\mathrm{T}} F_0(t, \boldsymbol{y}(t), \boldsymbol{y}(t-\tau_y), \boldsymbol{u}(t), \boldsymbol{u}(t-\tau_u)) \, \mathrm{d}t \tag{20}$$

with the given initial function for the state vector. Additional equality or inequality constraints can be imposed in terms of functionals $J_i(\boldsymbol{u})$. The Pontryagin maximum principle and the Bellman dynamic programming method are the frameworks being used to formulate computational approaches to time-delayed OC problems (see [1, Refs. 17,104,166]).

Delay equations were used by Buldaev [1, Ref. 37] to find the optimal control regimes of unfavourable infectious disease outcomes. The objective functional was expressed in terms of the virus population size, either at a given final time $t = T$, or a cumulative amount over $[0, T]$. Specific features of the studies are: (i) delays appear only in state variables; (ii) linear scalar control functions appearing additively or multiplicatively in one equation (for the virus) were considered; and (iii) unconstrained problems were treated. An algorithm based on nonclassical spike variations of the control function was developed, using a piecewise constant approximation of the control function on a uniform mesh with the step-size being an integer fraction of the delay. The model and adjoint system were solved by a fourth/fifth order RK method with Hermite interpolation for the delayed terms.

An optimal intravenous drug delivery in AIDS patients has been considered recently by Rundell et al. [1, Ref. 159]. The objective was to find a control strategy that minimizes the total drug administered, subject to the constraint that the patient recovers. The control function, appearing nonlinearly in the equations, was obtained numerically by applying convex minimization techniques based on linear matrix inequalities and the LMI toolbox from MATLAB was used to compute the optimizer. The time-delay was approximated by a fourth-order Bessel filter, whereas the nonlinearities were addressed by transforming the nonlinear model to a linear-fractional representation.

We refer in passing to another example of a formulation of a control problem with delay model inspired by recently proposed nonconventional approach (compared to standard pharmacokinetics models) to the anti-HIV drug administration by Beretta et al. [1, Ref. 22]. A cohort of drug-loaded red blood cells (RBC) with density function $u(t,a)$ at time t and age $a \in \mathbb{R}_+$ is injected at time $t=0$ into a patient. The cells with age $a \geqslant a^*$ ($a^* > 120$ days), called the senescent cells, are phagocytosed by macrophages thus causing the drug to be absorbed (x_4 stands for the average drug concentration in macrophages). The drug has therapeutic effect as long as $0 < m \leqslant x_4(t) < M$ for $t \in [t_1, t_2]$. The delay represents the digestion time, which can be described by a fixed or a distributed delay. The initial age distribution of the RBC can be experimentally preassigned, i.e., $u(0,a) = \phi(a)$, is a control variable, and only a fraction α of the total cell number $\int_0^{+\infty} u(t,a)\,\mathrm{d}a$ are senescent cells. The control function appears additively as a control function in the equation for RBC. The OC problem is stated as: *Choose the control function* $\phi(a)$ *in the interval* $[0, a^*]$ *and the parameter* α *such that* $\Delta t = t_2 - t_1 \to \max$, *subject to* (i) $x_4(t) < M$ *for all* $t > 0$, (ii) *the condition that* $u_0 = \int_0^{a^*} \phi(a)\,\mathrm{d}a/(1-\alpha)$, $u_0 \in [n_1, n_2]$, *be minimum*. A qualitative analysis of the problem suggested that a constant age distribution function should be a solution.

Further research in the numerical treatment of constrained nonlinear OC problems is needed to provide biomodellers with user-friendly adaptive packages.

Acknowledgements

The authors are most grateful to Professor C.T.H. Baker, who was a co-author of [1], for his valuable comments and numerous suggestions concerning the content of the paper. G. Bocharov was supported by the Alexander von Humboldt-Stiftung, Germany during a visit to the group of Professor K.P. Hadeler at Tübingen; Sections 1, 2 and 3 in particular reflect input (private communication) from Professor Hadeler during this visit, which is most appreciated. The comments by other colleagues of the authors (in particular Dr Buckwar and Dr Paul at Manchester) are also gratefully acknowledged. Any imperfections are, however, the responsibility of the authors.

References

[1] C.T.H. Baker, G.A. Bocharov, F.A. Rihan, A report on the use of delay differential equations in numerical modelling in the biosciences, MCCM Technical Report, Vol. 343, (1999), Manchester, ISSN 1360-1725. (Available over the World-Wide Web from URL: `http://www.ma.man.ac.uk/MCCM/MCCM.html`).

[2] C.T.H. Baker, E. Buckwar, Introduction to the numerical analysis of stochastic delay differential equations, MCCM Technical Report, Vol. 345, ISSN 1360-1725, University of Manchester, 1999.
[3] R.B. Banks, Growth and Diffusion Phenomena. Mathematical Frameworks and Applications, Texts in Applied Mathematics, Vol. 14, Springer, Berlin, 1994.
[4] R. Bellman, K.L. Cooke, Differential-difference Equations, Academic Press, New York, 1963.
[5] J.M. Cushing, Integrodifferential Equations and Delay Models in Population Dynamics, Lecture Notes in Biomathematics, Vol. 20, Springer, Berlin, 1977.
[6] O. Diekmann, S. van Gils, S. Verduyn Lunel, H.-O. Walter, Delay Equations, Functional-, Complex-, and Nonlinear Analysis, Springer, New York, 1995.
[7] R.D. Driver, Ordinary and Delay Differential Equations, Applied Mathematics Series, Vol. 20, Springer, Berlin, 1977.
[8] L.E. Elsgolt's, S.B. Norkin, Introduction to the Theory and Application of Differential Equations With Deviating Arguments, Academic Press, New York, 1973.
[9] K. Gopalsamy, Stability and Oscillations in Delay Differential Equations of Population Dynamics, Kluwer, Dordrecht, 1992.
[10] I. Györi, G. Ladas, Oscillation Theory of Delay Equations With Applications, Oxford Mathematical Monographs, Clarendon Press, Oxford, 1991.
[11] A. Halanay, Differential Equations, Stability, Oscillations, Time Lags, Academic Press, New York, 1966.
[12] J.K. Hale, Theory of Functional Differential Equations, Springer, New York, 1977.
[13] J.K. Hale, S.M. Verduyn Lunel, Introduction to Functional Differential Equations, Springer, New York, 1993.
[14] V.B. Kolmanovskii, A. Myshkis, Applied Theory of Functional Differential Equations, MIA, Vol. 85, Kluwer, Dordrecht, 1992.
[15] V.B. Kolmanovskii, V.R. Nosov, Stability of Functional Differential Equations, Academic Press, New York, 1986.
[16] Y. Kuang, Delay Differential Equations With Applications in Population Dynamics, Academic Press, Boston, 1993.
[17] N. MacDonald, Time-lags in Biological Models, Lecture Notes in Biomathematics, Vol. 27, Springer, Berlin, 1978.
[18] G.I. Marchuk, Mathematical Modelling of Immune Response in Infectious Diseases, Kluwer, Dordrecht, 1997.
[19] R. May, Stability and Complexity in Model Ecosystems, Princeton University Press, Princeton, NJ, 1974.
[20] J. Maynard Smith, Models in Ecology, Cambridge University Press, Cambridge, 1974.
[21] J.D. Murray, Mathematical Biology, Springer, Berlin, 1989.
[22] P. Waltman, Deterministic Threshold Models in the Theory of Epidemics, Lecture Notes in Biomathematics, Vol. 1, Springer, Berlin, 1974.

ELSEVIER

Journal of Computational and Applied Mathematics 125 (2000) 201–215

JOURNAL OF COMPUTATIONAL AND APPLIED MATHEMATICS

www.elsevier.nl/locate/cam

Dynamics of constrained differential delay equations

John Norbury[a, *], R. Eddie Wilson[b]

[a] Mathematical Institute, Oxford University, 27-29 St Giles', Oxford OX1 3LB, UK

[b] School of Computing and Mathematical Sciences, Oxford Brookes University, Headington, Oxford OX3 0BP, UK

Received 25 March 1999; received in revised form 27 July 1999

Abstract

A class of forced first-order differential delay equations with piecewise-affine right-hand sides is introduced, as a prototype model for the speed of a motor under control. A simple pure delay form is mainly considered. When forcing is zero, an exact stable periodic solution is exhibited. For large amplitude periodic forcing, existence of stable solutions, whose period is equal to that of the forcing function, is discussed, and these solutions are constructed for square wave forcing. Traditional numerical methods are discussed briefly, and a new approach based on piecewise-polynomial structure is introduced. Simulations are then presented showing a wide range of dynamics for intermediate values of forcing amplitude, when the natural period of the homogeneous equation and the period of the forcing function compete.

Keywords: Differential delay equations; Bifurcation theory; Control theory

1. Introduction

This paper announces our work on differential delay equations of the type

$$u'(t) = -M_{\text{dec}}, \qquad c\{f(t) - L(u,t)\} < -M_{\text{dec}}, \tag{1a}$$

$$u'(t) = c\{f(t) - L(u,t)\}, \quad -M_{\text{dec}} \leqslant c\{f(t) - L(u,t)\} \leqslant M_{\text{acc}}, \tag{1b}$$

$$u'(t) = M_{\text{acc}}, \qquad M_{\text{acc}} < c\{f(t) - L(u,t)\}. \tag{1c}$$

Here c, M_{dec}, $M_{\text{acc}} > 0$ are constants, and $L(\cdot,t)$ is a linear functional (e.g., see Eqs. (3)–(5)) which acts on the history $u(s)$, $s \leqslant t$, of u. For reasons of space, we concentrate here on the special case $L(u,t) = u(t-\tau)$ when (1), suitably rescaled, may be written in the form

$$\varepsilon u'(t) = \text{sgn}\{f_{A,T}(t) - u(t-1)\} \min\{1, |f_{A,T}(t) - u(t-1)|\}, \tag{2}$$

* Corresponding author.

0377-0427/00/$ - see front matter
PII: S0377-0427(00)00469-6

and prescribed with suitably smooth initial data $u_0(t)$, $t \in [-1, 0]$. Here $f_{A,T}$ is periodic with amplitude A and period T, and $\varepsilon > 0$. Section 2 considers periodic solutions of (2) when either $A = 0$ or A is large; Section 3 discusses numerical techniques (including a method which we believe is original), and Section 4 presents some numerical results, including interesting simulations for intermediate A, where the analytical results of Section 2 do not apply. In this section, we explain the model behind (1) and (2).

Eq. (1) has a control theory application where u represents, e.g., the speed of a motor which is trying to adjust to a prescribed (possibly time dependent) *target* speed f. The motor attempts this by acceleration/deceleration proportional to the difference of the target speed and an estimate of its own current speed (regime (1b)). In physical systems, delays are inevitable in the measurement process, so that the motor's measured speed is related to the recent history of its true speed. This justifies the presence of the delay operator L.

Because forces are limited in mechanical systems, one expects maximum rates at which the motor may speed up or slow down, so that if the difference of target and measured speeds becomes too large, then the acceleration/deceleration saturates, and one has either regime (1a) or (1c).

Note that (1) is a simplified system: we are not proposing it as an accurate model for any specific mechanical situation. However, we believe that it is a good prototype equation for lots of systems where there is delay in measurement and/or response, and some sort of saturation effect. Further, we believe that there might be applications in other areas, e.g., the right-hand side of (1) resembles the response of an electronic component known as the *operational amplifier* (see [6, Section 5.3, Fig. 5.7]), and we anticipate our theory having applications in modelling circuits which contain such components.

Although the right-hand side of (1) is linear in each of the regimes (1a)–(1c), it is a nonlinear equation taken as a whole, because of the nonsmooth swapping between regime (1b) (which we call *off the constraint*) and regimes (1a) or (1c) (which we call *on the constraint*), which depends on the solution itself. (Note that although nonsmooth, the right-hand side is continuous in u, so that if f is (piecewise) continuous, we can reasonably expect a (piecewise) C^1 solution for u for $t > 0$.)

To the best of our knowledge, only the theses of Allen (whose supervisor and co-worker is the first author) [1,2], have previously considered delay equations with this type of nonlinearity. There are analytical advantages to such piecewise-linearity: in Section 2 we are able to derive some *exact* nontrivial solutions, provided we track the points where the solution swaps between regimes (1a)–(1c). This is in marked contrast to other sorts of nonlinear delay equations, where, e.g., although one might be able to exhibit a Hopf bifurcation, one will not usually be able to construct finite-amplitude oscillatory solutions.

In practical situations, engineers have a choice of components which would change the way in which the motor measured its own speed (i.e., the delay operator L could be changed to some extent), and the responsiveness c might also be altered. The choice of c is a key design consideration – increasing c reduces the time scale on which the motor adapts to its target speed, but increasing c too far leads to oscillatory instability, because of the presence of delay.

It is convenient to normalise L so that constants are invariant, i.e., if $u(t) = 1$ for all t, then $L(u, t) = 1$ for all t. Simple examples of delay operators then include

$$L(u,t) = u(t-\tau), \quad \tau > 0, \quad \text{(the pure delay case) or more generally} \tag{3}$$

$$L(u,t) = \sum_{i=1}^{n} \alpha_i u(t-\tau_i) \quad \text{with} \quad \sum_{i=1}^{n} \alpha_i = 1, \quad \tau_i \geqslant 0, \tag{4}$$

or distributed forms such as

$$L(u,t) = \int_0^\infty K(s)u(t-s)\,\mathrm{d}s, \quad \text{where} \quad \int_0^\infty K(s)\,\mathrm{d}s = 1. \tag{5}$$

In the last case, it is well known (see e.g. [4, Section 1a]) that choosing $K(s) := p_n(s)\exp(-s)$, where p_n is a polynomial of degree n, allows (1) to be expanded to a system of $n+1$ *ordinary* differential equations. Apart from this case, differential delay equations such as (1) are infinite dimensional, and their general analysis has the technical complexity usually associated with partial differential equations.

In this paper, we consider the forward problem of fixing the delay functional, and determining the resulting dynamics. To simplify the analysis, we use only the pure delay form (3). (Some progress can be made in e.g. case (4) with $n=2$ and $\tau_1=0$, see [2, Chapters 4 and 5].) If we also make the symmetric choice $M_{\mathrm{dec}} = M_{\mathrm{acc}}$ (so that the maximum braking and acceleration are equal), then (1) may be written

$$u'(t) = c\,\mathrm{sgn}\{f(t) - u(t-\tau)\}\min\left\{\frac{M_{\mathrm{acc}}}{c}, |f(t) - u(t-\tau)|\right\}. \tag{6}$$

This can be nondimensionalised by writing $u(t) = M_{\mathrm{acc}}\tilde{u}(t)/c$, and $t = \tau\tilde{t}$, then $\tilde{\tilde{u}}(\tilde{t}) = \tilde{u}(\tau\tilde{t})$; with f suitably transformed, and with "tildes removed" we obtain (2), with $\varepsilon = 1/c\tau$. (Thus, ε becomes small as either the original responsiveness or the delay becomes large.)

To further simplify matters here, we consider only periodic target functions f. Noting the symmetry $u \mapsto u+k$, $f \mapsto f+k$ (k constant) of (2), we may assume without loss of generality that f has mean zero.

In the following sections we will refer to these special families of f:

(1) unbiased square waves

$$f(t) = \begin{cases} +A, & t \in [0, T/2) \\ -A, & t \in [T/2, T) \end{cases} \quad \text{and periodic extension;} \tag{7}$$

(2) and sine waves

$$f(t) = A\sin\left(\frac{2\pi t}{T}\right). \tag{8}$$

Throughout we use A and T, respectively, to denote the amplitude and period of f. Thus fixing the shape of f, (2) is a three parameter (ε, A, T) problem.

A key idea (not present in [2]) which we want to introduce, is the choice of piecewise-linear forcing functions such as (7) or, e.g.,

(3) symmetric, unbiased triangular waves

$$f(t) = \begin{cases} -A + \dfrac{4At}{T}, & t \in [0, T/2) \\ +A - \dfrac{4A(t - T/2)}{T}, & t \in [T/2, T) \end{cases} \quad \text{and periodic extension.} \tag{9}$$

These make available exact explicit solutions, and so simplify calculations. Further, they permit the use of a new numerical method, which we describe in Section 3.

2. Analytical results

2.1. Homogeneous equation: period 4 solutions

First, we consider (2) with $f \equiv 0$, i.e. how does the system respond to a constant target function? In this case, note that $u \equiv 0$ (corresponding to the system perfectly matching its target) is a solution.

The natural question is whether the zero solution is stable to small perturbations. The linearisation is trivial: if initial data is small (so that $|u_0(t)| < 1$ for $t \in [-1, 0))$, then up to the time when $|u|$ first exceeds one (if this is to happen) u is *off the constraint* and satisfies the linear equation

$$\varepsilon u'(t) = -u(t-1) \quad \text{exactly.} \tag{10}$$

Substituting $u = Re\{C \exp(\lambda + i\omega)t\}$, it may be shown that all modes decay for

$$\varepsilon > \varepsilon^* := 2/\pi. \tag{11}$$

In fact, (11) is enough to imply that $u \equiv 0$ is asymptotically stable, because by taking u_0 sufficiently small we may guarantee that u stays off the constraint, so that (10) holds for all $t > 0$.

Our numerical experiments indicate that $u \equiv 0$ is *globally* asymptotically stable for (11), however, a proof would be difficult. Because eigenvalues have nonzero imaginary parts, it does not follow that $|u(t)| \leqslant \sup |u_0|$ for all t: hence it is possible, even when $\sup |u_0| \leqslant 1$, that $|u|$ will exceed one and "hit the constraint", so that (10) no longer holds. When $\sup |u_0| > 1$, similar difficulties are encountered.

When $\varepsilon = \varepsilon^*$, the linear equation (10) has a family of oscillatory solutions $u = C \sin((\pi/2)t + t_0)$ of period 4. For these to solve (2) we require $|C| \leqslant 1$, as otherwise the constraint would be hit.

When $\varepsilon < \varepsilon^*$, the mode of period 4 is unstable (i.e., its λ is positive) and higher frequency modes lose stability as ε is decreased further. If this regime persisted, such modes would continue to grow exponentially without bound. However, for some t^*, $|u|$ must exceed one, so that at $t = t^* + 1$, the solution will hit the constraint. The solution cannot then blow up, because of the following result.

Lemma 1 (Coarse upper bound). *Given initial data u_0, there exists a $t^*_{u_0}$ such that*

$$|u(t)| < 1 + \frac{1}{\varepsilon} \quad for\ t > t^*. \tag{12}$$

Proof. First note that u must either enter $[-1, 1]$ and remain there, or must visit that interval infinitely often. To see this, suppose for a contradiction that $\exists t^\dagger$ so that for $\forall t > t^\dagger$, $u(t) \notin [-1, 1]$. By continuity of u, either (i) $u(t) > +1\ \forall t > t^\dagger$, or (ii) $u(t) < -1\ \forall t > t^\dagger$; assume (i) without loss of generality. In this case, $u'(t) = -1/\varepsilon$ for $t > t^\dagger + 1$, so after a further time $(u(t^\dagger + 1) - 1)\varepsilon$, $u(t)$ will enter $[-1, 1]$, and we have a contradiction.

Secondly, note that if $\exists t^*$ such that $u(t) \in [-1, 1]\ \forall t > t^*$, then the result follows immediately. Otherwise, $\exists t^*$ such that either (i) $u(t^*) = +1$, $u(t^* + \delta) > +1$ or (ii) $u(t^*) = -1$, $u(t^* + \delta) < -1$, for $\delta > 0$ sufficiently small; assume (i) without loss of generality. At time $t^* + 1$, the solution will hit the constraint and $u'(t)$ will become equal to $-1/\varepsilon$. Therefore, u has a maximum at $t \in (t^*, t^* + 1)$; u' is bounded above by $1/\varepsilon$ on this interval, so this maximum value is bounded strictly above by $1 + 1/\varepsilon$, and the lemma is proved.

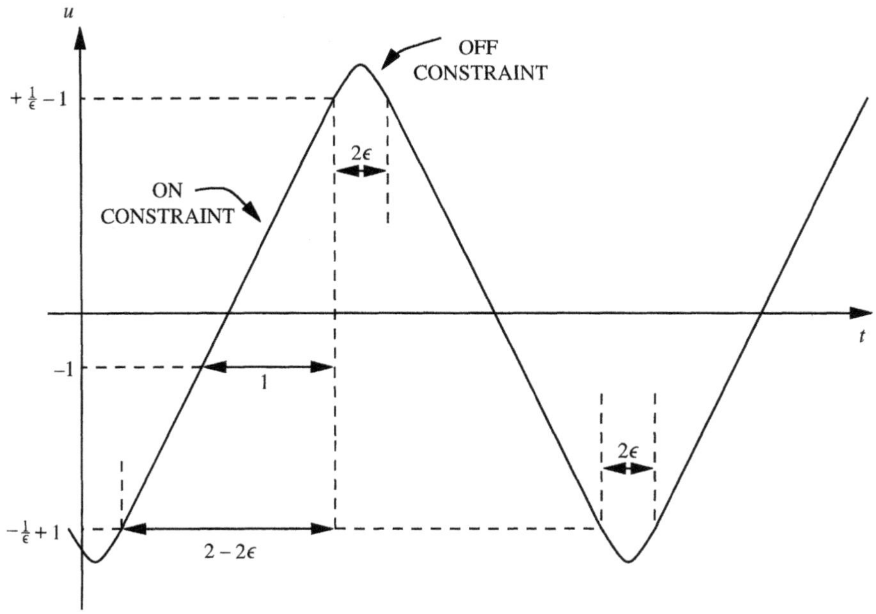

Fig. 1. Exact period 4 solutions of (2) in the homogeneous case $f \equiv 0$, when $\varepsilon \leqslant \frac{1}{2}$. See Eq. (13).

So for $\varepsilon < \varepsilon^*$, the solution is contained in $(-1-1/\varepsilon, +1+1/\varepsilon)$, and visits the unconstrained regime infinitely often. Numerical simulations indicate that the global attractor is a solution of period 4, and for $\varepsilon \leqslant \frac{1}{2}$, it is given by the piecewise-polynomial construction

$$u(t) = \begin{cases} \left(-\dfrac{1}{\varepsilon}+1\right)+\dfrac{t_1}{\varepsilon}, & 0 < t_1 := t - \text{const.} < 2-2\varepsilon, \\ \left(+\dfrac{1}{\varepsilon}-1\right)+\dfrac{t_2}{\varepsilon}-\dfrac{t_2^2}{2\varepsilon^2}, & 0 < t_2 := t_1 - (2-2\varepsilon) < 2\varepsilon, \\ \left(+\dfrac{1}{\varepsilon}-1\right)-\dfrac{t_3}{\varepsilon}, & 0 < t_3 := t_2 - 2\varepsilon < 2-2\varepsilon, \\ \left(-\dfrac{1}{\varepsilon}+1\right)-\dfrac{t_4}{\varepsilon}+\dfrac{t_4^2}{2\varepsilon^2}, & 0 < t_4 := t_3 - (2-2\varepsilon) < 2\varepsilon, \end{cases} \tag{13}$$

with periodic extension, which is shown in Fig. 1. Here t_i parametrise consecutive intervals of time on which the solution takes a different polynomial form. The key to (13) is to note that if the solution spends more than unit time (i.e., the rescaled delay) on the constraint, then the time at which the solution comes off the constraint is fully independent of the history prior to hitting the constraint. Some initial data are attracted to (13) in finite time, as a result of this *loss of memory* effect.

For $\frac{1}{2} < \varepsilon < \varepsilon^* \simeq 0.63661977$, there is also a stable period 4 attractor, but its form is more complicated, because time intervals of length more than one are spent off the constraint. This solution cannot be written as a piecewise-polynomial and must involve exponentials: we do not give details here.

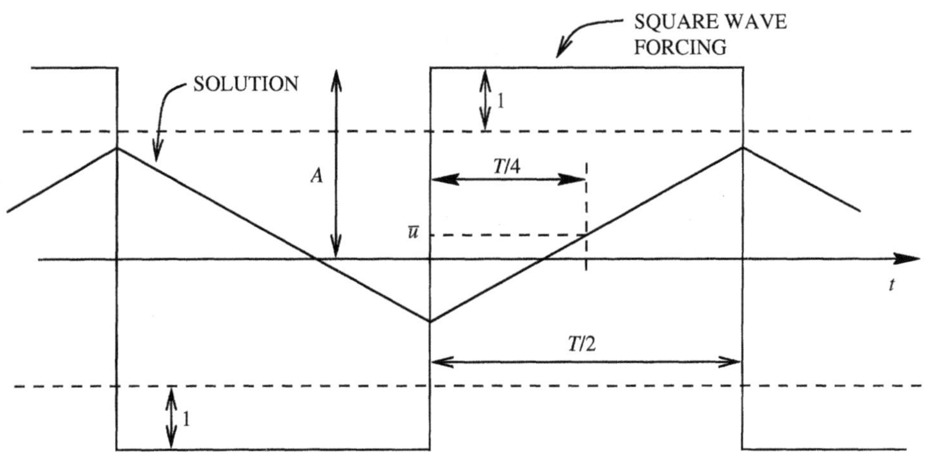

Fig. 2. Large amplitude square wave forcing: periodic solutions, wholly on the constraint, with the same period as the forcing.

2.2. Large amplitude forcing: periodic solutions

Now, we consider what happens to solutions of (2) when the amplitude A of the forcing function f is large. Numerics indicate that in this case, solutions with period T (equal to that of the forcing function) attract almost all initial data.

First, we consider square wave forcing (7). For sufficiently large A, we can derive piecewise-linear solutions with period T, which are on the constraint with positive gradient $1/\varepsilon$ when $f>0$, and which are on the constraint with negative gradient $-1/\varepsilon$ when $f<0$ (see Fig. 2). (The ability to swap between opposite constraints without going off the constraint is a special feature due to the forcing function being discontinuous.) Our numerical experiments indicate that solutions of this form are globally asymptotically stable.

Note that the piecewise-linear form shown in Fig. 2 is periodic, because equal times are spent in the positive and negative phases of the square wave (i.e. the square wave is unbiased). Further, if the putative solution fits between $-(A-1)$ and $+(A-1)$, it is always on the constraint (being always more than one different from the forcing function), and so indeed solves (2). Therefore, a *sufficient* condition for existence of such solutions is for there to be a constant $\bar{u}$ such that $\bar{u}+T/4\varepsilon\leqslant A-1$, and $\bar{u}-T/4\varepsilon\geqslant-(A-1)$. ($\bar{u}$ gives the mean of the solution u (see Fig. 2).) This is possible if

$$A\geqslant 1+\frac{T}{4\varepsilon}. \tag{14}$$

If (14) holds with strict inequality, then there is an interval of permissible $\bar{u}$, and hence a whole family of solutions of this type, for fixed A.

For period $T>2$, bound (14) is not necessary for solutions of this type (because it is actually not necessary to keep the solution inside $[-(A-1),A-1]$ to remain on the constraint). If $T>2$, a sharp bound

$$A \geqslant 1 + \frac{|T-4|}{4\varepsilon}, \tag{15}$$

for existence of this type of solution may be derived.

For other *shapes* of f, it is possible to construct (asymptotically, as $\varepsilon \to 0$) solutions of period T when there is large amplitude forcing, and these appear to be stable. See [2, Chapter 5], which considers in particular sine wave forcing (8). When f is continuous, such solutions have small $O(\varepsilon)$ intervals off the constraint. Note that for $\varepsilon \to 0$ asymptotics, it is convenient to write $u = v/\varepsilon$, and consider (2) in the rescaled form

$$v'(t) = \operatorname{sgn}\{f_{\tilde{A},T}(t) - v(t-1)\} \min\left\{1, \frac{1}{\varepsilon}|f_{\tilde{A},T}(t) - v(t-1)|\right\}, \tag{16}$$

where $\tilde{A} = \varepsilon A$. If $\varepsilon \to 0$ with $\tilde{A}$ fixed, then solutions v converge to a piecewise-linear "$\varepsilon = 0$" solution, whose gradient is ± 1. For continuous f and any one sufficiently large value of $\tilde{A}$, we usually find a *unique* symmetric (i.e., invariant under $t \mapsto t + T/2$, $u \mapsto -u$), globally asymptotically stable solution (in contrast to when f is a square wave, where we have shown that there is a whole interval of solutions). The symmetry is forced by higher-order matching conditions across the $O(\varepsilon)$ transition intervals. In Section 4 we exhibit these solutions numerically.

In [2, Chapter 4], it is also shown how the fixed-point method of Mawhin [5] may be used to prove the existence of the solutions of (2), with period T equal to that of the forcing function. However, this approach is nonconstructive and gives no information about the number or structure of such solutions.

3. Numerical techniques

Allen [2, Chapter 3] gives an account of finite-difference methods used to solve equations of type (1). At the most basic level, these are like ODE solvers, except the history of the solution up to that of the maximum delay must be stored. (In the case of a delay operator like (5), the history must be truncated, although it is easier to solve the equivalent ODE system by a standard method.)

However, naive solvers for (1) lose an order of accuracy because of nonsmoothness at points where the solution changes between regimes (1a), (1b) and (1c). These changeover points must be treated with extra care: strategies involve either (i) *tracking*, in which one aims to place a mesh point exactly on the changeover point, or (ii) *refinement*, in which the mesh size is locally reduced so that some error bound is satisfied.

The loss of smoothness in our equation is in addition to the well-known problem which one usually finds in differential delay equations: the solution of e.g. (2) is generically nonsmooth at $t = 0$ owing to the discrepancy between $u'_{-}(0) = u'_0(0)$ and $u'_{+}(0)$; the latter is given by the right-hand side, and involves the independent value $u_0(-1)$. This nonsmoothness propagates to $t = 1, 2, 3, \ldots$, although it becomes an order weaker at each subsequent time. The effect of this discontinuity, and methods for dealing with it, have been well documented by Baker and collaborators (see, e.g., [3]).

Our original contribution to the numerics is to note that if we take the pure delay equation (2) with piecewise-polynomial forcing f and piecewise-polynomial initial data u_0, then the solution u is also piecewise-polynomial. To see this, note that in (2) one is either integrating a constant (if on the constraint) to give a linear in t part to the solution, or one is integrating the difference of polynomial f and the history $u(t-1)$. If we assume that u is piecewise-polynomial at earlier times (to start the inductive process, use piecewise-polynomial u_0) then it follows that $u(t)$ is polynomial, being either linear or found by integrating the difference of two polynomials.

We have written a C++ solver for (2) based on this piecewise-polynomial approach (for details, see [7]). The scheme does not introduce rounding error, because time is not discretised. Instead, the code represents the solution (and forcing function, initial data) in terms of *pieces*, which each consist of a vector of polynomial coefficients and the length of the time interval for which they apply. Integration, time-translation, etc., can then be achieved by *exact* (up to rounding error) operations.

Note that it is necessary to find the knots which join different polynomial *pieces* of the solution. To do this, one must consider the location of knots in the history and in the forcing function. Further, to know when one goes on or off the constraint, we have written a scheme to find the first positive zero of the polynomial equations $f(t) - u(t-1) \pm 1 = 0$. This last method is a possible source of numerical error if the degree of the solution polynomial becomes too large. However, our code calculates the exact solutions of Section 2 to 14 significant figures, which is almost at the limit of double-precision arithmetic.

Unfortunately, our new method does not work if either the initial data or forcing function is not piecewise-polynomial – e.g. we cannot use it to tackle problems with sine wave forcing (8), although of course we can approximate the sine wave arbitrarily closely by a piecewise-polynomial.

Further, when we do have piecewise-polynomial forcing function and initial data, our scheme cannot cope with *all* solutions of (2) as $t \to \infty$. Although the solution is piecewise-polynomial, sometimes we cannot bound its degree as $t \to \infty$. This is most easily seen for $\varepsilon > \varepsilon^*$, when given some initial data, $\exists t^*$ so that the solution is off the constraint for all $t > t^*$. In this case, the degree of solution polynomial pieces will tend to increase by one for each unit in time, as they are always found by an integration of the history.

This last effect would also occur if our approach was used to solve smooth DDEs whose nonlinearity was (nonconstant) polynomial in u. The approach only works for (2), if a sufficient proportion of time is spent on the constraint (where the solution is linear) so that for all t, u is on the constraint at $t-n$, for $n \in \mathbb{N}$ uniformly bounded above. If this is not the case, the order of the solution polynomial can increase, and/or knot points can accumulate.

4. Simulations

In this section, we present the results of some simulations of (2) using the finite difference and exact polynomial schemes discussed in Section 3. The aim is not to give a comprehensive catalogue of solution behaviour – rather we give examples of some interesting parameter regimes. We have analytical constructions for some of the solutions shown, but for reasons of space, these will be left to a later paper.

All the solutions shown were for initial data of the simple type $u_0 \equiv$ const. When the initial data has not been shown on plots, we have considered it unimportant, because we found the same large-time behaviour for all the initial data that we tried. Thus, the solutions which are displayed without initial data may be considered attractors.

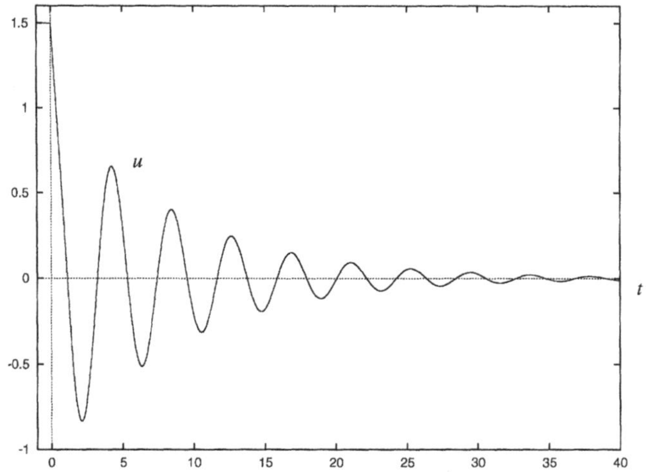

Fig. 3. Simulation with $f \equiv 0$ and $\varepsilon = \frac{3}{4} > \varepsilon^*$, showing exponential decay to zero.

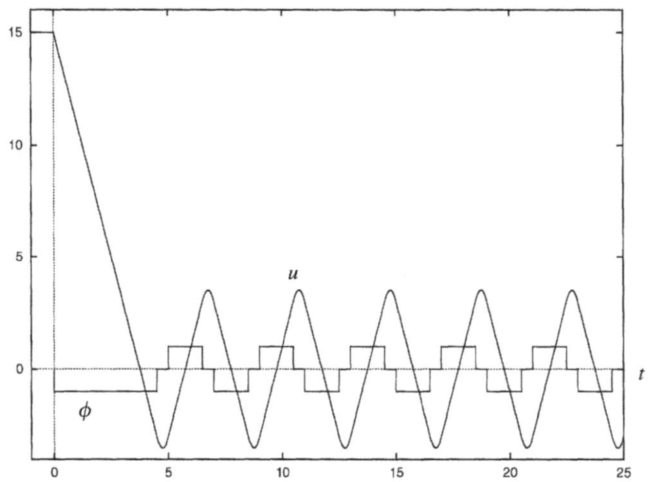

Fig. 4. Simulation with $f \equiv 0$, and $\varepsilon = \frac{1}{4} < \varepsilon^*$, showing (finite time) attraction to the period 4 solution given by (13) (see also Fig. 1). ϕ denotes the *state* of the solution: 0 for off the constraint, and ± 1 for on the constraint with positive or negative gradient, respectively.

First, Figs. 3 and 4 show the behaviour of the homogeneous equation where $f \equiv 0$, which was considered in Section 2.1. Depending on whether ε is greater or less than $2/\pi$, solutions either decay to zero or are attracted (possibly in finite time, depending on initial data) to a period 4 solution, which is given by (13) when $\varepsilon < \frac{1}{2}$. (The function ϕ plotted in Fig. 4 is a *state* variable, which equals 0 when the solution is off the constraint, and equals ± 1 when the solution is on the constraint with positive or negative gradient, respectively.)

Figs. 5–7 show solutions for "large" amplitude periodic forcing, which have the same period as the forcing function. (This situation was considered analytically in Section 2.2.) Fig. 5 is for square wave forcing, and shows four solutions with finite-time convergence to different members

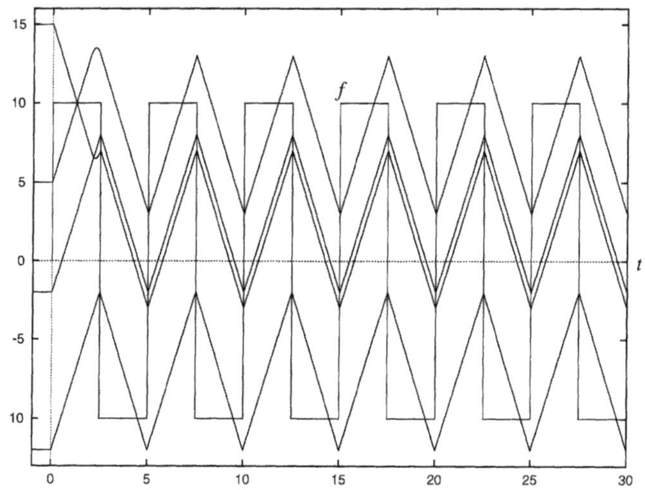

Fig. 5. Simulations with $\varepsilon=\frac{1}{4}$ and large amplitude $A=10$ square wave forcing f with period $T=5$. There is a one-parameter family of period 5 piecewise-linear solutions of the type constructed in Section 2.2 (see also Fig. 2). Finite-time attraction to four members of the solution family is shown.

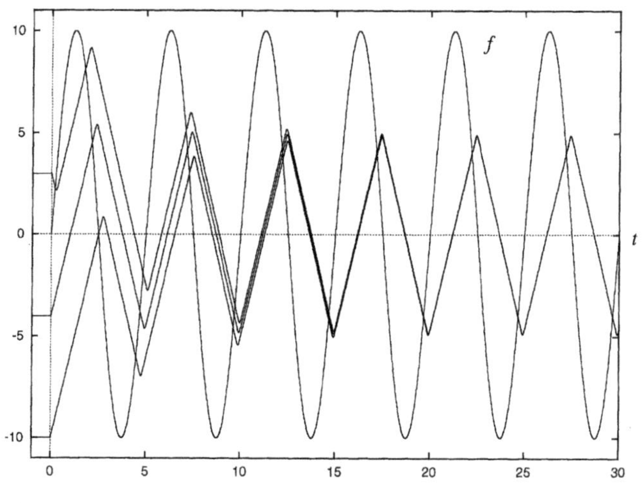

Fig. 6. Simulations with $\varepsilon=\frac{1}{4}$ and large amplitude $A=10$ sine wave forcing, with period $T=5$. Three solutions are shown converging to the unique, symmetric global attractor of period 5.

of the family constructed by Fig. 2. Fig. 6 is for sine wave forcing, and shows (in contrast to the square wave case) that there is a *unique* attracting periodic solution, which has $t\mapsto t+T/2$, $u\mapsto -u$ symmetry. We have found in simulations that the attracting periodic solution is symmetric and unique for other (large amplitude) *continuous* forcing functions.

Fig. 7 concerns the $\varepsilon\to 0$ limit of solutions u (with sine wave forcing) multiplied by ε. As claimed in Section 2.2, these converge to a piecewise-linear form, with gradient ± 1.

However, in our view, the most important parameter is the amplitude A of the forcing function f. For sufficiently large A, the global attractor appears to consist of solutions with period equal to

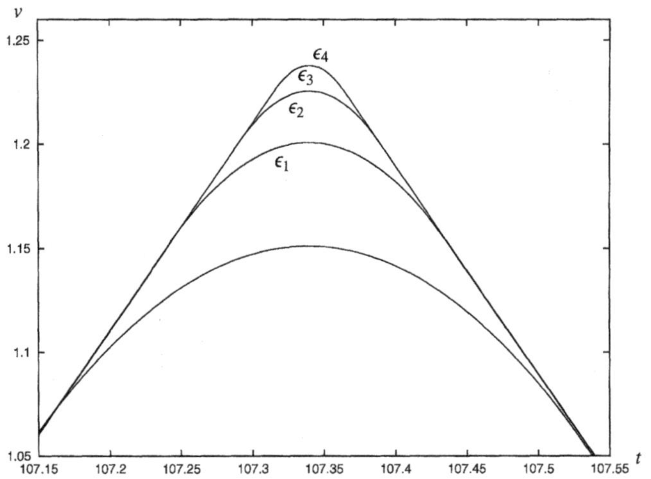

Fig. 7. Magnification of periodic solutions u as $\varepsilon \to 0$ with sine wave forcing, $T = 5$, and εA fixed (equal to 1.25); $v = \varepsilon u$ with $\varepsilon_1 = \frac{1}{2}$, $\varepsilon_2 = \frac{1}{4}$, $\varepsilon_3 = \frac{1}{8}$, and $\varepsilon_4 = \frac{1}{16}$ is shown. Transition intervals off the constraint are width $O(\varepsilon)$ so that v tends to a piecewise-linear form as $\varepsilon \to 0$.

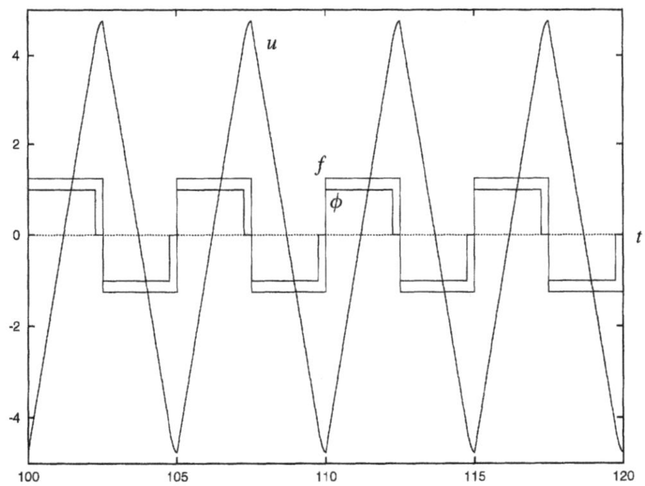

Fig. 8. Simulation with $\varepsilon = \frac{1}{4}$ and square wave forcing of moderate amplitude $A = 1.25$, and period $T = 5$. There is a unique symmetric global attractor, which spends the end of each up-/down-phase of the square wave *off* the constraint.

that of f. However, for $A = 0$, the attractor has period 4. If the period T of f is not equal to 4, then what form do the solutions take for intermediate values of A? Is the attractor periodic, and if so, does its period change in a smooth or discontinuous way as A is decreased? Figs. 8–14 address this question. We concentrate on the square wave-forcing case: Figs. 8–10 are for a *long-wavelength* situation, with $T = 5$; Figs. 11 and 12 are for a *resonant* case with $T = 1$. Throughout, we fix $\varepsilon = \frac{1}{4}$.

For $\varepsilon = \frac{1}{4}$ and square wave forcing with $T = 5$, solutions of the type of Fig. 5 (i.e., wholly on the constraint) cease to exist as A is decreased through 2. However, the attracting solution, shown in Fig. 8, remains periodic with period 5; but, it is unique, symmetric, and spends some time *off* the

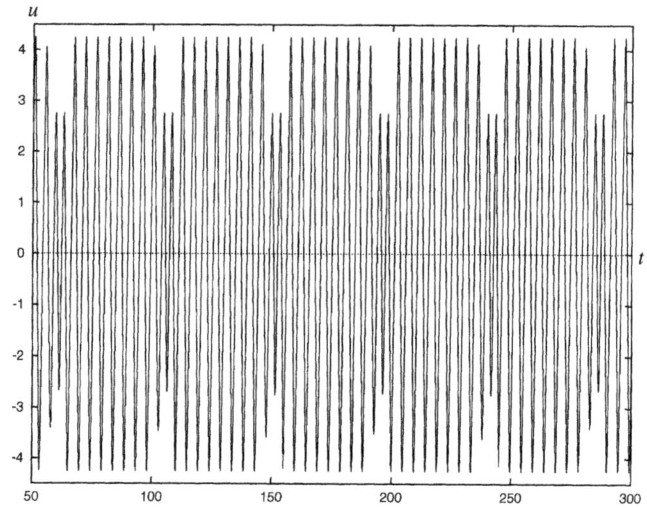

Fig. 9. Simulation with $\varepsilon = \frac{1}{4}$ and square wave forcing of small amplitude $A = 0.75$ and period $T = 5$. The solution appears to be approximately periodic with period 90.

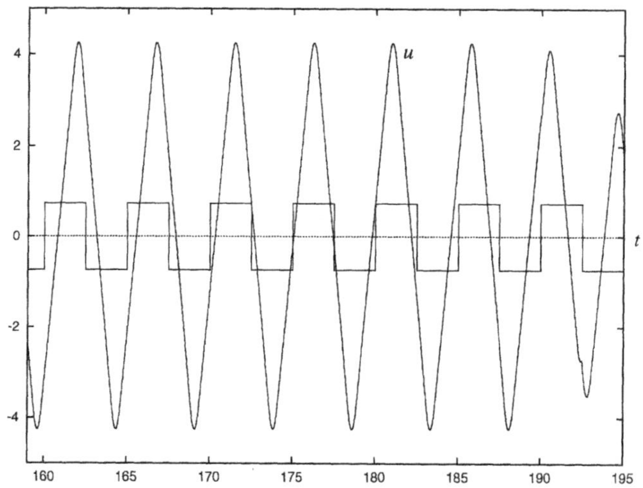

Fig. 10. Magnification of Fig. 9. The solution has features of period approximately 4.75, which drift in and out of phase with the square wave forcing.

constraint. We are able to construct the solution of Fig. 8 analytically (details will follow in another paper), and show (for these ε, T) that this form of solution cannot persist as A is decreased through 1. For $A < 1$, we have attracting solutions like that shown in Fig. 9. Depending on the exact value of A, we observe a solution which seems either periodic with very high period, or quasi-periodic. The solution (see Fig. 9) consists of regular oscillations which drift in and out of phase with the forcing function; at particular phase differences, these oscillations change form. As $A \to 0$, the solution remains quasi-periodic, but the modulation of the oscillations diminishes, so that the known period 4 solution is approached.

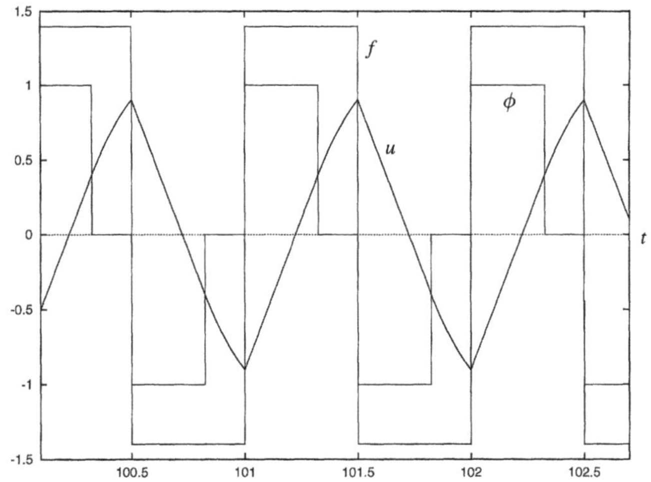

Fig. 11. Periodic solution with $\varepsilon = \frac{1}{4}$ and square wave forcing of amplitude $A = 1.4$ and period $T = 1$. The solution is symmetric and has period 1; the solution off the constraint is exponential, rather than polynomial, in t.

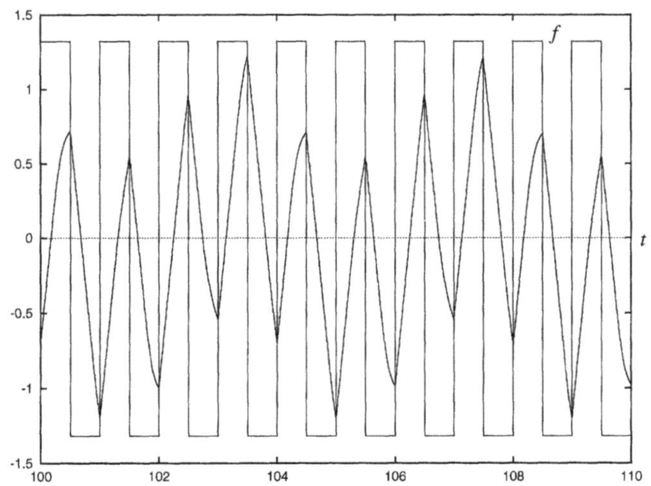

Fig. 12. Period 4 solution with $\varepsilon = \frac{1}{4}$ and square wave forcing of amplitude $A = 1.32$ and period $T = 1$.

Solutions where T is smaller can be more interesting, because of possible resonances between the forcing function and the delay (which is normalised as unity in (2)). In Figs. 11–14, solutions with $T = 1$ are shown, but similar sorts of effects may be achieved with $T = 2$, or with T close to either of these values.

Fig. 11 shows an attracting, symmetric solution of period 1, when f has intermediate amplitude $A = 1.4$. The solution has a similar form to that of Fig. 8 – however, note that the parts of the solution off the constraint are exponential, rather than polynomial, in t. (To see this, note that $u(t-1) = u(t)$, so that off the constraint $\varepsilon u'(t) = \pm A - u(t)$.) Hence, this is a nontrivial solution with polynomial forcing which cannot be found by the piecewise-polynomial scheme announced in Section 3.

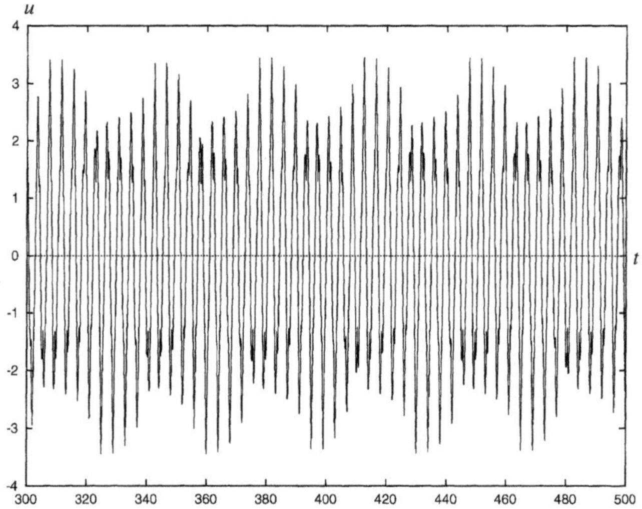

Fig. 13. Simulation with $\varepsilon = \frac{1}{4}$ and square wave forcing of amplitude $A = 1.22$ and period $T = 1$. The solution appears to be (approximately) periodic with period 32.

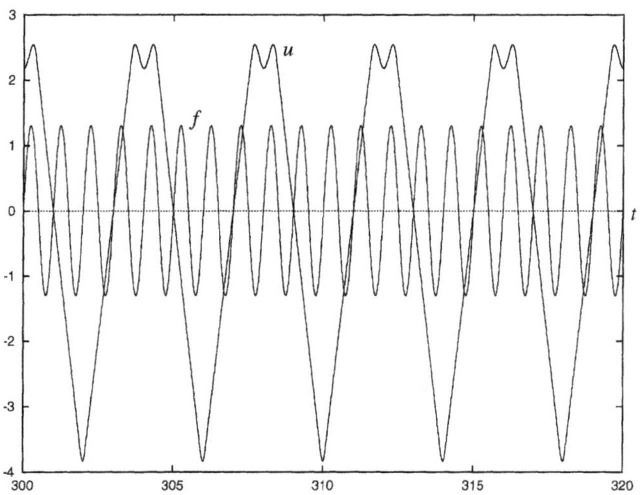

Fig. 14. Period 4 solution with $\varepsilon = \frac{1}{4}$, and sine wave forcing of amplitude $A = 1.3$ and period $T = 1$. Contrast with Fig. 12.

As A is reduced, there is a bifurcation where the period of the attractor jumps to 4 (see Fig. 12). Rather than continue with period 4 as $A \to 0$, there appears a sequence of period doubling and period halving bifurcations. E.g., Fig. 13 shows a complicated (apparently stable) solution with period 32. Other types of period 4 solution also seem possible (see Fig. 14), which is for sine wave forcing, and whose solution has a quite different form from that of Fig. 12.

Given the presence of period-doubling bifurcations, it is natural to ask if (2) can have a chaotic attractor for suitable f. We do not have a conclusive answer to this question – distinguishing between large period and chaotic solutions requires larger integration times than we have attempted so far. For square wave forcing, it might be possible to describe the dynamics of (2) with a family of maps, for which chaos could either be proved or shown not to occur. This is an open question which requires further work.

Acknowledgements

We would especially like to thank Brenda Allen, who in conjunction with the first author wrote down the differential delay equation considered here, and who motivated our interest in piecewise-affine delay equations more generally.

References

[1] B. Allen, Models of an engine controller, M.Sc. Thesis, Oxford Centre for Industrial and Applied Mathematics, Oxford University, September 1994.
[2] B. Allen, Non-smooth differential delay equations, D.Phil. Thesis, Faculty of Mathematical Sciences, Oxford Unive-rsity, July 1997.
[3] C.T.H. Baker, C.A.H. Paul, D.R. Wille, Issues in the numerical solution of evolutionary delay differential equations, Adv. Comput. Math. 3 (1995) 171–196.
[4] N. MacDonald, Time Lags in Biological Models, Lecture Notes in Biomathematics, Vol. 27, Springer, Berlin, 1978.
[5] J. Mawhin, Periodic solutions to nonlinear functional differential equations, J. Differential Equations 10 (1971) 240–261.
[6] L.A.A. Warnes, Analogue and Digital Electronics, MacMillan, New York, 1998.
[7] R.E. Wilson, Piecewise-polynomial solver for a class of delay differential equations, in preparation.

ELSEVIER

Journal of Computational and Applied Mathematics 125 (2000) 217–249

JOURNAL OF COMPUTATIONAL AND APPLIED MATHEMATICS

www.elsevier.nl/locate/cam

A perspective on the numerical treatment of Volterra equations

Christopher T.H. Baker [1]

Department of Mathematics, The University of Manchester, Oxford Road, Manchester M13 9PL, UK

Received 14 January 2000; received in revised form 27 March 2000

Abstract

We discuss the properties and numerical treatment of various types of Volterra and Abel–Volterra integral and integro-differential equations. © 2000 Elsevier Science B.V. All rights reserved.

Keywords: Volterra and Abel equations; Quadrature, Runge–Kutta and collocation methods; Convergence; Superconvergence; Stability

1. Introduction

Volterra's book *Leçons sur les équations intégrales et intégro-différentielles* appeared in 1913. Since then, a considerable literature on the theory and on applications (which include elasticity, semi-conductors, scattering theory, seismology, heat conduction, metallurgy, fluid flow, chemical reactions, population dynamics, etc.) — e.g., [52,55,57,58,76,77,95,108,111,117,118] — and on the numerics — e.g., [1,25,30,41,49,92,96,109] — has appeared.

The obligation to provide a perspective on the subject in the year 2000 has settled on this author. A paper such as this, in which we seek to convey basic theoretical and computational features, could have been written from many different viewpoints, including that of *mathematical modelling*, that of *robust numerical algorithms*, or that of the *mathematical analysis* of numerical methods. Each standpoint has its own relevance to the numerical simulation of real phenomena, and each relies on rather different foundations and methodologies. In particular, the mathematical theory of numerical formulae relies on detailed and sometimes intricate analytical arguments to generate a formal theory (cf. [49] and its extensive references and bibliographical notes) that has mathematical significance

[1] Professor of Mathematics, & Director, Manchester Center for Computational Mathematics (MCCM).
E-mail address: cthbaker@ma.man.ac.uk (C.T.H. Baker).

PII: S0377-0427(00)00470-2

in its own right but, at least in part, attempts to generate insight into the performance of real-life algorithms.

Within the space available, we give an introduction to the numerical treatment of classical Volterra equations. It is hoped that the author, while avoiding pedantry, is sufficiently careful to avoid mathematical pitfalls or ambiguities occasionally found in the literature that a reader may use the work as a basis for further critical reading and research. The mathematical style may be considered to be intermediate between that of the excellent introduction of Linz [96] (Math. Rev. 86m:65163) and the majestic detail provided by Brunner and van der Houwen [49] (Math. Rev. 88g:65136), which is currently out of print.

2. Basic theory

The classical forms of *Volterra integral equation of the first and second kind* and of *Volterra integro-differential equations* are, respectively,

$$\int_{t_0}^{t} K(t,s,y(s))\,\mathrm{d}s = g(t) \quad (t \in [t_0, T]), \tag{2.1a}$$

$$y(t) = g(t) + \int_{t_0}^{t} K(t,s,y(s))\,\mathrm{d}s \quad (t \in [t_0, T]), \tag{2.1b}$$

$$y'(t) = g(t) + \int_{t_0}^{t} K(t,s,y(s))\,\mathrm{d}s \quad (t \in [t_0, T]) \text{ with } y(t_0) = y_0 \tag{2.1c}$$

(given $g(\cdot)$ and $K(\cdot,\cdot,\cdot)$, find $y(\cdot)$) where the interval $[t_0, T]$ (interpreted as $[t_0, \infty)$ if T is unbounded) is prescribed, and where $g(\cdot) \in C[t_0, T]$ and

Hypothesis 1. (a) $K(t,s,v)$ is continuous for $t_0 \leqslant s \leqslant t \leqslant T$, $g(\cdot) \in C[t_0, T]$;
(b) $K(t,s,v)$ satisfies a uniform Lipschitz condition in v for $t_0 \leqslant s \leqslant t \leqslant T$.

As an alternative to (2.1c) we may encounter

$$y'(t) = F\left(t, y(t), \int_{t_0}^{t} K(t,s,y(s))\,\mathrm{d}s\right) \quad (t \in [t_0, T]) \quad \text{with } y(t_0) = y_0 \tag{2.1d}$$

(given $F(\cdot,\cdot,\cdot)$ and $K(\cdot,\cdot,\cdot)$, find $y(\cdot)$) where, in addition to Hypothesis 1 itself, $F(\cdot,\cdot,\cdot)$ satisfies appropriate Lipschitz conditions. The functions involved here can be real or complex valued. Higher-order integro-differential equations (say generalizing (2.1d)) also arise. Alternative assumptions to those in Hypothesis 1 allow us to consider equations of Abel type. Each of the Volterra integral equations (2.1a)–(2.1d), gives a corresponding *Abel* (or *Abel–Volterra*) equation when

Hypothesis 2. (a) $K(t,s,v) = (t-s)^{-\nu} H(t,s,v)$ where $0 < \nu < 1$ and $H(t,s,v)$ is continuous for $t_0 \leqslant s \leqslant t \leqslant T$, $g(\cdot) \in C[t_0, T]$; (b) $H(t,s,v)$ satisfies a uniform Lipschitz condition in v for $t_0 \leqslant s \leqslant t \leqslant T$.

Thus, the Abel equation of the second kind has the form

$$y(t) = g(t) + \int_{t_0}^{t} (t-s)^{-\nu} H(t,s,y(s))\, ds \quad (t \in [t_0, T)). \tag{2.1e}$$

The condition $0 < \nu < 1$ is customary when referring to Abel equations.

Remark 2.1. (a) In the theory, the uniformity of the Lipschitz conditions may be *relaxed.* However, to obtain the benefits of the numerical processes described later, one often requires continuity properties *stronger* than those required in Hypotheses 1 and 2 (e.g., higher-order differentiability). Moreover, one sometimes (e.g., in certain Runge–Kutta methods discussed later) requires a suitable smooth *extension* of $K(t,s,v)$ (or of $H(t,s,v)$) defined on an extended domain $t_0 \leqslant s \leqslant t + \Delta \leqslant T + \Delta$ with $\Delta > 0$:

$$K_{\text{ext}}(t,s,v) = \begin{cases} K(t,s,v) & \text{if } s \leqslant t, \\ K_{\text{new}}(t,s,v) & \text{if } s > t \end{cases} \tag{2.2}$$

or

$$H_{\text{ext}}(t,s,v) = \begin{cases} H(t,s,v) & \text{if } s \leqslant t, \\ H_{\text{new}}(t,s,v) & \text{if } s > t. \end{cases} \tag{2.3}$$

(b) If $K(t,s,v) = \mathsf{k}(t,s)v$, Eqs. (2.1a)–(2.1c) are *linear equations.*

(c) An integral of the form $\int_{t_0}^{t} k(t-s)\phi(s)\, ds$ is called a convolution of $k(\cdot)$ with $\phi(\cdot)$ — taking $k(t) \equiv 1$ is a special case. (Corresponding sums $\sum_{j=0}^{n} \omega_{n-j}\phi_j$, and the matrix–vector counterparts $\sum_{j=0}^{n} \boldsymbol{\Omega}_{n-j}\boldsymbol{\phi}_j$, are *discrete convolutions.*) If $K(t,s,v) = k(t-s)v$ or $K(t,s,v) = k(t-s)\varphi(v)$ for some continuous function $k(\cdot)$ and (usually Lipschitz–continuous) $\varphi(\cdot)$ the Volterra equations are (linear or nonlinear) *convolution equations.* Under additional conditions, the Laplace transform provides a tool for analysing linear convolution equations.

The ordinary differential equation (ODE) $y'(t) = f(t, y(t))$ $(t \geqslant t_0)$ can be re-written as a Volterra integral equation of the second kind:

$$y(t) = \int_{t_0}^{t} f(s, y(s))\, ds + y(t_0) \quad (t \geqslant t_0). \tag{2.4}$$

Furthermore, Eq. (2.1d) can be written as a system of equations, one of which is an ODE and the other a Volterra integral equation:

$$y'(t) = F(t, y(t), z(t)), \tag{2.5a}$$

$$z(t) = \int_{t_0}^{t} K(t,s,y(s))\, ds. \tag{2.5b}$$

Alternatively, by integrating (2.5a) and expressing it as

$$y(t) = y(t_0) + \int_{t_0}^{t} F(s, y(s), z(s))\, ds, \tag{2.5c}$$

we obtain the system of Volterra integral equations (2.5b), (2.5c). The intimate relation between systems of Volterra equations of the second kind and Volterra integro-differential equations encourages us (though sacrificing some insight and detail) to concentrate upon the former.

A starting point for any mathematical discussion is to ask what is known about the well-posedness of the problems under consideration. (Rigorous concepts of well-posedness depend upon choices of underlying function spaces. Sometimes, a solution is required to be *positive*.) If the Volterra equation of the first kind is to be interpreted as implying equality everywhere, we require $g(t_0)=0$ and, clearly, a solution does not exist for all $g(\cdot)$. If $g(t_0)=0$ and we have sufficient differentiability, we can deduce

$$K(t,t,y(t))+\int_{t_0}^{t} K_t(t,s,y(s))\,ds = g'(t).$$

In particular, if $K(t,s,v)=\mathsf{k}(t,s)v$ and $\mathsf{k}(t,t)$ is nonvanishing, we have

$$y(t)+\int_{t_0}^{t}\frac{\mathsf{k}_t(t,s)}{\mathsf{k}(t,t)}y(s)\,ds=\frac{g'(t)}{\mathsf{k}(t,t)}\quad(\text{for } t\geqslant t_0), \tag{2.6}$$

a linear Volterra equation of the second kind. On the other hand, if $\mathsf{k}(t,t)$ vanishes, but $g''(t)$ and $\mathsf{k}_{tt}(t,s)$ exist and $\mathsf{k}_t(t,t)$ is nonvanishing, we can obtain an equation of the second kind by a further differentiation. Our remarks serve to demonstrate that the first-kind equation can suffer various degrees of ill-posedness in the sense that if nonsmooth perturbations are made to $g(\cdot)$ a solution (other than in a generalized sense) may fail to exist. Inexact data therefore raises problems. Theoretical results on numerical procedures for first-kind equations usually assume a condition such as $\inf|\mathsf{k}(t,t)|>0$, or an analogous condition in the nonlinear case. Some numerical methods require one to supply the value $y(t_0)$ (when (2.6) is valid, $y(t_0)=g'(t_0)/\mathsf{k}(t_0,t_0)$).

Example 2.2. When $\int_{t_0}^{t} y(s)\,ds=g(t)$, the solution, if it exists, is $y(t)=g'(t)$. For Abel equations of the first kind, similar remarks are possible; e.g., if

$$\frac{1}{\Gamma(1-\nu)}\int_{t_0}^{t}(t-s)^{-\nu}y(s)\,ds=g(t)\quad \text{for } \nu\in(0,1) \tag{2.7a}$$

($\nu=\frac{1}{2}$ has special interest) then the solution $y(\cdot)$, if it exists, is expressible

$$y(t)=\frac{1}{\Gamma(\nu)}\int_{t_0}^{t}(t-s)^{\nu-1}g'(s)\,ds \tag{2.7b}$$

and $y(\cdot)$ is now a *fractional derivative* (or a *fractional integral*) of $g(\cdot)$ (cf. [49, p.8; 76, p.3; 96, p.73]). One can write (2.7) symbolically as $J^{(1-\nu)}y=g$, and $y=J^{\nu}Dg$ (where J is an operator of indefinite integration and D is the operator of differentiation and DJ is the identity).

Remark 2.3. On the face of it, there could be a link between the solution $y(\cdot)$ of the equation of the first kind (2.1a) and the solution $y_\varepsilon(\cdot)$ of the singularly perturbed equation $\varepsilon y_\varepsilon(t)=\int_{t_0}^{t}K(t,s,y_\varepsilon(s))\,ds-g(t)$ of the second kind (cf. [91]), either as $\varepsilon\nearrow 0$ or $\varepsilon\searrow 0$. However, such links can only be established under special hypotheses. This said, one can develop discretization formulae that are *candidates* for application to equations of the first kind by constructing them for the singularly perturbed equation of the second kind above and, formally, setting $\varepsilon=0$, provided one appreciates that whether or not these formulae are useful has to be investigated independently.

The theory for linear equations of the second kind is simpler than that of equations of the first kind. In particular, for the Volterra equation

$$y(t) = g(t) + \lambda \int_{t_0}^{t} \mathsf{k}(t,s)y(s)\,ds \quad (\text{for } t \geqslant t_0), \tag{2.8a}$$

we have

$$y(t) = g(t) + \lambda \int_{t_0}^{t} \mathsf{r}^{\{\lambda\}}(t,s)g(s)\,ds \quad (\text{for } t \geqslant t_0), \tag{2.8b}$$

where $\mathsf{r}^{\{\lambda\}}(\cdot,\cdot)$ is the *resolvent kernel* for (2.8a). The Neumann series gives the expression $\mathsf{r}^{\{\lambda\}}(t,s) = \sum_{r=1}^{\infty} \lambda^{r-1}\mathsf{k}^r(t,s)$ where $\mathsf{k}^r(t,s) = \int_{t_0}^{t} \mathsf{k}(t,\sigma)\mathsf{k}^{r-1}(\sigma,s)\,d\sigma$ $(r = 2,3,\ldots)$, and $\mathsf{k}^1(t,s) = \mathsf{k}(t,s)$. One has

$$\mathsf{r}^{\{\lambda\}}(t,\sigma) = \lambda \int_{t_0}^{t} \mathsf{k}(t,s)\mathsf{r}^{\{\lambda\}}(s,\sigma)\,ds + \mathsf{k}(t,\sigma) \quad (\text{for } t_0 \leqslant \sigma \leqslant t). \tag{2.9}$$

Since $\mathsf{r}^{\{\lambda\}}(s,\sigma) = 0$ if $\sigma > s$, the lower limit of integration in (2.9) can be replaced by σ. Smoothness properties of $\mathsf{r}^{\{\lambda\}}(t,s)$ follow from those of $\mathsf{k}(t,s)$, and the degree of smoothness of $y(\cdot)$ (which has an impact on the suitability of various numerical techniques) follows from the properties of $g(\cdot)$ and the resolvent. If $\mathsf{k}(t,s)$ is of Abel type, $\mathsf{k}(t,s) = (t-s)^{-\nu}\mathsf{h}(t,s)$, for $\nu \in (0,1)$, the resolvent is expressible [49, p.16] as $\mathsf{r}^{\{\lambda\}}(t,s) = (t-s)^{-\nu}\mathsf{p}^{\{\lambda,\nu\}}(t,s)$ with continuous $\mathsf{p}^{\{\lambda,\nu\}}(\cdot,\cdot)$. One deduces that the solution of an Abel equation of the second kind has unbounded derivatives at t_0 when $g(t)$ is smooth (asymptotic expansions as $t \to t_0$ can be found).

The Neumann series is related to an iteration that, for the more general nonlinear equation (2.1b), reads

$$y_{k+1}(t) = g(t) + \int_{t_0}^{t} K(t,s,y_k(s))\,ds \quad (k = 0,1,2,\ldots), \tag{2.10}$$

often with $y_0(t) = g(t)$. (The iteration may also be used to "refine" an approximation $\tilde{y}(t)$ by setting $y_0(\cdot) = \tilde{y}(\cdot)$ and computing the corresponding iterate $y_1(\cdot)$.) Iteration (2.10) collapses to the Picard iteration in the case of an integrated form of an ODE (2.4). It extends in an obvious manner to systems of Volterra integral equations, and, thence, through its application to (2.5b), (2.5c) to the integro-differential equation (2.1d).

Remark 2.4. For the linear equation $y'(t) = a(t)y(t) + \int_{t_0}^{t} \mathsf{k}(t,s)y(s)\,ds + g(t)$ $(t \geqslant t_0)$ subject to the initial condition $y(t_0) = y_0$, where $a(\cdot)$ is continuous, we can establish the existence of a *resolvent* $\mathsf{u}(t,s)$ such that $y(t) = \mathsf{u}(t,t_0)y_0 + \int_{t_0}^{t} \mathsf{u}(t,s)g(s)\,ds$; $(\partial/\partial s)\mathsf{u}(t,s) + a(t)\mathsf{u}(t,s) + \int_{s}^{t} \mathsf{k}(t,\sigma)\mathsf{u}(\sigma,s)\,d\sigma = 0$, $\mathsf{u}(t,t) = 1$.

We have introduced the resolvent kernels $\mathsf{r}^{\{\lambda\}}(t,s)$, and $\mathsf{u}(t,s)$, into our discussion because they have a rôle in a discussion of *stability of solutions* (see [49, p.493–494; 118]), and in an *error analysis of certain approximate solutions* — both studies involving aspects of perturbation theory. Clearly, if $g(\cdot)$ in (2.8a) is perturbed to $g(\cdot)+\delta g(\cdot)$ then, by (2.8b), $y(\cdot)$ suffers a consequent change

$$\delta y(t) = \lambda \int_{t_0}^{t} \mathsf{r}^{\{\lambda\}}(t,s)\delta g(s)\,ds + \delta g(t). \tag{2.11}$$

With appropriate conditions, we have a similar result[2] for the nonlinear equation (2.1b). Suppose that $y(\cdot)$ is the unique solution of (2.1b) and we have

$$u(t) = g(t) + \delta(t) + \int_{t_0}^{t} K(t,s;u(s))\,\mathrm{d}s, \tag{2.12a}$$

$$u(t) \equiv y(t) + \delta y(t) \tag{2.12b}$$

(where we assume that $\delta(t) \equiv \delta(u(\cdot);t)$). Then there exists a corresponding $R(t,s;u(s))$ such that

$$\delta y(t) = \delta(t) + \int_{t_0}^{t} R(t,s;u(s))\delta(s)\,\mathrm{d}s. \tag{2.13}$$

Let us obtain one such result, under assumptions that are obvious. Subtract from (2.12a) the unperturbed equation (2.1b) and we obtain

$$\delta y(t) = \delta(t) + \int_{t_0}^{t} \{K(t,s;u(s)) - K(t,s;y(s))\}\,\mathrm{d}s. \tag{2.14}$$

Hence, with $K_\natural(t,s;u(s)) = K_3(t,s;(1-w(s))y(s) + w(s)u(s))$ for some appropriate $w(s) \in [0,1]$, where $K_3(t,s;v) = (\partial/\partial v)K_3(t,s;v)$, we have

$$\delta y(t) = \delta(t) + \int_{t_0}^{t} K_\natural(t,s;u(s))\delta y(s)\,\mathrm{d}s, \tag{2.15}$$

which provides a result of the form (2.13) with $R(t,s;u(s))$ as the resolvent kernel satisfying $R(t,s;u(s)) = K_\natural(t,s;u(s)) + \int_{t_0}^{t} R(t,\sigma;u(\sigma))K_\natural(\sigma,s;u(s))\,\mathrm{d}\sigma$.

3. Some numerical methods

We review, selectively, numerical methods for Volterra integral and integro-differential equations, concentrating on the classical forms. Discrete methods for the solution of Volterra integral and integro-differential equations are based upon the development of a grid or mesh:

$$\mathcal{T} := \{t_0 < t_1 < t_2 < \cdots < t_{n-1} < t_n < \cdots\}, \quad h_n := t_{n+1} - t_n. \tag{3.1}$$

Remark 3.1. (a) The *width* (or *diameter*) of a grid $\mathcal{T}$ is $h(\mathcal{T}) := \sup\{h_n : t_n \in \mathcal{T}\}$. $\mathcal{T}$ is called *uniform* if $h_n \equiv h$ for all n, and *quasi-uniform* if there exists a finite κ such that $\sup h_n \leqslant \kappa \inf h_n$. If $t_N := \max_n t_n = T < \infty$ then $\mathcal{T}$ is a finite grid that provides a partition of the bounded interval $[t_0, T]$. Such a partition is called (*i*) *graded* with *grading exponent* α if $t_n - t_0 = (n/N)^\alpha$ [49, p.349 et seq.], or (*ii*) *geometric* [43] if $t_n - t_0 = \beta^{N-n}\{T - t_0\}$ for some $\beta \in (0,1)$. (b) Later, in connection with Runge–Kutta (RK) or collocation processes, we shall introduce a set of indexed abscissae $\{\vartheta_i\}_{i=1}^m$ and the points $t_{n,i} = t_n + \vartheta_i h_n$. In the case $0 \leqslant \vartheta_1 < \vartheta_2 < \cdots < \vartheta_m \leqslant 1$ these points define an

[2] Various similar formulae appear in the literature [11,24,94], sometimes under the heading of variation of constants formulae for integro-differential equations obtained by differentiating the integral equation and with restrictions on the form of perturbation. In this paper, we are trying to capture the spirit of the approach and we shall not need detailed information about $R(t,s;u(s))$ in (2.13), other than knowledge of its continuity properties, which are obvious in the linear case.

ordered set of abscissae $\mathscr{T}^{\#}(\boldsymbol{\vartheta}):=\{t_0 \leqslant t_0+\vartheta_1 h_0 < t_0+\vartheta_2 h_0 < \cdots < t_0+\vartheta_m h_0 \leqslant t_1 \leqslant t_1+\vartheta_1 h_1 < t_1+\vartheta_2 h_1 < \cdots < t_1+\vartheta_m h_1 \leqslant t_2 < \cdots\}$. We shall usually need to note whether $\vartheta_1=0$, $\vartheta_m=1$.

We now consider some *primitive quadrature methods*. A simple family of methods for classical Volterra equations can be built from the quadrature rule

$$\int_{t_j}^{t_{j+1}} \phi(s)\,\mathrm{d}s \approx h_j\{(1-\theta)\phi(t_j)+\theta\phi(t_{j+1})\} \tag{3.2a}$$

(where $\theta \in [0,1]$), which yield, in particular, the Euler rule ($\theta=0$), the backward Euler rule ($\theta=1$), and the trapezium rule ($\theta=\frac{1}{2}$). From the primitive rule (3.2a) one obtains the basic components for use with Volterra integral and integro-differential equations. In particular, one obtains by repeated application of the basic rules the approximations

$$\int_{t_0}^{t_n} \phi(s)\,\mathrm{d}s \approx \sum_{j=0}^{n-1} h_j\{(1-\theta)\phi(t_j)+\theta\phi(t_{j+1})\}. \tag{3.2b}$$

If we proceed formally, we can discretize (2.1b) using (for $n=0,1,2,\ldots$)

$$\tilde{y}(t_{n+1})=g(t_{n+1})+\sum_{j=0}^{n} h_j\{(1-\theta)K(t_{n+1},t_j,\tilde{y}(t_j))+\theta K(t_{n+1},t_{j+1},\tilde{y}(t_{j+1}))\}. \tag{3.3}$$

We need to establish that Eqs. (3.3) do have a solution and that it provides a good approximation to $y(\cdot)$. However, the basic idea underlying a wide class of numerical methods is already present since Eqs. (3.3) have the form

$$\tilde{y}(t_{n+1})=g(t_{n+1})+\sum_{j=0}^{n+1} \Omega_{n+1,j}K(t_{n+1},t_j,\tilde{y}(t_j)), \tag{3.4}$$

which are *discrete Volterra equations* that arise from a family of quadrature rules using weights $\{\Omega_{n,j}\}_{j=0}^{n}$ ($n=1,2,3,\ldots$) and abscissae $\{t_j\} \subset \mathscr{T}$.

Example 3.2. For the *test equation* $y(t)-\lambda\int_{t_0}^{t} y(s)\,\mathrm{d}s=g(t)$, the θ-rule with uniform step $h_r=h$ yields a recurrence that simplifies to

$$\frac{\tilde{y}(t_{n+1})-\tilde{y}(t_n)}{h}-\lambda\{\theta\tilde{y}(t_{n+1})+(1-\theta)\tilde{y}(t_n)\}=\frac{g(t_{n+1})-g(t_n)}{h} \tag{3.5}$$

(which is solvable if $\lambda\theta h \neq 1$) that simulates the corresponding analytical result $y'(t)-\lambda y(t)=g'(t)$. Those familiar with the numerics of ODEs can, in this example, readily infer properties from (3.5).

The θ-rules have relatively low order, and one may turn to alternatives. The set of weights that arise on taking $h_0=h_1=h'$, $h_2=h_3=h''$, $h_4=h_5=h'''$,$\ldots$ and combining Simpson's rule (repeated as necessary) with the trapezium rule is indicated in the following tableau:

Tableau of weights $\Omega_{n,j}$

$j=$	0	1	2	3	4	5	6	...
$n=1$	$\frac{h'}{2}$	$\frac{h'}{2}$						
$n=2$	$\frac{h'}{3}$	$\frac{4h'}{3}$	$\frac{h'}{3}$					
$n=3$	$\frac{h'}{3}$	$\frac{4h'}{3}$	$\frac{h'}{3}+\frac{h''}{2}$	$\frac{h''}{2}$				
$n=4$	$\frac{h'}{3}$	$\frac{4h'}{3}$	$\frac{h'}{3}+\frac{h''}{3}$	$\frac{4h''}{3}$	$\frac{h''}{3}$			
$n=5$	$\frac{h'}{3}$	$\frac{4h'}{3}$	$\frac{h'}{3}+\frac{h''}{3}$	$\frac{4h''}{3}$	$\frac{h''}{3}+\frac{h'''}{2}$	$\frac{h'''}{2}$		
		Simpson's rule		Simpson's rule	trapezium rule			
$n=6$	$\frac{h'}{3}$	$\frac{4h'}{3}$	$\frac{h'}{3}+\frac{h''}{3}$	$\frac{4h''}{3}$	$\frac{h''}{3}+\frac{h'''}{3}$	$\frac{4h'''}{3}$	$\frac{h'''}{3}$	
		Simpson's rule		Simpson's rule		Simpson's rule		
$\vdots$		...		...		...		

(3.6)

Consider the rather special case of a *uniform mesh* ($h_r \equiv h$): the corresponding quadrature weights $\{\Omega_{n,j}\}$ for $j \leqslant n$ is said to have a *repetition factor* r if there is a least integer r such that, for $n \geqslant n_0$, $\Omega_{n+r,j} = \Omega_{n,j}$ for $r_0 \leqslant j \leqslant n - r_1$ where n_0, r_0, r_1 are fixed. Where there is no exact repetition, the weights are said to have *asymptotic repetition factor* r' provided there exists a least integer r' such that, where r_0, r_1 are fixed, $\lim_{n\to\infty} \sup_{r_0 \leqslant j \leqslant n-r_1} |\Omega_{n+r',j} - \Omega_{n,j}| = 0$. With $h' = h'' = h''' = \cdots = h$, the weights in (3.6) have repetition factor unity.

The discretization in (3.6) may be thought of as based on a repeated application of Simpson's rule to approximate the "history" and an application of the trapezium rule or another Simpson's rule for the "local" or "incremental" term. This viewpoint is lost if, instead, we generate the weights $\Omega_{n,j}$ for odd values of n by applying the trapezium rule to the first sub-interval (where needed) rather than the last. Retain a uniform step h and we now have weights

$$
\begin{array}{lcccccc}
n=3 & \frac{h}{2} & \frac{h}{2}+\frac{h}{3} & \frac{4h}{3} & \frac{h}{3} & & \\
n=5 & \frac{h}{2} & \frac{h}{2}+\frac{h}{3} & \frac{4h}{3} & \frac{h}{3}+\frac{h}{3} & \frac{4h}{3} & \frac{h}{3} \\
 & \multicolumn{2}{c}{\text{trapezium rule}} & \multicolumn{2}{c}{\text{Simpson's rule}} & \multicolumn{2}{c}{\text{Simpson's rule}}
\end{array}, \qquad (3.7)
$$

with a repetition factor 2. *Although one can establish convergence results of the same order*, for both cases, the stability properties of weights (3.7) have led to them being rejected (see [110] for more insight) as unsuitable for numerical computation in the solution of arbitrary Volterra integral equations of the second kind; the scheme suffers the same defect as Simpson's rule for solving ODEs (under certain assumptions, it is weakly unstable).

As the last remark indicates, certain 'plausible' weights (that yield convergence) may be rejected for practical computation in the numerical solution of Volterra equations of the second kind. For equations of the first kind (2.1a), the situation is more delicate: Whilst the formal analogue of (3.4) reads

$$\sum_{r=0}^{n+1} \Omega_{n+1,j} K(t_{n+1}, t_j, \tilde{y}(t_j)) = g(t_{n+1}) \quad (n = 0, 1, 2, \ldots) \tag{3.8}$$

(cf. Remark 2.3) one has to be more circumspect in the choice of quadrature rules to obtain a satisfactory (or even convergent) scheme for (2.1a).

Example 3.3. Consider, for the case $\int_{t_0}^{t} y(s)\,ds = g(t)$ (where $g(t_0)=0$ and $g'(t)$ exists for $t \in [t_0, \infty)$), the application of a θ-rule with a uniform step $h_r \equiv h$:

$$(1-\theta)h\tilde{y}(t_0) + h\tilde{y}(t_1) + \cdots + h\tilde{y}(t_{n-1}) + \theta h\tilde{y}(t_n) = g(t_n) \quad (n = 1, 2, 3, \ldots).$$

For $\theta \in (0, 1]$, we deduce that (for $n = 0, 1, 2, \ldots$)

$$\tilde{y}(t_{n+1}) = \mu\tilde{y}(t_n) + \frac{g(t_{n+1}) - g(t_n)}{\theta h} \quad \text{where } \mu := \frac{\theta - 1}{\theta}, \tag{3.9}$$

which is an unstable recurrence if $|\mu| > 1$, namely if $\theta \in (0, \frac{1}{2})$. Unless $\theta = 1$ the value $\tilde{y}(t_0) \approx g'(t_0)$ in these equations has to be supplied independently. Clearly, if $\theta = 1$ (the backward or implicit Euler rule) then we deduce $\tilde{y}(t_{n+1}) = \{g(t_{n+1}) - g(t_n)\}/h$ which is a local approximation to $g'(t_{n+1})$ that clearly converges to the correct value as $h \to 0$. For $\theta \in (0, 1)$,

$$\tilde{y}(t_{n+1}) = \sum_{r=0}^{n} \mu^{n-r} \frac{g(t_{r+1}) - g(t_r)}{\theta h} + \mu^{n+1}\tilde{y}(t_0). \tag{3.10}$$

A difficulty with (3.10) is that it seeks to approximate $g'(t_{n+1})$ in terms of $g(t_0), g(t_1), g(t_2), \ldots, g(t_{n+1})$, and if $|\mu| > 1$ it gives increasing weight to $g(t_0)$ as n increases. In contrast, the value $g'(t_{n+1}) = \lim_{\delta \to 0} \{g(t_{n+1} + \delta) - g(t_{n+1})\}/\delta$ depends only upon limiting behaviour of $g(\cdot)$ in the neighbourhood of t_{n+1}. Given $g(t)$, one can (*without changing the derivative* $g'(t_{n+1})$) add an arbitrary multiple of any smooth function $\gamma(t)$ that has support (t_0, t_{n+1}), thereby *changing the computed approximation* (3.10) *by an arbitrary amount*. If $\theta = 0$, the coefficient matrix in our equations is formally singular, and to be useful the equations have to be re-interpreted:
$h\tilde{y}(t_1) = g(t_2) - h\tilde{y}(t_0)$; $h\tilde{y}(t_1) + h\tilde{y}(t_2) = g(t_3) - h\tilde{y}(t_0)$; $h\tilde{y}(t_1) + h\tilde{y}(t_2) + h\tilde{y}(t_3) = g(t_4) - h\tilde{y}(t_0), \ldots$ (where $\tilde{y}(t_0)$ is obtained independently).
We now find $\tilde{y}(t_n) = \{g(t_{n+1}) - g(t_n)\}/h$ which converges to $g'(t_\star)$ as $h \to 0$ and $n \to \infty$ with $t_n = t_\star \in (t_0, \infty)$ fixed.

One can generalize the composite Euler and trapezium rules in order to treat Abel equations; corresponding to (3.2b) one has, in particular, formulae

$$\int_{t_0}^{t_n} (t_n - s)^{-\nu}\phi(s)\,ds \approx \sum_{j=0}^{n-1} \left\{ \int_{t_j}^{t_{j+1}} (t_n - s)^{-\nu}\,ds \right\} \phi(t_j), \tag{3.11a}$$

$$\int_{t_0}^{t_n} (t_n - s)^{-\nu}\phi(s)\,ds \approx \sum_{j=0}^{n-1} \left\{ \int_{t_j}^{t_{j+1}} (t_n - s)^{-\nu}\,ds \right\} \phi(t_{j+1}), \tag{3.11b}$$

$$\int_{t_0}^{t_n} (t_n - s)^{-\nu}\phi(s)\,\mathrm{d}s$$
$$\approx \sum_{j=0}^{n-1}\left\{\int_{t_j}^{t_{j+1}} (t_n - s)^{-\nu}\frac{t_{j+1}-s}{h_j}\,\mathrm{d}s\;\phi(t_j) + \int_{t_j}^{t_{j+1}} (t_n - s)^{-\nu}\frac{s-t_j}{h_j}\,\mathrm{d}s\;\phi(t_{j+1})\right\} \tag{3.11c}$$

and the integrals that occur here as weights multiplying values $\phi(t_j)$ and $\phi(t_{j+1})$ can be written down very simply, given $\mathscr{T}$. Bounds on the error in such approximations are readily obtained in terms of the error in piecewise-constant or piecewise-linear interpolation to $\phi(\cdot)$. With a uniform grid $\mathscr{T}$, the above approximations yield (on summation) convolution sums approximating the convolution integrals $\int_{t_0}^{t_n}(t_n - s)^{-\nu}\phi(s)\,\mathrm{d}s$.

Remark 3.4. When solving Volterra integro-differential equations we can combine methods for ODEs and integral equations. Thus, if $\theta', \theta'' \in [0,1]$ we might write, discretizing (2.5a), (2.5b),

$$\tilde{y}(t_{n+1}) = \tilde{y}(t_n) + h_n\{(1-\theta')F(t_n, \tilde{y}(t_n), \tilde{z}(t_n)) + \theta' F(t_{n+1}, \tilde{y}(t_{n+1}), \tilde{z}(t_{n+1}))\},$$
$$\tilde{z}(t_k) := \sum_{r=0}^{k-1} h_r\{(1-\theta'')K(t_k, t_r, \tilde{y}(t_r)) + \theta'' K(t_k, t_{r+1}, \tilde{y}(t_{r+1}))\}, \quad k \in \{1,2,3,\ldots,(n+1)\}. \tag{3.12}$$

(We intend to indicate some flexibility of approach which might not be transparent if one viewed the integro-differential equation as a system of two integral equations: one may reasonably ask whether there is a purpose to selecting $\theta' \neq \theta''$.) If the integral term is, instead, of Abel type, there is a rôle for approximations (3.11) in the term for $\tilde{z}(t_r)$.

4. Relationship to ODE methods

The integrated form of the initial-value problem for the equation $y'(t) = f(t, y(t))$ (for $t \geqslant t_0$) is (2.4), and it is not surprising to find that (although Volterra equations have more complex character) there is a close connection between methods for the ODE and for classical Volterra integral equations of the second kind. We enumerate some features:

(1) For certain types of kernel $K(t,s,v)$ (in particular, $\sum_{j=1}^{N} T_j(t)S_j(s,v)$, or $(t-s)^n\exp\{-v(t-s)\}v)$ we find that (2.1b) and (2.1c), (2.1d) can be reduced to a system of ODEs (see [20] for an exploitation of such a result);

(2) An *embedding* relationship has a rôle in the theory of numerical methods. Consider $G(t,s)$ such that (one may compare with Eqs. (8.7)):

$$\frac{\partial}{\partial t}G(t,s) = K(t,s,G(t,t)) \quad (t_0 \leqslant s \leqslant t), \tag{4.1a}$$

$$G(t_0, s) = g(s) \quad (s \in [t_0, T]). \tag{4.1b}$$

Then the solution of (2.1b) satisfies $y(t) = G(t,t)$. We refer to [114,123].

(3) In principle, every ODE method (multistep, cyclic multistep, RK method, hybrid method, general linear method) generates a corresponding integral equation method for (2.1b), or (2.1d). By (2.4), every integral equation method for (2.1b) generates a (possibly novel) ODE method.

Remark 4.1. Jones and McKee [88] used a family of predictor–corrector methods of *variable order* and *variable step-size* to devise a method for nonlinear second kind Volterra integral equations with smooth kernels. Their strategy for changing the step size followed closely one employed for ODEs. (b) Bownds and Applebaum [21] have contributed a code based on separable kernels. (c) Shampine [116] addressed problems associated with the special nature of the ODE approach to Volterra equations.

Variable step ODE methods provide approximations to integrals over intervals $[t_0, t_n]$ in the form of sums, similar to (3.2b) (which originates with the θ-method). In particular, with a *uniform mesh*, the linear multistep method with first and second characteristic polynomials $\rho(\zeta){:=}\alpha_0\zeta^k + \alpha_1\zeta^{k-1} + \cdots + \alpha_k$, $\sigma(\zeta){:=}\beta_0\zeta^k + \beta_1\zeta^{k-1} + \cdots + \beta_k$ ($\alpha_0 \neq 0$) generates, with appropriate starting values, an approximation (termed a $\{\rho, \sigma\}$-reducible quadrature) of the form

$$\int_{t_0}^{t_n} \phi(s)\,\mathrm{d}s \approx \underbrace{h\sum_{j=0}^{n_0} w_{n,j}\phi(t_j)}_{\text{starting terms}} + \underbrace{h\sum_{j=n_0+1}^{n} \omega_{n-j}\phi(t_j)}_{\text{convolution sum}} \; (t_n \equiv t_0 + nh) \tag{4.2}$$

for $n \geqslant n_0$. An example arises if we take a uniform mesh in (3.2b). Formally,

$$\sum_{j=0}^{\infty} \omega_j \zeta^j = \frac{\zeta^k \sigma(\zeta^{-1})}{\zeta^k \rho(\zeta^{-1})} = \frac{\beta_0 + \beta_1\zeta + \cdots + \beta_k\zeta^k}{\alpha_0 + \alpha_1\zeta + \cdots + \alpha_k\zeta^k} \quad (\alpha_0 \neq 0) \tag{4.3}$$

is the *generating function* (related to the *Z-transform*) for the sequence of weights $\{\omega_\ell\}_{\ell=0}^{\infty}$. To ensure uniform bounds on the weights we require $\rho(\cdot)$ to be simple von Neumann (i.e., $\rho(\cdot)$ must have all its zeros on the closed unit disk centred on the origin, any on its boundary being simple). Adapting (4.2) to the discretization of nonlinear Volterra convolution integrals we obtain

$$\underbrace{\int_{t_0}^{t_n} k(nh - s)\varphi(y(s))\,\mathrm{d}s}_{\text{continuous convolution}} \approx h\sum_{j=0}^{n_0} \mathfrak{s}_{n,j}\varphi(\tilde{y}(jh)) + \underbrace{h\sum_{j=n_0+1}^{n} \mathfrak{w}_{n-j}\varphi(\tilde{y}(jh))}_{\text{discrete convolution}}, \tag{4.4}$$

where $\mathfrak{s}_{n,j} := k((n-j)h)w_{n,j}$, $\mathfrak{w}_{n-j} := k((n-j)h)\omega_{n-j}$. The *discrete convolution property* in (4.4) facilitates the use of FFT techniques (cf. [3,4,81,97,101]) in solving Volterra convolution integral and integro-differential equations. Gregory rules of a fixed order correspond to Adams-Moulton methods. The BDF formulae rules generate rules (4.2) with asymptotic repetition factor unity.

Remark 4.2. Suitable linear multistep formulae generate fractional integration rules for use with a uniform mesh in discretizing Abel equations:

$$\frac{1}{\Gamma(1-\nu)}\int_{t_0}^{t_n} (t_n - s)^{-\nu}\phi(s)\,\mathrm{d}s \approx h^{1-\nu}\sum_{j=0}^{n_0} w_{n,j}^{\{\nu\}}\phi(t_j) + h^{1-\nu}\sum_{j=n_0+1}^{n} \omega_{n-j}^{\{\nu\}}\phi(t_j) \; (t_n \equiv t_0 + nh). \tag{4.5}$$

We refer to [3,4,81,97,101]. It is noteworthy that the starting weights can be chosen so that (4.5) is exact for selected nonpolynomial terms (e.g., $\phi(s) = (s - t_0)^{\mu_i}$, with $\mu_i \in \mathbb{R}$ and where $y(t) \sim a_0 + a_1(t - t_0)^{\mu_1} + a_2(t - t_0)^{\mu_2} + a_3(t - t_0)^{\mu_3} + \cdots$, as $t \to t_0$, with $\mu_1 < \mu_2 < \mu_3 < \cdots < \mu_m$).

5. RK processes

We now consider *RK methods* with nonuniform meshes for Volterra equations of the second kind. At the risk of offending the purist, we develop these from a heuristic viewpoint. The early work was due to Pouzet [114] and exploited RK parameters arising for ODEs; Bel'tyukov [14] introduced schemes with additional parameters. For the rigorous foundations of RK-type methods in terms of order conditions and from a modern perspective, see [12,42]; for extensions to Abel equations see [100]; for integro-differential equations see [99].

The literature contains examples of *RK formulae for ODEs*, generated by the *RK triple* $(\boldsymbol{\vartheta}, \boldsymbol{A}, \boldsymbol{b})$ where $\boldsymbol{\vartheta} = [\vartheta_1, \vartheta_2, \ldots, \vartheta_m]^{\mathrm{T}} \in \mathbb{R}^m$, $\boldsymbol{A} = [a_{i,j}] \in \mathbb{R}^{m\times m}$, $\boldsymbol{b} = [b_1, b_2, \ldots, b_m]^{\mathrm{T}}$, commonly represented as an *RK* (or *Butcher*) *tableau* $\begin{array}{c|c} \boldsymbol{\vartheta} & \boldsymbol{A} \\ \hline & \boldsymbol{b}^{\mathrm{T}} \end{array}$. Examples of RK tableaux are

$$\begin{array}{c|ccc} 0 & 0 & & \\ \frac{1}{2} & \frac{1}{4} & \frac{1}{4} & \\ 1 & \frac{1}{6} & \frac{2}{3} & \frac{1}{6} \\ \hline (1) & \frac{1}{6} & \frac{2}{3} & \frac{1}{6} \end{array}, \quad \begin{array}{c|cc} \frac{1}{2} - \frac{\sqrt{3}}{6} & \frac{1}{4} & \frac{1}{4} - \frac{\sqrt{3}}{6} \\ \frac{1}{2} + \frac{\sqrt{3}}{6} & \frac{1}{4} + \frac{\sqrt{3}}{6} & \frac{1}{4} \\ \hline (1) & \frac{1}{2} & \frac{1}{2} \end{array}, \quad \begin{array}{c|cc} \frac{1}{3} & \frac{5}{12} & -\frac{1}{12} \\ 1 & \frac{3}{4} & \frac{1}{4} \\ \hline (1) & \frac{3}{4} & \frac{1}{4} \end{array}, \quad \begin{array}{c|ccc} 0 & 0 & & \\ \frac{1}{2} & \frac{5}{24} & \frac{1}{3} & -\frac{1}{24} \\ 1 & \frac{1}{6} & \frac{2}{3} & \frac{1}{6} \\ \hline (1) & \frac{1}{6} & \frac{2}{3} & \frac{1}{6} \end{array} \tag{5.1}$$

(a) Trapezium & Simpson (b) Gauss family (collocation RK) (c) Radau IIA family (collocation RK) (d) Lobatto IIA family (collocation RK)

(the parenthetical "(1)" denoting $\vartheta_{m+1} = 1$, that has been inserted here, is usually omitted).

Remark 5.1. (i) The choice of $\boldsymbol{\vartheta}$ can be indicated by a subscript, as in $\boldsymbol{\vartheta}_{[\mathrm{NC}_m]}$, $\boldsymbol{\vartheta}_{(\mathrm{Gauss}_m)}$, $\boldsymbol{\vartheta}_{(\mathrm{Radau}_m]}$, $\boldsymbol{\vartheta}_{[\mathrm{Lobatto}_m]}$ (the m quadrature points for closed Newton–Cotes formulae, open Gauss–Legendre formulae, Radau right-hand formulae, and closed Lobatto formulae, such as appear in (5.1)(a)–(d), respectively) to denote the abscissae. Note that (a) and (d) in (5.1) have the same $\boldsymbol{\vartheta}$. (ii) The RK parameters are *explicit* (or, more accurately, "*formally explicit*", since some reordering is possible) if $a_{i,j}=0$ for $j \geqslant i$, and *semi-implicit* if $a_{i,i} \neq 0$ for some i and $a_{i,j}=0$ for $j > i$. Other features to note are whether $\vartheta_i \in [0,1]$ for all i; whether $\{\vartheta_i\}_1^m$ are distinct; whether $0 \leqslant \vartheta_1 < \vartheta_2 < \cdots < \vartheta_m < \vartheta_m \leqslant 1$; whether $\boldsymbol{b}(\vartheta_i) = [a_{i,1}, a_{i,2}, \ldots, a_{i,m}]$; whether, if $\vartheta_m = 1$, $\boldsymbol{b} = [a_{m,1}, a_{m,2}, \ldots, a_{m,m}]$. (iii) The 'stability' and 'convergence' properties of the RK method when applied to an ODE have a bearing on our methods.

The abscissae ϑ_i and the points $t_n \in \mathscr{T}$ together define

$$\mathscr{T}^{\#}(\boldsymbol{\vartheta}) := \{t_{n,i} = t_n + \vartheta_i h_n\}, \tag{5.2}$$

the set of points anticipated in Remark 3.1. We shall wish to discretize the integrals $\int_{t_0}^{t_{n,i}} K(t_{n,i}, s, y(s))\,\mathrm{d}s$, and more generally $\int_{t_0}^{t_n+\sigma h_n} K(t_n + \sigma h_n, s, y(s))\,\mathrm{d}s$, in (2.1). Now any RK triple defines a family of quadrature rules

$$\int_0^{\vartheta_i} \phi(s)\,\mathrm{d}s \approx \sum_{j=1}^{m} a_{i,j}\phi(\vartheta_j). \tag{5.3a}$$

This family, if we set $\vartheta_{m+1}=1, a_{m+1,j}=b_j$, includes

$$\int_0^1 \phi(s)\,\mathrm{d}s \approx \sum_{j=1}^{m} b_j\phi(\vartheta_j). \tag{5.3b}$$

In addition, it is possible to provide a vector $\boldsymbol{b}(\sigma)=[b_1(\sigma), b_2(\sigma),\dots,b_m(\sigma)]^{\mathrm{T}}$, defined for $\sigma\in[0,1]$, that yields quadrature

$$\int_0^{\sigma} \phi(s)\,\mathrm{d}s \approx \sum_{j=1}^{m} b_j(\sigma)\phi(\vartheta_j) \quad \text{for all } \sigma\in[0,1], \tag{5.3c}$$

where $\boldsymbol{b}(1)\equiv\boldsymbol{b}$. The parameters $(\boldsymbol{\vartheta},\ A,\ \boldsymbol{b}(\sigma))$ define a *continuous RK process*. For those tableaux labelled "collocation-RK" in (5.1) (which are linked to piecewise-polynomial collocation processes) the polynomials $b_j(\cdot)$ can be obtained by indefinite integration of the Lagrangean form of the polynomial interpolating $\phi(\cdot)$ at $\vartheta_1,\dots,\vartheta_m$ and we find that $a_{i,j}=b_j(\theta_i)$ for $i,j=1,2,\dots,m$.

Remark 5.2. By an affine transformation, Eq. (5.3b) gives, as an alternative to (5.3a),

$$\int_0^{\vartheta_i} \phi(s)\,\mathrm{d}s \approx \sum_{j=1}^{m} \vartheta_i b_j\phi(\vartheta_i\vartheta_j) \tag{5.4}$$

that is useful when $\phi(\cdot)$ is smooth on $[0,\vartheta_i]$ but not on all of $[0,1]$.

We observe that

$$\int_{t_0}^{t_{n,i}} \phi(s)\,\mathrm{d}s = \sum_{j=0}^{n-1}\int_{t_j}^{t_{j+1}} \phi(s)\,\mathrm{d}s + \int_{t_n}^{t_{n,i}} \phi(s)\,\mathrm{d}s \tag{5.5a}$$

for continuous $\phi(\cdot)$, and the extended use of RK parameters yields

$$\int_{t_0}^{t_{n,i}} \phi(s)\,\mathrm{d}s \approx \sum_{j=0}^{n-1} h_j \sum_{k=1}^{m} b_k\phi(t_{j,k}) + h_n\sum_{k=1}^{m} a_{i,k}\phi(t_{n,k}) \tag{5.5b}$$

and, for $\sigma\in[0,1]$,

$$\int_{t_0}^{t_n+\sigma h_n} \phi(s)\,\mathrm{d}s \approx \sum_{j=0}^{n-1} h_j \sum_{k=1}^{m} b_k\phi(t_{j,k}) + h_n\sum_{k=1}^{m} b_k(\sigma)\phi(t_{n,k}). \tag{5.5c}$$

If $a_{i,j}\neq b_j(\vartheta_i)$, (5.5c) does *not* reduce to (5.5b) on setting σ to ϑ_i. Formally, the RK integration formulae (5.5b) allow one to discretize a Volterra integral or integro-differential equation with an "*extended RK method*". Thus, we may seek approximations $\tilde{y}_{n,i}\approx y(t_{n,i})$ to the solution $y(\cdot)$ of (2.1b), on $\mathscr{T}^{\#}(\boldsymbol{\vartheta})$, via the equations

$$\tilde{y}_{n,i} = \underbrace{g(t_{n,i}) + \sum_{j=0}^{n-1} h_j \sum_{k=1}^{m} b_k K(t_{n,i}, t_{j,k}, \tilde{y}_{j,k})}_{\text{history term ('lag' or 'tail' term)}} \; + \; \underbrace{h_n \sum_{k=1}^{m} a_{i,j} K(t_{n,i}, t_{n,k}, \tilde{y}_{n,k})}_{\text{'incremental term'}} \tag{5.6a}$$

$(i = 1, 2, \ldots, m, m+1)$ in which the equation with $i = m+1$ provides

$$\tilde{y}(t_{n+1}) = g(t_{n+1}) + \sum_{j=0}^{n} h_j \sum_{k=1}^{m} b_k K(t_{n+1}, t_{j,k}, \tilde{y}_{j,k}). \tag{5.6b}$$

Example 5.3. Compare the use of the left-most RK tableau in (5.1) with quadrature based on the weights in (3.6), taking steps $h_1 = 2h'$, $h_2 = 2h''$, $h_3 = 2h'''$, ...

The formulae $\tilde{y}_{n,i} \approx y(t_{n,i})$ have to be interpreted appropriately if one has an RK triple with $\vartheta_i = \vartheta_k$ for $i \neq k$ but (5.6) are essentially the *extended RK formulae* obtained in [114] using an RK triple $(\boldsymbol{\vartheta}, \boldsymbol{A}, \boldsymbol{b})$. Where the formulae involve $K(t_{n,i}, t_{n,j}, \tilde{y}_{n,j})$ with $t_{n,j} > t_{n,i}$, we are required to define a *smooth* extension (2.2) of $K(t,s,v)$ for values of $s > t$. Though this is not transparent from the above derivation, the order of convergence on $\mathscr{T}$ (defined in Section 7) is inherited from the ODE RK process *provided* $K(t,s,y(s))$ (or $K_{\text{ext}}(t,s,y(s))$ in (2.2), if it is required) is sufficiently smooth. Unless the RK formula is explicit (Remark 5.1), we have to solve for a block of unknown values $\{\tilde{y}_{n,i} : i = 1, 2, \ldots, m\}$, simultaneously. There is a unique solution of the equations given an appropriate Lipschitz condition (Remark 2.1) and a sufficiently small width of $\mathscr{T}$. An appropriate choice of nonlinear equation solver depends on various factors including the step size h_n; some questions remain open.

We can modify the above approach slightly and obtain an approximation $\tilde{y}(\cdot)$ defined on $[t_0, T]$, if we use continuous RK parameters $(\boldsymbol{\theta}, \boldsymbol{A}, \boldsymbol{b}(\sigma))$ and write

$$\tilde{y}(t_n + \sigma h_n) = g(t_n + \sigma h_n) + \sum_{j=0}^{n-1} h_j \sum_{k=1}^{m} b_k K(t_n + \sigma h_n, t_{j,k}, \tilde{y}_{j,k}) + h_n \sum_{k=1}^{m} b_k(\sigma) K(t_n + \sigma h_n, t_{n,k}, \tilde{y}_{n,k}), \tag{5.7}$$

which, having obtained $\{\tilde{y}_{n,i}\}$, yields a densely defined function $\tilde{y}(\cdot)$.

Evaluating the term $\sum_{j=0}^{n-1} h_j \sum_{k=1}^{m} b_k K(t_n + \sigma h_n, t_{j,k} \tilde{y}_{j,k})$ in (5.6) or (5.7) involves a considerable computational expense, which prompts a modification (a *mixed quadrature-RK method*) based upon the use of quadrature formulae of the type arising in (3.4). The discretized equations now read

$$\tilde{y}_{n,i} = g(t_{n,i}) + \underbrace{\sum_{j=0}^{n} \Omega_{n,j} K(t_{n,i}, t_j, \tilde{y}(t_j))}_{\text{quadrature history term}} + \underbrace{h_n \sum_{k=1}^{m} a_{i,k} K(t_{n,i}, t_{n,k}, \tilde{y}_{n,k})}_{\text{`incremental term'}}, \tag{5.8a}$$

from which we obtain

$$\tilde{y}(t_{n+1}) = g(t_{n+1}) + \sum_{j=0}^{n} \Omega_{n,j} K(t_{n+1}, t_j, \tilde{y}(t_j)) + h_n \sum_{k=1}^{m} b_k K(t_{n+1}, t_{n,k}, \tilde{y}_{n,k}) \tag{5.8b}$$

with a corresponding densely defined extension

$$\tilde{y}(t_{n+\sigma}) = g(t_{n+\sigma}) + \sum_{j=0}^{n} \Omega_{n,j} K(t_{n+\sigma}, t_j, \tilde{y}(t_j)) + h_n \sum_{k=1}^{m} b_k(\sigma) K(t_{n+\sigma}, t_{n,k}, \tilde{y}_{n,k})$$

for $\sigma \in (0,1]$. The incremental terms together define a one-step method with an increment function similar to that for an ODE. It will be clear that the history term and the increment function both contribute to the overall discretization error on $\mathscr{T}$. For variations on these methods, see [86]. Wolkenfelt [121] also discusses, *inter alia*, multilag variations of the mixed RK methods.

Remark 5.4. (a) Note that if one has constructed a dense-output approximation $\tilde{y}(t)$ for $t \in [t_0, t_n]$ then it is possible to compute the history terms $g(t)+\int_{t_0}^{t_n} K(t,s,\tilde{y}(s))\,\mathrm{d}s$ by adaptive quadrature. (b) In terms of controlling accuracy, a major problem to be faced is that the accuracy in approximating $y(\cdot)$ by $\tilde{y}(\cdot)$ 'in the past' may prove inadequate to allow a good approximation to $g(t)+\int_{t_0}^{t} K(t,s,y(s))\,\mathrm{d}s$ 'in the future'. (c) The structure of the extended and mixed (Pouzet) RK formulae discussed above makes them suitable for the application of FFT techniques in the case of convolution integral or integro-differential equations and a uniform mesh. (d) One may use embedded RK formulae and quadrature correction terms to try to control the discretization error. (e) The RK discretizations discussed here do not apply to Abel type equations, see [80] for a discussion of RK-type parameters for Abel equations. (f) We restricted attention to the classical (Pouzet) RK parameters. Some mathematical ingenuity has been invested into the derivation of Bel'tyukov RK-type formulae (these are generalizations [14,49, p.175] of the RK triples that were adapted by Pouzet from the ODE context).

For Volterra integro-differential equations, the extended use of $(\boldsymbol{\vartheta}, \boldsymbol{A}, \boldsymbol{b})$ for (2.5a),(2.5b), gives an approximating set of equations

$$\tilde{y}(t_{n,i}) = \tilde{y}(t_n) + h_n \sum_{j=1}^{m} a_{i,j} F(t_{n,j}, \tilde{y}(t_{n,j}),\ \tilde{z}(t_{n,j})), \tag{5.9a}$$

$$\tilde{z}(t_{n,r}) = \sum_{j=0}^{n-1} h_j \sum_{k=1}^{m} b_k K(t_{n,r}, t_{j,k}, \tilde{y}(t_{j,k})) + h_n \sum_{k=1}^{m} a_{i,j} K(t_{n,r}, t_{n,k}, \tilde{y}(t_{n,k})), \quad i,r \in \{1,2,\ldots,m\} \tag{5.9b}$$

with $\tilde{y}(t_{n+1}) = \tilde{y}(t_{n,m+1})$ there is a related mixed quadrature-RK formula.

6. Collocation and related methods

Collocation and Galerkin methods in the literature are, frequently though not exclusively, based upon polynomial spline or piecewise-polynomial densely defined approximations $\tilde{y}(t)$. In electronic databases for tracing the literature, perhaps a quarter of the citations on numerical methods for Volterra equations refer to collocation methods. This statistic may, however, be misleading, because the collocation methods described are often discretized to produce block-by-block methods that are intimately related to RK methods.

Given $\mathscr{T}$ in (3.1) we assume that the restriction of $\tilde{y}(\cdot)$ on $[t_0, t_1]$ and subsequently, for $n=1,2,\ldots$, each subinterval $(t_n, t_{n+1}]$ is a polynomial of degree q_n $(n = 0,1,2,\ldots)$. For simplicity we suppose $q_n = q$ for all n. Clearly, the restriction of $\tilde{y}(\cdot)$ on $(t_n, t_{n+1}]$ has associated limits $\tilde{y}(t_n+)$ and derivatives $\tilde{y}'(t_n+)$, $\tilde{y}''(t_n+), \ldots$, at each point t_n; the continuity class of $\tilde{y}(\cdot)$ for $t \geqslant t_0$ is determined by the requirement that left- and right-hand limits agree. To simplify, it is assumed that the order of the derivatives that exist (and are continuous) at t_n is independent of n, say $\kappa_n = \kappa$ ($\kappa = -1$ implies lack of continuity at points t_n), and

$$\tilde{y}(t_n+) = \tilde{y}(t_n-), \quad \tilde{y}'(t_n+) = \tilde{y}'(t_n-), \ldots, \tilde{y}^{(\kappa)}(t_n+) = \tilde{y}^{(\kappa)}(t_n-). \tag{6.1}$$

The approximation[3] is then termed a *polynomial spline of degree m* and of *continuity class* κ with *knots* $\mathcal{T}_N \equiv \{t_n\}_1^N \subset \mathcal{T}$. The linear space of such functions, denoted $S_m^\kappa(\mathcal{T}_N)$ has dimension $d = N(m-\kappa) + \kappa + 1$.

Any approximation to the solution of (2.1b), $\hat{y}(t) \approx y(t)$, gives rise to a corresponding *defect* $\delta(\hat{y}(\cdot); t)$ defined as

$$\delta(\hat{y}(\cdot); t) := \hat{y}(t) - \{g(t) + \int_{t_0}^{t} K(t, s, \hat{y}(s))\,ds\} \quad (t \in [t_0, T)). \tag{6.2}$$

Clearly, if Eq. (2.1b) has a unique solution, then $\hat{y}(\cdot) = y(\cdot)$ if and only if $\delta(\hat{y}(\cdot); t) \equiv 0$, and this suggests that one should seek an approximation that gives rise to a small defect. Analogous comments apply to (2.1d) and, provided that one interprets "smallness" of the defect in an appropriate sense, to some first-kind equations (2.1a). In collocation methods one seeks approximations of a particular type such that the defect vanishes at a set of points known as the *collocation points.* For Galerkin methods, one instead asks that the *moments of the defect*, taken with a prescribed set of functions, should vanish. "Refinement" (of the collocation or the Galerkin approximation) by iteration based on (2.10) can be useful.

We now return to the determination, through collocation-type techniques, of an approximation $\tilde{y}(\cdot) \in S_m^\kappa(\mathcal{T}_N)$. An arbitrary element of $S_m^\kappa(\mathcal{T}_N)$ is determined uniquely by d parameters where d is the dimension of the space $S_m^\kappa(\mathcal{T}_N)$. Given $\mathcal{T}^{\#}(\boldsymbol{\vartheta})$, we may therefore expect to be able to chose $\tilde{y}(\cdot)$ in order to satisfy the $\kappa+1$ continuity conditions (6.1) at each point t_n ($n = 1, 2, \ldots, N-1$) and an additional Nm collocation conditions of the type

$$\delta(\tilde{y}(\cdot); t_{n,r}) = 0, \qquad n = 0, 1, \ldots, N-1; \quad r \in \{1, 2, \ldots, m\}, \tag{6.3}$$

in which we ensure the defect vanishes at selected collocation points $t_{n,r} = t_n + \vartheta_r h_n$, here assumed to be distinct. From (6.2) we have, for $t \in [t_n, t_{n+1}]$,

$$\delta(\tilde{y}(\cdot); t) = \tilde{y}(t) - \left\{ g(t) + \sum_{j=0}^{n-1} \int_{t_j}^{t_{j+1}} K(t, s, \tilde{y}(s))\,ds + \int_{t_n}^{t} K(t, s, \tilde{y}(s))\,ds \right\}, \tag{6.4}$$

where, in each interval $(t_j, t_{j+1}]$, $\tilde{y}(\cdot)$ is a polynomial (of degree q, say).

We consider polynomial spline collocation with $\tilde{y}(\cdot) \in S_{m-1}^{-1}(\mathcal{T}_N)$, following the technique of Blom and Brunner [19]. Given $0 \leqslant \vartheta_1 < \vartheta_2 < \cdots < \vartheta_m \leqslant 1$, and the grid $\mathcal{T}$, $\tilde{y}(\cdot)$ is constructed to be a polynomial of degree $q = m-1$ in every interval $[t_n, t_{n+1}]$. Then $\tilde{y}(\cdot)$ is such that (for $\sigma \in (0,1)$)

$$\tilde{y}(t_n + \sigma h_n) = \sum_{i=1}^{m} \prod_{\substack{\ell \neq i \\ \ell = 1}}^{m} \frac{(\sigma - \vartheta_\ell)}{(\vartheta_i - \vartheta_\ell)} \tilde{y}(t_{n,i}) \quad \text{where } t_{n,i} = t_n + \vartheta_i h_n. \tag{6.5}$$

In general, $\tilde{y}(\cdot) \notin C[t_0, T]$ except when $\vartheta_1 = 0$ and $\vartheta_m = 1$.

[3] We have piecewise constant approximations if $m=1, \kappa=-1$; piecewise linear continuous approximations if $m=1, \kappa=0$; classical cubic splines if $m=3, \kappa=1$.

Taking the set $\{t_{n,r}\}$ as collocation points we have the *exact collocation equations* of the form

$$\tilde{y}(t_{n,r}) = g(t_{n,r}) + \sum_{j=0}^{n-1} h_j \int_0^1 K\left(t_{n,r}, t_j + \sigma h_j, \sum_i \prod_{\ell \neq i} \frac{(\sigma - \vartheta_\ell)}{(\vartheta_i - \vartheta_\ell)} \tilde{y}(t_{j,i})\right) \mathrm{d}\sigma$$

$$+ h_n \int_0^{\vartheta_r} K\left(t_{n,r}, t_n + \sigma h_n, \sum_i \prod_{\ell \neq i} \frac{(\sigma - \vartheta_\ell)}{(\vartheta_i - \vartheta_\ell)} \tilde{y}(t_{n,i})\right) \mathrm{d}\sigma. \tag{6.6}$$

For $n = 0, 1, 2, \ldots$, in turn, we fix n and run through the values of r in (6.6), to obtain sufficient equations to determine (if Hypothesis 1 holds and h_n is sufficiently small) the values $\{\tilde{y}(t_{n,r})\}$.

An unfortunate obstacle to implementation of the exact collocation scheme is the need to compute the integrals occurring in (6.6). We may therefore consider *discretized collocation equations*, in which the integrals are replaced by interpolatory quadrature and the formulae become more tractable. The presentation in [19] is based on the use of M-point quadrature, in the form

$$\int_0^1 \phi(s)\,\mathrm{d}s \approx \sum_{k=1}^{M} b_k \phi(\vartheta_k) \quad \text{and} \quad \int_0^{\vartheta_r} \phi(s)\,\mathrm{d}s \approx \sum_{k=1}^{M} \vartheta_r b_k \phi(\vartheta_r \vartheta_k) \tag{6.7a}$$

with $M = m - 1$ (if $\vartheta_m < 1$) or with $M = m$ (if $\vartheta_m = 1$), where

$$b_k = \int_0^1 \prod_{\substack{\ell \neq k \\ \ell = 1}}^{M} \frac{(s - \vartheta_\ell)}{(\vartheta_k - \vartheta_\ell)} \mathrm{d}s \quad (k = 1, 2, \ldots, M) \tag{6.7b}$$

(cf. Remark 5.2). Thus our discretized equations read

$$\tilde{y}(t_{n,r}) = g(t_{n,r}) + \sum_{j=0}^{n-1} h_j \sum_{k=1}^{M} b_k K(t_{n,r}, t_{j,k}, \tilde{y}(t_{j,k}))$$

$$+ h_n \vartheta_r \sum_{k=1}^{M} b_k K\left(t_{n,r}, t_n + \vartheta_k \vartheta_r h_n, \sum_{i=1}^{m} \prod_{\ell \neq i} \frac{(\vartheta_k \vartheta_r - \vartheta_\ell)}{(\vartheta_i - \vartheta_\ell)} \tilde{y}(t_{n,i})\right), \tag{6.8}$$

$$\tilde{y}(t_n + s h_n) = \sum_{i=1}^{m} \prod_{\substack{\ell \neq i \\ \ell = 1}}^{m} \frac{(s - \vartheta_\ell)}{(\vartheta_i - \vartheta_\ell)} \tilde{y}(t_{n,i}). \tag{6.9}$$

One may further refine the approximations $\tilde{y}(\cdot)$ by computing a discretized *iterated collocation approximation*, which can possess improved continuity and convergence properties. The abscissae $\boldsymbol{\vartheta}_{(\mathrm{Gauss}_m)}$, $\boldsymbol{\vartheta}_{(\mathrm{Radau}_m]}$, $\boldsymbol{\vartheta}_{[\mathrm{Lobatto}_m]}$ referred to earlier are candidates for use. Defining a new set of abscissae by supplementation of $\boldsymbol{\vartheta}_{(\mathrm{Gauss}_m)}$ with the points $\{0, 1\}$, or of $\boldsymbol{\vartheta}_{(\mathrm{Radau}_m]}$ with $\{0\}$, has also been suggested. The formulae above may be regarded as a generalization of the RK methods discussed previously.

Remark 6.1. (a) Hermann Brunner has authored or co-authored numerous papers in the extensive literature on collocation, including the recent report [43]. (b) For collocation methods for a wide class of integro-differential equations, see, for example, [36]. (c) Brunner [27] investigated the convergence rate of the numerical solution, by polynomial spline collocation, of the Abel integral equation $y(t) = g(t) + \int_0^t (t-s)^{-\nu} \mathsf{h}(t,s) y(s)\,\mathrm{d}s$, $t \in I := [0,T]$, $T < \infty$, $0 < \nu < 1$, assuming g and h to be smooth. Brunner showed that, independent of the degree of the polynomials, the convergence rate with respect to a *quasi-uniform mesh* is only $N^{\nu-1}$, where N denotes the number of subintervals on I. The optimal rate of convergence N^{-m}, where $m-1$ is the degree of the polynomials, is attainable using the special *graded mesh* $t_n = (n/N)^r T$, $r = m/(1-\nu)$, $n = 0, 1, \ldots, N$ (and similar results are valid for the approximate solution of the nonlinear equation $y(t) = g(t) + \int_0^t (t-s)^{-\nu} H(t,s,y(s))\,\mathrm{d}s$). (d) For a recent paper on collocation for Abel equations (which opens with a succinct review of previous results and of some practical difficulties), see [87]. (e) Brunner [26] introduced nonpolynomial spline functions in a discretized collocation method to obtain high-order approximations to the solutions of integral and integro-differential equations having Abel-type kernels. (f) For a paper that closes a number of long-standing gaps in our understanding of collocation for equations of the first kind, see [92].

7. Numerical analysis

Mathematical numerical analysis encompasses, *inter alia*, a study of convergence, order of convergence, and superconvergence of approximations, and their stability properties. There is a considerable literature in this area. A rigorous study must commence with the issue of existence and uniqueness of the true solution and of the approximate solution (or, in the case of nonuniqueness, clarification of which solution or approximation is being discussed).

The term *convergence* has to be interpreted in context. It can apply, with given assumptions (which may include specification of ϑ and restrictions on $\mathcal{T}$ — requiring it to be *uniform* or *quasi-uniform*, or a *graded partition* of $[t_0, T]$ – see Remark 3.1) to, e.g., convergence on an interval or on mesh points:

$$\sup_{t \in [t_0,T]} |y(t) - \tilde{y}(t)| \to 0 \quad \text{as } h(\mathcal{T}) \to 0 \quad \text{with } \mathcal{T} \in \mathfrak{T}, \tag{7.1a}$$

$$\sup_{t \in \mathcal{T}} |y(t) - \tilde{y}(t)| \to 0 \quad \text{as } h(\mathcal{T}) \to 0 \quad \text{with } \mathcal{T} \in \mathfrak{T}. \tag{7.1b}$$

Here $\mathcal{T} \in \mathfrak{T}$ denotes that relevant restrictions on $\mathcal{T}$ are satisfied, and $h(\mathcal{T}) := \sup_{t_r \in [t_0,T]} \{t_r - t_{r-1}\}$ is the width (or diameter) of $\mathcal{T}$. As for *order of convergence*, we may have, first (order of convergence of the densely defined approximation on the whole interval $[t_0, T]$)

$$\sup_{t \in [t_0,T]} |y(t) - \tilde{y}(t)| = \mathcal{O}(\{h(\mathcal{T})\}^{\rho}) \quad \text{with } \mathcal{T} \in \mathfrak{T}, \tag{7.1c}$$

second (order of convergence on the points $\{t_n\}$ and the points $\{t_{n,r}\}$)

$$\sup_{t \in \mathcal{T}^{\#}(\vartheta')} |y(t) - \tilde{y}(t)| = \mathcal{O}(\{h(\mathcal{T})\}^{\varrho}) \quad \text{with } \mathcal{T} \in \mathfrak{T}, \tag{7.1d}$$

third (order of convergence on the points $\{t_n\}$ of $\mathscr{T}$)

$$\sup_{t\in\mathscr{T}} |y(t) - \tilde{y}(t)| = \mathcal{O}(\{h(\mathscr{T})\}^{\varrho'}) \quad \text{with } \mathscr{T} \in \mathfrak{T}. \tag{7.1e}$$

If the value of ρ in (7.1c) is optimal, and $\varrho > \rho$, or if $\varrho' > \varrho$ where ϱ is optimal or $\varrho' > \rho$, we speak of *superconvergence* on $\mathscr{T}^{\#}(\vartheta')$ or on $\mathscr{T}$, respectively. Superconvergence results arise, in particular, in the context of RK methods (they should be familiar for RK methods for ODEs) and of collocation.

Remark 7.1. Statement (7.1c) is simply short hand for the claim that there exist $h^\star > 0$ and a constant M_ρ (that is finite, and independent of $h(\mathscr{T})$ and of $\mathscr{T} \in \mathfrak{T}$) such that

$$\sup_{t\in[t_0,T]} |y(t) - \tilde{y}(t)| \leqslant M_\rho \{h(\mathscr{T})\}^\rho \quad \text{when} \quad h(\mathscr{T}) \leqslant h^\star. \tag{7.2}$$

It does not follow from (7.2) that a reduction in the size of $h(\mathscr{T})$ will reduce $\sup_{t\in[t_0,T]}|y(t) - \tilde{y}(t)|$; nor can we infer that $\sup_{t\in[t_0,T]} |y(t) - \tilde{y}(t)|$ is proportional to $h(\mathscr{T})$. However, statements of orders of convergence can often be strengthened — so that, with a uniform mesh of width h, one may have, e.g.,

$$y(t) - \tilde{y}_h(t) = h^\rho e_\rho(h;t) + \mathcal{O}(h^{\rho^\star}), \quad \tilde{y}_h(t) \equiv \tilde{y}(t) \tag{7.3}$$

with $|e_\rho(h;t)|$ uniformly bounded for $0 < h \leqslant h^\star$, $t \in [t_0, T]$, where $\rho^\star > \rho$. The vanishing of the nonnull function $e_\rho(h;t)$ on a set of points $\mathscr{S}_h \subset [t_0, T]$ $(0 < h \leqslant h^\star)$ implies superconvergence on $\{\mathscr{S}_h\}$.

Occasionally, one sees estimates of the convergence rate at the point t of the form

$$\rho_{\text{computed}}^{h',\kappa h',\kappa_\star h'}(t) := \log_\kappa \left| \frac{y_{\kappa_\star h'}(t) - y_{\kappa h'}(t)}{y_{\kappa_\star h'}(t) - y_{h'}(t)} \right| \quad \text{where } \kappa_\star \ll \kappa < 1.$$

Such estimates may not have been reconciled with the theory. However, if, in (7.3), $e_\rho(h;t) \neq 0$ is independent of h ($e_\rho(h;t) \equiv e_\rho(t)$, say), then $y(t) - \tilde{y}_h(t)$ is, *for sufficiently small h*, monotonically decreasing as $h \to 0$ and *asymptotically proportional* to h^ρ (here, $|y(t) - \tilde{y}_h(t)|/h^\rho \to |e_\rho(t)|$ for some bounded nonnull $e_\rho(t)$, where $t \in [t_0, T]$, as $h \to 0$).

Whilst controlling the accuracy of numerical computation is, in fact, a complicated matter, the mathematical community finds delight in the precision of mathematical theorems and "concise" results (e.g., 'superconvergence occurs at the collocation points' or 'algebraic stability implies C-stability' — see below) have a particular appeal. Such theories are both elegant and rewarding and they can give confidence in the underlying algorithms, but real-life computational conditions and practical adaptive algorithms do not always conform to the theoretical conditions assumed and it is worthwhile reviewing, from time to time, the exact practical significance of the theory. As an example, in his paper on collocation for integro-differential equations with nonsmooth solutions, Brunner [29] remarked that even though the use of suitably graded meshes yields optimal convergence rates, such meshes have their limitations (e.g., they may demand a very small initial step size, and carry a risk of contamination by rounding error).

8. Some approaches to convergence analysis

We shall indicate some of the tools used in establishing convergence and orders of convergence. We commence with a *Gronwall-type inequality*, that follows from a result quoted in [49, p.41]:

Lemma 8.1. *For* $\Lambda > 0$ *and* $z_\ell > 0$ $(\ell = 0,1,2,\ldots)$, *suppose that*

$$|\xi_n| \leqslant \Lambda \sum_{\ell=0}^{n-1} z_\ell |\xi_\ell| + |\eta_n| \quad (for\ n = 0,1,\ldots), \tag{8.1}$$

then $|\xi_n| \leqslant |\eta_n| + \Lambda \sum_{r=0}^{n-1} \{z_r \prod_{q=r+1}^{n-1}(1+\Lambda z_q)\}|\eta_r|$ *for* $n = 0,1,2,\ldots$.

An example will indicate an application of (8.1). Assume that $\mathcal{T}$ is given (with a sequence of positive steps $\{h_n\}$ and $h(\mathcal{T}) = \sup h_r$), and consider a set of approximating formulae (3.4), which we assume have a unique solution $\{\tilde{y}(t_0),\ \tilde{y}(t_1),\ \tilde{y}(t_2),\ldots\}$. We find that the true solution of (2.1b) gives

$$y(t_{n+1}) = g(t_{n+1}) + \sum_{j=0}^{n+1} \Omega_{n+1,j} K(t_{n+1}, t_j, y(t_j)) + \eta_n, \tag{8.2}$$

where $\{\eta_n\}$ are the *local truncation errors*. If we subtract the corresponding equations (3.4) and write $e_n := \tilde{y}(t_n) - y(t_n)$ we have

$$e_{n+1} = \sum_{j=0}^{n+1} \Omega_{n+1,j}\{K(t_{n+1}, t_j, \tilde{y}(t_j)) - K(t_{n+1}, t_j, y(t_j))\} - \eta_n \tag{8.3}$$

and hence, by Hypothesis 1,

$$|e_{n+1}| \leqslant \sum_{j=0}^{n+1} \Omega_{n+1,j} L |e_j| + |\eta_n|. \tag{8.4}$$

Now an examination of the weights in (3.6) reveals a property

$$|\Omega_{n,j}| \leqslant W\{h_{j-1} + h_j\} \quad \text{for some finite absolute constant } W, \tag{8.5}$$

when $j = 0,1,2,\ldots,n$ that holds for more general weights, and we therefore take (8.5) as an assumption in the case of quadrature methods for Volterra equations. From Lemma 8.1 above, applying (8.1) with $z_\ell = h_\ell + h_{\ell+1}$, assuming $Wh(\mathcal{T}) \leqslant \frac{1}{4}$, we obtain without effort, on setting $\Lambda = 2LW$,

$$|e_n| \leqslant \sup_{r \in \{0,1,\ldots,n\}} |\eta_r| \{1 + 2\Lambda(T - t_0)\exp\{2\Lambda(T - t_0)\}\} \quad \text{for } t_n \in [t_0, T]. \tag{8.6}$$

This establishes that the order of convergence of quadrature approximations $\tilde{y}(t)$ to $y(t)$, for $t \in \mathcal{T}$, is determined by the local truncation errors. (Actually, this result can be refined to take account of starting procedures.)

Remark 8.2. For quadrature generated by cyclic multistep formulae and for RK formulae, *convergence* can be established by similar techniques though results on the order of accuracy tend to be pessimistic and a more refined analysis is required. For Abel equations of the second kind (2.1e) the weights analogous to $\Omega_{n,j}$ behave like $(n-j)^{-\nu}h^{1-\nu}$, and McKee [107] developed an appropriate Gronwall-type lemma for this case (see also [49, pp. 42–44]).

8.1. One-step methods

The discussion of one-step methods on a uniform mesh $\mathscr{T}$ by Hairer et al. [80] applies to various methods of RK type, and merits attention. To indicate something of their viewpoint, we re-write (2.1b) in the form

$$y(t) = G_n(t) + \int_{t_n}^{t} K(t,s,y(s))\,\mathrm{d}s \tag{8.7a}$$

for $t \geqslant t_n$, where $G_n(t) = g(t) + \int_{t_0}^{t_n} K(t,s,y(s))\,\mathrm{d}s$, so that

$$G_k(t) = G_{k-1}(t) + \int_{t_{k-1}}^{t_k} K(t,s,y(s))\,\mathrm{d}s \quad (k = 1,2,3,\ldots) \tag{8.7b}$$

with $G_0(t) = g(t)$. A discrete analogue of (8.7a) is provided by

$$\tilde{y}(t_{n+1}) = \tilde{G}_n(t_{n+1}) + h_n \Phi(t_{n+1}, \tilde{G}(\cdot); h_n). \tag{8.8a}$$

Here (cf. the theory of one-step methods for ODEs) $\Phi(\cdot,\cdot;\cdot)$ is an *increment function* (whose evaluation involves internal stage values) and $\{\tilde{G}_k(\cdot)\}$ are approximations to $\{G_k(\cdot)\}$. One assumes that the increment functions inherit, from the properties of $K(\cdot,\cdot,\cdot)$, a uniform Lipschitz condition with respect to its second argument (uniform with regard to the index and the step size).

For the development of this approach to a wide class of RK methods and various approximations to the history term, consult [80]. One should not forget that some[4] RK methods require a *smooth* extension such as (2.2) to achieve their potential in terms of accuracy.

For extended (Pouzet-)RK methods, $\tilde{G}_{n+1}(\cdot)$ is itself computed via the incremental function:

$$\tilde{G}_k(t) = \tilde{G}_{k-1}(t) + h_k \Phi(t, \tilde{G}_{k-1}(\cdot); h_k) \tag{8.8b}$$

in analogy with (8.7a), and moreover $\tilde{y}(t_{n+1}) = \tilde{G}_{n+1}(t_{n+1})$. In this case, the link with an embedding approach (cf. (4.1a)) is apparent.

8.2. Resolvents and variation of parameters in the analysis of collocation

There have been numerous results [49] on convergence and superconvergence properties of collocation-type methods. Some results [24] were achieved by relying on a (nonlinear) variation of constants formula (2.13) due to Brauer (the reader should also consult [11] which corrects errors in [15] where, in turn, a slip in [22] was noted) to relate the true and approximate solution. Our purpose here is to draw attention to the rôle of a suitable formula of type (2.13), without entering into the detail which has to be investigated to place the theory on a rigorous foundation.

[4] An extension of the values $K(t,s,v)$ for $s > t$ is *not required* in the Pouzet-type RK formulae if $a_{i,j} = 0$ when $\vartheta_j > \vartheta_i$ and $\vartheta_\ell \in [0,1]$ for all ℓ.

We appeal to (2.13) (with minimal assumptions concerning the nature of $R(t,s,u(s))$) to compare the solution $\tilde{y}(\cdot)$ of the collocation equations with the solution $y(\cdot)$ of (2.1b). We let $\delta(\tilde{y}(\cdot);t)$ in (6.4) denote the defect, which vanishes at the points $\{t_{n,i}\}_{i=1}^{m}$ for $n=0,1,2,\ldots,N-1$, and we have

$$\tilde{y}(t)-y(t)=\delta(\tilde{y}(\cdot);t)+\int_{t_0}^{t} R(t,s;\tilde{y}(s))\delta(\tilde{y}(\cdot);s)\,\mathrm{d}s \tag{8.9}$$

and, in particular,

$$\tilde{y}(t_n)-y(t_n)=\delta(\tilde{y}(\cdot);t_n)+\sum_{j=0}^{n-1}\int_{t_j}^{t_{j+1}} R(t_n,s,\tilde{y}(s))\delta(\tilde{y}(\cdot);s)\,\mathrm{d}s. \tag{8.10}$$

Using interpolatory quadrature with abscissae $t_{j,k}$ (the collocation points in the jth subinterval, or even a subset of these abscissae) and weights $h_j w_k$ we deduce that

$$\int_{t_j}^{t_{j+1}} R(t_n,s,\tilde{y}(s))\delta(\tilde{y}(\cdot);s)=h_j\sum_k w_k R(t_n,t_{j,k},\tilde{y}(t_{j,k}))\delta(\tilde{y}(\cdot);t_{j,k})+E_{n,j}, \tag{8.11}$$

where $E_{n,j}$ denotes the error in the quadrature, and the sum in (8.11) vanishes because the defect is zero at the collocation points. Thus,

$$\tilde{y}(t_n)-y(t_n)=\delta(\tilde{y}(\cdot);t_n)+\sum_{j=0}^{n-1}E_{n,j}.$$

To obtain high order of convergence on $\mathscr{T}$ under suitable smoothness conditions (the integrand in (8.11) should be shown to be smooth on $[t_j,t_{j+1}]$) one should therefore ensure (i) that $\delta(\tilde{y}(\cdot);t)$ vanishes on $\mathscr{T}$ (the points $\{t_n\}$ should be included in the collocation points), and (ii) that the points $t_{j,k}$ (or a subset thereof) give interpolatory quadrature with an error of high order (Gaussian-type quadrature, e.g., taking ϑ as $[\vartheta_{(\mathrm{Gauss}_m)},1]$, $\vartheta_{(\mathrm{Radau}_m]}$, or $\vartheta_{[\mathrm{Lobatto}_m]}$). For a complete analysis for linear equations see [49], cf. [23,51].

9. Stability

The stability theory for numerical methods for Volterra equations is still incomplete though considerable advances have been made. Stability is concerned with the effect on a solution (here defined over $[t_0,\infty)$) of perturbations in the problem, and differing definitions of stability arise when one considers different types of *admissible perturbations*. Let us therefore consider

$$y(t)=g(t)+\int_{t_0}^{t} K(t,s,y(s))\,\mathrm{d}s \quad (t\in[t_0,\infty)). \tag{9.1a}$$

With a choice of a class $\mathfrak{G}$ of admissible perturbations defined on $[t_0,\infty)$, consider the introduction of a perturbation $\delta g(\cdot)\in\mathfrak{G}$ to give

$$y(t)+\delta y(t)=g(t)+\delta g(t)+\int_{t_0}^{t} K(t,s,y(s)+\delta y(s))\,\mathrm{d}s \quad (t\in[t_0,\infty)). \tag{9.1b}$$

Possible choices of $\mathfrak{G}$ are

$$\mathfrak{G}:=\{\delta g(t)\in BC[t_0,\infty)\},\text{ i.e., } \delta g(t)\in C[t_0,\infty),\ \sup_{t\geqslant t_0}|\delta g(t)|<\infty, \tag{9.2a}$$

$$\mathfrak{G}:=\{\delta g(t) \in BC_0[t_0,\infty)\},\ \text{i.e.,}\ \delta g(t) \in BC[t_0,\infty),\ \lim_{t\to\infty} \delta g(t) = 0, \tag{9.2b}$$

$$\mathfrak{G}:=\{\delta g(t) \in \mathscr{L}_1[t_0,\infty)\},\ \text{i.e.,}\ \int_{t_0}^{\infty} |\delta g(s)|\,\mathrm{d}s < \infty. \tag{9.2c}$$

The notion of the stability of the solution $y(\cdot)$ of (9.1a) is related to the effect $\delta y(\cdot)$ that results from making the perturbation $\delta g(\cdot)$, e.g. *The solution $y(t)$ is globally stable with respect to perturbations $\delta g(\cdot) \in \mathfrak{G}$ in the case that $\sup_{t\in[t_0,\infty)}|\delta y(t)|$ is bounded whenever $\delta g(\cdot) \in \mathfrak{G}$; it is globally asymptotically stable if it is stable and in addition $\delta y(t) \to 0$ as $t \to \infty$ whenever $\delta g(\cdot) \in \mathfrak{G}$.*

Concepts of the stability of the *numerical solution* $\tilde{y}(t)$ follows analogous lines. These reflect whether $\tilde{y}(\cdot)$ is densely defined on $[t_0,\infty)$ or is regarded as a mesh function defined on $\mathscr{T}$ (or on $\mathscr{T}^{\#}(\vartheta)$ — for definiteness assume the former) and whether we take $\mathfrak{G}$ as in (9.2) or instead choose a corresponding $\tilde{\mathfrak{G}}$ comprising mesh functions, such as

$$\tilde{\mathfrak{G}}:=\{\{\delta g(t_0),\delta g(t_1),\delta g(t_2),\ldots\} \in \ell_\infty\},\ \text{i.e.,}\ \sup_{t_n\in\mathscr{T}}|\delta g(t)| < \infty, \tag{9.3a}$$

$$\tilde{\mathfrak{G}} := \{\{\delta g(t_0),\delta g(t_1),\delta g(t_2),\cdots\} \in \ell_\infty^0\},$$
$$\text{i.e.,}\quad \sup_{t_n\in\mathscr{T}} |\delta g(t_n)| < \infty, \quad \text{and} \quad \lim_{t_n\to\infty} \delta g(t_n) = 0, \tag{9.3b}$$

$$\tilde{\mathfrak{G}}:=\{\{\delta g(t_0),\delta g(t_1),\delta g(t_2),\ldots\} \in \ell_1\},\ \text{i.e.,}\ \sum_{n=0}^{\infty} |\delta g(t_n)| < \infty. \tag{9.3c}$$

We shall denote the choice by suffices ($\mathfrak{G}_{(9.2a)}$ through $\tilde{\mathfrak{G}}_{(9.3c)}$). Note the duplication in the sense that, e.g., $\delta g \in \mathfrak{G}_{(9.2a)}$ implies that $\delta g \in \tilde{\mathfrak{G}}_{(9.3a)}$ (but not vice versa). As an example of definitions of numerical stability we have

Suppose $\tilde{y}(\cdot)$ satisfies (3.4), *and*

$$\tilde{y}(t_{n+1}) + \delta\tilde{y}(t_{n+1}) = \sum_{r=0}^{n+1} \Omega_{n+1,j} K(t_{n+1}, t_j, \tilde{y}(t_j) + \delta\tilde{y}(t_j)) + g(t_{n+1}) + \delta g(t_n). \tag{9.4}$$

The solution $\tilde{y}(t)$ ($t \in \mathscr{T}$) of (3.8) *is globally stable with respect to perturbations $\delta g(\cdot) \in \tilde{\mathfrak{G}}$ in the case that $\sup_{t_n\in\mathscr{T}}|\delta\tilde{y}(t_n)|$ is bounded whenever $\delta g(\cdot) \in \tilde{\mathfrak{G}}$; it is globally asymptotically stable if it is stable and in addition $\delta\tilde{y}(t_n) \to 0$ as $t_n \to \infty$ with $t_n \in \mathscr{T}$ whenever $\delta g(\cdot) \in \tilde{\mathfrak{G}}$; it is* exponentially stable *if it is asymptotically stable and there exist $M, z > 0$ (corresponding to $\delta g(\cdot) \in \tilde{\mathfrak{G}}$) such that $\delta\tilde{y}(t_n) \leqslant M \exp\{-z(t_n - t_0)\}$ for $t_n \in \mathscr{T}$.*

With favourable assumptions (though *not* invariably), investigation of stability of a solution of a nonlinear equation can be reduced to that of the solutions of a corresponding linear equation (stability in the first approximation). Alternative approaches invoke Lyapunov theory or ad hoc qualitative arguments. The stability of any solution $y(t)$ of

$$y(t) = \lambda \int_{t_0}^{t} \mathsf{k}(t,s) y(s)\,\mathrm{d}s + g(t) \tag{9.5}$$

can be settled by investigation of

$$\delta y(t) = \lambda \int_{t_0}^{t} \mathsf{r}^{\{\lambda\}}(t,s) \delta g(s)\,\mathrm{d}s + \delta g(t) \quad (\text{for } t \geqslant t_0), \tag{9.6}$$

where $\mathsf{r}^{\{\lambda\}}(\cdot,\cdot)$ is the resolvent kernel. One has only to determine the boundedness and asymptotic behaviour of $\delta y(\cdot)$ given the condition $\delta g(\cdot) \in \mathfrak{G}$ and conditions on $\mathsf{r}^{\{\lambda\}}(t,s)$ that can be deduced from assumed conditions on $\mathsf{k}(t,s)$. E.g., boundedness of $\delta y(t)$ for all $t \geqslant t_0$ follows from the property $\sup_{t\geqslant t_0}|\delta g(t)| < \infty$ if the resolvent condition $\sup_{t\geqslant t_0} \int_{t_0}^{t} |\mathsf{r}^{\{\lambda\}}(t,s)|\,\mathrm{d}s < \infty$ is satisfied, and one has stability under this condition if $\delta g(\cdot) \in \mathfrak{G}_{(9.2a)}$.

Remark 9.1. In the integro-differential equation in Remark 2.4, the effect $\delta y(t)$ $(t > t_0)$ of perturbing the initial value $y(t_0)$ to $y(t_0)+\delta y(t_0)$ and of perturbing $g(\cdot)$ by a uniformly bounded function $\delta g(\cdot)$ to give $g(\cdot) + \delta g(\cdot)$ can be investigated in terms of the function $\mathsf{u}(\cdot,\cdot)$ discussed in that remark.

For discretized equations one has a similar result to (9.6). Consider, for example, *discrete Volterra equations* of the form

$$\tilde{y}_{n+1} = \lambda \sum_{i=0}^{n+1} \mathsf{a}_{n+1,j}\tilde{y}_j + g_{n+1} \quad (n = 0,1,\ldots) \tag{9.7}$$

with $\tilde{y}_0 = g_0$, that correspond when $\mathsf{a}_{n,j} = \Omega_{n,j}\mathsf{k}(t_n,t_j)$, to the application of the scheme in (3.4) to (9.5). Then (9.7) provides an infinite system of equations with a lower triangular coefficient matrix for which the solution (*a*) exists if $\lambda \mathsf{a}_{n,n} - 1 \neq 0$ for all n and (*b*) can then be expressed, if required, as

$$\tilde{y}_{n+1} = g_{n+1} + \lambda \sum_{j=0}^{n+1} \mathsf{b}_{n+1,j}^{\{\lambda\}} g_j \quad (n = 0,1,\ldots). \tag{9.8}$$

Thus, $\delta\tilde{y}_{n+1} = \delta g_{n+1} + \lambda \sum_{j=0}^{n+1} \mathsf{b}_{n+1,j}^{\{\lambda\}} \delta g_j$ and conditions for the stability of (9.7) with $\delta g \in \tilde{\mathfrak{G}}_{(9.3a)}$ can be expressed in terms of the condition $\sup_n \sum_{j=0}^{n+1} |\mathsf{b}_{n+1,j}^{\{\lambda\}}| < \infty$. (For RK and block-by-block formulae one may use vectors and submatrices in an analogous discussion.) In previous literature, this underlying approach is frequently concealed from view, because special structure of (9.7) is exploited to reduce Eqs. (9.7) to a finite-term recurrence relation or to one in which other structures can be exploited.

For equations of the second kind, numerical stability analysis focussed initially on methods with a uniform grid $\mathscr{T}$ applied to the equation

$$y(t) - \lambda \int_{t_0}^{t} y(s)\,\mathrm{d}s = g(t), \tag{9.9}$$

which (Example 3.2) reduces to an ODE if $g'(\cdot)$ exists. The corresponding stability results were therefore related to those for (possibly novel) ODE methods for $y'(t) = \lambda y(t) + g'(t)$, but of course a perturbation $\delta g(\cdot)$ in the integral equation corresponds to a change in the ODE of $\delta g'(\cdot)$ in $g'(\cdot)$ and of $\delta g(t_0)$ in the initial value. However, without assuming differentiability we can still introduce a continuous perturbation $\delta g(\cdot)$ in (9.9) and use the resolvent to write

$$\delta y(t) = \delta g(t) + \lambda \int_{t_0}^{t} \exp\{\lambda(t-s)\}\delta g(s)\,\mathrm{d}s. \tag{9.10}$$

Note that (i) when $\Re(\lambda) \leqslant 0$ then $\delta y(\cdot)$ is bounded when $\delta g(\cdot) \in \mathfrak{G}_{(9.2a)}$ (giving stability) and (ii) when $\Re(\lambda) < 0$ then $\delta y(t) \to 0$ as $t \to \infty$ when $\delta g(\cdot) \in \mathfrak{G}_{(9.2b)}$ (giving asymptotic stability). For the corresponding discrete equations based on an application of a $\{\rho,\sigma\}$-reducible quadrature (4.2) to (9.9) we find that perturbations satisfy

$$\delta\tilde{y}_n - \lambda h \sum_{j=n_0+1}^{n} \omega_{n-j}\delta\tilde{y}_j = \Delta_n \quad (n = n_0+1, n_0+2, \ldots), \tag{9.11}$$

where $\Delta_n := \delta g(nh) + h\sum_{j=0}^{n_0} w_{n,j}\delta\tilde{y}_j$ and where the boundedness of $\{w_{n,j}\}$ implies the boundedness of $\{\Delta_n\}$ when $\delta g(\cdot) \in \tilde{\mathfrak{G}}_{(9.3a)}$. Then, if $\lambda h\omega_0 \neq 1$,

$$\delta\tilde{y}_n = \Delta_n + \lambda h \sum_{j=n_0+1}^{n} \varpi_{n-j}^{\{\lambda h\}}\Delta_j \quad (n = n_0+1, n_0+2, \ldots), \tag{9.12}$$

where $(1 + \lambda h\{\varpi_0^{\{\lambda h\}} + \varpi_1^{\{\lambda h\}}\zeta + \varpi_2^{\{\lambda h\}}\zeta^2 + \cdots\})(1 - \lambda h\{\omega_0 + \omega_1\zeta + \omega_2\zeta^2 + \cdots\}) = 1$, and hence, by virtue of (4.3),

$$1 + \lambda h\{\varpi_0^{\{\lambda h\}} + \varpi_1^{\{\lambda h\}}\zeta + \varpi_2^{\{\lambda h\}}\zeta^2 + \cdots\} = \frac{\zeta^k\rho(\zeta^{-1})}{\zeta^k\rho(\zeta^{-1}) - \lambda h\zeta^k\sigma(\zeta^{-1})}. \tag{9.13}$$

Using partial fractions and expanding, it follows that if the polynomial $\rho(\zeta) - \lambda\sigma(\zeta)$ is *Schur* [5] then the coefficients $\{\varpi_n^{\{\lambda\}}\}$ satisfy $\sum_{n=0}^{\infty}|\varpi_n^{\{\lambda\}}| < \infty$, and we deduce *stability* for perturbations $\delta g(\cdot) \in \tilde{\mathfrak{G}}_{(9.3a)}$. If $\{\rho,\sigma\}$ is A-stable, this property holds whenever $\Re(\lambda) < 0$ for all h. Note, however, that when $\delta g \in \tilde{\mathfrak{G}}_{(9.3b)}$, we find $|\Delta_n| \nrightarrow 0$ and we do not deduce *asymptotic stability* without further restriction of the admissible δg; this minor difficulty disappears later when we introduce a suitable $\mathscr{L}_1$-kernel $k(\cdot)$ into the equation.

Remark 9.2. (a) Some of the techniques used for (9.9) can be modified for separable kernels, polynomial convolution kernels (e.g., $k(t) = \xi + \eta t$, for which (9.14) below reduces to $y''(t) = \xi y'(t) + \eta y(t) + g''(t)$), etc. (b) For integro-differential equations, $y'(t) = \xi y(t) + \eta\int_{t_0}^{t} y(s)\,ds + g(t)$ plays a similar rôle to (9.9). (c) Stability and stability regions for test equations are discussed in, for example, [8,7,10,50].

A natural question to ask is whether results for (9.9) provide any insight into results for more general equations; we indicate one approach to answering this question by considering the discretization of the linear convolution equation

$$y(t) - \lambda\int_{t_0}^{t} k(t-s)y(s)\,ds = g(t) \quad \text{with } \lambda \in \mathbb{C}, \tag{9.14}$$

Let us keep in mind the substitutions $\mathfrak{a}_n = hk(nh)\omega_n$, $\tilde{y}_j = \tilde{y}(jh)$, $g_n = g(nh)$, and consider the discrete convolution equations

$$\tilde{y}_{n+1} = \lambda\sum_{j=0}^{n+1}\mathfrak{a}_{n+1-j}\tilde{y}_j + g_{n+1} \quad (n = 0, 1, \ldots). \tag{9.15}$$

[5] All its zeros lie inside the unit circle centered on the origin, or, equivalently, $z^k\rho(z^{-1}) - \lambda hz^k\sigma(z^{-1}) = 0$ implies $|z| \geqslant 1$.

If $\mathfrak{a}(\zeta)$ denotes the formal power series $\sum_{j=0}^{\infty}\mathfrak{a}_j\zeta^j$, and if $\mathfrak{b}^{\{\lambda\}}(\zeta)$ denotes the formal power series $\sum_{j=0}^{\infty}\mathfrak{b}_j^{\{\lambda\}}\zeta^j$ where $(1-\lambda\mathfrak{a}(\zeta))(1+\lambda\mathfrak{b}^{\{\lambda\}}(\zeta))=1$ then

$$\tilde{y}_{n+1} = g_{n+1} + \lambda \sum_{j=0}^{n+1} \mathfrak{b}_{n+1-j}^{\{\lambda\}} g_{n+1} \quad (n=0,1,\ldots). \tag{9.16}$$

If, further,

$$\sum_{j=0}^{\infty} |\mathfrak{a}_j| < \infty \quad \text{and} \quad 1 \neq \lambda \sum_{j=0}^{\infty} \mathfrak{a}_j \zeta^j \text{ whenever } |\zeta| \leqslant 1, \tag{9.17}$$

then $\sum_{j=0}^{\infty}|\mathfrak{b}_j^{\{\lambda\}}|<\infty$ and hence the solution of (9.15) is stable for perturbations δg_n in $\tilde{\mathfrak{G}}_{(9.3a)}$ or in $\tilde{\mathfrak{G}}_{(9.3b)}$ and is asymptotically stable for perturbations δg_n in $\tilde{\mathfrak{G}}_{(9.3b)}$ or, indeed, in $\tilde{\mathfrak{G}}_{(9.3c)}$. Under assumptions on the 'starting weights' $\mathfrak{s}_{n+1,j}$ the analysis can be extended to discuss equations, cf. (4.2), of the form

$$\tilde{y}_{n+1} = \lambda \left\{ \sum_{j=0}^{n_0} \mathfrak{s}_{n+1,j} \tilde{y}_j + \sum_{j=n_0+1}^{n+1} \mathfrak{a}_{n+1-j} \tilde{y}_j \right\} + g_{n+1} \; (n \geqslant n_0). \tag{9.18}$$

Normally $\mathfrak{s}_{n,j}=w_{n,j}k((n-j)h)$, and $\mathfrak{a}_{n,j}=\omega_{n-j}k((n-j)h)$; hence if $k(\cdot)\in C[0,\infty)\cap L_1[0,\infty)$ (so that $k(t)\to 0$ as $t\to\infty$) and if the weights ω_ℓ are uniformly bounded and

$$\sum_{\ell=0}^{\infty} |k(\ell h)| < \infty, \tag{9.19}$$

we find $\mathfrak{s}_{n,j}=0$; $\sum_{\ell=0}^{\infty}|\mathfrak{a}_\ell|<\infty$. Condition (9.19) is a minor annoyance but should not be overlooked; it holds under our previous assumptions when $k(\cdot)$ is monotonic decreasing or is monotonic decreasing for sufficiently large arguments or when h is sufficiently small (say $h\leqslant h^\star(k(\cdot))$).

In our discussion of (9.9) we relied upon the simple nature of the kernel. Whilst (9.17) and (9.19) allow one to compute approximate stability regions for special $k(\cdot)$, what we seek now is a class of integral equations for which stability is readily analysed and classes of methods that simulate the stability properties of the true solution. A theory can be constructed if we consider *positive quadrature* and *completely monotone* or *positive definite functions*.

Remark 9.3. (a) A convolution quadrature is called *positive* (i) if and only if $\sum\sum\omega_{i-j}z_i\bar{z}_j\geqslant 0$ for all finite sequences of complex-valued z_ℓ. (ii) A quadrature is positive if it is a reducible quadrature generated by an A-stable LMF (ρ,σ). (b) A function $\psi:[0,\infty)\to\mathbb{C}$ is *positive definite* if (i) on defining $\phi(t)=\psi(t)$, $t\geqslant 0$, $\phi(t)=-\overline{\psi(-t)}$, $t\leqslant 0$, we have $\sum_i\sum_j\phi(t_i-t_j)z_i\overline{z_j}\geqslant 0$ for all t_ℓ and all finite sequences of complex valued z_ℓ, or (*ii*) equivalently, by a result of Bochner, if (a Laplace transform condition) $\Re\{\int_0^\infty\psi(s)\exp(-zs)\mathrm{d}s\}\geqslant 0$ if $\Re(z)>0$. (c) A function ψ that is *completely monotone* (that is, $\psi(\cdot)\in C^\infty[0,\infty)$ and $(-1)^jk^{(j)}(t)>0$ for $j=0,1,2,\ldots$) is positive definite. (d) When the convolution quadrature with weights ω_ℓ is positive and $k(\cdot)$ is a positive-definite function, the sequence $\{\mathfrak{a}_\ell\}$ (with $\mathfrak{a}_\ell=hk(\ell h)\omega_\ell$) is a positive definite sequence and the analogue of Bochner's Laplace transform result (in effect, a Z-transform condition) holds: $\Re\{\sum_{n=0}^{\infty}\mathfrak{a}_n\zeta^n\}\geqslant 0$ for $|\zeta|\leqslant 1$.

Consider the linear equation (9.14) for which a resolvent convolution kernel $r^{\{\lambda\}}(\cdot) \in C[0,\infty)$ provides the solution $y(t) = g(t) + \lambda \int_{t_0}^{t} r^{\{\lambda\}}(t-s)g(s)\,\mathrm{d}s$. If

Hypothesis 3. $k(\cdot) \in C[0,\infty) \cap \mathscr{L}_1[0,\infty)$, and $\Re(\lambda) < 0$, and either
(i) $k(\cdot)$ is *completely monotone*, or, more generally,
(ii) $k(\cdot)$ is *positive definite*,

then we have $\int_0^\infty |r^{\{\lambda\}}(s)\,\mathrm{d}s| < \infty$ and hence *all solutions of* (9.14) *are stable for perturbations* $\delta g(\cdot)$ in $\mathfrak{G}_{(9.2a)}$ (or a fortiori for perturbations $\delta g(\cdot)$ in $\mathfrak{G}_{(9.2b)}$), and asymptotically stable for perturbations $\delta g(\cdot)$ in $\mathfrak{G}_{(9.2b)}$ (or a fortiori for perturbations $\delta g(\cdot)$ in $\mathfrak{G}_{(9.2c)}$). The preceding remarks inspire the definition of a *C-stable* RK or quadrature method as one that, when applied to (9.14) with $\Re(\lambda) < 0$ and a bounded continuous *positive-definite* kernel $k(\cdot)$ yields *for all fixed h* approximations $\{\tilde{y}_n\}$ such that $\tilde{y}_n \to 0$ whenever $g(t_n) \to 0$ as $n \to \infty$. In fact, when using the concept of C-stability, we have to impose condition (9.19) or a similar condition, thereby restricting slightly the class of functions $k(\cdot)$. Expressed in our terms, the method is C-stable if, given Hypothesis 3 along with (9.19) (or along with an analogous assumption in the case of RK methods) the mesh-functions $\tilde{y}(t)$ ($t \in \mathscr{T}$) are asymptotically stable for all h with respect to perturbations in $\tilde{\mathfrak{G}}_{(9.3b)}$. It was Lubich who showed (*inter alia*) that A-stable $\{\rho,\sigma\}$-reducible quadrature formulae applied to (9.14) are C-stable, using assumption (9.19) (see [102]), in particular, therefore, under the assumption that $k(\cdot)$ is completely monotone. Given (9.19), $\{\mathfrak{a}_\ell\}_0^\infty$ is a positive definite ℓ_1 sequence, which (with our previous remarks) is sufficient to establish the claim.

Remark 9.4. (a) The work of Nevanlinna (cf. [112,113]) stimulated a strand of research in the stability analysis for numerical methods for Volterra equations. (b) Hairer and Lubich [78] showed that, provided one employs the correct extension of the positive-definite function $k(\cdot)$ to negative arguments, *algebraically stable* extended Pouzet–R–K methods are *C-stable under an assumption analogous to* (9.19). An order barrier was given. (c) The stability of collocation methods has been discussed in [17,18,65] etc. (d) Nonlinear equations were discussed in [72], where an approach of Corduneanu in [56] is exploited and analogous results are obtained. The discussion in [73] deals with a class of nonlinear convolution equations and refers to use that can be made of *strong* and *strict positivity*.

10. Concluding remarks

The term "Volterra equations" encompasses a variety of "nonanticipative" functional equations that generalize the classical equations we have studied here. It has been necessary to limit our discussion and to omit a number of growth areas, and we have concentrated on classical Volterra equations. This leaves some noteworthy omissions, including integro-differential equations with discretely distributed delays, such as

$$y'(t) = \lambda y(t)\left\{A - \int_{t_0}^{t} K(t,s,y(s))\,\mathrm{d}s - \sum_{i=1}^{N} B_i y(t-\tau_i(t))\right\},$$

where $\tau_i(t)\geqslant 0$ $(i=1,2,\dots,N)$ and partial integro-differential equations (see e.g., [106,117] and their references).

Equations with infinite memory that have the form

$$y(t)=\gamma(t)+\int_{-\infty}^{t} K(t,s,y(s))\,\mathrm{d}s \quad (t\in[t_0,T]), \tag{10.1a}$$

$$y(t)=\psi(t) \quad (-\infty<t\leqslant t_0), \tag{10.1b}$$

reduce, formally, to (2.1b) if, in the classical case, we set

$$g(t):=\gamma(t)+\int_{-\infty}^{t_0} K(t,s,\psi(s))\,\mathrm{d}s. \tag{10.2}$$

This begs the question of the accurate evaluation of (10.2) which can be computationally expensive. One can also consider equations with a lower limit of integration $\alpha(t)$ and an upper limit of integration $\beta(t)$ (where $\alpha(t)\leqslant\beta(t)<t$),

$$y(t)=\gamma(t)+\int_{\alpha(t)}^{\beta(t)} K(t,s,y(s))\,\mathrm{d}s \quad (t\in[t_0,T]), \tag{10.3}$$

$$y(t)=\psi(t) \quad \left(\inf_{t'\geqslant t_0}\alpha(t')\leqslant t\leqslant t_0\right), \tag{10.4}$$

if we define $K(t,s,u):=0$ for $s\notin[\alpha(t),\beta(t)]$. However, in order to treat "nonclassical" limits of integration, we must admit that $K(t,s,v)$ may suffer a jump discontinuity when $s=\alpha(t)$ or $\beta(t)$. It is clear that the form of $\alpha(t)$ has an impact upon the type of initial condition required to determine a unique solution. Whereas the analytical theories can to some extent unified, numerical practice should reflect the presence of discontinuities and it seems better to treat the original nonclassical equations directly. A number of papers in the literature do attempt that. Further variants include equations

$$y(t)=g(t)+\int_{\alpha(t)}^{\beta(t)} \mathscr{K}(t,s,y(t),y(s),y(a(t)),y(b(s)))\,\mathrm{d}s \quad (t\geqslant t_0) \tag{10.5}$$

with $\alpha(t)\leqslant\beta(t)\leqslant t$ and with $a(s)\leqslant t$, and $b(s)\leqslant t$ for $s\in[\alpha(t),\beta(t)]$.

As regards future work, there remain interesting theoretical questions about the numerical simulation of qualitative properties (e.g., the effect of different types of memory, the onset of periodicity through bifurcation, and [115] blow-up in solutions); there exist opportunities for the application of mathematical arguments to the design and validation of robust and efficient *adaptive codes* that are suited to numerical simulation of real-life models (particularly Volterra equations arising in the biosciences).

We conclude with some references to the literature. The material herein should provide an adequate base for the interested reader to pursue the above topics. The book of Brunner and van der Houwen [49] has an excellent bibliography to the date of its completion in 1985. Since the output referred to there is repeated here when we wish to draw particular attention to it, our choice of Refs. [1–123] may be regarded as reflecting idiosyncrasies of the author.

Acknowledgements

The author thanks Prof. Pieter van der Houwen (CWI., Amsterdam) and a number of his own students (in particular Yihong Song and Hongjiong Tian) and colleagues in MCCM for reading and commenting on drafts of this paper.

References

[1] C.T.H. Baker, Numerical analysis of Volterra functional and integral equations, in: I.S. Duff, G.A. Watson (Eds.), The State of the Art in Numerical Analysis, Oxford University Press, New York, 1997, pp. 193–222.
[2] C.T.H. Baker, M.S. Derakhshan, FFT techniques in the numerical solution of convolution equations, J. Comput. Appl. Math. 20 (1987) 5–24.
[3] C.T.H. Baker, M.S. Derakhshan, Computed approximations to some power series, Internat. Schriftenreihe Numer. Math. 81 (1986) 11–20.
[4] C.T.H. Baker, M.S. Derakhshan, Stability barriers to the construction of $\{\rho, \sigma\}$-reducible and fractional quadrature rules, Internat. Schriftenreihe Numer. Math. 85 (1988) 1–15.
[5] C.T.H. Baker, A. Tang, Stability analysis of continuous implicit Runge–Kutta methods for Volterra integro-differential systems with unbounded delay, Appl. Numer. Math. 24 (1997) 153–173.
[6] C.T.H. Baker, G.F. Miller (Eds.), Treatment of Integral Equations by Numerical Methods, Academic Press, London, 1982.
[7] C.T.H. Baker, M.S. Keech, Stability regions in the numerical treatment of Volterra integral equations, SIAM J. Numer. Anal. 15 (1978) 394–417.
[8] C.T.H. Baker, A. Makroglou, E. Short, Regions of stability in the numerical treatment of Volterra integro-differential equations, SIAM J. Numer. Anal. 16 (1979) 890–910.
[9] C.T.H. Baker, A. Tang, Numerical solution of Volterra integro-differential equations with an unbounded delay, Hellenic European Research on Mathematics and Informatics '94, Vols. 1, 2, Hellenic Mathematical Society, Athens, 1994, pp. 129–136.
[10] C.T.H. Baker, J.C. Wilkinson, Stability analysis of Runge–Kutta methods applied to a basic Volterra integral equation, J. Austral. Math. Soc. Ser. B 22 (1980/81) 515–538.
[11] P.R. Beesack, On some variation of parameter methods for integro-differential, integral, and quasilinear partial integro-differential equations, Appl. Math. Comput. 22 (1987) 189–215.
[12] A. Bellen, Z. Jackiewicz, R. Vermiglio, M. Zennaro, Natural continuous extensions of Runge–Kutta methods for Volterra integral equations of the second kind and their applications, Math. Comp. 52 (1989) 49–63.
[13] A. Bellen, Z. Jackiewicz, R. Vermiglio, M. Zennaro, A stability analysis of the trapezoidal method for Volterra integral-equations with completely positive kernels, J. Math. Anal. Appl. 152 (1990) 324–342.
[14] B.A. Bel'tyukov, Analogue of the Runge–Kutta method for the solution of nonlinear integral equations of Volterra type, Differential Equations 1 (1965) 417–433.
[15] S.H. Bernfeld, M.E. Lord, A nonlinear variation of constants method for integro differential and integral equations, Appl. Math. Comput. 4 (1978) 1–14.
[16] J.G. Blom, H. Brunner, The numerical solution of nonlinear Volterra integral equations of the second kind by collocation and iterated collocation methods, SIAM J. Sci. Statist. Comput. 8 (1987) 806–830.
[17] L. Blank, Stability of collocation for weakly singular Volterra equations, IMA J. Numer. Anal. 15 (1995) 357–375.
[18] L. Blank, Stability results for collocation methods for Volterra integral equations, Appl. Math. Comput. 79 (1996) 267–288.
[19] J.G. Blom, H. Brunner, Algorithm 689: collocation for nonlinear Volterra integral equations of the second kind, ACM Trans. Math. Software 17 (1991) 167–177.
[20] J.M. Bownds, Theory and performance of a subroutine for solving Volterra integral equations, Computing 28 (1982) 317–332.
[21] J.M. Bownds, L. Applebaum, Algorithm 627: A FORTRAN subroutine for solving Volterra integral equations, ACM Trans. Math. Software 11 (1985) 58–65.

[22] F. Brauer, A nonlinear variation of constants formula for Volterra equations, Math. Systems Theory 6 (1972) 226–234.
[23] H. Brunner, Superconvergence in collocation and implicit Runge–Kutta methods for Volterra-type integral equations of the second kind, Internat. Schriftenreihe Numer. Math. 53 (1980) 54–72.
[24] H. Brunner, The variation of constants formulas in the numerical analysis of integral and integro-differential equations, Utilitas Math. 19 (1980) 255–290.
[25] H. Brunner, A survey of recent advances in the numerical treatment of Volterra integral and integro-differential equations, J. Comput. Appl. Math. 8 (1982) 213–229.
[26] H. Brunner, Nonpolynomial spline collocation for Volterra equations with weakly singular kernels, SIAM J. Numer. Anal. 20 (1983) 1106–1119.
[27] H. Brunner, The numerical solution of weakly singular Volterra integral equations by collocation on graded meshes, Math. Comp. 45 (1985) 417–437.
[28] H. Brunner, The approximate solution of Volterra equations with nonsmooth solutions, Utilitas Math. 27 (1985) 57–95.
[29] H. Brunner, On the numerical solution by collocation of Volterra integro-differential equations with nonsmooth solutions, Internat. Schriftenreihe Numer. Math. 73 (1985) 74–92.
[30] H. Brunner, On the history of numerical methods for Volterra integral equations, CWI Newslett. 11 (1986) 3–20.
[31] H. Brunner, High-order methods for the numerical solution of Volterra integro-differential equations, J. Comput. Appl. Math. 15 (1986) 301–309.
[32] H. Brunner, Polynomial spline collocation methods for Volterra integro-differential equations with weakly singular kernels, IMA J. Numer. Anal. 6 (1986) 221–239.
[33] H. Brunner, Implicit Runge–Kutta–Nyström methods for general second-order Volterra integro-differential equations, Comput. Math. Appl. 14 (1987) 549–559.
[34] H. Brunner, The approximate solution of initial-value problems for general Volterra integro-differential equations, Computing 40 (1988) 125–137.
[35] H. Brunner, The numerical treatment of Volterra integro-differential equations with unbounded delay, J. Comput. Appl. Math. 28 (1989) 5–23.
[36] H. Brunner, Collocation methods for nonlinear Volterra integro-differential equations with infinite delay, Math. Comp. 53 (1989) 571–587.
[37] H. Brunner, Direct quadrature methods for nonlinear Volterra integro-differential equations with infinite delay, Utilitas Math. 40 (1991) 237–250.
[38] H. Brunner, Implicitly linear collocation methods for nonlinear Volterra equations, Appl. Numer. Math. 9 (1992) 235–247.
[39] H. Brunner, On discrete superconvergence properties of spline collocation methods for nonlinear Volterra integral equations, J. Comput. Math. 10 (1992) 348–357.
[40] H. Brunner, Open problems in the discretization of Volterra integral equations, Numer. Funct. Anal. Optim. 17 (1996) 717–736.
[41] H. Brunner, 1896–1996: one hundred years of Volterra integral equations of the first kind, Appl. Numer. Math. 24 (1997) 83–93.
[42] H. Brunner, E. Hairer, S.P. Nørsett, Runge–Kutta theory for Volterra integral equations of the second kind, Math. Comp. 39 (1982) 147–163.
[43] H. Brunner, Q. Hu, Q. Lin, Geometric meshes in collocation methods for Volterra integral equations with proportional time delays, Report 99-25 Seminar für Angewandte Mathematik, ETH, Zurich, 1999.
[44] H. Brunner, Y. Lin, S. Zhang, Higher accuracy methods for second-kind Volterra integral equations based on asymptotic expansions of iterated Galerkin methods, J. Integral Equations Appl. 10 (1998) 375–396.
[45] H. Brunner, A. Makroglou, R.K. Miller, On mixed collocation methods for Volterra integral equations with periodic solution, Appl. Numer. Math. 24 (1997) 115–130.
[46] H. Brunner, A. Pedas, G. Vainikko, The piecewise polynomial collocation method for nonlinear weakly singular Volterra equations, Math. Comp. 68 (1999) 1079–1095.
[47] H. Brunner, L. Qun, Y. Ningning, The iterative correction method for Volterra integral equations, BIT 36 (1996) 221–228.

[48] H. Brunner, Y. Ningning, On global superconvergence of iterated collocation solutions to linear second-kind Volterra integral equations, J. Comput. Appl. Math. 67 (1996) 185–189.
[49] H. Brunner, P.J. van der Houwen, The Numerical Solution of Volterra Equations, CWI Monographs, North-Holland, Amsterdam, 1986.
[50] H. Brunner, J.D. Lambert, Stability of numerical methods for Volterra integro-differential equations, Computing 12 (1974) 75–89.
[51] H. Brunner, S.P. Nørsett, Superconvergence of collocation methods for Volterra and Abel integral equations of the second kind, Numer. Math. 36 (1980/81) 347–358.
[52] T.A. Burton, Volterra Integral and Differential Equations, Academic Press, New York, 1983.
[53] B. Cahlon, Numerical solution of nonlinear Volterra integral equations, J. Comput. Appl. Math. 7 (1981) 121–128.
[54] B. Cahlon, On the stability of Volterra integral-equations with a lagging argument, BIT 35 (1995) 19–29.
[55] C. Corduneanu, Principles of Differential and Integral Operators, Chelsea, New York, 1988.
[56] C. Corduneanu, Integral Equations and Stability of Feedback Systems, Academic Press, New York, 1973.
[57] C. Corduneanu, Functional Equations with Causal Operators, Gordon and Breach, in preparation.
[58] C. Corduneanu, V. Lakshmikantham, Equations with unbounded delay: a survey, Nonlinear Anal. 4 (1980) 831–877.
[59] C. Corduneanu, I.W. Sandberg (Eds.), Volterra Equations and Applications, Gordon and Breach, Amsterdam, 2000.
[60] M.R. Crisci, V.B. Kolmanovskii, E. Russo, A. Vecchio, Stability of difference Volterra equations: direct Liapunov method and numerical procedure, Comput. Math. Appl. 36 (1998) 77–97.
[61] M.R. Crisci, V.B. Kolmanovskii, E. Russo, A. Vecchio, Boundedness of discrete Volterra equations, J. Math. Anal. Appl. 211 (1997) 106–130.
[62] M.R. Crisci, V.B. Kolmanovskii, E. Russo, A. Vecchio, Stability of continuous and discrete Volterra integro-differential equations by Liapunov approach, J. Integral Equations Appl. 7 (1995) 393–411.
[63] M.R. Crisci, E. Russo, A. Vecchio, Stability of collocation methods for Volterra integro-differential equations, J. Integral Equations Appl. 4 (1992) 491–507.
[64] M.R. Crisci, E. Russo, Z. Jackiewicz, A. Vecchio, Global stability of exact collocation methods for Volterra integro-differential equations, Atti Sem. Mat. Fis. Univ. Modena 39 (1991) 527–536.
[65] M.R. Crisci, E. Russo, A. Vecchio, Stability results for one-step discretized collocation methods in the numerical treatment of Volterra integral equations, Math. Comp. 58 (1992) 119–134.
[66] J. Dixon, S. McKee, R. Jeltsch, Convergence analysis of discretization methods for nonlinear first kind Volterra integral equations, Numer. Math. 49 (1986) 67–80.
[67] J. Dixon, S. McKee, A unified approach to convergence analysis of discretization methods for Volterra-type equations, IMA J. Numer. Anal. 5 (1985) 41–57.
[68] I.S. Duff, G.A. Watson (Eds.), The State of the Art in Numerical Analysis, Oxford University Press, New York, 1997.
[69] P.P.B. Eggermont, Improving the accuracy of collocation solutions of Volterra integral equations of the first kind by local interpolation, Numer. Math. 48 (1986) 263–279.
[70] P.P.B. Eggermont, On monotone Abel–Volterra integral equations on the half line, Numer. Math. 52 (1988) 65–79.
[71] P.P.B. Eggermont, Uniform error estimates of Galerkin methods for monotone Abel–Volterra integral equations on the half-line, Math. Comp. 53 (1989) 157–189.
[72] N.J. Ford, C.T.H. Baker, Qualitative behaviour and stability of solutions of discretised nonlinear Volterra integral equations of convolution type, J. Comput. Appl. Math. 66 (1996) 213–225.
[73] N.J. Ford, C.T.H. Baker, Preserving transient behaviour in numerical solutions of Volterra integral equations of convolution type, in: R. Agarwal, D. O'Regan (Eds.), Integral and Integro-Differential Equations, Gordon and Breach, Lausanne, 1999, pp. 77–89.
[74] N.J. Ford, C.T.H. Baker, J.A. Roberts, A. Nonlinear Volterra integro-differential equations — stability and numerical stability of θ-methods, J. Integral Equations Appl. 10 (1998) 397–416.
[75] R. Gorenflo, F. Mainardi, Fractional calculus: integral and differential equations of fractional order, in: A. Carpinteri, F. Mainardi (Eds.), Fractals and Fractional Calculus in Continuum Mechanics, CISM Courses and Lectures, Vol. 378, Springer, Vienna, 1997, pp. 223–276.
[76] R. Gorenflo, S. Vessela, Abel Integral Equations, Springer, Berlin, 1991.
[77] G. Gripenberg, S.-O. Londen, O. Staffans, Volterra Integral and Functional Equations, Cambridge University Press, Cambridge, 1990.

[78] E. Hairer, C. Lubich, On the stability of Volterra–Runge–Kutta methods, SIAM J. Numer. Anal. 21 (1984) 123–135.
[79] E. Hairer, C. Lubich, An extension of ODE-methods to Volterra integral equations, Teubner-Texte Math. 82 (1986) 55–63.
[80] E. Hairer, C. Lubich, S.P. Nørsett, Order of convergence of one-step methods for Volterra integral equations of the second kind, SIAM J. Numer. Anal. 20 (1983) 569–579.
[81] E. Hairer, C. Lubich, M. Schlichte, Fast numerical solution of weakly singular Volterra integral equations, J. Comput. Appl. Math. 23 (1988) 87–98.
[82] E. Hairer, C. Lubich, M. Schlichte, Fast numerical solution of nonlinear Volterra convolution equations, SIAM J. Sci. Statist. Comput. 6 (1985) 532–541.
[83] P.J. van der Houwen, Convergence and stability results in Runge–Kutta type methods for Volterra integral equations of the second kind, BIT 20 (1980) 375–377.
[84] P.J. van der Houwen, H.J.J. te Riele, Backward differentiation type formulas for Volterra integral equations of the second kind, Numer. Math. 37 (1981) 205–217.
[85] P.J. van der Houwen, P.H.M. Wolkenfelt, On the stability of multistep formulas for Volterra integral equations of the second kind, Computing 24 (1980) 341–347.
[86] P.J. van der Houwen, P.H.M. Wolkenfelt, C.T.H. Baker, Convergence and stability analysis for modified Runge–Kutta methods in the numerical treatment of second-kind Volterra integral equations, IMA J. Numer. Anal. 1 (1981) 303–328.
[87] Q. Hu, Superconvergence of numerical solutions to Volterra integral equations with singularities, SIAM J. Numer. Anal. 34 (1997) 1698–1707.
[88] H.M. Jones, S. McKee, Variable step size predictor–corrector schemes for second kind Volterra integral equations, Math. Comp. 44 (1985) 391–404.
[89] J.-P. Kauthen, Implicit Runge–Kutta methods for some integrodifferential-algebraic equations, Appl. Numer. Math. 13 (1993) 125–134.
[90] J.-P. Kauthen, Implicit Runge–Kutta methods for singularly perturbed integro-differential systems, Appl. Numer. Math. 18 (1995) 201–210.
[91] J.-P. Kauthen, A survey of singularly perturbed Volterra equations, Appl. Numer. Math. 24 (1997) 95–114.
[92] J.-P. Kauthen, H. Brunner, Continuous collocation approximations to solutions of first kind Volterra equations, Math. Comp. 66 (1997) 1441–1459.
[93] D. Kershaw, Some results for Abel–Volterra integral equations of the second kind, in: C.T.H. Baker, G.F. Miller (Eds.), Treatment of Integral equations by Numerical Methods, Academic Press, London, 1982, pp. 273–282.
[94] V. Lakshmikantham, S.G. Deo, Method of Variation of Parameters for Dynamic Systems, Gordon and Breach, Lausanne, 1998.
[95] V. Lakshmikantham, M. Rama Mohana Rao, Theory of Integro-differential Equations, Gordon and Breach, Lausanne, 1995.
[96] P. Linz, Analytical and Numerical Methods for Volterra Equations, SIAM, Philadelphia, PA, 1985.
[97] C. Lubich, Fractional linear multistep methods for Abel–Volterra integral equations of the first kind, IMA J. Numer. Anal. 7 (1987) 97–106.
[98] C. Lubich, A stability analysis of convolution quadratures for Abel–Volterra integral equations, IMA J. Numer. Anal. 6 (1986) 87–101.
[99] C. Lubich, Runge–Kutta theory for Volterra integro-differential equations, Numer. Math. 40 (1982) 119–135.
[100] C. Lubich, Runge–Kutta theory for Volterra and Abel integral equations of the second kind, Math. Comp. 41 (1983) 87–102.
[101] C. Lubich, Fractional linear multistep methods for Abel–Volterra integral equations of the second kind, Math. Comp. 45 (1985) 463–469.
[102] C. Lubich, On the stability of linear multistep methods for Volterra convolution equations, IMA J. Numer. Anal. 3 (1983) 439–465.
[103] C. Lubich, Runge–Kutta theory for Volterra and Abel integral equations of the second kind, Math. Comp. 41 (1983) 87–102.
[104] C. Lubich, Runge–Kutta theory for Volterra integro-differential equations, Numer. Math. 40 (1982) 119–135.
[105] C. Lubich, On the stability of linear multistep methods for Volterra integral equations of the second kind, in: C.T.H. Baker, G.F. Miller (Eds.), Treatment of Integral equations by Numerical Methods, Academic Press, London, 1982, pp. 233–238.

[106] C. Lubich, I.H. Sloan, V. Thomée, Nonsmooth data error estimates for approximations of an evolution equation with a positive-type memory term, Math. Comp. 65 (1996) 1–17.

[107] S. McKee, Generalized discrete Gronwall lemmas, Z. Angew. Math. Mech. 62 (1982) 429–434.

[108] S. McKee, Volterra integral and integro-differential equations arising from problems in engineering and science, Bull. Inst. Math. Appl. 24 (1988) 135–138.

[109] S. McKee, A review of linear multistep methods and product integration methods and their convergence analysis for first kind Volterra integral equations, in: C.T.H. Baker, G.F. Miller (Eds.), Treatment of Integral equations by Numerical Methods, Academic Press, London, 1982, pp. 153–161.

[110] S. McKee, H. Brunner, The repetition factor and numerical stability of Volterra integral equations, Comput. Math. Appl. 6 (1980) 339–347.

[111] R.K. Miller, Nonlinear Volterra Integral Equations, W.A. Benjamin, Menlo Park, CA, 1972.

[112] O. Nevanlinna, On the numerical solutions of some Volterra equations on infinite intervals, Rev. Anal. Numér. Theéorie Approx. 5 (1976) 31–57.

[113] O. Nevanlinna, On the stability of discrete Volterra equations, in: C.T.H. Baker, G.F. Miller (Eds.), Treatment of Integral equations by Numerical Methods, Academic Press, London, 1982, pp. 139–147.

[114] P. Pouzet, Etude, en vue de leur approximation numérique, des solutions d'équations intégrales et intégro-différentielles de type Volterra pour des problèmes de conditions initiales, Thése, Université de Strasbourg, 1962.

[115] A.A. Roberts, Analysis of explosion for nonlinear Volterra equations, J. Comput. Appl. Math. 97 (1998) 153–166.

[116] L.F. Shampine, Solving Volterra integral equations with ODE codes, IMA J. Numer. Anal. 8 (1988) 37–41.

[117] S. Shaw, J.R. Whiteman, Applications and numerical analysis of partial differential Volterra equations: a brief survey, Comput. Methods Appl. Mech. Eng. 150 (1997) 397–409.

[118] Z.B. Tsalyuk, Volterra integral equations, J. Soviet Math. 12 (1979) 715–758.

[119] A. Vecchio, Stability results on some direct quadrature methods for Volterra integro-differential equations, Dynamics Systems Appl. 7 (1998) 501–518.

[120] V. Volterra, Leçons sur les équations intégrales et intégro-différentielles, Gauthier-Villars, Paris, 1913.

[121] P.H.M. Wolkenfelt, The numerical analysis of reducible quadrature methods for Volterra integral and integro-differential equations — Academisch Proefschrift, Mathematisch Centrum, Amsterdam, 1981.

[122] P.H.M. Wolkenfelt, The construction of reducible quadrature-rules for Volterra integral and integro-differential equations, IMA J. Numer. Anal. 2 (1982) 131–152.

[123] P.H.M. Wolkenfelt, P.J. van der Houwen, C.T.H. Baker, Analysis of numerical methods for second kind Volterra equations by imbedding techniques, J. Integral Equations 3 (1981) 61–82.

JOURNAL OF COMPUTATIONAL AND APPLIED MATHEMATICS

ELSEVIER Journal of Computational and Applied Mathematics 125 (2000) 251–263

www.elsevier.nl/locate/cam

Numerical stability of nonlinear delay differential equations of neutral type ☆

Alfredo Bellen[a,] *, Nicola Guglielmi[b], Marino Zennaro[a]

[a] *Dipartimento di Scienze Matematiche, Università di Trieste, Piazzale Europa 1, I-34100 Trieste, Italy*

[b] *Dipartimento di Matematica Pura e Applicata, Università dell'Aquila, via Vetoio (Coppito), I-67010 L'Aquila, Italy*

Received 14 June 1999; received in revised form 2 September 1999

Abstract

This paper is devoted to the stability analysis of both the true solutions and the numerical approximations for nonlinear systems of neutral delay differential equations (NDDEs) of the form $y'(t) = F(t, y(t), G(t, y(t-\tau(t)), y'(t-\tau(t))))$. This work extends the results recently obtained by the authors Bellen et al. (BIT 39 (1999) 1–24) for the linear case. This is accomplished by considering a suitable reformulation of the given system, which transforms it into a nonlinear differential system coupled with an algebraic functional recursion. Numerical processes preserving the qualitative properties of the solutions are also investigated. © 2000 Elsevier Science B.V. All rights reserved.

Keywords: Neutral delay differential equations; Numerical stability; RK-methods for delay equations

1. Introduction

Delay differential equations (DDE) are assuming an increasingly important role in many applied disciplines. One of the reasons is that progress has been made in the mathematical understanding and theory of DDEs. Further, a multitude of different interesting problems lead to DDEs in different fields like biology, economy, circuit theory, control theory and electrodynamics (see, e.g., [5,7,10,14]). The mathematical theory necessary for the efficient solution and understanding of key issues like convergence and stability has been advanced especially in the last few years. A comprehensive introduction to the subject of DDEs and numerical solvers is given in a review paper by Baker et al. [1] and in a book chapter by Zennaro [19].

☆ This work was supported by the Italian M.U.R.S.T. and C.N.R.

* Corresponding author.

E-mail addresses: bellen@univ.trieste.it (A. Bellen), guglielm@univaq.it (N. Guglielmi), zennaro@univ.trieste.it (M. Zennaro).

PII: S0377-0427(00)00471-4

Our work was in part motivated by the solution of circuit theory time domain problems which lead to DDEs. Two important examples are the method of characteristics transmission line models [5] and the partial element equivalent circuit (PEEC) three-dimensional electromagnetic circuit model [14]. Recently, implementations of the PEEC approach have shown to be promising for the time domain analysis of electromagnetic interactions of packaged electronics [13]. The key feature of these problems is the fact that they involve delayed sources which are of the form $y(t-\tau)$ and $y'(t-\tau)$.

In this paper we investigate neutral delay differential equations (NDDEs) of the form

$$\begin{aligned} y'(t) &= F(t, y(t), G(t, y(t-\tau(t)), y'(t-\tau(t)))), \quad t \geqslant t_0, \\ y(t) &= g(t), \quad t \leqslant t_0, \end{aligned} \tag{1}$$

where F and G are complex continuous vector functions, $g(t)$ is a C^1-continuous complex-valued function and $\tau(t)$ is a continuous delay function such that

$$\text{(H1)} \quad \tau(t) \geqslant \tau_0 > 0 \quad \text{and} \quad \alpha(t) = t - \tau(t) \text{ is increasing } \forall t \geqslant t_0.$$

We first examine sufficient conditions for the contractivity and for the asymptotic stability of the true solutions of (1), which represents the model for the system to be studied. From a computational point of view, we are also interested in the qualitative behaviour of the numerical solution of the NDDE. For this, in the second part of the paper, we investigate both the contractivity and the asymptotic stability of the numerical solution furnished by RK-methods.

Confined to the linear case, several researchers have studied the asymptotic behaviour of the true solutions (see, e.g., [12]) as well as the numerical solutions (see, e.g., [6,9,11] and, quite recently, [3]). Relevant to the considered nonlinear test problem (1), instead, only very few results dealing with the qualitative behaviour of both the true and the numerical solutions have been published. Among them, we address the reader to the recent paper by Torelli and Vermiglio [17], where, however, both the model and the used approach are different with respect to those proposed here.

Our approach is based on the contractivity properties of the solutions of (1) and we extend the contractivity requirements to the numerical solutions. For the case of (nonneutral) nonlinear delay differential equations, this kind of methodology has been first introduced by Torelli [15,16] and then developed by Bellen and Zennaro [4], Bellen [2] and Zennaro [18,20].

2. The standard approach

The most standard approach considered in the literature consists in integrating (1) step-by-step. To this aim we need to determine the breaking points of the solution $y(t)$, which are points associated with discontinuities in the derivatives of $y(t)$, due to the presence of the functional argument $\alpha(t)$. By hypothesis (H1) on the delay function, we label these points as

$$\xi_0 = t_0 < \xi_1 < \cdots < \xi_n < \xi_{n+1} < \cdots,$$

ξ_{n+1} being the unique solution of $\alpha(\xi) = \xi_n$. Then we define the intervals $I_0 := [t_0 - \tau(t_0), \xi_0]$ and $I_n = [\xi_{n-1}, \xi_n]$, for $n \geqslant 1$. The analysis of the behaviour of the solutions can be done across the intervals I_n, by relating the solution in I_n with the one in I_{n-1}.

The classical way of numerically solving system (1) consists in making use of numerical methods for ODEs to integrate the system

$$y'(t) = F(t, y(t), G(t, \eta(\alpha(t)), \zeta(\alpha(t)))), \quad t \in I_n,$$

where $\eta(s)$ and $\zeta(s)$ are approximations of $y(s)$ and $y'(s)$, respectively, obtained by the numerical method itself in the previous interval I_{n-1}.

3. A suitable reformulation of the problem

In this work we propose a suitable reformulation of the problem which, apparently, does not require to approximate the derivative of the numerical solution. In order to do this, we rewrite problem (1) as follows:

$$\begin{aligned} y'(t) &= F(t, y(t), \Phi(t)), \quad t \geqslant t_0, \\ y(t) &= g(t), \quad t \leqslant t_0, \end{aligned} \tag{2}$$

where

$$\Phi(t) = \begin{cases} G(t, g(\alpha(t)), g'(\alpha(t))), & \text{if } t_0 \leqslant t < \xi_1, \\ G(t, y(\alpha(t)), F(\alpha(t), y(\alpha(t)), \Phi(\alpha(t)))) & \text{if } t \geqslant \xi_1. \end{cases} \tag{3}$$

Observe that $\Phi(t)$ is not continuous from the left at the breaking points ξ_k. In this approach the neutral system is transformed into an ordinary differential system plus an algebraic functional recursion. Consequently, new numerical schemes for the approximation of the solution are suggested.

Without loss of generality, we assume that

$$F(t, 0, 0) \equiv G(t, 0, 0) \equiv 0,$$

so that the system

$$\begin{aligned} y'(t) &= F(t, y(t), G(t, y(\alpha(t)), y'(\alpha(t)))), \quad t \geqslant t_0, \\ y(t) &= 0, \quad t \leqslant t_0, \end{aligned}$$

has the trivial solution $y(t) \equiv 0$. Our stability analysis is based on the contractivity properties of the solutions of the ordinary differential equation (ODE)

$$\begin{aligned} y'(t) &= F(t, y(t), \Phi(t)), \quad t \geqslant t_0, \\ y(t_0) &= y_0, \end{aligned} \tag{4}$$

with respect to the forcing term $\Phi(t)$. Preliminarily, we make some standard assumptions. Given an inner product $\langle \cdot, \cdot \rangle$ in $\mathbb{C}^m$ and the corresponding norm $||\cdot||$, we assume that $F : [t_0, +\infty) \times \mathbb{C}^m \times \mathbb{C}^m \to \mathbb{C}^m$ is continuous with respect to t and uniformly Lipschitz continuous with respect to the second and third variables, that is there exist two continuous functions $Y(t)$ and $W(t)$ such that

$$Y(t) \geqslant \sup_{x, y_1 \neq y_2} \frac{\Re\left[\langle F(t, y_1, x) - F(t, y_2, x), y_1 - y_2 \rangle\right]}{||y_1 - y_2||^2}, \tag{5}$$

$$W(t) \geqslant \sup_{y, x_1 \neq x_2} \frac{||F(t, y, x) - F(t, y, x)||}{||w_1 - w_2||}. \tag{6}$$

In the sequel it will be convenient to consider the function

$$H(t,x,z) = G(t,x,F(\alpha(t),x,z)). \tag{7}$$

Similar to what was done for F, we assume for H a continuous dependence with respect to t and a uniform Lipschitz continuity with respect to x and z, that is there exist two continuous functions $X(t)$ and $Z(t)$ such that

$$X(t) \geqslant \sup_{z,x_1 \neq x_2} \frac{||H(t,x_1,z) - H(t,x_2,z)||}{||x_1 - x_2||}, \tag{8}$$

$$Z(t) \geqslant \sup_{x,z_1 \neq z_2} \frac{||H(t,x,z_1) - H(t,x,z_2)||}{||z_1 - z_2||}. \tag{9}$$

Then recall the following result (see, for example [20]).

Lemma 1. *Consider system* (4) *where the components of the forcing term* $\Phi(t)$ *are assumed to be continuous functions. Moreover, chosen a suitable inner product* $\langle \cdot,\cdot \rangle$ *such that* (5) *and* (6) *hold, assume that*

$$Y(t) \leqslant 0, \quad t \geqslant t_0,$$

and for a bounded function $r(t) \geqslant 0$,

$$W(t) = -r(t)Y(t), \quad t \geqslant t_0.$$

Then, for all $t \geqslant t_0$, *the inequality*

$$||y(t)|| \leqslant E(t_0,t)||y(t_0)|| + (1 - E(t_0,t)) \sup_{t_0 \leqslant s \leqslant t} (r(s)\,||\Phi(s)||) \tag{10}$$

holds, where

$$E(t_1,t_2) = \exp\left(\int_{t_1}^{t_2} Y(s)\,\mathrm{d}s \right) \leqslant 1, \quad \forall t_2 \geqslant t_1.$$

By using the foregoing result, it is not difficult to modify the proofs of Theorems 3.1 and 3.2 in [3] in order to extend the contractivity and asymptotic stability results from the linear to the nonlinear case (1) considered here. In the sequel, we shall assume the function $\Phi(t)$ continuous in the closed interval $[\xi_{n-1}, \xi_n]$, $n \geqslant 1$, by considering

$$\Phi(\xi_n) = \lim_{t \to \xi_n^-} \Phi(t),$$

which exists. Therefore, there exists $\max_{\xi_{n-1} \leqslant s \leqslant \xi_n} ||\Phi(s)||$ for every $n \geqslant 1$. The following theorem concerns the contractivity properties of the solutions of (2) and (3). For technical reasons we shall assume $r(t) > 0$. In fact, we shall consider the ratio $r(t)/r(\alpha(t))$.

Theorem 2. *Assume that functions* (5), (6), (8) *and* (9) *fulfil the inequalities* $Y(t) \leqslant 0$, $r(t) > 0$ *and*

$$r(t)\left(X(t) + \frac{Z(t)}{r(\alpha(t))} \right) \leqslant 1, \quad \forall t \geqslant t_0. \tag{11}$$

Then the solution $y(t)$ *of* (1) *is such that*

$$||y(t)|| \leqslant \max\{||g(t_0)||, \kappa\}, \tag{12}$$

where

$$\kappa = \sup_{t_0 \leqslant s \leqslant \xi_1} ||r(s)\, G(s, g(\alpha(s)), g'(\alpha(s)))|| \tag{13}$$

for every initial function $g(t)$ *and for every delay* $\tau(t)$ *satisfying the assumption* (H1).

Proof. We proceed by developing a step-by-step analysis of system (2)–(3) over the intervals $\{I_n\}$. Thus consider the interval $I_n = [\xi_{n-1}, \xi_n]$ $(n \geqslant 1)$. By Lemma 1, for every $t \in I_n$, we have

$$||y(t)|| \leqslant E(\xi_{n-1}, t)||y(\xi_{n-1})|| + (1 - E(\xi_{n-1}, t)) \sup_{\xi_{n-1} \leqslant s \leqslant t} (r(s)||\Phi(s)||), \tag{14}$$

where $E(\xi_{n-1}, t) = \exp(\int_{\xi_{n-1}}^{t} Y(s)\,\mathrm{d}s) \leqslant 1$. Since the right-hand side of (14) is a convex combination, we immediately have

$$||y(t)|| \leqslant \max\left\{ ||y(\xi_{n-1})||, \sup_{\xi_{n-1} \leqslant s \leqslant t} (r(s)\, ||\Phi(s)||) \right\}. \tag{15}$$

Furthermore, by (3) and (7), it holds that

$$||\Phi(t)|| = \begin{cases} ||G(t, g(\alpha(t)), g'(\alpha(t)))||, & n = 1, \\ ||H(t, y(\alpha(t)), \Phi(\alpha(t)))||, & n \geqslant 2. \end{cases} \tag{16}$$

Therefore, for $n \geqslant 2$, since $H(t, 0, 0) \equiv 0$, we have

$$\begin{aligned} ||\Phi(t)|| &= ||H(t, y(\alpha(t)), \Phi(\alpha(t)))|| - ||H(t, 0, 0)|| \\ &= ||H(t, y(\alpha(t)), \Phi(\alpha(t)))|| - ||H(t, 0, \Phi(\alpha(t)))|| + ||H(t, 0, \Phi(\alpha(t)))|| - ||H(t, 0, 0)|| \\ &\leqslant X(t)||y(\alpha(t))|| + Z(t)||\Phi(\alpha(t))|| \end{aligned} \tag{17}$$

and, hence, assumption (11) implies

$$\begin{aligned} ||r(t)\Phi(t)|| &\leqslant (1 - v(t))||y(\alpha(t))|| + v(t)||r(\alpha(t))\Phi(\alpha(t))|| \\ &\leqslant \max\{||y(\alpha(t))||, ||r(\alpha(t))\Phi(\alpha(t))||\}, \end{aligned} \tag{18}$$

where $v(t) = (r(t)/r(\alpha(t)))Z(t) \leqslant 1$. Now, for any vector function $v(s)$ and any integer $l \geqslant 0$, set

$$|||v|||_l = \sup_{s \in I_l} ||v(s)||.$$

Therefore, routine calculations and (13) yield

$$|||r\,\Phi|||_1 = \kappa, \tag{19}$$

$$|||r\,\Phi|||_n \leqslant \max\{|||y|||_{n-1}, |||r\,\Phi|||_{n-1}\}, \quad n \geqslant 2. \tag{20}$$

Now, for all $n \geqslant 1$ define $\alpha_n = \max\{|||y|||_n, |||r\,\Phi|||_n\}$, so that (15) and (20) imply

$$|||y|||_n \leqslant \max\{|||y|||_{n-1}, |||r\,\Phi|||_{n-1}\}, \quad n \geqslant 1 \tag{21}$$

and $\alpha_n \leqslant \alpha_{n-1}$ for $n \geqslant 2$. Therefore,

$$|||y|||_n \leqslant \alpha_1 \quad \forall n \geqslant 2.$$

Finally, consider the case $n=1$, for which (15), (16) and (13) yield

$$|||y|||_1 \leqslant \max\{||g(t_0)||, \kappa\}.$$

Since $\alpha_1 = \max\{|||y|||_1, |||r\,\Phi|||_1\}$, the last inequality provides

$$\alpha_1 \leqslant \max\{||g(t_0)||, \kappa\}$$

and then (12) is proved. □

Remark 3. When G does not depend on $y'(\alpha(t))$, conditions (11) and (12) reduce to

$$r(t)X(t) \leqslant 1, \quad \forall t \geqslant t_0,$$

and

$$||y(t)|| \leqslant \max\left\{||g(t_0)||, \sup_{t_0 \leqslant s \leqslant \xi_1} ||r(s)\,G(s, g(\alpha(s)))||\right\}.$$

Remark 4. *In the fully linear nonautonomous case, that is*

$$\begin{aligned} y'(t) &= L(t)\,y(t) + M(t)\,y(\alpha(t)) + N(t)\,y'(\alpha(t)), & t \geqslant t_0, \\ y(t) &= g(t), & t \leqslant t_0, \end{aligned} \tag{22}$$

we can choose $Y(t) = \mu[L(t)]$, $\mu[\cdot]$ being the logarithmic norm, $W(t) = 1$, $r(t) = -1/\mu[L(t)]$, $X(t) = ||M(t) + N(t)L(\alpha(t))||$ and $Z(t) = ||N(t)||$. As a consequence, for the contractivity we require that $\mu[L(t)] < 0$ and

$$\frac{||M(t) + N(t)\,L(\alpha(t))||}{-\mu[L(t)]} \leqslant 1 - \frac{\mu[L(\alpha(t))]}{\mu[L(t)]}\,||N(t)||, \quad \forall t \geqslant t_0.$$

By slightly strengthening assumption (11), we can state a first result on the asymptotic stability of the solutions.

Theorem 5. *Assume that functions* (5), (6), (8) *and* (9) *fulfil the inequalities* $r(t) > 0$,

$$Y(t) \leqslant Y_0 < 0, \quad \forall t \geqslant t_0, \tag{23}$$

and

$$\begin{aligned} &\nu(t) \leqslant \tilde{\xi} < 1, \quad \forall t \geqslant t_0, \\ &X(t)\,r(t) \leqslant k(1 - \nu(t)), \quad \forall t \geqslant t_0, \ k < 1, \end{aligned} \tag{24}$$

where $\nu(t) = Z(t)\,r(t)/r(\alpha(t))$. *Then we have* $\lim_{t \to \infty} y(t) = 0$ *for every initial function* $g(t)$ *and for every delay* $\tau(t)$ *satisfying assumption* (H1) *and*

(H2) $\lim_{t \to +\infty} \alpha(t) = +\infty$.

Proof. Observe that, by (23), the function $r(t) = -(W(t)/Y(t))$ is continuous. In analogy to the previous theorem, we proceed by developing a step-by-step analysis over the intervals $\{I_n\}$. By (10), for every $t \in I_n$, we get

$$\begin{aligned} \|y(\xi_n)\| &\leqslant E(\xi_{n-1},\xi_n)\|y(\xi_{n-1})\| + (1 - E(\xi_{n-1},\xi_n)) \sup_{\xi_{n-1}\leqslant s\leqslant\xi_n} (r(s)\|\Phi(s)\|) \\ &\leqslant c_n |||y|||_n + (1-c_n)|||r\,\Phi|||_n, \end{aligned} \tag{25}$$

where $c_n = \exp(\int_{\xi_{n-1}}^{\xi_n} Y(s)\,ds)$ and, by (H1), $c_n < 1\,\forall n$. Furthermore, for the interval I_{n+1}, (10) yields

$$|||y|||_{n+1} \leqslant \max\{\|y(\xi_n)\|, |||r\,\Phi|||_{n+1}\}. \tag{26}$$

In turn, by (25), inequality (26) provides

$$|||y|||_{n+1} \leqslant \max\{c_n |||y|||_n + (1-c_n)|||r\,\Phi|||_n, |||r\,\Phi|||_{n+1}\}. \tag{27}$$

Now consider (3) and (7). By assumption (24), we have

$$\begin{aligned} |||r\,\Phi|||_{n+1} &\leqslant \max_{\xi_n\leqslant s\leqslant\xi_{n+1}} (v(s)|||r\,\Phi|||_n + k(1-v(s))|||y|||_n) \\ &= \zeta_{n+1}|||r\,\Phi|||_n + k(1-\zeta_{n+1})|||y|||_n, \quad n \geqslant 2, \end{aligned} \tag{28}$$

where $\zeta_{n+1} = v(s_{n+1})$, with $s_{n+1} \in [\xi_n, \xi_{n+1}]$ suitable point. For the sake of conciseness, we omit the rest of the proof, which is completely analogous to that given in Theorem 3.2 of Bellen et al. [3] for the linear case, and is based on a suitable analysis of relations (27) and (28). We remark that hypothesis (H2) is necessary. In fact, if it did not hold, there would exist only a finite number of breaking points ξ_n and we could not conclude with asymptotic stability. □

The following corollary is obtained by assuming that the delay function is bounded from above.

Corollary 6. *Assume the hypotheses of Theorem 5 and that*

(H3) $\tau(t) \leqslant \tau_M, \quad \forall t \geqslant t_0$ (*fading memory assumption*).

Then the solution $y(t)$ has an exponential asymptotic decay.

The proof is analogous to that given in [3], relevant to the linear case.

Remark 7. When G does not depend on $y'(\alpha(t))$, (24) yield the condition $X(t)r(t) \leqslant k < 1$.

4. The numerical scheme

Given a general ODE with forcing term $\Phi(t)$ of the form

$$y'(t) = f(t, y(t), \Phi(t)) \tag{29}$$

with initial condition $y(t_0) = y_0$ and a mesh $\Delta = \{t_0 < t_1 < \cdots < t_n < t_{n+1} < \cdots\}$, let us consider the s-stage RK-method

$$\begin{aligned} Y_{n+1}^i &= y_n + h_{n+1} \sum_{j=1}^{s} a_{ij} K_{n+1}^j, \quad i = 1,\ldots,s, \\ K_{n+1}^i &= f(t_{n+1}^i, Y_{n+1}^i, \Phi(t_{n+1}^i)), \\ y_{n+1} &= y_n + h_{n+1} \sum_{i=1}^{s} w_i K_{n+1}^i, \end{aligned} \tag{30}$$

where $t_{n+1}^j = t_n + c_j h_{n+1}$, $c_i = \sum_{j=1}^{s} a_{ij}$, $i = 1,\ldots,s$, and $h_{n+1} = t_{n+1} - t_n$.

A continuous extension of the s-stage RK-method (30) may be obtained by considering the following interpolation:

$$\eta(t_n + \theta h_{n+1}) = y_n + h_{n+1} \sum_{i=1}^{s} w_i(\theta) K_{n+1}^i, \quad 0 \leqslant \theta \leqslant 1, \tag{31}$$

where $w_i(\theta)$, $i = 1,\ldots,s$, are polynomials of given degree d such that $w_i(0) = 0$ and $w_i(1) = w_i$. Henceforth, we shall assume that $c_i \in [0,1]$ $\forall i$. In fact, if $c_i > 1$ for some i, there could be some problems with the considered method of steps, due to the possibility that $\eta(\alpha(t_n^i))$ could be not available from previous computations. A special instance, which will be referred to in the next sections, consists of linear interpolation. In this case we have $w_i(\theta) = \theta\, w_i$ and hence

$$\eta(t_n + \theta h_{n+1}) = (1-\theta)\, y_n + \theta\, y_{n+1}, \quad 0 \leqslant \theta \leqslant 1. \tag{32}$$

Now we propose a procedure for solving the test problem (1), which acts recursively on the intervals $I_j = [\xi_{j-1}, \xi_j]$, where the true solution $y(t)$ is regular. The procedure, also called *method of steps*, is summarized by the following scheme. On each interval I_j, $j \geqslant 1$, use a continuous RK-method on the mesh $\Delta_j = \{t_{j,0} \equiv \xi_{j-1} < t_{j,1} < \cdots < t_{j,N_j} \equiv \xi_j\}$ in I_j to solve the nonlinear system

$$\begin{aligned} y'(t) &= F(t, y(t), \bar{\Phi}(t)), \\ y(\xi_{j-1}) &= \eta(\xi_{j-1}), \end{aligned} \tag{33}$$

where $\eta(t)$ is the continuous approximation already computed by the numerical method in the previous intervals and the function $\bar{\Phi}(t)$ is given by one of the following schemes.

First scheme: *direct evaluation* (DE):

$$\bar{\Phi}(t) = \begin{cases} G(t, g(\alpha(t)), g'(\alpha(t))), \ t \in [t_0, \xi_1], \\ H(t, \eta(\alpha(t)), \bar{\Phi}(\alpha(t))), \ t \in [\xi_{j-1}, \xi_j],\ j \geqslant 2. \end{cases} \tag{34}$$

Second scheme: *piecewise polynomial approximation* (PA):

$$\bar{\Phi}(t) = \begin{cases} G(t, g(\alpha(t)), g'(\alpha(t))), & t \in [t_0, \xi_1], \\ H\left(t, \eta(\alpha(t)), \mathcal{Q}_{j-1}^d(\bar{\Phi}(\cdot))(\alpha(t))\right), & t \in [\xi_{j-1}, \xi_j],\ j \geqslant 2, \end{cases}$$

$$\mathcal{Q}_l^d : C([\xi_{l-1}, \xi_l], \mathbb{C}^m) \to \Pi_l^d([\xi_{l-1}, \xi_l], \mathbb{C}^m),$$

where d is the degree of the approximating polynomial and $\mathcal{Q}_l^d$ denotes a linear projector from the space of continuous vector functions $C([\xi_{l-1}, \xi_l], \mathbb{C}^m)$ into the space of piecewise vector polynomials

$\Pi_l^d([\xi_{l-1},\xi_l],\mathbb{C}^m)$. To be more precise, $\mathcal{Q}_l^d$ provides a vector polynomial $\bar{\Phi}$ of degree d on every subinterval $[t_{l,k},t_{l,k+1}]$ defined by the mesh Δ_l.

In the same way as in [3], we can prove the following result on the order of the proposed numerical schemes for the solution of (2) and (3).

Theorem 8. *For the solution of the NDDE* (1), *consider the DE-scheme, or the PA-scheme, with piecewise polynomial approximation of order p. Assume that the functions F and G are sufficiently smooth. Moreover, suppose to use a continuous RK-method of uniform global order p for the numerical solution of* (33) *on each interval I_n. Then the numerical solution has uniform global order p.*

5. Stability properties of the DE- and PA-scheme

In this section we analyze the contractivity and asymptotic stability properties of the DE- and PA-scheme proposed in Section 4.

Being the proposed numerical schemes based on the recursive solution of ordinary differential systems such as (33), we are interested in determining RK-methods which are contractive with respect to the test equation

$$\begin{aligned} y'(t) &= F(t,y(t),\Phi(t)), \\ y(t_n) &= y_n. \end{aligned} \tag{35}$$

To this purpose, we recall the following definition from [4].

Definition 9. A continuous RK-method is said to be BN_f-stable if the continuous numerical solution $\eta(t)$ of (35) satisfies

$$\max_{0\leqslant\theta\leqslant 1}\|\eta(t_n+\theta h)\|\leqslant\max\left\{\|y_n\|,\max_{1\leqslant j\leqslant s}\|r(t_n+c_j h)\,\Phi(t_n+c_j h)\|\right\} \tag{36}$$

for any stepsize $h>0$, $r(t)$ being the function considered in Lemma 1.

It has been proved [4] that Backward Euler ($p=1$) and 2-stage Lobatto III-C with linear interpolation ($p=2$) are BN_f-stable. These methods are given by the following Butcher tableaux:

$$\begin{array}{c|c} 1 & 1 \\ \hline & 1 \end{array} \qquad \begin{array}{c|cc} 0 & \frac{1}{2} & -\frac{1}{2} \\ 1 & \frac{1}{2} & \frac{1}{2} \\ \hline & \frac{1}{2} & \frac{1}{2} \end{array}.$$

As in Section 3, we shall assume the function $\bar{\Phi}(t)$ continuous in the closed interval $[\xi_{n-1},\xi_n]$, $n\geqslant 1$, by considering

$$\bar{\Phi}(\xi_n)=\lim_{t\to\xi_n^-}\bar{\Phi}(t),$$

which exists. The first result we are able to prove concerns the DE-scheme.

Theorem 10. *Consider the DE-scheme for the numerical solution of* (1), *and assume to use a* BN_f*-stable continuous RK-method for the solution of* (33). *If the hypotheses of Theorem* 2 *hold, then for any mesh* Δ *the numerical solution* η *satisfies the contractivity property*

$$||\eta(t)|| \leqslant \max\{||g(t_0)||, \kappa\}, \quad \forall t \geqslant t_0.$$

Proof. The proof is analogous to the one for the stability of the true solutions and presents several similarities to the proof of Theorem 5.1 in [3]. In fact, it is based on the property established by Definition 9 for the RK-method and on the assumption that we compute the function $\bar{\Phi}$ "exactly" (that is "recursively" from the previous steps). The first assumption allows to state an inequality for the numerical solution $\eta(t)$ analogous to (21) on each interval I_n, that is

$$|||\eta|||_n \leqslant \max\{|||\eta|||_{n-1}, |||r\bar{\Phi}|||_{n-1}\}. \tag{37}$$

In fact, denoting the mesh relevant to the interval I_n by

$$t_{n,0} \equiv \xi_{n-1} < t_{n,1} < \cdots < t_{n,N_n} \equiv \xi_n \tag{38}$$

and the corresponding stepsizes by $h_{n,k} = t_{n,k} - t_{n,k-1}$, $k \geqslant 1$, (36) yields

$$\max_{0 \leqslant \theta \leqslant 1} ||\eta(t_{n,0} + \theta h_{n,1})|| \leqslant \max\{||\eta(t_{n,0})||, |||r\bar{\Phi}|||_n\}.$$

By induction on the points $t_{n,k}$, $k = 1, \ldots, N_n - 1$, we easily get

$$\max_{0 \leqslant \theta \leqslant 1} ||\eta(t_{n,k} + \theta h_{n,k+1})|| \leqslant \max\{||\eta(t_{n,0})||, |||r\bar{\Phi}|||_n\}.$$

By definition of $\bar{\Phi}$, assumption (11) provides the inequalities

$$|||r\bar{\Phi}|||_1 \leqslant \kappa \quad \text{and} \quad |||r\bar{\Phi}|||_n \leqslant \max\{|||\eta|||_{n-1}, |||r\bar{\Phi}|||_{n-1}\}, \quad n \geqslant 2, \text{ and hence (37).}$$

From here on, the proof of contractivity is completely analogous to the one given in Theorem 2. □

The second result concerns instead the PA-scheme, where piecewise constant or piecewise linear interpolation is used for approximating the function $\bar{\Phi}$. If $t \in [\xi_{n-1}, \xi_n]$, there exists k such that $\alpha(t) \in [t_{n-1,k}, t_{n-1,k+1}] \subseteq [\xi_{n-2}, \xi_{n-1}]$ and therefore

$$\mathcal{Q}^0_{n-1}(\bar{\Phi}(\cdot))(\alpha(t)) = \bar{\Phi}(t_{n-1,k}),$$

$$\mathcal{Q}^1_{n-1}(\bar{\Phi}(\cdot))(\alpha(t)) = \frac{t_{n-1,k+1} - \alpha(t)}{h_{n-1,k+1}} \bar{\Phi}(t_{n-1,k}) + \frac{\alpha(t) - t_{n-1,k}}{h_{n-1,k+1}} \bar{\Phi}(t_{n-1,k+1}).$$

Theorem 11. *Consider the PA-scheme with constant or linear interpolant for the numerical solution of* (1) *and assume to use a* BN_f*-stable continuous RK-method for the solution of* (33). *If the hypotheses of Theorem* 2 *hold, then for every mesh* Δ *the numerical solution* η *satisfies the contractivity property*

$$||\eta(t)|| \leqslant \max\{||g(t_0)||, \kappa\}, \quad \forall t \geqslant t_0.$$

Proof. The proof proceeds as for the DE-case (Theorem 10), where the bound for $|||\bar{\Phi}|||_n$ remains the same because the piecewise constant or linear interpolation of $\bar{\Phi}$ is still bounded by the maximum value of $||\bar{\Phi}||$. Therefore, also in this case we can assert inequality (37) and then proceed as in the previous case. □

Now we pass to consider the asymptotic stability of the numerical schemes.

Lemma 12 (Zennaro [20]). *Let the continuous RK-method* (30)–(31) *be* BN_f*-stable and such that* $c_i \in [0,1]\ \forall i$. *Then, for any mesh* Δ, *under the condition* $Y(t) \leqslant Y_0 < 0$, the numerical solution $\{y_\ell\}$ *of* (35) *satisfies*

$$||y_{\ell+1}|| \leqslant \psi_{\ell+1}||y_\ell|| + (1-\psi_{\ell+1}) \max_{1\leqslant i\leqslant s} ||r(t_\ell + c_i h_{\ell+1})\Phi(t_\ell + c_i h_{\ell+1})||,$$

where $0 \leqslant \psi_{\ell+1} \leqslant \Theta(h_{\ell+1})$, $\Theta(h)$ *being the error growth function of the RK-method, which depends on* Y_0.

Hairer and Zennaro [8] proved that the error growth function $\Theta(h)$ is an asymptotically negative super-exponential function, that is

$$\begin{aligned}
&\Theta(0) = 1,\\
&\Theta(h) < 1, \quad \forall h > 0,\\
&\Theta(h')\,\Theta(h'') \leqslant \Theta(h' + h''), \quad \forall h', h'' \geqslant 0,\\
&\lim_{h\to\infty} \Theta(h) < 1.
\end{aligned} \tag{39}$$

Theorem 13. *Consider the DE-scheme for the numerical solution of* (1), *and assume to use a* BN_f*-stable continuous RK-method for the solution of* (33), *such that* $c_i \in [0,1]\ \forall i$. *If the hypotheses of Theorem 5 hold, then, for any mesh* Δ, *the numerical solution* η *asymptotically vanishes. Furthermore, if condition* (H3) *holds, the decay of* η *is exponential.*

Proof. With the previously introduced notation (38), as a consequence of the assumptions and by means of Lemma 12, we get

$$||\eta(t_{n,1})|| \leqslant \psi_{n,1}||\eta(t_{n,0})|| + (1-\psi_{n,1}) \max_{0\leqslant\theta\leqslant 1} ||r(t_{n,0} + \theta h_{n,1})\,\bar{\Phi}(t_{n,0} + \theta h_{n,1})||,$$

where

$$\psi_{l,k} \leqslant ||\Theta(h_{l,k})||, \quad l = 1,2,\ldots, \quad k = 1,\ldots,N_l.$$

With standard manipulations, by (38), we arrive at

$$\begin{aligned}
||\eta(\xi_n)|| &\leqslant (1-\Psi_n) \max_{0\leqslant\theta\leqslant 1} ||r(\xi_{n-1} + \theta\mathscr{H}_n)\bar{\Phi}(\xi_{n-1} + \theta\mathscr{H}_n)|| + \Psi_n||\eta(\xi_{n-1})||\\
&\leqslant \Psi_n|||\eta|||_n + (1-\Psi_n)|||r\bar{\Phi}|||_n,
\end{aligned} \tag{40}$$

where $\mathscr{H}_n = \sum_{i=1}^{N_n} h_{n,i} = \xi_n - \xi_{n-1}$ and $\Psi_n = \prod_{k=1}^{N_n} \psi_{n,k}$. By Lemma 12, we get

$$\Psi_n \leqslant \Theta(h_{n,1})\Theta(h_{n,2})\cdots\Theta(h_{n,N_n}) \leqslant \Theta(\mathscr{H}_n) \leqslant \Theta(\tau_0) < 1,$$

since $\mathscr{H}_n = \xi_n - \xi_{n-1} \geqslant \tau_0$ and $\Theta(h)$ is nonincreasing. We remark that formula (40) plays, in the numerical case, the role of (25). The method being BN_f-stable, for the interval I_{n+1} we get

$$|||\eta|||_{n+1} \leqslant \max\{||\eta(\xi_n)||, |||r\bar{\Phi}|||_{n+1}\}. \tag{41}$$

Hence, by applying estimate (40), inequality (41) yields

$$|||\eta|||_{n+1} \leqslant \max\{\Psi_n|||\eta|||_n + (1-\Psi_n)|||r\bar{\Phi}|||_n, |||r\bar{\Phi}|||_{n+1}\}. \tag{42}$$

By assumption (24) and by definition of DE-scheme, similar to (28) we have

$$|||r\bar{\Phi}|||_{n+1} \leqslant \bar{\zeta}_{n+1}|||r\bar{\Phi}|||_n + k(1-\bar{\zeta}_{n+1})|||\eta|||_n, \tag{43}$$

where $\bar{\zeta}_{n+1} = v(\bar{s}_{n+1})$ with $\bar{s}_{n+1} \in [\xi_n, \xi_{n+1}]$ suitable point. Formulae (41)–(43) play, in the numerical case, the role of formulae (26)–(28). Now, by virtue of this correspondence, the proof of the theorem proceeds in perfect analogy to that of Theorem 5 and Corollary 6. □

Similarly, relevant to the PA-scheme, we obtain the following result.

Theorem 14. *Consider the PA-scheme with constant or linear interpolant for the numerical solution of* (1) *and assume to use a* BN_f*-stable continuous RK-method for the solution of* (33). *If the hypotheses of Theorem* 5 *hold, then, for every mesh* Δ, *the numerical solution* η *asymptotically vanishes. Furthermore, if condition* (H3) *holds, the decay of* η *is exponential.*

We conclude the paper by giving an algorithmic description of both the classical method (see Section 2) and the new method (based on the PA-scheme) for integrating (1).

Scheme 1 (*classical method*).

(i) In the first interval $I_1 = [t_0, \xi_1]$:

(1) Compute

$$\bar{\Phi}(t_{n,i}) = G(t_{n,i}, g(\alpha(t_{n,i})), g'(\alpha(t_{n,i}))), \quad i = 1, \dots, s.$$

(2) Evaluate the stages of the RK-method

$$K_i = F(t_{n,i}, Y_i, \bar{\Phi}(t_{n,i})), \quad i = 1, \dots, s.$$

(3) Construct the polynomial approximations $\eta(t)$ of $y(t)$ and $\zeta(t)$ (possibly $\eta'(t)$) of $y'(t)$.

(ii) In the subsequent intervals $I_j = [\xi_{j-1}, \xi_j]$, $j \geqslant 2$:

(1) Compute the values

$$\bar{\Phi}(t_{n,i}) = G(t_{n,i}, \eta(\alpha(t_{n,i})), \zeta(\alpha(t_{n,i}))), \quad i = 1, \dots, s.$$

(2) As in (i)(2).

(3) As in (i)(3).

Scheme 2 (*novel method* (*PA*)).

(i) In the first interval $I_1 = [t_0, \xi_1]$:

(1) Compute

$$\bar{\Phi}(t_{n,i}) = G(t_{n,i}, g(\alpha(t_{n,i})), g'(\alpha(t_{n,i}))), \quad i = 1, \dots, s.$$

(2) Evaluate the stages of the RK-method

$$K_i = F(t_{n,i}, Y_i, \bar{\Phi}(t_{n,i})), \quad i = 1, \dots, s,$$

and then the continuous numerical approximation of the solution $\eta(t)$.

(3) Interpolate $\Phi(t)$ over the time points $\{t_{n,i}\}$ to obtain a polynomial approximation $\bar{\Phi}(t)$.

(ii) In the subsequent intervals $I_j = [\xi_{j-1}, \xi_j]$, $j \geqslant 2$:
   (1) Compute the values

$$\bar{\Phi}(t_{n,i}) = H(t_{n,i}, \eta(\alpha(t_{n,i})), \bar{\Phi}(\alpha(t_{n,i}))), \quad i = 1, \ldots, s.$$

   (2) As in (i)(2).
   (3) As in (i)(3).

References

[1] C.T.H. Baker, C.A.H. Paul, D.R. Willè, Issues in the numerical solution of evolutionary delay differential equations, Adv. Comput. Math. 3 (1995) 171–196.
[2] A. Bellen, Contractivity of continuous Runge–Kutta methods for delay differential equations, Appl. Numer. Math. 24 (1997) 219–232.
[3] A. Bellen, N. Guglielmi, M. Zennaro, On the contractivity and asymptotic stability of systems of delay differential equations of neutral type, BIT 39 (1999) 1–24.
[4] A. Bellen, M. Zennaro, Strong contractivity properties of numerical methods for ordinary and delay differential equations, Appl. Numer. Math. 9 (1992) 321–346.
[5] R.K. Brayton, Small signal stability criterion for networks containing lossless transmission lines, IBM J. Res. Develop. 12 (1968) 431–440.
[6] R.K. Brayton, R.A. Willoughby, On the numerical integration of a symmetric system of difference-differential equations of neutral type, J. Math. Anal. Appl. 18 (1967) 182–189.
[7] N. Guglielmi, Inexact Newton methods for the steady-state analysis of nonlinear circuits, Math. Models Methods Appl. Sci. 6 (1996) 43–57.
[8] E. Hairer, M. Zennaro, On error growth functions of Runge–Kutta methods, Appl. Numer. Math. 22 (1996) 205–216.
[9] G.-Da Hu, T. Mitsui, Stability analysis of numerical methods for systems of neutral delay-differential equations, BIT 35 (1995) 504–515.
[10] Y. Kuang, Delay Differential Equations with Application in Population Dynamics, Academic Press, Boston, 1993.
[11] J.X. Kuang, J.X. Xiang, H.J. Tian, The asymptotic stability of one-parameter methods for neutral differential equations, BIT 34 (1994) 400–408.
[12] L.M. Li, Stability of linear neutral delay-differential systems, Bull. Austral. Math. Soc. 38 (1988) 339–344.
[13] W. Pinello, A.C. Cangellaris, A. Ruehli, Hybrid electromagnetic modeling of noise interactions in packaged electronics based on the partial element equivalent circuit formulation, IEEE Trans. Microwave Theory Tech. 45 (1997) 1889–1896.
[14] A. Ruehli, U. Miekkala, A. Bellen, H. Heeb, Stable time domain solutions for EMC problems using PEEC circuit models, Proceedings of IEEE International Symposium on Electromagnetic Compatibility, Chicago, IL, 1994 pp. 371–376.
[15] L. Torelli, Stability of numerical methods for delay differential equations, J. Comput. Appl. Math. 25 (1989) 15–26.
[16] L. Torelli, A sufficient condition for GPN-stability for delay differential equations, Numer. Math. 59 (1991) 311–320.
[17] L. Torelli, R. Vermiglio, Stability of non-linear neutral delay differential equations, preprint, 1999.
[18] M. Zennaro, Contractivity of Runge–Kutta methods with respect to forcing terms, Appl. Numer. Math. 10 (1993) 321–345.
[19] M. Zennaro, Delay differential equations: theory and numerics, in: M. Ainsworth, W.A. Light, M. Marletta (Eds.), Theory and Numerics of Ordinary and Partial Differential Equations, Oxford University Press, Oxford, 1995 (Chapter 6).
[20] M. Zennaro, Asymptotic stability analysis of Runge–Kutta methods for nonlinear systems of delay differential equations, Numer. Math. 77 (1997) 549–563.

ELSEVIER
Journal of Computational and Applied Mathematics 125 (2000) 265–275

JOURNAL OF
COMPUTATIONAL AND
APPLIED MATHEMATICS

www.elsevier.nl/locate/cam

Numerical bifurcation analysis of delay differential equations

Koen Engelborghs, Tatyana Luzyanina [1], Dirk Roose *

Department of Computer Science, Katholieke Universiteit Leuven, Celestijnenlaan 200A, B-3001 Heverlee-Leuven, Belgium

Received 11 June 1999; received in revised form 6 January 2000

Abstract

Numerical methods for the bifurcation analysis of delay differential equations (DDEs) have only recently received much attention, partially because the theory of DDEs (smoothness, boundedness, stability of solutions) is more complicated and less established than the corresponding theory of ordinary differential equations. As a consequence, no established software packages exist at present for the bifurcation analysis of DDEs. We outline existing numerical methods for the computation and stability analysis of steady-state solutions and periodic solutions of systems of DDEs with several constant delays.

Keywords: Delay differential equations; Bifurcation analysis; Numerical methods

1. Delay differential equations

In this paper, we consider delay differential equations with multiple fixed discrete delays,

$$\dot{x}(t) = f(x(t), x(t-\tau_1), \ldots, x(t-\tau_m), \alpha), \tag{1}$$

where $x(t) \in \mathbb{R}^n$, $f : \mathbb{R}^{n\times(m+1)} \times \mathbb{R} \to \mathbb{R}^n$, $\alpha \in \mathbb{R}$, $\tau_j > 0$, $j = 1, \ldots, m$. Other types of DDEs exist and appear in applications. If, e.g., the right-hand side of (1) further depends on the derivative of $x(t)$ in the past,

$$\dot{x}(t) = f(x(t), x(t-\tau_1), \ldots, x(t-\tau_m), \dot{x}(t-\tau_1), \ldots, \dot{x}(t-\tau_m), \alpha), \tag{2}$$

the corresponding DDE is of *neutral* type. Delays can also be distributed (bounded or unbounded) or varying (time and/or state dependent). All these types of DDEs exhibit quite different theoretical properties, many of which are still under investigation. Here we restrict ourselves to DDEs of

* Corresponding author. Tel.: +32-16-20-10-15; fax: +32-16-20-53-08.
E-mail address: dirk.roose@cs.kuleuven.ac.be (D. Roose).
[1] On leave from the Institute of Mathematical Problems in Biology, RAS, Pushchino, Moscow Region 142292, Russia.

PII: S0377-0427(00)00472-6

type (1) and briefly mention some extra (and unsolved) difficulties occurring in the numerical analysis of (2).

A solution $x(t)$ of (1) and (2) on $t \in [0, \infty)$ is uniquely defined by specifying as initial condition a *function segment*, $x(\theta) = x_0(\theta)$, $-\tau \leqslant \theta \leqslant 0$, where $\tau = \max_i \tau_i$, $x_0 \in C$. Here $C = C([-\tau, 0], \mathbb{R}^n)$ is the space of continuous function segments mapping the delay interval into $\mathbb{R}^n$. Even if f and x_0 are infinitely smooth, a discontinuity in the first derivative of $x(t)$ generally appears at $t = 0$ which is propagated through time. The solution operator of (1), however, smoothes the initial data and discontinuities appear in higher derivatives as time increases, which is not the case for (2).

Numerical methods for time integration of (1) are indicated by C.T.H. Baker [26]. Here, we outline numerical methods for the bifurcation analysis of steady-state solutions and periodic solutions of (1). Since a DDE can be approximated by a system of ordinary differential equations (ODEs), this allows to use standard numerical methods for ODEs. However, an accurate approximation leads to a high-dimensional system of ODEs and thus to a very expensive method. In this paper we only present methods that are specifically developed for DDEs, which are both more efficient and more reliable.

2. Steady-state solutions

A steady-state solution (equilibrium) of (1) does not depend on the delays. Hence, (1) and the system of ODEs, obtained from (1) by putting all delays to zero, have the same steady-state solutions, $x^* \in \mathbb{R}^n$, which can be computed as zeros of the n-dimensional nonlinear system $f(x^*, x^*, \ldots, x^*, \alpha) = 0$.

However, the stability conditions are different. The stability of a steady-state solution x^* of (1) is determined by the roots of the characteristic equation. If we define the $n \times n$-dimensional characteristic matrix Δ as

$$\Delta(x^*, \lambda) = \lambda I - A_0(x^*) - \sum_{j=1}^{m} A_j(x^*)\,\mathrm{e}^{-\tau_j \lambda}, \tag{3}$$

with A_j the partial derivative of f with respect to its $(j+1)$th argument evaluated at x^*, then the characteristic equation reads

$$P(\lambda) = \det(\Delta(x^*, \lambda)) = 0. \tag{4}$$

The steady-state solution x^* is asymptotically stable provided that all the roots $\lambda \in \mathbb{C}$ of (4) have strict negative real part [14, p. 23]. Note that, in general, (4) has infinitely many roots.

If the right-hand side of the DDE depends on a physical parameter (like α in (1)) then a branch of steady-state solutions $(x^*(\alpha))$ can be computed as a function of the parameter using a continuation procedure. The stability of the steady-state can change whenever roots of (4) cross the imaginary axis. The corresponding bifurcation point is (generically) a turning point when a real root crosses through zero and a Hopf bifurcation when a complex pair of roots crosses the imaginary axis. In the latter case a branch of periodic solutions arises from the bifurcation point.

To determine the stability of x^*, a distinction can be made between two types of algorithms. The first type computes the *number* of unstable roots of Eq. (4), that is the number of roots λ with $\mathrm{Re}(\lambda) > 0$, without computing the roots themselves. The second type *locates* and computes the finite number of roots of (4) with real part greater than some negative constant, $\mathrm{Re}(\lambda) > -\beta$. Once detected, bifurcation points can be followed in two-parameter space using an appropriate determining system (see [20]).

2.1. Counting unstable roots

Since $P(\lambda)$ is an analytic function, the theorem on the logarithmic residue [4] holds. If $P(\lambda)$ is not zero on a closed contour $G \subset \mathbb{C}$ in the right-half plane, then the number of zeros of $P(\lambda)$ in the interior of G, counted according to their multiplicities, is

$$N_G = \frac{1}{2\pi i}\int_G \frac{P'(z)}{P(z)}\,dz. \tag{5}$$

In [16], (5) is approximated using numerical quadrature. Geometric interpretation of (5) leads to the argument principle which is used in [20],

$$N_G = \frac{1}{2\pi}\Delta_G \arg P(z), \tag{6}$$

where $\Delta_G \arg P(z)$ is the total increase of $\arg P(z)$ as z moves once around the contour G in the positive direction. The method suggested in [12] is based on a restatement of the argument principle which says that N_G equals the number of times the curve $J = P(G)$ (the mapping of G under $z \to P(z)$) encircles the origin.

In the papers mentioned above, different choices of the contour G are used. Two problems arise: roots in the right-half plane outside of G go unnoticed and (5) and (6) are ill-conditioned when $P(\lambda)$ has a root near G. The first problem is tackled in [16] using a Gershgorin-type bound to determine a region in the positive half plane, inside which all zeros (if any) of (4) must lie. In [20] the contour is user chosen. In [12] G consists of the semicircle $\lambda = R\exp(i\theta)$, $\theta \in [-\pi/2, \pi/2]$, and the segment $[-iR, iR]$ with $R \gg 0$. If an upper bound for R can be derived such that for $R > R_{crit}$, $\mathrm{Re}(P(iR)) < 0$, it is sufficient to consider the segment $[-iR_{crit}, iR_{crit}]$. The second problem can somewhat be alleviated using adaptivity in the stepsize of the numerical quadrature or in the determination of (6).

Recently, extensions of (5) have been used to compute not only the number but also the location of the roots inside a region in the complex plane (see [17] and the references therein). All the methods mentioned in this subsection suffer from the problem that they are computationally expensive for large n due to the computation of the $n \times n$ determinant in (4).

2.2. Locating and computing roots

Eq. (4) can be seen as a nonstandard, nonlinear eigenvalue problem with an infinite number of solutions of which only the rightmost solutions are of interest to determine the stability. Several algorithms are available for the efficient computation of selected eigenvalues of possibly very large matrices. Therefore, one transforms (4) to a standard eigenvalue problem. For this, we have to consider (1) in its proper state space, C.

Two different operators acting on C are related to (1) and have eigenvalues which correspond to the solutions of (4). First, consider the time integration operator $S(t_0)$ of the linearized equation of

(1), i.e. $S(t_0)y_0=y_{t_0}$ where y_0, y_{t_0} are solution segments of a solution $y(t)$ of the variational equation

$$\dot{y}(t)=A_0(x^*)y(t)+\sum_{i=1}^{m}A_i(x^*)y(t-\tau_i). \tag{7}$$

$S(t_0)$ is compact whenever $t_0\geqslant\tau$. Second, consider the infinitesimal generator $\mathscr{A}$ defined as

$$\mathscr{A}\phi=\lim_{t_0\to 0^+}\frac{S(t_0)\phi-\phi}{t_0},\qquad \phi\in\mathscr{D}(\mathscr{A}),$$

whose domain is dense in C. The eigenvalues of $\mathscr{A}$ coincide with the solutions of (4), while the eigenvalues of $S(t_0)$ are of the form $\exp(\lambda t_0)$, with λ a solution of (4), plus possibly zero [14]. Once an eigenvalue μ of $S(t_0)$ is found, the corresponding root of the characteristic equation can be extracted using $\mathrm{Re}(\lambda)=\log(|\mu|)/t_0$ and $\mathrm{Im}(\lambda)=\arcsin(\mathrm{Re}(\mu)/|\mu|)/t_0 \bmod \pi/t_0$. The full imaginary part can further be extracted from the corresponding eigenvector of $S(t_0)$.

We now indicate how both $\mathscr{A}$ and $S(t_0)$ can be discretized into large matrices J and $M(t_0)$ whose eigenvalues can be computed using established numerical algorithms like subspace iteration and Arnoldi's method (to compute dominant eigenvalues), if necessary combined with shift-invert or Cayley transformations (to compute rightmost eigenvalues) [24,21].

Let $\Pi=\{-\tau=t_0<t_1<\cdots<t_{L-1}<t_L=0\}$ denote a mesh on $[-\tau,0]$, $N=n\times(L+1)$. We denote the discrete initial condition by $\phi\in\mathbb{R}^N$ and introduce two new operators. Let Interp : $\mathbb{R}^N\to C$ interpolate the discrete representation ϕ on Π such that the resulting function segment is in C, while Discret : $C\to\mathbb{R}^N$ discretizes a function segment by taking its value on the mesh points of Π. Then $M(t_0)=\mathrm{Discret}(S(t_0)\mathrm{Interp}(\cdot))\in\mathbb{R}^{N\times N}$.

The construction of $M(t_0)$ can now be achieved as follows: the jth column of $M(t_0)$ ($M(t_0)e_j$ with e_j the jth unit vector) equals $\mathrm{Discret}(S(t_0)\mathrm{Interp}(e_j))$ which can be computed using one time integration. Hence the construction of $M(t_0)$ requires N time integrations, which is expensive for large N. In this situation it can be more efficient to approximate the eigenvalues of $M(t_0)$ without its full construction (see [6] for details).

An approximation for J is given by

$$J\approx\frac{M(h)-I}{h}, \tag{8}$$

where $0<h\ll 1$ denotes one time step of a given time integration scheme. In this case, the sparse matrix $M(h)$ can easily be written down explicitly. The eigenvalues of J approximate the eigenvalues of $\mathscr{A}$ only by $\mathscr{O}(h)$, even if the integration scheme is $\mathscr{O}(h^{k+1})$ because of the forward difference formula (8). More details about this second approach and a second-order alternative can be found in [10,11].

In [18] a third and quite different numerical approach is described. Using the end point values of the solution to a functional equation occurring in the Lyapunov theory of delay equations, a scalar function $v(d)$ is constructed which has a pole at every $d=\mathrm{Re}(\lambda)$ with λ a root of (4). The determination of $\mathrm{Re}(\lambda)$ is then reduced to finding the poles of $v(d)$. This method is, however, restricted to systems with commensurate delays.

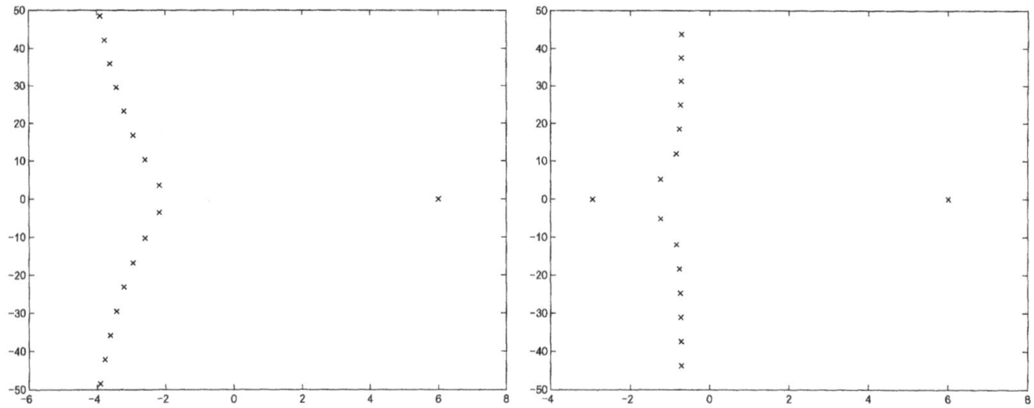

Fig. 1. Roots of the characteristic equation in a region of the complex plane for the delay equation $\dot{x}(t)=6x(t)+x(t-1)$ (left), respectively, the neutral equation $\dot{x}(t)=6x(t)+x(t-1)+\frac{1}{2}\dot{x}(t-1)$ (right).

2.3. Neutral equations

The time integration operator $S(t_0)$ of the linear neutral DDE,

$$\dot{y}(t)=A_0y(t)+\sum_{i=1}^{m}A_iy(t-\tau_i)+\sum_{i=1}^{m}B_i\dot{y}(t-\tau_i), \tag{9}$$

is no longer compact (even if $t_0 \geqslant \tau$). The spectrum of $S(t_0)$ consists of both a point spectrum and an essential spectrum. The point spectrum of $S(t_0)$ asymptotes to the essential spectrum and the real parts of the corresponding roots of the characteristic equation of (9) do not asymptote to $-\infty$. As a consequence, the stability-determining roots can have a very large imaginary part, see Fig. 1. Moreover, the radius of the essential spectrum can change discontinuously as a function of the delays [1,15,22]. Hence, the bifurcation analysis of neutral equations leads to many new and unsolved difficulties.

3. Periodic solutions

For notational convenience we restrict ourselves to equations with one delay,

$$\dot{x}(t)=f(x(t),x(t-\tau),\alpha), \tag{10}$$

but generalisation to multiple delays is straightforward. A periodic solution of (10) can be found as the solution of the following *two-point boundary value problem* (BVP)

$$\begin{aligned}&\dot{x}(t)=f(x(t),x(t-\tau),\alpha), \quad t\in[-\tau,T],\\&x_0=x_T,\\&s(x_0,x_T)=0,\end{aligned} \tag{11}$$

where T is the (unknown) period and $s:C\times C\to\mathbb{R}$ is a suitable phase condition to remove translational invariance. The periodic BVP (11) is an *infinite-dimensional* problem (a function segment $x_0\in C$ has to be determined), while it is a *finite-dimensional* problem in the case of ODEs.

Due to its periodicity, a periodic solution $x(t)$ is smooth for all t. The stability of a solution of (11) is determined by the spectrum of the linearized *monodromy operator*, $S(T,0) = \partial x_T(x_0)/\partial x_0$, where $x_T(x_0)$ is the result of integrating (10) over time T starting from initial condition x_0. Provided $T \geqslant \tau$, the operator $S(T,0)$ is compact and thus its spectrum, $\sigma(S(T,0))$, consists of a point spectrum with zero as its only cluster point [14, p. 191]. Any $\mu \neq 0$ in $\sigma(S(T,0))$ is called a characteristic (Floquet) multiplier. Furthermore, $\mu = 1$ is always an eigenvalue of $S(T,0)$ and it is referred to as the trivial Floquet multiplier.

In general, there are two approaches to compute periodic solutions of (11), i.e., shooting [13,19] and collocation [3,2,5] methods.

3.1. Shooting methods

System (11) can be rewritten in terms of the unknowns $x_0 \in C$ and $T \in \mathbb{R}$,

$$\begin{aligned} r(x_0, T) &:= x_T(x_0) - x_0 = 0, \\ s(x_0, T) &= 0. \end{aligned} \tag{12}$$

This nonlinear system can be solved iteratively using a Newton iteration,

$$\begin{bmatrix} \partial x_T/\partial x_0 - I & \partial x_T/\partial T \\ \partial s/\partial x_0 & \partial s/\partial T \end{bmatrix}\Bigg|_{(x_0^{(k)},T^{(k)})} \begin{bmatrix} \Delta x_0^{(k)} \\ \Delta T^{(k)} \end{bmatrix} = -\begin{bmatrix} r \\ s \end{bmatrix}\Bigg|_{(x_0^{(k)},T^{(k)})}, \tag{13}$$

$$x_0^{(k+1)} = x_0^{(k)} + \Delta x_0^{(k)}, \quad T^{(k+1)} = T^{(k)} + \Delta T^{(k)}.$$

At convergence, the Fréchet derivative $\partial x_T(x_0)/\partial x_0$ equals the linearized monodromy operator $S(T,0)$. After discretization, (13) is approximated by

$$\begin{bmatrix} M - I & g \\ c^{\mathrm{T}} & d \end{bmatrix} \begin{bmatrix} \Delta\phi \\ \Delta T \end{bmatrix} = -\begin{bmatrix} \tilde{r} \\ \tilde{s} \end{bmatrix}, \tag{14}$$

where

$$M = \text{Discret}\left(\frac{\partial x_T(\text{Interp}(\phi))}{\partial x_0}\Bigg|_{(\phi,T)=(\phi^{(k)},T^{(k)})} \text{Interp}(\cdot) \right) \in \mathbb{R}^{N\times N}, \tag{15}$$

and similar formulas hold for $g \in \mathbb{R}^N$, $c \in \mathbb{R}^N$, $d \in \mathbb{R}$, $\tilde{r} \in \mathbb{R}^N$ and $\tilde{s} \in \mathbb{R}$.

M is a high-dimensional dense matrix which requires at least N time integrations to construct (see Section 2.2 and [19] for details). This makes 'full' Newton [13] prohibitively expensive for large N. At convergence M is a discrete approximation of $S(T,0)$ and we call it the *monodromy matrix*.

The Newton–Picard approach [19] reduces the computational costs of the 'full' Newton iteration (14) by exploiting the fact that most multipliers lie close to zero. Suppose $\mu_1,\ldots,\mu_p$ are the p most dominant eigenvalues of M such that

$$|\mu_1| \geqslant |\mu_2| \geqslant \cdots \geqslant |\mu_p| > \rho > |\mu_{p+1}| \geqslant \cdots \geqslant |\mu_N| \tag{16}$$

with $p \ll N$ and $\rho < 1$. Suppose the columns of $V_p \in \mathbb{R}^{N\times p}$ form an orthonormal basis for the low-dimensional eigenspace U corresponding to $\mu_1,\ldots,\mu_p$. The number p is determined during the computation of V_p using subspace iteration [19]. Suppose further that the columns of $V_q \in \mathbb{R}^{N\times q}$,

$q = N - p$, form an orthonormal basis for $U^{\perp}$, the orthogonal complement of U in $\mathbb{R}^N$. The high-dimensional basis V_q is only used to derive the necessary formulas, it is not constructed in the actual implementation. The Newton correction $\Delta\phi \in \mathbb{R}^N$ has a unique decomposition

$$\Delta\phi = V_p \Delta\hat{\phi}_p + V_q \Delta\hat{\phi}_q, \quad \Delta\hat{\phi}_p \in \mathbb{R}^p, \ \Delta\hat{\phi}_q \in \mathbb{R}^q. \tag{17}$$

Projecting the first N equations of (14) on the two subspaces U and $U^{\perp}$ and substituting $\Delta\hat{\phi}_p$ and $\Delta\hat{\phi}_q$ for $\Delta\phi$ using (17) results in

$$\begin{bmatrix} V_q^{\mathrm{T}}(M-I)V_q & V_q^{\mathrm{T}}MV_p & V_q^{\mathrm{T}}g \\ V_p^{\mathrm{T}}MV_q & V_p^{\mathrm{T}}(M-I)V_p & V_p^{\mathrm{T}}g \\ c^{\mathrm{T}}V_q & c^{\mathrm{T}}V_p & d \end{bmatrix} \begin{bmatrix} \Delta\hat{\phi}_q \\ \Delta\hat{\phi}_p \\ \Delta T \end{bmatrix} = - \begin{bmatrix} V_q^{\mathrm{T}}\tilde{r} \\ V_p^{\mathrm{T}}\tilde{r} \\ \tilde{s} \end{bmatrix}. \tag{18}$$

Because U is an invariant subspace of M, we have $V_q^{\mathrm{T}}MV_p = 0_{q\times p}$. At the periodic solution g corresponds to the discretized eigenfunction of the trivial Floquet multiplier. The corresponding eigenvalue of M is approximately 1 and if ρ is not too close to 1, the corresponding eigenvector belongs to U. Hence, the term $V_q^{\mathrm{T}}g$ is very small at convergence and can be neglected near the periodic solution.

System (18) is now partially decoupled: one can first solve the large, $q \times q$ system

$$V_q^{\mathrm{T}}(M-I)V_q\Delta\hat{\phi}_q = -V_q^{\mathrm{T}}\tilde{r}, \tag{19}$$

iteratively with a Picard scheme (requiring one matrix–vector product with M per iteration) and, using its solution $\Delta\hat{\phi}_q$, solve the small $p \times p$ system

$$\begin{bmatrix} V_p^{\mathrm{T}}(M-I)V_p & V_p^{\mathrm{T}}g \\ c^{\mathrm{T}}V_p & d \end{bmatrix} \begin{bmatrix} \Delta\hat{\phi}_p \\ \Delta T \end{bmatrix} = - \begin{bmatrix} V_p^{\mathrm{T}}\tilde{r} \\ \tilde{s} \end{bmatrix} - \begin{bmatrix} V_p^{\mathrm{T}}M \\ c^{\mathrm{T}} \end{bmatrix} V_q\Delta\hat{\phi}_q \tag{20}$$

using a direct method. For (19) and (20), the matrix M is not constructed explicitly; instead, matrix–vector products with M are computed through time integration. More details can be found in [19,8].

3.2. Collocation methods

In this section we consider collocation methods based on piecewise polynomials [5]. Collocation based on truncated Fourier series [3,2] is quite similar.

We approximate the solution profile by a piecewise polynomial $u(t)$ on $[0,1]$ which is represented on each interval of a mesh $\Pi = \{0 = t_0 < t_1 < \cdots < t_L = 1\}$ as a polynomial of degree m,

$$u(t) = \sum_{j=0}^{m} u(t_{i+\frac{j}{m}})P_{i,j}(t), \quad t \in [t_i, t_{i+1}], \ i = 0,\ldots,L-1, \tag{21}$$

where $P_{i,j}(t)$ are the Lagrange polynomial basis functions, $t_{i+j/m} = t_i + jh_i/m$, $h_i = t_{i+1} - t_i$, $i = 0,\ldots,L-1$, $j=1,\ldots,m$. The approximation $u(t)$ is completely determined in terms of the coefficients $u_{i+j/m} = u(t_{i+j/m})$, $i = 0,\ldots,L-1$, $j = 0,\ldots,m-1$ and $u_L = u(t_L)$.

Let $X = \{c_{i,j} = t_i + c_j h_i,\ i = 0,1,\ldots,L-1,\ j = 1,\ldots,m\}$ be a given set of collocation points based on the collocation parameters $0 \leqslant c_1 < c_2 < \cdots < c_m \leqslant 1$. Then the idea of a collocation method is

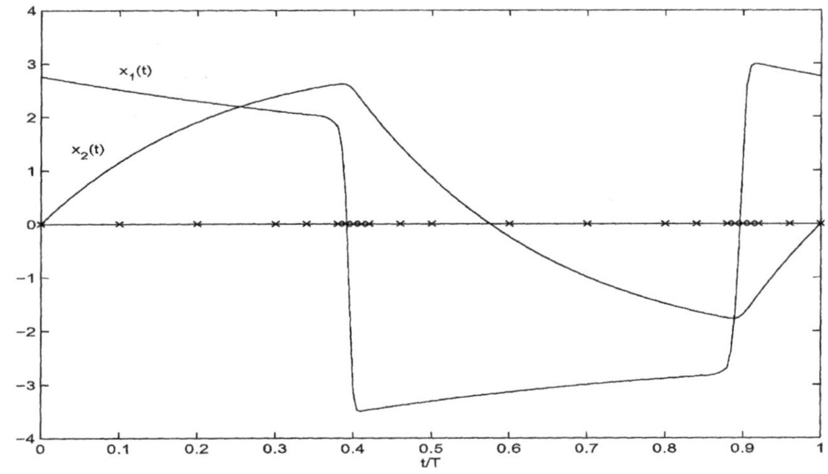

Fig. 2. Profile of the solution of the two-dimensional delay differential equation described in [23]. Mesh points (×) chosen by an adaptive mesh selection procedure.

to construct a (nonlinear) system of equations such that its solution $u(t)$ satisfies the time-rescaled original system of equations on the set X,

$$\dot{u}(c_{i,j}) = \begin{cases} Tf(u(c_{i,j}), u(c_{i,j} - \tau/T), \alpha) & \text{when } c_{i,j} - \tau/T \geqslant 0, \\ Tf(u(c_{i,j}), u(c_{i,j} - \tau/T + 1), \alpha) & \text{when } c_{i,j} - \tau/T < 0. \end{cases} \tag{22}$$

In (22) we used the periodicity condition to eliminate $u(t)$ for $t < 0$ and we assumed $T > \tau$. Using $c = c_{i,j}$, $\tilde{c} = (c - \tau/T) \bmod 1$, $t_k \leqslant \tilde{c} < t_{k+1}$, (22) has the following structure:

$$\sum_{j=0}^{m} u_{i+j/m} P'_{i,j}(c) = Tf\left(\sum_{j=0}^{m} u_{i+j/m} P_{i,j}(c), \quad \sum_{j=0}^{m} u_{k+j/m} P_{k,j}(\tilde{c}), \alpha\right), \tag{23}$$

where $P'(\cdot)$ is the derivative of $P(\cdot)$. The unknowns $u_{i+j/m}$ and T are computed by solving the linearization of (23) together with $u_0 = u_L$, $s(u, T) = 0$ iteratively. Instead of the collocation polynomial in the past (to evaluate $u(\tilde{c})$), special interpolating polynomials can be used which recover superconvergence results at mesh points. For difficult profiles adaptive mesh selection can be of interest, see Fig. 2. More details can be found in [5].

3.3. Stability and continuation

Once a periodic solution is found, its stability can be obtained by computing a discrete approximation M of the monodromy operator $S(T, 0)$ and its dominant eigenvalues (approximations to the stability determining Floquet multipliers). This is automatically done in shooting, whether implemented as a full Newton iteration or via the Newton–Picard approach. For the collocation methods, an approximation of $S(T, 0)$ can be constructed along the lines of Section 2.2 using the variational equation around the obtained periodic solution [2].

If (10) depends on a physical parameter α, branches of periodic solutions can be traced as a function of the parameter in a continuation procedure [3,2,19], see Fig. 3. Bifurcations occur whenever

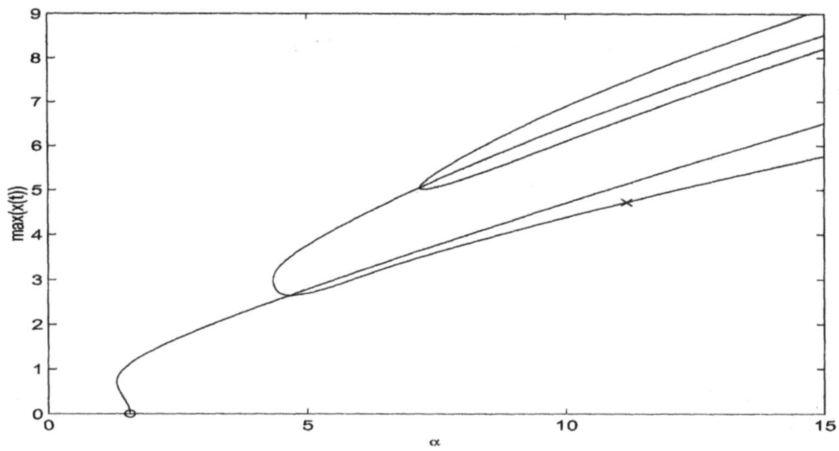

Fig. 3. Three branches of periodic solutions of the delay differential equation $\dot{x}(t) = -\alpha x(t-1)(1+x^2(t-1))/(1+x^4(t-1))$ [3,19]. The first branch emanates from a Hopf bifurcation (o), turns and undergoes a transcritical bifurcation. The intersecting branch undergoes a turning point, a symmetry-breaking pitchfork bifurcation and a torus bifurcation (x).

Floquet multipliers move into or out of the unit circle. Generically this is a turning point when a real multiplier crosses through 1, a period-doubling point when a real multiplier crosses through -1 and a torus bifurcation when a complex pair of multipliers cross the unit circle. Continuation can be started from a Hopf point or a (stable) solution profile obtained using simulation.

3.4. Neutral equations

No smoothing occurs when integrating neutral DDEs. If the neutral terms are linearly involved and the corresponding essential spectrum is stable, then smoothness of periodic solutions was proven in [14]. In [8] some branches of periodic solutions of a scalar one-delay neutral equation were computed using the method of Section 3.1. To use the spectral decomposition (16), it is required that the essential spectrum is well inside the unit circle. These numerical results indicate the sensitivity of the essential spectrum to both the delay and the period. For the same example it was proven that periodic solutions with discontinuous first derivative exist when the above-mentioned conditions are violated [7].

4. Conclusions

In this paper we outlined numerical methods for the computation and stability analysis of steady-state solutions and periodic solutions of systems of DDEs with several constant delays. These methods can be used in a continuation procedure to follow branches of solutions, to determine their stability and to locate bifurcation points. The methods could be adapted for DDEs with more general types of delays. The main problem here is that the theory is not fully developed for all types of delays. Equations with bounded distributed and state-dependent delays behave quite similar to equations of type (1) but larger differences occur when going to unbounded delays or neutral equations. For DDEs of neutral type, we briefly discussed some open problems and new difficulties.

While a number of software packages have been developed for time integration of DDEs, no software packages exist for the bifurcation analysis of DDEs. At present some (private) codes for stability analysis of steady states or for the computation and stability analysis of periodic solutions exist, but they are not yet available in a user friendly form. A Matlab package which implements some of the methods described in Sections 2.2 and 3.2 is in preparation by the authors [25]. The package XPP [9] contains an implementation of the method described in Section 2.1 for the stability analysis of steady-state solutions.

Acknowledgements

This research presents results of the research project OT/98/16, funded by the Research Council K.U. Leuven and of the research project IUAP P4/02 funded by the programme on Interuniversity Poles of Attraction, initiated by the Belgian State, Prime Minister's Office for Science, Technology and Culture. K. Engelborghs is a research assistant of the Fund for Scientific Research – Flanders (Belgium).

References

[1] C.E. Avellar, J.K. Hale, On the zeros of exponential polynomials, J. Math. Anal. Appl. 73 (1980) 434–452.
[2] A.M. Castelfranco, H.W. Stech, Periodic solutions in a model of recurrent neural feedback, SIAM J. Appl. Math. 47 (3) (1987) 573–588.
[3] E.J. Doedel, P.P.C. Leung, A numerical technique for bifurcation problems in delay differential equations, Congr. Num. 34 (1982) 225–237 (Proceedings of the 11th Manitoba Conference on Num. Math. Comput., University of Manitoba, Winnipeg, Canada).
[4] L.E. El'sgol'ts, H.W. Norkin, Introduction to the theory and application of differential equations with deviating arguments, Mathematics in Science and Engineering, Vol. 105, Academic Press, New York, 1973.
[5] K. Engelborghs, T. Luzyanina, K.J. in 't Hout, D. Roose, Collocation methods for the computation of periodic solutions of delay differential equations, Technical Report TW295, Department of Computer Science, K.U. Leuven, 1999.
[6] K. Engelborghs, D. Roose, Numerical computation of stability and detection of Hopf bifurcations of steady-state solutions of delay differential equations, Adv. Comput. Math. 10 (3–4) (1999) 271–289.
[7] K. Engelborghs, D. Roose, Smoothness loss of periodic solutions of a neutral functional differential equation: on a bifurcation of the essential spectrum, Dynamics Stability Systems 14 (3) (1999) 255–273.
[8] K. Engelborghs, D. Roose, T. Luzyanina, Bifurcation analysis of periodic solutions of neutral functional differential equations: a case study, Int. J. Bifurc. Chaos 8 (10) (1998) 1889–1905.
[9] B. Ermentrout, XPPAUT3.91 — The differential equations tool, University of Pittsburgh, Pittsburgh, 1998, (http://www.pitt.edu/~phase/).
[10] N.J. Ford, V. Wulf, Embedding of the numerical solution of a dde into the numerical solution of a system of odes, Technical Report, Manchester Centre for Computational Mathematics, University of Manchester, 1998.
[11] V. Wulf, N.J. Ford, Numerical Hopf bifurcation for a class of delay differential equations, J. Comput. Appl. Math. 115 (1–2) (2000) 601–616.
[12] S.A. Gourley, M.V. Bartuccelli, Parameter domains for instability of uniform states in systems with many delays, J. Math. Biol. 35 (1997) 843–867.
[13] K.P. Hadeler, Effective computation of periodic orbits and bifurcation diagrams in delay equations, Numer. Math. 34 (1980) 457–467.
[14] J.K. Hale, in: Theory of Functional Differential Equations, Applied Mathematical Sciences, Vol. 3, Springer, Berlin, 1977.

[15] J.K. Hale, S.M. Verduyn Lunel, in: Introduction to Functional Differential Equations, Applied Mathematical Sciences, Vol. 99, Springer, Berlin, 1993.
[16] B.D. Hassard, A code for Hopf bifurcation analysis of autonomous delay-differential systems, in: W.F. Langford, F.V. Atkinson, A.B. Mingarelli (Eds.), Oscillations, Bifurcations and Chaos, Canadian Mathematical Society Conference Proceedings, Vol. 8, Amer. Math. Soc., Providence, RI, 1987, pp. 447–463.
[17] P. Kravanja, M. Van Barel, A. Haegemans, Computing the zeros of analytic functions, Lecture Notes in Mathematics, Vol. 1727, Springer, Berlin, 2000.
[18] J. Louisell, in: L. Dugard, E.I. Verriest (Eds.), Numerics of the stability exponent and eigenvalue abscissas of a matrix delay system, Stability and Control of Time-Delay Systems, Lecture Notes in Control and Information Sciences, Vol. 228, Springer, Berlin, 1997.
[19] T. Luzyanina, K. Engelborghs, K. Lust, D. Roose, Computation, continuation and bifurcation analysis of periodic solutions of delay differential equations, Int. J. Bifurc. Chaos 7 (11) (1997) 2547–2560.
[20] T. Luzyanina, D. Roose, Numerical stability analysis and computation of Hopf bifurcation points for delay differential equations, J. Comput. Appl. Math. 72 (1996) 379–392.
[21] K. Meerbergen, D. Roose, Matrix transformations for computing rightmost eigenvalues of large sparse non-symmetric eigenvalue problems, IMA J. Numer. Anal. 16 (1996) 297–346.
[22] W. Michiels, K. Engelborghs, D. Roose, Sensitivity to delays in neutral equations, Technical Report TW286, Department of Computer Science, K.U. Leuven, Belgium, 1998.
[23] R.E. Plant, A FitzHugh differential-difference equation modeling recurrent neural feedback, SIAM J. Appl. Math. 40 (1) (1981) 150–162.
[24] Y. Saad, Numerical Methods for Large Eigenvalue Problems, Manchester University Press, 1992.
[25] K. Engelborghs, DDE-BIFTOOL: a Matlab package for bifurcation analysis of delay differential equations, Technical Report TW 305, K.U. Leuven, Belgium, May 2000.
[26] C.T.H. Baker, Retarded differential equations, J. Comput. Appl. Math. 125 (2000) 309–335.

ELSEVIER

Journal of Computational and Applied Mathematics 125 (2000) 277–285

JOURNAL OF COMPUTATIONAL AND APPLIED MATHEMATICS

www.elsevier.nl/locate/cam

How do numerical methods perform for delay differential equations undergoing a Hopf bifurcation?

Neville J. Ford *, Volker Wulf

Department of Mathematics, Chester College, Parkgate Road, Chester CH1 4BJ, UK

Received 23 July 1999; received in revised form 6 January 2000

Abstract

In this paper we consider the numerical solution of delay differential equations (DDEs) undergoing a Hopf bifurcation. Some authors use special methods to calculate bifurcating periodic solutions. We investigate what will happen when simple standard numerical methods (based on ODE methods) are used to obtain an approximate solution to the DDE. We want to establish whether the method will predict the true behaviour of the solution. We present three distinctive and complementary approaches to the analysis which together provide us with the result that ϑ-methods applied to a DDE will retain Hopf bifurcations and preserve their type, for sufficiently small $h > 0$. © 2000 Elsevier Science B.V. All rights reserved.

1. Introduction

A major concern of numerical analysts is the development of reliable algorithms to solve differential equations. The aim is to provide algorithms that consistently provide good-quality solutions to a wide class of equations. We want to be able to predict when the algorithms will perform well, and when they will fail.

If a differential equation has to be solved only over a short (finite) time interval, amongst the main issues are convergence of the numerical solution and the order of the method. On the other hand, if long-term behaviour of solutions (over infinite time intervals) is of more interest, then the errors may grow, and it may be impossible to prove that the numerical solution is close to the true solution. In this event, the desire to preserve qualitative behaviour may be more important. One might seek to show that both the exact solution to the problem and the numerical solution tend to zero as $t \to \infty$, that both exhibit the same stability properties for particular solutions, or that both exhibit periodic or even chaotic solutions. Unfortunately, convergence of a method over finite intervals does not

* Corresponding author. Tel.: +44-1244-392-748; fax: +44-1244-392-820.
E-mail address: njford@chester.ac.uk (N.J. Ford).

PII: S0377-0427(00)00473-8

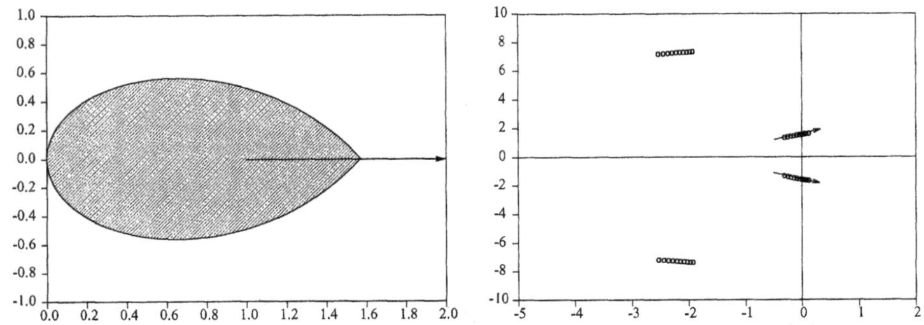

Fig. 1. Roots of the characteristic equation moving into the right half of the complex plane as the parameter λ changes. The shaded region in the left picture is the stability region of the zero solution of the delay logistic equation (here for all $\lambda \in \mathbb{C}$). The arrow depicts the path of $\lambda \in [1,2]$ and the right picture shows the corresponding path of some of the roots as they move into the right half plane.

guarantee persistence of long-term characteristics of solutions in the numerical approximation. The analysis of asymptotic stability of equilibrium solutions is a very well-established concern of the numerical analyst; the analysis of periodic solutions is less well understood and has been considered by several authors recently (see, for example, [13] and [18, Chapter 6 and references therein]).

In this paper we investigate the long term properties of numerical approximations to the solutions of the scalar *delay differential equation*

$$y'(t) = f(y(t), y(t-\tau), \lambda), \; t \geqslant 0, \qquad y(t) = \phi(t), \; -\tau \leqslant t \leqslant 0, \tag{1}$$

where $\tau > 0$ is a constant *time lag* and λ is a *real parameter* and we assume that $y(t) \equiv 0$ is an equilibrium solution for all λ, i.e. $f(0,0,\lambda) = 0$. A usual starting point for the analysis of long-term behaviour of solutions is to consider the values of λ for which the zero solution of (1) is asymptotically stable. This may be determined by looking at the roots of the *characteristic equation*

$$d(\mu, \lambda) = \mu - \alpha(\lambda) - \beta(\lambda)e^{-\tau\mu}, \tag{2}$$

where

$$\alpha(\lambda) = \frac{\partial}{\partial y(t)} f(0,0,\lambda), \qquad \beta(\lambda) = \frac{\partial}{\partial y(t-\tau)} f(0,0,\lambda). \tag{3}$$

If all the roots of (2) have negative real parts, then the zero solution of (1) is asymptotically stable. Obviously as the parameter λ varies some of the roots of (2) might leave the left half of the complex plane and $y(t) \equiv 0$ becomes unstable. For example, consider the *delay logistic equation*

$$y'(t) = -\lambda y(t-1)(1 + y(t)), \tag{4}$$

with characteristic equation

$$0 = \mu + \lambda e^{-\mu}. \tag{5}$$

It can be shown that for all $\lambda \in (0, \pi/2)$ all roots μ of (5) have negative real parts and consequently $y(t) \equiv 0$ is asymptotically stable (see, e.g., [11]). As λ moves beyond $\pi/2$ a pair of complex conjugate roots leaves the left half-plane (see Fig. 1). For $\lambda = \pi/2$ a pair of complex conjugate roots of (5) lies on the imaginary axis. The zero solution of the linear equation is stable but not asymptotically stable for this value of λ, but now the stability of the zero solution of (4) cannot be determined from looking at roots of (5) but one needs to take into account the nonlinear parts of Eq. (4); the linear stability theory breaks down here. For $\lambda > \pi/2$ the zero solution is unstable but it can be shown that neighbouring solutions remain bounded and become eventually periodic as $t \to \infty$. In fact, what happens is that in Eq. (4) a *Hopf bifurcation* occurs as λ passes through $\pi/2$. This is a genuinely nonlinear feature of (4). Hopf bifurcations are not found in linear problems (nor, indeed, in scalar ordinary differential equations). Further details can be found, for example, in [2,11].

Hopf bifurcations in equations of type (1) are quite well understood. On the other hand, the *behaviour* of standard numerical approximation methods (based on numerical methods for ordinary differential equations) applied to such equations has had little attention so far. The purpose of this paper is to provide some insight into what happens when numerical methods are applied to a problem that has a Hopf bifurcation: does the bifurcation persist in the approximation and can the numerical solution be relied upon in a neighbourhood of the bifurcation point?

We say Eq. (1) undergoes a Hopf bifurcation at the parameter value λ_* if the following conditions are met (see [2, Chapter X]):

H1: f is a C^k-smooth mapping from $\mathbb{R}^3$ into $\mathbb{R}$ and $f(0,0,\lambda)=0$ for all $\lambda \in \mathbb{R}$.
H2: $d(\mu,\lambda_*)$ has a pair of simple complex conjugate roots $\mu_{1,2} = \pm\,\mathrm{i}\omega_0$, $\omega_0 > 0$ and all other roots of (2) have negative real parts bounded away from zero.
H3: $\mathrm{Re}(\mu_1'(\lambda_*)) \neq 0$, where $\mu_1(\lambda)$ is the branch of roots of (2) with $\mu_1(\lambda_*) = \mathrm{i}\omega_0$.

If the above conditions hold then for all λ in a one-sided neighbourhood of λ_* there exists an invariant periodic orbit surrounding the origin. The periodic orbit is attractive (repelling) if the zero solution is asymptotically stable (unstable) for $\lambda=\lambda_*$. Since at λ_* the zero solution is a nonhyperbolic equilibrium its stability is determined by the nonlinear terms of (1). This leads to the definition of a *stability coefficient* (based upon the nonlinear terms) whose sign determines the stability of the periodic orbit. For a more in-depth discussion on Hopf bifurcations we refer to the relevant chapters in [2,12,15].

If one knew in advance that an equation had a periodic solution then there are special methods presented in the literature for finding a numerical approximation to the solution. (See, for example, the papers [3,4,17].) In practical applications this may not be realistic. One often applies a numerical method and finds an approximate solution without first investigating the dynamical behaviour of the exact solution. We adopt our approach here because we are interested in establishing how reliable are simple numerical schemes in this situation. The question we set out to investigate is: how does a variation of the parameter λ affect the numerical approximation of system (1) and does the Hopf bifurcation "persist" in some way? In other words, will a straightforward application of a simple numerical method display the true behaviour of the solution close to a Hopf bifurcation?

As an illustration we consider the ϑ-methods applied to (1) with stepsize $h = \tau/m$, $m \in \mathbb{N}$ given by

$$y_{n+1} = y_n + h\{(1-\vartheta)f(y_n, y_{n-m}, \lambda) + \vartheta f(y_{n+1}, y_{n+1-m})\}, \quad n \geqslant 1,$$
$$y_n = \phi(nh), \quad -m \leqslant n \leqslant 0, \tag{6}$$

where $\vartheta \in [0,1]$. This is the natural extension of the ϑ-methods for ODEs to the DDE case. We restrict the stepsize to integer fractions of the delay in order to avoid the need to interpolate lagged values. We want to establish whether the numerical scheme exhibits a Hopf bifurcation, and if so, at what value of the parameter the bifurcation arises. Finally we want to consider whether the nature of the bifurcation (subcritical or supercritical) is preserved in the approximation. This type of approach has been considered by [14] and in our previous work ([6,5] etc.).

In the remainder of the paper we present three different approaches to the above problem.

(1) First, we consider how far the existing (linear) stability analysis can help. In Section 2 we use the boundary locus technique, which is familiar from the linear stability analysis of numerical methods, and illustrate how we can identify parameter points at which a Hopf bifurcation could take place.
(2) Second (see Section 3), we undertake a direct bifurcation analysis of the difference equation (6) with fixed h and varying λ. This leads to the task of checking Neimark–Sacker bifurcation conditions in Eq. (6). We illustrate how this can be done and conclude that, while the results we have obtained in this way are entirely satisfactory, the approach leads to complicated algebra which means that it is hard to obtain good general results. This motivates our introduction of an alternative approach.
(3) A third approach (in Section 4) tries to avoid this algebraic complexity by using a projection method. We "project" the DDE onto a system of ODEs and are able to make use of known results on the preservation of Hopf bifurcations in ODEs in the context of numerical approximation of DDEs.

Finally, we indicate how the results we have obtained for simple numerical methods can be generalised to apply to much wider classes of method. It turns out that our calculations, although confined to the simplest prototype numerical schemes, provide the evidence needed to ensure that Hopf bifurcations are preserved in other numerical schemes.

2. Approach 1: application of the boundary locus method

In this section we seek to obtain information based on the existing linear stability theory. We will apply the boundary locus method (see [1,16]) to plot the boundaries of the regions (in the parameter space) for which the equilibrium solution is asymptotically stable. It is known that Hopf bifurcations arise on the boundary of the stability region at those points where two characteristic values simultaneously leave the left half plane. The boundary locus of (1) is the set

$$\partial\mathscr{D} = \{(\alpha,\beta) : \exists \nu \in \mathbb{R}, \mathrm{i}\nu - \alpha - \beta \mathrm{e}^{-\mathrm{i}\nu\tau} = 0\}, \tag{7}$$

where α and β are in the parameter space of the equation. If we are interested in a system that depends only on one parameter λ, one usually assumes $\lambda \in \mathbb{C}$. For instance for the delay logistic equation we have $\alpha(\lambda) = -\lambda$ and $\beta(\lambda) = 0$ and

$$\partial\mathscr{D}_{(4)} = \{\lambda \in \mathbb{C} : \lambda = -\mathrm{i}\nu \mathrm{e}^{\mathrm{i}\nu}, -\infty < \nu < \infty\}. \tag{8}$$

For delay equations the boundary locus $\partial\mathscr{D}$ subdivides the parameter plane into open sets. For all parameter values in one open set the number of roots of the characteristic equation with positive real part is constant. In particular where this number is zero the parameters lie in the region of asymptotic stability of the equation (see Fig. 1). Similarly, a boundary locus can be defined for the

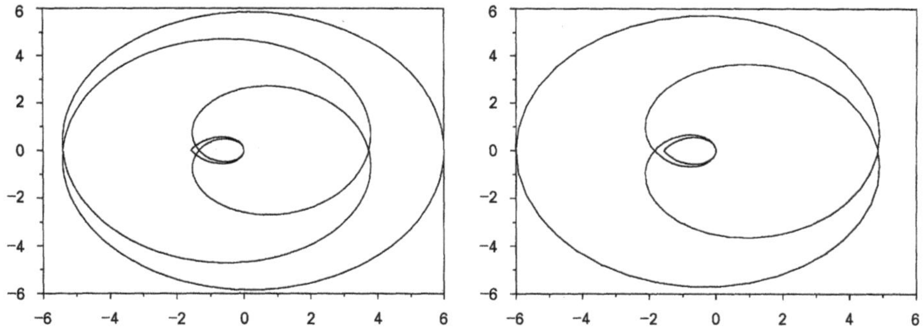

Fig. 2. The boundary loci for Euler forward (left) and Euler backward (right) with the stability region for the true delay equation superimposed.

discretization (6).

$$\partial\mathscr{D}_m = \{(\alpha, \beta) : \exists\theta \in [-\pi, \pi], z = e^{i\theta},$$
$$z^{m+1} - z^m - h(1-\vartheta)(\alpha z^m + \beta) - h\vartheta(\alpha z^{m+1} + \beta z) = 0\}. \tag{9}$$

For each m, $\partial\mathscr{D}_m$ partitions the parameter space into open sets so that the number of roots of modulus greater than one of the characteristic polynomial corresponding to (6) is constant in each open set. The region of asymptotic stability of the problem is therefore the union of the open sets where the number of such roots is zero.

Every point on the boundary locus corresponds to a parameter value at which there is a root of the characteristic equation exactly on the imaginary axis (in the case of the DDE) or on the unit circle (in the case of the discrete scheme). For a Hopf bifurcation, we require two such roots to arise for the same parameter value, and this can be observed as a parameter value where the boundary locus curve crosses itself (see Fig. 2).

In [6] we showed that for certain numerical methods the two curves approximate each other for all equations where the linearization yields a pure delay equation. As a consequence we were able to prove that if a Hopf bifurcation in the DDE occurs for some parameter value λ_* then there exists a nearby parameter value λ_m where a Hopf bifurcation occurs in the numerical approximation. For strongly stable linear multistep methods we have $\lambda_m = \lambda_* + \mathcal{O}(h^p)$, where p is the order of the method.

The approach is as follows: we assume that the original equation has a Hopf bifurcation for a particular parameter value λ_*. In this case, the boundary locus for the DDE crosses itself on the x-axis at λ_*. We establish that the boundary locus for the numerical scheme also crosses itself at λ_m close to λ_*, that the point of intersection lies on the real axis, and that adjacent to the value λ_m on the real axis is an interval of stability and an interval of instability for the numerical scheme.

For a linear problem the boundary locus is sufficient to determine the stability behaviour of solutions of the equation at particular parameter values. However, for nonlinear problems, the boundary locus provides us with the parameter values at which the linearised equation loses its stability. What happens to the nonlinear problem at this point requires analysis of the nonlinear parts of equation.

Remark 1. One could extend the analysis in a natural way to consider other classes of nonlinear delay differential equation.

Remark 2. The authors have provided a similar analysis for the application of Runge–Kutta methods to delay differential equations (see [8]).

3. Approach 2: a Neimark–Sacker bifurcation analysis

In this section we consider how one would perform a direct bifurcation analysis of difference equation (6). This will enable us to determine how the *nonlinearity* of the problem affects its behaviour at a Hopf bifurcation. For further details of the definitions and background, we refer the reader to [2,11,12,15].

For each $m \in \mathbb{N}$ we can define a map $F_m : \mathbb{R}^{m+1} \times \mathbb{R} \to \mathbb{R}^{m+1}$ by

$$Y^{n+1} = F_m(Y^n, \lambda), \tag{10}$$

where $Y_i^{n+1} = Y_{i-1}^n$, $i = 1, \ldots, m$ and Y_0^{n+1} is defined as the solution of

$$Y_0^{n+1} = Y_0^n + h\{(1-\vartheta)f(Y_0^n, Y_m^n, \lambda) + \vartheta f(Y_0^{n+1}, Y_{m-1}^n, \lambda)\}. \tag{11}$$

With $h = \tau/m$ we have $Y_i^n = y_{n-i}$, where $\{y_n\}$ is the solution of (6) and is therefore equivalent to iterating (10). For $\vartheta = 0$ (Explicit Euler) the authors showed in [7] the following:

Theorem 3. *Assume that the differential equation* (1) *undergoes a supercritical* (*subcritical*) *Hopf bifurcation at the parameter value* λ_*, *then for sufficiently small step sizes the map* (10) *undergoes a supercritical* (*subcritical*) *Neimark–Sacker bifurcation at a parameter value* $\lambda_h = \lambda_* + \mathcal{O}(h)$.

A Neimark–Sacker bifurcation or Hopf bifurcation for maps is characterized by the following: given a parameter dependent map on $\mathbb{R}^n$, $n \geqslant 2$,

$$x \mapsto A(\lambda)x + G(x, \lambda) \tag{12}$$

with

N1: G is a C^k-smooth mapping, $k \geqslant 2$, from $\mathbb{R}^n \times \mathbb{R}$ into $\mathbb{R}^n$, $G(0, \lambda) = 0$, $G_x(0, \lambda) = 0$, $\lambda \in \mathbb{R}$.

N2: $A(\lambda^*)$ has a complex conjugate pair of eigenvalues $\gamma_{1,2} = e^{\pm i\theta_0}$, while all other eigenvalues have modulus strictly less than one.

N3: $r'(\lambda^*) \neq 0$, where $r(\lambda)$ is the modulus of the branch of eigenvalues with $r(\lambda^*) = 1$, i.e. $r(\lambda) = |\gamma_{1,2}(\lambda)|$.

Under the above hypotheses the map (12) has an invariant closed curve of radius $\mathcal{O}(\sqrt{(|\lambda^* - \lambda|)})$ surrounding the origin for all λ in a one-sided neighbourhood of λ^*. The closed curve is attracting (repelling) if zero is an asymptotically stable (unstable) fixed point of (12) at $\lambda = \lambda^*$. Since at $\lambda = \lambda^*$ zero is a nonhyperbolic fixed point the nonlinear part $G(\cdot, \lambda^*)$ determines the attractivity of the bifurcating invariant curve (see, e.g., [15]). Theorem 3 shows that for m large enough each map (10) undergoes a Neimark–Sacker bifurcation with $\lambda^* = \lambda_h$.

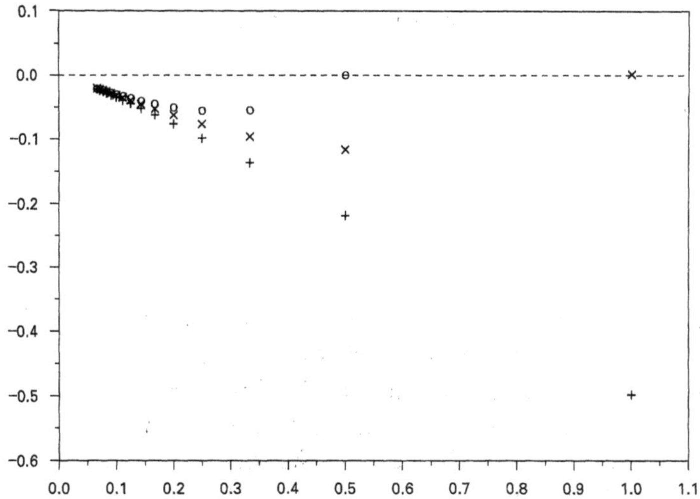

Fig. 3. Plots of stability coefficient against step size for $\vartheta = 0(+), \frac{1}{2}(\times)$ and $1(\circ)$ applied to the delay logistic equation (see [19,9]).

Remark 4. Theorem 3 applies only to the Euler forward method. The main difficulty in extending the result to more general methods lies in determining the stability of the bifurcating closed curves. This requires us to determine the sign of quite complicated expressions involving the nonlinear part G. For the Euler forward method, F_m is explicitly given but for $\vartheta \neq 0$ say, F_m, and therefore G, is known only implicitly making the analysis even more complicated.

We have shown (see [5]) that the approach can be extended for specific methods applied to particular problems, although the calculations remain complicated. In the examples we have calculated explicitly, the sign of the stability coefficient is preserved for sufficiently small $h > 0$ (Fig. 3).

The computational complexity of these calculations leads us to consider whether improved insight can be obtained through a more innovative approach.

4. Approach 3: the use of a projection method

The numerical solution of a scalar DDE yields a discrete system of the same general form as arises in the numerical solution of a system of ordinary differential equations. We aim to use this property and known analysis for ODEs under discretisation to derive results that were difficult to derive directly for the DDE.

We start with map (10) for the Euler Forward method which has the form

$$\begin{aligned} Y_0^{n+1} &= Y_0^n + hf(Y_0^n, Y_m^n, \lambda), \\ Y_1^{n+1} &= Y_0^n, \\ &\vdots \\ Y_m^{n+1} &= Y_{m-1}^n. \end{aligned} \tag{13}$$

We seek a system of ODEs for which the same system (13) is the Euler forward discretization. One easily finds that the system

$$\begin{aligned} Y_0' &= f(Y_0, Y_m), \\ Y_1' &= h^{-1}(Y_0 - Y_1), \\ &\vdots \\ Y_m' &= h^{-1}(Y_{m-1} - Y_m), \end{aligned} \tag{14}$$

discretized with stepsize h yields exactly (13). As we have shown in [20] the system (14) undergoes a Hopf bifurcation of the same type as the DDE (1) at some parameter value $\lambda_m = \lambda_* + \mathcal{O}(h)$. We can now use known results on the persistence of Hopf bifurcations in ODEs under approximations to obtain the result that (13) undergoes a Neimark–Sacker bifurcation at $\lambda_h = \lambda_m + \mathcal{O}(h)$ (see, e.g., [10]). We therefore have reconfirmed Theorem 3 using known results from ODE theory and without recourse to complicated calculations.

We seek to generalise this result to more realistic methods, and employ a broadly similar approach. Unfortunately it turns out that, even for ϑ-methods apart from Euler forward, it is not possible to derive an exactly equivalent system of ODEs and therefore one must consider the DDE method as a perturbation of the corresponding ODE method. The method we have used proceeds as follows:

(1) We wish to establish the *sign* of the stability coefficient of the approximate scheme and our existing analysis (in this section and in the previous one) has shown that the sign of the stability coefficient (for small enough $h > 0$) is *correct* for the Euler forward method.
(2) We write some other numerical scheme as a perturbation of the Euler forward scheme and we consider the stability coefficient of the perturbed scheme.
(3) We show (see [7] for the analysis) that, for all ϑ-methods, and for certain other numerical methods, the sign of the stability coefficient (as $h \to 0$) is unchanged.
(4) We conclude that the numerical approximation will display a Hopf bifurcation of corresponding type to the one found in the original DDE.

Acknowledgements

We are pleased to acknowledge the helpful comments from the referees and the editor. Volker Wulf was supported in this work by a Chester College Research Student Bursary.

References

[1] C.T.H. Baker, N.J. Ford, Some applications of the boundary locus method and the method of D-partitions, IMA J. Numer. Anal. 11 (1991) 143–158.
[2] O. Diekman, S.A. van Gils, S.M. Verduyn Lunel, H.-O. Walther, Delay Equations: Functional-, Complex- and Nonlinear Analysis, Springer, New York, 1995.
[3] K. Engelborghs, K. Lust, D. Roose, A Newton Picard method for accurate computation of period doubling bifurcation points of large-scale systems of ODEs, Report TW251, Leuven, 1996.
[4] K. Engelborghs, D. Roose, Numerical computation of stability and detection of Hopf bifurcations of steady state solutions of delay differential equations, Report TW274, Leuven, 1998.

[5] N.J. Ford, V. Wulf, Numerical Hopf bifurcation for the delay logistic equation, Technical Report 323, Manchester Centre for Computational Mathematics (ISSN 1360 1725), 1998.
[6] N.J. Ford, V. Wulf, The use of boundary locus plots in the identification of bifurcation points in the numerical approximation of delay differential equations, J. Comput. Appl. Math. 111 (1999) 153–162.
[7] N.J. Ford, V. Wulf, Numerical Hopf bifurcation for a class of delay differential equations, J. Comput. Appl. Math. 115 (2000) 601–616.
[8] N.J. Ford, Volker Wulf, A note on the approximation of Hopf bifurcations in DDEs by Runge–Kutta methods, Technical Report (to appear), Manchester Centre for Computational Mathematics (ISSN 1360 1725), 2000.
[9] N.J. Ford, V. Wulf, Hopf bifurcation for numerical approximations to the delay logistic equation, in: S.K. Dey (Ed.), Proceedings of the IMACS Congr. on Computer Simulation and Mathematical Modelling, Illinois, 1998, Institute of Applied Science and Computations, pp. 249–254.
[10] E. Hairer, Ch. Lubich, The life-span of backward error analysis for numerical integrators, Numer. Math. 76 (1997) 441–462.
[11] J.K. Hale, S.M. Verdyn Lunel, Introduction to Functional Differential Equations, Springer, Berlin, New York, 1993.
[12] B.D. Hassard, N.D. Kazarnikoff, Y.-H. Wan, Theory of Hopf Bifurcation, Cambridge University Press, Cambridge, 1981.
[13] K.J. in't Hout, Ch. Lubich, Periodic solutions of delay differential equations under discretization, BIT 38 (1998) 72–91.
[14] T. Koto, Naimark–Sacker bifurcations in the Euler method for a delay differential equation, BIT 39 (1999) 110–115.
[15] Y.A. Kuznetsov, Elements of Applied Bifurcation Theory, Springer, New York, 1995.
[16] J.D. Lambert, Numerical Methods for Ordinary Differential Equations, Wiley, Chichester, 1991.
[17] T. Luzyanina, K. Engelborghs, K. Lust, D. Roose, Computation, continuation and bifurcation analysis of periodic solutions of delay differential equations, Report TW252, Leuven, 1997.
[18] A.M. Stuart, A.R. Humphries, Dynamical Systems and Numerical Analysis, Cambridge University Press, Cambridge, 1996.
[19] V. Wulf, Numerical analysis of delay differential equations undergoing a Hopf bifurcation, Ph.D. Thesis, University of Liverpool, Liverpool, 1999.
[20] V. Wulf, N.J. Ford, Insight into the qualitative behaviour of numerical solutions to some delay differential equations, in: E.A. Lipitakis (Ed.), Proceedings of HERCMA 1998, Athens, 1999, L.E.A, pp. 629–636.

ELSEVIER

Journal of Computational and Applied Mathematics 125 (2000) 287–295

JOURNAL OF COMPUTATIONAL AND APPLIED MATHEMATICS

www.elsevier.nl/locate/cam

Designing efficient software for solving delay differential equations

Christopher A.H. Paul[1]

Mathematics Department, Manchester University, Manchester M13 9PL, United Kingdom

Received 19 June 1999; received in revised form 7 October 1999

Abstract

In this paper, the efficient implementation of numerical software for solving delay differential equations is addressed. Several strategies that have been developed over the past 25 years for improving the efficiency of delay differential equation solvers are described. Of particular interest is a new method of automatically constructing the network dependency graph used in tracking derivative discontinuities.

Keywords: Numerical software; Delay differential equations

1. Introduction

The past 10 years have seen a substantial increase in both the research into and the use of delay differential equations (DDEs), such as

$$y'(t) = f(t, y(t), y(t-\tau(t))), \quad \text{where } \tau(t) \geqslant 0. \tag{1}$$

The increased use of DDEs in mathematical modelling has been fuelled by the availability of efficient and robust software for solving such types of problem.

In this paper, some aspects of the design of efficient numerical software for solving DDEs are discussed. In particular, several methods of *automatically* obtaining useful information about a DDE are described and then used to improve the efficiency with which DDE is solved.

DDE (1) may be solved numerically by combining an "interpolation" method (for evaluating delayed solution values) with an ordinary differential equation (ODE) integration method (for solving the resulting "ODE"). However, there are certain features of DDEs, such as the propagation of derivative discontinuities [4,6] and the possibility of vanishing delays [1], that make such an over-simplification unhelpful when writing efficient and robust software, unless measures are taken to address them.

[1] Funded by the U.K. Engineering & Physical Science Research Council.

0377-0427/00/$ - see front matter
PII: S0377-0427(00)00474-X

2. Evaluating delayed solution values

In order to evaluate delayed solution values, it is necessary to construct a suitable continuous solution for each accepted step and store this solution for later use. Whilst there are several possible continuous extensions for Runge–Kutta methods (see [2, p. 186]), the most flexible are those one-step natural continuous extensions [7] that are also Hermite interpolants: They can be used to provide defect error control [3], to evaluate delayed *derivative* values for NDDEs, and to construct "smooth" graphical output (where appropriate).

2.1. Storage of the past solution

A key feature of any DDE solver is the ability to evaluate delayed solution values efficiently. This feature is related to, *although not necessarily the same as*, the ability to provide a dense output of the solution. Consider the DDE $y'(t) = y(t - \tau(t))$. If the *delay function* $\tau(t)$ is bounded, then the oldest solution information may eventually be safely overwritten. However, if $\tau(t)$ is unbounded, then it is necessary to keep as much of the past solution as possible. This is because once a delayed solution value that is not available is required, the DDE cannot be solved any further. Given that the storage available for the past solution is limited, this can best be achieved by using a *cyclic* storage structure — once a cyclic storage structure is "full", the oldest solution information is overwritten. This means that some DDEs may be solved over much larger ranges of integration than would otherwise be possible.

2.2. Dense output

One of the requirements of any good DDE solver is the ability to produce a dense output (continuous extension) of the solution. Using a cyclic storage structure for storing the past solution means that it is possible that, even after successfully solving a DDE, not all of the solution is available for dense output. However, if the dense output points are known *before* the DDE is solved, then they can be output as soon as the appropriate continuous solution becomes available. This means that graphical output can be provided, and parameter fitting problems can be solved, over much larger ranges of integration.

2.3. Retrieving delayed solution values

Efficient implementation of the routine that calculates delayed solution values is essential, otherwise the time taken to solve simple DDEs may be dominated by "housekeeping". When a delayed solution value is required, it is necessary to locate the correct previously stored solution — if a fixed stepsize is used then the location of the stored solution can be easily computed, otherwise it is necessary to search the solution history. For a DDE with more than one delay function, if only one pointer is used to search the solution history, considerable time can be spent searching backwards and forwards between calculating different delayed solution values [5, p. 200]. For a DDE with more than three delay functions, such as

$$y'(t) = y(t-1) - y^2(t-7) + y(t-3) + y(t-5), \tag{2}$$

Table 1
Archi [5] timings for solving Eq. (2) over the range [0, 10] using a fixed stepsize

Stepsize	Four pointers Worst ordering	Best ordering	One pointer Worst ordering	Best ordering
0.001	0.89 s	0.89 s	89.79 s	74.89 s
0.0005	1.78 s	1.78 s	356.48 s	297.35 s
0.00025	3.50 s	3.50 s	1442.13 s	1185.42 s

there is also the problem of the most efficient order for evaluating the delayed solution values — a question that should not really have to concern the user. The best ordering of the delayed arguments for the above example is

$$y'(t) = y(t-1) + y(t-3) + y(t-5) - y^2(t-7),$$

because then the delayed solution values are evaluated in strictly increasing order. The answer to both problems is simply to have separate pointers for each delay function; then the ordering of delayed arguments is irrelevant and the time spent searching for the solution information is minimised (see Table 1).

3. Derivative discontinuities

In order to treat derivative discontinuities (from now on referred to as discontinuities) efficiently and effectively, it is necessary to know how they arise and how they are propagated. It is also useful to understand what impact discontinuities have on the numerical solution of DDEs. There are two ways of improving the efficiency of a DDE solver in the presence of discontinuities, *defect control* [3] and *discontinuity tracking* [6]. How discontinuities are treated determines how some parts of the user-interface should be specified.

3.1. Origin and propagation of derivative discontinuities

Consider the scalar DDE

$$y'(t) = y(t-\tau(t)) \; (t \geqslant t_0), \quad y(t) = \Psi(t) \; (t < t_0), \quad y(t_0) = y_0. \tag{3}$$

Discontinuities can originate in the *initial function* $\Psi(t)$, from the *initial point* t_0 when $\Psi^{(k)}(t_0-) \neq y^{(k)}(t_0+)$ for some integer $k \geqslant 0$, and from discontinuities in the DDE itself and the delay functions. Suppose that $t_0 = 0$, $\tau(t) = 1$ and $\Psi(t_0-) = y_0$ but that $\Psi'(t_0-) \neq y'(t_0+)$, then there is a discontinuity in $y'(t)$ at $t = 0$. This discontinuity is propagated by the *lag function* $\alpha(t) = t - \tau(t)$ so that $y''(1-) = \Psi'(0-) \neq y'(0+) = y''(1+)$.

Note that when a discontinuity is propagated in a DDE it is also *smoothed*, that is to say, it occurs in a higher derivative. For a general lag function $\alpha(t)$, a discontinuity in $y^{(r)}(t)$ typically propagates into $y^{(r+1)}(t)$. However if σ_k is an odd-multiple zero of $\alpha(t) = \sigma_i$, where σ_i is some previous discontinuity, then σ_k occurs in a much higher derivative [4, p. 847].

3.2. Impact of derivative discontinuities

A fundamental assumption made when deriving integration methods is that the solution is sufficiently smooth (over each step). If the solution is not sufficiently smooth, a high-order integration method collapses to lower order and consequently, the reliability of the local error estimator will be affected. When this happens, the extra computational cost involved in using a high-order method can be wasted because a cheaper, lower order method could have been used to achieve the same order with better error control. Thus when a DDE is not sufficiently smooth, unless appropriate measures are taken, the efficiency of a high-order integration method can be severely compromised.

3.3. Defect control of derivative discontinuities

One way in which discontinuities can be detected and located is by using the *defect* $\delta(t)$ [3]; in the case of Eq. (1),

$$\delta(t) = f(t, y(t), y(t - \tau(t))) - y'(t).$$

An advantage of defect control (over discontinuity tracking) is that no extra information about the DDE is required in order to locate discontinuities. However, in practice, an estimate of the maximum size of $|\delta(t)|$ on each step is used, and this estimate is based on asymptotic arguments that are no longer valid if a low-order discontinuity occurs in the step.

"Significant discontinuities" [2] are detected by monitoring the sequence of rejected steps and their associated defects. Once detected, it is usually necessary to locate the position of the discontinuity more precisely. However, because defect control is concerned with the accuracy of the solution and not its order, it is not generally necessary to locate the discontinuity precisely.

Having detected a discontinuity in the interval $(t_n, t_n + h)$, a bisection method can be used to locate it within a smaller interval $(r, s) \in (t_n, t_n + h)$, where the defect over the interval (r, s) is acceptably small. (In general, the exact location of a discontinuity cannot be found because there is no precise event function, and thus high-order event location methods cannot be used.) Each bisection iteration requires a defect evaluation, so that repeated derivative evaluations may be required. However, in systems of DDEs only one solution component can have the "most significant" discontinuity at any one point. Thus, evaluating all the derivative components can be very inefficient. This problem may be overcome by specifying the derivatives so that they can be evaluated separately. (However, this approach may severely affect the efficiency of the numerical integration method.) Consider the DDE

$$y_1'(t) = y_1(t)y_2(t-1), \quad y_2'(t) = y_2(t^2) - y_3(y_1(t)), \quad y_3'(t) = y_2(t-3). \tag{4}$$

[2] A "significant discontinuity" is any discontinuity that gives rise to such a large defect (*estimate*) that a step is rejected. Thus, not all low-order discontinuities are "significant" and, in fact, the treatment of discontinuities relies on a discontinuity becoming "less significant" as the size of the step in which it occurs decreases.

The corresponding FORTRAN 90 code could be written as

```
REAL FUNCTION DERIVATIVE(COMPONENT,T)
USE DELAY_SOLVER
INTEGER, INTENT(IN) :: COMPONENT
REAL, INTENT(IN) :: T
SELECT CASE (COMPONENT)
CASE(1)
  DERIVATIVE=DELAY(1,T)*DELAY(2,T-1)
CASE(2)
  DERIVATIVE=DELAY(2,T*T)-DELAY(3,DELAY(1,T))
CASE(3)
  DERIVATIVE=DELAY(2,T-3)
END SELECT
END FUNCTION DERIVATIVE
```

where the module DELAY_SOLVER contains a function DELAY(i,t) that evaluates the ith solution component at the argument t.

3.4. Tracking derivative discontinuities

Discontinuity tracking maintains the order of a numerical solution by including low-order discontinuities in the meshpoints. It is much more intimately connected with the structure of the DDE being solved than defect control. Software that tracks discontinuities à la Willé and Baker [6] requires the user to specify the lag functions and derivative functions separately, as well as specifying the initial discontinuities and the network dependency graph (Section 3.4.3).

3.4.1. The derivative discontinuity tracking equations

Reconsider Eq. (3) and assume that $\Psi(t_0-)\neq y(t_0+)$. The discontinuity in $y(t)$ at $t=t_0$ is propagated to the point $t=\sigma_1$ when the lag function *crosses* the position of the discontinuity, that is to say

$$(\alpha(\sigma_1+)-t_0)\times(\alpha(\sigma_1-)-t_0)<0. \tag{5}$$

Condition (5) is "almost equivalent" to requiring $\alpha(\sigma_1)=t_0$, except that this does not exclude even-multiplicity zeros, whereas only odd-multiplicity zeros satisfy the "switching" condition (5). However, since multiple zeros are numerically ill-conditioned [4], *in practice* they are given no special treatment.

The discontinuity at $t=\sigma_1$ may itself be propagated to a point $t=\sigma_2$, where $\alpha(\sigma_2)=\sigma_1$. It is also possible, depending on the lag function, that the discontinuity at $t=t_0$ is again propagated to a point $t=\sigma_2$, where $\sigma_2>\sigma_1$ can be arbitrarily large. Thus, $y(t)$ cannot be guaranteed to be ultimately smooth.

Thus, discontinuity tracking in a *scalar* DDE involves the evaluation of the lag functions and a record of the (initial) discontinuities and how they have been propagated. The need to evaluate the lag functions means that they must be specified separately from the DDE (Section 3.4.3).

3.4.2. Tracking derivative discontinuities in systems of DDEs

In a system of DDEs discontinuity tracking can be complicated by discontinuities being propagated between solution components. This fact gives rise to the concept of *strong* and *weak coupling* and *network dependency graphs* (NDGs) [6]. Strong coupling describes the propagation of discontinuities between different solution components by an ODE term, that is to say $\tau(t) \equiv 0$. Weak coupling describes the propagation of discontinuities within the same solution component and between different solution components by a DDE term, that is to say $\tau(t) \not\equiv 0$. Given a system of DDEs, the NDG is key to tracking the propagation of discontinuities. For the system of DDEs (4), the NDG is

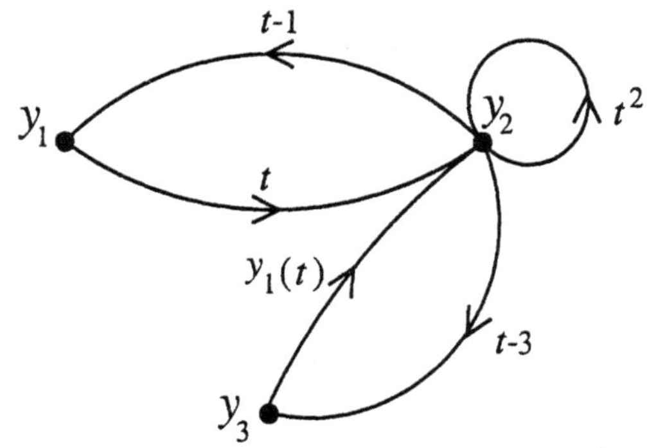

Thus, tracking discontinuities requires the following *extra* information about the DDE to be specified: (i) the initial discontinuities and (ii) the network dependency graph. Whilst the initial discontinuities cannot be automatically determined, it is possible to construct the NDG automatically.

3.4.3. Constructing the network dependency graph

Reconsider the DDE system (4). Unlike using defect control for treating discontinuities (Section 3.3), the lag functions have to be specified separately from the derivative functions. Thus, the DDE could be specified using FORTRAN 90 as

```
REAL FUNCTION DERIVATIVE(COMPONENT,T)
USE DELAY_SOLVER
INTEGER, INTENT(IN) :: COMPONENT
REAL, INTENT(IN) :: T
SELECT CASE (COMPONENT)
CASE(1)
  DERIVATIVE=DELAY(1,1)*DELAY(2,2)
CASE(2)
  DERIVATIVE=DELAY(2,3)-DELAY(3,4)
CASE(3)
  DERIVATIVE=DELAY(2,5)
END SELECT
END FUNCTION DERIVATIVE
```

where the function DELAY(i,j) evaluates the ith solution component using the jth lag function. The corresponding lag functions would then be given as

```
REAL FUNCTION LAG(COMPONENT,T)
USE DELAY_SOLVER
INTEGER, INTENT(IN) :: COMPONENT
REAL, INTENT(IN) :: T
SELECT CASE (COMPONENT)
CASE(1)
  LAG=T
CASE(2)
  LAG=T-1
CASE(3)
  LAG=T*T
CASE(4)
  LAG=DELAY(1,1)
CASE(5)
  LAG=T-3
END SELECT
END FUNCTION LAG
```

By evaluating each of the lag functions in turn and monitoring calls to the DELAY function, it is possible to determine which lag functions are state-independent (do not depend on the solution) and which are state-dependent (and on which solution components and other lag functions they depend). This information is necessary for constructing the NDG, as well as being useful when tracking discontinuities because there are additional difficulties associated with discontinuities propagated by state-dependent lag functions.

Next, by evaluating each derivative component in turn and monitoring calls to the DELAY function, it is possible to determine the links in the NDG and which lag functions are associated with each link. For state-dependent lag functions there is also the question of the propagation of discontinuities in the lag functions, which gives rise to additional links in the NDG. This then completes the construction of the NDG.

3.4.4. *Efficient tracking of derivative discontinuities*

The implementation of discontinuity tracking is even more complicated than already suggested. Having specified the initial discontinuities and constructed the NDG, it is necessary to calculate ("predict") where discontinuities will be propagated to *before* attempting to solve the DDE over an interval in which a discontinuity occurs. This can be achieved efficiently by determining whether an interval contains a discontinuity just before attempting to solve over it.

Testing the switching condition

$$(\alpha(t_n + h) - \sigma_i) \times (\alpha(t_n) - \sigma_i) < 0$$

on each step $[t_n, t_n+h]$ might appear to be sufficient for detecting if the discontinuity σ_i is propagated into the interval $[t_n, t_n + h]$. However, if the lag function $\alpha(t)$ propagates the discontinuity σ_i into the interval an even number of times, then the propagated discontinuities will not be detected (*see below*).

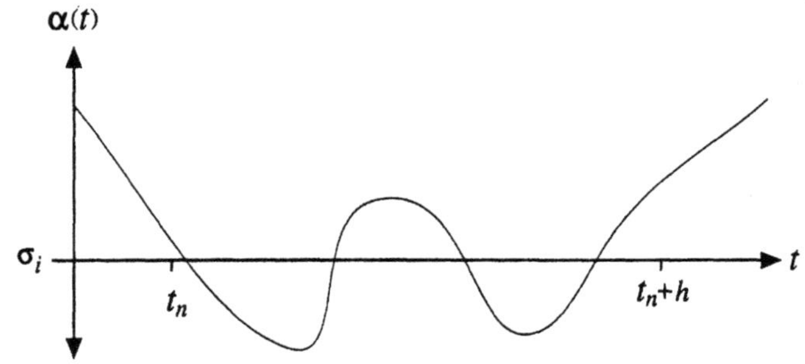

Once detected, the location of the discontinuity must be found more precisely by solving the equation $\alpha(t) = \sigma_i$. However, the convergence test is usually of the form $|\alpha(t) - \sigma_i| < v|\sigma_i|\varepsilon$, where $v > 1$ and ε is the unit-roundoff. For some DDEs, for example, $y'(t) = y(t^2)y\left(t - \frac{1}{10}\right)$ with $t_0 = 0$, this gives rise to numerous spurious discontinuities [5, p. 86]. A strategy for reducing the impact of this problem is to require discontinuities in the same solution component and derivative to be at least a distance $\delta \gg \varepsilon$ apart before they are considered to be distinct.

Having located a discontinuity at the point $t = \sigma_k$, it still remains to advance the solution up to $t = \sigma_k$. For DDEs in which $y(t)$ is discontinuous at the point $t = t_0$, it is particularly important to evaluate delayed solution values correctly [5, p. 87]: If $\alpha(t) \to t_0+$ as $t \to \sigma_k-$, then $y(\alpha(\sigma_k-))$ should be evaluated as $y(t_0+)$ and not $y(t_0-)$. The correct evaluation of delayed solution values can be achieved by monitoring whether the current interval is $[t_n, \sigma_k)$ or $[\sigma_k, t_{n+1}]$ and by knowing whether $\alpha(t_n) < t_0$ or $\alpha(t_n) > t_0$. (For continuous solutions this problem does not arise because $|y(t_0 + \varepsilon) - y(t_0 - \varepsilon)| = O(\varepsilon)$.)

Discontinuity tracking maintains the order of the solution by including discontinuities in the meshpoints, whilst attempting to avoid unnecessarily small and inefficient stepsizes. However for discontinuities propagated by state-dependent lag functions, very small stepsizes may still arise because the discontinuities may "move" slightly as the solution advances [5, p. 150]. This problem has still to be adequately addressed by codes that track discontinuities.

Tracking discontinuities can be computationally expensive, and it becomes more expensive as the number of discontinuities that need to be tracked increases [5, p. 177]. However, it is not necessary to track every discontinuity: The smoothing of discontinuities when they are propagated means that they eventually occur in a sufficiently high derivative that they can be ignored. Also, although it is necessary to track every low-order discontinuity (for a general lag function), if the user can specify a bound on the size of the delays then the oldest discontinuities can eventually be safely ignored.

4. Conclusion

The analytical theory underlying the numerical solution of DDEs has advanced considerably over the past decade. However, the number of available DDE solvers has remained somewhat limited and has thus delayed the widespread use of DDEs by non-mathematicians. Whilst there are a number of theoretical areas in the numerical solution of DDEs, convergence, stability, bifurcations, oscillations, etc., this paper has aimed at shedding light on some of the practical issues in writing an efficient and robust code for solving DDEs. Having identified some of the obvious and less obvious design problems, the next generation of DDE solvers should hopefully be more efficient and easier to use.

References

[1] C.T.H. Baker, C.A.H. Paul, Parallel continuous Runge–Kutta methods and vanishing lag delay differential equations, Adv. Comput. Math. 1 (1993) 367–394.

[2] C.T.H. Baker, C.A.H. Paul, D.R. Willé, Issues in the numerical solution of evolutionary delay differential equations, Adv. Comput. Math. 3 (1995) 171–196.

[3] W.H. Enright, H. Hayashi, A delay differential equation solver based on a continuous Runge–Kutta method with defect control, Numer. Algorithms 16 (1997) 349–364.

[4] M.A. Feldstein, K.W. Neves, High-order methods for state-dependent delay differential equations with non-smooth solutions, SIAM J. Numer. Anal. 21 (1984) 844–863.

[5] C.A.H. Paul, Runge–Kutta methods for functional differential equations, Ph.D. Thesis, Department of Mathematics, University of Manchester, 1992.

[6] D.R. Willé, C.T.H. Baker, The tracking of derivative discontinuities in systems of delay differential equations, Appl. Numer. Math. 9 (1992) 209–222.

[7] M. Zennaro, Natural continuous extensions of Runge–Kutta methods, Math. Comp. 46 (1986) 119–133.

ELSEVIER Journal of Computational and Applied Mathematics 125 (2000) 297–307

JOURNAL OF COMPUTATIONAL AND APPLIED MATHEMATICS

www.elsevier.nl/locate/cam

Introduction to the numerical analysis of stochastic delay differential equations

Evelyn Buckwar [1]

Department of Mathematics, The Victoria University of Manchester, Manchester M13 9PL, UK

Received 29 July 1999; received in revised form 17 February 2000

Abstract

We consider the problem of the numerical solution of stochastic delay differential equations of Itô form

$$\mathrm{d}X(t) = f(X(t), X(t-\tau))\mathrm{d}t + g(X(t), X(t-\tau))\mathrm{d}W(t), \quad t \in [0, T]$$

and $X(t) = \Psi(t)$ for $t \in [-\tau, 0]$, with given f, g, Wiener noise W and given $\tau > 0$, with a prescribed initial function Ψ. We indicate the nature of the equations of interest and give a convergence proof for explicit single-step methods. Some illustrative numerical examples using a strong Euler–Maruyama scheme are provided. © 2000 Elsevier Science B.V. All rights reserved.

1. Introduction

We are concerned here with the evolutionary problem for Itô stochastic delay differential equations or SDDEs. SDDEs generalise both deterministic delay differential equations (DDEs) and stochastic ordinary differential equations (SODEs). One might therefore expect the numerical analysis of DDEs and the numerical analysis of SODEs to have some bearing upon the problems that concern us here. We refer to [14] for an overview of the issues in the numerical treatment of DDEs. For a reprise of the basic issues in the numerical treatment of SODEs, see [4]; for more extensive treatments see [8,11]. In this article we will be interested in obtaining approximations to strong solutions of an SDDE. One reason to be interested in this kind of approximation is to examine the dependence of the solution on the initial function or on parameters that are contained in the definition of the SDDE. The article is based on [2].

[1] Supported by TMR Grant No. ERBFMBICT983282.

PII: S0377-0427(00)00475-1

We shall use a brief discussion of some model problems to introduce SDDEs to the reader.

- (*Cell population growth*) Consider a large (in order to justify continuous as opposed to discrete growth models) population $N(t)$ of cells at time t evolving with a proportionate rate $\rho_0 > 0$ of 'instantaneous' and a proportionate rate ρ_1 of 'delayed' cell growth. By 'instantaneous' cell growth, we mean that the rate of growth is dependent on the *current* cell population, and by 'delayed' cell growth, we mean that the rate of growth is dependent on some *previous* cell population. If the number $\tau > 0$ denotes the average cell-division time, the following equation provides a model
$$N'(t) = \rho_0 N(t) + \rho_1 N(t-\tau), \quad t \geqslant 0, \qquad N(t) = \Psi(t), \quad t \in [-\tau, 0].$$
Now assume that these biological systems operate in a noisy environment whose overall noise rate is distributed like white noise $\beta\, \mathrm{d}W(t)$. Then we will have a population $X(t)$, now a random process, whose growth is described by the SDDE
$$\mathrm{d}X(t) = (\rho_0 X(t) + \rho_1 X(t-\tau))\mathrm{d}t + \beta\, \mathrm{d}W(t), \quad t > 0,$$
with $X(t) = \Psi(t)$ for $-\tau \leqslant t < 0$. This is a constant delay equation with additive noise (the delay is only in the drift term).
- (*Population growth again*) Assume now that in the above equation we want to model noisy behaviour in the system itself, e.g. the intrinsic variability of the cell proliferation or other individual differences and the interaction between individuals. This leads to the multiplicative noise term, as in
$$\mathrm{d}X(t) = (\rho_0 X(t) + \rho_1 X(t-\tau))\,\mathrm{d}t + \beta X(t)\,\mathrm{d}W(t), \quad t > 0,$$
with $X(t) = \Psi(t)$ for $-\tau \leqslant t < 0$.
- (*More examples*) For some additional examples we can refer to examples in neural control mechanisms: neurological diseases [3], pupil light reflex [9] and human postural sway [6].

2. General formulation

Let $(\Omega, \mathscr{A}, P)$ be a complete probability space with a filtration $(\mathscr{A}_t)$ satisfying the usual conditions, i.e. the filtration $(\mathscr{A}_t)_{t\geqslant 0}$ is right-continuous, and each $\mathscr{A}_t$, $t \geqslant 0$, contains all P-null sets in $\mathscr{A}$. In this article we will prove convergence of a numerical method in the mean-square-sense, i.e. we say that $X \in \mathscr{L}^2 = \mathscr{L}^2(\Omega, \mathscr{A}, P)$ if $\mathscr{E}(|X|^2) < \infty$ and we define the norm $\|X\|_2 = (\mathscr{E}(|X|^2))^{1/2}$. We refer to [13] for the background on probability theory and to [1,7] for properties of a Wiener process and stochastic differential equations.

Let $0 = t_0 < T < \infty$. Let $W(t)$ be a one-dimensional Brownian motion given on the filtered probability space $(\Omega, \mathscr{A}, P)$. We consider the scalar autonomous stochastic delay differential equation (SDDE)

$$\begin{aligned} \mathrm{d}X(t) &= \overbrace{f(X(t), X(t-\tau))}^{\text{drift coefficient}}\, \mathrm{d}t + \overbrace{g(X(t), X(t-\tau))}^{\text{diffusion coefficient}}\, \mathrm{d}W(t), \quad t \in [0, T] \\ X(t) &= \Psi(t), \quad t \in [-\tau, 0] \end{aligned} \tag{1}$$

with one fixed delay, where $\Psi(t)$ is an $\mathscr{A}_{t_0}$-measurable $C([-\tau, 0], \mathbb{R})$-valued random variable such that $\mathscr{E}\|\Psi\|^2 < \infty$ ($C([-\tau, 0], \mathbb{R})$ is the Banach space of all continuous paths from $[-\tau, 0] \to \mathbb{R}$

equipped with the supremum norm). The functions $f:\mathbb{R}\times\mathbb{R}\to\mathbb{R}$ and $g:\mathbb{R}\times\mathbb{R}\to\mathbb{R}$ are assumed to be continuous. Eq. (1) can be formulated rigorously as

$$X(t)=X(0)+\int_0^t f(X(s),X(s-\tau))\,\mathrm{d}s+\int_0^t g(X(s),X(s-\tau))\,\mathrm{d}W(s) \tag{2}$$

for $t\in[0,T]$ and with $X(t)=\Psi(t)$, for $t\in[-\tau,0]$. The second integral in (2) is a stochastic integral, which is to be interpreted in the Itô sense. If g does not depend on X the equation has *additive noise*, otherwise the equation has *multiplicative noise*. We refer to [10,12] for the following definition and a proof of Theorem 2.

Definition 1. An $\mathbb{R}$-valued stochastic process $X(t):[-\tau,T]\times\Omega\to\mathbb{R}$ is called a strong solution of (1), if it is a measurable, sample-continuous process such that $X|[0,T]$ is $(\mathscr{A}_t)_{0\leqslant t\leqslant T}$-adapted and X satisfies (1) or (2), almost surely, and satisfies the initial condition $X(t)=\Psi(t)$ ($t\in[-\tau,0]$)). A solution $X(t)$ is said to be path-wise unique if any other solution $\hat{X}(t)$ is stochastically indistinguishable from it, that is $P(X(t)=\hat{X}(t)$ for all $-\tau\leqslant t\leqslant T)=1$.

Theorem 2. *Assume that there exist positive constants $L_{f,i}$, $i=1,2$ and K_f, such that both the functions f and g satisfy a uniform Lipschitz condition and a linear growth bound of the following form: For all $\xi_1,\xi_2,\eta_1,\eta_2,\xi,\eta\in\mathbb{R}$ and $t\in[0,T]$*

$$|f(\xi_1,\eta_1)-f(\xi_2,\eta_2)|\leqslant L_{f,1}|\xi_1-\xi_2|+L_{f,2}|\eta_1-\eta_2|,$$
$$|f(\xi,\eta)|^2\leqslant K_f(1+|\xi|^2+|\eta|^2)$$

and likewise for g with constants $L_{g,i}$, $i=1,2$, and K_g. Then there exists a path-wise unique strong solution to Eq. (1).

3. Numerical analysis for an autonomous SDDE

Define a mesh with a uniform step h on the interval $[0,T]$ and $h=T/N, t_n=n\cdot h, n=0,\dots,N$. We assume that there is an integer number N_τ such that the delay can be expressed in terms of the stepsize as $\tau=N_\tau\cdot h$. We consider strong approximations $\tilde{X}_n$ of the solution to (1), using a stochastic explicit single-step method with an increment function ϕ incorporating increments $\Delta W_{n+1}:=W_{(n+1)h}-W_{nh}$ of the driving Wiener process. For all indices $n-N_\tau\leqslant 0$ define $\tilde{X}_{n-N_\tau}:=\Psi(t_n-\tau)$, otherwise

$$\tilde{X}_{n+1}=\tilde{X}_n+\phi(h,\tilde{X}_n,\tilde{X}_{n-N_\tau},\Delta W_{n+1}),\quad n=0,\dots,N-1. \tag{3}$$

Notation 1. We denote by $X(t_{n+1})$ the value of the exact solution of Eq. (1) at the meshpoint t_{n+1} and by $\tilde{X}_{n+1}$ the value of the approximate solution using (3) and by $\tilde{X}(t_{n+1})$ the value obtained after just one step of (3), i.e., $\tilde{X}(t_{n+1})=X(t_n)+\phi(h,X(t_n),X(t_n-\tau),\Delta W_{n+1})$.

With this notation we can give the following definitions.

Definition 3. The *local error* of the above approximation $\{\tilde{X}(t_n)\}$ is the sequence of random variables $\delta_{n+1}=X(t_{n+1})-\tilde{X}(t_{n+1})$, $n=0,\dots,N-1$. The global error of the above approximation $\{\tilde{X}_n\}$ is the sequence of random variables $\varepsilon_n:=X(t_n)-\tilde{X}_n$, $n=1,\dots,N$.

Note that ε_n is $\mathscr{A}_{t_n}$-measurable since both $X(t_n)$ and $\tilde{X}_n$ are $\mathscr{A}_{t_n}$-measurable random variables.

Definition 4. Method (3) is *consistent with order* p_1 *in the mean and with order* p_2 *in the mean-square sense* if the following estimates hold with $p_2 \geqslant \frac{1}{2}$ and $p_1 \geqslant p_2 + \frac{1}{2}$:

$$\max_{0 \leqslant n \leqslant N-1} |\mathscr{E}(\delta_{n+1})| \leqslant Ch^{p_1} \quad \text{as } h \to 0, \tag{4}$$

$$\max_{0 \leqslant n \leqslant N-1} (\mathscr{E}|\delta_{n+1}|^2)^{1/2} \leqslant Ch^{p_2} \quad \text{as } h \to 0, \tag{5}$$

where the (generic) constant C does not depend on h, but may depend on T and the initial data.

We also will assume the following properties of the increment function ϕ: assume there exist positive constants C_1, C_2 such that for all $\xi, \xi', \eta, \eta' \in \mathbb{R}$

$$|\mathscr{E}(\phi(h, \xi, \eta, \Delta W_{n+1}) - \phi(h, \xi', \eta', \Delta W_{n+1}))| \leqslant C_1 h(|\xi - \xi'| + |\eta - \eta'|), \tag{6}$$

$$\mathscr{E}(|\phi(h, \xi, \eta, \Delta W_{n+1}) - \phi(h, \xi', \eta', \Delta W_{n+1})|^2) \leqslant C_2 h(|\xi - \xi'|^2 + |\eta - \eta'|^2). \tag{7}$$

We now state the main theorem of this article.

Theorem 5. *We assume that the conditions of Theorem* 2 *are fulfilled and that the increment function* ϕ *in* (3) *satisfies estimates* (6) *and* (7). *Suppose the method defined by* (3) *is consistent with order* p_1 *in the mean and order* p_2 *in the mean-square sense, so that* (4) *and* (5) *hold* (*where the constant* C *does not depend on* h). *Then, approximation* (3) *for Eq.* (1) *is convergent in* $\mathscr{L}^2$ (*as* $h \to 0$ *with* $\tau/h \in \mathbb{N}$) *with order* $p = p_2 - \frac{1}{2}$. *That is, convergence is in the mean-square sense and*

$$\max_{1 \leqslant n \leqslant N} (\mathscr{E}|\varepsilon_n|^2)^{1/2} \leqslant Ch^p \quad \text{as } h \to 0, \text{ where } p = p_2 - \tfrac{1}{2}. \tag{8}$$

Proof. Since we have exact initial values we set $\varepsilon_n = 0$ for $n = -N_\tau, \ldots, 0$. Now beginning with $\varepsilon_{n+1} = X(t_{n+1}) - \tilde{X}_{n+1}$, using Notation 1, adding and subtracting $X(t_n)$ and $\phi(h, X(t_n), X(t_n - \tau), \Delta W_{n+1})$ and rearranging we obtain $\varepsilon_{n+1} \leqslant \varepsilon_n + \delta_{n+1} + u_n$, where

$$u_n := \phi(h, X(t_n), X(t_n - \tau), \Delta W_{n+1}) - \phi(h, \tilde{X}_n, \tilde{X}_{n-N_\tau}, \Delta W_{n+1}).$$

Squaring both sides, employing the conditional mean with respect to the σ-algebra $\mathscr{A}_{t_0}$, and taking absolute values, we obtain

$$\begin{aligned} \mathscr{E}(\varepsilon_{n+1}^2 | \mathscr{A}_{t_0}) \leqslant{} & \mathscr{E}(\varepsilon_n^2 | \mathscr{A}_{t_0}) + \underbrace{\mathscr{E}(|\delta_{n+1}|^2 | \mathscr{A}_{t_0})}_{1)} + \underbrace{\mathscr{E}(|u_n|^2 | \mathscr{A}_{t_0})}_{2)} \\ & + \underbrace{2|\mathscr{E}(\delta_{n+1} \cdot u_n | \mathscr{A}_{t_0})|}_{3)} + \underbrace{2|\mathscr{E}(\delta_{n+1} \cdot \varepsilon_n | \mathscr{A}_{t_0})|}_{4)} + \underbrace{2|\mathscr{E}(\varepsilon_n \cdot u_n | \mathscr{A}_{t_0})|}_{5)} \end{aligned} \tag{9}$$

which holds almost surely. We will now estimate the separate terms in (9) individually and in sequence; all the estimates hold almost surely. We will frequently use the Hölder inequality, the inequality $2ab \leqslant a^2 + b^2$ and properties of conditional expectation.

- For the term labelled 1) in (9) we have, due to the assumed consistency in the mean-square sense of the method,

$$\mathscr{E}(|\delta_{n+1}|^2|\mathscr{A}_{t_0}) = \mathscr{E}(\mathscr{E}(|\delta_{n+1}|^2|\mathscr{A}_{t_n})|\mathscr{A}_{t_0}) \leqslant c_1 h^{2p_2}.$$

- For the term labelled 2) in (9) we have, due to property (7) of the increment function,

$$\mathscr{E}(|u_n|^2|\mathscr{A}_{t_0}) \leqslant c_2 h \mathscr{E}(|\varepsilon_n|^2|\mathscr{A}_{t_0}) + c_2 h \mathscr{E}(|\varepsilon_{n-N_\tau}|^2|\mathscr{A}_{t_0}).$$

- For the term labelled 3) in (9) we obtain, by employing the consistency condition and property (7) of the increment function ϕ,

$$\begin{aligned} 2|\mathscr{E}(\delta_{n+1}\cdot u_n|\mathscr{A}_{t_0})| &\leqslant 2(\mathscr{E}(|\delta_{n+1}|^2|\mathscr{A}_{t_0}))^{1/2}(\mathscr{E}(|u_n|^2|\mathscr{A}_{t_0}))^{1/2} \\ &\leqslant \mathscr{E}(\mathscr{E}(|\delta_{n+1}|^2|\mathscr{A}_{t_n})|\mathscr{A}_{t_0}) + \mathscr{E}(|u_n|^2|\mathscr{A}_{t_0}) \\ &\leqslant c_3 h^{2p_2} + h c_6 \mathscr{E}(\varepsilon_n^2|\mathscr{A}_{t_0}) + h c_6 \mathscr{E}(\varepsilon_{n-N_\tau}^2|\mathscr{A}_{t_0}). \end{aligned}$$

- For the term labelled 4) we have, due to the consistency condition,

$$\begin{aligned} 2|\mathscr{E}(\delta_{n+1}\cdot \varepsilon_n|\mathscr{A}_{t_0})| &\leqslant 2\mathscr{E}(|\mathscr{E}(\delta_{n+1}|\mathscr{A}_{t_n})|\cdot|\varepsilon_n||\mathscr{A}_{t_0}) \\ &\leqslant 2(\mathscr{E}|\mathscr{E}(\delta_{n+1}|\mathscr{A}_{t_n})|^2)^{1/2}\cdot(\mathscr{E}(|\varepsilon_n|^2|\mathscr{A}_{t_0}))^{1/2} \\ &\leqslant 2(\mathscr{E}(c_5 h^{p_1})^2)^{1/2}\cdot(\mathscr{E}(|\varepsilon_n|^2|\mathscr{A}_{t_0}))^{1/2} \\ &= 2(\mathscr{E}(c_5^2 h^{2p_1-1}))^{1/2}\cdot(h\mathscr{E}(|\varepsilon_n|^2|\mathscr{A}_{t_0}))^{1/2} \\ &\leqslant c_5^2 h^{2p_1-1} + h\mathscr{E}(|\varepsilon_n|^2|\mathscr{A}_{t_0}). \end{aligned}$$

- For the term labelled 5) in (9) we have, using property (6) of the increment function ϕ,

$$\begin{aligned} 2|\mathscr{E}(\varepsilon_n\cdot u_n|\mathscr{A}_{t_0})| &\leqslant 2\mathscr{E}(|\mathscr{E}(u_n|\mathscr{A}_{t_n})||\varepsilon_n||\mathscr{A}_{t_0}) \\ &\leqslant c_6 h\mathscr{E}(|\varepsilon_n|^2|\mathscr{A}_{t_0}) + 2c_6 h\mathscr{E}(|\varepsilon_n||\varepsilon_{n-N_\tau}||\mathscr{A}_{t_0}) \\ &\leqslant c_6 h\mathscr{E}(|\varepsilon_n|^2|\mathscr{A}_{t_0}) + c_6 h 2(\mathscr{E}(|\varepsilon_n|^2|\mathscr{A}_{t_0}))^{1/2}\cdot(\mathscr{E}(|\varepsilon_{n-N_\tau}|^2|\mathscr{A}_{t_0}))^{1/2} \\ &\leqslant c_6 h\mathscr{E}(|\varepsilon_n|^2|\mathscr{A}_{t_0}) + c_6 h\mathscr{E}(|\varepsilon_n|^2|\mathscr{A}_{t_0}) + c_6 h\mathscr{E}(|\varepsilon_{n-N_\tau}|^2|\mathscr{A}_{t_0}) \\ &\leqslant c_6 h\mathscr{E}(|\varepsilon_n|^2|\mathscr{A}_{t_0}) + c_6 h\mathscr{E}(|\varepsilon_{n-N_\tau}|^2|\mathscr{A}_{t_0}). \end{aligned}$$

Combining these results, we obtain, with $2p_2 \leqslant 2p_1 - 1$,

$$\mathscr{E}(\varepsilon_{n+1}^2|\mathscr{A}_{t_0}) \leqslant (1+c_7 h)\mathscr{E}(\varepsilon_n^2|\mathscr{A}_{t_0}) + c_7 h^{2p_2} + c_8 h\mathscr{E}(|\varepsilon_{n-N_\tau}|^2|\mathscr{A}_{t_0}).$$

Now we will proceed by using an induction argument over consecutive intervals of length τ up to the end of the interval $[0, T]$.

Step 1: $t_n \in [0, \tau]$, i.e., $n = 1, \ldots, N_\tau$ and $\varepsilon_{n-N_\tau} = 0$.

$$\begin{aligned} \mathscr{E}(\varepsilon_{n+1}^2|\mathscr{A}_{t_0}) &\leqslant (1+c_7 h)\mathscr{E}(\varepsilon_n^2|\mathscr{A}_{t_0}) + c_7 h^{2p_2} \\ &\leqslant c_7 h^{2p_2}\sum_{k=0}^{n}(1+c_6 h)^k = c_7 h^{2p_2}\frac{(1+c_6 h)^{n+1}-1}{(1+c_6 h)-1} \\ &\leqslant c_9 h^{2p_2-1}((e^{c_6 h})^{n+1}-1) \leqslant c_9 h^{2p_2-1}(e^{c_6 T}-1). \end{aligned}$$

Step 2: $t_n \in [k\tau, (k+1)\tau]$ and we make the assumption $\mathscr{E}(|\varepsilon_{n-N_\tau}|^2|\mathscr{A}_{t_0}) \leqslant c_{10} h^{2p_2-1}$.

$$\begin{aligned} \mathscr{E}(\varepsilon_{n+1}^2|\mathscr{A}_{t_0}) &\leqslant (1+c_7 h)\mathscr{E}(\varepsilon_n^2|\mathscr{A}_{t_0}) + c_7 h^{2p_2} + c_8 h\mathscr{E}(|\varepsilon_{n-N_\tau}|^2|\mathscr{A}_{t_0}) \\ &\leqslant (1+c_7 h)\mathscr{E}(\varepsilon_n^2|\mathscr{A}_{t_0}) + c_7 h^{2p_2} + h c_{10} h^{2p_2-1} \end{aligned}$$

$$= (1 + c_7 h)\mathscr{E}(\varepsilon_n^2|\mathscr{A}_{t_0}) + c_{11}h^{2p_2}$$
$$\leqslant c_{12}h^{2p_2-1}(e^{c_6 T} - 1),$$

by the same arguments as above. This implies, almost surely,

$$(\mathscr{E}(\varepsilon_{n+1}^2|\mathscr{A}_{t_0}))^{1/2} \leqslant c_9 h^{p_2-1/2},$$

which proves the theorem. □

Remark 6. Assumption (6) reduces to the condition of Lipschitz-continuity for the increment function ϕ in the deterministic setting, i.e., without noise. This is a standard assumption for convergence in the theory of numerical analysis for deterministic ordinary differential equations, as it implies the zero-stability of the numerical method.

4. The Euler–Maruyama scheme

As a start we have considered strong Euler–Maruyama approximations with a fixed stepsize on the interval $[0,T]$, i.e., $h = T/N$, $t_n = n \cdot h$, $n = 0,\ldots,N$. In addition we have assumed that there is an integer number N_τ such that the delay can be expressed in terms of the stepsize as $\tau = N_\tau h$.

For Eq. (1) the increment function ϕ_{EM} of the Euler–Maruyama scheme has the following form in the method (3):

$$\phi_{\mathrm{EM}}(h, \tilde{X}_n, \tilde{X}_{n-N_\tau}, \Delta W_{n+1}) = hf(\tilde{X}_n, \tilde{X}_{n-N_\tau}) + g(\tilde{X}_n, \tilde{X}_{n-N_\tau})\Delta W_{n+1} \tag{10}$$

for $0 \leqslant n \leqslant N-1$ and with $\Delta W_{n+1} := W_{(n+1)h} - W_{nh}$, denoting independent $N(0,h)$-distributed Gaussian random variables.

Theorem 7. *If the functions f and g in Eq.* (1) *satisfy the conditions of Theorem* 2, *then the Euler–Maruyama approximation is consistent with order $p_1 = 2$ in the mean and order $p_2 = 1$ in the mean square.*

We gave a complete proof in [2], based on the consistency analysis given in [11] for SODEs and using a theorem from Mao [10, Lemma 5.5.2], which provides the necessary moment inequalities for the solution of (1).

Lemma 8. *If the functions f and g in Eq.* (1) *satisfy the conditions of Theorem* 2, *then the increment function ϕ_{EM} of the Euler–Maruyama scheme (given by* (10)) *satisfies estimates* (6) *and* (7) *for all $\xi, \xi', \eta, \eta' \in \mathbb{R}$.*

$$|\mathscr{E}(\phi_{\mathrm{EM}}(h, \xi, \eta, \Delta W_{n+1}) - \phi_{\mathrm{EM}}(h, \xi', \eta', \Delta W_{n+1}))|$$

$$= |\mathscr{E}(hf(\xi,\eta) + g(\xi,\eta)\Delta W_{n+1} - hf(\xi',\eta') - g(\xi',\eta')\Delta W_{n+1})|$$

$$\leqslant h|f(\xi,\eta) - f(\xi',\eta')| + |g(\xi,\eta) - g(\xi',\eta')||\mathscr{E}(\Delta W_{n+1})|$$

$$\leqslant h(L_1|\xi - \xi'| + L_2|\eta - \eta'|)$$

Table 1

Time step	0.25	0.125	0.0625	0.03125
I ε	0.0184	0.00404	0.000973	0.000244
II ε	0.1088654	0.04912833	0.02437045	0.01213507

$$\begin{aligned}
&\mathscr{E}(|\phi_{\mathrm{EM}}(h,\xi,\eta,\Delta W_{n+1})-\phi_{\mathrm{EM}}(h,\xi',\eta',\Delta W_{n+1})|^2)\\
&=\mathscr{E}(|hf(\xi,\eta)+g(\xi,\eta)\Delta W_{n+1}-hf(\xi',\eta')-g(\xi',\eta')\Delta W_{n+1}|^2)\\
&\leqslant h^2|f(\xi,\eta)-f(\xi',\eta')|^2+|g(\xi,\eta)-g(\xi',\eta')|^2\mathscr{E}|\Delta W_{n+1}|^2\\
&\leqslant h^2(L_1^2|\xi-\xi'|^2+L_2^2|\eta-\eta'|^2)+h(L_3^2|\xi-\xi'|^2+L_4^2|\eta-\eta'|^2),
\end{aligned}$$

from which the estimates follow.

Remark 9. Theorem 7 and the last lemma imply that for the Euler–Maruyama method Theorem 5 is valid, with order of convergence $p=\frac{1}{2}$ in the mean-square-sense. If Eq. (1) has additive noise, then the Euler–Maruyama approximation is consistent with order $p_1=2$ in the mean and order $p_2=\frac{3}{2}$ in the mean square, which implies an order of convergence $p=1$ in the mean-square-sense.

5. Numerical experiments

We have used the equation

$$\mathrm{d}X(t)=\{aX(t)+bX(t-1)\}\,\mathrm{d}t+\{\beta_1+\beta_2X(t)+\beta_3X(t-1)\}\,\mathrm{d}W(t)$$

as a test equation for our method. In the case of additive noise ($\beta_2=\beta_3=0$) we have calculated an explicit solution on the first interval $[0,\tau]$ by the method of steps (see, e.g., [5]), using $\Psi(t)=1+t$ for $t\in[-1,0]$ as an initial function. The solution on $t\in[0,1]$ is given by

$$X(t)=\mathrm{e}^{at}\left(1+\frac{b}{a^2}\right)-\frac{b}{a}t-\frac{b}{a^2}+\beta\mathrm{e}^{at}\int_0^t \mathrm{e}^{-as}\,\mathrm{d}W(s).$$

We have then used this solution as a starting function to compute an 'explicit solution' on the second interval $[\tau,2\tau]$ with a standard SODE-method and a small stepsize. In the case of multiplicative noise we have computed an 'explicit solution' on a very fine grid (2048 steps) with the Euler–Maruyama scheme.

One of our tests concerned the illustration of the theoretical order of convergence. In this case the mean-square error $\mathscr{E}|X(T)-\tilde{X}_N|^2$ at the final time $T=2\tau$ was estimated in the following way. A set of 20 blocks each containing 100 outcomes ($\omega_{i,j}$; $1\leqslant i\leqslant 20, 1\leqslant j\leqslant 100$), were simulated and for each block the estimator $\varepsilon_i=\frac{1}{100}\sum_{j=1}^{100}|X(T,\omega_{i,j})-\tilde{X}_N(\omega_{i,j})|^2$ was formed. In Table 1 ε denotes the mean of this estimator, which was itself estimated in the usual way: $\varepsilon=\frac{1}{20}\sum_{i=1}^{20}\varepsilon_i$.

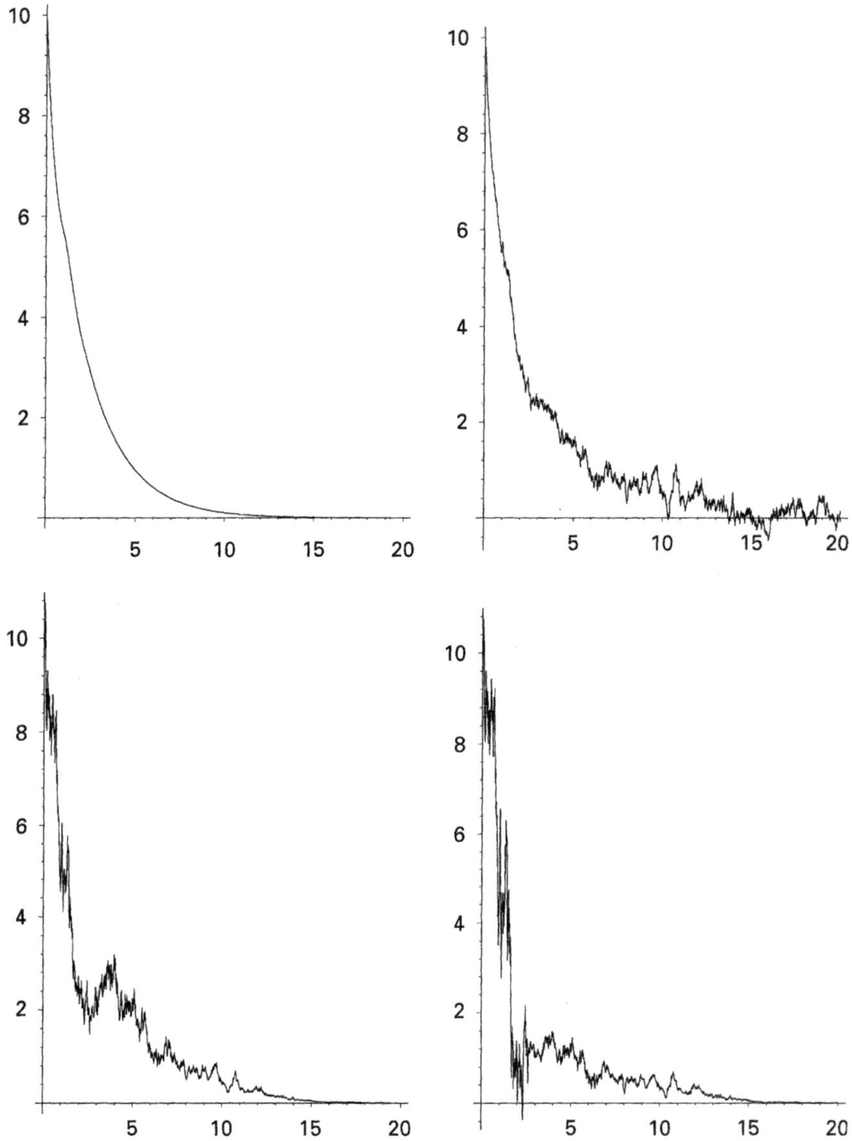

Fig. 1. Upper left: $\beta_i = 0,\ i = 1, \dots, 3$, upper right: $\beta_1 = 0.5,\ \beta_i = 0,\ i \neq 1$, lower left: $\beta_2 = 0.5,\ \beta_i = 0,\ i \neq 2$, lower right: $\beta_3 = 0.5,\ \beta_i = 0,\ i \neq 3$.

We have used the set of coefficients I $a = -2, b = 0.1, \beta_1 = 1$ and II $a = -2, b = 0.1, \beta_2 = 1$ (the other coefficients in the diffusion term are set to 0). The figures display $\max_{1 \leqslant n \leqslant N} \mathscr{E}|X(T) - \tilde{X}_N|^2$, which according to (8) in Theorem 5 is bounded by $c^2 h^{2p}$, and they are compatible with the results given in Remark 9, i.e. $p = 1$ in (I), the example with additive noise, and $2p = \frac{1}{2}$ in (II), an example with multiplicative noise.

One may consider Eq. (1) as a deterministic delay equation perturbed by white noise. In this context Figs. 1 and 2 show the influence of the parameters β_i on the solution of the deterministic

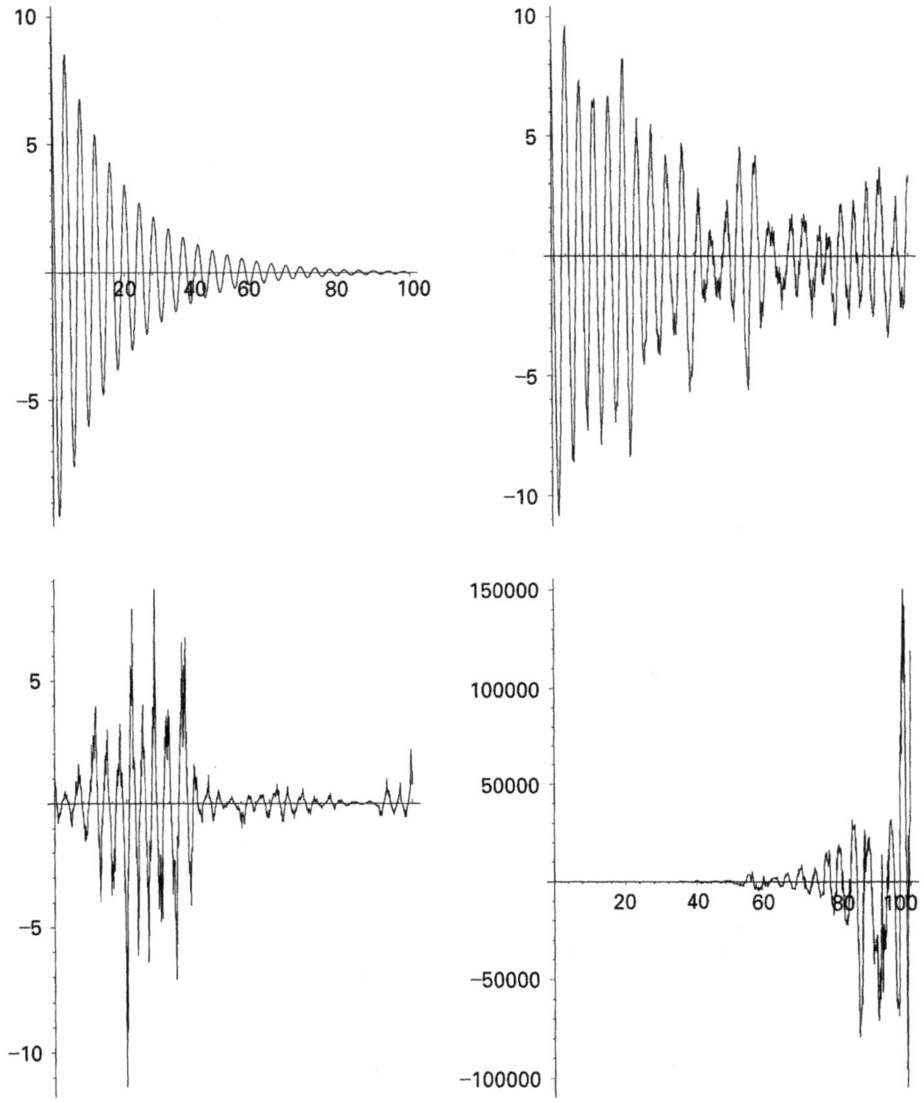

Fig. 2. Upper left: $\beta_i = 0,\ i = 1,\ldots,3$, upper right: $\beta_1 = 1,\ \beta_i = 0,\ i \neq 1$, lower left: $\beta_2 = 1,\ \beta_i = 0,\ i \neq 2$, lower right: $\beta_3 = 1,\ \beta_i = 0,\ i \neq 3$.

test equation $x'(t) = ax(t) + bx(t - \tau)$. In the first four pictures $a = -2, b = 1$, in the second four pictures $a = 0, b = 1.45$.

As a last experiment we varied the stepsize in order to observe some stability behaviour of the Euler–Maruyama method. Using the coefficients $a = -16, b = 1$ and two stepsizes: $h = \frac{1}{16}$ (left figure) and $h = \frac{1}{32}$ (right figure), we observe the same stability behaviour as for the deterministic equation, i.e., a change from unstable to stable, when varying the coefficients of the diffusion term. In the pictures we have $\beta_1 = 0.5$ ($\beta_i = 0, i \neq 1$), $\beta_2 = 0.5$ ($\beta_i = 0, i \neq 2$), $\beta_3 = 0.5$ ($\beta_i = 0, i \neq 3$), respectively.

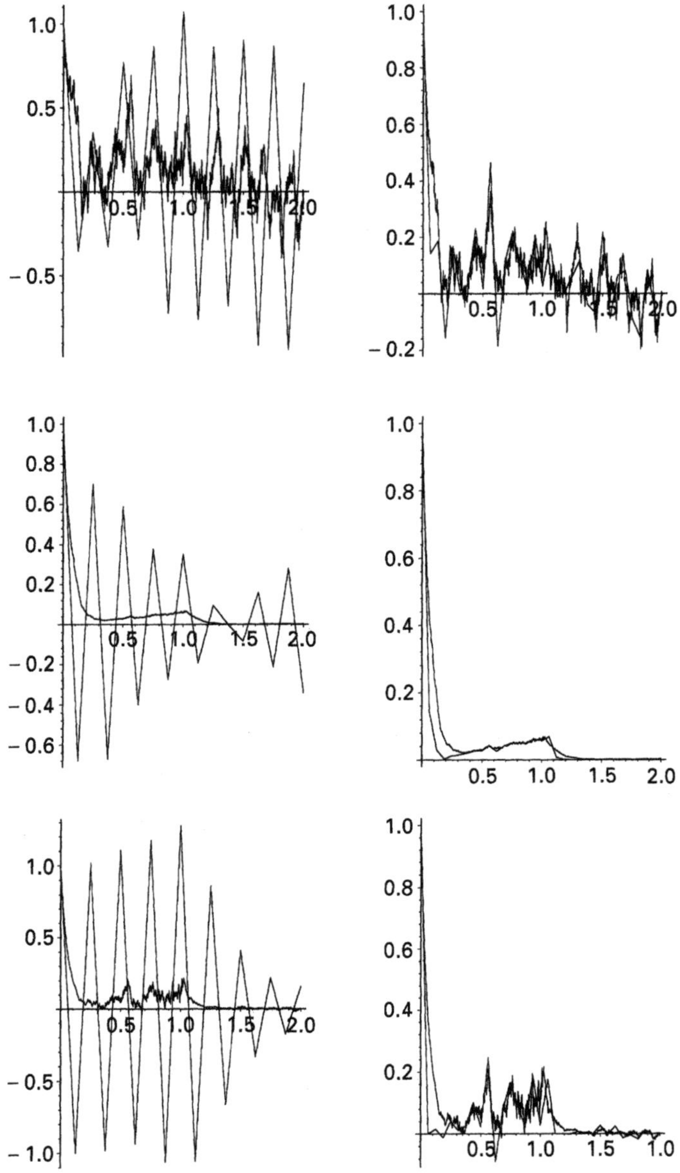

6. Conclusions

This article provides an introduction to the numerical analysis of stochastic delay differential equations. When one seeks to advance the study further, one sees open a number of unanswered questions, involving (for example)

- the design of numerical methods for more general kinds of memory (e.g., time or state dependent time lags);
- the stability and dynamical properties of the numerical methods;
- the design of numerical methods for more general problems (e.g., stochastic integrodifferential equations).

We hope that such issues will be addressed in sequels to this report.

References

[1] L. Arnold, Stochastic Differential Equations: Theory and Applications, Wiley-Interscience, New York, 1974.

[2] C.T.H. Baker, E. Buckwar, Introduction to the numerical analysis of stochastic delay differential equations, MCCM Numerical Analysis Technical Report, Manchester University, ISSSN 1360–1725, 1999.

[3] A. Beuter, J. Bélair, Feedback and delays in neurological diseases: a modelling study using dynamical systems, Bull. Math. Biol. 55 (3) (1993) 525–541.

[4] J.M.C. Clark, The discretization of stochastic differential equations: a primer, in: H. Neunzert (Ed.), Road Vehicle Systems and Related Mathematics; Proceedings of the second DMV-GAMM Workshop, Torino, 1987, Teubner, Stuttgart, and Kluwer Academic Publishers, Amsterdam, pp. 163–179.

[5] R.D. Driver, Ordinary and Delay Differential Equations, Applied Mathematical Sciences, Vol. 20, Springer, New York, 1977.

[6] C.W. Eurich, J.G. Milton, Noise-induced transitions in human postural sway, Phys. Rev. E 54 (6) (1996) 6681–6684.

[7] I. Karatzas, S.E. Shreve, Brownian Motion and Stochastic Calculus, Springer, New York, 1991.

[8] P.E. Kloeden, E. Platen, Numerical Solution of Stochastic Differential Equations, Springer, Berlin, 1992.

[9] M.C. Mackey, A. Longtin, J.G. Milton, J.E. Bos, Noise and critical behaviour of the pupil light reflex at oscillation onset, Phys. Rev. A 41 (12) (1990) 6992–7005.

[10] X. Mao, Stochastic Differential Equations and their Applications, Horwood Publishing Limited, Chichester, 1997.

[11] G.N. Milstein, Numerical Integration of Stochastic Differential Equations, Kluwer Academic Publishers, Dordrecht, 1995 (Translated and revised from the 1988 Russian original.)

[12] S.E.A. Mohammed, Stochastic Functional Differential Equations, Pitman (Advanced Publishing Program), Boston, MA, 1984.

[13] D. Williams, Probability with Martingales, Cambridge University Press, Cambridge, 1991.

[14] M. Zennaro, Delay differential equations: theory and numerics, in: M. Ainsworth, J. Levesley, W.A. Light, M. Marletta (Eds.), Theory and Numerics of Ordinary and Partial Differential Equations (Leicester, 1994), Oxford University Press, New York, 1995, pp. 291–333.

ELSEVIER Journal of Computational and Applied Mathematics 125 (2000) 309–335

JOURNAL OF COMPUTATIONAL AND APPLIED MATHEMATICS

www.elsevier.nl/locate/cam

Retarded differential equations

Christopher T.H. Baker [1]

Department of Mathematics, The University of Manchester, Oxford Road, Manchester M13 9PL, UK

Received 5 October 1999; received in revised form 4 April 2000

Abstract

Retarded differential equations (RDEs) are differential equations having retarded arguments. They arise in many realistic models of problems in science, engineering, and medicine, where there is a time lag or after-effect. Numerical techniques for such problems may be regarded as extensions of dense-output methods for ordinary differential equations (ODEs), but scalar RDEs are inherently infinite dimensional with a richer structure than their ODE counterparts. We give background material, develop a theoretical foundation for the basic numerics, and give some results not previously published. © 2000 Elsevier Science B.V. All rights reserved.

Keywords: Retarded differential equations; Delay and neutral delay differential equations; Continuity and stability; Numerics; Mesh and densely defined approximations; Convergence; Order of convergence; Numerical stability

1. Introduction

Many real-life problems that have, in the past, sometimes been modelled by initial-value problems for differential equations actually involve a significant memory effect that can be represented in a more refined model, using a differential equation incorporating retarded or delayed arguments (arguments that 'lag behind' the current value). The last few decades have seen an expanding interest in problems variously classified as delay differential equations (DDEs), retarded differential equations (RDEs), retarded functional differential equations (RFDEs), or neutral delay differential equations (NDDEs). (Stochastic DDEs, whose basic numerics are addressed in [6], also arise.)

Amongst the application areas are the biosciences, economics, materials science, medicine, public health, and robotics; in a number of these there is an underlying problem in control theory. Regarding the independent variable (t, say) as representing "time" in an evolutionary problem, the significance

E-mail address: cthbaker@maths.man.ac.uk (C.T.H. Baker).

[1] Professor of Mathematics & Director, Manchester Centre for Computational Mathematics (MCCM).

PII: S0377-0427(00)00476-3

of any "time-lag" is in a sense determined by its size relative to the underlying "time-scales". Parameter estimation [3,4,9,11] in prospective models can assist in assessing the need to incorporate time-lags. To the extent that numerical analysts should be influenced by applications (see, for example, [5]) and by the theoretical background, the starting point for a study of numerical methods lies with mathematical modelling and in dynamical systems defined by Volterra (that is, nonanticipative) equations. The advent of robust general numerical routines for DDEs and NDDEs then changes the opportunities for mathematical modelling (through numerical simulation) using equations for which closed-form solutions do not exist.

Our main task is to convey the essence of the subject to the reader by means of a rigorous presentation that foregoes any attempt at a complete discussion (this would require a book rather than a paper), and the structure of this paper is as follows. Following this introduction, we provide some background theory (i) concerning the functional equations and (ii) concerning the background numerical analysis. We then discuss (emphasizing relatively simple formulae to advance the development, but indicating extensions) the major theoretical issues: *existence of an approximate solution*, *convergence to the true solution*, and *numerical stability*; the aim, always, is to adopt a mathematically sound viewpoint whilst noting the limitations of the discussion. We conclude with a mention of some further important issues. Though the material is to a large extent available in the literature, our personal viewpoint pervades the presentation — and we were unable to resist the temptation to include new material.

To illustrate problems of the type that interest us, we may consider

$$y'(t) = f(t, y(t), y(\alpha(t))), \quad t \geqslant t_0 \quad \text{where } \alpha(t) \leqslant t, \tag{1.1a}$$

$$y(t) = \psi(t), \quad t \in [t_{\min}, t_0] \quad \text{where } t_{\min} := \inf_{t \geqslant t_0} \alpha(t) \tag{1.1b}$$

($t_{\min}$ denotes a generic value, dependent on the problem) with one retarded or 'lagging' argument $\alpha(t)$. In general, the equations of interest present an *initial function* problem rather than an initial-value problem familiar in ordinary differential equations (ODEs): the solution $y(t)$ of (1.1) is defined by $\psi(t)$ on an initial interval depending on the initial point t_0. Thus, $y(t) = y(\psi, t_0; t)$.

Eq. (1.1a) is an example of a DDE. By way of illustration, one may take $\alpha(t) = t - \tau_\star$ where the *lag* $\tau_\star > 0$ is fixed and $t_{\min} = t_0 - \tau_\star$. That this form of lag is common in the modelling literature may owe more to the difficulty of treating more general equations analytically than to the realism of the model. In any event, (1.1a) may itself be generalized. Systems with multiple delays e.g., $y'(t) = G(t, y(t), y(\alpha_1(t)), y(\alpha_2(t)), \ldots, y(\alpha_m(t)))$, $t \geqslant t_0$, $y(t) \in \mathbb{R}^N$, $(\alpha_i(t) \leqslant t;\ i = 1, 2, \ldots, m)$ also occur. We may also consider a system of *neutral* differential equations, or NDDEs, say (y, F vector-valued)

$$y'(t) = F(t, y(t), y(\alpha(t)), y'(\beta(t))), \quad t \geqslant t_0 \tag{1.2a}$$

with $\alpha(t), \beta(t) \leqslant t$; if $t^0_{\min} = \inf_{t \geqslant t_0} \alpha(t)$ and $t^1_{\min} = \inf_{t \geqslant t_0} \beta(t)$ then the initial conditions are that

$$y(t) = \psi_0(t), \quad t \in [t^0_{\min}, t_0] \quad \text{and} \quad y'(t) = \psi_1(t),\ t \in [t^1_{\min}, t_0]. \tag{1.2b}$$

We may set $t_{\min} := \min\{t^0_{\min}, t^1_{\min}\}$. We term $\alpha(t)$, $\beta(t)$ the *delayed arguments*, and $\tau(t) := t - \alpha(t) \geqslant 0$, $\zeta(t) := t - \beta(t) \geqslant 0$ the corresponding *lags*. An NDDE (1.2a) is characterized by the dependence of $y'(t)$ on $y'(\beta(t))$, as an argument of F. Frequently, but not always, $\psi_1(t) = \psi_0'(t)$. We concentrate for simplicity on DDEs and NDDEs. The theory of NDDEs is rather less straightforward than that of DDEs, and it is commonplace to impose sufficient, rather than necessary, conditions on F for existence and uniqueness. We observe, in passing, that an alternative standard type of NDDE is represented by

$$\{y(t) - \gamma(t, y(\beta(t)))\}' = \Gamma(t, y(t), y(\alpha(t))), \tag{1.3}$$

which with suitable assumptions is in "Hale's form". The theory of (1.3) seems to be better developed than that for (1.2a), and its numerical treatment can merit separate investigation.

A classification of the types of delayed argument is important in the modelling, the analysis, and the numerics. We refer to $\alpha(t) = t - \tau(t)$ in (1.2a), and by analogy to $\beta(t) = t - \zeta(t)$, to distinguish (i) *fading* or *persistent* memory, respectively, $\alpha(t) \to \infty$ as $t \to \infty$ or $\alpha(t) \not\to \infty$ as $t \to \infty$; (ii) *bounded lag* if $\sup \tau(t) < \infty$; (iii) *constant* and *state-dependent* lag if, respectively, $\alpha(t) = t - \tau_\star$, with $\tau_\star$ fixed, or $\alpha(t) \equiv \alpha(t, y(t)) := t - \tau(t, y(t))$. Finally, we have (iv) a *vanishing lag* if $\alpha(t_\star) = t_\star$. (Analogously, (i) $\beta(t) \to \infty$ or $\beta(t) \not\to \infty$ as $t \to \infty$, (ii) $\sup \zeta(t) < \infty$, (iii) $\zeta(t) \equiv \zeta_\star$, (iv) $\beta(t_\star) = t_\star$.)

Remark 1.1. Where left- and right-hand derivatives of the solution $y(\cdot)$ do not agree, in an RDE such as that in (1.1) or (1.2), $y'(t)$ is interpreted as the right-hand derivative. The right- and left-hand derivatives of a function $\phi(\cdot)$, are (provided the required limit exists), respectively

$$\phi'_+(t) = \lim_{\delta \to 0} \frac{\phi(t + |\delta|) - \phi(t)}{|\delta|} \quad \text{and} \quad \phi'_-(t) = \lim_{\delta \to 0} \frac{\phi(t) - \phi(t - |\delta|)}{|\delta|}. \tag{1.4}$$

2. Background theory

We touch on some theoretical issues, asking the reader to consult the literature [2] (see, e.g., citations [16, 78, 83, 102, 104, 125, 151, 153, 200] in [30] and [59]) for more detail. We give some very simple DDEs to illustrate interesting features. Our first example is the scalar, linear, DDE

$$y'(t) = \lambda_\star y(t) + \mu_\star y(t - \tau_\star), \quad t \geqslant t_0 \quad \text{with} \quad y(t) = \psi_\star(t),\ t \in [t_0 - \tau_\star, t_0]. \tag{2.1a}$$

By a simple change of variable, one can (if appropriate) normalize the lag $\tau_\star$ to unity and obtain the equation

$$u'(t) = \lambda u(t) + \mu u(t - 1), \quad t \geqslant t_0, \quad \text{with} \quad u(t) = \psi(t), \quad t \in [t_0 - 1, t_0], \tag{2.1b}$$

where $u(t) := y(\tau_\star t)$, $\psi(t) = \psi_\star(\tau_\star t)$, $\lambda = \lambda_\star \tau_\star$, $\mu = \mu_\star \tau_\star$. On the other hand, the further substitution $v(t) = \exp(-\lambda t)u(t)$ then gives, with $\hat{\lambda} := \mu \exp(-\lambda)$, the equation $v'(t) = \hat{\lambda} v(t - 1)$ $(t \geqslant t_0)$, a "*pure delay equation*". This can be solved, in principle, by repeated integration: we have $v(t) = \hat{\lambda} \int_{t_0+(n-1)}^{t-1} v(s)\,ds + v(t_0 + n)$, with $t \in [t_0 + n, t_0 + (n+1)]$, for $n = 0, 1, 2, \ldots$.

Our second example is the popular *delayed logistic equation*

$$y'(t) = \Lambda_\star y(t)\{1 - y(t - \tau_\star)\}, \quad \tau_\star > 0 \tag{2.2}$$

[2] We are constrained in this presentation by pressures on space, and this limits the comprehensiveness of our references (the references [1–77] that are listed here carry additional citations), and limits our discussion (our main aim is to convey, without over-simplification, a rigorous underlying approach to theory and practice). For future reading, we refer to a forthcoming book in [26].

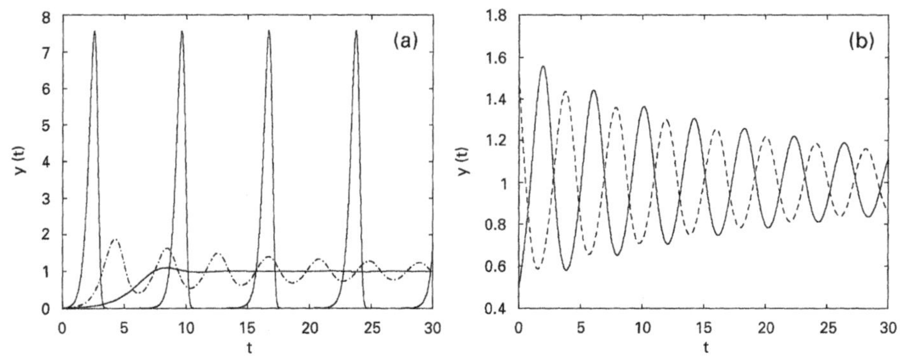

Fig. 1. Some solutions of $u'(t) = \Lambda u(t)\{1 - u(t-1)\}$ in (2.3): (a) for $\Lambda = 0.75, 1.5$ and 3.0, and a fixed $\psi(\cdot)$; (b) for $\Lambda = 1.5$; $\psi(t) = 0.5$ and $\psi(t) = 1.5$ $(t \in [-1, 0])$.

with an initial function defined on $[t_0 - \tau_\star, t_0]$, or (normalizing $\tau_\star$ to unity):

$$u'(t) = \Lambda u(t)\{1 - u(t-1)\} \quad (t \geqslant t_0), \tag{2.3a}$$

$$u(t) = \psi(t), \quad t \in [t_0 - 1, t_0], \quad \Lambda = \Lambda_\star \tau_\star \in \mathbb{R}. \tag{2.3b}$$

For $\Lambda > 0$, a positive initial function $\psi(t) > 0$ gives a positive solution $u(t)$ $(t \geqslant t_0)$ and with $w(t) := \ln u(t)$ one has the DDE $w'(t) = \Lambda\{1 - \exp(w(t-1))\}$, another "pure delay equation". Numerical solutions of (2.3) are plotted in Fig. 1. The realism of the model is open to question but it has some typical features: (a) The equation has a positive equilibrium solution $u(t) = 1$ for all Λ, but the qualitative behaviour of solutions of the scalar equation depends upon Λ and chaotic behaviour can arise. For large Λ, a solution can remain small for a substantial interval in t before increasing to a large value and then decaying again (cf. the case $\Lambda = 3$ which has high peaks, in Fig. 1(a)). The work [37] yields asymptotic expressions (valid as Λ becomes large) for maxima and minima and ultimate periodicity. (b) Solutions corresponding to different *initial functions* can assume the same value at a point $t = t_\star$ though they are not identical; see Fig. 1(b). For ODEs where the solution is uniquely determined by initial data at an arbitrary initial point this cannot happen. A further example is the *pantograph equation* $y'(t) = y(\gamma t)$ $(t \geqslant t_0 \geqslant 0)$; $y(t) = \psi(t)$, $t \in [\gamma t_0, t_0]$, $\gamma \in (0, 1)$. If $t_0 = 0$ this is an *initial-value* problem (cf. [29] for an extension). In certain limiting cases (when $\tau_\star = 0$ or $\gamma = 0$), equations mentioned above reduce to ODEs, e.g., $y'(t) = \Lambda_\star y(t)\{1 - y(t)\}$, the logistic ODE. If we set $\gamma = 0$ in the pantograph equation we obtain $y'(t) = y(0)$ which displays a persistent memory ($y(0)$ is never "forgotten").

Remark 2.1. Eq. (2.3a) arises from $x'(t) = \rho x(t)\{1 - K^{-1}x(t - \tau_\star)\}$, on a change of variables. Sir Robert May, FRS (at the time of writing, the Chief Scientific Adviser to HM Government) proposed, in 1973, study of a related *system* (with $\kappa, \xi, \rho, \omega, \tau_\star > 0$, $K > 0$) $y_1'(t) = \rho y_1(t)\{1 - K^{-1}y_1(t - \tau_\star)\} - \kappa y_1(t)y_2(t)$; $y_2'(t) = -\omega y_2(t) + \xi y_1(t)y_2(t)$. Other similar systems arise.

Theorem 2.2 (Existence; cf. [57]). *With $\delta > 0$, $t_0 \in [t_{\min}, t^{\max}]$, there is a unique solution of* (1.2a) *on $[t_{\min}, t^{\max}]$ if, whenever $u, U \in C^1[t_{\min}, t^{\max}]$ and $v, V \in C[t_{\min}, t^{\max}]$,* (i) *$F(t, u(t), u(\alpha(t)), u'(\beta(t)))$ is* *continuous on* $[t_{\min}, t^{\max}]$ *and, further,* (ii) $\|F(t,u(t),u(\alpha(t)),v(\beta(t))) - F(t,U(t),U(\alpha(t)),V(\beta(t)))\| \leqslant L_1(\sup_{s\in[t_{\min},t]}|u(s) - U(s)| + \sup_{s\in[t_{\min},t-\delta]}|v(s) - V(s)|) + L_2 \sup_{s\in[t-\delta,t]}|v(s) - V(s)|$ *holds for* $t \in [t_0, t^{\max}]$ *with* $L_{1,2} \geqslant 0$ *and* $L_2 < 1$.

One of the important characteristics of an RDE is the *sensitivity of a particular solution* to changes in the problem. This may be sensitivity to changes in the parameters [11] in the equation ("structural stability"), or sensitivity to changes in the initial conditions, or to persistently acting disturbances, and each gives rise to a definition of *stability*. To give definitions we shall refer to a solution of the NDDE problem (1.2) (then the DDE problem (1.1) is a special case), and we consider the perturbed problem

$$y'(t) + \delta y'(t)$$
$$= F(t, y(t) + \delta y(t), y(\alpha(t)) + \delta y(\alpha(t)), y'(\beta(t)) + \delta y'(\beta(t))) + \delta F(t) \quad (t \geqslant t_0), \tag{2.4a}$$

$$y(t) + \delta y(t) = \psi_0(t) + \delta\psi_0(t), \quad y'(t) + \delta y'(t) = \psi_1(t) + \delta\psi_1(t) \quad (t \leqslant t_0), \tag{2.4b}$$

conditions (2.4b) being valid on the appropriate initial intervals. We can measure the overall size of the input perturbations by, for example,

$$\Delta = \sup_{t\geqslant t_0} \|\delta F(t)\| + \max\left\{\sup_{t\leqslant t_0} \|\delta\psi_0(t)\|, \sup_{t\leqslant t_0} \|\delta\psi_1(t)\|\right\}, \tag{2.5}$$

where $\|\cdot\|$ is a vector norm. It may be appropriate to restrict the classes of perturbations (the *admissible perturbations*); for example, we may require $\delta F(t) \to 0$ as $t \to \infty$ or $\delta F(\cdot) \in \mathscr{L}[t_0, \infty)$, or (if $t_{\min}$ is $-\infty$) $\delta\psi(t) \to 0$ as $t \to -\infty$. For nonneutral equations, $\psi_1(\cdot)$ is absent and $\delta\psi_1(\cdot)$ vanishes. Stability concerns the boundedness of $\delta y(\cdot)$ on $[0, \infty)$, and the limiting behaviour of $\delta y(t)$ as $t \to \infty$, when the problem is perturbed; note that a solution of a *nonlinear* problem may be *unbounded but stable*. In consequence, to attempt (without further comment) to define stability of a solution of a *linear homogeneous* equation in terms of boundedness of its solutions $y(\cdot)$ can lead to misunderstanding of the nonlinear case. The unqualified term *stability* often refers to *stability with respect to perturbed initial conditions* in which we require that $\delta F(\cdot)$ vanishes, so the only perturbations are in $\psi_0(\cdot)$, $\psi_1(\cdot)$. We now give formal definitions of selected concepts.

Definition 2.3 (Stability). (a) Given t_0 and a class of admissible perturbations satisfying (2.5), a solution $y(\cdot)$ of the neutral equation (1.2) is (i) stable, if for each $\varepsilon > 0$ there exists $\Delta^* = \Delta^*(\varepsilon, t_0) > 0$ such that $\|\delta y(t)\| < \varepsilon$, when $t \geqslant t_0$ whenever $\Delta < \Delta^*$; (ii) asymptotically stable, if it is stable and there exists $\Delta^\dagger = \Delta^\dagger(t_0)$ such that $\delta y(t) \to 0$ as $t \to \infty$ whenever $\Delta < \Delta^\dagger$.

There are many complementary theoretical tools for analysing stability; stability theory for bounded lags may not extend to unbounded lags and theory for DDEs does not always extend to NDDEs.

Remark 2.4. (a) For linear equations (2.1), some results are well known. Thus, the use of Laplace transforms and an investigation of the zeros of the quasi-polynomial that serves as a stability function allows one to establish that *all solutions of the DDE* (2.1a) *are stable* with respect to perturbed initial conditions *when the point* $(\lambda, \mu) := (\lambda_\star \tau_\star, \mu_\star \tau_\star)$ *lies in the stability region* Σ which is the

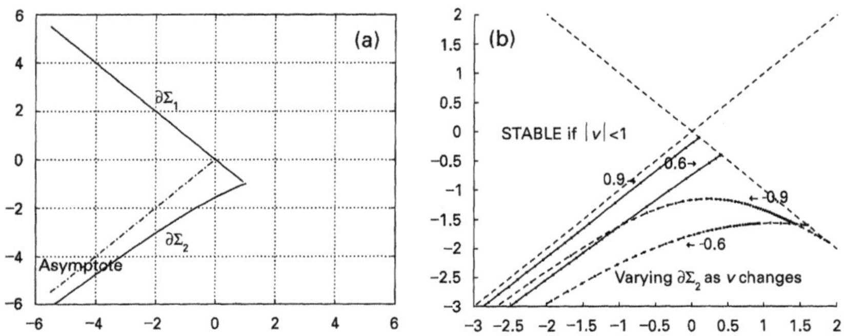

Fig. 2. Representation of the boundary $\partial\Sigma_1 \cup \partial\Sigma_2$ of the unbounded exact stability region Σ in the (λ, μ)-plane: (a) as given in (2.6), showing that Σ contains the wedge $\mathscr{W}$ of points (λ, μ) such that $|\mu| < -\lambda$ and $\tau_\star > 0$; $\lambda = \mu$ is an asymptote to $\partial\Sigma_2$; (b) showing for comparison the corresponding boundary (2.10) for the NDDE (2.9) in the case $v = -0.9, -0.6, 0.6, 0.9$; the stability regions again include $\mathscr{W}$ (for $|v| < 1$).

region of the (λ, μ)-plane that includes the half-line $\lambda < 0$, $\mu = 0$ and whose boundary

$$\partial\Sigma = \partial\Sigma_1 \cup \partial\Sigma_2 \text{ is formed by the loci} \tag{2.6a}$$

$$\partial\Sigma_1 := \{\mu = -\lambda\}, \tag{2.6b}$$

$$\partial\Sigma_2 := \{(\lambda = \omega \cot \omega,\ \mu = -\omega \operatorname{cosec} \omega);\ 0 < \omega < \pi\}. \tag{2.6c}$$

In particular, therefore, we obtain a stability condition

$$|\mu_\star| \leqslant -\lambda_\star \tag{2.7}$$

that is independent of $\tau_\star > 0$ from the observation $\Sigma \supset \mathscr{W} := \{(\lambda, \mu) \text{ such that } |\mu| < -\lambda\}$. For *complex-valued* $\lambda_\star, \mu_\star$, an analogue of (2.7) is that solutions are stable if $|\mu_\star| < -\Re\,\lambda_\star$; Guglielmi [41, p. 409] indicates the region $\Sigma^\dagger$ of complex parameters $(\lambda_\star\tau_\star, \mu_\star\tau_\star)$ for which all solutions of (2.1a) are stable.

(b) The above results extend after some precautions to certain NDDEs such as

$$y'(t) = \lambda_\star y(t) + \mu_\star y(t - \tau_\star) + v_\star y'(t - \tau_\star) \quad (\lambda_\star, \mu_\star, v_\star \in \mathbb{C}) \text{ with } |v_\star| < 1. \tag{2.8}$$

For the linear neutral delay differential equation, (2.8) a sufficient condition for stability is $|\lambda_\star \bar{v}_\star - \bar{\mu}_\star| + |\lambda_\star v_\star - \mu_\star| < -2\Re\,\lambda_\star$; if $\lambda_\star, \mu_\star, v_\star \in \mathbb{R}$, it suffices that $|\mu_\star| < -\lambda_\star$ and $|v_\star| < 1$. The lag can be normalized to unity if we consider

$$y'(t) = \lambda y(t) + \mu y(t-1) + v y'(t-1) \quad (\lambda := \lambda_\star\tau_\star, \mu := \mu_\star\tau_\star, v = v_\star) \tag{2.9}$$

Boundaries of the stability region are presented graphically (after some re-labelling) in [60, p. 117]. We here note that the locus corresponding to $\partial\Sigma_2$ in (2.6c), for v not necessarily 0, is the set of parameterized points

$$\{(\lambda, \mu)\colon \lambda = \omega \cot \omega - v\omega \operatorname{cosec}(\omega), \quad \mu = -\omega \operatorname{cosec}(\omega) + v\omega \cot(\omega)\}, \tag{2.10}$$

and sample loci are drawn in Fig. 2(b) for $v = -0.9, -0.6, 0.6$, and 0.9.

(c) For systems of ODEs, the test equation $y'(t) = \Lambda y(t)$ with $\Lambda \in \mathbb{C}$ (rather than with $\Lambda \in \mathbb{R}$) can be motivated by consideration of the equations $y_i'(t) = \sum a_{i,j} y_j(t)$, or, in compact notation,

$\boldsymbol{y}'(t) = \boldsymbol{A}\boldsymbol{y}(t)$, where the real matrix $\boldsymbol{A} = [a_{i,j}]$ is supposed to be similar to the diagonal matrix $\boldsymbol{diag}(\Lambda_1, \Lambda_2, \ldots, \Lambda_m)$ with complex $\{\Lambda_\ell\}$. The corresponding test equation (2.8) with complex coefficients should probably be considered less significant than in the complex-coefficient ODE case, because (i) it will be a rare occasion for three matrices $\boldsymbol{L} = [\lambda_{i,j}]$, $\boldsymbol{M} = [\mu_{i,j}]$, $\boldsymbol{N} = [\nu_{i,j}]$ to be simultaneously diagonalizable — i.e., under the same similarity transform — and (ii) the system of NDDEs that one would prefer to consider has the form

$$y_i'(t) = \sum_j \lambda_{i,j} y_j(t) + \sum_j \mu_{i,j} y_j(t - \tau_{i,j}) + \sum_j \nu_{i,j} y_j'(t - \zeta_{i,j}) \tag{2.11}$$

where the positive values $\{\tau_{i,j}, \zeta_{i,j}\}$ are generally all *different* (and may be t– or state–dependent).

If we turn to nonlinear equations, the solution $y(t) \equiv 1$ of (2.3) is readily shown to have different stability properties from those of the solution $y(t) \equiv 0$ and this illustrates the general observation that stability properties attach to a particular solution $y(t) \equiv y(\psi, t_0; t)$ of a nonlinear equation. Let us observe also that there are conditions in the literature under which stability with respect to certain types of persistent perturbations may be deduced from asymptotic stability with respect to initial perturbations.

Emphasis on test equations, such as (2.1) has limited interest unless one can relate the results to more general equations. In this respect, Baker and Tang [12] established results on stability with respect to initial perturbations, that include the following as a special case, and they investigated analogues for some numerical approximations.

Theorem 2.5. *Define* $||u(\cdot)||^{[t-k\tau_\star, t]} := \sup_{\sigma \in [t-k\tau_\star, t]} |u(\sigma)|$. *Suppose* $\alpha(t) \equiv t - \tau(t, u(t)) \leqslant t$ $(t \geqslant t_0)$ *and* $|\tau(t, v) - \tau_\star| \to 0$ *uniformly as* $t \to \infty$, $v \to 0$, *and, for some finite* $k \geqslant 1$, $|f(t, u(t), u(\alpha(t, u(t)))) - \{\lambda_\star u(t) + \nu_\star u(t - \tau_\star)\}| = o(||u(\cdot)||^{[t-k\tau_\star, t]})$ *uniformly as* $||u(\cdot)||^{[t-k\tau_\star, t]} \to 0, t \to \infty$. *Then, the zero solution of* (1.1) *is asymptotically stable if the zero solution of* (2.1*a*) *is asymptotically stable* (*and unstable if the zero solution of* (1.1) *is unstable*).

We turn to some further key features that are of interest in the context of numerical analysis for RDEs. These include (i) propagated discontinuities in the solution or its derivatives, or *large* derivative values; (ii) delay terms that can *stabilize* or can *destabilize* solutions of the problem; (iii) onset of periodicity or chaotic behaviour as a parameter in the defining equations is varied. We can illustrate these by reference to scalar equations, such as (2.3).

Firstly, consider the question of smoothness. Suppose that the initial function $\psi(t) \in C^1[t_0 - \tau_\star, t_0]$ in (2.2) is such that its (left-hand) derivative $\psi_\star'(t_0)$ is not equal to $\Lambda_\star \psi_\star(t_0)\{1 - \psi_\star(t_0 - \tau_\star)\}$. Then (since $y(\cdot)$ and $\psi_\star(\cdot)$ coincide on $[t_0 - \tau_\star, t_0]$), it follows that the right-hand derivative $y_+'(t_0)$ does not agree with the left-hand derivative $y_-'(t_0)$. In this case, we say that the derivative of y has a jump at t_0. We deduce, from (2.2), that $y''(t)$ has a similar jump at $t_0 + \tau_\star$, $y'''(t)$ has a jump at $t_0 + 2\tau_\star$, and so on. Here, the solution becomes smoother as t increases. There exists a well-developed theory of tracking of the points where derivatives of the solution have jump discontinuities, see [69,75] and other citations in [10]. The reader may consider a modification of (2.2),

$$y'(t) = \Lambda_\star y(t)\{1 - y(t) - \tau_\star y'(t - \tau_\star)\} \qquad (t \geqslant 0) \tag{2.12}$$

which is an NDDE, as an example where smoothing does not take place. Note that the type of approximation that purports to relate (2.2)–(2.12) is most suspect!

Secondly, we address stability. Consider the equilibrium $y(t) \equiv 1$ for the ODE $y'(t) = \Lambda^\star y(t)\{1 - y(t)\}$ which, by 'stability in the first approximation', is stable with respect to initial perturbations if $\Lambda^\star > 0$. However, $y(t) \equiv 1$ is a stable solution of $y'(t) = \Lambda_\star y(t)\{1 - y(t - \tau_\star)\}$ under the introduction of (small) perturbations $\varepsilon(t)$ for $t \in [-\tau_\star, 0]$ if (by a refined version of 'stability in the first approximation') corresponding solutions of $\varepsilon'(t) = -\Lambda_\star \varepsilon(t - \tau_\star)\{1 + \varepsilon(t)\}$ are bounded. A *stability condition* now is that $\Lambda_\star \tau_\star \in (0, \frac{1}{2}\pi)$ ($y(t) \equiv 1$ is unstable for $\Lambda_\star \tau_\star > \frac{1}{2}\pi$). Note the continuous dependence on $\tau_\star$ as $\tau_\star \searrow 0$.

Finally, consider, in association with bifurcation theory (see [30, p. 301]), the qualitative behaviour of the solutions of the delayed logistic equation as $\Lambda_\star \tau_\star$ varies. The solutions of (2.2) with $\Lambda_\star \tau_\star > 0$ and a positive $\psi(t)$ ($t \in [-\tau_\star, 0]$) are positive, converge monotonically to 1 if $0 \leqslant \Lambda_\star \tau_\star \leqslant e^{-1}$, converge to 1 in an oscillatory fashion if $e^{-1} \leqslant \Lambda_\star \tau_\star < \pi/2$, and display oscillatory behaviour for $\Lambda_\star \tau_\star \geqslant \pi/2$ (a Hopf bifurcation point — it is not a coincidence that the locus $\partial \Sigma_2$ in Fig. 2(a) crosses $\lambda = 0$ where $\mu = -\pi/2$). The numerical analysis of Hopf bifurcation is addressed in the literature (cf. [32,38], where differing perspectives are presented, and [34,35,64,65,76] along with the citations given therein). This area merits the reader's further consideration: Bifurcation, the onset of periodicity, and the behaviour of periodic solutions, are of practical interest.

3. Background numerical analysis

Numerical computation is designed to give quantitative insight (and, thereby, qualitative insight) into the solution of various mathematical models. As in the numerical solution of ODEs, there are two inter-related types of question that concern us: the one relates to the design of algorithms, and the other relates to what can be established rigorously about their properties and performance. This requirement of rigour imposes assumptions that in general reduce the degree of realism but nevertheless (one hopes) allows some practical insight to be gained. Additional valuable citations will be found in [3,10,26,43,77].

The choice of numerical techniques for the treatment of RDEs (DDEs and NDDEs) relies heavily on the construction of densely-defined *continuous extensions*. (Similar constructions produce dense-output in the numerics of ODEs.) The method of defining the continuous extension can, in addition to the properties of the other components of our methods, *affect both the accuracy and the stability of the numerical method* [10,26,77]. We concentrate upon the scalar version of (1.2a), for definiteness, so that

$$y'(t) = F(t, y(t), y(\alpha(t)), y'(\beta(t))) \quad \text{for } t \geqslant t_0, \tag{3.1}$$

(with $y(t) = \psi_0(t)$, $y'(t) = \psi_1(t)$, $t \leqslant t_0$). The restriction to DDEs, and the extension to systems, will be obvious.

Using the initial conditions $\tilde{y}(t) = \psi_0(t)$, $\widetilde{y'}(t) = \psi_1(t)$ for $t \leqslant t_0$, our numerical solution proceeds, usually with adaptive step sizes $\{h_n\}$, with a mesh

$$\mathscr{T} := \{t_0 < t_1 < t_2 < \cdots < t_{n-1} < t_n < \cdots\}, \quad h_n := t_{n+1} - t_n, \tag{3.2}$$

by obtaining $\tilde{y}(t), \widetilde{y'}(t)$ for $t \geqslant t_0$. The *width* of $\mathscr{T}$ is $\max_n h_n$. One advances, from t_n to t_{n+1} at the nth stage, in 'evolutionary' mode. The choice of $\mathscr{T}$ is clearly important and should take account of the nature of the solution. (Some of the published analysis treats the uniform step-size case, $h_n \equiv h$.) Where one seeks the solution on $[t_0, t^{\max}]$, we suppose that $t^{\max}$ is the only possible point of accumulation of $\{t_\ell\}$. In practice, the steps h_n are chosen on the basis of one or more of the following: (a) knowledge (to within a given tolerance) of the points where the solution or a "significant" derivative has a jump discontinuity (or is relatively large), (b) estimates of the *local truncation error*, (c) estimates of the *defect* (see [36,45]; and citation [46] of [10]). Determination of the points referred to in (a) can be attempted using *tracking* theory or by using indicators from estimates obtained under (b) or (c). For some problems, the solution becomes smoother as t increases, and the problem of discontinuities is transient; for certain NDDEs, and certain DDEs with unbounded lag, the problem of lack of smoothness persists. Where discontinuities cause a problem, the author's view is that modest-order one-step methods, such as those based on Runge–Kutta (RK) processes with RK abscissae in $[0,1]$, and *local* approximation techniques for providing a densely defined approximation, together with control of the step sizes, have great appeal (they are self-starting and can more readily avoid derivative discontinuities). Some RK methods are equivalent to a form of collocation. RK methods with abscissae outside $[0,1]$, linear multistep methods, and nonlocal extensions, can have rôles in the particular case of problems with smooth (or ultimately smooth) solutions.

3.1. The construction of some approximating formulae

For state-dependent $\alpha(t) \equiv \alpha(t, y(t)), \beta(t) \equiv \beta(t, y(t))$ we write

$$\tilde{\alpha}(t) = \hat{\alpha}(t, \tilde{y}(t)) \equiv \min(t, \alpha(t, \tilde{y}(t))), \tag{3.3a}$$

$$\tilde{\beta}(t) = \hat{\beta}(t, \tilde{y}(t)) \equiv \min(t, \beta(t, \tilde{y}(t))). \tag{3.3b}$$

Write $t_n + sh_n$ as t_{n+s}; as the calculation proceeds, one retains information to generate approximations $\tilde{y}(t_{n+s})$ (and $\widetilde{y'}(t_{n+s})$), $s \in (0,1)$, when required.

We orientate the reader by considering *the θ-method* applied, with fixed $\theta \in [0,1]$, to (3.1); later, we consider RK methods. With $\tilde{y}_k := \tilde{y}(t_k) \approx y(t_k)$, $\tilde{F}_k := F(t_k, \tilde{y}(t_k), \tilde{y}(\tilde{\alpha}(t_k)), \widetilde{y'}(\tilde{\beta}(t_k)))$, suppose we have computed and stored

$$\tilde{y}_0, \tilde{F}_0; \quad \tilde{y}_1, \tilde{F}_1; \quad \tilde{y}_2, \tilde{F}_2; \quad \cdots \quad \tilde{y}_n, \tilde{F}_n \tag{3.4}$$

(see Remark 3.2 for a generalization) and require to advance from t_n to t_{n+1}. The θ-formula for $\tilde{y}_{n+1}$ reads

$$\tilde{y}_{n+1} := \tilde{y}_n + (1-\theta)h_n\tilde{F}_n + \theta h_n F(t_{n+1}, \tilde{y}_{n+1}, \tilde{y}(\tilde{\alpha}(t_{n+1})), \widetilde{y'}(\tilde{\beta}(t_{n+1}))). \tag{3.5}$$

Of course, $\widetilde{y'_k} = \widetilde{y'}(t_k) := \tilde{F}_k$ approximates $y'(t_k)$. The θ-formula is implicit if $\theta \neq 0$ in which case one needs the delayed function value $\tilde{y}(\tilde{\alpha}(t_{n+1}))$ or $\widetilde{y'}(\tilde{\beta}(t_{n+1}))$. Taking $\theta = \frac{1}{2}$ defines *the trapezium rule*

$$\tilde{y}_{n+1} := \tilde{y}_n + \tfrac{1}{2}h_n\tilde{F}_n + \tfrac{1}{2}h_n F(t_{n+1}, \tilde{y}_{n+1}, \tilde{y}(\tilde{\alpha}(t_{n+1})), \widetilde{y'}(\tilde{\beta}(t_{n+1}))). \tag{3.6}$$

In the *special case* where $\alpha(t) = t - \tau_\star$, $\beta(t) = t - \zeta_\star$ and h can be (and is) fixed so that $\tau_\star/h = M \in \mathbb{N}$ and $\zeta_\star/h = M' \in \mathbb{N}$, the natural complete definition is

$$\tilde{y}_{n+1} := \tilde{y}_n + \tfrac{1}{2}h_n\tilde{F}_n + \tfrac{1}{2}h_nF(t_{n+1}, \tilde{y}_{n+1}, \tilde{y}_{n+1-M}, \tilde{F}_{n+1-M'}), \tag{3.7a}$$

$$\widetilde{y'}_{n+1} \equiv \tilde{F}_{n+1} := F(t_{n+1}, \tilde{y}_{n+1}, \tilde{y}_{n+1-M}, \tilde{F}_{n+1-M'}). \tag{3.7b}$$

In general, however, $\tilde{\alpha}(t_{n+1})$ or $\tilde{\beta}(t_{n+1}) \notin \mathscr{T}$. One then needs approximation formulae that extend the solution defined on the mesh $\mathscr{T}$, to compute $\tilde{y}(\tilde{\alpha}(t))$ at $\tilde{\alpha}(t) = t_\kappa + sh_\kappa$ for $s = s(\mathscr{T};t)$, and to compute $\widetilde{y'}(\tilde{\beta}(t))$ with $\tilde{\beta}(t) = t_{\kappa'} + qh_{\kappa'}$ for $q = q(\mathscr{T};t)$ $(s, q \in (0,1))$. Generic relationships of the type

$$\varphi(t_{k+s}) = A(s)\varphi(t_k) + B(s)\varphi(t_{k+1}) + h_k\{C(s)\varphi'(t_k) + D(s)\varphi'(t_{k+1})\} + \mathcal{O}(h_k^{\varrho_k}),$$
$$\varphi'(t_{k+s}) = \frac{1}{h_k}\{A'(s)\varphi(t_k) + B'(s)\varphi(t_{k+1})\} + C'(s)\varphi'(t_k) + D'(s)\varphi'(t_{k+1}) + \mathcal{O}(h_k^{\varrho_k'})$$

(where ϱ_k, ϱ_k' depend on the smoothness of φ on $[t_k, t_{k+1}]$) generate such approximations on omitting the Landau order terms. We obtain piecewise-constant, piecewise-linear, piecewise-quadratic and piecewise-cubic approximations interpolating $\varphi(t)$ at $\{t_k\}$ if, respectively, $A(s) = 1, B(s) = C(s) = D(s) = 0$; $A(s) = 1 - s, B(s) = s, C(s) = D(s) = 0$;

$$A(s) = 1, \quad B(s) = 0, \quad C(s) = (s - \tfrac{1}{2}s^2), \quad D(s) = \tfrac{1}{2}s^2 \tag{3.8}$$

and (the piecewise cubic case)

$$A(s) = \{1 + 2s\}(1 - s)^2, \qquad B(s) = 1 - A(s),$$
$$C(s) = s(1 - s)^2, \qquad D(s) = s - C(s). \tag{3.9}$$

As a simple low-order method, we might employ the Euler formula ($\theta = 0$ in the θ-method) with piecewise-linear interpolation for both $\tilde{y}(t_{k+s})$ and $\widetilde{y'}(t_{k+s})$. In this manner, we obtain

$$\tilde{y}(t_{n+1}) := \tilde{y}_n + h_n\tilde{F}_n, \tag{3.10a}$$

$$\begin{aligned}\tilde{y}(t_{k+s}) &:= \tilde{y}_k + h_n\{(1-s)\tilde{F}_k + (1-s)\tilde{F}_{k+1}\}\\ &= (1-s)\tilde{y}_k + s\tilde{y}_{k+1} \quad (k \leqslant n,\ s \in [0,1]),\end{aligned} \tag{3.10b}$$

$$\widetilde{y'}(t_{k+s}) := (1-s)\tilde{F}_k + s\tilde{F}_{k+1} \quad (k \leqslant n,\ s \in [0,1]). \tag{3.10c}$$

An alternative to the expression in (3.10c) is, of course, the derivative obtained from (3.10b), namely

$$\{\tilde{y}_{k+1} - \tilde{y}_k\}/h, \quad s \in [0,1). \tag{3.10d}$$

A method based on (3.10d) will, in general, have different properties from one based on (3.10c).

For each choice of $\{A(s), B(s), C(s), D(s)\}$, we obtain extensions of the form $\tilde{y}(t_{k+s}) = A(s)\tilde{y}_k + B(s)\tilde{y}_{k+1} + h_k\{C(s)\widetilde{y'}_k + D(s)\widetilde{y'}_{k+1}\}$ (and similarly for $\widetilde{y'}(t_{k+s})$) if we set $\varphi(t_\ell)$ to $\tilde{y}_\ell$, and $\varphi'(t_\ell)$ to $\tilde{F}_\ell$. By (3.5), $\tilde{y}_{k+1}$ can be eliminated and (since $A(s)+B(s)=1$) we deduce, given θ, approximations of the type

$$\tilde{y}(t_k + sh_k) \equiv \tilde{y}(t_{k+s}) = \tilde{y}_k + h_k\{\mathfrak{b}_1(s)\tilde{F}_k + \mathfrak{b}_2(s)\tilde{F}_{k+1}\}, \tag{3.11a}$$

$$\widetilde{y'}(t_k + sh_k) \equiv \widetilde{y'}(t_{k+s}) = \mathfrak{c}_1(s)\tilde{F}_k + \mathfrak{c}_2(s)\tilde{F}_{k+1}. \tag{3.11b}$$

All choices above give *local* approximations on $[t_k, t_{k+1}]$ using computed values of $\tilde{y}(\cdot)$ and $\widetilde{y'}(\cdot)$ at arguments in $[t_k, t_{k+1}]$. For convenience, we suppose we take one type of approximation consistently for all k.

If we combine (3.9) and (3.6), we find $\mathfrak{b}_1(s) = s - \frac{1}{2}s^2$, $\mathfrak{b}_2(s) = \frac{1}{2}s^2$. The same result arises on using (3.8). If we differentiate (3.9) and use (3.6), we find $\mathfrak{c}_1(s) = 1 - s$, $\mathfrak{c}_2(s) = s$. Here, $\mathfrak{c}_\ell(s) = \mathfrak{b}'_\ell(s)$, but this is not (cf. (3.10b)–(3.10c)) essential in general. The trapezium rule gives, with the chosen extensions,

$$\tilde{y}(t_{n+s}) := \tilde{y}_n + h_n\{(s - \tfrac{1}{2}s^2)\tilde{F}_n + \tfrac{1}{2}s^2 F(t_{n+1}, \tilde{y}_{n+1}, \tilde{y}(\tilde{\alpha}(t_{n+1})), \tilde{y}'(\tilde{\beta}(t_{n+1})))\}, \tag{3.12a}$$

$$\widetilde{y'}(t_{k+s}) := (1-s)\tilde{F}_k + s\tilde{F}_{k+1} \quad (k = 0, 1, \ldots, n). \tag{3.12b}$$

Note that we have $\tilde{y}(t_\ell) = \tilde{y}_\ell$, $\tilde{y}'(t_\ell) = \tilde{F}_\ell$, for $\ell \in \{k, k+1\}$. Approximation theory yields (*inter alia*) the following results related to (3.12).

Lemma 3.1. *If, for* $r \in \{0,1,2\}$, $y(\cdot) \in C^{2-r}[t_k, t_{k+1}]$ *is Lipschitz continuous* (*in particular if* $y(\cdot) \in C^{3-r}[t_k, t_{k+1}]$) *and if*

$$y(t_{k+s}) = y(t_k) + h_k\{(s - \tfrac{1}{2}s^2)y'(t_k) + \tfrac{1}{2}s^2 y'(t_{k+1})\} + \eta_{h_k}(y; s), \tag{3.13a}$$

$$y'(t_{k+s}) = (1-s)y'(t_k) + sy'(t_{k+1}) + \eta^\natural_{h_k}(y; s), \tag{3.13b}$$

then $\sup_{s\in[0,1]} |\eta_{h_k}(y;s)| = \mathcal{O}(h_k^{3-r})$, $\sup_{s\in[0,1]} |\eta^\natural_{h_k}(y;s)| = \mathcal{O}(h_k^{2-r})$.

Our discussion illustrates the general approach to adapting an ODE solver to treat a DDE or NDDE using auxiliary approximations to compute solution or derivative values at retarded arguments and there exists a wide choice for the latter. We progress from the θ-methods to consider *RK methods*. The ODE literature contains examples of *continuous RK formulae* that incorporate an inbuilt method for generating dense output. Such a formula is generated by the *continuous RK triple* $(\boldsymbol{\vartheta}, \boldsymbol{A}, \boldsymbol{b}(s))$ featured, with an example, in the tableau in (3.14):

$$\begin{array}{c|c} \boldsymbol{\vartheta} & \boldsymbol{A} \\ \hline s & \boldsymbol{b}^{\mathrm{T}}(s), \end{array} \quad \text{for example,} \quad \begin{array}{c|cc} 0 & 0 & \\ 1 & \frac{1}{2} & \frac{1}{2} \\ \hline s & s - \frac{1}{2}s^2 & \frac{1}{2}s^2 \end{array}, \quad s \geqslant 0. \tag{3.14}$$

We have $A = [a_{i,j}] \in \mathbb{R}^{m\times m}$, $\boldsymbol{b}(s) = [b_1(s), b_2(s), \ldots, b_m(s)]^{\mathrm{T}}$ and $\boldsymbol{\vartheta} = [\vartheta_1, \vartheta_2, \ldots, \vartheta_m]^{\mathrm{T}} \in \mathbb{R}^m$. The RK parameters are "*formally explicit*" if $a_{i,j} = 0$ for $j \geqslant i$ and will be called "*local*" if $\vartheta_i \in [0,1]$ for all i. An RK triple allows one to obtain a formula for the numerical solution of a DDE.

For an NDDE one requires a further vector of weights $\boldsymbol{c}^{\mathrm{T}}(s)$, conveniently but not necessarily taken as the derivative of $\boldsymbol{b}^{\mathrm{T}}(s)$ (cf. (3.11), but note (3.10b), (3.10c)). Such parameters define a *RK-quadruple* $(\boldsymbol{\vartheta}, \boldsymbol{A}, \boldsymbol{b}(s), \boldsymbol{c}(s))$ corresponding to an augmented tableau:

$$\begin{array}{c|c} \boldsymbol{\vartheta} & \boldsymbol{A} \\ \hline s & \boldsymbol{b}^{\mathrm{T}}(s) \\ \hline & \boldsymbol{c}^{\mathrm{T}}(s), \end{array} \quad \text{for example,} \quad \begin{array}{c|cc} 0 & 0 & \\ 1 & \frac{1}{2} & \frac{1}{2} \\ \hline s & s - \frac{1}{2}s^2 & \frac{1}{2}s^2 \\ \hline & 1-s & s \end{array}, \quad s \geqslant 0. \tag{3.15}$$

We write $t_{n,i} := t_n + \vartheta_i h_n$, $\tilde{y}_{n,i} := \tilde{y}(t_{n,i})$ (also, as above, $t_{n+s} = t_n + sh_n$ for $s \in [0,1]$), and the continuous RK discretization of (3.1) is given by

$$\tilde{y}(t_{n+s}) := \tilde{y}_n + h_n \sum_{j=1}^{m} b_j(s)\tilde{F}_{n,j}, \tag{3.16a}$$

$$\widetilde{y'}(t_{n+s}) := \sum_{j=1}^{m} c_j(s)\tilde{F}_{n,j}, \tag{3.16b}$$

$$\tilde{y}_{n,i} = \tilde{y}_n + h_n \sum_{j=1}^{m} a_{i,j}\tilde{F}_{n,j}, \qquad \tilde{F}_{n,j} = F(t_{n,j}, \tilde{y}_{n,j}, \tilde{y}(\tilde{\alpha}(t_{n,j})), \widetilde{y'}(\tilde{\beta}(t_{n,j}))). \tag{3.16c}$$

Eqs. (3.16c) together constitute equations to be solved for $\{\tilde{F}_{n,i}\}_{i=1}^{m}$ for substitution in (3.16a), (3.16b). For compatibility, we evaluate past values using

$$\tilde{y}(\tilde{\alpha}(t_{n,j})) = \tilde{y}_{k_{n,j}} + h_{k_{n,j}} \sum_{\ell=1}^{m} b_\ell(s_{n,j})\tilde{F}_{k_{n,j},\ell} \quad (\tilde{\alpha}(t_{n,j}) = t_{k_{n,j}} + s_{n,j}h_{k_{n,j}}), \tag{3.16d}$$

$$\widetilde{y'}(\tilde{\beta}(t_{n,j})) = \sum_{\ell=1}^{m} c_\ell(q_{n,j})\tilde{F}_{k'_{n,j},\ell} \quad (\tilde{\beta}(t_{n,j}) = t_{k'_{n,j}} + q_{n,j}h_{k'_{n,j}}). \tag{3.16e}$$

Alongside the general form in (3.15), we gave an illustration (the trapezium rule). That tableau produces a continuous RK method equivalent to Eqs. (3.12a)–(3.12b). A Dormand & Prince RK tableau and an extension due to Shampine are employed in the code `Archi` [71].

There may be complications in the use of (3.16). Eqs. (3.16a), (3.16b) may become implicit if $\tau(t) < h_n$ for some $t \in [t_n, t_{n+1}]$ (*e.g.*, for vanishing lag), even if the RK parameters are formally explicit. If α (or β) is state-dependent ($\tilde{\alpha}(t_{n,j})$ signifies $\hat{\alpha}(t_{n,j}, \tilde{y}(t_{n,j}))$), (3.16) are implicit and must be solved iteratively.

Remark 3.2. For RK methods, one may store $\tilde{y}_0, \{\tilde{F}_{0,j}\}_{j=1}^{m}$; $\tilde{y}_1, \{\tilde{F}_{1,j}\}_{j=1}^{m}$; $\tilde{y}_2, \{\tilde{F}_{2,j}\}_{j=1}^{m}; \cdots; \tilde{y}_n$, $\{\tilde{F}_{n,j}\}_{j=1}^{m}; \ldots$. If the equations are solved iteratively, one records $\tilde{y}_n^{[r]}$ and $\tilde{F}_{n,j}^{[r]}$ where $\tilde{F}_{n,j}^{[r]} := F(t_{n,j}, \tilde{y}_{n,j}^{[r]}, \tilde{y}(\tilde{\alpha}^{[r]}(t_{n,j})), \widetilde{y'}(\tilde{\beta}^{[r]}(t_{n,j})))$ (or alternatively $\tilde{y}_n^{[r+1]}$ and $\tilde{F}_{n,j}^{[r]}$), with $\tilde{\alpha}^{[r]}(t_{n,k}) = \hat{\alpha}(t, \tilde{y}^{[r]}(t_{n,k}))$, and $\tilde{\beta}^{[r]}(t_{n,k}) = \hat{\beta}(t, \tilde{y}^{[r]}(t_{n,k}))$ $(r \equiv r_n)$.

4. Approximate solutions — existence, uniqueness, convergence

Consider a sequence $\mathfrak{T} = \{\mathcal{T}^{[0]}, \mathcal{T}^{[1]}, \mathcal{T}^{[2]}, \ldots\}$ of meshes (3.2) whose widths $\{h_{\max}^{[0]} \geqslant h_{\max}^{[1]} \geqslant h_{\max}^{[2]}, \ldots\}$ tend to zero. The approximation $\tilde{y}(t)$, if it exists, denotes $\tilde{y}(\mathcal{T}^{[m]}; t)$ for some $\mathcal{T}^{[m]} \in \mathfrak{T}$ with width $h_{\max}^{[m]} \equiv h_{\max}(\mathcal{T}^{[m]})$.

Definition 4.1. Given $\tilde{y}(t) \equiv \tilde{y}(\mathscr{T}^{[m]};t)$, the error on the grid points and the global error, are, respectively,

$$e^{[m]}([t_0,t^{\max}]) := \sup\{|y(t_n) - \tilde{y}(t_n)| \text{ for } t_n \in \mathscr{T}^{[m]} \cap [t_0,t^{\max}]\},$$

$$\mathfrak{e}^{[m]}([t_0,t^{\max}]) := \sup\{|y(t) - \tilde{y}(t)| \text{ for } t \in [t_0,t^{\max}]\}.$$

For fixed $t^{\max} < \infty$, the approximations $\tilde{y}(\cdot)$ are (i) *convergent on grid-points in* $[t_0,t^{\max}]$ if $\lim_{m\to\infty} e^{[m]}([t_0,t^{\max}]) = 0$, (ii) *convergent of order* p *on grid-points in* $[t_0,t^{\max}]$ if $e^{[m]}([t_0,t^{\max}]) = \mathcal{O}((h^{[m]}_{\max})^p)$ as $m \to \infty$, and (iii) *convergent of order* ϱ *on the interval* $[t_0,t^{\max}]$ if $\mathfrak{e}^{[m]}([t_0,t^{\max}]) = \mathcal{O}((h^{[m]}_{\max})^\varrho)$ as $m \to \infty$.

Clearly, $\varrho \leqslant p$; there are cases where $\varrho < p$. Further, there are cases where the order of convergence depends upon the sequence of meshes $\{\mathscr{T}^{[m]}\}$. (Consider the method defined by (3.6) together with the low-order extension $\tilde{y}(t_{n+s}) := \tilde{y}(t_n)$ for $s \in (0,1)$. Apply it to $y'(t) = y(t-1)$ with $t_i = t_0 + ih$ and uniform step h: compare the case $\mathscr{T}^{[m]} \in \mathfrak{T}$ if $h = 1/m$ ($m \in \mathbb{N}$), where no extension is needed, and the case $\mathscr{T}^{[m]} \in \mathfrak{T}$ if $h = 1/(\sqrt{2}m)$, $m \in \mathbb{N}$.) One *caveat* is in order: theories that apply for small discretization parameters ("as $h \to 0$") may not give the desired insight in real life (e.g., if h is large relative to the time-scale of the problem). The concept of stiff order in the study of ODEs (where practical step sizes do not produce asymptotically correct convergence behaviour) reflects this observation.

We illustrate some more general results by reference to method (3.12) for (3.1). Recall that $t_{n+s} = t_n + sh_n$, for $s \in (0,1]$. We use (if the solution $y(\cdot)$ and the approximation $\tilde{y}(\cdot)$ exist and are unambiguously defined) the notation

$$F_{n+s} = F(t_{n+s}, y(t_{n+s}), y(\alpha(t_{n+s})), y'(\beta(t_{n+s}))), \tag{4.1a}$$

$$\tilde{F}_{n+s} = F(t_{n+s}, \tilde{y}(t_{n+s}), \tilde{y}(\tilde{\alpha}(t_{n+s})), \widetilde{y'}(\tilde{\beta}(t_{n+s})))). \tag{4.1b}$$

Henceforth, we presume the existence of a unique true solution $y(t) \equiv y(\psi, t_0; t)$ but discuss existence of $\tilde{y}(\cdot)$. We detect a difficulty on considering (2.12) with an initial function $\psi_\star(t)=1$, $\psi'_\star(t)=0$ and solution $y(t) = 1$. Taking a uniform mesh with $h_n = h$ and setting $\Lambda := \Lambda_\star h$, (3.12) gives, with $\tilde{y}(0) = 1$, a quadratic equation for $\tilde{y}(h)$ of which one solution is $\tilde{y}_1 = 1$ corresponding to which $\tilde{y}(t) = 1$, $\widetilde{y'}(t) = 0$ for $t \leqslant h$. By consistently taking the "appropriate" root, we have as one solution $\tilde{y}(t)=1$, $\widetilde{y'}(t)=0$ for $t \leqslant nh$. At the nth stage the equation for $\tilde{y}_{n+1}$ is then $\frac{1}{2}\Lambda\tilde{y}^2_{n+1} + (1-\Lambda)\tilde{y}_{n+1} - 1 = 0$. (The path from $\tilde{y}_n$ to $\tilde{y}_{n+1}$ is in general multi-valued though the existence of a real sequence $\{\tilde{y}_\ell\}_{\ell\geqslant 0}^{\infty}$ on every path is not guaranteed.) The preceding quadratic equation has as its solutions $\tilde{y}^{\pm}_{n+1} = \{\Lambda - 1 \pm |1+\Lambda|\}/\{2\Lambda\}$, namely the value 1 (the true solution) and the value $-1/\Lambda$ (which becomes infinite as $h \to 0$). This example suggests that *an implicit formula, such as* (3.12), *if applied to a nonlinear NDDE, may give rise to a multi-valued approximation* $\tilde{y}(t)$ *of which only some branches are defined for all* $t \geqslant t_0$ *and also satisfy* $\lim_{h\to 0} \sup_{t\in[t_0,t^{\max}]} |\tilde{y}(t) - y(t)| = 0$.

Suppose $\tilde{y}(t)$ exists for $t \in [t_{\min}, t_n]$. Now, $\tilde{y}_{n+1} = \tilde{y}(t_{n+1})$, if it exists, satisfies $\tilde{y}_{n+1} = \varphi_n(\tilde{y}_{n+1})$ where (with notation (3.3))

$$\varphi_n(x) := \tilde{y}_n + \tfrac{1}{2}h_n\tilde{F}_n + \tfrac{1}{2}h_n F(t_{n+1}, x, \tilde{y}(\hat{\alpha}(t_{n+1},x)), \widetilde{y'}(\hat{\beta}(t_{n+1},x))). \tag{4.2}$$

The function $\varphi_n(\cdot)$ satisfies a global Lipschitz condition with constant $\frac{1}{2}h_n L$ if, with $\Phi(x) := F(t_{n+1}, x, \tilde{y}(\hat{\alpha}(t_{n+1},x)), \widetilde{y'}(\hat{\beta}(t_{n+1},x)))$, we have $|\Phi(x') - \Phi(x'')| \leqslant L|x' - x''|$ uniformly for all x', x''.

A line of enquiry is to discover reasonable conditions that determine a suitable L; we can then use the following lemma which may be found in a student text.

Lemma 4.2 (Fixed-point iteration). *Consider the general fixed-point iteration* $x_{k+1}=\phi(x_k)$. *Suppose (given* $\varepsilon>0$) *that* $\phi(\cdot)$ *continuous on* $[x_\star-\varepsilon,x_\star+\varepsilon]$ *and* $|\phi(x')-\phi(x'')|\leqslant\Lambda|x'-x''|$ *for all* $x',x''\in[x_\star-\varepsilon,x_\star+\varepsilon]$ *where* $0\leqslant\Lambda<1$ *and let* $x_0\in[x_\star-\varepsilon,x_\star+\varepsilon]$ *be such that* $|x_0-\phi(x_0)|\leqslant(1-\Lambda)\varepsilon$. *Then* $\lim_{r\to\infty}x_r$ *exists and is the only fixed point of* $\phi(\cdot)$ *lying in* $[x_\star-\varepsilon,x_\star+\varepsilon]$.

If the *global* Lipschitz condition holds, we can apply this result to the iteration $x_{r+1}=\varphi_n(x_r)$, with arbitrary x_0 and ε as large as necessary, by ensuring (say) that $h_nL<1$, so there exists a unique value $\tilde{y}(t_{n+1})$ for all h_n sufficiently small, and a unique $\tilde{y}(t)$ exists for $t\in[t_{\min},t_{n+1}]$. We have (when $h^{[m]}_{\max}$ is sufficiently small) the basis of a proof, by induction, of the existence of a unique $\tilde{y}(t)$ for all $t\in[t_{\min},t^{\max}]$. To bound the error, compare (3.12) with

$$y(t_{n+s})=y(t_n)+h_n\{(s-\tfrac{1}{2}s^2)F_n+\tfrac{1}{2}s^2F_{n+1}\}+\eta_{h_n}(y;s), \tag{4.3a}$$

$$y'(t_{k+s})=(1-s)F_k+sF_{k+1}+\eta^{\natural}_{h_k}(y;s)\quad(k=0,1,\ldots,n), \tag{4.3b}$$

use (cf. Eq. (3.13)) theoretical bounds on $\{\eta_{h_n}(y;s),\eta^{\natural}_{h_k}(y;s)\}$, apply Lipschitz conditions and use inequalities familiar in the numerics of ODEs.

4.1. Theory based on local Lipschitz conditions

Global conditions are not always realistic and we will request local Lipschitz conditions (this material appears to be a novel extension of previously published results). We sacrifice generality by imposing conditions on the neutral term: in particular, we suppose β to be state-independent and consider

$$y'(t)=F(t,y(t),y(\alpha(t,y(t))),y'(\beta(t)))\quad\text{for } t\geqslant t_0. \tag{4.4}$$

Thus, *scalar state-dependent DDEs* and *state-independent NDDEs* are covered by our discussion. (The rather intricate details that we supply allow the reader to construct the extension to state-dependent $\beta(t,y(t))$ if Lipschitz conditions apply globally.)

Our main task is to use conditions on the discretization error to establish, for $h^{[m]}_{\max}$ sufficiently small, the existence of $\tilde{y}(t)$, $\widetilde{y'}(t)$ in a region in which local Lipschitz conditions apply. We focus on formulae (3.12), and use the notation $\eta_{h_k}(y;s)$, $\eta^{\natural}_{h_k}(y;s)$ in Lemma 3.1 and

$$\eta_n(y):=\sup_{\ell\in\{0,1,\ldots,(n-1)\}}|\eta_{h_\ell}(y)|,\qquad \eta_{h_\ell}(y)=\sup_{s\in[0,1]}|\eta_{h_\ell}(y;s)|, \tag{4.5a}$$

$$\eta^{\natural}_n(y):=\sup_{\ell\in\{0,1,\ldots,(n-1)\}}|\eta^{\natural}_{h_\ell}(y)|,\qquad \eta^{\natural}_{h_\ell}(y)=\sup_{s\in[0,1]}|\eta^{\natural}_{h_\ell}(y;s)| \tag{4.5b}$$

and, for a given $\omega\in(0,\infty)$,

$$\eta^{\omega}_{h_\ell}(y):=|\eta_{h_\ell}(y)|+h_\ell\omega|\eta^{\natural}_\ell(y)|. \tag{4.5c}$$

To impose *local* conditions for u, w we define the neighbourhoods

$$\mathscr{D}(\varepsilon,\varepsilon_1;s):=\{(s,u,w);\ |u-y(s)|\leqslant\varepsilon(s),|w-y'(s)|\leqslant\varepsilon_1(s)\} \tag{4.6}$$

for $s \in [t_0, t^{\max}]$ ($\varepsilon(\cdot)$, $\varepsilon_1(\cdot)$ continuous and strictly positive on $[t_0, t^{\max}]$). We ask (i) that $\sup_{t\in[t_{\min},t^{\max}]} |y(t)| \leqslant M$, $\sup_{t\in[t_{\min},t^{\max}]} |y'(t)| \leqslant M$, (ii) that $y'(\cdot)$ should be Lipschitz-continuous on $[t_{\min}, t^{\max}]$ with Lipschitz constant M; (iii) (uniform boundedness of F) that

$$|F(t, u(t), u(\alpha(t, u(t))), w(\beta(t)))| \leqslant M \tag{4.7}$$

for $t \in [t_0, t^{\max}]$ when $|u(t) - y(t)| \leqslant \varepsilon(t)$ and $|w(\sigma) - y'(\sigma)| \leqslant \varepsilon_1(\sigma)$ for $\sigma \in [t_0, t]$, $t_0 \leqslant t \leqslant t^{\max}$. We require (iv) Lipschitz conditions $|\alpha(\sigma, u) - \alpha(\sigma, U)| \leqslant \mu_\alpha |u - U|$, $|F(\sigma, U, v, w) - F(\sigma, u, v, w)| \leqslant \mu_2 |U - u|$, $|F(\sigma, u, V, w) - F(\sigma, u, v, w)| \leqslant \mu_3 |V - v|$, $|F(\sigma, u, v, W) - F(\sigma, u, v, w)| \leqslant \mu_4 |W - w|$. *We ask that* $\mu_4 < 1$ (*cf. Theorem* 2.2, *with the condition* $L_2 < 1$). The conditions hold for $u, U \in [y(\sigma) - \varepsilon(\sigma), y(\sigma) + \varepsilon(\sigma)]$, $v, V \in [y(\sigma_\alpha) - \varepsilon_\alpha, y(\sigma_\alpha) + \varepsilon_\alpha]$ ($\varepsilon_\alpha = \varepsilon(s_\alpha)$), and $w, W \in [y'(\sigma_\beta) - \varepsilon_\beta, y'(\sigma_\beta) + \varepsilon_\beta]$, with $\varepsilon_\beta = \varepsilon_1(\sigma_\beta)$, whenever $\sigma_\alpha, \sigma_\beta \leqslant \sigma$, and $\sigma \in [t_0, t^{\max}]$. We ask (if necessary by a redefinition) that when $t \in [t_n, t_{n+1}]$,

$$\varepsilon_1(t) \geqslant \mu \sup_{\sigma\in[t_0,t_{n+1}]} \varepsilon(\sigma) + \varepsilon_\star, \quad \text{where } \varepsilon_\star > 0, \tag{4.8}$$

where $\mu = \omega\mu_\star$, with $\mu_\star = \mu_2 + \mu_3 + M\mu_3\mu_\alpha$, $\omega = (1 - \mu_4)^{-1}$. In relation to $\mathscr{T}$, we also ask for continuity of $y''(t)$ on each interval $[t_k, t_{k+1}]$ and that $h^{[m]}_{\max}\mu < 1$, $h^{[m]}_{\max}L < 1$, where $L = \mu_2 + \mu_3\mu_\alpha M$. Setting $\hat{\mu} := \mu_3 + \mu(1 + \mu_4)$ and $\delta_n = \{h_n \eta^\natural_{h_n}(y) + \eta_{h_n}(y; 1)\}$ we require that, for $n = 0, 1, \ldots,$

$$2\exp\{2\mu(t_n - t_0)\} \sum_{\ell=0}^{n} \eta^\omega_{h_\ell} \leqslant \varepsilon(t) \quad \text{when } t \in [t_n, t_{n+1}], \tag{4.9a}$$

$$\left(1 - \frac{1}{2}h_n L\right) \left\{(1 + \hat{\mu}h_n)\exp\{2\mu(t_n - t_0)\} \sum_{k=0}^{n-1} \eta_{h_k}(y) + |\delta_n|\right\} \leqslant \varepsilon(t_{n+1}) \tag{4.9b}$$

and

$$\omega \sup_{t_\ell \leqslant t^{\max}} |\eta^\natural_{h_\ell}(y)| \leqslant \varepsilon_\star. \tag{4.9c}$$

Conditions (4.9) are satisfied on taking $h^{[m]}_{\max}$ sufficiently small, since $\eta^\omega_{h_\ell}(y) = \mathcal{O}(h_\ell h_{\max})$, and $\sum_k \eta^\omega_{h_k} = \{\sum_k h_k\}\,\mathcal{O}(h^{[m]}_{\max})$. To simplify, we assume $h^{[m]}_{\max}$ is smaller than a "minimum possible lag":

$$0 < \max_\ell h_\ell < \inf_{t\in[t_0,t^{\max}]} \min\left(\inf_{|u-y(t)|\leqslant\varepsilon(t)} (t - \alpha(t, u)), t - \beta(t)\right). \tag{4.10}$$

Then, $\tilde{\alpha}(t_{\ell+1}) \leqslant t_\ell$, $\tilde{\beta}(t_{\ell+1}) \leqslant t_\ell$ for $\ell \in \mathbb{N}$. (If $\alpha(t) = t - \tau(t)$, $\beta(t) = t - \zeta(t)$ with $\xi := \min\{\inf_t \tau(t), \inf_t \zeta(t)\} > 0$ it suffices to take $h^{[m]}_{\max} < \xi$.)

Theorem 4.3. *If all the preceding assumptions hold, there is a unique approximate solution* $\tilde{y}(t)$ *defined on* $[t_0, t^{\max}]$ *and satisfying* $|y(t) - \tilde{y}(t)| \leqslant \varepsilon(t)$. *Furthermore,* $\sup_{[t_0,t^{\max}]} |y(t) - \tilde{y}(t)| \to 0$ *as* $h^{[m]}_{\max} \to 0$ (*that is, as* $m \to \infty$).

4.1.1. The proof of Theorem 4.3

We shall indicate a proof of the above, via some lemmas. When brevity requires, we write

$$e(t) = y(t) - \tilde{y}(t) \quad \text{and} \quad e_1(t) = y'(t) - \tilde{y}'(t), \tag{4.11a}$$

if defined. For $r \in \mathbb{N}$, $\mathfrak{I}_r$ will denote the set

$$\mathfrak{I}_r := [t_{\min}, t_r] \cup \{t_{r+1}\}. \tag{4.11b}$$

Lemma 4.4. *Suppose, for some* $r \in \mathbb{N}$, *that there exists a unique* $\tilde{y}(t)$ *that satisfies* $|y(t)-\tilde{y}(t)| \leqslant \varepsilon(t)$ *for* $t \in \mathfrak{I}_r$, *and* $|y'(t) - \tilde{y}'(t)| \leqslant \varepsilon_1(t)$ *for* $t \in [t_{\min}, t_r]$. *Then* (a) $\tilde{y}(t)$, $\widetilde{y'}(t)$ *exist on* $[t_{\min}, t_{r+1}]$, *and* $|\tilde{y}'(t)| \leqslant M$ *for* $t \in [t_{\min}, t_{r+1}]$. (b) *With* $\mu = \omega\mu_\star$ *defined above, and* $\ell = 0, 1, \ldots, r$,

$$\sup_{t\in[t_0,t_{\ell+1}]} |y'(t) - \tilde{y}'(t)| \leqslant \mu \sup_{t\in[t_0,t_{\ell+1}]} |y(t) - \tilde{y}(t)| + \omega|\eta^{\natural}_{\ell+1}(y)| \tag{4.12}$$

and, uniformly for $\ell \in \{0, 1, \ldots, r\}$,

$$\sup_{t\in[t_0,t_{\ell+1}]} |y(t) - \tilde{y}(t)| \leqslant 2\exp\{2\mu(t_\ell - t_0)\} \sum_{k=0}^{\ell} \eta^{\omega}_{h_\ell}(y) = \mathcal{O}(h^{[m]}_{\max}). \tag{4.13}$$

Finally, $|y(t) - \tilde{y}(t)| \leqslant \varepsilon(t)$, *and* $|y'(t) - \widetilde{y'}(t)| \leqslant \varepsilon_1(t)$ *for* $t \in [t_0, t_{r+1}]$.

Proof. We outline a proof of the above, but first note that a similar result (the value of μ is changed) holds even if β is state-dependent. By assumption, the trapezium-rule approximation $\tilde{y}(t)$ is unambiguously defined for $t \in [t_0, t_r]$ and at t_{r+1}, so $\tilde{F}_\ell$ is defined for $\ell = 0, 1, \ldots, (r+1)$ and hence there exists a unique twice-differentiable approximation $\tilde{y}(t)$ on $[t_{\min}, t_{r+1}]$ satisfying $|y(t)-\tilde{y}(t)| \leqslant \varepsilon(t)$ on $[t_0, t_r] \cup \{t_{r+1}\}$, and $\widetilde{y'}(t_\ell) = \tilde{F}_\ell \equiv F(t_\ell, \tilde{y}(t_\ell), \tilde{y}(\hat{\alpha}(t_\ell, \tilde{y}(t_\ell))), \widetilde{y'}(\beta(t_\ell)))$ if $\ell \leqslant r+1$ (cf. (3.3)). By (3.12b) with (4.7), $|\widetilde{y'}(t)| \leqslant M$ $(t \in [t_0, t_{r+1}])$ and (a) follows.

Now $y'(t_\ell) = F_\ell \equiv F(t_\ell, y(t_\ell), y(\alpha(t_\ell, y(t_\ell))), y'(\beta(t_\ell)))$, and $\widetilde{y'}(t_\ell) = \tilde{F}_\ell$ as above, and $|e_1(t_\ell)| = |F_\ell - \tilde{F}_\ell|$. If we use $|y(\alpha(t_\ell, y(t_\ell))) - \tilde{y}(\hat{\alpha}(t_\ell, \tilde{y}(t_\ell)))| \leqslant |y(\alpha(t_\ell, y(t_\ell))) - y(\hat{\alpha}(t_\ell, \tilde{y}(t_\ell)))| + |y(\hat{\alpha}(t_\ell, \tilde{y}(t_\ell))) - \tilde{y}(\hat{\alpha}(t_\ell, \tilde{y}(t_\ell)))| \leqslant M|\alpha(t_\ell, y(t_\ell)) - \alpha(t_\ell, \tilde{y}(t_\ell))| + |e(\hat{\alpha}(t_\ell, \tilde{y}(t_\ell))|$ we find

$$|e_1(t_\ell)| \leqslant \mu_2|e(t_\ell)| + \mu_3\left\{\sup_{t\in[t_0,t_{\ell-1}]} |e(t)| + \mu_\alpha M|e(t_\ell)|\right\} + \mu_4 \sup_{t\in[t_0,t_{\ell-1}]} |e_1(t)|. \tag{4.14}$$

If $s \in [0, 1]$, then $|e_1(t_{\ell+s})| \leqslant (1-s)|e_1(t_\ell)| + s|e_1(t_{\ell+1})| + |\eta^{\natural}_\ell(y; s)| \leqslant \max_{j\in\{\ell,\ell+1\}} |e_1(t_j)| + |\eta^{\natural}_{h_\ell}(y; s)|$. Thus, with $\ell \in \{0, 1, \ldots, r\}$, $\mathfrak{I}_\ell \equiv [t_0, t_\ell] \cup \{t_{\ell+1}\}$, $\mu_\star = \mu_2 + \mu_3 + \mu_3\mu_\alpha M$, $0 < \mu_4 < 1$,

$$|e_1(t_{\ell+s})| \leqslant \mu_\star \sup_{t\in\mathfrak{I}_\ell} |e(t)| + \mu_4 \sup_{t\in[t_0,t_\ell]} |e_1(t)| + |\eta^{\natural}_{h_\ell}(y; s)|. \tag{4.15}$$

From (4.15) we deduce $\sup_{t\in[t_0,t_{\ell+1}]} |e_1(t)| \leqslant \mu_4 \sup_{t\in[t_0,t_{\ell+1}]} |e_1(t)| + \mu_\star \sup_{t\in\mathfrak{I}_\ell} |e(t)| + |\eta^{\natural}_{\ell+1}(y)| \leqslant \mu_4 \sup_{t\in[t_0,t_{\ell+1}]} |e_1(t)| + \mu_\star \sup_{t\in[t_0,t_{\ell+1}]} |e(t)| + |\eta^{\natural}_{\ell+1}(y)|$, and so (4.12) follows.

For $\ell = 0, 1, \ldots, r$, $|y(t_{\ell+s}) - \tilde{y}(t_{\ell+s})| \leqslant |y(t_\ell) - \tilde{y}(t_\ell)| + h_\ell\{(s - \frac{1}{2}s^2)|e_1(t_\ell)| + \frac{1}{2}s^2|e_1(t_{\ell+1})|\} + |\eta_{h_\ell}(y)| \leqslant |y(t_\ell) - \tilde{y}(t_\ell)| + \mu h_\ell\{(s - \frac{1}{2}s^2) \sup_{t\in[t_0,t_\ell]} |e(t)| + \frac{1}{2}s^2 \sup_{t\in[t_0,t_{\ell+1}]} |e(t)|\} + |\eta^{\omega}_{h_\ell}(y)|$. We deduce that $(1 - \frac{1}{2}\mu h_\ell) \sup_{t\in[t_\ell,t_{\ell+1}]} |e(t)| \leqslant (1 + \frac{1}{2}\mu h_\ell) \sup_{t\in[t_0,t_\ell]} |e(t)| + \eta^{\omega}_{h_\ell}(y)$, and, if $\mu h_\ell < 1$,

$$\sup_{t\in[t_0,t_{\ell+1}]} |e(t)| \leqslant (1 + 2\mu h_\ell) \sup_{t\in[t_0,t_\ell]} |e(t)| + 2\eta^{\omega}_{h_\ell}(y), \tag{4.16}$$

where, by definition, $\eta^{\omega}_{h_\ell}(y) = |\eta_{h_\ell}(y)| + \omega h_\ell|\eta^{\natural}_\ell(y)| = \mathcal{O}(h_\ell h^{[m]}_{\max})$. By induction, since $1 + 2\mu_\star h_\ell \leqslant \exp(2\mu_\star h_\ell)$, result (4.13) follows. The remainder of the lemma follows by reference to (4.9a), (4.9c). □

Consider, with notation (3.3) the Lipschitz-continuity of the functions

$$\varphi_n(x) := \tilde{y}_n + \tfrac{1}{2}h_n\tilde{F}_n + \tfrac{1}{2}h_nF(t_{n+1}, x, \tilde{y}(\hat{\alpha}(t_{n+1}, x)), \widetilde{y'}(\hat{\beta}(t_{n+1}))). \tag{4.17}$$

Lemma 4.5. *Suppose $|y(t)-\tilde{y}(t)|\leqslant\varepsilon(t)$ for $t\in\mathfrak{I}_{n-1}$, and $|y'(t)-\widetilde{y'}(t)|\leqslant\varepsilon_1(t)$ for $t\in[t_{\min},t_{n-1}]$. Then there exists a finite L such that $|\varphi_n(x')-\varphi_n(x'')|\leqslant\frac{1}{2}h_nL|x'-x''|$, whenever $x',x''\in[y(t_{n+1})-\varepsilon(t_{n+1}),y(t_{n+1})+\varepsilon(t_{n+1})]$, $t_{n+1}\in[t_1,t^{\max}]$.*

Proof. We indicate a proof, taking advantage of the state-independence of β, and invoking (4.10). By Lemma 4.4 $|e(t)|\leqslant\varepsilon(t)$, $|e_1(t)|\leqslant\varepsilon_1(t)$, and $|\widetilde{y'}(t)|\leqslant M$, $t\in[t_{\min},t_n]$. Now, $|\tilde{y}(\hat{\alpha}(t_{n+1},x'))-\tilde{y}(\hat{\alpha}(t_{n+1},x''))|\leqslant|x'-x''|\mu_\alpha\sup_{\sigma\leqslant t_n}|\widetilde{y'}(\sigma)|$ giving the bound $\mu_\alpha M|x'-x''|$ for this quantity. By the repeated use of the triangle inequality, we have the stated result for $\varphi_n(\cdot)$ in (4.17), with $L=\mu_2+\mu_3\mu_\alpha M$.

We outline a proof of Theorem 4.3. We consider the stage at which we compute $\tilde{y}(t)$ on $[t_n,t_{n+1}]$ and wish to validate the conditions of Lemma 4.4 with $r=n$, assuming that they (and hence the conclusions of the Lemma) hold with $r=n-1$. To achieve our objective we have to establish the existence of $\tilde{y}(t_{n+1})$ such that $|y(t)-\tilde{y}(t)|\leqslant\varepsilon(t)$ for $t\in\mathfrak{I}_n$, and we apply Lemma 4.2 to $x_{k+1}=\varphi_n(x_k)$. For theoretical purposes, take $x_\star=x_0=y(t_{n+1})$ (which exists by assumption). Now $|e(t)|\leqslant\varepsilon(t)$, $|e_1(t)|\leqslant\varepsilon_1(t)$ $(t\leqslant t_n)$, $\alpha(t_{n+1},y(t_{n+1}))$, $\beta(t_{n+1})\leqslant t_n$; if we set $E_n:=F(t_{n+1},y(t_{n+1}),\tilde{y}(\alpha(t_{n+1},y(t_{n+1}))),\widetilde{y'}(\beta(t_{n+1})))-F(t_{n+1},y(t_{n+1}),y(\alpha(t_{n+1},y(t_{n+1}))),y'(\beta(t_{n+1})))$ then

$$|E_n|\leqslant\left\{\mu_3\sup_{t\in[t_0,t_n]}|y(t)-\tilde{y}(t)|+\mu_4\sup_{t\in[t_0,t_n]}|y'(t)-\widetilde{y'}(t)|\right\},\tag{4.18}$$

Lemma 4.4 allows us to bound (4.18) in terms of $\sup_{t\in[t_0,t_n]}|y(t)-\tilde{y}(t)|$; we have $|E_n|\leqslant\{\mu_3+\mu\mu_4\}\sup_{t\in[t_0,t_n]}|e(t)|+\omega\mu_4|\eta_n^\natural(y)|$. The value $\varphi_n(y(t_{n+1}))$ is

$$\tilde{y}_n+\tfrac{1}{2}h_n\tilde{F}_n+\tfrac{1}{2}h_nF(t_{n+1},y(t_{n+1}),\tilde{y}(\hat{\alpha}(t_{n+1},y(t_{n+1}))),\widetilde{y'}(\hat{\beta}(t_{n+1},y(t_{n+1}))))$$

$$=y(t_n)+\tfrac{1}{2}h_n\{F_n+F_{n+1}\}-e(t_n)+\tfrac{1}{2}h_n\{\tilde{F}_n-F_n+E_n\}\tag{4.19}$$

$$=y(t_n)+\int_{t_n}^{t_{n+1}}y'(s)\,\mathrm{d}s+\eta_{h_n}(y;1)-e(t_n)+\tfrac{1}{2}h_n\{\tilde{F}_n-F_n+E_n\}.$$

$$=y(t_{n+1})+\mathscr{E}_{n+1}\tag{4.20}$$

with $\mathscr{E}_{n+1}=\eta_{h_n}(y;1)-e(t_n)+\frac{1}{2}h_n\{\tilde{F}_n-F_n+E_n\}$. We have, with $\delta_n=\{h_n\eta_{h_n}^\natural(y)+\eta_{h_n}(y;1)\}$ and $\hat{\mu}:=\mu_3+\mu(1+\mu_4)$,

$$|\mathscr{E}_{n+1}|\leqslant|e(t_n)|+\tfrac{1}{2}h_n\{\mu_3+\mu(1+\mu_4)\}\sup_{t\in[t_0,t_n]}|e(t)|+\{\tfrac{1}{2}h_n(1+\mu_4)|\eta_n^\natural(y)|+|\eta_{h_n}(y;1)|\},\tag{4.21}$$

$$\leqslant(1+\tfrac{1}{2}\hat{\mu}h_n)\sup_{t\in[t_0,t_n]}|e(t)|+\{\tfrac{1}{2}h_n(1+\mu_4)\eta_{h_n}^\natural(y)+|\eta_{h_n}(y;1)|\}$$

$$\leqslant(1+\hat{\mu}h_n)\exp\{2\mu(t_{n-1}-t_0)\}\sum_{k=0}^{n-1}\eta_{h_k}^\omega(y)+|\delta_n|,\tag{4.22}$$

$$=\mathscr{O}(h_{\max}^{[m]}).\tag{4.23}$$

Now $|y(t_{n+1})-\varphi_n(y(t_{n+1}))|=|\mathscr{E}_{n+1}|$. In Lemma 4.2, concerning fixed-point iterations, set $x_\star=x_0=y(t_{n+1})$ and $\Lambda=\frac{1}{2}h_nL$ (where $h_n<1/L$ and where L is given by Lemma 4.5), and we determine that there is a unique value $\tilde{y}_{n+1}$ with $\tilde{y}_{n+1}=\varphi_n(\tilde{y}_{n+1})$ and satisfying

$$|\tilde{y}_{n+1}-y(t_{n+1})|\leqslant\varepsilon(t_{n+1})\quad\text{in the case }|\mathscr{E}_{n+1}|\leqslant\{1-\tfrac{1}{2}h_nL\}^{-1}\varepsilon(t_{n+1}).\tag{4.24}$$

We then have $|\tilde{y}(t_{n+s}) - y(t_{n+s})| \leqslant \varepsilon(t_{n+s})$ if, in particular, Eq. (4.9b) holds and, by (4.8), (4.9c) and (4.12), $|\widetilde{y'}(t_{n+s}) - y(t_{n+s})| \leqslant \varepsilon_1(t_{n+s})$, $s \in [0,1]$. □

Theorem 4.3 now follows by induction, since the assumptions in Lemma 4.4(c) follow from our discussion. We note that our results provide a rate of convergence for the error on the mesh-points (Definition 4.1), $\sup_n |y(t_{n+1}) - \tilde{y}(t_{n+1})| = \mathcal{O}(h_{\max}^{[m]2})$, in terms of the global error $\sup_t |y(t) - \tilde{y}(t)| = \mathcal{O}(h_{\max}^{[m]})$. Note that it is useful to have asymptotically correct expansions of the error, in addition to orders of convergence, but convergence can be provable with relaxed assumptions, using Lemma 3.1, with $r \in \{1,2\}$, or a suitable extension.

5. Numerical stability

Stability (with or without qualifying adjectives or acronyms — e.g., "absolute stability", "relative stability", "stiff stability", "A-stability"), and its opposite instability, are amongst the most over-used (perhaps misused) terms in the literature on evolutionary problems. Numerical stability, applied to a class of discretization methods, can refer to *zero stability* as in "stability plus consistency implies convergence". In *strong stability*, dominant terms in the error in $\tilde{y}(t)$ satisfy an equation naturally associated with perturbations in the original problem. In contrast, Simpson's rule for $y'(t) = \lambda y(t)$ with constant step h is "unstable" as it has a spurious "growth parameter" that gives rise to a contribution of the form $-(1 - \frac{1}{3}\lambda h + \mathcal{O}(h^2))^n$ in $\tilde{y}(t_n)$; this clearly has no association with the original problem (and can grow in magnitude for negative λ). For selected λ, μ, it can readily be shown that this analysis extends to Simpson's rule for $y'(t) = \lambda y(t) + \mu y(t - \tau_\star)$ where $h = 1/N$, $N \in \mathbb{N}$; this does not concern us further.

The preponderance of work on numerical stability has related to numerical solutions of DDEs rather than NDDEs. We will provide a definition of stability that parallels Definition 2.3. Concerning initial perturbations one addresses the case where the initial functions are perturbed to $\psi_0(t) + \delta\psi_0(t)$ and $\psi_1(t) + \delta\psi_1(t)$. To simplify, we consider the trapezium rule formulated as in (3.12a)–(3.12b) and leave the reader to generalize. For persistent (or steady acting) perturbations $\varDelta_{0,1}(t)$ we use the notation

$$\tilde{\mathfrak{F}}_k := F(t_k, \tilde{y}_k + \delta\tilde{y}_k, \tilde{y}(\tilde{\alpha}(t_k)) + \delta\tilde{y}(\tilde{\alpha}(t_k)), \widetilde{y'}(\tilde{\beta}(t_k)) + \delta\widetilde{y'}(\tilde{\beta}(t_k)))$$

and consider

$$\tilde{y}(t_{n+s}) + \delta\tilde{y}(t_{n+s}) = \tilde{y}_n + \delta\tilde{y}_n + h_n\{(s - \tfrac{1}{2}s^2)\tilde{\mathfrak{F}}_n + \tfrac{1}{2}s^2\tilde{\mathfrak{F}}_{n+1}\} + \varDelta_0(t_{n+s}), \tag{5.1a}$$

$$\tilde{y}'(t_k + sh_k) = (1-s)\tilde{\mathfrak{F}}_k + s\tilde{\mathfrak{F}}_{k+1} + \varDelta_1(t_{n+s}). \tag{5.1b}$$

Given a vector norm $\|\cdot\|$, we can measure the input perturbations by

$$\tilde{\varDelta} = \sup_{t \geqslant t_0} \max\{\|\varDelta_0(t)\|, \|\varDelta_1(t)\|\} + \max\left\{\sup_{t \leqslant t_0} \|\delta\psi_0(t)\|, \sup_{t \leqslant t_0} \|\delta\psi_1(t)\|\right\}. \tag{5.2}$$

For nonneutral DDEs, $\varDelta_1$ is absent (in general, $\varDelta_1$ could be eliminated by incorporation into a revised definition of $\varDelta_0$). If one wishes to cover numerical perturbations, one does not require the derivative of $\varDelta_0$ to equal $\varDelta_1$ (an unsuitable requirement if $\widetilde{y'}(\cdot) \neq \tilde{y}'(\cdot)$ in the defining extensions), nor even continuity of $\varDelta_0(\cdot), \varDelta_1(\cdot)$. However, one can by suitable choices attempt to simulate the

Table 1
Some of the stability terms in previous literature

Test equation (subject to conditions:)	Conditions ($h_n \equiv h$ fixed)	Type of stability	Reduces to the ODE concept
$y'(t) = \lambda_\star y(t) + \mu_\star y(t-\tau_\star)$ $\lvert\mu_\star\rvert < -\Re(\lambda_\star)$ cf. (2.7)	$\tau_\star/h \in \mathbb{N}$ arbitrary h	P-stability GP-stability	A-stability
$y'(t) = \lambda_\star(t)y(t) + \mu_\star(t)y(t-\tau_\star)$ $\lvert\mu_\star(t)\rvert < -\Re(\lambda_\star(t))$	$\tau_\star/h \in \mathbb{N}$ arbitrary h	PN-stability GPN-stability	AN-stability
A system $y'(t) = \mathfrak{f}(t, y(t), y(t-\tau_\star))$ subject to (5.3)	$\tau_\star/h \in \mathbb{N}$ arbitrary h	RN-stability GRN-stability See (5.4).	BN-stability

perturbations $\delta F(t)$ in Definition 2.3 The accompanying Table 1 presents some of the terms that appear in the literature. The natural definitions of numerical stability are analogues of Definition 2.3, and concern boundedness of $\delta\tilde{y}(\cdot)$ on $[0,\infty)$, and behaviour of $\delta\tilde{y}(t)$ as $t \to \infty$.

Definition 5.1 (Numerical stability). (a) For given admissible perturbations, a solution $\tilde{y}(\cdot)$ of (3.12) is termed (i) *stable*, if for each $\varepsilon > 0$ there exists $\tilde{\Delta}^* = \tilde{\Delta}^*(\varepsilon, t_0) > 0$ such that $\|\delta y(t)\| < \varepsilon$ when $t \geqslant t_0$ for any $\tilde{\Delta} < \tilde{\Delta}^*$; (ii) *asymptotically stable*, if it is stable and, given t_0, there exists $\tilde{\Delta}^\dagger = \tilde{\Delta}^\dagger(t_0)$ such that $\delta\tilde{y}(t) \to 0$ as $t \to \infty$ when $\tilde{\Delta} < \tilde{\Delta}^\dagger$; (iii) *uniformly asymptotically stable*, if the number $\tilde{\Delta}^\dagger$ in definition (ii) is independent of t_0; (iv) *ξ-exponentially stable*, if it is *asymptotically stable* and, given t_0, there exist finite $K = K(t_0)$ and $\tilde{\Delta}^\ddagger = \tilde{\Delta}^\ddagger(t_0)$ such that $\delta\tilde{y}(t) \leqslant K\exp\{-\xi(t-t_0)\}$ (with $\xi > 0$) for $t \geqslant t_0$ when $\tilde{\Delta} < \tilde{\Delta}^\ddagger$. (b) The stability is "*stability with respect to perturbed initial conditions*" if we require that $\delta F(\cdot)$ vanishes, so that the only perturbations are in $\psi_0(\cdot)$, $\psi_1(\cdot)$.

The condition $|\mu_\star(t)| < -\Re(\lambda_\star(t))$ is sufficient to ensure *asymptotic stability* (with respect to perturbed initial perturbations) of every solution of $y'(t) = \lambda_\star(t)y(t) + \mu_\star(t)y(t-\tau_\star)$ for *every* $\tau_\star > 0$. Recalling the concept of contractivity when studying ODEs, we note that nonuniqueness of solutions passing through a point (cf. Fig. 1(b)) makes the definition of contractivity for ODEs inappropriate for DDEs. The condition imposed on $\mathfrak{f}$ in $y'(t) = \mathfrak{f}(t, y(t), y(t-\tau_\star))$ for RN- and GRN-stability is (compare with $|\mu_\star(t)| < -\Re(\lambda_\star(t))$, used above)

$$\sup_{y,x_1 \neq x_2} \frac{\|\mathfrak{f}(t,y,x_1) - \mathfrak{f}(t,y,x_2)\|}{\|x_1 - x_2\|} \leqslant - \sup_{x,y_1 \neq y_2} \frac{\Re\langle(\mathfrak{f}(t,y_1,x) - \mathfrak{f}(t,y_2,x), y_1 - y_2)\rangle}{\|y_1 - y_2\|^2}, \tag{5.3}$$

where $\langle\cdot,\cdot\rangle$ is a vector inner product and $\|u\| := \langle u,u\rangle^{1/2}$. This ensures that

$$\|y(\psi;t) - y(\psi + \delta\psi;t)\| \leqslant \sup_{t \leqslant t_0} \|\delta\psi(t)\| \tag{5.4}$$

and RN- and GRN-stable methods preserve property (5.4) under discretization, with the respective constraints on h; clearly, $\tilde{y}(\cdot)$ is then *stable with respect to initial perturbations*.

Some theoretical tools for analysing stability are Z-transform theory (in the case of constant-coefficient constant-lag DDEs), boundary-locus techniques, linearized stability, the fundamental matrix, comparison theorems, Lyapunov functions and functionals (the work [62] echos the analytical theory in an attractive manner), and Halanay inequalities. It cannot be overemphasized that *such tools complement each other* in providing insight. The stability properties attach to a particular solution $\tilde{y}(t) \equiv \tilde{y}(\mathscr{T}, \psi, t_0; t)$, and as in the analytic case a solution of a nonlinear problem may be unbounded but stable. It is often assumed, without loss of generality, that the null function is a solution so that stability can be defined in terms of stability of the null solution (one considers the effect of "small nonnull initial functions").

For *linear* equations all solutions corresponding to a given $\mathscr{T}$ simultaneously have the same stability properties. We indicate some features by considering the case of the test equation

$$y'(t) = \lambda_\star y(t) + \mu_\star y(t - \tau_\star) + v_\star y'(t - \zeta_\star) \quad \text{with } \tau_\star = Nh,\ \zeta_\star = N'h \tag{5.5}$$

(where $N, N' \in \mathbb{N}$). The general θ-method yields

$$\begin{aligned}(1 - \theta\lambda_\star h)\tilde{y}_{n+1} &= (1 + (1-\theta)\lambda_\star h)\tilde{y}_n + \theta\mu_\star h\tilde{y}_{n+1-N} \\ &\quad + (1-\theta)\mu_\star h\tilde{y}_{n-N} + \theta v_\star h\widetilde{y'}_{n+1-N'} + (1-\theta)v_\star h\widetilde{y'}_{n-N'},\end{aligned} \tag{5.6a}$$

$$\widetilde{y'}_{n+1} = \lambda_\star \tilde{y}_{n+1} + \mu_\star \tilde{y}_{n+1-N} + v_\star \widetilde{y'}_{n+1-N'} \tag{5.6b}$$

and in the present case we avoid the need for interpolation for past values and past derivative values. With $\theta = \frac{1}{2}$ we obtain an example of (3.7).

We can now derive (if $\theta\lambda_\star h \neq 1$) a vector recurrence between the vectors $\boldsymbol{\phi}_k = [\tilde{y}_k, \widetilde{y'_k}]^{\mathrm{T}}$ $(k \in \mathbb{N})$ of the form

$$\boldsymbol{\phi}_{n+1} + \boldsymbol{A}_1\boldsymbol{\phi}_n + \boldsymbol{A}_N\boldsymbol{\phi}_{n-N} + \boldsymbol{A}_{N'}\boldsymbol{\phi}_{n-N'} = 0. \tag{5.7}$$

Here, $\boldsymbol{A}_\ell \equiv \boldsymbol{A}_\ell(\lambda_\star, \mu_\star, v_\star; h)$, for $\ell = 1, 2, \ldots, \max(N, N')$. A slightly different recurrence is obtained if the derivative values $\widetilde{y'_\ell}$ are obtained using numerical differentiation

$$\begin{aligned}(1 - \theta\lambda h)\tilde{y}_{n+1} &= (1 + (1-\theta)\lambda h)\tilde{y}_n + \theta\mu h\tilde{y}_{n+1-N} + (1-\theta)\mu h\tilde{y}_{n-N} \\ &\quad + \theta v h\widetilde{y'}_{n+1-N'} + (1-\theta)v h\widetilde{y'}_{n-N'},\end{aligned} \tag{5.8a}$$

$$\widetilde{y'}_{n+1-N} = \{\tilde{y}_{n+1-N} - \tilde{y}_{n-N}\}/h. \tag{5.8b}$$

If desired, a vector recurrence such as (5.7) can be re-expressed, using a block companion matrix, as a two-term autonomous vector recurrence of the form $\boldsymbol{\varphi}_{n+1} = \boldsymbol{M}\boldsymbol{\varphi}_n$ with an amplification matrix $\boldsymbol{M} \equiv \boldsymbol{M}(\lambda_\star, \mu_\star, v_\star; h)$ of order $2 \times \max\{N, N'\}$ — e.g., if $N' > N$:

$$\underbrace{\begin{pmatrix} \boldsymbol{\phi}_{n+1} \\ \boldsymbol{\phi}_n \\ \vdots \\ \boldsymbol{\phi}_{n-N'+1} \end{pmatrix}}_{\varphi_{n+1}} = \underbrace{\begin{pmatrix} -\boldsymbol{A}_1 & -\boldsymbol{A}_2 & \cdots & -\boldsymbol{A}_N & \cdots & -\boldsymbol{A}_{N'} \\ \boldsymbol{0} & \boldsymbol{I} & \cdots & \boldsymbol{0} & \cdots & \boldsymbol{0} \\ \vdots & & & & & \\ \boldsymbol{0} & \boldsymbol{0} & \cdots & \cdots & \boldsymbol{I} & \boldsymbol{0} \end{pmatrix}}_{\text{amplification matrix } \boldsymbol{M}} \times \underbrace{\begin{pmatrix} \boldsymbol{\phi}_n \\ \boldsymbol{\phi}_{n-1} \\ \vdots \\ \boldsymbol{\phi}_{n-N'} \end{pmatrix}}_{\varphi_n}. \tag{5.9}$$

Solutions of the latter recurrence are (uniformly) *stable with respect to initial perturbations* if $\rho(\boldsymbol{M}) \leqslant 1$ and any eigenvalues of modulus unity are semi-simple; uniformly *asymptotically stable with respect to initial perturbations* if $\rho(\boldsymbol{M}) < 1$ (and *ξ-exponentially stable* if $\rho(\boldsymbol{M}) \leqslant 1 - \xi h$, $\xi > 0$). *Stability with respect to uniformly bounded persistent perturbations* is assured when

$\rho(M) < 1$. *Note that since the dimensionality of $\boldsymbol{M}$ varies with h, stability for $h = h'$ and for h'' need not be quantitatively comparable properties.* The eigenvalues of $\boldsymbol{M}$ are, of course, the zeros of the characteristic polynomial

$$\begin{aligned}&\mathscr{P}(z) := \det\{z^{N^*+1}\boldsymbol{I} + z^{N^*}\boldsymbol{A}_1 + z^{N^*-N}\boldsymbol{A}_N + z^{N^*-N'}\boldsymbol{A}_{N'}\},\\ &\mathscr{P}(z) \equiv \mathscr{P}(\lambda, \mu, v, h; z), \quad N^* = \max\{N, N'\}.\end{aligned} \tag{5.10}$$

Regions of stability can be computed using the boundary-locus technique. To illustrate, such a region can be associated with the parameters of the test equation (2.1a) if we fix $v_\star = 0$ in the above, and the region of stability $\mathscr{S}(h)$ can then be defined[3] as

$$\mathscr{S}(h) := (\lambda = \lambda_\star \tau_\star, \mu = \mu_\star \tau_\star) \in \mathbb{R} \times \mathbb{R} \tag{5.11a}$$

such that

$$\mathscr{P}(z) = 0 \text{ implies } \quad |z| < 1 \quad \text{or} \quad |z| = 1 \text{ and } z \text{ is semi-simple,} \tag{5.11b}$$

with a corresponding definition of $\mathscr{S}_\star(h)$ when λ, $\mu \in \mathbb{C}$. Guglielmi [41] offered the following definitions in terms of the exact stability regions Σ and $\Sigma^\dagger$ referred to in Remark 2.4:

Definition 5.2. A numerical method is called (i) $\tau(0)$-*stable* if, for $v = 0$,

$$S(\tau_\star/N) \supseteq \Sigma \text{ whenever } 1 \leqslant N \in \mathbb{N}; \tag{5.12}$$

(ii) τ-*stable* if, for $v = 0$, $\mathscr{S}(\tau_\star/N) \supseteq \Sigma^\dagger$ whenever $1 \leqslant N \in \mathbb{N}$.

Guglielmi [41] showed that the trapezium rule is $\tau(0)$-stable but not τ-stable. We may contrast the rather strict requirements of τ- or $\tau(0)$-stability with the observation that, even for the Euler rule, given arbitrary fixed $(\lambda, \mu) \in \Sigma$ there exists a corresponding $M = M(\lambda, \mu) \in \mathbb{N}$ such that $(\lambda, \mu) \in \mathscr{S}(\tau_\star/N)$ for all $N \geqslant M$ (here, $v = 0$).

Remark 5.3. (a) Regions of stability can be computed using the boundary-locus technique, in which one seeks the loci on which $\mathscr{P}(\cdot)$ has a zero of modulus unity. Such a region can be obtained in (λ, μ)-plane for each parameter v of the test equation (2.1a) or of (2.8) or (5.5). Further, one can use this approach for the cases where $\tau_\star/h$ or $\zeta_\star/h$ are noninteger, so that interpolation is required for the lagging value, and its effect on stability observed. The bounded stability regions in Fig. 3 correspond to the use of the Euler method. From Fig. 3(a) it may be observed that for $h = 1/(m + \varsigma)$ with modest $m \in \mathbb{N}$, $\varsigma \in (0, 1)$, the stability region for the test equation

$$y'(t) = \lambda y(t) + \mu y(t-1) \tag{5.13}$$

depends in a pronounced manner on ς. The effect is much less pronounced for implicit formulae such as the trapezium rule.

(b) In Fig. 3(b) the stability regions are those for the Euler method applied with $h = \frac{1}{25}$ to the test equation

$$y'(t) = \lambda y(t) + \mu y(t-1) + v y'(t-1). \tag{5.14}$$

---

[3] A zero z of $\mathscr{P}(\cdot)$ defined by (5.10) is semi-simple if $\mathscr{P}'(z) \neq 0$ (i.e., if it is simple) or if its geometric and algebraic multiplicities, as an eigenvalue of $\boldsymbol{M}$ are the same.

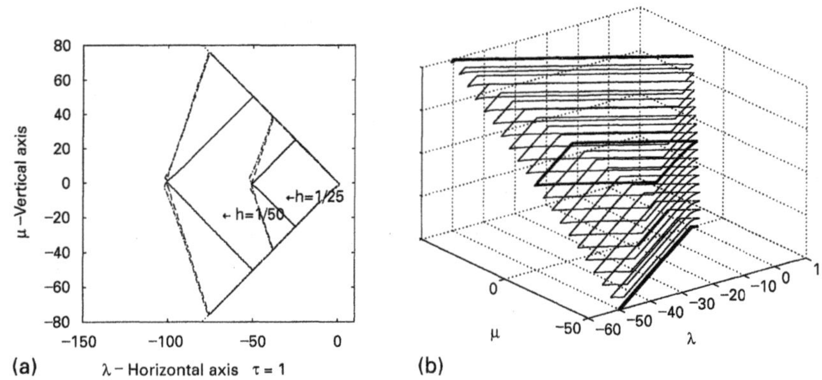

Fig. 3. Boundaries of stability regions for the explicit Euler method: (a) applied to the DDE (2.1a) with step size h ($h=\frac{1}{25}$ and $h=\frac{1}{50}$ — unbroken lines, $h=\frac{3}{72}$, $\frac{3}{69}$ and $\frac{3}{153}$, $\frac{3}{156}$ — broken lines), in the (λ,μ)-plane; (b) applied to the NDDE (2.8) with step $h=\frac{1}{25}$, in (λ,μ,v)-space with $\lambda\in(-60,0)$, $\mu\in(-60,60)$ and (vertical axis) $v\in(-1,1)$. See Remark 5.3.

The "lozenge" shown emphasized for $v=0$ in Fig. 3(b) corresponds to the case shown in Fig. 3(a) with $h=\frac{1}{25}$. As v varies from -1 to $+1$ in Fig. 3(b) the stability region in the (λ,μ)-plane first expands (until $v=0$) and then contracts, and as it does so it moves from proximity to $\lambda=-\mu$ towards proximity to $\lambda=\mu$. The exact stability regions were indicated in Fig. 2(b), and it seems that it has to be concluded that *the Euler formulae given above seem unsuitable for NDDEs.* This and similar observations are being pursued elsewhere.

In an elegant paper, Guglielmi and Hairer [42] develop the relationship between stability analysis and order stars.

Clearly, some interesting results can be obtained with the approaches indicated above but it is already clear that difficulties will be encountered if (i) step-sizes h_n are non-constant, and (ii) the lags $\tau_\star,\zeta_\star$ are replaced by time- or state-dependent values. In general, the two-term recurrence $\boldsymbol{\varphi}_{n+1}=\boldsymbol{M}\boldsymbol{\varphi}_{n+1}$ introduced above has to be replaced by a local recurrence $\boldsymbol{\psi}_{n,n+1}=\boldsymbol{M}_n\boldsymbol{\psi}_{n,n+1}$ where the dimension of the vectors $\boldsymbol{\psi}_{n,n+1},\boldsymbol{\psi}_{n,n}$ and the order of $\boldsymbol{M}_n$ changes with n, and the stability analysis is less straightforward.

A rather different approach relies upon a generalization of an inequality of Halanay (see [44, pp. 377–378 et seq.]) in the study of stability of DDEs, to which we restrict ourselves, though extensions are possible. The basic result of Halanay states that if $p(t)$ is a positive scalar-valued function satisfying $p'(t)\leqslant -Ap(t)+B\sup_{s\in -\tau_\star\leqslant s\leqslant t}p(t)$ for $t\geqslant t_0$, where $A>B>0$ then there exist positive k,ξ such that $p(t)\leqslant k\exp\{-\xi(t-t_0)\}$. In applications, this result may be used to establish exponential stability. A useful extension is to the case where $\alpha(t)\to\infty$ as $t\to\infty$ and (with $A>B>0$)

$$p'(t)\leqslant -Ap(t)+B\sup_{s\in[\alpha(t),t]}p(s)\quad\text{for } t\geqslant t_0 \tag{5.15}$$

whereupon we have (a result helpful in establishing asymptotic stability, but not exponential stability) $p(t)\to 0$ as $t\to\infty$. We note a further extension [13] to a *nonlinear inequality of Halanay type.* For the analysis of numerical stability, one can seek *discrete analogues* of the Halanay inequalities indicated above, in which a right-hand derivative is replaced by a difference $\Delta p(t)$. The approach is suggested by comparison of $y'(t) = \lambda y(t) + \mu y(t - \tau(t))$ with the trapezium rule applied with arbitrary step h, which yields expressions of the form $\Delta \tilde{y}_n = \frac{1}{2}\lambda h \tilde{y}_{n+1} + \frac{1}{2}\lambda h \tilde{y}_n + \frac{1}{2}\mu h\{\gamma_{n+1,\ell_n} \tilde{y}_{\ell_{n+1,n}} + \gamma^*_{n+1,\ell_n} \tilde{y}_{\ell_{n+1,n}+1}\} + \frac{1}{2}\mu h\{\gamma_{n,\ell_n} \tilde{y}_{\ell_{n,n}} + \gamma^*_{n,\ell_n} \tilde{y}^*_{\ell_{n,n}+1}\}$ where $\Delta \tilde{y}_n = \tilde{y}_{n+1} - \tilde{y}_n$ and where $\ell_{n+1,n}$ and $\ell_{n,n} \to \infty$ as $n \to \infty$.

6. Further issues and concluding remarks

Here, we shall address some problems that are in our view ongoing and therefore merit further attention, and greater space than we can afford here.

We turn first to stiffness in retarded equations, on which further study is required. The concept of stiffness in numerical analysis is, whilst the terminology is widely used in ODEs, open to varying (frequently controversial) mathematical interpretations [1]. For systems of ODEs, some would define stiffness in terms of the occurrence of different time-scales. Since a scalar DDE is infinite dimensional and in some sense incorporates a countable set of time-scales (the solution of many delayed equations can be expressed in the form of an infinite sum of exponential terms with differing time-scales) generalization of this idea seems unfruitful. Stiffness is sometimes related to the situation where explicit methods do not work, or where stability rather than accuracy constrains the step size. A *particular solution* $y(t)$ of a given DDE or NDDE will be regarded by this author as *stiff in the neighbourhood of a point* $t_\star$ when (i) $y(t)$ is smooth in the neighbourhood of $t_\star$, but (ii) accurate numerical simulation of the behaviour of $y(\cdot)$ can only be achieved by constraining the step size or the choice of method so that local errors are not amplified. In colloquial terms, (ii) relates to stability, but stability is defined formally in terms of behaviour of perturbations as $t \to \infty$ rather than in the shorter term as implied here. One naturally turns to experience with ODEs on detection of stiffness and for treatment using numerical methods based upon highly stable implicit formulae. However, some care has to be exercised here, because including delay terms modifies the behaviour of the solution and the numerical solution. For example, a solution of the equation $y'(t) = \lambda y(t)$ with $\lambda \ll 0$ is generally regarded as stiff, but solutions of $y'(t) = \lambda y(t) - 2\lambda y(t-1)$ (with $\lambda \ll 0$) are not even stable. The design of methods for detecting stiffness, and methods that switch in or out of stiff mode, therefore merit further examination.

We introduce a class of problem not discussed above, *constrained DDEs*, or *delay differential algebraic equations* (DDAEs) of the form (say)

$$\begin{aligned} u'(t) &= f(t, u(t), v(t), u(\alpha_1(t)), v(\alpha_1(t)), \ldots, u(\alpha_q(t)), v(\alpha_q(t))), \\ 0 &= g(t, u(t), v(t), u(\alpha_1(t)), v(\alpha_1(t)), \ldots, u(\alpha_q(t)), v(\alpha_q(t))), \end{aligned} \tag{6.1}$$

that is, a system of DDEs coupled with constraints. Such systems are modifications, incorporating delayed arguments, of differential algebraic equations — hence, the description DDAEs. The constraints need not be algebraic constraints, and problems with inequality rather than equality constraints also arise. For the numerics of DDAEs, see [2] and its citations, and [46,63]; for neutral DDAEs see [58]. One difficulty is that DDAEs can be equivalent to neutral differential equations with deviating arguments in which some arguments are advanced ("$\alpha(t) > t$, $\beta(t) > t$") rather than delayed. If one excludes this possibility, it appears that concerns should focus on the numerics of the related NDDEs, the problem of overcoming apparent "stiffness" (depending on the local index of the solution) and the question of error control in the presence of poorly behaved derivatives.

Closely allied to DDAEs are *singular perturbation problems*, such as

$$\begin{aligned} u_\varepsilon'(t) &= f(t, u_\varepsilon(t), v_\varepsilon(t)), \\ \varepsilon v_\varepsilon'(t) &= g(t, u_\varepsilon(t), v_\varepsilon(t), u_\varepsilon(\alpha_1(t)), v_\varepsilon(\alpha_1(t)), \ldots, u_\varepsilon(\alpha_q(t)), v_\varepsilon(\alpha_q(t))) \end{aligned} \tag{6.2}$$

or, a simpler example, $\varepsilon u_\varepsilon'(t) = f(t, u_\varepsilon(t), u_\varepsilon(t-\tau_\star))$. For a preview of some results on the numerics see [3]. Finally, we concentrated on DDEs and NDDEs, but one can discuss Volterra integro-differential equations with delays.

We conclude with a reference to a different viewpoint, recently introduced in [23] in their analysis of a linear system of DDEs with a fixed lag $\tau_\star$. The basis of the work [23] is the observation that the problem

$$y'(t) = f(t, y(t), y(t-\tau_\star)), \quad t \in [-\tau, \infty), \tag{6.3a}$$

$$y(t) = \psi(t), \quad t \in [t_0 - \tau_\star, t_0] \tag{6.3b}$$

may be solved — in particular if

$$\psi'(t_0) = f(t_0, \psi(t_0), \psi(t_0 - \tau_\star)), \tag{6.3c}$$

by constructing the solution of the PDE

$$\frac{\partial}{\partial t} u(t,s) = \frac{\partial}{\partial s} u(t,s), \quad t \geqslant t_0, \quad s \in [t_0 - \tau_\star, t_0], \tag{6.4a}$$

$$\frac{\partial}{\partial s} u(t, t_0) = f(t, u(t, t_0), u(T, t_0 - \tau_\star)), \tag{6.4b}$$

$$u(t_0, s) = \psi(s), \quad s \in [t_0 - \tau_\star, t_0], \tag{6.4c}$$

on setting $y(t+s) = u(t,s)$. Bellen and Maset relate this to an abstract Cauchy problem that can be treated numerically, and consider stability and convergence for linear systems ($y'(t) = My(t) + Ny(t-\tau_\star)$); the approach has further potential.

Acknowledgements

The author acknowledges funding from EPSRC, the London Mathematical Society, NAG Ltd. and the University of Manchester, which supported fruitful collaboration with co-workers and former and current PhD students, a number of whom kindly commented on a draft of this work. He also gives thanks to F. Rihan for assisting in preparation of some of the figures, and to an anonymous referee for helpful remarks.

References

[1] R.C. Aitken (Ed.), Stiff Computation, Oxford University Press, Oxford, 1985.

[2] U.M. Ascher, L.R. Petzold, The numerical solution of delay–differential–algebraic equations of retarded and neutral type, SIAM J. Numer. Anal. 32 (1995) 1635–1657.
[3] C.T.H. Baker, G.A. Bocharov, A. Filiz, N.J. Ford, C.A.H. Paul, F.A. Rihan, A. Tang, R.M. Thomas, H. Tian, D.R. Willé, Numerical modelling by delay and Volterra functional differential equations, in: E.A. Lipitakis (Ed.), Topics in Computer Mathematics and its Applications, LEA Athens, Hellas, 1999, pp. 113–137. ISBN 960-85176-6-4.
[4] C.T.H. Baker, G.A. Bocharov, C.A.H. Paul, F.A. Rihan, Modelling and analysis of time-lags in cell proliferation, J. Math. Biol. 37 (1998) 341–371.
[5] C.T.H. Baker, G.A. Bocharov, F.A. Rihan, A report on the use of delay differential equations in numerical modelling in the biosciences, MCCM Technical Report 343, 1999. ISSN 1360–1725.
[6] C.T.H. Baker, E. Buckwar, Introduction to the numerical analysis of stochastic delay differential equations. MCCM Technical Report 345, 1999. ISSN 1360–1725.
[7] C.T.H. Baker, C.A.H. Paul, Computing stability regions — Runge–Kutta methods for delay differential equations, IMA J. Numer. Anal. 14 (1994) 347–362.
[8] C.T.H. Baker, C.A.H. Paul, A global convergence theorem for a class of parallel continuous explicit Runge–Kutta methods and vanishing lag delay differential equations, SIAM J. Numer. Anal. 33 (1996) 1559–1576.
[9] C.T.H. Baker, C.A.H. Paul, Pitfalls in parameter estimation for delay differential equations, SIAM J. Sci. Comput. 18 (1997) 305–314.
[10] C.T.H. Baker, C.A.H. Paul, D.R. Willé, Issues in the numerical solution of evolutionary delay differential equations, Adv. Comput. Math. 3 (1995) 171–196.
[11] C.T.H. Baker, F.A. Rihan, Sensitivity analysis of parameters in modelling with delay-differential equations. MCCM Technical Report 349, 1999 ISSN 1360–1725.
[12] C.T.H. Baker, A. Tang, Stability of non-linear delay integro-differential equations via approximating equations, MCCM Technical Report 327, 1998. ISSN 1360–1725.
[13] C.T.H. Baker, A. Tang, Generalized Halanay inequalities for Volterra functional differential equations and discretized versions, in: C. Corduneanu, I.W. Sandberg, (Eds.), Volterra Equations and Applications, Gordon and Breach, Amsterdam, 1999, pp. 39–55. ISBN 1023-6155.
[14] V.L. Bakke, Z. Jackiewicz, Stability analysis of linear multistep methods for delay differential equations, Internat. J. Math. Math. Sci. 9 (1986) 447–458.
[15] V.K. Barwell, Special stability problems for functional differential equations, BIT 15 (1975) 130–135.
[16] A. Bellen, One-step collocation for delay differential equations, J. Comput. Appl. Math. 10 (1984) 275–283.
[17] A. Bellen, Constrained mesh methods for functional differential equations, Internat. Schriftenreihe Numer. Math. 74 (1985) 52–70.
[18] A. Bellen, N. Guglielmi, A. Ruehli, On a class of stable methods for linear systems of delay differential equations of neutral type, IEEE Trans. Circuits Systems 46 (1999) 212–216.
[19] A. Bellen, N. Guglielmi, L. Torelli, Asymptotic stability properties of Theta-methods for the pantograph equation, Appl. Numer. Math. 24 (1997) 279–293.
[20] A. Bellen, N. Guglielmi, M. Zennaro, On the contractivity and asymptotic stability of systems of delay differential equations of neutral type, BIT 39 (1999) 1–24.
[21] A. Bellen, N. Guglielmi, M. Zennaro, Numerical stability of nonlinear delay differential equations of neutral type, this issue, J. Comput. Appl. Math. 125 (2000) 251–263.
[22] A. Bellen, Z. Jackiewicz, M. Zennaro, Stability analysis of one-step methods for neutral delay-differential equations, Numer. Math. 52 (1988) 605–619.
[23] A. Bellen, S. Maset, Numerical solution of constant coefficient linear delay differential equations as abstract Cauchy problems Numer. Math. 84 (2000) 351–374.
[24] A. Bellen, M. Zennaro, Numerical solution of delay differential equations by uniform corrections to an implicit Runge–Kutta method, Numer. Math. 47 (1985) 301–316.
[25] A. Bellen, M. Zennaro, Strong contractivity properties of numerical methods for ordinary and delay differential equations, Appl. Numer. Math. 9 (1992) 321–346.
[26] A. Bellen, M. Zennaro, Numerical Methods for Delay Differential Equations, Oxford University Press, Oxford, in preparation.
[27] H. Brunner, Collocation Methods for Volterra Integral and Related Functional Differential Equations, book in preparation.

[28] M.D. Buhmann, A. Iserles, Numerical analysis of functional equations with a variable delay, Pitman Research Notes Mathematics Series, vol. 260, (1991), pp. 17–33, Longman Sci. Tech., Harlow, 1992.
[29] M.D. Buhmann, A. Iserles, S.P. Nørsett, Runge–Kutta methods for neutral differential equations, WSSIAA 2 (1993) 85–98.
[30] O. Diekmann, S.A. van Gils, S.M. Verduyn Lunel, H-O. Walther, Delay Equations, Springer, New York, 1994. ISBN 0-387-94416-8.
[31] L.E. El'sgolts, S.B. Norkin, Introduction to the Theory and Application of Differential Equations with Deviating Arguments, Academic Press, New York, 1973. LIBR. OF CONGRESS CAT. NO. 73-811.
[32] K. Engelborghs, T. Luzyanina, D. Roose, Numerical bifurcation analysis of delay differential equations, this issue, J. Comput. Appl. Math. 125 (2000) 265–275.
[33] K. Engelborghs, D. Roose, Smoothness loss of periodic solutions of a neutral functional-differential equation: on a bifurcation of the essential spectrum, Dynamics Stability Systems 14 (1999) 255–273.
[34] K. Engelborghs, D. Roose, Numerical computation of stability and detection of Hopf bifurcations of steady state solutions of delay differential equations, Adv. Comput. Math. 10 (1999) 271–289.
[35] K. Engelborghs, D. Roose, T. Luzyanina, Bifurcation analysis of periodic solutions of neutral functional-differential equations: a case study, Internat. J. Bifur. Chaos Appl. Sci. Eng. 8 (1998) 1889–1905.
[36] W.H. Enright, H. Hayashi, Convergence analysis of the solution of retarded and neutral delay differential equations by continuous numerical methods, SIAM J. Numer. Anal. 35 572–585.
[37] A.C. Fowler, An asymptotic analysis of the delayed logistic equation when the delay is large, IMA J. Appl. Math. 28 (1982) 41–49.
[38] N.J. Ford, V. Wulf, How do numerical methods perform for delay differential equations undergoing a Hopf bifurcation? this issue, J. Comput. Appl. Math. 125 (2000) 277–285.
[39] K. Gopalsamy, Stability and Oscillations in Delay Differential Equations of Population Dynamics, Kluwer Academic, London, 1992. ISBN 0-7923-1594-4.
[40] N. Guglielmi, On the asymptotic stability properties of Runge–Kutta methods for delay differential equations, Numer. Math. 77 (1997) 467–485.
[41] N. Guglielmi, Delay dependent stability regions of theta-methods for delay differential equations, IMA J. Numer. Anal. 18 (1998) 399–418.
[42] N. Guglielmi, E. Hairer, Order stars and stability for delay differential equations, Numer. Math. 83 (1999) 371–383.
[43] E. Hairer, S.P. Nørsett, G. Wanner, Solving Ordinary Differential Equations – I, Springer, Berlin, 1987. ISBN 3-540-17145-2.
[44] A. Halanay, Differential Equations — Stability, Oscillations, Time Lags Academic Press, New York, 1966. LIBR. OF CONGRESS CAT. NO. 65-25005.
[45] H. Hayashi, Numerical solution of delay differential equations, Ph.D. Thesis, University of Toronto, 1995.
[46] R. Hauber, Numerische Behandlung von retardierten differential-algebraischen Gleichungen, Doctoral Thesis, Universität München, 1994.
[47] K.J. in 't Hout, The stability of a class of Runge–Kutta methods for delay differential equations, Appl. Numer. Math. 9 (1992) 347–355.
[48] K.J. in 't Hout, A new interpolation procedure for adapting Runge–Kutta methods to delay differential equations, BIT 32 (1992) 634–649.
[49] K.J. in 't Hout, Runge–Kutta methods in the numerical solution of delay differential equations, Ph.D. Thesis, The University of Leiden, 1992.
[50] K.J. in 't Hout, C. Lubich, Periodic solutions of delay differential equations under discretization, BIT 38 (1998) 72–91.
[51] K.J. in 't Hout, M.N. Spijker, Stability analysis of numerical methods for delay differential equations, Numer. Math. 59 (1991) 807–814.
[52] C. Huang, Dissipativity of Runge–Kutta methods for dynamical systems with delay, IMA J. Numer. Anal. 20 (2000) 153–166.
[53] C. Huang, H. Fu, S. Li, G. Chen, Stability analysis of Runge–Kutta methods for non-linear delay differential equations, BIT 39 (1999) 270–280.
[54] C. Huang, S. Li, H. Fu, G. Chen, Stability and error analysis of one-leg methods for nonlinear delay differential equations, J. Comput. Appl. Math. 103 (1999) 263–279.

[55] A. Iserles, Numerical analysis of delay differential equations with variable delays, Ann. Numer. Math. 1 (1994) 133–152.
[56] A. Iserles, Exact and discretized stability of the pantograph equation, Appl. Numer. Math. 24 (1997) 295–308.
[57] Z. Jackiewicz, One-step methods of any order for neutral functional differential equations, SIAM J. Numer. Anal. 21 (1984) 486–511.
[58] T. Jankowski, M. Kwapisz, Convergence of numerical methods for systems of neutral functional-differential-algebraic equations, Appl. Math. 40 (1995) 457–472.
[59] V.B. Kolmanovskii, A. Myskhis, Applied Theory of Functional Differential Equations, Kluwer Academic, Dordrecht, 1992. ISBN 0-7923-2013-1.
[60] V.B. Kolmanovskii, V.R. Nosov, Stability of Functional Differential Equations, Academic, London, 1986. ISBN 0-12-417941-X.
[61] T. Koto, A stability property of A-stable collocation-based Runge–Kutta methods for neutral delay differential equations, BIT 36 (4) (1996) 855–859.
[62] T. Koto, The stability of natural Runge–Kutta methods for nonlinear delay differential equations, Japan J. Ind. Appl. Math. 14 (1997) 111–123.
[63] Y. Liu, Numerical solution of implicit neutral functional-differential equations, SIAM J. Numer. Anal. 36 (1999) 516–528.
[64] T. Luzyanina, K. Engelborghs, K. Lust, D. Roose, Computation, continuation and bifurcation analysis of periodic solutions of delay differential equations, Internat. J. Bifur. Chaos Appl. Sci. Eng. 7 (1997) 2547–2560.
[65] T. Luzyanina, D. Roose, Numerical stability analysis and computation of Hopf bifurcation points for delay differential equations, J. Comput. Appl. Math. 72 (1996) 379–392. 6
[66] G. Meinardus, G. Nürnberger, approximation theory and numerical methods for delay differential equations, Internat. Schriftenreihe Numer. Math. 74 (1985) 13–40.
[67] K.W. Neves, Automatic integration of functional differential equations: an approach, ACM Trans. Math. Software 1 (1975) 357–368.
[68] K.W. Neves, Algorithm 497 — automatic integration of functional differential equations, ACM Trans. Math. Software 1 (1975) 369–371.
[69] K.W. Neves, A. Feldstein, Characterization of jump discontinuities for state dependent delay differential equations, J. Math. Anal. Appl. 56 (1976) 689–707.
[70] K.W. Neves, S. Thompson, Software for the numerical solution of systems of functional differential equations with state dependent delays, Appl. Numer. Math. 9 (1992) 385–401.
[71] C.A.H. Paul, A User Guide to ARCHI, MCCM Technical Report 283, 1995. ISSN 1360–1725.
[72] L. Qiu, B. Yang, J. Kuang, The NGP-stability of Runge–Kutta methods for systems of neutral delay differential equations, Numer. Math. 81 (1999) 451–459.
[73] L. Torelli, Stability of numerical methods for delay differential equations, J. Comput. Appl. Math. 25 (1989) 15–26.
[74] L. Torelli, A sufficient condition for GPN-stability for delay differential equations, Numer. Math. 59 (1991) 311–320.
[75] D.R. Willé, C.T.H. Baker, The tracking of derivative discontinuities in systems of delay-differential equations, Appl. Num. Math. 9 (1992) 209–222.
[76] V. Wulf, Numerical analysis of delay differential equations undergoing a Hopf bifurcation Ph.D. Thesis, University of Liverpool, 1999.
[77] M. Zennaro, Delay differential equations: theory and numerics, in: M. Ainsworth, L. Levesley, W.A. Light, M. Marletta (Eds.), Theory and Numerics of Ordinary and Partial Differential Equations, Oxford University Press, Oxford, 1995. ISBN 0-19-851193-0.

ELSEVIER

Journal of Computational and Applied Mathematics 125 (2000) 337–345

JOURNAL OF COMPUTATIONAL AND APPLIED MATHEMATICS

www.elsevier.nl/locate/cam

Adaptive space–time finite element solution for Volterra equations arising in viscoelasticity problems

Simon Shaw, J.R. Whiteman*

BICOM, Institute of Computational Mathematics, Brunel University, Uxbridge, Middlesex UB8 3PH, UK

Received 20 February 2000

Abstract

We give a short overview of our recent efforts towards constructing adaptive space–time finite element solvers for some partial differential Volterra equations arising in viscoelasticity theory. © 2000 Elsevier Science B.V. All rights reserved.

Keywords: Volterra equation; Viscoelasticity; Finite element method; Adaptivity

MSC: 73F15; 45D05; 65M60

1. Introduction

Viscoelastic materials, such as polymers, under external loading exhibit features typical of both elastic solids (instantaneous elastic deformation) and viscous fluids (creep, or flow, over long times).

In the classical theory of linear viscoelasticity the stress tensor, $\underline{\sigma} := \{\sigma_{ij}\}_{i,j=1}^{n}$, in a viscoelastic body $\Omega \subset \mathbb{R}^n$ is related to the current strain tensor, $\underline{\varepsilon} := \{\varepsilon_{ij}\}_{i,j=1}^{n}$, and the strain history through a hereditary integral (see [4]). Typically one has either of the constitutive relationships

$$\underline{\sigma}(t) = E(0)\underline{\boldsymbol{D}}\,\underline{\varepsilon}(t) - \int_0^t E_s(t-s)\underline{\boldsymbol{D}}\,\underline{\varepsilon}(s)\,\mathrm{d}s, \tag{1}$$

$$\underline{\sigma}(t) = E(t)\underline{\boldsymbol{D}}\,\underline{\varepsilon}(0) + \int_0^t E(t-s)\underline{\boldsymbol{D}}\,\underline{\varepsilon}_s(s)\,\mathrm{d}s. \tag{2}$$

Here t is the time variable, with $t = 0$ a datum prior to which it is assumed that $\underline{\varepsilon} = \underline{\boldsymbol{0}}$, the subscript denotes partial differentiation, and $\underline{\boldsymbol{D}}$ is a fourth-order tensor of elastic coefficients. The *stress*

* Corresponding author. Tel.:+44-1-895-203270; fax:+44-1-895-203303.
E-mail address: john.whiteman@brunel.ac.uk (J.R. Whiteman).

PII: S0377-0427(00)00477-5

relaxation function, $E(t)$, is non-negative and monotone decreasing and we invariably assume that it has the form

$$E(t)=E_0+\sum_{i=1}^{N}E_i\mathrm{e}^{-\alpha_i t} \tag{3}$$

for $E_0 \geqslant 0$, $E_i > 0$ and $\alpha_i > 0$ for each i. This form is typical when the viscoelasticity is modelled by springs and dashpots after the manner of Maxwell, Kelvin and Voigt (see for example, [5]), but other forms have also been used (see the review by Johnson in [8]).

When the viscoelastic body occupies an open domain $\Omega \subset \mathbb{R}^n$ and is acted upon by a system of body forces, $\boldsymbol{f} := (f_i)_{i=1}^n$, one obtains an initial value problem for the displacement, $\boldsymbol{u} := (u_i)_{i=1}^n$, by merging either of (1) or (2) with Newton's second law: for $i = 1,\dots,n$

$$\begin{aligned}
&\varrho u_i''(t) - \sigma_{ij,j} = f_i && \text{in } (0,T)\times\Omega,\\
&\boldsymbol{u} = \boldsymbol{0} && \text{on } (0,T)\times\partial\Omega,\\
&\boldsymbol{u}(\boldsymbol{x},0) = \boldsymbol{u}_0(\boldsymbol{x}) && \text{in } \Omega,\\
&\boldsymbol{u}'(\boldsymbol{x},0) = \boldsymbol{u}_1(\boldsymbol{x}) && \text{in } \Omega.
\end{aligned} \tag{4}$$

Here, ϱ is the mass density of the body; the primes denote time differentiation; the σ_{ij} are components of the stress tensor, $\underline{\boldsymbol{\sigma}}$, with $\sigma_{ij,j} := \partial\sigma_{ij}/\partial x_j$; the summation convention is in force; T is some positive final time; and, more general boundary conditions are possible (see [22]). The equations are closed with the linear strain–displacement map,

$$\varepsilon_{ij}(\boldsymbol{u}) := \frac{1}{2}\left(\frac{\partial u_i}{\partial x_j}+\frac{\partial u_j}{\partial x_i}\right).$$

Using this in (4) along with either of the constitutive laws results in a partial differential equation with memory: a *partial differential Volterra equation*.

For example, substituting for the strain in (1), and using the result in the first of (4) yields (with summation implied), for each $i = 1,\dots,n$

$$\begin{aligned}
&\varrho u_i''(t) - \frac{\partial}{\partial x_j}\left[\frac{E(0)D_{ijkl}}{2}\left(\frac{\partial u_k(t)}{\partial x_l}+\frac{\partial u_l(t)}{\partial x_k}\right)\right]\\
&\quad + \frac{\partial}{\partial x_j}\int_0^t \frac{E_s(t-s)D_{ijkl}}{2}\left(\frac{\partial u_k(s)}{\partial x_l}+\frac{\partial u_l(s)}{\partial x_k}\right)\mathrm{d}s = f_i.
\end{aligned}$$

These are the usual elastodynamic equations but augmented with a Volterra integral. With appropriately defined partial differential operators A and B we can write this in the more compact form

$$\varrho\boldsymbol{u}''(t)+A\boldsymbol{u}(t)=\boldsymbol{f}(t)+\int_0^t B(t-s)\boldsymbol{u}(s)\,\mathrm{d}s, \tag{5}$$

which we refer to as a *hyperbolic Volterra equation*. Note that A and B are closely related to the linear elasticity operator, and that here, and below, we are using (1).

Frequently, engineers assume that the deformation is *quasistatic*, wherein $\varrho \mathbf{u}''(t) \approx \mathbf{0}$, and so arrive at the *elliptic Volterra equation*

$$A\mathbf{u}(t) = \mathbf{f}(t) + \int_0^t B(t-s)\mathbf{u}(s)\,\mathrm{d}s. \tag{6}$$

This, essentially, is a Volterra equation of the second kind for the displacement $\mathbf{u}$, and we will return to it below.

In, for example, [1] Cohen et al. suggest that viscoelastic stress is also an important factor influencing diffusion processes in polymers. To model this they suggest introducing a viscoelastic stress dependence into the classical Fick's law and arrive at the *parabolic Volterra equation*

$$u'(t) + Au(t) = f(t) + \int_0^t B(t-s)u(s)\,\mathrm{d}s. \tag{7}$$

Here u is the concentration and A and B are now Laplacian-type diffusion operators. (Actually the theory developed by Cohen et al. in [1] incorporates a crucially important nonlinearity.)

The numerical analysis of the pure-time versions of these problems using classical methods (e.g. finite difference, quadrature, collocation) is extensive and we have attempted a brief survey in [17]. The space–time problems have also received some attention with (7) apparently dominating. Thomée et al. are major contributors to this area (see for example [24]) where spatial discretization is effected with the finite element method but, usually, the temporal discretization is based on classical methods.

In this paper we discuss temporal finite element discretizations of (6) and (7) for pure-time and space–time problems, and demonstrate how duality and the Galerkin framework can be exploited to provide a posteriori error estimates suitable for adaptive error control. Our work is related to, and motivated by, the techniques of Eriksson et al. described in, for example, [2].

In Section 2 below we consider the pure-time versions of (6) and (7), and we describe the finite element approximation and a posteriori error estimation technique. We also discuss the related issue of data-stability estimates. In Section 3 we outline our results so far for (6), in the space–time context, and then finish in Section 4 by describing our aims to extend this work.

2. Pure-time problems

2.1. Second kind Volterra equations

The simplest pure-time problem related to these partial differential Volterra (PDV) equations is the scalar Volterra equation of the second kind associated with (6). We consider this problem as: find $u \in L_p(0,T)$ such that

$$u(t) = f(t) + \int_0^t \phi(t-s)u(s)\,\mathrm{d}s \tag{8}$$

for given data T, $f \in L_p(0,T)$, $\phi \in L_1(0,T)$ and some $p \in (1,\infty]$. Consider $L_p(0,T)$ as the dual of $L_q(0,T)$ (so that $p^{-1}+q^{-1}=1$ for $p>1$), and let $(\cdot,\cdot)$ denote the $L_2(0,T)$ inner product. The "variational" form of (8) is: find $u \in L_p(0,T)$ such that

$$(u,v) = (f,v) + (\Lambda u, v) \quad \forall v \in L_q(0,T), \tag{9}$$

where we define

$$\Lambda u(t) := \int_0^t \phi(t-s)u(s)\,\mathrm{d}s$$

for convenience and brevity.

Discretizing $(0, T)$ into $0 = t_0 < t_1 < \cdots < t_i < \cdots < t_N = T$, and defining the time steps, $k_i := t_i - t_{i-1}$, and subintervals, $\mathscr{I}_i := [t_{i-1}, t_i]$, we now let $V^k \subset L_\infty(0, T)$ denote the space of piecewise constant functions with respect to this partition (mesh). The piecewise constant finite element approximation to (9) is then: find $U \in V^k$ such that

$$(U, v) = (f, v) + (\Lambda U, v) \quad \forall v \in V^k. \tag{10}$$

Here of course U is the piecewise constant approximation to u. Choosing v to be the characteristic function of each $\mathscr{I}_i$ in turn, then yields a time stepping scheme for U similar to that produced by a standard quadrature approximation. We refer to [16] for full details.

To derive an a posteriori error estimate for the Galerkin error $e := u - U$ we first subtract (10) from (9) and obtain the *Galerkin orthogonality* relationship

$$(e - \Lambda e, v) = 0 \quad \forall v \in V^k. \tag{11}$$

The next step is to introduce a dual backward problem: find $\chi \in L_q(0, T)$ such that

$$\chi(t) = g(t) + \Lambda^* \chi(t).$$

Here $g \in L_q(0, T)$ is arbitrary and Λ^* is dual to Λ in that for all $v \in L_p(0, T)$

$$(v, \Lambda^* \chi) := \int_0^T \int_t^T \phi(s-t)\chi(s)v(t)\,\mathrm{d}s\,\mathrm{d}t = (\Lambda v, \chi)$$

by interchanging the order of integration. This dual problem can also be given by a variational formulation

$$(v, \chi) = (v, g) + (v, \Lambda^* \chi) \quad \forall v \in L_p(0, T)$$

and choosing $v = e \in L_p(0, T)$ in this we get

$$(e, g) = (e, \chi) - (e, \Lambda^* \chi) = (e - \Lambda e, \chi).$$

Introducing an interpolant, $\pi\chi \in V^k$, to χ and assuming the estimate

$$\|\chi - \pi\chi\|_{L_q(0,T)} \leqslant C_\pi \|\chi\|_{L_q(0,T)}, \tag{12}$$

(for C_π a positive constant) we can use Galerkin orthogonality to write

$$(e, g) = (R(U), \chi - \pi\chi), \tag{13}$$

where $R(U) := e - \Lambda e \equiv f - U + \Lambda U$ is the *residual*, and is computable. Assuming now the existence of a stability factor $S(T)$ such that

$$\|\chi\|_{L_q(0,T)} \leqslant S(T)\|g\|_{L_q(0,T)}, \tag{14}$$

we have by Hölder's inequality,

$$|(e, g)| = |(R(U), \chi - \pi\chi)| \leqslant C_\pi S(T)\|R(U)\|_{L_p(0,T)}\|g\|_{L_q(0,T)}.$$

The a posteriori error estimate now follows by duality since for $p>1$,

$$\|u-U\|_{L_p(0,T)} := \sup\left\{\frac{|(e,g)|}{\|g\|_{L_q(0,T)}} : g\in L_q(0,T)\setminus\{0\}\right\}$$

giving

$$\|u-U\|_{L_p(0,T)} \leqslant C_\pi S(T)\|R(U)\|_{L_p(0,T)} \quad \text{for } p>1. \tag{15}$$

Notice that (12) actually imparts no useful information, and so setting $\pi\chi := 0$ we can take $C_\pi = 1$ in the above equation. Below we illustrate a different approach.

This estimate is computable in terms of the data f and ϕ, the finite element solution U, and the stability factor $S(T)$. Full details of this result (for $p=1$ also) are given in [16]. Also given there are an a priori error estimate and an upper bound on the residual.

Clearly, it is important to have a high-quality estimate for the stability factor $S(T)$. In general, one would have to either approximate this numerically or use the (usually) non-optimal result from Gronwall's lemma. However, for the viscoelasticity problem as described earlier it is possible to derive a sharp estimate for $S(T)$ by exploiting the fading memory of the problem. In particular, for viscoelastic solids, under physically reasonable assumptions, it can be proven that $S(T)=O(1)$ independently of T. Full details of this are in [21].

A major problem with (15) is that the time steps $\{k_i\}$ do not explicitly appear, and this means that we cannot use the estimate to provide an adaptive time step controller. To incorporate the time steps into an a posteriori error estimate we replace the non-optimal interpolation-error estimate, (12), with the sharper

$$\|\kappa^{-1}(\chi-\pi\chi)\|_{L_q(0,T)} \leqslant c_\pi \|\chi'\|_{L_q(0,T)},$$

where $\kappa|_{\mathscr{I}_i} := k_i$, for each i, is the piecewise constant time step function. In place of (13) we can now write

$$|(e,g)| = |(\kappa R(U), \kappa^{-1}(\chi-\pi\chi))|.$$

Now, in [18] we demonstrate (for convolution equations) that if (14) holds then so too does,

$$\|\chi'\|_{L_q(0,T)} \leqslant S(T)\|g'\|_{L_q(0,T)} \quad \forall g\in \mathring{W}_q^1(0,T).$$

Here we recall that the Sobolev space $\mathring{W}_q^1(0,T)$ contains all functions with first derivative in $L_q(0,T)$ which vanish at $t=0$ and $t=T$. Thus,

$$|(e,g)| \leqslant c_\pi S(T)\|\kappa R(U)\|_{L_p(0,T)}\|g'\|_{L_q(0,T)}$$

and the residual is now weighted by the time steps. The price we pay for this is that the argument used above to estimate $\|u-U\|_{L_p(0,T)}$ no longer holds because g' and not g appears on the right. So, to "remove" g' we must estimate $u-U$ in a weaker norm.

Specializing to the case $p=\infty$ and $q=1$ we first recall the negative Sobolev space $W_\infty^{-1}(0,T)$ with norm

$$\|w\|_{W_\infty^{-1}(0,T)} := \sup\left\{\frac{|(w,v)|}{\|v'\|_{L_1(0,T)}} : v\in \mathring{W}_1^1(0,T)\setminus\{0\}\right\}.$$

Using this weaker norm we then have another a posteriori error estimate,

$$\|u - U\|_{W_\infty^{-1}(0,T)} \leqslant c_\pi S(T) \|\kappa R(U)\|_{L_\infty(0,T)}.$$

The presence of the time steps on the right now allows for adaptive time step control. For example, if we require that $\|u - U\|_{W_\infty^{-1}(0,T)} \leqslant \mathrm{TOL}$, where $\mathrm{TOL} > 0$ is a user-specified tolerance level, then it is sufficient to iteratively choose each time step so as to satisfy,

$$\|k_i R(U)\|_{L_\infty(t_{i-1},t_i)} \leqslant \frac{\mathrm{TOL}}{c_\pi S(T)}.$$

More details on this, and some numerical experiments are given in [18].

2.2. An ODE with memory

It was necessary to introduce the negative norm above because one cannot expect to bound the derivative of the solution to the dual problem, χ', in terms of g alone. Loosely speaking, this is because the original problem, (8), contains no time derivative. On the other hand, the problem: find $u \in W_p^1(0,T)$ such that,

$$u'(t) + \sigma(t)u(t) = f(t) + \Lambda u(t) \quad \text{in } (0,T)$$

with $U(0) = u_0$ and $\sigma > 0$ in $(0,T)$ given, does contain a time derivative and one can avoid all mention of negative norms when deriving error estimates.

This problem was studied in [15] as a pure-time prototype for the non-Fickian diffusion equation (7). In essentially the same way as described above (with a distributional interpretation of the time derivative), we can formulate a piecewise constant finite element approximation to this problem. The resulting time stepping algorithm is then

$$U_i - U_{i-1} + U_i \int_{t_{i-1}}^{t_i} \sigma(t)\,\mathrm{d}t = \int_{t_{i-1}}^{t_i} f(t)\,\mathrm{d}t + \int_{t_{i-1}}^{t_i} \Lambda U(t)\,\mathrm{d}t,$$

where U_i is the constant approximation to u on (t_{i-1}, t_i) and $U_0 = u_0$. A similar duality argument to that illustrated in the previous subsection leads to the a posteriori error estimate

$$\|u - U\|_{L_\infty(0,t_i)} \leqslant C_s \max_{1 \leqslant j \leqslant i} \{k_j \|R(U)\|_{L_\infty(t_{i-1},t_i)} + |U_j - U_{j-1}|\},$$

where C_s is a stability constant, and the residual is now defined by

$$R(U(t)) := f(t) - \sigma(t)U(t) + \Lambda U(t)$$

in each subinterval. Full details of this result, along with an a priori error estimate, numerical experiments, quadrature error estimates, and an alternative solution algorithm using a continuous piecewise linear approximation, are given in [15].

We turn now briefly to space–time problems.

3. Space–time problems

Although it is instructive to study these pure-time prototypes, adaptive solution algorithms for the viscoelasticity problems described earlier must be based on a posteriori error estimates for space–

time discretizations. Our work in this direction has so far been confined to the quasistatic problem represented by (6) and we give only a short description of our results here.

For this problem we have a space–time finite element method based upon a continuous piecewise linear approximation in the space variables, and a discontinuous piecewise constant, or linear, approximation in the time variable. Using $\boldsymbol{U}$ to denote the approximation to $\boldsymbol{u}$ we give in [22] (see also the more detailed [19,20]) the following a posteriori error estimate:

$$\|\boldsymbol{u} - \boldsymbol{U}\|_{L_\infty(0,T;H)} \leqslant S(T)(\mathscr{E}_\Omega(\boldsymbol{U}) + \mathscr{E}_\mathscr{I}(\boldsymbol{U}) + \mathscr{E}_V(\boldsymbol{U})), \qquad (16)$$

where $H \subset \boldsymbol{H}^1(\Omega)$ is the natural Hilbert energy–space for the problem. In this estimate, $\mathscr{E}_\Omega$, $\mathscr{E}_\mathscr{I}$ and $\mathscr{E}_V$ are residuals which are computable in terms of the data and the finite element solution $\boldsymbol{U}$, and $S(t)$ is the stability factor introduced before in (14).

The term $\mathscr{E}_\Omega$ contains the spatial discretization error and can be used to guide adaptive space mesh refinement. It is essentially identical to the residual derived for linear elasticity by Johnson and Hansbo in [10]. Some numerical experiments demonstrating such adaptive spatial error control using this estimate are given in the report [23], which also contains details of how the algorithm can be written in terms of internal variables, as used by Johnson and Tessler in, for example, [11]. The paper [22] also contains an a priori error estimate, discrete data-stability estimates and upper bounds on the residuals (where appropriate – see below). The stability factor is again given in [21].

The residual $\mathscr{E}_\mathscr{I}$ is either unstable as $h \to 0$ (i.e. useless) or – when written in a different form – prohibitively expensive to implement. We are currently working on a remedy for this involving weaker norms, as described above for the second kind Volterra equation.

It is the term $\mathscr{E}_V$ that causes a novel difficulty in this estimate. The spatial residuals in $\mathscr{E}_\Omega$ are constructed by integrating the discrete solution by parts over each element to arrive at a *distributional* divergence of the discrete stress. This divergence comprises two parts: the smooth function inside the element (which is zero in our case of piecewise linear approximation), and the stress jumps across inter-element boundaries. The difficulty arises because the stress is history dependent. This means that we have to integrate by parts over not just the elements in the current mesh, but also over all elements in all previous meshes. The internal edges that appeared in previous meshes but are no longer present in the current mesh (e.g. due to derefinement) are therefore "left behind" when forming the standard residual $\boldsymbol{f} + \nabla \cdot \underline{\boldsymbol{\sigma}}^h$ (which constitutes $\mathscr{E}_\Omega$), and so we consign the stress jumps across these edges to the term $\mathscr{E}_V$. In the particular case where only nested refinements in the space mesh are permitted, no edges are left behind in this way and we have $\mathscr{E}_V \equiv 0$.

To deal with mesh derefinement would appear to require fairly complex data structures in the computer code in order to track all these resulting previous edges. Also, it is not likely that $\mathscr{E}_V$ will act in any way other than to degrade the quality of the estimate since it contains historical contributions to the current stress. These can then only act to reinforce one another in the estimate when in fact the residual could be much smaller due to cancellation. Our feeling at the moment is that a representation of the algorithm in terms of internal variables could go some way toward removing the $\mathscr{E}_V$ residual, since then all hereditary information is automatically represented on the current mesh. The price of this is that the error estimates will then be restricted to viscoelasticity problems for which Prony series relaxation functions (as in (3)) are appropriate. This does not seem to be an unreasonable restriction.

4. Future work

Clearly there is enormous scope for further development of this work, and we close with just a few of the more obvious suggestions.

- Incorporate the time step into (16) by measuring the error in a suitable negative norm.
- Extend the work of Eriksson and Johnson in, for example, [3] on parabolic problems to include (7).
- Extend the discontinuous Galerkin approximation described above to the dynamic problem (5) building on the work of Hulbert and Hughes in [6,7], and Johnson in [9].
- Incorporate physically important nonlinear effects such as the reduced time model discussed by Knauss and Emri in, for example, [12,13]. Some early numerical computations based on this type of constitutive nonlinearity are given in [14].

Acknowledgements

The research summarized herein has been sponsored in part by the United States Army through its European Research Office (Contract No. N68171-97-M-5763), and in part by the Brunel University Research Committee: 96–97 QR funding scheme.

References

[1] D.S. Cohen, A.B. White Jr., T.P. Witelski, Shock formation in a multidimensional viscoelastic diffusive system, SIAM J. Appl. Math. 55 (1995) 348–368.
[2] K. Eriksson, D. Estep, P. Hansbo, C. Johnson, Introduction to adaptive methods for differential equations. Acta Numer. Cambridge University Press, Cambridge, 1995, 105–158.
[3] K. Eriksson, C. Johnson, Adaptive finite element methods for parabolic problems, I: a linear model problem, SIAM J. Numer. Anal. 28 (1991) 43–77.
[4] J.D. Ferry, Viscoelastic Properties of Polymers, Wiley, New York, 1970.
[5] W.N. Findley, J.S. Lai, K. Onaran, Creep and relaxation of nonlinear viscoelastic materials (with an introduction to linear viscoelasticity), in: H.A. Lauwerier, W.T. Koiter (Eds.), Applied Mathematics and Mechanics, North-Holland, Amsterdam, 1976.
[6] T.J.R. Hughes, G.M. Hulbert, Space-time finite element methods for elastodynamics: formulations and error estimates, Comput. Methods Appl. Mech. Eng. 66 (1988) 339–363.
[7] G.M. Hulbert, T.J.R. Hughes, Space-time finite element methods for second-order hyperbolic equations, Comput. Methods Appl. Mech. Eng. 84 (1990) 327–348.
[8] A.R. Johnson, Modeling viscoelastic materials using internal variables, Shock Vibr. Digest 31 (1999) 91–100.
[9] C. Johnson, Discontinuous Galerkin finite element methods for second order hyperbolic problems, Comput. Methods Appl. Mech. Eng. 107 (1993) 117–129.
[10] C. Johnson, P. Hansbo, Adaptive finite element methods in computational mechanics, Comput. Methods Appl. Mech. Eng. 101 (1992) 143–181.
[11] A.R. Johnson, A. Tessler, A viscoelastic high order beam finite element, in: J.R. Whiteman (Ed.), The Mathematics of Finite Elements and Applications. MAFELAP 1996, Wiley, Chichester, 1997, pp. 333–345.
[12] W.G. Knauss, I.J. Emri, Non-linear viscoelasticity based on free volume consideration, Comput. Struct. 13 (1981) 123–128.
[13] W.G. Knauss, I.J. Emri, Volume change and the nonlinearly thermo-viscoelastic constitution of polymers, Polym. Eng. Sci. 27 (1987) 86–100.

[14] S. Shaw, M.K. Warby, J.R. Whiteman, Numerical techniques for problems of quasistatic and dynamic viscoelasticity, in: J.R. Whiteman (Ed.), The Mathematics of Finite Elements and Applications. MAFELAP 1993, Wiley, Chichester, 1994, pp. 45–68.

[15] S. Shaw, J.R. Whiteman, Backward Euler and Crank–Nicolson finite element variants with rational adaptivity and a posteriori error estimates for an integrodifferential equation. Math. Comput., submitted for publication, see www.brunel.ac.uk/~icsrbicm, 1996.

[16] S. Shaw, J.R. Whiteman, Discontinuous Galerkin method with a posteriori $L_p(0,t_i)$ error estimate for second-kind Volterra problems, Numer. Math. 74 (1996) 361–383.

[17] S. Shaw, J.R. Whiteman, Applications and numerical analysis of partial differential Volterra equations: a brief survey, Comput. Methods Appl. Mech. Eng. 150 (1997) 397–409.

[18] S. Shaw, J.R. Whiteman, Negative norm error control for second-kind convolution Volterra equations, Numer. Math. 85 (2000) 329–341; BICOM Technical Report 98/6, see www.brunel.ac.uk/~icsrbicm, 1998.

[19] S. Shaw, J.R. Whiteman, Numerical solution of linear quasistatic hereditary viscoelasticity problems I: a priori estimates, BICOM Technical Report 98/2, see www.brunel.ac.uk/~icsrbicm, 1998.

[20] S. Shaw, J.R. Whiteman, Numerical solution of linear quasistatic hereditary viscoelasticity problems II: a posteriori estimates, BICOM Technical Report 98/3, see www.brunel.ac.uk/~icsrbicm, 1998.

[21] S. Shaw, J.R. Whiteman, Optimal long-time $L_p(0, T)$ data stability and semidiscrete error estimates for the Volterra formulation of the linear quasistatic viscoelasticity problem, Numer. Math., to appear; BICOM Technical Report 98/7, see www.brunel.ac.uk/~icsrbicm., 1998.

[22] S. Shaw, J.R. Whiteman, Numerical solution of linear quasistatic hereditary viscoelasticity problems, SIAM J. Numer. Anal (1999), to appear.

[23] S. Shaw, J.R. Whiteman, Robust adaptive finite element schemes for viscoelastic solid deformation: an investigative study, Technical Report, BICOM, Brunel University, Uxbridge, England, 1999, TR99/1 (US Army ERO Seed Project Report).

[24] V. Thomée, L.B. Wahlbin, Long-time numerical solution of a parabolic equation with memory, Math. Comput. 62 (1994) 477–496.

ELSEVIER

Journal of Computational and Applied Mathematics 125 (2000) 347–357

JOURNAL OF COMPUTATIONAL AND APPLIED MATHEMATICS

www.elsevier.nl/locate/cam

CP methods for the Schrödinger equation

L.Gr. Ixaru

Department of Theoretical Physics, Institute of Physics and Nuclear Engineering, P.O. Box MG - 6, Bucharest, Romania

Received 5 May 1999

Abstract

After a short survey over the efforts in the direction of solving the Schrödinger equation by using piecewise approximations of the potential function, the paper focuses on the piecewise perturbation methods in their CP implementation. The presentation includes a short list of problems for which CP versions are available, a sketch of the derivation of the CPM formulae, a description of various ways to construct or identify a certain version and also the main results of the error analysis. One of the most relevant results of the latter is that the energy dependence of the error is bounded, a fact which places these methods on a special position among the numerical methods for differential equations. A numerical illustration is also included in which a CPM based code for the regular Sturm–Liouville problem is compared with some other, well-established codes. © 2000 Elsevier Science B.V. All rights reserved.

1. Introduction

The piecewise perturbation methods are a class of numerical methods specially devised for the solution of the Schrödinger equation. The CP methods form a subclass whose algorithms are both easy to construct and very convenient for applications.

Given the one-dimensional Schrödinger equation,

$$y'' = (V(x) - E)y, \quad a \leqslant x \leqslant b, \tag{1.1}$$

where the potential $V(x)$ is a well-behaved function and the energy E is a free parameter, one can formulate either an initial-value (IV) or a boundary-value (BV) problem. The latter typically takes the form of a Sturm–Liouville (SL) problem in which the eigenvalues and the associated eigenfunctions are required.

Analytic solutions of this equation, either for the IV or for the SL problem, are known only for a restricted number of expressions for the function $V(x)$, let such functions be denoted by $\bar{V}(x)$, and for many years the physicists used to select that $\bar{V}(x)$ which is the closest to the given $V(x)$ and

E-mail address: ixaru@theor1.theory.nipne.ro (L.Gr. Ixaru).

PII: S0377-0427(00)00478-7

to accept that the analytical solution of the equation with the chosen $\bar{V}(x)$ is satisfactory for their investigations.

It was also obvious that a closer approximation may be achieved if the replacement of $V(x)$ by $\bar{V}(x)$ is made piecewise, i.e., if the interval $I_{a,b}=[a,b]$ is first conveniently partitioned, $r_0 = a, r_1, r_2, \ldots, r_{k_{\max}} = b$, and a suitable $\bar{V}(x)$ is introduced on each elementary interval $I_k=[r_{k-1},r_k]$. If so is done, the solution on the whole $I_{a,b}$ requires an extra algebraic manipulation of the piecewise analytic solutions, but such a task can be accomplished efficiently only by a computer. This is why the first systematic investigations along these lines are relatively recent [3–5,17]. In [3,5,17] the piecewise $\bar{V}(x)$ was a constant, the average value of $V(x)$ over I_k, while in [4] it was a straight-line segment which is tangent to $V(x)$. In the first case, the two linear independent solutions are given by trignometric or hyperbolic functions while in the second by the Airy functions. However, the error analysis of the two versions [6] has shown that both of them produce second-order methods, although the latter corresponds to a potentially better approximation.

Two directions of investigation were adopted: The first consists on assuming a piecewise polynomial form for $\bar{V}(x)$ and asking for the best fit of the polynomial coefficients on each I_k. The most important result is due to Pruess [12] and it says in essence that the best fit is by developing $V(x)$ over shifted Legendre polynomials. If, over each I_k, $\bar{V}(x)$ is chosen as an Nth degree polynomial obtained in this way, then the error is $O(h^{2N+2})$, see also [11,14]. However, seen from the perspective of a practical application, the use of polynomials of a degree higher than one is problematic in so much the accurate computation of the two linear independent solutions is difficult.

The other direction takes for $\bar{V}(x)$ only the potentials for which the two independent solutions have known, analytic forms which can be calculated efficiently. To further improve the accuracy, the corrections from the perturbation $\Delta V(x)=V(x)-\bar{V}(x)$ are added on each I_k, see Chapter 3 of [7]. In this way a general family of piecewise perturbation methods (PPM) is delimited.

The success of the latter approach depends, on course, on the correctness and efficiency when calculating the perturbation corrections. As shown in [7], if on each I_k, $\bar{V}(x)$ is taken as a constant and $\Delta V(x)$ is a polynomial, then the perturbation corrections have simple analytic forms. The numerical methods obtained on this basis are referred to as forming the CPM (short for constant (based) perturbation method) family. Each member of the family is identified by the degree N of the polynomial $\Delta V(x)$ and by the number of perturbation corrections Q retained in the algorithm. Clearly, the version CPM[N,Q] with a conveniently chosen value for Q will furnish the same accuracy as a method based just on a piecewise Nth degree polynomial for $\bar{V}(x)$. However, as said, while the algorithm of the former is at hand, writing an algorithm for the latter is not simple at all. The simplest version, in which $V(x)$ is approximated piecewise by a constant but no correction is introduced, is identified either as CPM[0,0] or directly as CPM(0).

CPM versions were also formulated for the case when the independent variable x, $V(x)$ and E are complex (see [10]) and for solving systems of coupled Schrödinger equations [7]. An extension of the latter in complex is now in final tests. A highly accurate version for Eq. (1.1) was obtained recently in [8] and a computer program based on it for the solution of the regular Sturm–Liouville problem is published in [9].

Among the salient advantages of the CP algorithms we mention: (i) the accuracy is uniform with respect to E, a feature unparalleled by any other numerical method; (ii) there is an easy control of the error; (iii) the step widths are unusually big and the computation is fast; (iv) the form of the algorithm allows a direct evaluation of the Prüfer phase and of the derivative of the solution with respect to E. When a SL problem is solved these data make the search of the eigenvalues very efficient. Finally, (v), the algorithms are of a form which allows using parallel computation.

It is also important to mention that the CP represents only one possible way of implementing a PPM. One could well take a piecewise line for $\bar{V}(x)$ and thus an LP implementation would result. Some work in this direction is also mentioned in [7] but up to now there is no general rule for constructing the corrections. Moreover, there are hints that the very construction of the successive corrections is problematic, if not just impossible, for the most challenging case of the systems of coupled equations. Another case of interest refers to the situation when $V(x)$ contains a centrifugal term, $l(l+1)/x^2$. The potential is then singular at the origin and, on a short interval around the origin, a specially tuned implementation should be used, with $\bar{V}(x)=l(l+1)/x^2$ for the reference potential. In principle, the generation of the formulae for the corrections is possible for this case. As from the practical point of view, the expressions of the first- and second-order corrections can be generated without difficulty but more performant packages are perhaps needed for the higher-order corrections.

Seen from this perspective the existing PPM-based programs may seem rather poor and rigid in so much each of them uses only one and the same preset version on all steps. Flexible programs, i.e., programs instructed to automatically select the most convenient version on each I_k, can be written. A substantial enhancement in efficiency is expected from such flexible programs and this is of particular interest for the solution of large systems of coupled equations.

2. The CPM algorithm

We focus on the initial-value problem for the Schrödinger equation (2.1) with $y(a)=y_0$, $y'(a)=y'_0$. The currrent interval I_k of the partition is denoted generically by $I=[X, X+h]$.

On I the solution is advanced by the so-called propagation matrix algorithm

$$\begin{bmatrix} y(X+h) \\ y'(X+h) \end{bmatrix} = \begin{bmatrix} u(h) & v(h) \\ u'(h) & v'(h) \end{bmatrix} \begin{bmatrix} y(X) \\ y'(X) \end{bmatrix}. \tag{2.1}$$

Functions $u(\delta)$ and $v(\delta)$, where $\delta=x-X$, called propagators, are the solutions of the local problem

$$y''(\delta)=(V(X+\delta)-E)y(\delta), \quad \delta\in[0,h] \tag{2.2}$$

with the initial values $y(0)=1$, $y'(0)=0$ for u and $y(0)=0$, $y'(0)=1$ for v. The one-step propagation matrix is

$$\boldsymbol{P}(\delta)=\begin{bmatrix} u(\delta) & v(\delta) \\ u'(\delta) & v'(\delta) \end{bmatrix} \tag{2.3}$$

and its inverse reads

$$\boldsymbol{P}^{-1}(\delta)=\begin{bmatrix} v'(\delta) & -v(\delta) \\ -u'(\delta) & u(\delta) \end{bmatrix}, \tag{2.4}$$

because $\mathrm{Det}[\boldsymbol{P}(\delta)]=1$. It follows that the knowledge of $u(h)$, $v(h)$, $u'(h)$ and $v'(h)$ is sufficient to advance the solutions in both directions.

To construct the propagators $v(\delta)$ and $u(\delta)$ the pertubation approach is used. Let

$$\bar{V} = \frac{1}{h}\int_0^h V(X+\delta)\,\mathrm{d}\delta, \qquad \Delta V(\delta) = V(X+\delta) - \bar{V}. \tag{2.5}$$

The original potential then reads $V(X+\delta) = \bar{V} + \Delta V(\delta)$, where $\bar{V}$ is a constant.

The procedure consists in taking $\bar{V}$ as the reference potential and $\Delta V(\delta)$ as a perturbation.

As explained in [7], each of $u(\delta)$ and $v(\delta)$, denoted generically by $p(\delta)$, is written as a perturbation series:

$$p(\delta) = p_0(\delta) + p_1(\delta) + p_2(\delta) + p_3(\delta) + \cdots, \tag{2.6}$$

where the zeroth-order term $p_0(\delta)$ is the solution of $p_0'' = (\bar{V} - E)p_0$, with $p_0(0) = 1$, $p_0'(0) = 0$ for u_0 and $p_0(0) = 0$, $p_0'(0) = 1$ for v_0. The correction p_q, $q = 1, 2, \ldots$, obeys the equation

$$p_q'' = (\bar{V} - E)p_q + \Delta V(\delta)p_{q-1}, \quad p_q(0) = p_q'(0) = 0. \tag{2.7}$$

With $Z(\delta) = (\bar{V} - E)\delta^2$ and functions $\xi(Z), \eta_0(Z), \eta_1(Z), \ldots$, defined in Appendix A, the zeroth-order propagators are

$$u_0(\delta) = \xi(Z(\delta)), \qquad v_0(\delta) = \delta\eta_0(Z(\delta)) \tag{2.8}$$

and the following iteration procedure exists to construct the corrections.

Correction p_{q-1} is assumed as known and of such a form that the product $\Delta V(\delta)p_{q-1}$ reads

$$\Delta V(\delta)p_{q-1}(\delta) = Q(\delta)\xi(Z(\delta)) + \sum_{m=0}^{\infty} R_m(\delta)\delta^{2m+1}\eta_m(Z(\delta)). \tag{2.9}$$

Then $p_q(\delta)$ and $p_q'(\delta)$ are of the form

$$p_q(\delta) = \sum_{m=0}^{\infty} C_m(\delta)\delta^{2m+1}\eta_m(Z(\delta)), \tag{2.10}$$

$$p_q'(\delta) = C_0(\delta)\xi(Z(\delta)) + \sum_{m=0}^{\infty}(C_m(\delta) + \delta C_{m+1}(\delta))\delta^{2m+1}\eta_m(Z(\delta)), \tag{2.11}$$

where $C_0(\delta), C_1(\delta), \ldots$ are given by quadrature (see again [7]):

$$C_0(\delta) = \frac{1}{2}\int_0^\delta Q(\delta_1)\,\mathrm{d}\delta_1, \tag{2.12}$$

$$C_m(\delta) = \frac{1}{2}\delta^{-m}\int_0^\delta \delta_1^{m-1}[R_{m-1}(\delta_1) - C_{m-1}''(\delta_1)]\,\mathrm{d}\delta_1, \quad m = 1, 2, \ldots. \tag{2.13}$$

To calculate successive corrections for u, the starting functions in $\Delta V(\delta)p_0(\delta)$ are $Q(\delta) = \Delta V(\delta)$, $R_0(\delta) = R_1(\delta) = \cdots = 0$, while for v they are $Q(\delta) = 0$, $R_0(\delta) = \Delta V(\delta)$, $R_1(\delta) = R_2(\delta) = \cdots = 0$.

The practical inconvenience is that successive quadratures starting from an arbitrary $\Delta V(\delta)$ are difficult to manipulate. For this reason, there is an intermediate stage in the procedure in which $V(X+\delta)$ is approximated by a polynomial in δ. More exactly, it is assumed that $V(X+\delta)$ can be written as a series over shifted Legendre polynomials $P_n^*(\delta/h)$ in the following way:

$$V(X+\delta) = \sum_{n=0} V_n h^n P_n^*(\delta/h). \tag{2.14}$$

The expressions of several $P_n^*(\gamma)$ polynomials, $\gamma \in [0,1]$, are as follows (see [1]):

$$P_0^*(\gamma)=1, \qquad P_1^*(\gamma)=-1+2\gamma,$$

$$P_2^*(\gamma)=1-6\gamma+6\gamma^2, \qquad P_3^*(\gamma)=-1+12\gamma-30\gamma^2+20\gamma^3.$$

The original $V(X+\delta)$ is then approximated by the truncated series

$$V^{(N)}(X+\delta)=\sum_{n=0}^{N} V_n h^n P_n^*(\delta/h). \tag{2.15}$$

As a matter of fact, the option for the shifted Legendre polynomials relies on the fact that such $V^{(N)}$ represents the best approximation to V in $L^2(X,X+h)$ by a polynomial of degree $\leqslant N$; the choice is consistent to that in [12].

Then the equation

$$y^{(N)''}=(V^{(N)}(X+\delta)-E)y^{(N)}, \quad \delta \in [0,h] \tag{2.16}$$

is the one whose propagators are actually constructed via CPM. With

$$\bar{V}=V_0, \quad \Delta V(\delta)=\Delta V^{(N)}(\delta)=\sum_{n=1}^{N} V_n h^n P_n^*(\delta/h), \tag{2.17}$$

integrals (2.12) and (2.13) can be solved analytically. Each $C_m(\delta)$ is a polynomial and the series (2.10) and (2.11) are finite.

Each value for N and for the maximal number of perturbation corrections Q would result in a version identified as CPM[N,Q]. The versions described in [7] take either $N=Q=0$ or $N=2$ as a default value and $Q=1,2$. These versions are there denoted as CPM(Q), $Q=0,1,2$, respectively.

The existence of powerful packages for analytic computation enabled to recently obtaining expressions of $u(h), hu'(h), v(h)/h$ and $v'(h)$ with more terms. In Appendix B we collect the expressions obtained in [8] by MATHEMATICA. The number of terms is large enough to be transparent for a pertinent error analysis and also to generate an algorithm of order 12 when $Z(h)=(V_0-E)h^2 \to 0$ and of order 10 when $-Z(h) \to +\infty$.

3. Error analysis of CPM[N, Q]

As explained above, CPM[N,Q] consists of two stages to be performed at each step. The first consists of approximating $V(X+\delta)$ by $V^{(N)}(X+\delta)$. This approximation gives rise to the errors

$$\varepsilon_k^{(N)}=\max\{|y(x_k)-y^{(N)}(x_k)|, |y'(x_k)-y^{(N)'}(x_k)|\}, \quad k=1,2,\ldots,k_{\max}. \tag{3.1}$$

The second stage consists of solving (16) by the perturbation technique with Q corrections included. The associated errors are

$$\bar{\varepsilon}_k^{[N,Q]}=\max\{|y^{(N)}(x_k)-\bar{y}(x_k)|, |y^{(N)'}(x_k)-\bar{y}'(x_k)|\}, \quad k=1,2,\ldots,k_{\max}, \tag{3.2}$$

where $\bar{y}(x_k)$ and $\bar{y}'(x_k)$ are the numerical values obtained by propagating the solution along the step intervals by using CPM[N,Q]. The error of the whole procedure, that is,

$$\varepsilon_k^{[N,Q]}=\max\{|y(x_k)-\bar{y}(x_k)|, |y'(x_k)-\bar{y}'(x_k)|\}, \quad k=1,2,\ldots,k_{\max}, \tag{3.3}$$

is bounded by the sum of both errors, namely

$$\varepsilon_k^{[N,Q]} \leqslant \bar{\varepsilon}_k^{[N,Q]} + \varepsilon_k^{(N)}. \tag{3.4}$$

As shown in [8], for each CPM$[N,Q]$ a $\bar{h}$ exists such that

Theorem. *If* CPM$[N,Q]$ *is applied to propagate the solution on an equidistant partition with* $h \leqslant \bar{h}$, *then*

— *if the energy* E *is such that* $|Z(h)|$ *is small in all intervals, a constant* C_N *exists such that*

$$\varepsilon_k^{[N,Q]} < C_N h^{2N+2}, \quad k = 1,2,\ldots,k_{\max}, \tag{3.5}$$

provided $Q \geqslant \lfloor \frac{2}{3}N \rfloor + 1$, $N = 1,2,\ldots,5$ *and* $Q = 0$ *for* $N = 0$. *The energy dependence of* C_N *is stronger and stronger as* N *increases.*

— *if* E *is such that* $Z(h) \ll 0$ *in all intervals, an* energy independent *constant* $\bar{C}_N^{\text{as}}$ *exists such that*

$$\varepsilon_k^{[N,Q]} < \bar{C}_N^{\text{as}} h^N / \sqrt{E}, \quad k = 1,2,\ldots,k_{\max}, \tag{3.6}$$

provided $Q \geqslant 1$ *if* $N = 1,2,\ldots,$ *and again* $Q = 0$ *if* $N = 0$.

The limitation of N up to 5 in the first part of the theorem is caused by the limited number of terms in Eqs. (B.5)–(B.8). However, there are reasons to think that the corresponding claim remains valid at any N. As a matter of fact it is of the same type as a result obtained for Sturm–Liouville problems, Theorem 9.5 in [14].

The theorem suggests that, for one and the same partition, the value of the energy E dictates two different behaviours of the error. If E is close enough to $V(x)$, specifically such that $|Z(h)|$ is small in each interval of the partition, then the method behaves as a method of order $n_0 = 2N + 2$. In contrast, when E is so high that $Z(h)$ is large and negative, it is another, asymptotic order $n_{\text{as}} = N$ which is appropriate. The theorem also shows that there is a damping of the error when E is increased.

The existence of two distinct orders allows an alternative way of formulating and identifying a CPM version. We can just ask to retain in the algorithm only the terms consistent with some input values for n_0 and n_{as}. This will lead to a unique N but to a sum over incomplete perturbations. The version corresponding to such a requirement is denoted as CPM$\{n_0, n_{\text{as}}\}$. The version corresponding to all terms given in Appendix B is identified as CPM$\{12, 10\}$.

The possibility of introducing an asymptotic order is unparalleled among the usual numerical methods for differential equations (Runge–Kutta or multistep, say). For these methods the accuracy quickly deteriorates when E is increased. A direct consequence is that the CPM's are the only methods for which the partition of the integration interval can be formulated from the very beginning of the run and never altered again, no matter how small or big is the energy. The E independent coefficients $C_m^{(u)}, C_m^{(u')}, C_m^{(v)}$ and $C_m^{(v')}$ (see Eqs. (B.5)–(B.8) in Appendix B, for version CPM$\{12, 10\}$) are also computed once on each step and stored. When the solution for a given E is advanced on successive steps, only the E dependent ξ and η_m remain to be calculated (this implies the computation of the pair of the trignometric or hyperbolic sine and cosine functions plus a few arithmetic operations) and they are introduced in the simple formulae (B.1)–(B.4) for the propagators. (The sums effectively consist of only a few terms; for CPM$\{12, 10\}$ the maximal m is five.) This very possibility of separating the relatively time consuming task of generating the partition and of calculating the quantities to be used later on, from

Table 1
Comparison of the four codes for problem 7 of Appendix B in [14] at $\beta = 20$

s	Reference E_s	ΔE_s			
		SLCPM12	SLEDGE	SLEIGN	SL02F
0	0.0000000000000245	−6.9(−10)	−4.5(−10)	1.8(−09)	−4.9(−11)
1	77.9169567714434	−1.3(−09)	6.5(−08)	1.1(−08)	−1.4(−09)
2	151.4627783464567	2.0(−10)	−4.2(−04)?	−1.1(−08)	−2.4(−04)
3	151.4632236576587	−3.6(−10)	6.0(−08)?	−3.0(−08)	−3.3(−09)
4	151.4636689883517	2.0(−10)	3.5(−04)	−1.3(−08)	2.3(−04)?
5	220.1542298352598	1.2(−09)	−1.3(−08)	2.2(−08)	−5.4(−09)
CPU time (s)		11.1	30.5	334.4	260.5

the repeatedly asked but fast executable task of integrating the equation at various values for E, represents perhaps the most important factor which makes the run with the CPM algorithms so efficient.

4. A numerical illustration

The results of the CPM$\{12, 10\}$ based code SLCPM12 [9] are now compared with the results of the codes SLEDGE [13], SLEIGN [2] and SL02F [16] for problem 7 of Appendix B of [14]. The potential function has the Coffey–Evans form $V(x) = -2\beta\cos(2x) + \beta^2\sin^2(2x)$ with $\beta = 20$, $a = -\pi/2$, $b = \pi/2$ and the boundary conditions are $y(a) = 0$, $y(b) = 0$.

The codes SLEDGE, SLEIGN and SL02F were accessed through SLDRIVER [15]. In all programs one and the same value for the tolerance is imposed, $\mathrm{TOL} = 10^{-8}$, and the first six eigenvalues are required.

By its very construction, the program SLCPM12 furnishes two approximations for each eigenvalue. The first, called the basic eigenvalue, is the result of the computation on the very partition consistent with the imposed TOL. The second, called the reference eigenvalue, results by repeating the run on a partition in which each step of the former is halved. Since the smallest of the two orders of the method is ten, the reference eigenvalue is expected to be by three orders of magnitude more accurate than the basic eigenvalue and therefore, the difference of the two produces an accurate estimation of the error in the latter. The knowledge of the reference eigenvalues is also used for the evaluation of the errors in the eigenvalues produced by the other codes.

In Table 1 we give the reference E_s and the deviations $\Delta E_s = E_s - \bar{E}_s$ where $\bar{E}_s$ are the energies furnished by SLCPM12 on its basic partition (this consisted of only 38 steps), and the usual outputs for the other three programs. Question marks were placed on the cases for which warnings have been mentioned during the computation. The associated CPU times from a PC with a 386 processor are also reported. We see that the results from the SLCPM12 are substantially more accurate than from the other codes and also that the computation is faster. The capacity of SLCPM12 of simultaneously producing highly accurate reference eigenvalues represents an additional advantage.

Appendix A

Functions $\xi(Z), \eta_0(Z), \eta_1(Z), \ldots$, originally introduced in [7] (they are denoted there as $\bar{\xi}(Z), \bar{\eta}_0(Z), \bar{\eta}_1(Z), \ldots$), are defined as follows:

$$\xi(Z) = \begin{cases} \cos(|Z|^{1/2}) & \text{if } Z \leqslant 0, \\ \cosh(Z^{1/2}) & \text{if } Z > 0, \end{cases} \tag{A.1}$$

$$\eta_0(Z) = \begin{cases} \sin(|Z|^{1/2})/|Z|^{1/2} & \text{if } Z < 0, \\ 1 & \text{if } Z = 0, \\ \sinh(Z^{1/2})/Z^{1/2} & \text{if } Z > 0, \end{cases} \tag{A.2}$$

while $\eta_m(Z)$ with $m > 0$ are further generated by recurrence

$$\eta_1(Z) = [\xi(Z) - \eta_0(Z)]/Z, \tag{A.3}$$

$$\eta_m(Z) = [\eta_{m-2}(Z) - (2m-1)\eta_{m-1}(Z)]/Z, \quad m = 2, 3, 4, \ldots, \tag{A.4}$$

if $Z \neq 0$, and by following values at $Z = 0$:

$$\eta_m(0) = 1/(2m+1)!!, \quad m = 1, 2, 3, 4, \ldots . \tag{A.5}$$

Some useful properties are as follows:

(i) Series expansion:

$$\eta_m(Z) = 2^m \sum_{q=0}^{\infty} \frac{g_{mq} Z^q}{(2q + 2m + 1)!}, \tag{A.6}$$

with

$$g_{mq} = \begin{cases} 1, & \text{if } m = 0, \\ (q+1)(q+2)\ldots(q+m) & \text{if } m > 0. \end{cases} \tag{A.7}$$

(ii) Asymptotic behaviour at large $|Z|$:

$$\eta_m(Z) \approx \begin{cases} \xi(Z)/Z^{(m+1)/2} & \text{for odd } m, \\ \eta_0(Z)/Z^{m/2} & \text{for even } m. \end{cases} \tag{A.8}$$

(iii) Differentiation properties:

$$\xi'(Z) = \tfrac{1}{2}\eta_0(Z), \qquad \eta'_m(Z) = \tfrac{1}{2}\eta_{m+1}(Z), \quad m = 0, 1, 2, \ldots . \tag{A.9}$$

(iv) Generating differential equation: $\eta_m(Z)$, $m = 0, 1, \ldots$ is the regular solution of

$$Zw'' + \tfrac{1}{2}(2m+3)w' - \tfrac{1}{4}w = 0. \tag{A.10}$$

(v) Relation with the spherical Bessel functions:

$$\eta_m(-x^2) = \sqrt{\pi/2}\, x^{-(m+1/2)} J_{m+1/2}(x), \quad m = 0, 1, \ldots . \tag{A.11}$$

Appendix B

The CPM standard form of the four elements of the propagation matrix $\boldsymbol{P}(\delta)$, Eq. (2.3), at $\delta=h$ is

$$u(h)=\xi(Z)+\sum_{m=1}^{\infty}C_m^{(u)}\eta_m(Z), \tag{B.1}$$

$$hu'(h)=Z\eta_0(Z)+\sum_{m=0}^{\infty}C_m^{(u')}\eta_m(Z), \tag{B.2}$$

$$v(h)/h=\eta_0(Z)+\sum_{m=2}^{\infty}C_m^{(v)}\eta_m(Z), \tag{B.3}$$

$$v'(h)=\xi(Z)+\sum_{m=1}^{\infty}C_m^{(v')}\eta_m(Z), \tag{B.4}$$

where the C coefficients depend only on the perturbation while the energy dependence is absorbed entirely in the Z-dependent functions ξ and η_m.

We give below the expressions of the coefficients as obtained in [8] by MATHEMATICA. With $V_0, V_1, V_2, \ldots$ defined in Eq. (2.14), $Z=(V_0-E)h^2$ and $\bar{V}_n=V_nh^{n+2}$, $n=1,2,\ldots$ these expressions are as follows:

$$\begin{aligned}
C_1^{(u)}&=-[\bar{V}_1+\bar{V}_3+\bar{V}_5+\bar{V}_7+\bar{V}_9]/2+\mathrm{O}(h^{13}),\\
C_2^{(u)}&=[5\bar{V}_3+14\bar{V}_5+27\bar{V}_7+44\bar{V}_9]/2-[105\bar{V}_1^2+63\bar{V}_2^2+45\bar{V}_3^2+35\bar{V}_4^2]/2520+\mathrm{O}(h^{13}),\\
C_3^{(u)}&=-3[21\bar{V}_5+99\bar{V}_7+286\bar{V}_9]/2+[2\bar{V}_1\bar{V}_2+\bar{V}_1\bar{V}_3+2\bar{V}_1\bar{V}_4+2\bar{V}_2\bar{V}_3\\
&\quad+\bar{V}_1\bar{V}_5+2\bar{V}_1\bar{V}_6+2\bar{V}_2\bar{V}_5+2\bar{V}_3\bar{V}_4+\bar{V}_1\bar{V}_7+\bar{V}_3\bar{V}_5]/4\\
&\quad-[63\bar{V}_2^2-60\bar{V}_3^2+35\bar{V}_4^2]/840+\mathrm{O}(h^{13}),\\
C_4^{(u)}&=3[429\bar{V}_7+2860\bar{V}_9]/2+[-9\bar{V}_1\bar{V}_5+3\bar{V}_2\bar{V}_4-54\bar{V}_1\bar{V}_6-42\bar{V}_2\bar{V}_5-36\bar{V}_3\bar{V}_4\\
&\quad-22\bar{V}_1\bar{V}_7+3\bar{V}_2\bar{V}_6-9\bar{V}_3\bar{V}_5]/4+5[-9\bar{V}_3^2+35\bar{V}_4^2]/168\\
&\quad+[35\bar{V}_1^3+42\bar{V}_1^2\bar{V}_2+35\bar{V}_1^2\bar{V}_3+21\bar{V}_1\bar{V}_2^2+54\bar{V}_1\bar{V}_2\bar{V}_3+6\bar{V}_2^3]/1680+\mathrm{O}(h^{13}),\\
C_5^{(u)}&=-36465\bar{V}_9/2+[396\bar{V}_1\bar{V}_6+252\bar{V}_2\bar{V}_5+210\bar{V}_3\bar{V}_4+143\bar{V}_1\bar{V}_7-33\bar{V}_2\bar{V}_6\\
&\quad+15\bar{V}_3\bar{V}_5-210\bar{V}_4^2]/4-[805\bar{V}_1^2\bar{V}_3+651\bar{V}_1\bar{V}_2^2+420\bar{V}_1^2\bar{V}_4+300\bar{V}_1\bar{V}_2\bar{V}_3-\bar{V}_2^3]/1680\\
&\quad+\bar{V}_1^4/1152+\mathrm{O}(h^{13}),\\
C_m^{(u)}&=0+\mathrm{O}(h^{t(m)})\quad\text{with } t(m)\geqslant 13\quad \forall m\geqslant 6,\\
C_0^{(u')}&=[\bar{V}_2+\bar{V}_4+\bar{V}_6+\bar{V}_8+\bar{V}_{10}]/2+\mathrm{O}(h^{14}),\\
C_1^{(u')}&=-[3\bar{V}_2+10\bar{V}_4+21\bar{V}_6+36\bar{V}_8+55\bar{V}_{10}]/2-[105\bar{V}_1^2+63\bar{V}_2^2+45\bar{V}_3^2+35\bar{V}_4^2]/2520\\
&\quad+\mathrm{O}(h^{14}),
\end{aligned} \tag{B.5}$$

$$C_2^{(u')} = [35\bar{V}_4 + 189\bar{V}_6 + 594\bar{V}_8 + 1430\bar{V}_{10}]/2 - [2\bar{V}_1\bar{V}_3 + 2\bar{V}_1\bar{V}_5 + \bar{V}_2\bar{V}_4 + 2\bar{V}_1\bar{V}_7 + \bar{V}_2\bar{V}_6 + 2\bar{V}_3\bar{V}_5]/4 - [735\bar{V}_1^2 + 378\bar{V}_2^2 + 675\bar{V}_3^2 + 350\bar{V}_4^2]/2520 + \mathrm{O}(h^{14}),$$

$$C_3^{(u')} = -[693\bar{V}_6 + 5148\bar{V}_8 + 21450\bar{V}_{10}]/2 + [16\bar{V}_1\bar{V}_3 + 43\bar{V}_1\bar{V}_5 + 26\bar{V}_2\bar{V}_4 + 82\bar{V}_1\bar{V}_7 + 48\bar{V}_2\bar{V}_6 + 63\bar{V}_3\bar{V}_5]/4 + [1197\bar{V}_2^2 + 3735\bar{V}_3^2 + 4900\bar{V}_4^2]/840 + [7\bar{V}_1^2\bar{V}_2 - 35\bar{V}_1^2\bar{V}_4 + 54\bar{V}_1\bar{V}_2\bar{V}_3 - 15\bar{V}_2^3]/1680 + \mathrm{O}(h^{14}),$$

$$C_4^{(u')} = [19305\bar{V}_8 + 182325\bar{V}_{10}]/2 - [261\bar{V}_1\bar{V}_5 + 165\bar{V}_2\bar{V}_4 + 1210\bar{V}_1\bar{V}_7 + 726\bar{V}_2\bar{V}_6 + 726\bar{V}_3\bar{V}_5]/4 - 5[1917\bar{V}_3^2 + 7392\bar{V}_4^2]/504 + [2331\bar{V}_1^2\bar{V}_2 + 2310\bar{V}_1^2\bar{V}_4 + 4752\bar{V}_1\bar{V}_2\bar{V}_3 + 234\bar{V}_2^3]/5040 + \bar{V}_1^4/1152 + \mathrm{O}(h^{14}),$$

$$C_5^{(u')} = -692835\bar{V}_{10}/2 + [6578\bar{V}_1\bar{V}_7 + 3927\bar{V}_2\bar{V}_6 + 3165\bar{V}_3\bar{V}_5]/4 + 8785\bar{V}_4^2/24 - [31395\bar{V}_1^2\bar{V}_4 + 44370\bar{V}_1\bar{V}_2\bar{V}_3 + 5679\bar{V}_2^3]/5040 + 13\bar{V}_1^4/1152 + \mathrm{O}(h^{14}),$$

$$C_m^{(u')} = 0 + \mathrm{O}(h^{t(m)}) \quad \text{with } t(m) \geqslant 14 \quad \forall m \geqslant 6, \tag{B.6}$$

$$C_2^{(v)} = -[\bar{V}_2 + \bar{V}_4 + \bar{V}_6 + \bar{V}_8 + \bar{V}_{10}]/2 + \mathrm{O}(h^{14}),$$

$$C_3^{(v)} = [7\bar{V}_4 + 18\bar{V}_6 + 33\bar{V}_8]/2 - [35\bar{V}_1^2 + 21\bar{V}_2^2 + 15\bar{V}_3^2]/840 + \mathrm{O}(h^{12}),$$

$$C_4^{(v)} = -[99\bar{V}_6 + 429\bar{V}_8]/2 + [2\bar{V}_1\bar{V}_3 + 2\bar{V}_1\bar{V}_5 + 3\bar{V}_2\bar{V}_4]/4 + [63\bar{V}_2^2 + 40\bar{V}_3^2]/280 + \mathrm{O}(h^{12}),$$

$$C_5^{(v)} = 2145\bar{V}_8/2 - [27\bar{V}_1\bar{V}_5 + 19\bar{V}_2\bar{V}_4]/4 - 115\bar{V}_3^2/56 + 11\bar{V}_1^2\bar{V}_2/240 + \mathrm{O}(h^{12}),$$

$$C_m^{(v)} = 0 + \mathrm{O}(h^{t(m)}) \quad \text{with } t(m) \geqslant 12 \quad \forall m \geqslant 6, \tag{B.7}$$

$$C_1^{(v')} = [\bar{V}_1 + \bar{V}_3 + \bar{V}_5 + \bar{V}_7 + \bar{V}_9]/2 + \mathrm{O}(h^{13}),$$

$$C_2^{(v')} = -[5\bar{V}_3 + 14\bar{V}_5 + 27\bar{V}_7 + 44\bar{V}_9]/2 - [105\bar{V}_1^2 + 63\bar{V}_2^2 + 45\bar{V}_3^2 + 35\bar{V}_4^2]/2520 + \mathrm{O}(h^{13}),$$

$$C_3^{(v')} = 3[21\bar{V}_5 + 99\bar{V}_7 + 286\bar{V}_9]/2 - [2\bar{V}_1\bar{V}_2 - \bar{V}_1\bar{V}_3 + 2\bar{V}_1\bar{V}_4 + 2\bar{V}_2\bar{V}_3 - \bar{V}_1\bar{V}_5 + 2\bar{V}_1\bar{V}_6 + 2\bar{V}_2\bar{V}_5 + 2\bar{V}_3\bar{V}_4 - \bar{V}_1\bar{V}_7 - \bar{V}_3\bar{V}_5]/4 - [63\bar{V}_2^2 - 60\bar{V}_3^2 + 35\bar{V}_4^2]/840 + \mathrm{O}(h^{13}),$$

$$C_4^{(v')} = -3[429\bar{V}_7 + 2860\bar{V}_9]/2 - [9\bar{V}_1\bar{V}_5 - 3\bar{V}_2\bar{V}_4 - 54\bar{V}_1\bar{V}_6 - 42\bar{V}_2\bar{V}_5 - 36\bar{V}_3\bar{V}_4$$
$$+ 22\bar{V}_1\bar{V}_7 - 3\bar{V}_2\bar{V}_6 + 9\bar{V}_3\bar{V}_5]/4 + 5[9\bar{V}_3^2 + 35\bar{V}_4^2]/168 - [35\bar{V}_1^3 - 42\bar{V}_1^2\bar{V}_2$$
$$+ 35\bar{V}_1^2\bar{V}_3 + 21\bar{V}_1\bar{V}_2^2 - 54\bar{V}_1\bar{V}_2\bar{V}_3 - 6\bar{V}_2^3]/1680 + O(h^{13}),$$
$$C_5^{(v')} = 36465\bar{V}_9/2 + [-396\bar{V}_1\bar{V}_6 - 252\bar{V}_2\bar{V}_5 - 210\bar{V}_3\bar{V}_4 + 143\bar{V}_1\bar{V}_7 - 33\bar{V}_2\bar{V}_6$$
$$+ 15\bar{V}_3\bar{V}_5 - 210\bar{V}_4^2]/4 + [805\bar{V}_1^2\bar{V}_3 + 651\bar{V}_1\bar{V}_2^2 - 420\bar{V}_1^2\bar{V}_4 - 300\bar{V}_1\bar{V}_2\bar{V}_3$$
$$+ 60\bar{V}_2^3]/1680 + \bar{V}_1^4/1152 + O(h^{13}),$$
$$C_m^{(v')} = 0 + O(h^{t(m)}) \quad \text{with } t(m) \geqslant 13 \quad \forall m \geqslant 6. \tag{B.8}$$

References

[1] M. Abramowitz, I.A. Stegun, Handbook of Mathematical Functions, 8th Edition, Dover, New York, 1972.

[2] P.B. Bailey, M.K. Gordon, L.F. Shampine, Automatic solution of the Sturm–Liouville problem, ACM Trans. Math. Software 4 (1978) 193–207.

[3] J. Canosa, R. Gomes de Oliveira, A new method for the solution of the Schrödinger equation, J. Comput. Phys. 5 (1970) 188–207.

[4] R.G. Gordon, A new way for constructing wave functions for bound states and scattering, J. Chem. Phys. 51 (1969) 14–25.

[5] L.Gr. Ixaru, An algebraic solution of the Schrödinger equation, Internal Report IC/69/6, International Centre for Theoretical Physics, Trieste, 1969.

[6] L.Gr. Ixaru, The error analysis of the algebraic method to solve the Schrödinger equation, J. Comput. Phys. 9 (1972) 159–163.

[7] L.Gr. Ixaru, Numerical Methods for Differential Equations and Applications, Reidel, Dordrecht, 1984.

[8] L.Gr. Ixaru, H. De Meyer, G. Vanden Berghe, CP methods for the Schrödinger equation revisited, J. Comput. Appl. Math. 88 (1998) 289.

[9] L.Gr. Ixaru, H. De Meyer, G. Vanden Berghe, SLCPM12, a program for the solution of regular Sturm–Liouville problems, Comput. Phys. Comm. 118 (1999) 259.

[10] L.Gr. Ixaru, M. Rizea, T. Vertse, Piecewise perturbation methods for calculating eigensolutions of a complex optical potential, Comput. Phys. Comm. 85 (1995) 217–230.

[11] J.W. Paine, F.R. de Hoog, Uniform estimation of the eigenvalues of Sturm–Liouville problem, J. Austral. Math. Soc. Ser. B 21 (1980) 365–383.

[12] S. Pruess, Estimating the eigenvalues of Sturm–Liouville problems by approximating the differential equation, SIAM J. Numer. Anal. 10 (1973) 55–68.

[13] S. Pruess, C.T. Fulton, Mathematical software for Sturm–Liouville problems, ACM Trans. Math. Software 19 (1993) 360–376.

[14] J.D. Pryce, Numerical Solution of Sturm–Liouville Problems, Oxford University Press, Oxford, 1993.

[15] J.D. Pryce, SLDRIVER installation guide and user guide, Technical Report SEAS/CISE/JDP03/96, Royal Military College of Sciences, Shrivenham, UK, 1996.

[16] J.D. Pryce, M. Marletta, Automatic solution of Sturm–Liouville problems using the Pruess method, J. Comput. Appl. Math. 39 (1992) 57–78.

[17] L.Gr. Ixaru, The algebraic approach to the scattering problem, Internal Report IC/69/7, International Centre for Theoretical Physics, Trieste, 1969.

JOURNAL OF
COMPUTATIONAL AND
APPLIED MATHEMATICS

ELSEVIER Journal of Computational and Applied Mathematics 125 (2000) 359–366

www.elsevier.nl/locate/cam

Asymptotic correction of Numerov's eigenvalue estimates with natural boundary conditions

Alan L. Andrew

Mathematics Department, La Trobe University, Bundoora, Victoria 3083, Australia

Received 18 November 1999; received in revised form 14 February 2000

Abstract

Asymptotic correction, at negligible extra cost, greatly improves the accuracy of higher eigenvalues computed by finite difference or finite element methods, and generally increases the accuracy of the lower ones as well. This paper gives a brief overview of the technique and describes how its previous use with Numerov's method may be extended to problems with natural boundary conditions. Numerical results indicate that it is just as successful as with Dirichlet boundary conditions.

Keywords: Sturm–Liouville problems; Eigenvalues; Numerov's method

1. Asymptotic correction

The term "asymptotic correction" is used here, following [1], to describe a technique first studied [19] in connection with the computation of the eigenvalues, $\lambda_1^{(0)} < \lambda_2^{(0)} < \cdots$, of the regular Sturm–Liouville problem

$$-y'' + qy = \lambda y, \tag{1}$$

$$y(0) = y(\pi) = 0. \tag{2}$$

The key idea of asymptotic correction is that, at least for sufficiently smooth q, the leading asymptotic term in the error in the computed eigenvalue is independent of q. Moreover, when q is constant, the error in the estimate of $\lambda_k^{(0)}$, obtained by the classical second-order centred finite difference method with the interval $[0,\pi]$ divided into n equal subintervals, is known to be exactly $\varepsilon_1(k,h)$, where $h := \pi/n$ and, following [3], we use the notation

$$\varepsilon_r(k,h) := k^2 - \frac{12\sin^2(kh/2)}{h^2[3 + (1-r)\sin^2(kh/2)]}. \tag{3}$$

0377-0427/00/$ - see front matter
PII: S0377-0427(00)00479-9

Hence, the known error for the case of constant q can be used to "correct" the estimate obtained for general q. Indeed, it was shown in [19] that, at least for "sufficiently small" kh, this correction reduced the $O(k^4h^2)$ error in the estimate for $\lambda_k^{(0)}$ obtained by the classical finite difference estimate to one of $O(kh^2)$. An alternative form of the error, which does not include the restriction to "sufficiently small" kh, is suggested in [9] and discussed further in [1]. Sometimes the technique has been given other names, including "algebraic correction" [13] and "the AAdHP correction" [20].

Anderssen and de Hoog [8] extended the results of [19] to problems involving general separated boundary conditions. For non-Dirichlet boundary conditions, they showed, using asymptotic formulae for eigenvalues of the discrete problem [12], that, for "sufficiently small" kh, the error in the corrected eigenvalues is $O(h^2)$. With these boundary conditions there is usually no simple closed form expression for the error in the case of constant q, but they gave a simple numerical method for approximating this error. Fortunately, as noted in [4], in the three important special cases

$$y'(0) = y'(\pi) = 0, \tag{4}$$

$$y(0) = y'(\pi) = 0 \tag{5}$$

and

$$y'(0) = y(\pi) = 0, \tag{6}$$

the exact errors obtained by the classical second-order centered finite difference method for constant q have the closed forms $\varepsilon_1(k-1,h)$, $\varepsilon_1(k-\frac{1}{2},h)$ and $\varepsilon_1(k-\frac{1}{2},h)$, respectively. Asymptotic correction has been shown to be especially effective for the finite element method [2,10,17]. For finite element eigenvalues with linear hat coordinate functions, the appropriate corrections for (1) with boundary conditions (2), (4), (5) and (6) are $\varepsilon_3(k,h)$, $\varepsilon_3(k-1,h)$, $\varepsilon_3(k-\frac{1}{2},h)$ and $\varepsilon_3(k-\frac{1}{2},h)$, respectively. The method is extended to periodic and semiperiodic boundary conditions in [2,4]. An alternative to the classical difference approximations used in [4] is suggested in [21], and analyzed in [11]. Numerical evidence shows that asymptotic correction can also be useful for some methods for which a complete theory is still lacking [5,15], including the computation of eigenvalues of certain partial differential operators [1,13].

A survey of results on asymptotic correction up to 1992, including a discussion of some open questions, is given in [5], while more recent developments, including progress on the problems listed in [5], are considered in [7]. Asymptotic correction is particularly useful for the computation of a substantial number of eigenvalues of regular Sturm–Liouville problems to moderate accuracy (say 8 significant figures). For singular (or nearly singular) problems, for the computation of eigenfunctions, and usually even for the highly accurate computation of just the first eigenvalue, there are better methods [20]. Partly because it is so efficient for dealing simultaneously with many eigenvalues, asymptotic correction has proved especially useful in the solution of inverse eigenvalue problems [6,13,14,17,18]. Many authors have suggested that inverse Sturm–Liouville problems be solved numerically by using algorithms for matrix inverse eigenvalue problems, but, as shown in [17,18], appropriate use of asymptotic correction is crucial to the viability of such methods. Before solving the corresponding discrete inverse eigenvalue problem, the correction which would be added to the discrete eigenvalues in the forward problem must be *subtracted* from the observed eigenvalues of the continuous problem.

The rest of this paper concerns the use of asymptotic correction with the deservedly popular Numerov method, which gives fourth-order accuracy while using only tridiagonal matrices. Previous work on asymptotic correction with Numerov's method has considered only the boundary conditions (2). Here an extension to the boundary conditions (4), (5) and (6) is proposed and tested. As shown in [9] for (2), the corrections required for these boundary conditions have the same form as for the second-order finite difference or finite element methods, but with ε_2 instead of ε_1 or ε_3. As with second-order methods [2], results given here for $[0,\pi]$ may be generalized to an arbitrary finite interval.

2. Application to Numerov's method

Numerov's method approximates (1) by the three term recurrence relation

$$-[12 - h^2q(x_{i-1})]y_{i-1} + [24 + 10h^2q(x_i)]y_i - [12 - h^2q(x_{i+1})]y_{i+1} = h^2\Lambda[y_{i-1} + 10y_i + y_{i+1}], \tag{7}$$

where $x_i := ih$, and Λ and y_i are the Numerov approximations of λ and $y(x_i)$. With the boundary conditions $y_0 = y_n = 0$ corresponding to (2), this gives a matrix eigenvalue problem $A^{(0,n)}\boldsymbol{x} = \Lambda B^{(0,n)}\boldsymbol{x}$, whose eigenvalues, $\Lambda_1^{(0,n)} < \Lambda_2^{(0,n)} < \cdots < \Lambda_{n-1}^{(0,n)}$ are the Numerov approximations of $\lambda_1^{(0)} < \lambda_2^{(0)} < \cdots < \lambda_{n-1}^{(0)}$. (Note $A^{(0,n)}$ and $B^{(0,n)}$ are $(n-1)\times(n-1)$ and tridiagonal.) It is known that $\lambda_k^{(0)} - \Lambda_k^{(0,n)} = \mathrm{O}(k^6h^4)$, this estimate being sharp. In particular, when q is constant, $\lambda_k^{(0)} - \Lambda_k^{(0,n)} = \varepsilon_2(k,h) = k^6h^4/240 + \mathrm{O}(k^8h^6)$. Consequently, the estimate for $\lambda_k^{(0)}$ produced by asymptotic correction is $\tilde{\Lambda}_k^{(0,n)} := \Lambda_k^{(0,n)} + \varepsilon_2(k,h)$. It was shown in [9] that, if $q \in C^4[0,\pi]$, then

$$\tilde{\Lambda}_k^{(0,n)} - \lambda_k^{(0)} = \mathrm{O}(k^4h^5/\sin(kh)), \tag{8}$$

so that $\tilde{\Lambda}_{n-1}^{(0,n)} - \lambda_{n-1}^{(0)} = \mathrm{O}(1)$, which is similar to a result found to be sharp for certain second-order methods [2,4,10]. However, numerical results [1,5,9] suggest that (8) can be sharpened to $\tilde{\Lambda}_k^{(0,n)} - \lambda_k^{(0)} = \mathrm{O}(k^3h^5/\sin(kh))$, this latter estimate being sharp.

By Taylor's theorem, it follows from differentiating (1) that, if $y'(a) = \alpha y(a)$, then

$$y(a+h) - y(a-h) = h[\alpha(2 + (q(a) - \lambda)h^2/3) + h^2q'(a)/3]y(a) + \mathrm{O}(h^5). \tag{9}$$

Consequently, a fourth-order approximation of the condition $y'(0) = 0$ is

$$3(y_1 - y_{-1}) = h^3q'(x_0)y_0, \tag{10}$$

while a fourth-order approximation of $y'(\pi) = 0$ is

$$3(y_{n+1} - y_{n-1}) = h^3q'(x_n)y_n. \tag{11}$$

Eliminating y_{-1} between (10) and (7) with $i = 0$ gives the equation

$$[24 + 10h^2q(x_0) + h^3q'(x_0)(4 - h^2q(x_{-1})/3)]y_0 - [24 - h^2(q(x_{-1}) + q(x_1))]y_1 = h^2\Lambda[(10 - h^3q'(x_0)/3)y_0 + 2y_1], \tag{12}$$

while eliminating y_{n+1} between (11) and (7) with $i = n$ gives

$$-[24 - h^2(q(x_{n-1}) + q(x_{n+1}))]y_{n-1} + [24 + 10h^2q(x_n) - h^3q'(x_n)(4 - h^2q(x_{n+1})/3)]y_n = h^2\Lambda[2y_{n-1} + (10 + h^3q'(x_n)/3)y_n]. \tag{13}$$

Let $\lambda_k^{(1)}$, $\lambda_k^{(2)}$ and $\lambda_k^{(3)}$ denote the kth eigenvalues of (1) with boundary conditions (4), (5) and (6), respectively. Using (12) and (13) with (7), again for $i=1,\dots,n-1$, gives, as Numerov approximations of $\lambda_1^{(1)} < \lambda_2^{(1)} < \cdots < \lambda_{n+1}^{(1)}$, the eigenvalues, $\Lambda_1^{(1,n)} < \Lambda_2^{(1,n)} < \cdots < \Lambda_{n+1}^{(1,n)}$, of a matrix eigenvalue problem $A^{(1,n)}\boldsymbol{x}=\Lambda B^{(1,n)}\boldsymbol{x}$, where the matrices $A^{(1,n)}$ and $B^{(1,n)}$ so defined are $(n+1)\times(n+1)$ and again tridiagonal. Similarly, Numerov approximations of $\lambda_1^{(2)} < \lambda_2^{(2)} < \cdots < \lambda_n^{(2)}$ and $\lambda_1^{(3)} < \lambda_2^{(3)} < \cdots < \lambda_n^{(3)}$ are given by the eigenvalues $\Lambda_1^{(2,n)} < \Lambda_2^{(2,n)} < \cdots < \Lambda_n^{(2,n)}$ and $\Lambda_1^{(3,n)} < \Lambda_2^{(3,n)} < \cdots < \Lambda_n^{(3,n)}$ of the matrix eigenvalue problems $A^{(2,n)}\boldsymbol{x} = \Lambda B^{(2,n)}\boldsymbol{x}$ and $A^{(3,n)}\boldsymbol{x} = \Lambda B^{(3,n)}\boldsymbol{x}$, respectively, the former being defined by combining (7), for $i=1,\dots,n-1$ with (13) and $y_0=0$, and the latter by combining (7), for $i=1,\dots,n-1$ with (12) and $y_n=0$. The $n\times n$ matrices $A^{(2,n)}, B^{(2,n)}, A^{(3,n)}$ and $B^{(3,n)}$ so defined are again tridiagonal. Indeed $A^{(0,n)}$ and $B^{(0,n)}$ are obtained from $A^{(1,n)}$ and $B^{(1,n)}$ by deleting the first and last row and column, $A^{(2,n)}$ and $B^{(2,n)}$ by deleting the first row and column, and $A^{(3,n)}$ and $B^{(3,n)}$ by deleting the last row and column.

When $q=0$, the kth eigenvalues of (1), with boundary conditions (2),(4),(5) and (6), are k^2, $(k-1)^2$, $(k-\frac{1}{2})^2$ and $(k-\frac{1}{2})^2$, respectively. Moreover, when $q=0$, (7), (10) and (11) reduce to

$$-12y_{i-1} + 24y_i - 12y_{i+1} = h^2\Lambda[y_{i-1} + 10y_i + y_{i+1}], \tag{14}$$

$y_{-1} = y_1$ and $y_{n+1} = y_{n-1}$, respectively. It is readily checked that, for all real numbers θ and β, $y_i = \sin(i\theta+\beta)$ satisfies (14) for all integers i, when

$$\Lambda = \frac{12[\sin^2(\theta/2)]}{h^2[3-\sin^2(\theta/2)]}.$$

The boundary condition $y_0=0$ requires $\beta=m\pi$ for some integer m, while $y_{-1}=y_1$ requires $\beta=(m+\frac{1}{2})\pi$ or $\sin(\theta)=0$. Thus possible values of θ, and hence the complete set of eigenvalues and corresponding eigenvectors of the matrix equations $A^{(r,n)}\boldsymbol{x} = \Lambda B^{(r,n)}\boldsymbol{x}$, $r=0,1,2,3$, are readily determined by the remaining boundary condition. Hence, since the corrected estimate, $\tilde{\Lambda}_k^{(r,n)}$, of the kth eigenvalue of (1) with appropriate boundary conditions is obtained by adding to $\Lambda_k^{(r,n)}$ a quantity which would give the exact answer when $q=0$, it is readily deduced, using (3), that

$$\tilde{\Lambda}_k^{(0,n)} := \Lambda_k^{(0,n)} + \varepsilon_2(k,h),$$

$$\tilde{\Lambda}_k^{(1,n)} := \Lambda_k^{(1,n)} + \varepsilon_2(k-1,h),$$

$$\tilde{\Lambda}_k^{(2,n)} := \Lambda_k^{(2,n)} + \varepsilon_2(k-\tfrac{1}{2},h),$$

$$\tilde{\Lambda}_k^{(3,n)} := \Lambda_k^{(3,n)} + \varepsilon_2(k-\tfrac{1}{2},h).$$

The denominator $\sin(kh)$ in (8) is no problem with (2) as the matrices $A^{(0,n)}$ and $B^{(0,n)}$ are $(n-1)\times(n-1)$ and $h=\pi/n$. However, because of the larger dimension of the matrices associated with (4)–(6), any extension of (8) to these boundary conditions must allow for the possibility that $\sin(kh)=0$. For second-order methods, a related result [2] for (5) and (6) has denominator $\sin((k-\frac{1}{2})h)$, and an alternative approach [4] shows that with (4) the error in the highest eigenvalue computed using asymptotic correction is again O(1), compared with the $O(n^2)$ error in the corresponding uncorrected eigenvalues. Note also that (8) implies that, when $r=0$,

$$\tilde{\Lambda}_k^{(r,n)} - \lambda_k^{(r)} = O(k^4h^5/\sin((k-\tfrac{1}{2})h)). \tag{15}$$

Table 1
Errors in computed solutions of (1), (4) with $q(x) = 10\cos(2x)$

k	$\lambda_k - \Lambda_k^{(40)}$	$\lambda_k - \tilde{\Lambda}_k^{(40)}$	$\lambda_k - \tilde{\Lambda}_k^{(20)}$	$(\lambda_k - \tilde{\Lambda}_k^{(n)})\sin((k-\frac{1}{2})h)/k^4h^5$			
				$n = 10$	20	40	80
1	2.07E–5	2.07E–5	3.35E–4	0.288	0.275	0.272	0.271
2	9.92E–5	9.91E–5	1.61E–3	0.254	0.245	0.243	0.243
3	1.68E–4	1.58E–4	2.57E–3	0.128	0.127	0.127	0.127
4	2.80E–4	1.64E–4	2.70E–3	0.057	0.058	0.058	0.059
5	9.16E–4	2.64E–4	4.41E–3	0.046	0.048	0.049	0.049
6	2.95E–3	4.61E–4	7.81E–3	0.042	0.048	0.050	0.050
7	8.13E–3	6.77E–4	1.17E–2	0.049	0.044	0.046	0.047
8	1.98E–2	9.24E–4	1.66E–2	0.011	0.039	0.042	0.043
9	4.34E–2	1.21E–3	2.26E–2	0.068	0.035	0.038	0.039
10	8.74E–2	1.54E–3	3.04E–2	−0.031	0.032	0.035	0.036
11	1.64E–1	1.93E–3	4.06E–2	0.014	0.029	0.032	0.034
14	7.98E–1	3.40E–3	1.01E–1		0.023	0.026	0.027
16	1.90E0	4.72E–3	2.17E–1		0.022	0.023	0.024
18	4.06E0	6.41E–3	3.20E–1		0.012	0.020	0.022
19	5.76E0	7.42E–3	3.07E0		0.057	0.019	0.021
20	8.00E0	8.58E–3	−6.06E0		−0.031	0.018	0.020
21	1.09E+1	9.91E–3	−4.06E0		0.017	0.017	0.019

Moreover, numerical results reported below suggest that, for $k = n-1, n, n+1$, $\tilde{\Lambda}_k^{(1,n)} - \lambda_k^{(1)} = O(1)$, as would also follow from (15). The author conjectures that, if $q \in C^4[0,\pi]$, then (15) is true for $r = 0, 1, 2, 3$. He hopes to return to this question in a later paper.

Fourth-order difference approximations to the more general boundary conditions $y'(0) = \alpha_1 y(0)$ and $y'(\pi) = \alpha_2 y(\pi)$ are easily derived from (9), and in principle asymptotic correction can also be used with Numerov's method in this more general case, but it is less efficient than in the cases considered here, for two reasons. (i) Elimination of y_{-1} and y_{n+1} between these equations and (7), produces a quadratic eigenvalue problem $A\mathbf{x} = \Lambda B\mathbf{x} + \Lambda^2 C\mathbf{x}$ which requires more work to solve than the simpler $A\mathbf{x} = \Lambda B\mathbf{x}$ obtained with (2), (4), (5) or (6). (Although the difference equations can also be solved by shooting [16], this is less satisfactory when a large number of eigenvalues are required.) (ii) There is no longer a simple closed-form solution for the case of constant q, and we must resort to a numerical procedure for this, as in [8].

Table 1 shows some results obtained by applying asymptotic correction, as described above, to the Numerov estimates given by (7), (12), (13) for (1), (4) in the case $q(x) = 10\cos(2x)$ (Mathieu's equation). The last four columns check (15) for $n = 10$, 20, 40 and 80. To reduce clutter in the headings of both tables, superscript (1) is omitted, with $\lambda_k^{(1)}, \Lambda_k^{(1,n)}$, etc. written as $\lambda_k, \Lambda_k^{(n)}$ etc. The "exact" eigenvalues, $\lambda_k^{(1)}$, were computed as $C_k(160, 120)$, using the extrapolation formula

$$C_k(n,m) := \frac{n^5 \sin((k-\frac{1}{2})\pi/n)\tilde{\Lambda}_k^{(1,n)} - m^5 \sin((k-\frac{1}{2})\pi/m)\tilde{\Lambda}_k^{(1,m)}}{n^5 \sin((k-\frac{1}{2})\pi/n) - m^5 \sin((k-\frac{1}{2})\pi/m)} \tag{16}$$

suggested by (15). Comparison with $C_k(120,80)$ suggests that all listed results are correct to within one in the least significant figure shown. Our results also suggest that, as with (2), the growth of the error with k was initially slower than that allowed by (15). However, whereas with (2) this was true for all k, our results showed a sharp increase from $\lambda_{n-2}^{(1)} - \tilde{\Lambda}_{n-2}^{(1,n)}$ to $\lambda_{n-1}^{(1)} - \tilde{\Lambda}_{n-1}^{(1,n)}$. Moreover, for $n = 10, 40, 80$ and 120, the values of $\lambda_{n-1}^{(1)} - \tilde{\Lambda}_{n-1}^{(1,n)}$, $\lambda_{n}^{(1)} - \tilde{\Lambda}_{n}^{(1,n)}$ and $\lambda_{n+1}^{(1)} - \tilde{\Lambda}_{n+1}^{(1,n)}$ all differed from the values for $n=20$ shown in Table 1 by less than 1%, showing the O(1) estimate given by (15) for these three quantities to be sharp. With (4), asymptotic correction produces no change for $k=1$, but for all $k>1$ and all $n \geqslant 10$ it produced an improvement. However, although a choice of $n<10$ is unlikely in practice, (15) does not ensure that asymptotic correction will always produce an improvement for very small n. In fact for $n=5$ it produced improvement only for $k=2$, 4 and 6. Limitations of asymptotic correction with very small n are discussed further in [1,5].

The analysis of [9] shows that $|\lambda_k^{(0)} - \tilde{\Lambda}_k^{(0,n)}|$ generally increases as the norms of the first four derivatives, $q^{(j)}$, of q increase. To investigate the variation of $|\lambda_k^{(1)} - \tilde{\Lambda}_k^{(1,n)}|$ with q, the above calculations were repeated with $q = 2\cos(2x)$. Qualitatively, the results were very similar to those reported above. Again $\lambda_{n-1}^{(1)} - \tilde{\Lambda}_{n-1}^{(1,n)}$, $\lambda_{n}^{(1)} - \tilde{\Lambda}_{n}^{(1,n)}$ and $\lambda_{n+1}^{(1)} - \tilde{\Lambda}_{n+1}^{(1,n)}$ changed very little with n (their values being approximately 0.18, −1.05 and −0.22, respectively) and were much larger in magnitude than $\lambda_{n-2}^{(1)} - \tilde{\Lambda}_{n-2}^{(1,n)}$. All errors were less than $\frac{1}{5}$ (and most between $\frac{1}{15}$ and $\frac{1}{30}$) of their value when $q(x) = 10\cos(2x)$, thus indicating that the dependence on the $q^{(j)}$ of the error in the results obtained with asymptotic correction is mildly superlinear. The numerical results of [19] suggest that the errors in the corrected estimates obtained by the second order method considered there also have mildly superlinear dependence on the $q^{(j)}$. This offers an explanation of the very high accuracy of the results obtained for the first example ((4.1),(4.2)) of [19]: for that example all $q^{(j)}$ are bounded above by e, compared with the larger least upper bound e^{π} for the next example ((4.3),(4.4)) of that paper. Although significantly larger values of the $q^{(j)}$ would require a finer mesh to obtain accuracy comparable to that reported for our examples, the tridiagonal structure of the matrices allows quite fine meshes to be handled efficiently. However, one attraction of asymptotic correction is that, in the common case in which the $q^{(j)}$ are of modest size, excellent accuracy may be obtained with n only 100 or so. In this case, it may be convenient to use a general purpose method, such as the MATLAB command "eig(A,B)", which was used in the calculations reported here.

Asymptotic correction removes only the leading term (as $k \to \infty$) of the asymptotic expansion for the error. This dramatically improves estimates of the higher eigenvalues when the derivatives, $q^{(j)}$, are not too large, but is no substitute for mesh refinement when traditional methods strike trouble with the lowest eigenvalues, especially when the $q^{(j)}$ are large. Asymptotic correction produces only small changes in the lowest eigenvalues and none in $\Lambda_1^{(1,n)}$. Nevertheless, to assess the performance of our method with more difficult problems, we tested it on the Coffey–Evans potential [20], which on $[0,\pi]$ takes the form

$$q(x) = 2\beta\cos(2x) + \beta^2\sin^2(2x). \tag{17}$$

Results were computed for boundary conditions (4), with $\beta=10$ and 20, for $n=20, 40, 80, 150, 180, 200$ and 240. The lower eigenvalues of this problem are tightly clustered, especially for larger values of β, and this causes difficulties with many methods. For all n when $\beta=10$, and for all $n>20$ when $\beta=20$, asymptotic correction produced an improvement for *all* $k>1$, most improvements being

Table 2
Results for (1), (4) with $q(x) = 2\beta\cos(2x) + \beta^2\sin^2(2x)$

	$\beta = 10$			$\beta = 20$		
k	λ_k	$\lambda_k - \tilde{\Lambda}_k^{(80)}$	$\lambda_k - \tilde{\Lambda}_k^{(40)}$	λ_k	$\lambda_k - \tilde{\Lambda}_k^{(80)}$	$\lambda_k - \tilde{\Lambda}_k^{(40)}$
1	0.0000000	1.43E–4	2.32E–3	0.0000000	1.17E–3	1.93E–2
2	37.7596285	4.42E–4	7.52E–3	77.9161943	7.38E–3	1.22E–1
3	37.8059002	9.93E–5	1.61E–3	77.9161957	9.95E–4	1.63E–2
4	37.8525995	4.43E–4	6.88E–3	77.9161972	9.89E–4	1.62E–2
5	70.5475097	1.97E–3	3.22E–2	151.463224	2.33E–2	3.90E–1
6	92.6538177	1.37E–3	2.25E–2	220.143526	4.31E–2	8.48E–1
7	96.2058159	7.28E–4	1.20E–2	220.154230	1.85E–2	3.09E–1
8	102.254347	2.09E–3	3.45E–2	220.164945	2.52E–2	3.08E–1
38	1419.26453	5.79E–2	–5.71E–2	1572.80102	8.32E–1	–5.57E+1
39	1494.25080	6.23E–2	1.02E+1	1647.60306	8.95E–1	–4.48E+1
40	1571.23811	6.69E–2	–2.63E+1	1724.42022	9.62E–1	–1.44E+2
41	1650.22636	7.19E–2	–2.42E+1	1803.25098	1.03E0	–1.42E+2

substantial. Even for $n = \beta = 20$ it produced an improvement for most k. Comparing results for $\beta = 10$ and $\beta = 20$ again shows a mildly superlinear growth of errors with $\|q^{(j)}\|$. For both values of β our results again supported (15), the largest value of

$$(\lambda_k^{(1)} - \tilde{\Lambda}_k^{(1,n)})\sin((k - \tfrac{1}{2})h)/k^4h^5$$

occurring at $k = 1$ and the lowest tested value of n. As shown in Table 2, which tabulates some of our results for (1),(4),(17), even $|\lambda_k^{(1)} - \tilde{\Lambda}_k^{(1,n)}|$ often decreased when k was increased. The values shown as the "exact" λ_k in Table 2 were computed as $C_k(240, 180)$ using (16). They are labelled "λ_k" rather than "$C_k(240, 180)$" to avoid clutter in the headings, not because it is claimed that the least significant figures shown are all correct. Nevertheless, comparison with $C_k(200, 150)$ suggests that all figures shown for $\lambda_k - \tilde{\Lambda}^{(40)}$ and $\lambda_k - \tilde{\Lambda}^{(80)}$ are correct. This indicates that our method achieves quite good accuracy with a fairly coarse mesh for this problem also. Table 2 shows results for the lower eigenvalues (where the clustering of eigenvalues can cause problems) and also those terms (all near $k = n$) for which $\lambda_k^{(1)} - \tilde{\Lambda}_k^{(1,40)}$ is negative. Despite significant fluctuations in the magnitude of the error near $k = n$, even the larger errors were smaller than the errors in the corresponding uncorrected estimates, except for $n = 20$ (and presumably for some lower n as well). With both Mathieu's equation and the Coffey–Evans equation, q' vanishes at both boundaries, and this simplifies (12) and (13). Results for some examples with $q'(0)q'(\pi) \neq 0$ and with boundary conditions (5) are given in [7]. These results also satisfy (15).

References

[1] A.L. Andrew, Asymptotic correction of finite difference eigenvalues, in: J. Noye, R. May (Eds.), Computational Techniques and Applications: CTAC-85, North-Holland, Amsterdam, 1986, pp. 333–341.

[2] A.L. Andrew, Correction of finite element eigenvalues for problems with natural or periodic boundary conditions, BIT 28 (1988) 254–269.

[3] A.L. Andrew, Efficient computation of higher Sturm–Liouville eigenvalues, in: R.P. Agarwal, Y.M. Chow, S.J. Wilson (Eds.), Numerical Mathematics, Singapore, 1988, International Series of Numerical Mathematics, 86, Birkhäuser, Basel, 1988, pp. 1–9.
[4] A.L. Andrew, Correction of finite difference eigenvalues of periodic Sturm–Liouville problems, J. Austral. Math. Soc. Ser. B 30 (1989) 460–469.
[5] A.L. Andrew, Asymptotic correction of computed eigenvalues of differential equations, Ann. Numer. Math. 1 (1994) 41–51.
[6] A.L. Andrew, Some recent developments in inverse eigenvalue problems, in: D. Stewart, D. Singleton, H. Gardner (Eds.), Computational Techniques and Applications: CTAC-93, World Scientific, Singapore, 1994, pp. 94–102.
[7] A.L. Andrew, Twenty years of asymptotic correction for eigenvalue computation, J. Austral. Math. Sci. Soc. Ser. B(E), to appear.
[8] R.S. Anderssen, F.R. de Hoog, On the correction of finite difference eigenvalue approximations for Sturm–Liouville problems with general boundary conditions, BIT 24 (1984) 401–412.
[9] A.L. Andrew, J.W. Paine, Correction of Numerov's eigenvalue estimates, Numer. Math. 47 (1985) 289–300.
[10] A.L. Andrew, J.W. Paine, Correction of finite element estimates for Sturm–Liouville eigenvalues, Numer. Math. 50 (1986) 205–215.
[11] D.J. Condon, Corrected finite difference eigenvalues of periodic Sturm–Liouville problems, Appl. Numer. Math. 30 (1999) 393–401.
[12] F.R. de Hoog, R.S. Anderssen, Asymptotic formulas for discrete eigenvalue problems in Liouville form, Math. Models Methods Appl. Sci., to appear.
[13] C.R. Dun, Algebraic correction methods for two-dimensional eigenvalue problems, PhD Thesis, Australian National University, 1995.
[14] R.H. Fabiano, R. Knobel, B.D. Lowe, A finite difference algorithm for an inverse Sturm–Liouville problem, IMA J. Numer. Anal. 15 (1995) 75–88.
[15] P. Ghelardoni, Approximations of Sturm–Liouville eigenvalues using boundary value methods, Appl. Numer. Math. 23 (1997) 311–325.
[16] J.P. Leroy, R. Wallace, Renormalized Numerov method applied to eigenvalue equations: extensions to include single derivative terms and a variety of boundary conditions, J. Phys. Chem. 89 (1985) 1928–1932.
[17] J.T. Marti, Small potential corrections for the discrete eigenvalues of the Sturm–Liouville problem, Numer. Math. 57 (1990) 51–62.
[18] J. Paine, A numerical method for the inverse Sturm–Liouville problem, SIAM J. Sci. Statist. Comput. 5 (1984) 149–156.
[19] J.W. Paine, F.R. de Hoog, R.S. Anderssen, On the correction of finite difference eigenvalue approximations for Sturm–Liouville problems, Computing 26 (1981) 123–139.
[20] J.D. Pryce, Numerical Solution of Sturm–Liouville Problems, Oxford University Press, London, 1993.
[21] G. Vanden Berghe, M. Van Daele, H. De Meyer, A modified difference scheme for periodic and semiperiodic Sturm–Liouville problems, Appl. Numer. Math. 18 (1995) 69–78.

ELSEVIER

Journal of Computational and Applied Mathematics 125 (2000) 367–383

JOURNAL OF
COMPUTATIONAL AND
APPLIED MATHEMATICS

www.elsevier.nl/locate/cam

Numerical methods for higher order Sturm–Liouville problems

Leon Greenberg[a,*], Marco Marletta[b]

[a] *Department of Mathematics, University of Maryland, College Park, Maryland, MD 20740, USA*

[b] *Department of Mathematics and Computer Science, University of Leicester, University Road, Leicester LE1 7RH, UK*

Received 1 December 1999; received in revised form 21 February 2000

Abstract

We review some numerical methods for self-adjoint and non-self-adjoint boundary eigenvalue problems.

MSC: 34A12; 34B; 65L

Keywords: Self-adjoint; Eigenvalue; Oscillation theory; Miss-distance function

1. Introduction

Spectral problems for differential equations arise in many different physical applications. Perhaps quantum mechanics is the richest source of self-adjoint problems, while non-self-adjoint problems arise in hydrodynamic and magnetohydrodynamic stability theory. The problems in hydrodynamic and MHD stability are almost always of 'higher order', either because they involve a coupled system of ordinary differential equations, or because they have been reduced to a single equation of differential order $2m$, $m > 1$. Self-adjoint problems may also be of higher order: in particular, as mentioned in [21], certain quantum-mechanical partial differential eigenproblems can be reduced to systems of ordinary differential eigenproblems.

The solution of ODE eigenproblems presents particular difficulties to the numerical analyst who wants to construct library quality software. General purpose boundary value problem codes do not generally cope well with eigenproblems. Fortunately an increasing number of pure spectral theorists have brought their skills to bear on the numerical solution of these problems. Because of the sheer size of the literature, in this paper we restrict ourselves to a very brief summary of our own

* Corresponding author.

0377-0427/00/$ - see front matter
PII: S0377-0427(00)00480-5

work. A larger bibliography, which gives more (but still inadequate) credit to some of the other mathematicians involved in this area, may be found in [20].

2. Self-adjoint problems

In this section we shall consider a $2m$th order, nonsingular, self-adjoint problem of the form:

$$\begin{aligned}&(-1)^m(p_m(x)y^{(m)})^{(m)}+(-1)^{m-1}(p_{m-1}(x)y^{(m-1)})^{(m-1)}\\&\quad+\cdots+(p_2(x)y'')''-(p_1(x)y')'+p_0(x)y=\lambda w(x)y,\quad a<x<b,\end{aligned}\tag{2.1}$$

together with separated, self-adjoint boundary conditions. (The precise form of the boundary conditions will be given below.) We assume that all coefficient functions are real valued. The technical conditions for the problem to be nonsingular are: the interval (a,b) is finite; the coefficient functions p_k $(0\leqslant k\leqslant m-1)$, w and $1/p_m$ are in $L^1(a,b)$; and the essential infima of p_m and w are both positive. Under these assumptions, the eigenvalues are bounded below. (This is proved, for example, in [11], where the proof shows that the Rayleigh quotient is bounded below.) For good numerical performance however, the coefficients need to be piecewise smooth (where the degree of smoothness depends on the order of the numerical method used).

The eigenvalues can be ordered: $\lambda_0\leqslant\lambda_1\leqslant\lambda_2\leqslant\cdots$, where $\lim_{k\to\infty}\lambda_k=+\infty$ and where each eigenvalue has multiplicity at most m (so $\lambda_{k+m}>\lambda_k$ for all k). The restriction on the multiplicity arises from the fact that for each λ there are at most m linearly independent solutions of the differential equation satisfying either of the endpoint conditions which we shall describe below. The numerical methods discussed in this section are based on calculation of the following counting function:

$$N(\lambda)=\text{The number of eigenvalues of (2.1) (together with boundary conditions) that are}<\lambda.\tag{2.2}$$

We shall give two formulas for $N(\lambda)$ below, and indicate some methods to calculate it. If we can calculate $N(\lambda)$, then we can approximate eigenvalues. If $\lambda'<\lambda''$ are two values such that $N(\lambda')\leqslant j$ and $N(\lambda'')\geqslant j+1$, then the jth eigenvalue λ_j lies in the interval $\lambda'\leqslant\lambda_j<\lambda''$. Now λ_j can be approximated by applying the bisection method to $N(\lambda)$ (accelerated by an iterative rootfinder applied to various continuous functions associated with the eigenvalues).

Although the solutions of (2.1) depend on (x,λ), we shall often suppress λ in the notation. Corresponding to a solution $y(x)$ of (2.1), we define quasiderivatives:

$$\begin{aligned}u_k&=y^{(k-1)},\quad 1\leqslant k\leqslant m,\\v_1&=p_1y'-(p_2y'')'+(p_3y''')''+\cdots+(-1)^{m-1}(p_my^{(m)})^{(m-1)},\\v_2&=p_2y''-(p_3y''')'+(p_4y^{(4)})''+\cdots+(-1)^{m-2}(p_my^{(m)})^{(m-2)},\\&\vdots\\v_k&=p_ky^{(k)}-(p_{k+1}y^{(k+1)})'+(p_{k+2}y^{(k+2)})''+\cdots+(-1)^{m-k}(p_my^{(m)})^{(m-k)},\\&\vdots\\v_m&=p_my^{(m)}.\end{aligned}\tag{2.3}$$

Consider the column vector functions: $u(x)=(u_1,u_2,\ldots,u_m)^{\mathrm{T}}$, $v(x)=(v_1,v_2,\ldots,v_m)^{\mathrm{T}}$, $z(x)=(u_1,u_2,\ldots,u_m,v_1,v_2,\ldots,v_m)^{\mathrm{T}}$. Let S be the $2m\times 2m$ symmetric matrix

$$S(x,\lambda)=\begin{pmatrix}
(\lambda w-p_0) & 0 & & & & & & 0 & & & & & \\
 & -p_1 & 0 & & & & & 1 & 0 & & & & \\
 & & -p_2 & & & & & & 1 & 0 & & & \\
 & & & \ddots & \ddots & & & & & \ddots & \ddots & & \\
 & & & & -p_{m-2} & 0 & & & & & 1 & 0 & \\
 & & & & & -p_{m-1} & & & & & & 1 & 0 \\
0 & 1 & & & & & & & & & & & \\
 & 0 & 1 & & & & & & & & & & \\
 & & 0 & 1 & & & & & & & & & \\
 & & & \ddots & \ddots & & & & & & & & \\
0 & 0 & 0 & \cdot & 0 & 1 & 0 & \cdots & & & & & 0 \\
0 & 0 & 0 & \cdot & 0 & 0 & 0 & \cdots & & & & & 1/p_m
\end{pmatrix}. \tag{2.4}$$

and let J be the $2m\times 2m$ symplectic matrix

$$J=\begin{pmatrix} 0 & -I \\ I & 0 \end{pmatrix}.$$

Then Eq. (2.1) is equivalent to

$$Jz'=S(x,\lambda)z. \tag{2.5}$$

General, separated, self-adjoint boundary conditions for (2.1) are of the form

$$A_1u(a)+A_2v(a)=0, \qquad B_1u(b)+B_2v(b)=0, \tag{2.6}$$

where A_1,A_2,B_1,B_2 are $m\times m$ real matrices, such that $A_1A_2^{\mathrm{T}}=A_2A_1^{\mathrm{T}}$, $B_1B_2^{\mathrm{T}}=B_2B_1^{\mathrm{T}}$, and the $m\times 2m$ matrices (A_1A_2) and (B_1B_2) have rank m.

We now consider $2m\times m$ matrices

$$Z(x)=\begin{pmatrix} U(x) \\ V(x) \end{pmatrix}$$

that are solutions of the extended Hamiltonian system

$$JZ' = S(x,\lambda)Z. \tag{2.7}$$

The (linearly independent) column vectors of $Z(x)$ are solutions of (2.5).

2.1. The unitary marix $\Theta(x,\lambda)$

It can be shown that the matrix function $U^{\mathrm{T}}(x)V(x) - V^{\mathrm{T}}(x)U(x)$ is constant, and this constant equals 0 if Z satisfies either of the boundary conditions (2.6). If $U^{\mathrm{T}}(x)V(x) - V^{\mathrm{T}}(x)U(x) = 0$, (and $Z = (U^{\mathrm{T}}, V^{\mathrm{T}})^{\mathrm{T}}$ has rank m, as we suppose), then the $m \times m$ matrix $V - \mathrm{i}U$ is invertible and the matrix

$$\Theta(x) = (V + \mathrm{i}U)(V - \mathrm{i}U)^{-1} \tag{2.8}$$

is unitary. The matrix $\Theta(x)$ and its phase angles were introduced into oscillation theory by Atkinson [1] and Reid [24].

We now integrate (2.7) from the left and right endpoints toward a chosen point $c \in [a,b]$. Let

$$Z_{\mathrm{L}}(x) = \begin{pmatrix} U_{\mathrm{L}}(x) \\ V_{\mathrm{L}}(x) \end{pmatrix}, \qquad Z_{\mathrm{R}}(x) = \begin{pmatrix} U_{\mathrm{R}}(x) \\ V_{\mathrm{R}}(x) \end{pmatrix}$$

be the solutions of (2.7) with initial conditions $Z_{\mathrm{L}}(a) = (A_2, -A_1)^{\mathrm{T}}$, $Z_{\mathrm{R}}(b) = (B_2, -B_1)^{\mathrm{T}}$. Let $\Theta_{\mathrm{L}}(x)$ and $\Theta_{\mathrm{R}}(x)$ be the unitary matrices obtained from $Z_{\mathrm{L}}(x)$ and $Z_{\mathrm{R}}(x)$ by formula (2.8). The eigenvalues of $\Theta_{\mathrm{L}}(x)$ and $\Theta_{\mathrm{R}}(x)$ are $\{\exp(\mathrm{i}\theta_j^{\mathrm{L}}(x)); 1 \leqslant j \leqslant m\}$ and $\{\exp(\mathrm{i}\theta_j^{\mathrm{R}}(x)); 1 \leqslant j \leqslant m\}$, respectively. The phase angles $\theta_j^{\mathrm{L}}(x), \theta_j^{\mathrm{R}}(x)$ are uniquely determined continuous functions when normalized by the conditions:

$$\theta_1^{\mathrm{L}}(x) \leqslant \theta_2^{\mathrm{L}}(x) \leqslant \cdots \leqslant \theta_m^{\mathrm{L}}(x) \leqslant \theta_1^{\mathrm{L}}(x) + 2\pi,$$

$$\theta_1^{\mathrm{R}}(x) \leqslant \theta_2^{\mathrm{R}}(x) \leqslant \cdots \leqslant \theta_m^{\mathrm{R}}(x) \leqslant \theta_1^{\mathrm{R}}(x) + 2\pi,$$

$$0 \leqslant \theta_j^{\mathrm{L}}(a) < 2\pi, \quad 0 < \theta_j^{\mathrm{R}}(b) \leqslant 2\pi.$$

At a given point $c \in [a,b]$, let

$$\Theta_{\mathrm{LR}}(c) = \Theta_{\mathrm{L}}^*(c)\Theta_{\mathrm{R}}(c), \tag{2.9}$$

and let $\{\exp(\mathrm{i}\omega_j); 1 \leqslant j \leqslant m\}$ be the eigenvalues of $\Theta_{\mathrm{LR}}(c)$, where the ω_j are normalized by the condition

$$0 \leqslant \omega_j < 2\pi. \tag{2.10}$$

It is known that when $0 < \omega_j(\lambda) < 2\pi$, $\omega_j(\lambda)$ is a strictly decreasing function of λ. The normalization (2.10) ensures that $N(\lambda)$ is continuous from the left.

Recalling that all of the functions arising from (2.1) depend on (x,λ), we shall use the following notations:

$$\operatorname{Argdet} \Theta_{\mathrm{L}}(x,\lambda) = \theta_1^{\mathrm{L}}(x) + \theta_2^{\mathrm{L}}(x) + \cdots + \theta_m^{\mathrm{R}}(x),$$

$$\operatorname{Argdet} \Theta_{\mathrm{R}}(x,\lambda) = \theta_1^{\mathrm{R}}(x) + \theta_2^{\mathrm{R}}(x) + \cdots + \theta_m^{\mathrm{R}}(x),$$

$$\overline{\operatorname{Argdet}}\, \Theta_{\mathrm{LR}}(c,\lambda) = \omega_1 + \omega_2 + \cdots + \omega_m. \tag{2.11}$$

The overbar on $\overline{\text{Argdet}}\,\Theta_{LR}(c,\lambda)$ indicates that the angles are normalized to lie in the interval $[0,2\pi)$. We can now give the first formula for the function $N(\lambda)$, which is the number of eigenvalues of (2.1) and (2.6) that are less than λ. The following is proved in [11].

Theorem 1. *For any* $c \in [a,b]$,

$$N(\lambda) = \frac{1}{2\pi}\,(\text{Argdet}\,\Theta_L(c,\lambda) + \overline{\text{Argdet}}\,\Theta_{LR}(c,\lambda) - \text{Argdet}\,\Theta_R(c,\lambda)). \tag{2.12}$$

The matrix $S(x,\lambda)$ in (2.4) can be partitioned into $m \times m$ submatrices:

$$S = \begin{pmatrix} S_{11} & S_{12} \\ S_{21} & S_{22} \end{pmatrix}.$$

The differential equation (2.7) then translates into a differential equation for $\Theta(x,\lambda)$:

$$\Theta' = \mathrm{i}\Theta\Omega, \quad a < x < b, \tag{2.13}$$

where Ω is the Hermitian matrix given by

$$2\Omega = (\Theta^* - I)S_{11}(\Theta - I) + \mathrm{i}(\Theta^* - I)S_{12}(\Theta + I)$$
$$- \mathrm{i}(\Theta^* + I)S_{21}(\Theta - I) + (\Theta^* + I)S_{22}(\Theta + I). \tag{2.14}$$

At the same time, $A = \text{Argdet}\,\Theta$ satisfies the equation

$$A' = \text{trace}\,\Omega. \tag{2.15}$$

There are some existing specialized codes that can integrate the system consisting of Eqs. (2.13)–(2.15). For example, the code by Dieci et al. [7] is constructed specifically for (2.13). Marletta's code [21] for solving Hamiltonian systems works by solving (2.13). One can use these or more general initial value solvers to calculate $\text{Argdet}\,\Theta_L(c,\lambda)$ and $\text{Argdet}\,\Theta_R(c,\lambda)$. $N(\lambda)$ can then be calculated by formula (2.12). Note that we need only know $\Theta_L(c,\lambda)$ and $\Theta_R(c,\lambda)$ to calculate $\overline{\text{Argdet}}\,\Theta_{LR}(c,\lambda)$, since the angles ω_j are normalized to lie in the interval $[0,2\pi)$. This is not the case for $\text{Argdet}\,\Theta_L(c,\lambda)$ or $\text{Argdet}\,\Theta_R(c,\lambda)$. This is probably the best one can do for general self-adjoint $2m$th-order problems. However for 4th and 6th-order problems, there are faster, more efficient, and more elegant methods. These will be discussed below.

2.2. The symmetric matrix W and correction parameter σ

In order to develop another formula for $N(\lambda)$, we return to the matrices $U_L(x)$ and $U_R(x)$, and we define the following integer-valued functions:

$$\nu_L(x) = \text{nullity}\,U_L(x) = m - \text{rank}\,U_L(x) \quad \text{for } a < x < c,$$
$$\nu_R(x) = \text{nullity}\,U_R(x) = m - \text{rank}\,U_R(x) \quad \text{for } c < x < b,$$

$$N_L(c,\lambda) = \sum_{a<x<c} \nu_L(x), \qquad N_R(c,\lambda) = \sum_{c<x<b} \nu_R(x). \tag{2.16}$$

It is shown in [11] that $\nu_{\mathrm{L}}(x)$ and $\nu_{\mathrm{R}}(x)$ can differ from zero at only finitely many points x; therefore the sums in (2.16) are finite. If $U_{\mathrm{L}}(x)$ and $U_{\mathrm{R}}(x)$ are nonsingular, we define

$$W_{\mathrm{L}}(x) = V_{\mathrm{L}}(x)U_{\mathrm{L}}(x)^{-1}, \quad W_{\mathrm{R}}(x) = V_{\mathrm{R}}(x)U_{\mathrm{R}}(x)^{-1}. \tag{2.17}$$

It is known that $W_{\mathrm{L}}(x)$ and $W_{\mathrm{R}}(x)$ are symmetric matrices. (This follows from the fact that $U^{\mathrm{T}}(x)V(x) = V^{\mathrm{T}}(x)U(x)$). For any symmetric matrix W, let $\nu(W)$ be the negative index of inertia (number of negative eigenvalues) of W. We can now give a second formula for $N(\lambda)$. The following theorem is proved in [11].

Theorem 2. *If $U_{\mathrm{L}}(c,\lambda)$ and $U_{\mathrm{R}}(c,\lambda)$ are nonsingular, then*

$$N(\lambda) = N_{\mathrm{L}}(c,\lambda) + N_{\mathrm{R}}(c,\lambda) + \nu(W_{\mathrm{L}}(c,\lambda) - W_{\mathrm{R}}(c,\lambda)). \tag{2.18}$$

There is a more general formula:

$$N(\lambda) = N_{\mathrm{L}}(c,\lambda) + N_{\mathrm{R}}(c,\lambda) + \sigma(c,\lambda). \tag{2.19}$$

If $\det U_{\mathrm{L}}(c,\lambda) \neq 0 \neq \det U_{\mathrm{R}}(c,\lambda)$, Eq. (2.18) implies $\sigma(c,\lambda) = \nu(W_{\mathrm{L}}(c,\lambda) - W_{\mathrm{R}}(c,\lambda))$. More generally,

$$\sigma(c,\lambda) = \frac{1}{2\pi}(\overline{\operatorname{Argdet} \Theta_{\mathrm{L}}(c,\lambda)} + \overline{\operatorname{Argdet} \Theta_{\mathrm{LR}}(c,\lambda)} - \overline{\operatorname{Argdet} \Theta_{\mathrm{R}}(c,\lambda)}), \tag{2.20}$$

where the overbars indicate normalized angles:

$$\overline{\operatorname{Argdet} \Theta_{\mathrm{L}}(c,\lambda)} = \sum_{i=1}^{m} \bar{\theta}_i^{\mathrm{L}}, \qquad \overline{\operatorname{Argdet} \Theta_{\mathrm{R}}(c,\lambda)} = \sum_{i=1}^{m} \bar{\theta}_i^{\mathrm{R}},$$

$$\theta_i^{\mathrm{L}} = 2\pi n_i^{\mathrm{L}} + \bar{\theta}_i^{\mathrm{L}}, \qquad \theta_i^{\mathrm{R}} = -2\pi n_i^{\mathrm{R}} + \bar{\theta}_i^{\mathrm{R}},$$

where n_i^{L} and n_i^{R} are nonnegative integers, and

$$0 \leqslant \bar{\theta}_i^{\mathrm{L}} < 2\pi, \qquad 0 < \bar{\theta}_i^{\mathrm{R}} \leqslant 2\pi.$$

Numerical methods for problems of order 4 and 6 are given in [13–15], using coefficient approximation. The coefficient functions are approximated by piecewise-constant functions (equal to their values at the centers of the mesh intervals). This gives an $O(h^2)$ approximation to the original problem. It turns out that for orders 4 and 6, $N(\lambda)$ can be calculated exactly for the approximate problems, using formulas (2.18)–(2.20). On each mesh interval, the approximate ODE has constant coefficients, and the exact solutions can be found. Nevertheless, it is still difficult to calculate the contribution $N(x_{i-1},x_i)$ of a mesh interval to $N_{\mathrm{L}}(c,\lambda)$ or $N_{\mathrm{R}}(c,\lambda)$. Fortunately, it turns out that there is a simple relation between $N(x_{i-1},x_i)$ and the oscillation number $N_0(x_{i-1},x_i)$ corresponding to any other solution $Z_0(x) = (U_0^{\mathrm{T}}(x), V_0^{\mathrm{T}}(x))^{\mathrm{T}}$ of the approximate problem on $[x_{i-1},x_i]$. For 4th-order problems, $N_0(x_{i-1},x_i)$ can be calculated for the solution $Z_0(x)$ satisfying Dirichlet conditions at x_i: $U_0(x_i) = 0$, $V_0(x_i) = I$. For 6th-order problems, a special solution $Z_0(x)$ is devised for each case, depending on the number of real and purely imaginary roots of the characteristic equation. The case with 6 purely imaginary roots is still too difficult to calculate directly, and requires a homotopy theorem to show that they all have the same behavior. In these problems, the integration of the extended Hamiltonian system (2.7) is stabilized by using Ricatti variables, and the error is controlled by Richardson extrapolation.

3. Non-self-adjoint problems

While our numerical methods for self-adjoint problems have all been based on the well-developed oscillation theory for such problems, no such theory exists for non-self-adjoint problems. Numerical methods for such problems have tended to be more ad-hoc: one typical approach is to adjoin to the differential equation an additional equation $\mathrm{d}\lambda/\mathrm{d}x = 0$ plus an additional boundary condition determining the normalization and sign of the eigenfunction; this gives a boundary value problem which can be solved with a boundary value code. Finite difference and finite element methods have also been used, but perhaps the most popular method involving the representation of the eigenfunctions by a finite basis set has been the Chebychev τ method, which has been extensively developed and used by many authors including Straughan and Walker [26].

Although there is no oscillation theory for non-self-adjoint problems there is nevertheless a rich literature on the analytical aspects of these problems, including the classical works of Naimark [22] and Gohberg and Krein [10]. Many non-self-adjoint operators which arise in applications (see, e.g., all of the examples of Chandrasekhar [5]) are relatively compact perturbations of self-adjoint operators and are therefore unlikely to exhibit the extreme ill-conditioning of eigenvalues observed by Davies [6] and Trefethen [27]. Birkhoff [3] was perhaps the first person to obtain the asymptotic distribution of the eigenvalues for a general class of nth-order problems with this property, which we term *Birkhoff regularity*. For numerical methods based on shooting, Birkhoff regularity has important consequences: for example, it allows one to develop very efficient methods for counting the number of eigenvalues of a problem in a half-plane $\mathrm{Re}\,\lambda < s$.

3.1. Asymptotics and Birkhoff regularity

We consider a differential equation of even order $n = 2m$ of the form

$$y^{(n)} + p_{n-2}(x)y^{(n-2)} + \cdots + p_0(x)y = \lambda y, \quad x \in [0,1], \tag{3.1}$$

in which the coefficients p_k are smooth, together with $2m$ evenly separated boundary conditions normalized to the form

$$U_{0\nu}(y) := y_{(0)}^{(j_\nu)} + \sum_{i=0}^{j_\nu - 1} \alpha_{i\nu} y^{(i)}(0) = 0 \quad (\nu = 1,2,\ldots,m), \tag{3.2}$$

$$U_{1\nu}(y) := y_{(1)}^{(k_\nu)} + \sum_{i=0}^{k_\nu - 1} \beta_{i\nu} y^{(i)}(1) = 0 \quad (\nu = 1,2,\ldots,m). \tag{3.3}$$

Here the integers j_ν and k_ν satisfy $2m-1 \geqslant j_1 > j_2 > \cdots > j_m \geqslant 0$ and $2m-1 \geqslant k_1 > k_2 > \cdots > k_m \geqslant 0$. We require asymptotic information about the behavior of the solutions of (3.1) for large $|\lambda|$. Put $\lambda = -\rho^n$ in (3.1) and consider the sectors $S_k = \{\rho \in \mathbb{C} \mid k\pi/n \leqslant \arg\rho \leqslant (k+1)\pi/n\}$, $k = 0,1,\ldots,2n-1$. Let $\omega_1,\ldots,\omega_n$ be the nth roots of unity.

Theorem 3 (Birkhoff [2]). *Suppose that the coefficients in* (3.1) *are continuous in* [0,1]. *Then in each sector* S_k *the equation*

$$y^{(n)} + p_{n-2}(x)y^{(n-2)} + \cdots + p_0(x)y = -\rho^n y \tag{3.4}$$

has n linearly independent solutions $y_1(x,\rho),\dots,y_n(x,\rho)$ which are analytic functions of $\rho\in S_k$ for all sufficiently large $|\rho|$ and which have the asymptotic properties

$$y_k = \mathrm{e}^{\rho\omega_k x}(1+\mathrm{O}(1/\rho)), \tag{3.5}$$

$$\frac{\mathrm{d}^j y_k}{\mathrm{d}x^j} = \rho^j \mathrm{e}^{\rho\omega_k x}(\omega_k^j+\mathrm{O}(1/\rho)), \quad j=1,\dots,n-1. \tag{3.6}$$

Now consider the sector S_0, and suppose $\omega_1,\dots,\omega_n$ are ordered so that

$$\mathrm{Re}\,(\rho\omega_1)\leqslant \mathrm{Re}\,(\rho\omega_2)\leqslant\cdots\leqslant \mathrm{Re}\,(\rho\omega_n), \quad \rho\in S_0. \tag{3.7}$$

Let $j_1,\dots,j_m$ and $k_1,\dots,k_m$ be the integers in (3.2) and (3.3) and consider

$$\begin{vmatrix} \omega_1^{j_1} & \cdots & \omega_{m-1}^{j_1} & \omega_m^{j_1} & \omega_{m+1}^{j_1} & 0 & \cdots & 0 \\ \vdots & & \vdots & \vdots & \vdots & \vdots & & \vdots \\ \omega_1^{j_m} & \cdots & \omega_{m-1}^{j_m} & \omega_m^{j_m} & \omega_{m+1}^{j_m} & 0 & \cdots & 0 \\ 0 & \cdots & 0 & s\omega_m^{k_1} & \frac{1}{s}\omega_{m+1}^{k_1} & \omega_{m+2}^{k_1} & \cdots & \omega_n^{k_1} \\ \vdots & & \vdots & \vdots & \vdots & \vdots & & \vdots \\ 0 & \cdots & 0 & s\omega_m^{k_m} & \frac{1}{s}\omega_{m+1}^{k_m} & \omega_{m+2}^{k_m} & \cdots & \omega_n^{k_m} \end{vmatrix} = \frac{\theta_{-1}}{s} - s\theta_1, \tag{3.8}$$

where

$$\theta_{-1} = \begin{vmatrix} \omega_1^{j_1} & \cdots & \omega_m^{j_1} \\ \vdots & & \vdots \\ \omega_1^{j_m} & \cdots & \omega_m^{j_m} \end{vmatrix} \begin{vmatrix} \omega_{m+1}^{k_1} & \cdots & \omega_n^{k_1} \\ \vdots & & \vdots \\ \omega_{m+1}^{k_m} & \cdots & \omega_n^{k_m} \end{vmatrix}, \quad \theta_1 = \begin{vmatrix} \omega_1^{j_1} & \cdots & \omega_{m-1}^{j_1} & \omega_{m+1}^{j_1} \\ \vdots & & & \vdots \\ \omega_1^{j_m} & \cdots & \omega_{m-1}^{j_m} & \omega_{m+1}^{j_m} \end{vmatrix} \begin{vmatrix} \omega_m^{k_1} & \omega_{m+2}^{k_1} & \cdots & \omega_n^{k_1} \\ \vdots & & & \vdots \\ \omega_m^{k_m} & \omega_{m+2}^{k_m} & \cdots & \omega_n^{k_m} \end{vmatrix}. \tag{3.9}$$

Definition 4. The boundary conditions are Birkhoff regular if $\theta_{-1}\theta_1\neq 0$.

Although we have stated this definition for the ordering (3.7) for $\rho\in S_0$, it is easily seen that the definition does not depend on the sector chosen. Moreover, the following result has been proved recently in [16].

Theorem 5. *For even order $n=2m$ all evenly divided, separated, λ-independent boundary conditions* (3.2) *and* (3.3) *are Birkhoff regular.*

Birkhoff regularity has two important consequences. Firstly, asymptotic expressions for the eigenvalues were proved by Birkhoff [3] (see Theorem 6); secondly, an asymptotic expression can be obtained for a certain analytic *miss-distance function* $f(\lambda)$ whose zeros are the eigenvalues (see Section 3.2 below).

Theorem 6. *For $n=2m$, Eq.* (3.1) *with evenly separated λ-independent boundary conditions has precisely two sequences of eigenvalues λ_k^+ and λ_k^- given for large k by*

$$\lambda_k^{\pm} = (-1)^m(2k\pi)^n[1-(-1)^m m\log\xi^{\pm}/(k\pi\mathrm{i})+\mathrm{O}(1/k^2)],$$

where ξ^+ and ξ^- are the distinct roots of the equation $\theta_1\xi^2=\theta_{-1}$ for the sector S_0 and log *is any fixed branch of the natural logarithm.*

3.2. The miss-distance function and the characteristic determinant

Eq. (3.1) can be transformed to a 1st-order equation in n variables by many methods. If the coefficients are sufficiently smooth then we can first write it in the form

$$y^{(n)}+(q_{n-2}(x)y^{(m-1)})^{(m-1)}+(q_{n-3}(x)y^{(m-1)})^{(m-2)}+(q_{n-4}(x)y^{(m-2)})^{(m-2)}$$
$$+\cdots+(q_2(x)y')'+q_1(x)y'+q_0(x)y=\lambda y, \quad x\in(0,1). \tag{3.10}$$

We then consider new variables defined by

$$u_k=y^{(k-1)}, \quad k=1,\dots,m, \tag{3.11}$$

$$v_k=(-1)^{k-1}[y^{(n-k)}+(q_{n-2}y^{(m-1)})^{(m-k-1)}+(q_{n-3}y^{(m-1)})^{(m-k-2)}$$
$$+\cdots+(q_{2k+2}y^{(k+1)})'+q_{2k+1}y^{(k+1)}+q_{2k}y^{(k)}], \quad k=1,\dots,m. \tag{3.12}$$

Let $u=(u_1,\dots,u_m)^{\mathrm T}$, $v=(v_1,\dots,v_m)^{\mathrm T}$ and $z=(u^{\mathrm T},v^{\mathrm T})^{\mathrm T}$. Eq. (3.1) becomes

$$Jz'=S(x,\lambda)z, \tag{3.13}$$

where

$$J=\begin{pmatrix}0 & -I\\ I & 0\end{pmatrix}, \qquad S(x,\lambda)=\begin{pmatrix}S_{11}(x,\lambda) & S_{12}\\ S_{21} & S_{22}\end{pmatrix},$$

the $m\times m$ matrices S_{12}, S_{21} and S_{22} being independent of x and λ. Likewise the boundary conditions (3.2) and (3.3) can be expressed in the form

$$A_1u(0)+A_2v(0)=\mathbf{0}=B_1u(1)+B_2v(1). \tag{3.14}$$

Now let $Z_{\mathrm L}=(U_{\mathrm L}^{\mathrm T},V_{\mathrm L}^{\mathrm T})^{\mathrm T}$ and $Z_{\mathrm R}=(U_{\mathrm R}^{\mathrm T},V_{\mathrm R}^{\mathrm T})^{\mathrm T}$ be $2m\times m$ solution matrices of (3.13) of full rank m, such that each column of $Z_{\mathrm L}$ satisfies the boundary condition at $x=0$ and each column of $Z_{\mathrm R}$ satisfies the boundary condition at $x=1$: in particular,

$$A_1U_{\mathrm L}(0)+A_2V_{\mathrm L}(0)=\mathbf{0}=B_1U_{\mathrm R}(1)+B_2V_{\mathrm R}(1). \tag{3.15}$$

Fix $c\in[0,1]$. Then λ is an eigenvalue if and only if there exist nonzero vectors ξ and ζ such that $Z_{\mathrm L}(c,\lambda)\xi=Z_{\mathrm R}(c,\lambda)\zeta$; the corresponding eigenfunction z of (3.13) is then given by

$$z(x)=Z_{\mathrm L}(x,\lambda)\xi, \quad 0\leqslant x\leqslant c; \quad Z_{\mathrm R}(x,\lambda)\zeta, \quad c\leqslant x\leqslant 1.$$

The existence of ξ and ζ to satisfy $Z_{\mathrm L}(c,\lambda)\xi=Z_{\mathrm R}(c,\lambda)\zeta$ is equivalent to the condition

$$f(\lambda):=\det(Z_{\mathrm L}(c,\lambda),Z_{\mathrm R}(c,\lambda))=0. \tag{3.16}$$

This equation defines our miss-distance function $f(\lambda)$.

The more commonly used miss distance is the *characteristic determinant* (see [22]) defined in terms of the boundary operators $U_{0\nu}$ and $U_{1\nu}$. Let $y_1(x,\lambda),\dots,y_{2m}(x,\lambda)$ be any $2m=n$ linearly independent solutions of (3.10) which are also analytic functions of λ in some domain $\Omega \subseteq \mathbb{C}$. Then the characteristic determinant is

$$\Delta(\lambda) = \begin{vmatrix} U_{01}(y_1) & U_{01}(y_2) & \cdots & U_{01}(y_n) \\ \vdots & & & \vdots \\ U_{0m}(y_1) & U_{0m}(y_2) & \cdots & U_{0m}(y_n) \\ U_{11}(y_1) & U_{11}(y_2) & \cdots & U_{11}(y_n) \\ \vdots & & & \vdots \\ U_{1m}(y_1) & U_{1m}(y_2) & \cdots & U_{1m}(y_n) \end{vmatrix}. \tag{3.17}$$

It is known that for $\lambda \in \Omega$, the zeros of $\Delta(\lambda)$ are precisely the eigenvalues in Ω; moreover Keldysh has shown that the order of a zero λ_* of Δ is precisely the algebraic multiplicity [1] of λ_* as an eigenvalue [22]. Since so much is known about $\Delta(\lambda)$ it is obviously important to know the relationship between $\Delta(\lambda)$ and $f(\lambda)$: the following result is proved in [16].

Theorem 7. *Let $u_{1i},\dots,u_{mi}$, $v_{1i},\dots,v_{mi}$ be the quasiderivatives for the solution $y_i(x,\lambda)$, for $i = 1,\dots,n$. Let*

$$Y_{12}(x,\lambda) = \begin{pmatrix} u_{11} & \cdots & u_{1n} \\ \vdots & & \vdots \\ u_{m1} & \cdots & u_{mn} \\ v_{11} & \cdots & v_{1n} \\ \vdots & & \vdots \\ v_{m1} & \cdots & v_{mn} \end{pmatrix}, \tag{3.18}$$

which is a fundamental matrix for (3.13). *Let $W_L = Z_L^* Z_L$ and $W_R = Z_R^* Z_R$, which are Gram matrices and nonsingular. Let*

$$A = \begin{pmatrix} A_1 & A_2 \\ U_L^*(0) & V_L^*(0) \end{pmatrix}, \qquad B = \begin{pmatrix} U_R^*(1) & V_R^*(1) \\ B_1 & B_2 \end{pmatrix}. \tag{3.19}$$

Then A and B are invertible and

$$f(\lambda) = (-1)^m \frac{\det W_L(0) \det W_R(1)}{\det A \det B} \frac{\Delta(\lambda) \det Y_{12}(c,\lambda)}{\det Y_{12}(0,\lambda) \det Y_{12}(1,\lambda)}. \tag{3.20}$$

Since $f(\lambda)$ is an entire function this result, combined with the known properties of $\Delta(\lambda)$, imply that the order of a point λ_* as a zero of f is the algebraic multiplicity of λ_* as an eigenvalue of the problem. Moreover, by choosing for $y_1,\dots,y_n$ the n solutions whose asymptotics are described in

[1] For non-self-adjoint problems the algebraic and geometric multiplicities of an eigenvalue may be different. An eigenvalue λ_* of a general non-self-adjoint operator L has not only eigenfunctions, but additional *associated functions* which are elements of the null-spaces of the operators $(L-\lambda_*)^p$, $p = 1,2,3,\dots$. The algebraic multiplicity is the dimension of the sum of all these null spaces.

Theorem 3, we can obtain the asymptotics for $f(\lambda)$ for large $|\lambda|$. We shall see in the next section how important this can be.

3.3. λ-Integration

For self-adjoint problems all the eigenvalues are real, and there is a monotone increasing miss-distance function which takes prescribed values at the eigenvalues. For non-self-adjoint problems one has the harder problem of finding the zeros of an entire function $f(\lambda)$ in the complex plane, already addressed by many authors, e.g., [18,28]. An often used approach is based on the argument principle: the number of zeros of f inside a closed contour Γ is $(1/2\pi i)\int_\Gamma f'(\lambda)/f(\lambda)\,d\lambda$. The integral is computed by splitting up Γ into a number of segments $[z_j, z_{j+1}]$ such that for each j, for $z \in [z_j, z_{j+1}]$, $w_j(z) := f(z)/f(z_j)$ traces out a curve which lies entirely in the right half-plane $\mathrm{Re}(w_j) > 0$. The integral is then equal to $\sum_j \log(f(z_{j+1})/f(z_j))$. In practice it is usually impossible to verify the condition $\mathrm{Re}(w_j(z)) > 0$ for all $z \in [z_j, z_{j+1}]$, and so one replaces this by a heuristic such as $|\arg w_j(z_j)| < \pi/4$, where arg is the branch of the argument taking values in $(-\pi, \pi]$. Various strategies have been proposed for choosing the points z_j.

Knowing the number of zeros of f in, say, a rectangle in $\mathbb{C}$, one can set up a recursive bisection procedure to home in on individual zeros. For simple zeros it is usually possible, when the rectangles become sufficiently small, to switch to a quasi-Newton method based on finite difference approximation of the derivative, and converge rapidly to the zero.

In applications related to linear stability analysis it is often important to know whether or not any eigenvalues of a problem lie in a half-plane. From Theorem 3 there will be infinitely many eigenvalues in the half-plane $(-1)^m \mathrm{Re}(\lambda) > 0$, so the question is: how many eigenvalues lie in the half-plane $(-1)^m \mathrm{Re}(\lambda) < 0$? Ideally one would like the answer to be given by the integral $\int_{-i\infty}^{+i\infty} f'(\lambda)/f(\lambda)\,d\lambda$, but for the function f defined by (3.16) this integral does not converge. To circumvent this we use (3.20). The asymptotics of $\Delta(\lambda)$ are known [22, p. 60], as are those of the solutions $y_1, \ldots, y_n$, so the asymptotics of the terms $\det Y_{12}(0,\lambda)$, $\det Y_{12}(1,\lambda)$ and $\det Y_{12}(c,\lambda)$ can also be computed, all in terms of analytic functions of λ. One is then able to find a function $g(\lambda)$ which is (a) analytic in the half-plane $(-1)^m \mathrm{Re}(\lambda) \leqslant 0$, with no zeros there, (b) such that as $|\lambda| \to \infty$ in this half-plane, $f(\lambda)/g(\lambda) \to 1$. Defining a new miss-distance $\hat{f}$ by $\hat{f} = f/g$, the number of eigenvalues in the half-plane $(-1)^m \mathrm{Re}(\lambda) < 0$, counted according to algebraic multiplicity, is given by $\int_{-i\infty}^{+i\infty} \hat{f}'(\lambda)/\hat{f}(\lambda)\,d\lambda$.

3.4. x-Integration

Evaluating $f(\lambda)$ defined by (3.16) involves integrating the differential system in some form. Because λ may be large for evaluating some of the integrals mentioned at the end of Section 3.3, one should perhaps reformulate the system in a more stable set of variables; ideally one should also use a special numerical method capable of integrating the system for large $|\lambda|$ at a reasonable cost. One method of achieving these ends is to use the *compound matrix method*, described in Drazin and Reid [8, p. 311]. This involves using variables closely related to Riccati variables but satisfying a linear system of ODEs instead of the usual nonlinear system. The linearity can be exploited by using a special integrator for linear ODEs, e.g., a method [4,17] based on the Magnus series [19].

Unfortunately the compound matrix method involves an ODE in binomial $(2n, n)$ variables and is therefore impractical for equations of order > 6. However, many high-order problems actually originated from *systems* of equations of order 2. (This is true of the Orr–Sommerfeld equation, for example.) In terms of the matrices $Z_{\mathrm{L}} = (U_{\mathrm{L}}^{\mathrm{T}}, V_{\mathrm{L}}^{\mathrm{T}})^{\mathrm{T}}$ and $Z_{\mathrm{R}} = (U_{\mathrm{R}}^{\mathrm{T}}, V_{\mathrm{R}}^{\mathrm{T}})^{\mathrm{T}}$, let $U = U_{\mathrm{L}}$ (or U_{R}) and let $V = V_{\mathrm{L}}$ (resp. V_{R}); then these equations may be written as

$$-U'' + Q(x,\lambda)U = \mathbf{0}, \tag{3.21}$$

with $V = U'$. Eq. (3.21) can be solved for each fixed λ by replacing the $m \times m$ coefficient matrix $Q(x,\lambda)$ by a matrix $\hat{Q}(x,\lambda)$ which is piecewise constant on mesh intervals $(x_{j-1}, x_j]$, $j=0,\dots,N$, $x_0=0$, $x_N = 1$. On each mesh interval one can solve this approximate equation 'exactly' (i.e., symbolically) and hence obtain a symbolic expression for the Riccati variables associated with the system. Evaluated in the correct way, this symbolic expression gives a stable way of finding the Riccati variables for the approximated system. This method has the disadvantage that the error is at best $O(h^2)$, where h is a typical steplength; however it has the advantage that for a given mesh, the relative accuracy of $f(\lambda)$ often does not deteriorate as $|\lambda|$ increases. The $O(h^2)$ can be improved to higher order by Richardson extrapolation.

4. Numerical examples

We shall give two examples each of self-adjoint and non-self-adjoint problems. We begin with the self-adjoint examples.

(1) Consider the so-called modified harmonic oscillator, which consists of the equation

$$\ell(y) = -y'' + (x^2 + x^4)y = \lambda y$$

on the interval $(-\infty, \infty)$. No boundary conditions are needed because the problem is of limit-point type: the requirement that the eigenfunctions be square integrable suffices as a boundary condition. We truncate this problem to the interval $(-100, 100)$, and impose the boundary conditions $y(-100) = 0 = y(100)$. Now consider the square $L = \ell^2$ of the above operator on the interval $(-100, 100)$. Thus the fourth-order problem is

$$L(y) = y^{(\mathrm{iv})} - 2((x^2 + x^4)y')' + (x^8 + 2x^6 + x^4 - 12x - 2)y = \lambda y,$$

with boundary conditions $y(c) = y''(c) = 0$, for $c = \pm 100$. The eigenvalues of L are the squares of the eigenvalues of ℓ. Clearly the coefficients become quite large at the endpoints, so this problem tests how well the code SLEUTH can cope with stiffness.

(2) Self-adjoint fourth-order problems often arise in the study of vibration and buckling of beams. For example, Roseau [25, p. 141] analyzes vibrations in a turbine blade. By looking for normal modes of transverse vibration in the associated wave equation, he obtains the eigenproblem consisting of the differential equation

$$(EIy'')'' - ((F - \omega^2 I\rho)y')' - \omega^2 \rho y = 0, \quad 0 < x < \ell,$$

subject to the boundary conditions

$$y(0) = y'(0) = 0, \qquad EIy''(\ell) = (EIy'')'(\ell) - (F - \omega^2 I\rho)y(\ell) = 0.$$

Table 1
Eigenvalues found by SLEUTH at $TOL = 10^{-6}$

Problem number	Eigenvalue index	Eigenvalue approximation	Code relative error estimate	"True" relative error	CPU (secs)	Number of extrapolations, number of mesh points
	0	1.9386467	1E − 6	2E − 6	8.4	4,320
1	9	2205.7105	5E − 8	9E − 6	12.7	5,176
	99	1044329.235	9E − 9	1E − 9	113.7	3,504
	0	1.8115460	* * *	2E − 7	* * *	2,320
2	1	5.9067512	* * *	3E − 8	* * *	3,640
	2	10.8786209	* * *	2E − 9	* * *	5,2560

Here ω is the vibrational frequency; y is the displacement perpendicular to the blade; E is the Young's modulus; I is the moment of inertia of a cross-section of the blade; ρ is the linear density of the blade; and F is the (variable) centrifugal force:

$$F(x) = \Omega^2 \int_x^{\ell} \rho A(s)(r+s)\,\mathrm{d}s,$$

where Ω is the angular velocity, $A(\cdot)$ is the cross-sectional area of the blade, and r is the radius of the turbine.

We took $E = I = A(x) = \Omega = \ell = 1$ and $r = 2/3$. With the cross-sectional area constant we chose $\rho(x) = x$, corresponding to a blade made of a material of nonuniform density. Then $F(x) = (1/3)(2+2x+x^2)(1-x)$. We converted the problem to a standard eigenproblem by introducing a new eigenparameter λ:

$$y^{(\mathrm{iv})} - (((1/3)(2+2x+x^2)(1-x) - \omega^2 x)y')' - \omega^2 xy = \lambda y, \quad 0 < x < 1;$$

the boundary conditions are actually just Dirichlet conditions $u_1 = u_2 = 0$ at $x = 0$ and Neumann conditions $v_1 = v_2 = 0$ at $x = 1$ (see (2.3) for definitions of u_1, u_2, v_1 and v_2). For each value $\omega > 0$ this problem has an infinite sequence of λ-eigenvalues

$$\lambda_0(\omega) \leqslant \lambda_1(\omega) \leqslant \lambda_2(\omega) \leqslant \cdots.$$

The results in Greenberg [12] imply that $\lambda_k(\omega)$ is a strictly decreasing function of ω; the kth eigenvalue ω_k of the original nonlinear problem is defined by $\lambda_k(\omega_k) = 0$. Using a simple rootfinding process, we determined ω_0, ω_1 and ω_2.

The results are shown in Table 1. (We do not quote CPU times for Problem 2 as these depend very strongly on the rootfinding method used and the quality of the initial approximation.)

The two non-self-adjoint problems we shall consider both involve the Orr–Sommerfeld equation for plane laminar flow:

$$(-D^2+\alpha^2)^2 y + \mathrm{i}\alpha R(U(x)(-D^2+\alpha^2)y + U''(x)y) = \lambda(-D^2+\alpha^2)y, \tag{4.1}$$

where $D = \mathrm{d}/\mathrm{d}x$, and $U(x)$ is a flow profile whose stability is in question. The parameters α and R are the wave number and Reynolds number, respectively.

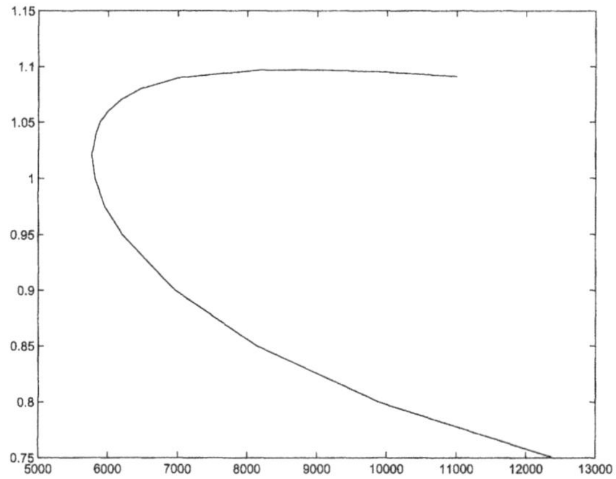

Fig. 1. Marginal curve for plane Poiseuille flow

(3) In this first example, we sketch the marginal curve for the Poiseuille profile: $U(x)=1-x^2$ on the interval $[-1,1]$, with Dirichlet boundary conditions $y(c)=y'(c)=0$ for $c=\pm 1$. (By symmetry, this reduces to the equivalent problem on $[0,1]$, with boundary conditions $y'(0)=y'''(0)=0$ and Dirichlet conditions at $x=1$.) If for some (R,α), the problem has an eigenvalue λ with $\mathrm{Re}(\lambda)<0$, then the flow is unstable under a perturbation with wave number α and Reynolds number R. If all eigenvalues have $\mathrm{Re}(\lambda)>0$, then the perturbed flow is stable. The pair (R,α) is *marginal* if all eigenvalues satisfy $\mathrm{Re}(\lambda)\geqslant 0$ and there is at least one eigenvalue with $\mathrm{Re}(\lambda)=0$. The minimum R on the marginal curve is the *critical Reynolds number*. This value R_{crit} has the property that any pair (R,α) is stable if $R<R_{\mathrm{crit}}$. Using the code SLNSA [16], a nonlinear solver and a path-following procedure, we sketched the marginal curve for plane Poiseuille flow (see Fig. 1). We used only 100 meshpoints, yet found the critical Reynolds number $R_{\mathrm{crit}}=5771.8$, which compares well with the result of Orszag [23] of $R_{\mathrm{crit}}=5772.2$.

(4) Gheorghiu and Pop [9] considered the Orr–Sommerfeld equation for a liquid film flowing over an inclined plane, with a surface tension gradient. In Eq. (4.1) they replace λ by $\alpha R\lambda$ on the right-hand side, a rescaling which does not change the stability criteria. The problem then consists of the differential equation on the interval $[0,1]$ and the following λ-dependent boundary conditions:

$$(-\mathrm{i}\lambda-U(0))(y''(0)+\alpha^2y(0))+U''(0)y(0)=0,$$

$$U''(0)y'''(0)+\mathrm{i}\alpha\{R(-\mathrm{i}\lambda-U(0))+3\mathrm{i}\alpha\}U''(0)y'(0)$$

$$-\mathrm{i}\alpha\{2\cot\beta+\alpha^2C_a+(-\mathrm{i}\lambda-U(0))RU'(0)\}\{y''(0)+\alpha^2y(0)\}=0,$$

$$y(1)=0,\quad y'(1)=0.$$

Using the flow profile

$$U(x)=(1-x)(1+x+\tau),$$

Gheorghiu and Pop calculate the critical Reynolds number for $\tau=\pm 1.75$ and $\cot\beta=1.19175$. We took $C_a=1$ (as the results seem independent of C_a). For this problem, R_{crit} is obtained as the

Table 2
Leftmost eigenvalue for Problem (4)

α	R	τ	Magnus λ	Vector SL λ
40 *meshpoints* (20/40 *extrap. for* Vector SL)				
10^{-2}	0.7947	1.75	$(5.7\times10^{-7}, 3.754980)$	$(5.7\times10^{-7}, 3.754980)$
10^{-2}	0.7949	1.75	$(-1.4\times10^{-6}, 3.754980)$	$(5.7\times10^{-7}, 3.754980)$
80 *meshpoints* (40/80 *extrap. for* Vector SL)				
10^{-2}	0.7947	1.75	$(5.7\times10^{-7}, 3.754980)$	$(5.7\times10^{-7}, 3.754980)$
10^{-2}	0.7949	1.75	$(-1.4\times10^{-6}, 3.754980)$	$(5.7\times10^{-7}, 3.754980)$
40 *meshpoints* (20/40 *extrap. for* Vector SL)				
10^{-4}	0.7945	1.75	$(1.2\times10^{-11}, 3.750000)$	$(1.4\times10^{-8}, 3.750000)$
10^{-4}	0.7946	1.75	$(-1.0\times10^{-8}, 3.750000)$	$(-3.2\times10^{-9}, 3.750000)$
80 *meshpoints* (40/80 *extrap. for* Vector SL)				
10^{-4}	0.7944	1.75		$(8.6\times10^{-9}, 3.750000)$
10^{-4}	0.7945	1.75	$(1.6\times10^{-11}, 3.750000)$	$(-1.0\times10^{-8}, 3.750000)$
10^{-4}	0.7946	1.75	$(-1.0\times10^{-8}, 3.750000)$	
40 *meshpoints* (20/40 *extrap. for* Vector SL)				
10^{-2}	11.9300	−1.75	$(2.8\times10^{-6}, 0.249792)$	$(2.8\times10^{-6}, 0.249792)$
10^{-2}	11.9400	−1.75	$(-3.8\times10^{-6}, 0.249792)$	$(-3.8\times10^{-6}, 0.249792)$
80 *meshpoints* (40/80 *extrap. for* Vector SL)				
10^{-2}	11.9300	−1.75	$(2.8\times10^{-6}, 0.249792)$	$(2.8\times10^{-6}, 0.249792)$
10^{-2}	11.9400	−1.75	$(-3.8\times10^{-6}, 0.249792)$	$(-3.8\times10^{-6}, 0.249792)$
40 *meshpoints* (20/40 *extrap. for* Vector SL)				
10^{-4}	11.9173	−1.75		$(2.4\times10^{-8}, 0.250000)$
10^{-4}	11.9174	−1.75	$(6.6\times10^{-10}, 0.250000)$	$(-2.0\times10^{-8}, 0.250000)$
10^{-4}	11.9175	−1.75	$(-3.9\times10^{-12}, 0.250000)$	
80 *meshpoints* (40/80 *extrap. for* Vector SL)				
10^{-4}	11.9174	−1.75	$(6.6\times10^{-10}, 0.250000)$	
10^{-4}	11.9175	−1.75	$(-2.9\times10^{-13}, 0.250000)$	$(5.3\times10^{-9}, 0.250000)$
10^{-4}	11.9176	−1.75		$(-8.3\times10^{-9}, 0.250000)$

limiting case as $\alpha\searrow 0$. In Table 2 we show the left-most eigenvalue in the complex plane for various values of R and α, and for the two different values of τ. We compare the Magnus method with the coefficient approximation vector Sturm–Liouville method. The values of R are chosen close to the stability/instability boundary predicted by Gheorgiu and Pop for the case $\alpha\searrow 0$, which are $R=0.7945$ in the case $\tau=1.75$ and $R=11.9175$ in the case $\tau=-1.75$. Both methods show the sign of the real part of the left-most eigenvalue changing at values of the Reynolds number close to these predicted values, for small α, even though the number of meshpoints used is very modest. It is particularly interesting to note the exceptional accuracy of both methods when $\alpha=10^{-2}$: they agree to all digits quoted, even using just 40 mesh intervals.

5. Conclusions

We have discussed some numerical methods for self-adjoint and non-self-adjoint Sturm–Liouville problems. We have concentrated on our own work because of space limitations, and we apologize to the many authors whose important contributions have not been included. The methods discussed here for self-adjoint problems not only approximate the eigenvalues and eigenvectors, but by approximating the counting function $N(\lambda)$, they also find the eigenvalue index (and in fact can aim for an eigenvalue with given index). For high eigenvalues, the ODE methods discussed here are usually more accurate and less costly than Galerkin or finite difference methods.

For self-adjoint problems of orders 4 and 6, coefficient approximation together with the W-matrix method (as discussed in [13–15]) is the cheapest method we know with high accuracy. Self-adjoint problems of order greater than 6 require Θ-matrices, and solution of the equation $\Theta' = \mathrm{i}\Theta\Omega$. Two methods for this are Marletta's method [21] using the Magnus series (which keeps Θ unitary) and the method of Dieci et al. [7] (which projects to unitary matrices). The computational costs of these methods seem to be remarkably similar (see [21] for a comparison). These methods can be quite expensive for high-order problems; and finding new, accurate methods with lower cost is an important and challenging problem.

For non-self-adjoint problems we have discussed the methods given in [16], using the argument principle. The code described in [16] can find the eigenvalues in a rectangle, left half-plane, or vertical strip. It can find the kth eigenvalue as ordered by the real part. The x-integration is carried out using compound matrices (which can be quite expensive) or, when possible, by transformation to a 2nd-order vector Sturm–Liouville problem (which is considerably cheaper). Some further problems and future directions are:

- methods for singular problems, including the approximation of essential spectra,
- analysis and codes for systems of mixed order (or block operators), and the associated problems with rational coefficients,
- applications of the various codes discussed here to physical problems, especially in hydrodynamics and magnetohydrodynamics.

Acknowledgements

The authors are grateful for the support of the EPSRC (through grants GR/J99438, GR/K59620, GR/L74354) which enabled them to work closely together on these projects.

References

[1] F. Atkinson, Discrete and Continuous Boundary Problems, Academic Press, New York, 1964.
[2] G.D. Birkhoff, On the asymptotic character of the solutions of certain linear differential equations containing a parameter, Trans. AMS 9 (1908) 219–223.
[3] G.D. Birkhoff, Boundary value problems and expansion problems of ordinary differential equations, Trans. AMS 9 (1908) 373–395.
[4] S. Blanes, F. Casas, J.A. Oteo, J. Ros, Magnus and Fer expansions for matrix differential equations: the convergence problem, J. Phys. A: Math. Gen. 31 (1998) 259–268.

[5] S. Chandrasekhar, On characteristic value problems in high order differential equations which arise in studies on hydrodynamic and hydromagnetic stability, in Proceedings of the Symposium on Special Topics in Applied Mathematics, Amer. Math. Monthly 61 (2) (1955) 32–45.
[6] E.B. Davies, Pseudospectra: the harmonic oscillator and complex resonances. Preprint, Department of Mathematics, King's College London, 1998.
[7] L. Dieci, R.D. Russell, E.S. Van Vleck, Unitary integrators and applications to continuous orthonormalization techniques, SIAM J. Numer. Anal. 31 (1994) 261–281.
[8] P.G. Drazin, W.H. Reid, Hydrodynamic Stability, Cambridge University Press, Cambridge, 1981.
[9] C.I. Gheorghiu, I.S. Pop, A modified Chebyshev-tau method for a hydrodynamic stability problem. Approximation and Optimization: Proceedings of the International Conference on Approximation and Optimization (Romania) – ICAOR, Cluj-Napoca, July 29–August 1, 1996. Vol. II, 1997, pp. 119–126.
[10] I.C. Gohberg, M.G. Krein, Introduction to the Theory of Linear Nonselfadjoint Operators, Trans. Math. Monographs 18 (1969).
[11] L. Greenberg, A. Prüfer method for calculating eigenvalues of selfadjoint systems of ordinary differential equations, Parts 1 and 2, University of Maryland Technical Report TR91-24, 1991.
[12] L. Greenberg, An oscillation method for fourth-order selfadjoint two-point boundary value problems with nonlinear eigenvalues, SIAM J. Math. Anal. 22 (1991) 1021–1042.
[13] L. Greenberg, M. Marletta, Oscillation theory and numerical solution of fourth order Sturm-Liouville problems, IMA J. Numer. Anal. 15 (1995) 319–356.
[14] L. Greenberg, M. Marletta, The code SLEUTH for solving fourth order Sturm-Liouville problems, ACM Trans. Math. Software 23 (1997) 453–493.
[15] L. Greenberg, M. Marletta, Oscillation theory and numerical solution of sixth order Sturm-Liouville problems, SIAM J. Numer. Anal. 35 (1998) 2070–2098.
[16] L. Greenberg, M. Marletta, Numerical solution of nonselfadjoint Sturm-Liouville problems and related systems, SIAM J. Numer. Anal., submitted.
[17] A. Iserles, S.P. Nørsett, On the solution of linear differential equations in Lie groups, Philos. Trans. Roy. Soc. London A 357 (1999) 983–1019.
[18] P. Kravanja, T. Sakurai, M. Van Barel, On locating clusters of zeros of analytic functions, Report TW280, Dept. of Computer Science, Katholieke Universiteit Leuven, Belgium, 1998.
[19] W. Magnus, On the exponential solution of differential equations for a linear operator, Comm. Pure Appl. Math. 7 (1954) 649–673.
[20] C.R. Maple, M. Marletta, Algorithms and software for selfadjoint ODE eigenproblems, in: D. Bainov (Ed.), Proceedings of the 8th International Colloquium on Differential Equations, VSP, Utrecht, 1998.
[21] M. Marletta, Numerical solution of eigenvalue problems for Hamiltonian systems, Adv. Comput. Math. 2 (1994) 155–184.
[22] M.A. Naimark, Linear Differential Operators, Vol. 1, Ungar, New York, 1968.
[23] S.A. Orszag, Accurate solution of the Orr-Sommerfeld stability equation, J. Fluid Mech. 50 (1971) 689–703.
[24] W.T. Reid, Sturmian Theory for Ordinary Differential Equations, Springer, New York, 1980.
[25] M. Roseau, Vibrations in Mechanical Systems, Springer, Berlin, 1987.
[26] B. Straughan, D.W. Walker, Two very accurate and efficient methods for computing eigenvalues and eigenfunctions in porous convection problems, J. Comput. Phys. 127 (1996) 128–141.
[27] L.N. Trefethen, Pseudospectra of linear operators, SIAM Rev. 39 (1997) 383–406.
[28] X. Ying, I.N. Katz, A reliable argument principle algorithm to find the number of zeros of an analytic function in a bounded domain, Numer. Math. 53 (1988) 143–163.

JOURNAL OF COMPUTATIONAL AND APPLIED MATHEMATICS

ELSEVIER Journal of Computational and Applied Mathematics 125 (2000) 385–393

www.elsevier.nl/locate/cam

On a computer assisted proof of the existence of eigenvalues below the essential spectrum of the Sturm–Liouville problem

B.M. Brown[a,*], D.K.R. McCormack[a], A. Zettl[b]

[a]*Department of Computer Science, Cardiff University of Wales, Newport Road, P.O. Box 916, Cardiff CF2 3XF, UK*
[b]*Department of Mathematical Sciences, Northern Illinois University, Dekalb, Illinois 60115, USA*

Received 11 October 1999; received in revised form 20 March 2000

Abstract

There is considerable interest in determining the existence of eigenvalues of the Sturm–Liouville problem

$$-(py')' + qy = \lambda wy,$$

where the independent variable $x \in [0,\infty)$ and p, q and w are real-valued functions, and λ is the spectral parameter. In general, an analytic attack on this problem is quite difficult and usually requires the use of the variational principal together with choice of suitable test functions. We show how results from functional analysis together with interval analysis and interval arithmetic can be used, not only to determine the existence of such eigenvalues, but also to compute provably correct bounds on their values.

1. Introduction

This paper is a follow up of [7]. In [7] the authors presented a new method for proving the existence of an eigenvalue below the essential spectrum of Sturm–Liouville problems defined by

$$-y'' + qy = \lambda y, \quad \text{on } J = [0,\infty), \quad y(0) = 0, \tag{1}$$

where q is a real $L^1(0,\infty)$ perturbation[1] of a real-valued periodic function p on J. This method combines operator theory and "standard" numerical analysis with interval analysis and interval

* Correspondence address: Department of Computer Science, Cardiff University of Wales, Newport Road, P.O. Box 916, Cardiff CF2 3XF, UK.

E-mail address: malcolm.brown@cs.cf.ac.uk (B.M. Brown).

[1] In this paper 'function g is an X perturbation of function f' means that $g = f + \Delta f$ where Δf is a function of kind X: that is, 'perturbation' means the result g, not the perturbing function Δf.

PII: S0377-0427(00)00481-7

arithmetic, and is illustrated by showing that there is at least one eigenvalue for (1) with

$$q(x)=\sin\left(x+\frac{1}{1+x^2}\right) \tag{2}$$

which lies below the essential spectrum.

In this paper we extend and develop further the notions in [7] to cover additional classes of problems. In Section 2 we introduce the relevant notation and review the approach taken in [7]. In Section 3 we show how some of the restrictions required by the approach in [7] may be removed enabling us to prove the existence of several eigenvalues below the essential spectrum of (1). For completeness, we include in Section 3 a short review of the interval analytic background that is relevant to our work. Section 4 contains examples, some of which are of physical interest, which illustrate our method.

2. Mathematical formulation of the method

Of particular interest are the two cases (i) $q\in L^1(0,\infty)$ or (ii) $q=p+\hat{q}$ where $\hat{q}$ is in $L^1(0,\infty)$ and p is a periodic function with fundamental periodic interval $[a,b]$. We remark that if p is a constant then this case is covered by (i). It is well known that in both cases q is in the limit-point case at infinity and that therefore (1) determines a unique self-adjoint operator S in $L^2[0,\infty)$ with domain

$$\{y\in L^2(0,\infty)\colon y,\ y'\in \mathrm{AC}_{\mathrm{loc}}[0,\infty),\ -y''+qy\in L^2(0,\infty),\ y(0)=0\}.$$

$\mathrm{AC}_{\mathrm{loc}}$ being the set of functions that are locally absolutely continuous. For both cases (i) and (ii), the potential q ensures that S has an essential spectrum $\sigma_{\mathrm{e}}(q)$, bounded below and unbounded above. Indeed in case (i), $\sigma_{\mathrm{e}}(q)$ occupies the nonnegative real axis, while for case (ii), it is known that $\sigma_{\mathrm{e}}(q)=\sigma_{\mathrm{e}}(p)$ and $\sigma_{\mathrm{e}}(q)$ lies in bands, bounded below and extending to infinity. We denote by $\sigma(q)$ the spectrum of S and write

$$\sigma_0(q)=\inf\sigma_{\mathrm{e}}(q). \tag{3}$$

In order to describe our approach to proving the existence of eigenvalues λ_n below the essential spectrum of the operator S we introduce the following notation. Denote by $\lambda_j^{\mathrm{N}}([a,b],q)$, $\lambda_j^{\mathrm{D}}([a,b],q)$, $\lambda_j^{\mathrm{P}}([a,b],q)$, $(0\leqslant j\leqslant k-1)$ the first k Neumann, Dirichlet and periodic eigenvalues, respectively, of the regular Sturm–Liouville problems (SLP) consisting of the left-hand side of (1) together with either Neumann $y'(a)=0=y'(b)$, Dirichlet $y(a)=0=y(b)$ or periodic $y(a)=y(b)$, $y'(a)=y'(b)$ boundary conditions.

The well-known inequality

$$\lambda_0^{\mathrm{N}}([a,b],q)\leqslant\lambda_0^{\mathrm{P}}([a,b],q)<\lambda_0^{\mathrm{D}}([a,b],q)<\lambda_1^{\mathrm{P}}([a,b],q) \tag{4}$$

may be found in [11, Theorem 13.10, pp. 209–212].

Our proof of the existence of eigenvalues below the essential spectrum depends on the following:

Theorem 2.1 (Bailey et al. [2]). *If*

$$\lambda_j^{\mathrm{D}}([0,b],q)<\sigma_0(q) \tag{5}$$

for some b, $0<b<\infty$, and some $j\in\{0,1,2,\ldots\}$, then S has at least $j+1$ eigenvalues $<\sigma_0(q)$.

Proof. (see Bailey et al. [2]). (We remark that this result could also be obtained from the min–max characterization of eigenvalues.)

Our approach to proving the existence of j eigenvalues below $\sigma_0(q)$ is first to compute a verified lower bound l for $\sigma_0(q)$ and a verified upper bound u for $\lambda_j^D([0,b],q)$ for some b and some j such that

$$u < l.$$

We recall that $\lambda_j^D([0,b],q)$ is a decreasing function of b and, by [2], $\lambda_j^D([0,b],q) \to \lambda_j < \sigma_0(q)$ implies that $\lambda_j \in \sigma(q)$.

In [7] we dealt with the case when q is an $L^1(0,\infty)$ perturbation of a periodic potential p. Here it follows that $\sigma_e(p)=\sigma_e(q)$ and further from the Floquet theory [8,11] that $\inf \sigma_e(p)=\lambda_0^P([a,b],q)$. Thus, it follows from (4) that we can take $l=\lambda_0^N([a,b],q)$. We remark that when $q \in L^1(0,\infty)$, $\sigma_0(q)=0$ we can take $l=0$.

In [7] we illustrated the above method by taking $q(x)=\sin(x+1/(1+x^2))$. This is an $L^1(0,\infty)$ perturbation of $\sin(x)$ and since the essential spectrum is invariant under a unitary map $\sigma_e(\sin)=\sigma_e(\cos)$. It follows then that we can take $l=\lambda_0^N([0,2\pi],\cos)$. We remark that for this problem, since cos is an even function, $\lambda_0^N([0,2\pi],\cos)=\lambda_0^P([0,2\pi],\cos)$ giving in this example the optimal value for l. However, we cannot hope for this to occur in more general examples.

In this paper we turn to wider considerations and address the problem of computing upper bounds for several eigenvalues below $\sigma_e(q)$ both when $q \in L^1(0,\infty)$ and when q is an $L^1(0,\infty)$ perturbation of a periodic potential p. It is clear that in both of these classes of examples the method that we have outlined above has two principal components to proving the existence of $\lambda_j < \sigma_e(q)$, $j \geqslant 0$, viz:

(1) finding a lower bound l for $\sigma_e(q)$;
(2) finding an upper bound u for λ_j satisfying $u < l$.

We discuss a new method to achieve this in the next subsection.

We remark that once a lower bound for $\sigma_0(p)$ is known upper bounds for eigenvalues below the essential spectrum can be obtained from the min–max characterization. However, this requires use of test functions which have to be constructed for each problem. By using SLEIGN2 our method does not require the construction of test functions, in effect these are constructed automatically by SLEIGN2. Further, while this paper is restricted to obtaining upper bounds for eigenvalues below σ_0, the methods of [6], under appropriate smoothness conditions on q allow enclosures for these eigenvalues to be determined.

2.1. A new algorithm to prove the existence of eigenvalues below σ_0

Upper bounds for eigenvalues of S may be obtained from the BEWZ approximation [2]. This shows that for operators S with exactly k eigenvalues below $\sigma_0(q)$

$$\lambda_j^D([0,X],q) \to \lambda_j([0,\infty],q), \quad (0 \leqslant j \leqslant k-1)$$
$$\lambda_j^D([0,X],q) \to \sigma_0(q) \quad (k \leqslant j)$$

as $X \to \infty$, the convergence being from above.

As we remarked above when $q \in L^1(0,\infty)$ we have $\sigma_0(q) = 0$ and so in this case the problem of determining a lower bound for the number of eigenvalues below the essential spectrum is to find an integer j such that for some $X > 0$,

$$\lambda_j^{\mathrm{D}}([0,X],q) < 0.$$

When q is an $L^1(0,\infty)$ perturbation of p, our previous method used $\lambda_0^{\mathrm{N}}([a,b],p)$ as a lower bound for $\sigma_0(q)$. This has the disadvantage that eigenvalues $\lambda_j([0,\infty],q)$ with

$$\lambda_0^{\mathrm{N}}([a,b],p) \leqslant \lambda_j([0,\infty],q) < \sigma_0(q)$$

fail to be detected. In this paper we use a lower bound for $\sigma_0(q)$ which is better than $\lambda_0^{\mathrm{N}}([a,b],p)$ and thus are able to find eigenvalues of S above $\lambda_0^{\mathrm{N}}([a,b],p)$.

We first introduce some notation. Assume q is an $L^1(a,\infty)$ perturbation of a periodic function p with fundamental periodic interval $[a,b]$. Let $\phi_1(x,\lambda)$ be the solution of

$$-y'' + qy = \lambda y$$

on $[a,b]$ determined by the initial conditions $y(a) = 0$, $y'(a) = 1$ and let $\phi_2(x,\lambda)$ be the solution determined by $y(a) = 1$, $y'(a) = 0$. Then defining

$$D(\lambda) = \phi_1'(b,\lambda) + \phi_2(b,\lambda), \tag{6}$$

where the differentiation is with respect to x it follows that the periodic eigenvalues $\lambda_n^{\mathrm{P}}([a,b],p)$ are the roots of the equation

$$D(\lambda) = 2. \tag{7}$$

Our method of obtaining an interval enclosure for $\sigma_0(q)$, is to obtain an enclosure $[\mu]$ i.e. an interval of the real line which contains μ, for the solution μ of (7) such that

$$[\mu] \leqslant [\lambda_0^{\mathrm{D}}([a,b],p)]. \tag{8}$$

Any verified computation of the above inequality, in view of (4), will exclude all the higher periodic eigenvalues. Our algorithm for establishing (8) is to first use SLEIGN2, a "standard numerical analysis" Sturm–Liouville eigenvalue solver to obtain numerical estimates for both $\lambda_0^{\mathrm{N}}([a,b],p)$ and $\lambda_0^{\mathrm{D}}([a,b],p)$. We then use the algorithm and code reported on in [6] to obtain enclosures of the true eigenvalues. Next, SLEIGN2 is used to find both a numerical approximation of $\lambda_0^{\mathrm{P}}([a,b],p)$ and some ε determined by the error tolerance returned by SLEIGN2. Again the methods of [6] are used to verify that the interval $[\lambda_0^{\mathrm{P}}([a,b],p) - \varepsilon,\ \lambda_0^{\mathrm{P}}([a,b],p) + \varepsilon]$ contains the approximation $\hat{\lambda}_0^{\mathrm{P}}([a,b],p)$ of $\lambda_0^{\mathrm{P}}([a,b],p)$. This is achieved by computing enclosures for both of

$$D(\hat{\lambda}_0^{\mathrm{P}}([a,b],p) \pm \varepsilon) - 2.$$

Provided these enclosures are of different sign, since $D(\lambda)$ is continuous, this establishes the result. Higher eigenvalues λ_j are found similarly. By sign $[J]$ for some interval J of the real line, we mean the sign of any member of the interval J. This is defined only when all members of J have the same sign.

3. The numerical method

In this section we give a brief overview of the concepts of interval arithmetic that are needed in this work together with both a short account of Lohner's AWA algorithm and an algorithm to enclose eigenvalues. A fuller discussion of these relevant concepts may be found in [7]. An in-depth discussion of interval computation, can be found in [1], Lohner's AWA algorithm is discussed in [5,9] and the enclosure algorithm for eigenvalues can be found in [6].

All computer realisations of algorithms consist of finitely many instances of the four basic operations of arithmetic. When these are applied to real numbers, modeled in a finite number of bits, rounding errors can occur. Interval arithmetic seeks to provide safe upper and lower bounds on a calculation which take these into account. A simple-minded implementation of this concept would lead to an explosion in the interval width and many sophisticated techniques are available to control this problem [1].

Most algorithms involve other approximation errors which must also contribute to the final enclosure. An example of this is the numerical solution of an initial value problem (IVP)

$$u' = f(x,u), \quad u(0) = u_0, \tag{9}$$

where $f : [0,\infty) \times R^n \to R^n$ is sufficiently smooth. In addition, we shall assume that a solution is known at $x = x_0$. The approach developed by Lohner to enclose the solution of the IVP uses the Taylor method to determine the solution at $x_0 + h$ from its known value at x_0, viz.

$$u(x_0 + h) = u(x_0) + h\phi(x_0, h) + z_{x_0+h}, \tag{10}$$

where $u(x_0) + h\phi(x_0, h)$ is the $(r-1)$th degree Taylor polynomial of u expanded about x_0 and z_{x_0+h} is the associated local error. The error term is not known exactly since the standard formula gives, for some unknown τ,

$$z_{x_0+h} = u^{(r)}(\tau)h^r/r!, \quad \tau \in [x_0, x_0 + h]. \tag{11}$$

Lohner's algorithm uses Banach's fixed-point theorem to compute in interval arithmetic a bound for this error term. We refer the reader to [5,9] for a complete discussion of the method.

The enclosures for the eigenvalues that we need are computed using the methods reported on in [6]. Briefly, Eq. (1) and boundary conditions

$$a_1 y(a) + a_2 y'(a) = 0 \quad \text{and} \quad b_1 y(b) + b_2 y'(b) = 0$$

for real a and b with $b > a$ together with Prüfer transformation

$$y = \rho \sin\theta, \qquad y' = \rho\cos\theta \tag{12}$$

yields

$$\frac{\mathrm{d}\theta}{\mathrm{d}x} = (\lambda - q(x))\sin^2\theta + \cos^2\theta \tag{13}$$

with initial condition $\theta(0;\lambda) = \theta(0) = \alpha \in [0,\pi)$ where $\tan\alpha = a_2/a_1$, $(a_1 \neq 0)$ and $\alpha = \pi/2$ otherwise. Standard results in the spectral theory of SLP allow us to classify the nth eigenvalue of the SLP, starting counting at $n = 0$, with the stated separated boundary conditions as that unique λ which is such that Eq. (13) has a solution θ with

$$\theta(0,\lambda) = \alpha, \quad \theta(b,\lambda) = \beta + n\pi, \quad \beta \in (0,\pi],$$

where $\tan\beta = b_2/b_1$, $(b_1 \neq 0)$ and $\beta = \pi/2$ otherwise. However, it is numerically more convenient to work with a pair of initial value problems for θ_L and θ_R defined by (13). The solutions $\theta_L(x,\lambda)$ and $\theta_R(x,\lambda)$ satisfy the initial conditions (14) and (15), respectively,

$$\theta_L(0) = \alpha, \tag{14}$$

$$\theta_R(b) = \beta + n\pi. \tag{15}$$

We next choose a point c, with $0 < c < b$ and define the miss-match distance

$$D(\lambda) \equiv \theta_L(c,\lambda) - \theta_R(c,\lambda). \tag{16}$$

The nth eigenvalue is the unique value λ_n with $D(\lambda_n) = 0$. By continuity we have for $\mu_1 < \mu_2$ that if $\mathrm{sign}[D(\mu_1)] = -\mathrm{sign}[D(\mu_2)]$ then $\lambda_n \in [\mu_1, \mu_2]$.

Our algorithm for enclosing the nth eigenvalue λ_n proceeds as follows:

(1) Obtain an estimate, $\hat{\lambda}_n$, for λ_n with a "standard" numerical Sturm–Liouville solver together with an error estimate ε (we use SLEIGN2 for this);

(2) Form the quantities

$$\mu_1 = \hat{\lambda}_n - \varepsilon, \quad \mu_2 = \hat{\lambda}_n + \varepsilon.$$

(3) Use the AWA algorithm to compute enclosures for

$$D(\mu_1), \quad D(\mu_2).$$

(4) If (the interval) $\mathrm{sign}[D(\mu_1)] = -\mathrm{sign}[D(\mu_2)]$ then $\lambda_n \in [\mu_1, \mu_2]$ otherwise increase ε and recompute $D(\mu_j)$, $j = 1, 2$.

4. Numerical examples

In this section we show how the theory and algorithms that we have developed in this paper may be applied to prove results above eigenvalues below the essential spectrum of a number of Sturm–Liouville problems.

4.1. $q \in L^1(0,\infty)$

We commence by proving the existence of several eigenvalues below the essential spectrum for a number of problems with $q \in L^1(0,\infty)$.

4.1.1. $q(x) = -c\exp(-x/4)\cos(x)$

Here we take $q(x) = -c\exp(-x/4)\cos(x)$ where c is some positive constant. This example has been discussed by Brown et al. [3] in relation to problems of spectral concentration. Since $\sigma_0(q) = 0$ we take $l = 0$ and obtain a lower bound on the number of negative eigenvalues. In Table 1 above we give for different values of c the largest eigenvalue together with its index, that we have been able to approximate using SLEIGN2. We also give the safe upper bound for it obtained by our method.

Table 1
Eigenvalues below the essential spectrum for $q(x) = -c\exp(-x/4)\cos(x)$

c	Eigenvalue index	Eigenvalue approximation	Upper bound
26	4	−0.453059	−0.45305
19	3	−0.098782	−0.0987
16	2	−0.181076	−0.181
5	1	−0.215400	−0.216
4	0	−0.264342	−0.264
1.31	0	−0.000451	−0.000451

Table 2
Eigenvalues below essential spectrum for $q(x) = -c\exp(-x^2)$

c	Eigenvalue index	Eigenvalue approximation	Upper bound
50	2	−0.232229	−0.232
40	1	−6.214214	−6.213
20	1	−0.122240	−0.122
10	0	−2.543410	−2.541
2.85	0	−0.000375	−0.00038

4.1.2. $q(x) = -c\exp(-x^2)$

For $c = 1$ this example has been discussed in connection with resonances by Siedentop [10] and Brown et al. [4]. Again $\sigma_0(q) = 0 = l$ and we find a lower bound for the number of negative eigenvalues. The results are contained in Table 2 above.

4.2. Periodic potentials

The examples given here illustrate the power of the methods of this paper to establish the existence of eigenvalues below $\sigma_0(p)$ which we could not reach in [7].

4.2.1. $q(x) = c\sin(x + 1/(1 + x^2))$

The lowest point of the essential spectrum is $\sigma_0 = \lambda_0^{\mathrm{P}}([0, 2\pi], c\sin(x))$. In Table 3 we give enclosures for $\lambda_0^{\mathrm{N}}([0, 2\pi], c\sin)$, $\lambda_0^{\mathrm{P}}([0, 2\pi], c\sin)$ and $\lambda_0^{\mathrm{D}}([0, 2\pi], c\sin)$ for a selection of differing values of c while in Table 4 we give numerical estimates, obtained from SLEIGN2, together with the highest eigenvalue index that we are able to determine and enclosures for the eigenvalues. Thus, we see that when $c = 8$ there are at least 3 eigenvalues below σ_0. We remark that since

$$[\lambda_2^{\mathrm{D}}([0, 20], 8\sin(x + 1/(1 + x^2)))] > [\lambda_0^{\mathrm{N}}([0, 2\pi], 8\sin(x))]$$

(see Tables 3 and 4) this result could not have been obtained using the methods of [7].

Table 3
Neumann, periodic and Dirichlet eigenvalues of $q(x) = c\sin(x)$

c	Eigenvalue	Approximation	Enclosure
1	$\lambda_0^{\mathrm{N}}([0,2\pi],\sin)$	−0.53370249	$[-0.53370^{2}_{3}]$
	$\lambda_0^{\mathrm{P}}([0,2\pi],\sin)$	−0.378489	$[-0.3784^{89}_{90}]$
	$\lambda_0^{\mathrm{D}}([0,2\pi],\sin)$	−0.18339010	$[-0.1833^{89}_{90}]$
2	$\lambda_0^{\mathrm{N}}([0,2\pi],\sin)$	−1.23567617	$[-1.23567^{6}_{7}]$
	$\lambda_0^{\mathrm{P}}([0,2\pi],\sin)$	−1.0701309	$[-1.0701^{2}_{3}]$
	$\lambda_0^{\mathrm{D}}([0,2\pi],\sin)$	−0.92090643	$[-0.92090^{3}_{4}]$
4	$\lambda_0^{\mathrm{N}}([0,2\pi],\sin)$	−2.77020288	$[-2.77020^{2}_{3}]$
	$\lambda_0^{\mathrm{P}}([0,2\pi],\sin)$	−2.6516838	$[-2.651682^{0}_{5}]$
	$\lambda_0^{\mathrm{D}}([0,2\pi],\sin)$	−2.55827832	$[-2.55827^{6}_{8}]$
6	$\lambda_0^{\mathrm{N}}([0,2\pi],\sin)$	−4.41135359	$[-4.41135^{3}_{4}]$
	$\lambda_0^{\mathrm{P}}([0,2\pi],\sin)$	−4.3330181	$[-4.333016^{5}_{6}]$
	$\lambda_0^{\mathrm{D}}([0,2\pi],\sin)$	−4.27201653	$[-4.27201^{6}_{7}]$
8	$\lambda_0^{\mathrm{N}}([0,2\pi],\sin)$	−6.11676788	$[-6.11676^{7}_{8}]$
	$\lambda_0^{\mathrm{P}}([0,2\pi],\sin)$	−6.06466963	$[-6.0646698^{6}_{7}]$
	$\lambda_0^{\mathrm{D}}([0,2\pi],\sin)$	−6.02358961	$[-6.0235^{89}_{90}]$

Table 4
Eigenvalues for the perturbed problem $q(x) = c\sin(x + 1/(1 + x^2))$

c	σ_0	Eigenvlaue index	Eigenvalue	Enclosure $\lambda_2^D([0,20],q)$
1	$[-0.3784^{89}_{90}]$	2	−0.34114653	$[-0.34114^{6}_{7}]$
2	$[-1.0701^{2}_{3}]$	2	−1.06208432	$[-1.0620^{8}_{9}]$
4	$[-2.651682^{0}_{5}]$	2	−2.65129447	$[-2.651^{3}_{4}]$
6	$[-4.333016^{5}_{6}]$	2	−4.33348751	$[-4.333^{5}_{487}]$
8	$[-6.0646698^{6}_{7}]$	2	−6.06536102	$[-6.06536^{0}_{1}]$

Acknowledgements

We are indebted to the referee for correcting an error in the statement of our main result and for numerous suggestions which have significantly improved the paper.

References

[1] G. Alefeld, J. Herzberger, Introduction to interval computations, Academic Press, New York, 1983.

[2] P.B. Bailey, W.N. Everitt, J. Weidman, A. Zettl, Regular approximations of singular Sturm–Liouville problems, Results Math. 23 (1993) 3–22.

[3] B.M. Brown, M.S.P. Eastham, D.K.R. McCormack, Spectral concentration and rapidly decaying potentials, J. Comput. Appl. Math. 81 (1997) 333–348.
[4] B.M. Brown, M.S.P. Eastham, D.K.R. McCormack, Resonances and analytic continuation for exponentially decaying Sturm–Liouville potentials, 1999, to appear.
[5] B.M. Brown, W.D. Evans, V.G. Kirby, M. Plum, Safe numerical bounds for the Titchmarsh-Weyl $m(\lambda)$-function, Math. Proc. Camb. Phil. Soc. 113 (1993) 583–599.
[6] B.M. Brown, D.K.R. McCormack, M. Marletta, On computing enclosures for the eigenvalues of Sturm–Liouville problems, Math. Nach. 213 (1999) 17–33.
[7] B. Brown, D. McCormack, A. Zettl, On the existence of an eigenvalue below the essential spectrum, Proc. Roy. Soc. London A 455 (1999) 2229–2234.
[8] M.S.P. Eastham, The Spectral Theory of Periodic Differential Equations, Scottish Academic Press, Edinburgh, 1973.
[9] R. Lohner, Einschliessung der Lösung gewöhnlicher Anfangs- und Randwertaufgaben und Anwendungen, Doctoral Thesis, Universität Karlsruhe, 1988.
[10] H. Siedentop, A generalization of Rouché's theorem with application to resonances, in: Resonances (Lertorpet, 1987), Springer, Berlin, 1989, pp. 77–85.
[11] J. Weidmann, in: Spectral theory of ordinary differential operators, Lecture Notes in Mathematics, Vol. 1258, Springer, Berlin, 1987.

JOURNAL OF COMPUTATIONAL AND APPLIED MATHEMATICS

ELSEVIER Journal of Computational and Applied Mathematics 125 (2000) 395–421

www.elsevier.nl/locate/cam

Numerical analysis for one-dimensional Cauchy singular integral equations

Peter Junghanns *, Bernd Silbermann

Fakultät für Mathematik, Technische Universität Chemnitz, D-09107 Chemnitz, Germany

Received 8 September 1999; received in revised form 19 November 1999

Abstract

The paper presents a selection of modern results concerning the numerical analysis of one-dimensional Cauchy singular integral equations, in particular the stability of operator sequences associated with different projection methods. The aim of the paper is to show the main ideas and approaches, such as the concept of transforming the question of the stability of an operator sequence into an invertibility problem in a certain Banach algebra or the concept of certain scales of weighted Besov spaces to prove convergence rates of the sequence of the approximate solutions. Moreover, computational aspects, in particular the construction of fast algorithms, are discussed.

MSC: 45E05; 45F15; 45G05; 45L05; 45L10; 65R20

Keywords: Cauchy singular integral equation; Stability of operator sequences; Projection methods

1. Introduction

The present paper is mainly devoted to the numerical solution of Cauchy singular integral equations (CSIEs) of the form

$$a(t)u(t) + \frac{1}{\pi i}\int_\Gamma \left[\frac{b(t)}{s-t} + h(s,t)\right] u(s)\,ds = f(t), \quad t \in \Gamma, \tag{1.1}$$

where $a, b, f : \Gamma \to \mathbb{C}$ and $h : \Gamma \times \Gamma \to \mathbb{C}$ are given functions and $u : \Gamma \to \mathbb{C}$ is sought. Concerning the integration curve Γ we will restrict ourselves to the unit circle $\Gamma = \mathbb{T} = \{t \in \mathbb{C}: |t| = 1\}$ (the periodic case) and the interval $\Gamma = (-1, 1)$ (the nonperiodic case). To be able to write Eq. (1.1) in

* Corresponding author.

E-mail addresses: peter.junghanns@mathematik.tu-chemnitz.de (P. Junghanns), bernd.silbermann@mathematik.tu-chemnitz.de (B. Silbermann).

0377-0427/00/$ - see front matter
PII: S0377-0427(00)00482-9

short form, we define the multiplication operator aI, the Cauchy singular integral operator (CSIO) S_Γ, and the integral operator H_Γ by

$$u \mapsto au, \quad u \mapsto \frac{1}{\pi \mathrm{i}} \int_\Gamma \frac{u(s)}{s-\cdot}\,\mathrm{d}s \quad \text{and} \quad u \mapsto \frac{1}{\pi \mathrm{i}} \int_\Gamma h(s,\cdot)u(s)\,\mathrm{d}s,$$

respectively. In case of $\Gamma = (-1,1)$ we will shortly write S and H instead of $S_{(-1,1)}$ and $H_{(-1,1)}$, respectively. The operator equations and the approximate methods under consideration here can be described in the following way. Let $\mathscr{H}_1$, $\mathscr{H}_2$ be Hilbert spaces and $P_n^j\colon \mathscr{H}_j \to \mathscr{H}_j$, $j = 1,2$, $n = 1,2,\ldots$, be two sequences of self-adjoint projections converging strongly to the identity operator. By $\mathscr{L}(\mathscr{H}_1, \mathscr{H}_2)$ denote the Banach space of all linear and bounded operators from $\mathscr{H}_1$ into $\mathscr{H}_2$. In the case of $\mathscr{H}_1 = \mathscr{H}_2 =: \mathscr{H}$ we will shortly write $\mathscr{L}(\mathscr{H})$ instead of $\mathscr{L}(\mathscr{H}, \mathscr{H})$. For given $A \in \mathscr{L}(\mathscr{H}_1, \mathscr{H}_2)$, $f \in \mathscr{H}_2$, and $A_n \in \mathscr{L}(\operatorname{im} P_n^1, \operatorname{im} P_n^2)$, $n = 1,2,\ldots$, consider the operator equation

$$Au = f, \quad u \in \mathscr{H}_1, \tag{1.2}$$

together with the approximate equations

$$A_n u_n = f_n, \quad u_n \in \operatorname{im} P_n^1, \tag{1.3}$$

where $f_n \in \operatorname{im} P_n^2$ is some approximation of f.

Our main concern is the so-called stability of the sequence $\{A_n\} = \{A_n\}_{n=1}^\infty$, which means by definition that there exists an n_0 such that $A_n\colon \operatorname{im} P_n^1 \to \operatorname{im} P_n^2$ is invertible for all $n \geqslant n_0$ and that the inverses A_n^{-1} are uniformly bounded, i.e.

$$\sup \left\{ \|A_n^{-1} P_n^2\|_{\mathscr{H}_2 \to \mathscr{H}_1} \colon n \geqslant n_0 \right\} < \infty.$$

If the sequence $\{A_n\}$ is stable and if u^* is a solution of (1.2) and u_n^* are the solutions of (1.3), then the estimation

$$\|P_n^1 u^* - u_n^*\|_{\mathscr{H}_1} \leqslant \|A_n^{-1} P_n^2\|_{\mathscr{H}_2 \to \mathscr{H}_1} \|A_n P_n^1 u^* - f_n\|_{\mathscr{H}_2}$$

shows that u_n^* converges to u^* in the norm of $\mathscr{H}_1$ if $A_n P_n^1 \to A$ (strong convergence) and $f_n \to f$ (in $\mathscr{H}_2$).

In the present paper two general approaches for investigating the stability of operator sequences are considered. The first one is a C^*-algebra approach and is restricted to the case $\mathscr{H}_1 = \mathscr{H}_2 =: \mathscr{H}$ and will be introduced to study Galerkin (finite section) and collocation methods in both the periodic and the nonperiodic case. The second approach uses a decomposition of the operator $A = B + T$ into a so-called dominant part B and a compact perturbation T together with the respective approximating operators B_n and T_n. The main tool is that B and B_n are chosen in such a way that $B_n u_n = B u_n$ for all $u_n \in \operatorname{im} P_n^1$. This second approach we will demonstrate in the nonperiodic case, but we should remark that, in various concrete situations the application of the first approach desires for considerations which are due to the second approach (for example, the investigation of local representatives, when local principles are applied).

We want to point out that, with the aim of limiting the bibliography, in general we will not refer to the original papers, but if possible, to textbooks or monographs, where the interested reader can also find the original references.

2. The periodic case

2.1. Finite section and collocation methods

Let $\mathscr{H}_1=\mathscr{H}_2=\mathscr{H}:=\boldsymbol{L}^2(\mathbb{T})$ be the Hilbert space of all square integrable (complex valued) functions on the unit circle $\mathbb{T}$ equipped with the inner product

$$\langle u,v\rangle_{\mathbb{T}}:=\frac{1}{2\pi}\int_0^{2\pi}u(\mathrm{e}^{\mathrm{i}s})\overline{v(\mathrm{e}^{\mathrm{i}s})}\,\mathrm{d}s.$$

It is well known that $\{e_n\}_{n=-\infty}^{\infty}$ with $e_n(t)=t^n$ forms a complete orthonormal system in $\boldsymbol{L}^2(\mathbb{T})$. Thus, we use the orthoprojections $P_n^1=P_n^2=P_n^{\mathbb{T}}$ defined by $P_n^{\mathbb{T}}u=\sum_{k=-n}^{n}\langle u,e_k\rangle_{\mathbb{T}}e_k$ and look for an approximate solution u_n of the equation

$$Au:=au+bS_{\mathbb{T}}u+H_{\mathbb{T}}u=f, \tag{2.1}$$

in the form $u_n=\sum_{k=-n}^{n}\xi_{kn}e_k$. We assume that a and b are piecewise continuous functions (i.e. $a,b\in\boldsymbol{PC}(\mathbb{T})$), which means that they have at each point of $\mathbb{T}$ one-sided limits and that (without loss of generality) the left-sided limits coincide with the funtion values. The finite section method consists in finding the unknowns ξ_{kn} by solving

$$\langle Au_n,e_k\rangle_{\mathbb{T}}=\langle f,e_k\rangle_{\mathbb{T}},\quad k=-n,\ldots,n. \tag{2.2}$$

As an example of a collocation method we consider

$$(Au_n)(t_{jn}^{\mathrm{L}})=f(t_{jn}^{\mathrm{L}}),\quad j=-n,\ldots,n, \tag{2.3}$$

where $t_{jn}^{\mathrm{L}}=\mathrm{e}^{2\pi\mathrm{i}j/(2n+1)}$. Obviously, the finite section method (2.2) can be written in the form (1.3) with $A_n=P_n^{\mathbb{T}}AP_n^{\mathbb{T}}$ and $f_n=P_n^{\mathbb{T}}f$. By $L_n^{\mathbb{T}}$ we will refer to the interpolation operator

$$L_n^{\mathbb{T}}a=\sum_{k=-n}^{n}\alpha_k e_k\quad\text{with }\alpha_k=\frac{1}{2n+1}\sum_{j=-n}^{n}a(t_{jn}^{\mathrm{L}})(t_{jn}^{\mathrm{L}})^{-k},$$

which has the property $(L_n^{\mathbb{T}}a)(t_{jn}^{\mathrm{L}})=a(t_{jn}^{\mathrm{L}})$, $j=-n,\ldots,n$. Then, the collocation method (2.3) can also be written in the form (1.3), where $A_n=L_n^{\mathbb{T}}AP_n^{\mathbb{T}}$ and $f_n=L_n^{\mathbb{T}}f$. To find necessary and sufficient conditions for the stability of the sequences $\{P_n^{\mathbb{T}}AP_n^{\mathbb{T}}\}$ and $\{L_n^{\mathbb{T}}AP_n^{\mathbb{T}}\}$ we use the following algebraization of the stability problem.

2.2. Algebraization

By $\mathscr{F}$ we denote the C^*-algebra of all sequences $\{A_n\}$ of linear operators $A_n:\operatorname{im}P_n^{\mathbb{T}}\to\operatorname{im}P_n^{\mathbb{T}}$, for which the strong limits

$$A=s-\lim_{n\to\infty}A_nP_n^{\mathbb{T}}\in\mathscr{L}(\mathscr{H})\quad\text{and}\quad A^*=s-\lim_{n\to\infty}A_n^*P_n^{\mathbb{T}},$$

exist. The norm in $\mathscr{F}$ is defined by $\|\{A_n\}\|_{\mathscr{F}}:=\sup\{\|A_nP_n\|_{\mathscr{H}\to\mathscr{H}}\colon n=1,2,\ldots\}$ and the algebraic operations by $\{A_n\}+\{B_n\}:=\{A_n+B_n\}$, $\{A_n\}\{B_n\}:=\{A_nB_n\}$, and $\{A_n\}^*:=\{A_n^*\}$. Let $\mathscr{N}$ be the two-sided closed ideal of $\mathscr{F}$ consisting of all sequences $\{C_n\}$ of operators $C_n:\operatorname{im}P_n^{\mathbb{T}}\to\operatorname{im}P_n^{\mathbb{T}}$ such that $\lim_{n\to\infty}\|C_nP_n^{\mathbb{T}}\|_{\mathscr{H}\to\mathscr{H}}=0$. Then, the stability of $\{A_n\}\in\mathscr{F}$ is equivalent to the invertibility of the coset $\{A_n\}+\mathscr{N}\in\mathscr{F}/\mathscr{N}$ (cf. [21, Proposition 7.5]). The task of finding necessary and sufficient conditions for the invertibility of the coset $\{A_n\}+\mathcal{N}$ in the quotient algebra $\mathcal{F}/\mathcal{N}$ is, in general, too complicated, because of the algebra $\mathcal{F}$ is to large and the ideal $\mathcal{N}$ is too small. For this reason we define a further sequence $\{W_n^{\mathbb{T}}\}$ of operators $W_n:\mathcal{H}\to\mathcal{H}$ by

$$W_n^{\mathbb{T}}u=\sum_{k=-n}^{-1}\langle u,e_{-n-1-k}\rangle_{\mathbb{T}}e_k+\sum_{k=0}^{n}\langle u,e_{n-k}\rangle_{\mathbb{T}}e_k$$

and the C^*-subalgebra $\mathcal{F}^W$ of $\mathcal{F}$ consisting of all sequences $\{A_n\}\in\mathcal{F}$, for which the strong limits

$$\tilde{A}=s-\lim_{n\to\infty}W_n^{\mathbb{T}}A_nW_n^{\mathbb{T}}\quad\text{and}\quad\tilde{A}^*=s-\lim_{n\to\infty}(W_n^{\mathbb{T}}A_nW_n^{\mathbb{T}})^*P_n^{\mathbb{T}},$$

exist. Furthermore, by $\mathcal{J}$ we refer to the collection of all sequences $\{A_n\}$ of the form

$$A_n=P_n^{\mathbb{T}}K_1P_n^{\mathbb{T}}+W_n^{\mathbb{T}}K_2W_n^{\mathbb{T}}+C_n\quad\text{with } K_j\in\mathcal{K}(\mathcal{H}),\quad\{C_n\}\in\mathcal{N},$$

where $\mathcal{K}(\mathcal{H})\subset\mathcal{L}(\mathcal{H})$ denotes the ideal of all compact operators. Then, $\mathcal{J}$ is a two-sided closed ideal of $\mathcal{F}^W$ and the stability of a sequence $\{A_n\}\in\mathcal{F}^W$ is equivalent to the invertibility of the operators $A,\tilde{A}:\mathcal{H}\to\mathcal{H}$ and the coset $\{A_n\}+\mathcal{J}\in\mathcal{F}^W/\mathcal{J}$ (see [21, Proposition 7.6, Theorem 7.7]).

To prove the invertibility of the above-mentioned cosets one can apply local principles, for example the local principles of Gohberg and Krupnik and of Allan and Douglas, which we will shortly describe in the following. Let $\mathcal{B}$ be a unital Banach algebra. A subset $\mathcal{M}\subset\mathcal{B}$ is called a localizing class, if $0\notin\mathcal{M}$ and if for any two elements $a_1,a_2\in\mathcal{M}$ there exists an $a\in\mathcal{M}$ such that $a_ja=aa_j=a$ for $j=1,2$. Two elements $x,y\in\mathcal{B}$ are called $\mathcal{M}$-equivalent if

$$\inf\{\|(x-y)a\|_{\mathcal{B}}:a\in\mathcal{M}\}=\inf\{\|a(x-y)\|_{\mathcal{B}}:a\in\mathcal{M}\}=0.$$

An element $x\in\mathcal{B}$ is said to be $\mathcal{M}$-invertible if there exist $z_j\in\mathcal{B}$ and $a_j\in\mathcal{M}$, $j=1,2$, such that $z_1xa_1=a_1$ and $a_2xz_2=a_2$. A system $\{\mathcal{M}_\tau\}_{\tau\in\mathcal{I}}$ of localizing classes $\mathcal{M}_\tau$ of $\mathcal{B}$ is called a covering system if, for any set $\{a_\tau\}_{\tau\in\mathcal{I}}$ with $a_\tau\in\mathcal{M}_\tau$, one can choose a finite number of elements $a_{\tau_1},\dots,a_{\tau_m}$ the sum of which is invertible in $\mathcal{B}$.

Local principle of Gohberg and Krupnik (see [8, Theorem XII.1.1]). Let $\{\mathcal{M}_\tau\}_{\tau\in\mathcal{I}}$ be a covering system of localizing classes of the unital Banach algebra $\mathcal{B}$. If $x\in\mathcal{B}$ commutes with all elements of $\bigcup_{\tau\in\mathcal{I}}\mathcal{M}_\tau$ and if $x_\tau\in\mathcal{B}$ is $\mathcal{M}_\tau$-equivalent to x for all $\tau\in\mathcal{I}$, then x is invertible in $\mathcal{B}$ if and only if x_τ is $\mathcal{M}_\tau$-invertible for all $\tau\in\mathcal{I}$.

Let $\mathcal{B}$ be a unital Banach algebra and $\mathcal{B}_c$ a closed subalgebra of the center of $\mathcal{B}$ containing the identity element. For every maximal ideal $x\in M(\mathcal{B}_c)$, let $\mathcal{J}_x$ denote the smallest closed ideal of $\mathcal{B}$ which contains x, i.e.

$$\mathcal{J}_x=\operatorname{clos}_{\mathcal{B}}\left\{\sum_{j=1}^{m}a_jc_j:a_j\in\mathcal{B},\ c_j\in x,\ m=1,2,\dots,\right\}.$$

Local principle of Allan and Douglas (comp. [9, Sections 1.4.4, 1.4.6]). An element $a\in\mathcal{B}$ is invertible in $\mathcal{B}$ if and only if $a+\mathcal{J}_x$ is invertible in $\mathcal{B}/\mathcal{J}_x$ for all $x\in M(\mathcal{B}_c)$. (In case $\mathcal{J}_x=\mathcal{B}$ we define that $a+\mathcal{J}_x$ is invertible.)

2.3. Realization

In this section we describe how the local principles can be applied to investigate stability of the finite section and the collocation method presented in Section 2.1, and we formulate the main results.

In the case of the finite section method one can choose as a covering system of localizing classes in the quotient algebra $\mathscr{F}^W/\mathscr{J}$ the system $\{\mathscr{M}_\tau\}_{\tau\in\mathbb{T}}$ with

$$\mathscr{M}_\tau = \{\{P_n^{\mathbb{T}} P_{\mathbb{T}} f P_{\mathbb{T}} P_n^{\mathbb{T}} + P_n^{\mathbb{T}} Q_{\mathbb{T}} f Q_{\mathbb{T}} P_n^{\mathbb{T}}\} + \mathscr{J} : f \in m_\tau\},$$

where $P_{\mathbb{T}} = \frac{1}{2}(I + S_{\mathbb{T}})$, $Q_{\mathbb{T}} = \frac{1}{2}(I - S_{\mathbb{T}})$, and m_τ denotes the collection of all continuous functions $f : \mathbb{T} \to [0,1]$ for which there exists a neighborhood U_f of τ such that $f(t) = 1$ for all $t \in U_f$.

Theorem 2.1 (Prössdorf and Silbermann [21, Theorem 7.14]). *Let $a, b \in \boldsymbol{PC}(\mathbb{T})$, $H_{\mathbb{T}} \in \mathscr{K}(\boldsymbol{L}^2(\mathbb{T}))$, and $c_\pm := a \pm b$. Then the sequence $\{P_n^{\mathbb{T}}(aI + bS_{\mathbb{T}} + H_{\mathbb{T}})P_n^{\mathbb{T}}\}$ is stable in $\boldsymbol{L}^2(\mathbb{T})$ if and only if the conditions*

(a) *the operators $aI + bS_{\mathbb{T}} + H_{\mathbb{T}}$, $c_\pm P_{\mathbb{T}} + Q_{\mathbb{T}} : \boldsymbol{L}^2(\mathbb{T}) \to \boldsymbol{L}^2(\mathbb{T})$ are invertible,*
(b) *for all $t \in \mathbb{T}$ and for all $\mu \in [0,1]$,*

$$\frac{c_+(t+0)}{c_+(t-0)}\mu + \frac{c_-(t+0)}{c_-(t-0)}(1-\mu) \notin (-\infty, 0]$$

are satisfied.

In the case of the collocation method one can use the localizing classes

$$\mathscr{M}_\tau = \{\{L_n^{\mathbb{T}} f P_n^{\mathbb{T}}\} + \mathscr{J} : f \in m_\tau\}, \quad \tau \in \mathbb{T},$$

which also form a covering system of localizing classes in $\mathscr{F}^W/\mathscr{J}$.

Theorem 2.2 (Prössdorf and Silbermann [21, Theorem 7.19]). *Let $a, b \in \boldsymbol{PC}(\mathbb{T})$ and $h : [-1,1]^2 \to \mathbb{C}$ be continuous. Then the sequence $\{L_n^{\mathbb{T}}(aI + bS_{\mathbb{T}} + H_{\mathbb{T}})P_n^{\mathbb{T}}\}$ is stable in $\boldsymbol{L}^2(\mathbb{T})$ if and only if the operator $aI + bS_{\mathbb{T}} + H_{\mathbb{T}} : \boldsymbol{L}^2(\mathbb{T}) \to \boldsymbol{L}^2(\mathbb{T})$ is invertible.*

Theorem 2.2 remains true if the interpolation operator $L_n^{\mathbb{T}}$ and the projection $P_n^{\mathbb{T}}$ are replaced by $M_n^{\mathbb{T}}$ and $Q_n^{\mathbb{T}}$, respectively, where

$$M_n^{\mathbb{T}} b = \sum_{k=-n-1}^{n} \beta_k e_k \quad \text{with } \beta_k = \frac{1}{2n+2} \sum_{j=-n-1}^{n} b(t_{jn}^M)(t_{jn}^M)^{-k}, \quad t_{jn}^M = \mathrm{e}^{\pi \mathrm{i} j/(n+1)}$$

and

$$Q_n^{\mathbb{T}} u = \sum_{k=-n-1}^{n} \langle u, e_k \rangle_{\mathbb{T}} e_k.$$

Let us ask for stability conditions for sequences $\{A_n\}$ belonging to the smallest closed subalgebra $\mathscr{A}$ of $\mathscr{F}$, which contains all sequences of the form $\{L_n^{\mathbb{T}}(aI + bS)P_n^{\mathbb{T}}\}$ with $a, b \in \boldsymbol{PC}(\mathbb{T})$ as well as all sequences belonging to $\mathscr{J}$. We remark that the mappings $W\{L_n^{\mathbb{T}}(aI + bS)P_n^{\mathbb{T}}\} = s - \lim L_n^{\mathbb{T}}(aI + bS)P_n^{\mathbb{T}} = aI + bS$ and $\tilde{W}\{L_n^{\mathbb{T}}(aI + bS)P_n^{\mathbb{T}}\} = s - \lim W_n^{\mathbb{T}} L_n^{\mathbb{T}}(aI + bS)W_n^{\mathbb{T}} = \tilde{a}I + \tilde{b}S$, $\tilde{a}(t) := a(t^{-1})$, extend to *-homomorphisms $W, \tilde{W} : \mathscr{A} \to \mathscr{L}(\boldsymbol{L}^2(\mathbb{T}))$. For the ideas of the proof of the following result, which is essentially based on the application of the local principle of Allan and Douglas, we refer to [23, Section 4].

Theorem 2.3. *A sequence* $\{A_n\} \in \mathscr{A}$ *is stable if and only if the operators* $W\{A_n\}, \tilde{W}\{A_n\} : \boldsymbol{L}^2(\mathbb{T}) \to \boldsymbol{L}^2(\mathbb{T})$ *are invertible.*

2.4. Convergence rates and corrected collocation methods

The scale of periodic Sobolev spaces $\boldsymbol{H}^s(\mathbb{T})$, $s \in \mathbb{R}$, defined, for example, as the completion of $\boldsymbol{C}^\infty(\mathbb{T})$ w.r.t. the norm

$$\|u\|_{\mathbb{T},s} = \left(\sum_{k=-\infty}^{\infty} (1+|k|)^{2s} |\langle u, e_k \rangle_{\mathbb{T}}|^2 \right)^{1/2},$$

is a powerful tool to consider convergence rates of the approximation error for various numerical methods. That the approximation error of the finite section and the collocation method will behave differently, is already suggested by the following result.

(A) If $f \in \boldsymbol{H}^s(\mathbb{T})$ then

$$\|f - P_n^{\mathbb{T}} f\|_{\mathbb{T},t} \leqslant \text{const}\, n^{t-s} \|f\|_{\mathbb{T},s} \quad \text{for all } t \leqslant s$$

and, if $s > \frac{1}{2}$,

$$\|f - L_n^{\mathbb{T}} f\|_{\mathbb{T},t} \leqslant \text{const}\, n^{t-s} \|f\|_{\mathbb{T},s} \quad \text{for all } t \text{ with } 0 \leqslant t \leqslant s,$$

where the constants are independent of n, t, and f.

Let $a, b \in \boldsymbol{H}^r(\mathbb{T})$ for some $r > \frac{1}{2}$ and $A = aI + bS_{\mathbb{T}} + H_{\mathbb{T}}$. Assume that $\mathbb{H}_{\mathbb{T}} \in \mathscr{L}(\boldsymbol{L}^2(\mathbb{T}), \boldsymbol{H}^r(\mathbb{T}))$ and that the respective sequence $\{P_n^{\mathbb{T}} A P_n^{\mathbb{T}}\}$ or $\{L_n^{\mathbb{T}} A P_n^{\mathbb{T}}\}$ is stable in $\boldsymbol{L}^2(\mathbb{T})$. Let $u^* \in \boldsymbol{L}^2(\mathbb{T})$ denote the solution of (2.1), where $f \in \boldsymbol{H}^s(\mathbb{T})$ for some $s \leqslant r$ and, in case of the collocation method, $\frac{1}{2} < s$. Then assertion (A) leads to the following.

(B) If $u_n^* \in \operatorname{im} P_n^{\mathbb{T}}$ is the solution of the finite section method (2.2) or the collocation method (2.3), respectively, then

$$\|u^* - u_n^*\|_{\mathbb{T},t} \leqslant \text{const}\, n^{t-s} \|f\|_{\mathbb{T},s}$$

for all $t \leqslant s$ in case of the finite section method and for all $t \in [0, s]$ in case of the collocation method.

The error estimate in (B) remains true if we consider instead of the collocation method (2.3) the collocation–quadrature method

$$a(t_{jn}^{\mathrm{L}}) u_n(t_{jn}^{\mathrm{L}}) + b(t_{jn}^{\mathrm{L}})(S_{\mathbb{T}} u_n)(t_{jn}^{\mathrm{L}}) + (H_{\mathbb{T},n} u_n)(t_{jn}^{\mathrm{L}}) = f(t_{jn}^{\mathrm{L}}), \quad j = -n, \ldots, n, \tag{2.4}$$

where

$$(H_{\mathbb{T},n} u)(t) = \frac{2}{2n+1} \sum_{j=-n}^{n} h(t_{jn}^{\mathrm{L}}, t) t_{jn}^{\mathrm{L}} u(t_{jn}^{\mathrm{L}})$$

and $h(s,t)$ is assumed to be sufficiently smooth.

In what follows we describe an idea for the construction of methods (different from the qualocation method proposed by I. Sloan), whose numerical realization is nearly as simple as that of the collocation method and whose convergence rates in negative Sobolev norms are nearly as good as that of the finite section method. Let $\{m_n\}$ and $\{M_n\}$ be two strictly monotonically increasing sequences of natural numbers satisfying $m_n \leqslant M_n \leqslant n$ and $\lim_{n\to\infty}(M_n - m_n) = \infty$. Define

$$u_n = u_n^{\mathrm{L}} + P_{m_n}^{\mathbb{T}}(u_{M_n}^{P} - u_n^{\mathrm{L}}), \tag{2.5}$$

where u_n^{P} and u_n^{L} denote the solutions of the finite section method (2.2) and the collocation method (2.3), respectively.

Theorem 2.4 (Berthold and Silbermann [3, Theorem 3.4]). *Let the above assumptions on a, b, and f be fulfilled and let $-r \leqslant t < -\frac{1}{2}$. Assume additionally that the $\mathbf{L}^2$-adjoint H^* of the operator H belongs to $\mathscr{L}(\boldsymbol{H}^{|t|}(\mathbb{T}), \boldsymbol{H}^r(\mathbb{T}))$. If both the finite section and the collocation method are stable in $\mathbf{L}^2(\mathbb{T})$ and if $f \in \boldsymbol{H}^s(\mathbb{T})$ for some s with $\frac{1}{2} < s \leqslant r$, then the corrected approximation defined by (2.5) satisfies the error estimate*

$$\|u^* - u_n\|_{\mathbb{T},t} \leqslant \mathrm{const}(m_n^t n^{-s} + (M_n - m_n)^{-r} + M_n^{-r-s})\|f\|_{\mathbb{T},s}. \tag{2.6}$$

If, for example, we choose $M_n = 2m_n$ and $m_n \sim \sqrt{n}$ then we get

$$\|u^* - u_n\|_{\mathbb{T},t} \leqslant \mathrm{const}\begin{cases} n^{t/2-s}, & r \geqslant 2s + |t|, \\ n^{-r/2}, & r < 2s + |t|, \end{cases}$$

which is better than $\mathrm{O}(n^{-s})$ if $r > 2s$.

2.5. Fast algorithms

In general the matrices of the algebraic systems resulting from the above considered methods are not structured. Nevertheless it is possible to construct fast algorithms based on these methods and having a complexity of $\mathrm{O}(n \log n)$. These algorithms are based on the decomposition of the operator into a simply structured dominant part $aI + bS_{\mathbb{T}} = cP_{\mathbb{T}} + dQ_{\mathbb{T}}$ ($c = a + b$, $d = a - b$) and a smoothing part $H_{\mathbb{T}}$ and on the observation that the behaviour of the Fourier coefficients of high order of the approximate solution is essentially determined by the dominant part of the operator. A further aim besides the low complexity is to preserve the convergence rate of the error for a possibly wide range of Sobolev norms.

Firstly we demonstrate Amosov's idea for the example of the collocation method (2.3), and secondly we describe shortly how it is possible to combine this idea with the concept of the corrected collocation method (2.5). Amosov's method or the so-called parametrix–quadrature method for (2.4) consists in determining an approximation by

$$u_n = P_m^{\mathbb{T}} u_m^{\mathrm{L}} + Q_m^{\mathbb{T}} z_n, \tag{2.7}$$

where $z_n = L_n^{\mathbb{T}} B L_n^{\mathbb{T}} f$ and B denotes the parametrix $c^{-1}P_{\mathbb{T}} + d^{-1}Q_{\mathbb{T}}$, $Q_m^{\mathbb{T}} = I - P_m^{\mathbb{T}}$ is the complementary projection of $P_m^{\mathbb{T}}$, and u_m^{L} is the solution of (2.4) with m instead of n and $g = f - (cP_{\mathbb{T}} + dQ_{\mathbb{T}})Q_m^{\mathbb{T}} z_n$ instead of f. The positive integer $m < n$ is chosen in such a way that $(2n+1)/(2m+1)$ is also an integer. The algorithm can be devided into three steps.

Step 1: The computation of $z_n = L_n^{\mathbb{T}} B L_n^{\mathbb{T}} f$. Let $\tilde{f}_n = [f(t_{jn}^{\mathrm{L}})]_{j=-n}^{n}$ and $\hat{f}_n = [\langle f, e_j \rangle_{\mathbb{T}}]_{j=-n}^{n}$ denote the vectors of the function values at the collocation points and the vector of the Fourier coefficients of f with indices $-n, \dots, n$, respectively. Then

$$\hat{z}_n = F_n^{-1}(C_n^{-1} F_n I_n^+ + D_n^{-1} F_n I_n^-) F_n^{-1} \tilde{f}_n,$$

where $F_n = [(t_{jn}^{\mathrm{L}})^k]_{j,k=-n}^{n}$ and $F_n^{-1} = [(1/(2n+1))(t_{jn}^{\mathrm{L}})^{-k}]_{j,k=-n}^{n}$ are the Fourier matrix of order $2n+1$ and its inverse, respectively. The other matrices are defined as

$$I_n^+ = \begin{bmatrix} 0 & 0 \\ 0 & I_{n+1} \end{bmatrix} \in \mathbb{C}^{(2n+1)\times(2n+1)}, \quad I_n^- = I_{2n+1} - I_n^+, \quad C_n = \operatorname{diag} \tilde{c}_n, \quad D_n = \operatorname{diag} \tilde{d}_n$$

(I_n is the unit matrix of order n).

Step 2: Calculation of $L_m^{\mathbb{T}} g$. We have

$$\tilde{g}_m = \tilde{f}_m - (C_m I_{m,n} F_n I_n^+ + D_m I_{m,n} F_n I_n^-) I_n^m \hat{z}_n,$$

where I_n^m is a diagonal matrix

$$\operatorname{diag}[\underbrace{1,\dots,1}_{n-m}, 0, \dots, 0, \underbrace{1,\dots,1}_{n-m}]$$

of order $2n+1$ and $I_{m,n}$ is the $(2m+1)\times(2n+1)$-matrix with the entries $[I_{m,n}]_{jk} = 1$ if $t_{jm}^{\mathrm{L}} = t_{kn}^{\mathrm{L}}$ and $[I_{m,n}]_{jk} = 0$ otherwise, $j = -m, \dots, m$, $k = -n, \dots, n$.

Step 3: Solution of the $m \times m$ algebraic system corresponding to

$$L_n^{\mathbb{T}}(cP_{\mathbb{T}} + dQ_{\mathbb{T}} + H_{\mathbb{T}}) u_m^{\mathrm{L}} = L_m^{\mathbb{T}} g.$$

The complexity of the first two steps is $\mathrm{O}(n \log n)$, that one of the third step equals $\mathrm{O}(m^3)$. But, choosing $m \sim n^{1/3}$ then the calculation of u_n needs only $\mathrm{O}(n \log n)$ operations. If $f \in \boldsymbol{H}^s(\mathbb{T})$ for some $s > \frac{1}{2}$ and $a, b \in \boldsymbol{H}^r(\mathbb{T})$, $H \in \mathscr{L}(\boldsymbol{L}^2(\mathbb{T}), \boldsymbol{H}^r(\mathbb{T}))$ for some $r > s$ and if the collocation method (2.3) is stable in $\boldsymbol{L}^2(\mathbb{T})$, then, for all sufficiently large n and $m \sim n^{1/3}$, we have the estimate

$$\|u^* - u_n\|_{\mathbb{T},t} \leqslant \mathrm{const}\, n^{t-s} \|f\|_{\mathbb{T},s}$$

for all $t > \frac{1}{2}$ with $s - (r-s)/2 \leqslant t \leqslant s$, where u_n is the approximation defined by (2.7) (see [1]).

The parametrix-Galerkin method combines the Amosov algorithm with the corrected collocation method. Here the approximation u_n is defined by $u_n = P_{m_n}^{\mathbb{T}} y_{M_n}^{\mathrm{P}} + Q_{m_n}^{\mathbb{T}} z_n$, where we use the same notations as in Section 2.4, and it is important that for this approximation u_n the same error estimate (2.6) as for the pure corrected collocation method holds true (see [4]). A generalization of this approach, i.e. the combination of Amosov's idea with the idea of corrected collocation methods, to a wider class of pseudodifferential equations on closed curves can be found in [24], where additionally a two-grid iteration method is used for the solution of the Galerkin equation.

2.6. *Spline approximation methods*

Here we like to demonstrate how spline approximation methods for CSIEs on closed curves, especially on the unit circle, can be investigated using a generalization of the C^*-algebra approach described in Section 2.2. Via the parametrization $s \mapsto \mathrm{e}^{2\pi \mathrm{i} x}$, $0 \leqslant x < 1$, we have a one-to-one correspondence between functions u on $\mathbb{T}$ and 1-periodic functions $\tilde{u}$ on $\mathbb{R}$, where $\tilde{u}(x) = u(e^{2\pi i x})$. Throughout this section we shall identify the functions u and $\tilde{u}$.

Let δ be a nonnegative integer and n a natural number, and let $\mathscr{S}_n^\delta := \mathscr{S}_n^\delta(\Delta)$ denote the space of smooth 1-periodic splines of degree δ on the uniform mesh $\Delta := \{k/n : k = 0, \ldots, n-1\}$. Thus, $\mathscr{S}_n^\delta$ ($\delta \geqslant 1$) consists of periodic $\boldsymbol{C}^{\delta-1}$ piecewise polynomials of degree δ and has dimension n. $\mathscr{S}_n^0$ is defined as the corresponding space of piecewise constant functions. Let $\varepsilon \in [0, 1)$ be a fixed number, where $\varepsilon > 0$ if $\delta = 0$. The ε-collocation method for the equation

$$Au = f, \tag{2.8}$$

which we consider in $\boldsymbol{L}^2(\mathbb{T})$ for $A = aI + bS_{\mathbb{T}}$, determines an approximate solution $u_n \in \mathscr{S}_n^\delta$ by

$$(Au_n)\left(\frac{k+\varepsilon}{n}\right) = f\left(\frac{k+\varepsilon}{n}\right), \quad k = 0, \ldots, n-1. \tag{2.9}$$

It is easily seen that the left-hand side of (2.9) makes sense in case of $\delta = 0$ if and only if $0 < \varepsilon < 1$. The Galerkin method defines $u_n \in \mathscr{S}_n^\delta$ by

$$\langle Au_n, \varphi_n \rangle_{\mathbb{T}} = \langle f, \varphi_n \rangle_{\mathbb{T}} \quad \text{for all } \varphi_n \in \mathscr{S}_n^\delta. \tag{2.10}$$

The simplest quadrature method for Eq. (2.8) is the so-called method of "discrete vortices", which reads as follows. Let $t_k = e^{(\pi i(2k+1))/n}$, $s_k = e^{2\pi i k/n}$, $k = 0, \ldots, n-1$, and determine the approximate values ξ_k of $u(s_k)$ by the system

$$a(t_j)\frac{\xi_{j+1} + \xi_j}{2} + b(t_j)\sum_{k=0}^{n-1} \frac{2\xi_k}{s_k - t_j}\frac{s_k}{n} = f(t_j), \quad j = 0, \ldots, n-1. \tag{2.11}$$

Once a suitable basis of $\mathscr{S}_n^\delta$ is chosen, the approximate equation is reduced to an $n \times n$ linear system for the unknown coefficients of u_n, whose matrix has a special structure, namely the structure of a paired circulant. This property owns also many other spline approximation methods. So we have to study the stability of some sequences constituted by paired circulants.

Let $\ell^2(n)$ denote the n-dimensional complex Hilbert space $\mathbb{C}^n$ provided with the inner product

$$\langle \xi, \eta \rangle = \sum_{k=0}^{n-1} \xi_k \overline{\eta_k}, \quad \xi = [\xi_k]_{k=0}^{n-1}, \; \eta = [\eta_k]_{k=0}^{n-1} \in \mathbb{C}^n$$

and the norm $\|\xi\| = \sqrt{\langle \xi, \xi \rangle}$. In what follows, each operator $\boldsymbol{A}_n \in \mathscr{L}(\ell^2(n))$ will be identified with the corresponding matrix in $\mathbb{C}^{n \times n}$ (w.r.t. the standard basis in $\mathbb{C}^n$). Introduce the unitary operators $\boldsymbol{U}_n, \boldsymbol{U}_n^{-1} \in \mathscr{L}(\ell^2(n))$ by

$$\boldsymbol{U}_n \xi := \left[\frac{1}{\sqrt{n}} \sum_{j=0}^{n-1} e^{2\pi i kj/n} \xi_j\right]_{k=0}^{n-1}, \qquad \boldsymbol{U}_n^{-1} \xi := \left[\frac{1}{\sqrt{n}} \sum_{j=0}^{n-1} e^{-2\pi i kj/n} \xi_j\right]_{k=0}^{n-1}.$$

A finite Toeplitz matrix $\boldsymbol{T}_n = [a_{j-k}]_{j,k=0}^{n-1}$ is said to be a circulant if $a_{-k} = a_{n-k}$, $k = 1, \ldots, n-1$. It is easily seen that $\boldsymbol{A}_n \in \mathscr{L}(\ell^2(n))$ is a circulant if and only if there exists a vector $\zeta = [\zeta_k]_{k=0}^{n-1} \in \mathbb{C}^n$ such that $\boldsymbol{A}_n = \boldsymbol{U}_n \boldsymbol{M}_\zeta \boldsymbol{U}_n^{-1}$, where $\boldsymbol{M}_\zeta$ is the diagonal matrix diag $[\zeta_0, \ldots, \zeta_{n-1}]$. Obviously, the numbers ζ_k are just the eigenvalues of $\boldsymbol{A}_n$ and $\zeta^{(k)} = [(1/\sqrt{n})\, e^{2\pi i jk/n}]_{j=0}^{n-1}$, $k = 0, \ldots, n-1$, are the corresponding eigenvectors.

Given a bounded function $\rho:\mathbb{T}\to\mathbb{C}$, let ρ_n and $\tilde{\rho}_n$ denote the diagonal matrices

$$\operatorname{diag}[\rho(t_0),\dots,\rho(t_{n-1})] \quad \text{and} \quad \operatorname{diag}[\rho(\tau_0),\dots,\rho(\tau_{n-1})],$$

respectively, where $t_k=\mathrm{e}^{2\pi \mathrm{i}k/n}$ and $\tau_k=\mathrm{e}^{2\pi\mathrm{i}(k+\varepsilon)/n}$. Consider the circulant $\hat{\rho}_n:=\boldsymbol{U}_n\rho_n\boldsymbol{U}_n^{-1}$. Obviously,

$$\|\hat{\rho}_n\|=\|\rho_n\|\leqslant \sup_{t\in\mathbb{T}}|\rho(t)| \quad \text{and} \quad \|\tilde{\rho}_n\|\leqslant \sup_{t\in\mathbb{T}}|\rho(t)|.$$

Let $\boldsymbol{C}_n\in\mathscr{L}(\ell^2(n))$ satisfy $\lim_{n\to\infty}\|\boldsymbol{C}_n\|=0$. For given $a,b\in\boldsymbol{C}(\mathbb{T})$ and $\alpha,\beta\in\boldsymbol{PC}(\mathbb{T})$ we introduce the paired circulants

$$\boldsymbol{B}_n=\tilde{a}_n\hat{\alpha}_n+\tilde{b}_n\hat{\beta}_n+\boldsymbol{C}_n. \tag{2.12}$$

Since $\sup\{\|\boldsymbol{B}_n\|:n=1,2,\dots\}<\infty$, it makes sense to consider the smallest Banach algebra $\mathscr{A}$ (w.r.t. the sup-norm and componentwise operations), which contains all sequences of the form (2.12). Define the functions $g_j^{(n)}=\sqrt{n}\chi_j^{(n)}$, where $\chi_j^{(n)}$ denotes the characteristic function of the arc $[\mathrm{e}^{2\pi\mathrm{i}j/n},\mathrm{e}^{2\pi\mathrm{i}(j+1)/n})\subset\mathbb{T}$. By L_n we denote the orthogonal projection from $\boldsymbol{L}^2(\mathbb{T})$ onto $\operatorname{span}\{g_j^{(n)}: j=0,\dots,n-1\}$. If we identify the operators of $\mathscr{L}(\operatorname{im}L_n)$ with their matrices corresponding to the basis $\{g_j^{(n)}: j=0,\dots,n-1\}$, we obtain that $\mathscr{L}(\operatorname{im}L_n)=\mathscr{L}(\ell^2(n))$. In particular, the operators $\boldsymbol{U}_n$ as well as $\tilde{a}_n,\tilde{b}_n,\hat{\alpha}_n,\hat{\beta}_n,\boldsymbol{C}_n$, and $\boldsymbol{B}_n=\tilde{a}_n\hat{\alpha}_n+\tilde{b}_n\hat{\beta}_n+\boldsymbol{C}_n$ will be thought of as belonging to $\mathscr{L}(\operatorname{im}L_n)$. Moreover, setting $K_n^\varepsilon f=\sum_{j=0}^{n-1}f(\mathrm{e}^{2\pi\mathrm{i}(j+\varepsilon)/n})\chi_j^{(n)}$ we get

$$\boldsymbol{B}_n=K_n^\varepsilon aL_n\boldsymbol{U}_nK_n^0\alpha\boldsymbol{U}_n^{-1}+K_n^\varepsilon bL_n\boldsymbol{U}_nK_n^0\beta\boldsymbol{U}_n^{-1}+\boldsymbol{C}_n.$$

Furthermore, the system $\{g_j^{(n)}: j=0,\dots,n-1\}$ forms an orthogonal basis. Consequently, for any $\boldsymbol{D}_n\in\mathscr{L}(\operatorname{im}L_n)$, the matrix of the adjoint operator $\boldsymbol{D}_n^*$ corresponding to this basis is exactly the adjoint matrix of $\boldsymbol{D}_n$. We conclude that $\boldsymbol{B}_n^*=\widehat{(\bar{\alpha})}_n\widetilde{(\bar{a})}_n+\widehat{(\bar{\beta})}_n\widetilde{(\bar{b})}_n+\boldsymbol{C}_n^*$. Now it is easy to see that $\mathscr{A}$ actually forms a C^*-algebra. So we are interested in whether elements of $\mathscr{A}$ are invertible modulo the ideal $\mathscr{G}$ of all sequences tending in norm to zero. The solution of this problem is rather involved. We try to point out the main steps (for further details see [21, Chapter 10]).

(A) We define two families of C^*-homomorphisms from $\mathscr{A}$ into $\mathscr{L}(\boldsymbol{L}^2(\mathbb{T}))$, which reflect the needed information, as we will see later on. For $\tau\in\mathbb{T}$ define

$$k_E=k_E(\varepsilon,\tau,n)\in\{0,\dots,n-1\} \quad \text{by } \tau\in(\mathrm{e}^{2\pi\mathrm{i}(k_E+\varepsilon-1)/n},\mathrm{e}^{2\pi\mathrm{i}(k_E+\varepsilon)/n}].$$

We set

$$\boldsymbol{T}_n^{(\tau,\varepsilon)}: \operatorname{im}L_n\to\operatorname{im}L_n,\quad \sum_{j=0}^{n-1}\xi_j g_j^{(n)}\mapsto\sum_{j=0}^{n-1}\xi_j g_{j-k_E}^{(n)} \quad (g_{-k}^{(n)}:=g_{n-k}^{(n)}).$$

Now define $\boldsymbol{E}_n^{(\tau,1)}:\operatorname{im}L_n\to\operatorname{im}L_n$ by $\boldsymbol{E}_n^{(\tau,1)}=\boldsymbol{U}_n\boldsymbol{T}^{(\tau,0)}\boldsymbol{U}_n^{-1}$. Notice that

$$\boldsymbol{E}_n^{(\tau,1)}\tilde{a}_n(\boldsymbol{E}_n^{(\tau,1)})^{-1}=\tilde{a}_n \quad \text{for } a\in\boldsymbol{C}(\mathbb{T}) \quad \text{and} \quad \boldsymbol{E}_n^{(\tau,1)}\hat{\alpha}_n(\boldsymbol{E}_n^{(\tau,1)})^{-1}=\hat{\gamma}_n,$$

where $\gamma(t)=\alpha(t\cdot t_E)$ with $t_E=\mathrm{e}^{2\pi\mathrm{i}j_E/n}$ defined by $j_E=j_E(\tau,n)\in\{0,\dots,n-1\}$ and $\tau\in(\mathrm{e}^{2\pi\mathrm{i}(j_E-1)/n},\mathrm{e}^{2\pi\mathrm{i}j_E/n}]$. One can prove that, for any sequence $\{\boldsymbol{A}_n\}\in\mathscr{A}$ and for any $\tau\in\mathbb{T}$, there exist the strong limits

$$W_{(\tau,1)}\{\boldsymbol{A}_n\}=s-\lim_{n\to\infty}\boldsymbol{E}_n^{(\tau,1)}\boldsymbol{A}_n(\boldsymbol{E}_n^{(\tau,1)})^{-1}$$

and

$$W_{(\tau,1)}\{A_n\}^* = s - \lim_{n\to\infty} [E_n^{(\tau,1)} A_n (E_n^{(\tau,1)})^{-1}]^*.$$

In particular, for the sequences B_n in (2.12) one obtains

$$W_{(\tau,1)}\{B_n\} = [a\alpha(\tau+0) + b\beta(\tau+0)]P_{\mathbb{T}} + [a\alpha(\tau-0) + b\beta(\tau-0)]Q_{\mathbb{T}}.$$

It follows that $W_{(\tau,1)} : \mathscr{A} \to \mathscr{L}(L^2(\mathbb{T}))$ is a C^*-homomorphism.

(B) Introduce the family of operators $P_n \in \mathscr{L}(L^2(\mathbb{T}))$,

$$(P_n f)(t) := \sum_{k=-[n/2]}^{[(n-1)/2]} f_k t^k, \qquad f_k := \frac{1}{2\pi \mathrm{i}} \int_{\mathbb{T}} f(t) t^{-k-1}\, \mathrm{d}t.$$

Then there is an isometric isomorphism $E_n : \operatorname{im} L_n \to \operatorname{im} P_n$,

$$\sum_{j=0}^{n-1} \xi_j g_j^{(n)}(t) \mapsto \sum_{j=0}^{[(n-1)/2]} \xi_j t^j + \sum_{j=[(n-1)/2]+1}^{n-1} \xi_j t^{j-n}.$$

Finally we set $E_n^{(\tau,2)} := E_n T_n^{(\tau,\varepsilon)} : \operatorname{im} L_n \to \operatorname{im} P_n$. Again one can prove that, for any sequence $\{A_n\} \in \mathscr{A}$ and any $\tau \in \mathbb{T}$ there exist the strong limits

$$W_{(\tau,2)}\{A_n\} := s - \lim_{n\to\infty} E_n^{(\tau,2)} A_n (E_n^{(\tau,2)})^{-1}$$

and

$$W_{(\tau,2)}\{A_n\}^* = s - \lim_{n\to\infty} [E_n^{(\tau,2)} A_n (E_n^{(\tau,2)})^{-1}]^*.$$

In particular, for B_n from (2.12) we get

$$W_{(\tau,2)}\{B_n\} = a(\tau)\tilde{\alpha} + b(\tau)\tilde{\beta},$$

where $\tilde{\alpha}(t) = \alpha(1/t)$. Thus, $W_{(\tau,2)} : \mathscr{A} \to \mathscr{L}(L^2(\mathbb{T}))$ is also a C^*-homomorphism.

(C) Let $\mathscr{J}$ be the closure of

$$\left\{ \sum_{k=1}^{r} \{(E_n^{(\omega_k,1)})^{-1} L_n T_k E_n^{(\omega_k,1)}\} + \{C_n\} \right\},$$

$\omega_k \in \mathbb{T}$, $T_k \in \mathscr{K}(L^2(\mathbb{T}))$, $\|C_n\| \to 0$, $r \in \{1,2,\ldots\}$, and consider the smallest C^*-algebra $\mathscr{A}_0$ containing $\mathscr{A}$ and $\mathscr{J}$. Then it turns out that the above defined two families of C^*-homomorphisms are also defined on $\mathscr{A}_0$ and that $\mathscr{J}$ forms a two-sided closed ideal in $\mathscr{A}_0$.

(D) The cosets $\{\tilde{a}_n\} + \mathscr{J}$, $a \in C(\mathbb{T})$, form a C^*-subalgebra $\mathbb{C}$ of the center of $\mathscr{A}_0/\mathscr{J}$, which is *-isomorphic to $C(\mathbb{T})$. Let $\mathscr{J}_\tau$ denote the smallest closed ideal in $\mathscr{A}_0/\mathscr{J}$ which contains the maximal ideal $\tau \in \mathbb{T}$ of $\mathbb{C}$. Then, by $(\{A_n\}+\mathscr{J})+\mathscr{J}_\tau \mapsto W_{(\tau,2)}\{A_n\}$ there is generated a homomorphism $(\mathscr{A}_0/\mathscr{J})/\mathscr{J}_\tau \to \mathscr{L}(\boldsymbol{L}^2(\mathbb{T}))$, which we also denote by $W_{(\tau,2)}$. The image of $(\mathscr{A}_0/\mathscr{J})/\mathscr{J}_\tau$ under this homomorphism is exactly $\boldsymbol{PC}(\mathbb{T})$, and $\alpha \mapsto (\{\widehat{(\alpha)}_n+\mathscr{J})+\mathscr{J}_\tau\}$ describes the inverse homomorphism $W_{(\tau,2)}^{-1}: \boldsymbol{PC}(\mathbb{T}) \to (\mathscr{A}_0/\mathscr{J})/\mathscr{J}_\tau$. These arguments entail that, for $\{A_n\} \in \mathscr{A}_0$, the invertibility of $W_{(\tau,2)}\{A_n\}$ for all $\tau \in \mathbb{T}$ ensure the invertibility of $\{A_n\}+\mathscr{J}$ in $\mathscr{A}_0/\mathscr{J}$, and this invertibility leads to the Fredholmness of all $W_{(\tau,1)}\{A_n\}$. Conversely, a little thought shows that the Fredholmness of all $W_{(\tau,1)}\{A_n\}$ implies the invertibility of all $W_{(\tau,2)}\{A_n\}$.

We end up with the following stability result.

Theorem 2.5. *A sequence $\{A_n\} \in \mathscr{A}_0$ is stable if and only if $W_{(\tau,1)}\{A_n\}$ is invertible for all $\tau \in \mathbb{T}$.*

The determination of the functions α and β for methods (2.9)–(2.11) is by no means simple. They are computed in [21, Chapter 10]. Using their concrete nature and Theorem 2.5 one obtains the following.

Theorem 2.6. *Assume the operator $A=aI+bS_{\mathbb{T}}$ with continuous coefficients a and b to be invertible in $\boldsymbol{L}^2(\mathbb{T})$.*

(a1) *If δ is odd and $\varepsilon=0$ or if δ is even and $\varepsilon=\frac{1}{2}$ (if δ is odd and $\varepsilon=\frac{1}{2}$ or δ is even and $\varepsilon=0$), then the collocation method* (2.9) *is stable if and only if*

$$a(\tau)+\mu b(\tau) \neq 0\,(\mu a(\tau)+b(\tau)\neq 0) \quad \text{for all } \mu \in [-1,1],\ \tau \in \mathbb{T}.$$

(a2) *If $0<\varepsilon<1$ and $\varepsilon \neq \frac{1}{2}$, then the ε-colloccation method* (2.9) *is stable if and only if*

$$a(\tau)+\mu b(\tau) \neq 0 \quad \text{for all } \mu \in [-1,1],\ \tau \in \mathbb{T}.$$

(b) *The Galerkin method* (2.10) *is stable if and only if*

$$a(\tau)+\mu b(\tau) \neq 0 \quad \text{for all } \mu \in [-1,1],\ \tau \in \mathbb{T}.$$

(c) *The quadrature method* (2.11) *is stable if and only if*

$$\mu a(\tau)+b(\tau) \neq 0 \quad \text{for all } \mu \in [-1,1],\ \tau \in \mathbb{T}.$$

Again convergence rates are available in the scale of the periodic Sobolev spaces $\boldsymbol{H}^s(\mathbb{T})$ (see [21, Chapter 10]). For further extensions (piecewise continuous coeffcients, curves with corners or even composed curves) see [9,21]. Let us finally remark that the stability of the qualocation method proposed by I. Sloan can be studied completely with the help of Theorem 2.5.

3. The nonperiodic case

3.1. The classical collocation method

To describe the main ideas concerning the construction of the classical collocation method and its investigation it is sufficient to consider equations of type (1.1), where the coefficients a and b are constants, since the main tool is the mapping property (3.1) which can be generalized to the case of Hölder-continuous coefficients a and b (cf. [21, Theorem 9.14]). Beside the weighted $\boldsymbol{L}^2$-convergence of the classical collocation method for CSIEs, in this section we also consider the weighted uniform convergence based on the application of special weighted Besov norms. Moreover, we will shortly discuss the application of the classical collocation method to some classes of CSIEs with weakly singular perturbation kernels and of hypersingular integral equations of Prandtl's type. Finally we present some results concerning fast algorithms based on Amosov's idea.

3.1.1. CSIOs, Jacobi polynomials, and the classical collocation method

Let α and β be real numbers greater than -1. Then the classical Jacobi polynomials $P_n^{\alpha,\beta}(x)$, $n=0,1,2,\dots$, are defined, for example, by the generalized Rodrigues' formula

$$(1-x)^\alpha(1+x)^\beta P_n^{\alpha,\beta}(x)=\frac{(-1)^n}{2^n n!}\left(\frac{\mathrm{d}}{\mathrm{d}x}\right)^n\left\{(1-x)^{n+\alpha}(1+x)^{n+\beta}\right\}.$$

It is well known that these polynomials satisfy the orthogonality relations

$$\int_{-1}^1 P_n^{\alpha,\beta}(x)P_k^{\alpha,\beta}(x)(1-x)^\alpha(1+x)^\beta\,\mathrm{d}x=\delta_{nk}h_n^{\alpha,\beta},$$

where δ_{nk} denotes the Kronecker delta and

$$h_n^{\alpha,\beta}=\begin{cases}\dfrac{2^{\alpha+\beta+1}}{2n+\alpha+\beta+1}\,\dfrac{\Gamma(n+\alpha+1)\Gamma(n+\beta+1)}{n!\Gamma(n+\alpha+\beta+1)} & \text{if } \alpha+\beta\neq-1 \text{ or } n=1,2,\dots,\\ \Gamma(\alpha+1)\Gamma(\beta+1) & \text{if } \alpha+\beta=-1 \text{ and } n=0.\end{cases}$$

Thus, the normalized Jacobi polynomials are given by

$$p_n^{\alpha,\beta}(x)=[h_n^{\alpha,\beta}]^{-1/2}P_n^{\alpha,\beta}(x),\quad n=0,1,2,\dots\,.$$

Now, let a and $b<0$ be real numbers with $a^2+b^2=1$ and let β_0 with $0<\beta_0<1$ be defined by $a-\mathrm{i}b=\mathrm{e}^{\mathrm{i}\pi\beta_0}$. Choose integers λ and ν such that $\alpha:=\lambda+\beta_0$ and $\beta:=\nu-\beta_0$ lie between -1 and 1. Then, for the operator $A_0=(aI+\mathrm{i}bS)v^{\alpha,\beta}I$ with $v^{\alpha,\beta}(t)=(1-t)^\alpha(1+t)^\beta$, the very important relation

$$(A_0p_n^{\alpha,\beta})(t)=(-1)^\lambda p_{n-\kappa}^{-\alpha,-\beta}(t),\quad -1<t<1,\ n=0,1,2,\dots, \tag{3.1}$$

holds true, where $\kappa=-\alpha-\beta=-\lambda-\nu$ and $p_{-1}^{-\alpha,-\beta}(t)\equiv0$. This relation suggests to consider Eq. (1.2) with $A=A_0+Hv^{\alpha,\beta}I$ in the pair $(\mathscr{H}_1,\mathscr{H}_2)$ of Hilbert spaces $\mathscr{H}_1=\boldsymbol{L}^2_{\alpha,\beta}$ and $\mathscr{H}_2=\boldsymbol{L}^2_{-\alpha,-\beta}$, where, for $\gamma>-1$ and $\delta>-1$, $\boldsymbol{L}^2_{\gamma,\delta}$ denotes the Hilbert space of all w.r.t. the weight $v^{\gamma,\delta}(t)$ square integrable functions with the inner product

$$\langle u,v\rangle_{\gamma,\delta}:=\int_{-1}^1 u(t)\overline{v(t)}v^{\gamma,\delta}(t)\,\mathrm{d}t.$$

Moreover, to prove convergence rates the following Hilbert scale of weighted Sobolev spaces is very convenient. Let $s\geqslant0$ and define

$$\boldsymbol{L}^{2,s}_{\gamma,\delta}:=\left\{u\in\boldsymbol{L}^2_{\gamma,\delta}:\sum_{n=0}^\infty(1+n)^{2s}|\langle u,p_n^{\gamma,\delta}\rangle_{\gamma,\delta}|^2<\infty\right\}.$$

Equipped with the scalar product

$$\langle u, v\rangle_{\gamma,\delta,s} := \sum_{n=0}^{\infty}(1+n)^{2s}\langle u, p_n^{\gamma,\delta}\rangle_{\gamma,\delta}\overline{\langle v, p_n^{\gamma,\delta}\rangle_{\gamma,\delta}},$$

$\{L^{2,s}_{\gamma,\delta}\}_{s\geqslant 0}$ is a Hilbert scale generated by the operator $\mathscr{B}u := \sum_{n=0}^{\infty}(1+n)\langle u, p_n^{\gamma,\delta}\rangle_{\gamma,\delta}p_n^{\gamma,\delta}$ with the domain $D(\mathscr{B}) = L^{2,1}_{\gamma,\delta}$. As an immediate consequence of (3.1) we get that the operator A_0 can be extended to a bounded linear operator $A_0 : L^2_{\alpha,\beta} \to L^2_{-\alpha,-\beta}$ and that $A_0 : L^{2,s}_{\alpha,\beta} \to L^{2,s}_{-\alpha,-\beta}$ is a one-sided invertible Fredholm operator with index κ, where $\ker A_0 = \operatorname{span}\{p_0^{\alpha,\beta}\}$ in case $\kappa = 1$ and $\operatorname{im} A_0 = \{f \in L^{2,s}_{-\alpha,-\beta} : \langle f, p_0^{-\alpha,-\beta}\rangle_{-\alpha,-\beta} = 0\}$ in case $\kappa = -1$. If by $S_n^{\gamma,\delta} : L^2_{\gamma,\delta} \to L^2_{\gamma,\delta}$ we denote the Fourier projection

$$S_n^{\gamma,\delta}u := \sum_{k=0}^{n-1}\langle u, p_k^{\gamma,\delta}\rangle_{\gamma,\delta}p_k^{\gamma,\delta}$$

then, for all $s \geqslant 0$ and for all $u \in L^{2,s}_{\gamma,\delta}$, we have

(A) $\lim_{n\to\infty}\|u - S_n^{\gamma,\delta}u\|_{L^{2,s}_{\gamma,\delta}} = 0$,
(B) $\|u - S_n^{\gamma,\delta}u\|_{L^{2,t}_{\gamma,\delta}} \leqslant n^{t-s}\|u\|_{L^{2,s}_{\gamma,\delta}}$, $0 \leqslant t \leqslant s$,
(C) $\|S_n^{\gamma,\delta}u\|_{L^{2,t}_{\gamma,\delta}} \leqslant n^{t-s}\|u\|_{L^{2,s}_{\gamma,\delta}}$, $t \geqslant s$.

Moreover,

(D) for $s > \frac{1}{2}$, the space $L^{2,s}_{\gamma,\delta}$ is continuously embedded into the space $C_{\tilde{\gamma},\tilde{\delta}}$ with $\tilde{\gamma} = \frac{1}{2}\max\{0, \gamma + \frac{1}{2}\}$ and $\tilde{\delta} = \frac{1}{2}\max\{0, \delta + \frac{1}{2}\}$.

Here, for nonnegative real numbers ρ and τ, by $C_{\rho,\tau}$ we refer to the Banach space of all continuous functions $u : (-1, 1) \to \mathbb{C}$, for which $v^{\rho,\tau}u$ is continuous on $[-1, 1]$, equipped with the norm $\|u\|_{\infty,\rho,\tau} = \sup\{v^{\rho,\tau}(t)|u(t)| : t \in [-1, 1]\}$.

To approximate the integral operator $H_0 = Hv^{\alpha,\beta}I$ Gaussian rules are often used. The Gaussian rule w.r.t. the Jacobi weight $v^{\gamma,\delta}(t)$ is given by

$$G_n^{\gamma,\delta}(u) := \frac{1}{\pi}\int_{-1}^{1}(L_n^{\gamma,\delta}u)(t)v^{\gamma,\delta}(t)\,\mathrm{d}t = \sum_{k=1}^{n}\lambda_{kn}^{\gamma,\delta}u(t_{kn}^{\gamma,\delta}),$$

where the interpolation operator $L_n^{\gamma,\delta}$ w.r.t. the zeros $t_{kn}^{\gamma,\delta}$, $t_{1n}^{\gamma,\delta} < \cdots < t_{nn}^{\gamma,\delta}$, of $p_n^{\gamma,\delta}(t)$ is defined by

$$L_n^{\gamma,\delta}u := \sum_{k=1}^{n}u(t_{kn}^{\gamma,\delta})\ell_{kn}^{\gamma,\delta}, \qquad \ell_{kn}^{\gamma,\delta}(t) = \prod_{j=1, j\neq k}^{n}\frac{t - t_{jn}^{\gamma,\delta}}{t_{kn}^{\gamma,\delta} - t_{jn}^{\gamma,\delta}},$$

and $\lambda_{kn}^{\gamma,\delta}$ are the Christoffel numbers $\lambda_{kn}^{\gamma,\delta} = \frac{1}{\pi}\int_{-1}^{1}\ell_{kn}^{\gamma,\delta}(t)v^{\gamma,\delta}(t)\,\mathrm{d}t$. For $s > \frac{1}{2}$ and $u \in L^{2,s}_{\gamma,\delta}$, the relations

(E) $\lim_{n\to\infty}\|u - L_n^{\gamma,\delta}u\|_{L^{2,s}_{\gamma,\delta}} = 0$,
(F) $\|u - L_n^{\gamma,\delta}u\|_{L^{2,t}_{\gamma,\delta}} \leqslant \text{const}\, n^{t-s}\|u\|_{L^{2,s}_{\gamma,\delta}}$, $0 \leqslant t \leqslant s$,

hold true, where the constant does not depend on n, t, and u. With the help of (3.1) one can show that

$$(A_0 u_n)(t_{j,n-\kappa}^{-\alpha,-\beta}) = b \sum_{k=1}^{n} \frac{\lambda_{kn}^{\alpha,\beta}}{t_{kn}^{\alpha,\beta} - t_{j,n-\kappa}^{-\alpha,-\beta}} u_n(t_{kn}^{\alpha,\beta}), \quad j = 1, \ldots, n-\kappa,$$

for each algebraic polynomial $u_n(t)$ of degree less than or equal to $2n$. Now, the classical collocation method for the equation

$$a v^{\alpha,\beta}(t) u(t) + \frac{1}{\pi} \int_{-1}^{1} \left[\frac{b}{s-t} + h(s,t) \right] v^{\alpha,\beta}(s) u(s)\, ds = f(t), \tag{3.2}$$

consists in the determination of an approximate solution $u_n(t) = \sum_{k=1}^{n} \xi_{kn} \ell_{kn}^{\alpha,\beta}$ by solving the system

$$\sum_{k=1}^{n} \lambda_{kn}^{\alpha,\beta} \left[\frac{b}{t_{kn}^{\alpha,\beta} - t_{j,n-\kappa}^{-\alpha,-\beta}} + h(t_{kn}^{\alpha,\beta}, t_{j,n-\kappa}^{-\alpha,-\beta}) \right] \xi_{kn} = f(t_{j,n-\kappa}^{-\alpha,-\beta}), \quad j = 1, \ldots, n-\kappa. \tag{3.3}$$

In case of $\kappa = 1$ an additional condition, for example $\sum_{k=1}^{n} \lambda_{kn}^{\gamma,\delta} \xi_{kn} = 0$, has to be considered. Of course, taking into account again relation (3.1) system (3.3) is equivalent to

$$(A_0 + L_{n-\kappa}^{-\alpha,-\beta} H_n) u_n = L_{n-\kappa}^{-\alpha,-\beta} f, \tag{3.4}$$

where, for continuous functions $u : (-1,1) \to \mathbb{C}$ the operator H_n is defined by

$$(H_n u)(t) = \sum_{k=1}^{n} \lambda_{kn}^{\alpha,\beta} h(t_{kn}^{\alpha,\beta}, t) u(t_{kn}^{\alpha,\beta}). \tag{3.5}$$

3.1.2. Weighted $\boldsymbol{L}^2$ -convergence

By $\boldsymbol{C}_\varphi^r$, $r \geqslant 0$ an integer and $\varphi(t) = \sqrt{1-t^2}$, we denote the space of all r times differentiable functions $u : (-1,1) \to \mathbb{C}$ satisfying the conditions $u^{(k)} \varphi^k \in \boldsymbol{C}[-1,1]$ for $k = 0, 1, \ldots, r$. Then the operators H_n defined in (3.5) have the following approximation property:

(G) If $h(.,t) \in \boldsymbol{C}_\varphi^r$ for some integer $r \geqslant s > \frac{1}{2}$ uniformly w.r.t. $t \in [-1,1]$ then, for $0 \leqslant \tau \leqslant s$ and $u \in \boldsymbol{L}_{\alpha,\beta}^{2,s}$,

$$\|L_m^{-\alpha,-\beta}(H_n - H v^{\alpha,\beta}) u\|_{\boldsymbol{L}_{-\alpha,-\beta}^{2,\tau}} \leqslant \text{const}\, m^\tau n^{-s} \|u\|_{\boldsymbol{L}_{\alpha,\beta}^{2,s}}.$$

In the following theorem, which can be proved with the help of relations (A)–(G) for simplicity we restrict the formulations to the case of index $\kappa = 0$.

Theorem 3.1 (cf. Berthold et al. [2]). *Let $\alpha + \beta = 0$ and assume that Eq.* (3.2) *has a unique solution $u^* \in \boldsymbol{L}_{\alpha,\beta}^2$. If, for some $s > \frac{1}{2}$ and some integer $r \geqslant s$, $h(.,t) \in \boldsymbol{C}_\varphi^r$ uniformly w.r.t. $t \in [-1,1]$, $h(t,.) \in \boldsymbol{L}_{-\alpha,-\beta}^{2,s}$ uniformly w.r.t. $t \in [-1,1]$, and $f \in \boldsymbol{L}_{-\alpha,-\beta}^{2,s}$, then $u^* \in \boldsymbol{L}_{\alpha,\beta}^{2,s}$ and the approximate*

Eqs. (3.4) *possess for all sufficiently large n a unique solution* $u_n^* = \sum_{k=1}^n \xi_{kn}^* \ell_{kn}^{\alpha,\beta}$, *where*

$$\|u_n^* - u^*\|_{L^{2,\tau}_{\alpha,\beta}} \leqslant \text{const}\, n^{\tau-s}\|u^*\|_{L^{2,s}_{\alpha,\beta}}, \quad 0 \leqslant \tau \leqslant s \tag{3.6}$$

and the constant does not depend on n, τ, and u^*.

3.1.3. Weighted Besov spaces and weighted uniform convergence

This section presents a concept for studying weighted uniform convergence of the classical collocation method. The main tool of this approach is to define a scale of subspaces of the weighted space $\boldsymbol{C}_{\rho,\tau}$ of continuous functions depending on the order of best weighted uniform approximation by algebraic polynomials and to study the mapping properties of the Cauchy singular and regular integral operators as well as integral operators with weakly singular kernels w.r.t. this scale of subspaces. For details we refer the interested reader to [12].

Let $\mathbb{P}_n$ denote the set of algebraic polynomials of degree less than n. For constants $\rho, \tau \geqslant 0$ and a function $f \in \boldsymbol{C}_{\rho,\tau}$ we denote by $E_n^{\rho,\tau}(f)$ the best weighted uniform approximation of f by polynomials belonging to $\mathbb{P}_n$, i.e. $E_n^{\rho,\tau}(f) = \inf\{\|f - p\|_{\infty,\rho,\tau} : p \in \mathbb{P}_n\}$, $E_0^{\rho,\tau}(f) = \|f\|_{\infty,\rho,\tau}$. For a sequence $\{b_n\}$ of positive real numbers with $\lim_{n\to\infty} b_n = 0$ we define the weighted Besov space

$$\boldsymbol{C}_{\rho,\tau}^{\{b_n\}} = \left\{ u \in \boldsymbol{C}_{\rho,\tau} : \|u\|_{\rho,\tau,\{b_n\}} := \sup_{n=0,1,2,\ldots} b_n^{-1} E_n^{\rho,\tau}(u) < \infty \right\}.$$

Then $\boldsymbol{C}_{\rho,\tau}^{\{b_n\}}$ is a Banach space compactly embedded into $\boldsymbol{C}_{\rho,\tau}$, and the embedding $\boldsymbol{C}_{\rho,\tau}^{\{b_n\}} \subset \boldsymbol{C}_{\rho,\tau}^{\{c_n\}}$ is compact if $\lim_{n\to\infty} b_n/c_n = 0$. In case of $b_n = (n+1)^{-\gamma}\log^{\delta}(n+2)$ with $\gamma > 0$ and $\delta \geqslant 0$ we write $\boldsymbol{C}_{\rho,\tau}^{\gamma,\delta}$ instead of $\boldsymbol{C}_{\rho,\tau}^{\{b_n\}}$.

Let $A_0 = (aI + \mathrm{i}bS)v^{\alpha,\beta}I : \boldsymbol{L}^2_{\alpha,\beta} \to \boldsymbol{L}^2_{-\alpha,-\beta}$ be the operator considered in Section 3.1.1 (cf. (3.1)) and define $\hat{A}_0 = (aI - \mathrm{i}bS)v^{-\alpha,-\beta}I : \boldsymbol{L}^2_{-\alpha,-\beta} \to \boldsymbol{L}^2_{\alpha,\beta}$, which is an at least one-sided inverse of A_0. Moreover, let

$$\alpha = \alpha^+ - \alpha^-, \quad \beta = \beta^+ - \beta^-, \quad 0 \leqslant \alpha^\pm, \beta^\pm < 1.$$

Then we have the following important property of the operators A_0 and $\hat{A}_0$:

(A) $A_0 \in \mathscr{L}(\boldsymbol{C}^{\gamma,\delta}_{\alpha^+,\beta^+}, \boldsymbol{C}^{\gamma,\delta+1}_{\alpha^-,\beta^-})$ and $\hat{A}_0 \in \mathscr{L}(\boldsymbol{C}^{\gamma,\delta}_{\alpha^-,\beta^-}, \boldsymbol{C}^{\gamma,\delta+1}_{\alpha^+,\beta^+})$.

By $h \in \boldsymbol{C}_{\nu,\zeta,s} \cap \boldsymbol{C}^{\{b_n\}}_{\rho,\tau,t}$ we mean that the function $h(s,t)v^{\nu,\zeta}(s)v^{\rho,\tau}(t)$ is continuous on $[-1,1]^2$ and that $h_s^{\nu,\zeta} \in \boldsymbol{C}^{\{b_n\}}_{\rho,\tau}$ uniformly w.r.t. $s \in [-1,1]$, where $h_s^{\nu,\zeta}(t) = h(s,t)v^{\nu,\zeta}(s)$.

(B) If $h \in \boldsymbol{C}_{\nu,\zeta,s} \cap \boldsymbol{C}^{\{b_n\}}_{\rho,\tau,t}$ and $\nu + \alpha^- < 1$, $\zeta + \beta^- < 1$, then $H_0 \in \mathscr{L}(\boldsymbol{C}_{\alpha^+,\beta^+}, \boldsymbol{C}^{\{b_n\}}_{\rho,\tau})$, where $H_0 = Hv^{\alpha,\beta}I$.

In what follows, for simplicity we again assume that $\kappa = 0$ and consider for the approximate solution of (3.2) instead of the classical collocation method (3.3) the more general collocation method

$$(A_0 u_n + H_n u_n)(t_{jn}) = f(t_{jn}), \quad j = 1, \ldots, n, \tag{3.7}$$

where $\{t_{1n}, \ldots, t_{nn}\}$ is a sequence of partitions of the interval $[-1,1]$ with $-1 < t_{1n} < \cdots < t_{nn} < 1$. By L_n we denote the respective interpolation operator and by $\|L_n\|_{\rho,\tau}$ the weighted Lebesque constant defined by

$$\|L_n\|_{\rho,\tau} = \sup\{\|L_n f\|_{\infty,\rho,\tau} : f \in \mathbf{C}_{\rho,\tau}, \|f\|_{\infty,\rho,\tau} = 1\}.$$

Theorem 3.2 (Junghanns and Luther [12, Theorem 6.8]). *Let $\alpha+\beta=0$ and assume that Eq. (3.2) has a unique solution $u^* \in \mathbf{C}_{\alpha^+,\beta^+}$. If $h \in \mathbf{C}^{\varepsilon,\eta}_{v,\zeta,s} \cap \mathbf{C}^{\gamma,\delta}_{\rho,\tau,t}$ with some nonnegative constants ρ,τ,v,ζ satisfying $\rho \leqslant \alpha^-$, $\tau \leqslant \beta^-$, $v+\alpha^- < 1$, $\zeta+\beta^- < 1$, if $f \in \mathbf{C}^{\gamma,\delta}_{\rho,\tau}$, and if $\lim_{n\to\infty} n^{-\gamma}\log^{\delta+1} n\|L_n\|_{\rho,\tau}=0$, then, for all, sufficiently large n Eqs. (3.7) are uniquely solvable and the respective polynomials u_n^* converge in the norm of $\mathbf{C}_{\alpha^+,\beta^+}$ to u^*, where*

$$\|u_n^* - u^*\|_{\infty,\alpha^+,\beta^+} \leqslant \text{const}\left(\frac{\log^{\delta+1} n}{n^\gamma}\|L_n\|_{\rho,\tau} + \frac{\log^{\eta+1} n}{n^\varepsilon}\right)\|f\|_{\mathbf{C}^{\gamma,\delta}_{\rho,\tau}}$$

with a constant independent of n and f.

3.1.4. Fast algorithms

In case $|\alpha| = |\beta| = \frac{1}{2}$, i.e. $a=0$, $b=-1$, we can adopt Amosov's idea to the collocation method (3.3) in order to get a fast algorithm for the numerical solution of (3.2) with $O(n\log n)$ complexity, which retains the convergence rate (3.6) of the classical collocation method for a scale of weighted Sobolev norms. We demonstrate the approach for the example of the equation

$$\frac{1}{\pi}\int_{-1}^{1}\left[\frac{1}{t-\sigma} + h(\sigma,t)\right]\sqrt{\frac{1+\sigma}{1-\sigma}}u(\sigma)\,d\sigma = f(t). \tag{3.8}$$

In what follows we assume that, for some $s > \frac{1}{2}$ and some $r > s$, $f \in \mathbf{L}^{2,s}_{\rho^{-1}} = \mathbf{L}^{2,s}_{1/2,-1/2}$, $\rho(t) = \sqrt{(1+t)/(1-t)}$, $h(.,t) \in \mathbf{L}^{2,r}_\rho$ uniformly w.r.t. $t \in [-1,1]$, and $h(\sigma,.) \in \mathbf{L}^{2,r}_{\rho^{-1}}$ uniformly w.r.t. $\sigma \in [-1,1]$. Moreover, assume that Eq. (3.8) has a unique solution $u \in \mathbf{L}^2_\rho$. We choose integers m and n, such that $0<m<n$ and $d:=(2n+1)/(2m+1)$ is an integer, and define an approximation $u_n = w_m + (I - S^\rho_m)v_n$, where $v_n \in \operatorname{im} S^\rho_n$ and $w_m \in \operatorname{im} S^\rho_m$ are the solutions of

$$\frac{1}{\pi}\int_{-1}^{1}\frac{v_n(\sigma)}{t-\sigma}\rho(\sigma)\,d\sigma = (L_n^{\rho^{-1}} f)(t) \tag{3.9}$$

and

$$\frac{1}{\pi}\int_{-1}^{1}\frac{w_m(\sigma)}{t-\sigma}\rho(\sigma)\,d\sigma + (L_m^{\rho^{-1}} H_m w_m)(t) = (L_m^{\rho^{-1}} g)(t) \tag{3.10}$$

with

$$g(t) = f(t) - \frac{1}{\pi}\int_{-1}^{1}\frac{(S^\rho_m v_n)(\sigma)}{t-\sigma}\rho(\sigma)\,d\sigma,$$

respectively. Define matrices

$$\mathbf{U}^\rho_n = \left[\frac{\cos((j+1/2)(2k-1)/(2n+1))\pi}{\cos((k-1/2)\pi/(2n+1))}\right]_{j=0,\,k=1}^{n-1,\,n} \quad \text{and} \quad \mathbf{U}^{\rho^{-1}}_n = \left[\frac{\sin((2j+1)k\pi/(2n+1))}{\sin(k\pi/(2n+1))}\right]_{j=0,\,k=1}^{n-1,\,n}$$

as well as $\mathbf{H}_n = [h(t^\rho_{kn}, t^{\rho^{-1}}_{jn})]_{j,k=1}^n$ and $\mathbf{\Lambda}^\rho_n = \operatorname{diag}[\lambda^\rho_{1n},\ldots,\lambda^\rho_{nn}]$. Let $\hat{v}_n = [\langle v_n, p^\rho_j\rangle_\rho]_{j=0}^{n-1}$ and $\tilde{f}_n = [f(t^{\rho^{-1}}_{jn})]_{j=1}^n$. Then Eq. (39) is equivalent to

$$\hat{v}_n = -\mathbf{U}^{\rho^{-1}}_n \mathbf{\Lambda}^{\rho^{-1}}_n \tilde{f}_n. \tag{3.11}$$

Eq. (3.10) can be written as

$$[(-U_m^{\rho^{-1}})^{\mathrm{T}} U_m^{\rho} + H_m](U_m^{\rho})^{\mathrm{T}} \hat{w}_m = \tilde{g}_m. \tag{3.12}$$

Since $U_n^{\rho^{-1}}$ can be handled as a sine transform, (3.11) can be realized with $\mathrm{O}(n\log n)$ complexity. Moreover, because of $\{t_{jm}^{\rho^{-1}}\}_{j=1}^m \subset \{t_{jn}^{\rho^{-1}}\}_{j=1}^n$ the right-hand side of (3.12) is equal to $\tilde{f}_m - r_{nm}(U_n^{\rho^{-1}})^{\mathrm{T}}[0,\dots,0,[\hat{v}_n]_m,\dots,[\hat{v}_n]_{n-1}]^{\mathrm{T}}$ with $r_{nm}[\xi_k]_{k=1}^n = [\xi_{d\cdot k}]_{k=1}^m$ and can also be computed with $\mathrm{O}(n\log n)$ complexity. Thus, if $m \sim n^{1/3}$ the complexity is $\mathrm{O}(n\log n)$ in order to determine u_n provided the values $f(t_{jn}^{\rho^{-1}})$ and $h(t_{km}^{\rho}, t_{jm}^{\rho^{-1}})$ are given. Moreover, under the above assumptions one can prove that $\|u^* - u_n\|_{\rho,t} \leqslant \mathrm{const}\, n^{t-s}\|u^*\|_{\rho,s}$ for all $t > \frac{1}{2}$ with $s-(r-s)/2 \leqslant t \leqslant s$. This approach, i.e. the application of Amosov's idea to the nonperiodic case, was demonstrated in [2] for the first time. In [13] there is shown, how it is possible to apply the concept of weighted Besov spaces (see Section 3.1.3) to prove convergence rates in weighted uniform norms for this fast algorithm.

3.1.5. Operators related to CSIOs

Here we study two types of SIOs which are closely related to the CSIO $A_0 = (aI + \mathrm{i}bS)v^{\alpha,\beta}I$ considered in Sections 3.1.1 and 3.1.3. The first operator is the hypersingular integral operator DA_0, where $D = \mathrm{d}/\mathrm{d}t$ denotes the operator of generalized differentiation and which can be written as a finite part integral operator

$$(DA_0u)(t) = a\frac{\mathrm{d}}{\mathrm{d}t}[v^{\alpha,\beta}(t)u(t)] + \frac{b}{\pi}\int_{-1}^{1}\frac{v^{\alpha,\beta}(\sigma)}{(\sigma-t)^2}u(\sigma)\,\mathrm{d}\sigma, \quad t\in(-1,1). \tag{3.13}$$

The second one is the weakly SIO

$$(Wu)(t) = a\int_{-1}^{t} v^{\alpha,\beta}(\sigma)u(\sigma)\,\mathrm{d}\sigma - \frac{b}{\pi}\int_{-1}^{1} v^{\alpha,\beta}(\sigma)\ln|\sigma-t|\,u(\sigma)\,\mathrm{d}\sigma, \quad t\in(-1,1). \tag{3.14}$$

The operator D of generalized differentiation is a continuous isomorphism between the spaces $\boldsymbol{L}_{\gamma,\delta}^{2,s+1,0}$ and $\boldsymbol{L}_{1+\gamma,1+\delta}^{2,s}$ for all $s \geqslant 0$ and all $\gamma,\delta > -1$, where $\boldsymbol{L}_{\gamma,\delta}^{2,s,r}$ denotes the subspace of $\boldsymbol{L}_{\gamma,\delta}^{2,s}$ of those functions f, for which $\langle f, p_k^{\gamma,\delta}\rangle_{\gamma,\delta} = 0$ for $k = 0,\dots,r$, and $r \geqslant 0$ and $\boldsymbol{L}_{\gamma,\delta}^{2,s,r} = \boldsymbol{L}_{\gamma,\delta}^{2,s}$ for $r < 0$. From this and the mapping property (3.1) one can conclude that DA_0 belongs to $\boldsymbol{L}(\boldsymbol{L}_{\alpha,\beta}^{2,s+1}, \boldsymbol{L}_{1-\alpha,1-\beta}^{2,s})$ for all $s \geqslant 0$ and that

$$DA_0u = (-1)^{\lambda}\sum_{n=0}^{\infty}\sqrt{(n-\kappa)(n+1)}\langle u, p_n^{\alpha,\beta}\rangle_{\alpha,\beta}\, p_{n-\kappa-1}^{1-\alpha,1-\beta}$$

for all $u \in \boldsymbol{L}_{\alpha,\beta}^{2,s+1}$. Consequently, $DA_0 : \boldsymbol{L}_{\alpha,\beta}^{2,s+1,\kappa} \to \boldsymbol{L}_{1-\alpha,1-\beta}^{2,s}$ is a continuous isomorphism. With these mapping properties one can investigate collocation and collocation–quadrature methods for example for integro-differential equations of Prandtl's type and construct fast algorithms using the concept presented in Section 3.1.4 (see, for example, [5,6]).

For the weakly SIO from (3.14), in case $\alpha+\beta=-1$ one can use the relations

$$\int_{-1}^{t} v^{\alpha,\beta}(\sigma)p_n^{\alpha,\beta}(\sigma)\,\mathrm{d}\sigma = -\frac{1}{n}v^{\alpha+1,\beta+1}(t)p_{n-1}^{\alpha+1,\beta+1}(t), \quad t\in(-1,1)$$

and

$$\int_{-1}^{1} v^{\alpha,\beta}(\sigma)\ln|\sigma-t|\,p_n^{\alpha,\beta}(\sigma)\,\mathrm{d}\sigma = \frac{1}{n}\int_{-1}^{1}\frac{v^{\alpha+1,\beta+1}(\sigma)}{\sigma-t}p_{n-1}^{\alpha+1,\beta+1}(\sigma)\,\mathrm{d}\sigma, \quad t\in(-1,1),$$

for $n=1,2,\ldots$, which lead to

$$Wp_n^{\alpha,\beta} = -\frac{1}{n}p_n^{-\alpha-1,-\beta-1}, \quad n=1,2,\ldots .$$

Analogous considerations are possible for $\alpha+\beta=0$ and 1. In all cases we get that $W: \boldsymbol{L}^{2,s}_{\alpha,\beta} \to \boldsymbol{L}^{2,s+1}_{\beta,\alpha}$ is a continuous operator.

In particular, in the case of the generalized airfoil equation

$$\frac{1}{\pi}\int_{-1}^{1}\left[\frac{1}{t-\sigma}+h_1(\sigma,t)\ln|\sigma-t|+h_2(\sigma,t)\right]\sqrt{\frac{1+\sigma}{1-\sigma}}u(\sigma)\,\mathrm{d}\sigma = f(t),$$

these mapping properties can be used to investigate collocation and collocation–quadrature methods in scales of weighted Sobolev spaces and to use Amosov's idea for the construction of fast algorithms (cf. [2,11,20]).

3.1.6. SIEs with weakly singular perturbation kernels

Here we show how it is possible to use the concept of weighted Besov spaces presented in Section 3.1.3 for investigating collocation methods for SIEs with weakly singular perturbation kernels of the form $h(s,t)=(s-t)^{-1}[k(s,t)-k(s,s)]$. We again consider the case $\kappa=0$ and the collocation–quadrature method (3.7). But, to get sufficiently good convergence rates it seems to be necessary to choose special collocation points $\{t_{1n},\ldots,t_{nn}\}$. For nonnegative constants ρ and τ with $\rho\leqslant\alpha^-$ and $\tau\leqslant\beta^-$ choose numbers $r,s\in\{0,1\}$ in such a way that

$$-\frac{\alpha}{2}+\frac{1}{4}-\rho\leqslant r\leqslant -\frac{\alpha}{2}+\frac{5}{4}-\rho \quad\text{and}\quad -\frac{\beta}{2}+\frac{1}{4}-\tau\leqslant r\leqslant -\frac{\beta}{2}+\frac{5}{4}-\tau. \tag{3.15}$$

Then, let $m=n-r-s$ and $\{t_{jn}\}_{j=1}^{n}=\{t_{jm}^{-\alpha,-\beta}\}_{j=1-s}^{m+r}$, where $t_{0m}^{-\alpha,-\beta}\in(-1,\min\{t_{1m}^{-\alpha,-\beta},t_{1n}^{\alpha,\beta}\})$ and $t_{m+1,m}\in(\max\{t_{mm}^{-\alpha,-\beta},t_{nn}^{\alpha,\beta}\},1)$ are knots satisfying

$$1+t_{0m}^{-\alpha,-\beta}\sim n^{-2} \ \text{ if } \tau>0, \quad 1-t_{m+1,m}^{-\alpha,-\beta}\sim n^{-2} \ \text{ if } \rho>0$$

and

$$\min\{t_{1m}^{-\alpha,-\beta},t_{1n}^{\alpha,\beta}\}-t_{0m}^{-\alpha,-\beta}\sim n^{-2}, \qquad t_{m+1,m}^{-\alpha,-\beta}-\max\{t_{mm}^{-\alpha,-\beta},t_{nn}^{\alpha,\beta}\}\sim n^{-2}.$$

If L_n denotes the Lagrange interpolation operator w.r.t. these nodes $\{t_{jm}^{-\alpha,-\beta}\}_{j=1-s}^{m+r}$ then, due to [7, Theorem 4.1], $\|L_n\|_{\rho,\tau}=\mathrm{O}(\log n)$.

Theorem 3.3 (Junghanns and Luther [14, Theorem 4.3]). *Let $\alpha+\beta=0$ and assume that Eq.* (3.2) *has a unique solution in $u^*\in\boldsymbol{C}_{\alpha^+,\beta^+}$. In* (3.15) *choose $\rho=\alpha^-$, $\tau=\beta^-$ and the collocation nodes t_{jn} as above. If $h(s,t)=(s-t)^{-1}[k(s,t)-k(s,s)]$ with $k\in\boldsymbol{C}^{\gamma,\delta}_{0,0,s}\cap\boldsymbol{C}^{\gamma,\delta}_{0,0,t}$, $f\in\boldsymbol{C}^{\gamma,\delta+1}_{\alpha^-,\beta^-}$, then, for all sufficiently large n, Eqs.* (3.7) *are uniquely solvable and the solutions u_n^* satisfy*

$$\|u_n^*-u^*\|_{\infty,\alpha^+,\beta^+}\leqslant \text{const}\,\frac{\log^{\delta+3}n}{n^{\gamma}}\|f\|_{\boldsymbol{C}^{\gamma,\,\delta+1}_{\alpha^-,\beta^-}}$$

with a constant independent of n and f.

3.2. Finite section and collocation methods w.r.t. weighted Chebyshev polynomials

If the ideas of Section 3.1 are applied to the case of variable coefficients $a(t)$ and $b(t)$ then a lot of difficulties occur. A first essential drawback is the fact that then in general one has to determine the parameters of Gaussian rules associated to generalized Jacobi weights, which are related to the coefficients a and b of the operator equation. This in turn requires the computation of the recurrence coefficients of the orthogonal polynomials w.r.t. such weights. Thus, one has a considerable computational complexity in the preprocessing, which, in particular, renders more difficult the application of these methods to nonlinear CSIEs. A second disadvantage is the fact that the mapping properties, on which the classical collocation method is based, require a certain Hölder continuity of the coefficients a and b. Last but not least, the classical collocation method is more or less restricted to the scalar case and in general not applicable to the case of a system of CSIEs.

The approach we present in this section is only based on the parameters of classical Chebyshev polynomials, which implies a very cheap preprocessing, and is applicable both to the case of piecewise continuous coefficients and to the case of a system of CSIEs. Although technical details as well as results are different the theoretical investigations, especially for the collocation method, follow the same lines as in Sections 2.2 and 2.3.

3.2.1. Stability of a finite section and a collocation method

Let $\sigma(t) = (1-t^2)^{-1/2}$ denote the Chebyshev weight of the first kind. For $\Gamma = (-1,1)$ and piecewise continuous coefficients $a,b \in \boldsymbol{PC}[-1,1]$, we consider Eq. (1.1) in the Hilbert space $\mathscr{H} = \boldsymbol{L}^2_\sigma := \boldsymbol{L}^2_{-1/2,-1/2}$, where, for simplicity, we restrict ourselve to the case $h \equiv 0$. For definiteness we agree that $a(-1+0) = a(-1)$ for each $a \in \boldsymbol{PC}[-1,1]$. The CSIO $S : \boldsymbol{L}^2_\sigma \to \boldsymbol{L}^2_\sigma$ is bounded, i.e. $S \in \mathscr{L}(\boldsymbol{L}^2_\sigma)$ (see [8, Theorem I.4.1]). By $\varphi(t) = \sqrt{1-t^2}$ we denote the Chebyshev weight of second kind and by $U_n(t)$ the respective orthonormal polynomial of degree n with positive leading coefficient, i.e.

$$U_n(\cos x) = \sqrt{\frac{2}{\pi}}\,\frac{\sin(n+1)x}{\sin x}, \quad x \in [0,\pi], \quad n = 0,1,2,\dots .$$

Then the system $\{\tilde{u}_n\}_{n=0}^\infty$ with $\tilde{u}_n := \varphi U_n$ is an orthonormal basis in $\boldsymbol{L}^2_\sigma$. We define the respective Fourier projections P_n by

$$P_n u = \sum_{k=0}^{n-1} \langle u, \tilde{u}_k \rangle_\sigma \tilde{u}_k$$

and, for the operator $A = aI + bS$, we are interested in the stability of the sequence $\{P_nAP_n\}$ of the respective finite sections corresponding to the Galerkin method. Moreover, if t^φ_{jn} are the Chebyshev nodes of second kind, $t^\varphi_{jn} = \cos(j\pi/(n+1))$, $j = 1,\dots,n$, we investigate the collocation method

$$(Au_n)(t^\varphi_{jn}) = f(t^\varphi_{jn}), \quad j = 1,\dots,n, \quad u_n \in \operatorname{im} P_n. \tag{3.16}$$

Of course, this collocation method can be written equivalently as

$$M_nAu_n = M_nf, \quad u_n \in \operatorname{im} P_n,$$

where M_n denotes the (weighted) interpolation operator $M_n = \varphi L^\varphi_n \varphi^{-1} I$ with the usual Lagrange interpolation operator L^φ_n w.r.t. the nodes t^φ_{jn}. Hence, concerning the collocation method (3.16), our aim is to study the stability of the sequence $\{M_nAP_n\}$ in $\boldsymbol{L}^2_\sigma$.

We introduce the operator $J:\mathbf{L}^2_\sigma \to \mathbf{L}^2(\mathbb{T})$, $u \mapsto \sum_{n=0}^{\infty}\langle u,\tilde{u}_n\rangle_\sigma e_n$. Of course, $J:\mathbf{L}^2_\sigma \to \mathbf{H}(\mathbb{T})$ is an isometric isomorphism, where $\mathbf{H}(\mathbb{T}):=\{f \in \mathbf{L}^2(\mathbb{T}){:}\ \langle f,e_{-n}\rangle = 0,\ n=1,2,\ldots\}$ is the Hardy space. Define $\chi \in \mathbf{PC}(\mathbb{T})$ by

$$\chi(t)=\begin{cases} 1, & \Im t>0,\\ -1, & \Im t<0\end{cases}$$

and, for a function $u:[-1,1]\to\mathbb{C}$, $\hat{u}(t)=u(\Re\, t)$. Defining $W_\mathbb{T}:\mathbf{L}^2(\mathbb{T})\to\mathbf{L}^2(\mathbb{T})$ by $(W_\mathbb{T}f)(t):=f(\bar{t})$ and, for $a\in\mathbf{L}^\infty(\mathbb{T})$, the Toeplitz operator $T(a):\mathbf{H}(\mathbb{T})\to\mathbf{H}(\mathbb{T})$ and the Hankel operator $H(a)$: $\mathbf{H}(\mathbb{T})\to\mathbf{H}(\mathbb{T})$ by $T(a):=P_\mathbb{T}aP_\mathbb{T}$ and $H(a):=P_\mathbb{T}ae_{-1}W_\mathbb{T}P_\mathbb{T}$, respectively, we get

(A) $aI=J^{-1}[T(\hat{a})-H(\hat{a}e_{-1})]J$,
(B) $S=-J^{-1}[T(\chi)+H(\chi e_{-1})]J$.

The proof of the stability criterion for the finite section method (Theorem 3.4) is based on a result presented in [22] (cf. also [9, Section 6.2]) for the stability of a sequence $\{B_n\}$ of finite sections $B_n=P_n^\mathbb{T}BP_n^\mathbb{T}+I-P_n^\mathbb{T}$, where $B\in\mathscr{L}(\mathbf{L}^2(\mathbb{T}))$ and $\{B_n\}$ belongs to the smallest closed subalgebra of the algebra of all bounded sequences $\{A_n\}$, $A_n\in\mathscr{L}(\mathbf{L}^2(\mathbb{T}))$ (equipped with component-wise algebraic operations and the supremum norm, cf. Section 2.2), containing the constant sequences $\{P_\mathbb{T}\}$, $\{e_{-1}W_\mathbb{T}\}$, and $\{aI\}$ for $a\in\mathbf{PC}(\mathbb{T})$, as well as the sequences $\{P_{kn}^\mathbb{T}\}$ for every positive integer k. This is possible because of the following two observations. First, the sequence $\{P_nAP_n\}$ is stable in $\mathbf{L}^2_\sigma$ if and only if the sequence $\{P_n^\mathbb{T}(JAJ^{-1}P_\mathbb{T}+Q_\mathbb{T})P_n^\mathbb{T}\}$ is stable in $\mathbf{L}^2(\mathbb{T})$. Second, in view of (A) and (B) we have

$$JAJ^{-1}=T(\hat{a})-H(\hat{a}e_{-1})-[T(\hat{b})-H(\hat{b}e_{-1})][T(\chi)+H(\chi e_{-1})]$$

(for details, see [19, Section 5]).

Theorem 3.4. *Let $a,b\in\mathbf{PC}[-1,1]$ and $c_\pm=a\pm b$. Then the sequence $\{P_n(aI+bS)P_n\}$ is stable in $\mathbf{L}^2_\sigma$ if and only if the following four conditions are fulfilled:*

(a) *The operator $aI+bS:\mathbf{L}^2_\sigma\to\mathbf{L}^2_\sigma$ is invertible.*
(b) *The operator $(\hat{a}-\hat{b}\chi)P_\mathbb{T}+Q_\mathbb{T}:\mathbf{L}^2(\mathbb{T})\to\mathbf{L}^2(\mathbb{T})$ is invertible,*
(c) *The point 0 lies outside the half-circle, which is formed by the segment $[c_+(1),c_-(1)]$ and the half-circle line from $c_-(1)$ to $c_+(1)$ that lies to the left of the line from $c_-(1)$ to $c_+(1)$, and outside the half-circle, which is formed by the segment $[c_-(-1),c_+(-1)]$ and the half-circle line from $c_+(-1)$ to $c_-(-1)$ that lies to the left of the line from $c_+(-1)$* to *$c_-(-1)$.*
(d) *For all $t\in(-1,1)$ and all $\mu\in[0,1]$, the relation*

$$\frac{c_+(t+0)}{c_+(t-0)}\mu+\frac{c_-(t+0)}{c_-(t-0)}(1-\mu)\notin(-\infty,0]$$

holds true.

To study the stability of the collocation method we use the concept explained in Section 2.2, where the C^*-algebra $\mathscr{F}^W$ is constructed in the same way, but now w.r.t. the space $\mathscr{H}=\mathbf{L}^2_\sigma$, the projections P_n instead of $P_n^{\mathbb{T}}$, and the operators W_n instead of $W_n^{\mathbb{T}}$, where $W_n u = \sum_{k=0}^{n-1} \langle u, \tilde{u}_{n-1-k} \rangle_\sigma \tilde{u}_k$. The covering system of localizing classes can be chosen in the same manner as in Section 2.3 for the collocation method, but with the interpolation operator M_n instead of $L_n^{\mathbb{T}}$. One can show that the sequence $\{M_n(aI + bS)P_n\}$ belongs to the algebra $\mathscr{F}^W$ if $a, b \in \boldsymbol{PC}$ and that the respective coset commutes with all elements from the localizing classes. The strong convergence

$$s - \lim_{n\to\infty} W_n M_n(aI + bS)W_n = aI - bS \quad \text{in } \boldsymbol{L}_\sigma^2$$

leads to the following result (see [19, Section 6]).

Theorem 3.5. *For $a, b \in \boldsymbol{PC}[-1, 1]$, the sequence $\{M_n(aI + bS)P_n\}$ is stable in $\boldsymbol{L}_\sigma^2$ if and only if the operators $aI \pm bS : \boldsymbol{L}_\sigma^2 \to \boldsymbol{L}_\sigma^2$ are invertible.*

Now let us investigate the stability of sequences $\{A_n\}$ belonging to the smallest closed subalgebra $\mathscr{A}$ of $\mathscr{F}$, which contains all sequences of the form $\{M_n(aI + bS)P_n\}$ with $a, b \in \boldsymbol{PC}[-1, 1]$ and the ideal $\mathscr{J}$. Although this algebra is not a C^*-algebra it turns out that $\mathscr{A}$ is an inverse closed subalgebra of $\mathscr{F}^W$. For the proof of the following result, which is essentially based on the application of the local principle of Allan and Douglas, and to some generalizations we refer to the forthcoming papers [15,17].

Theorem 3.6. *There is an isomorphism η from $\mathscr{A}/\mathscr{J}$ onto an algebra of continuous functions living on $((-1, 1) \times [0, 1]) \cup (\{\pm 1\} \times \bar{\mathbb{D}})$. This isomorphism sends $\{M_n a P_n\} + \mathscr{J}$ into*

$$(t, \mu) \mapsto \begin{bmatrix} a(t+0)\mu + a(t)(1-\mu) & (a(t+0) - a(t))\sqrt{\mu(1-\mu)} \\ (a(t+0) - a(t))\sqrt{\mu(1-\mu)} & a(t+0)(1-\mu) + a(t)\mu \end{bmatrix}$$

for $(t, \mu) \in (-1, 1) \times [0, 1]$ and $(t, z) \mapsto a(t)$ for $(t, z) \in \{\pm 1\} \times \bar{\mathbb{D}}$. The coset $\{M_n S P_n\} + \mathscr{J}$ is sent into

$$(t, \mu) \mapsto \begin{bmatrix} 1 & 0 \\ 0 & -1 \end{bmatrix} \quad \textit{for } (t, \mu) \in (-1, 1) \times [0, 1]$$

and $(t, z) \mapsto z$ for $(t, z) \in \{\pm 1\} \times \bar{\mathbb{D}}$. The sequence $\{A_n\} \in \mathscr{A}$ is stable if and only if the operators $W\{A_n\}, \tilde{W}\{A_n\} : \boldsymbol{L}_\sigma^2 \to \boldsymbol{L}_\sigma^2$ ($W\{A_n\} = s - \lim A_n P_n$, $\tilde{W}\{A_n\} := s - \lim W_n A_n W_n$) are invertible and $\eta\{A_n\}(t, z) \neq 0$ for all $(t, z) \in \{\pm 1\} \times \bar{\mathbb{D}}$.

We remark that the invertibility of $W\{A_n\}$ already implies $\det \eta(\{A_n\} + \mathscr{J})(t, \mu) \neq 0$ for all $(t, \mu) \in (-1, 1) \times [0, 1]$.

3.2.2. Convergence rates

The aim of this section is to introduce a suitable scale of Sobolev spaces in order to give a convergence rate for the error of the collocation method. The definition of such spaces is suggested by the orthonormal system $\{\tilde{u}_n\}$ in $\boldsymbol{L}^2_\sigma$, which we used as ansatz functions. Thus, we define

$$\tilde{\boldsymbol{L}}^{2,s}_\sigma := \left\{ u \in \boldsymbol{L}^2_\sigma \colon \|u\|_{s,\sim} := \left(\sum_{n=0}^{\infty} (1+n)^{2s} |\langle u, \tilde{u}_n \rangle_\sigma|^2 \right)^{1/2} < \infty \right\}, \quad s \geqslant 0.$$

Since $\langle u, \tilde{u}_n \rangle_\sigma = \langle \sigma u, U_n \rangle_\varphi$ we have $\|u\|_{s,\sim} = \|\sigma u\|_{\boldsymbol{L}^{2,s}_\varphi}$, where $\boldsymbol{L}^{2,s}_\varphi = \boldsymbol{L}^{2,s}_{1/2,1/2}$ is a special case of the Sobolev spaces introduced in Section 3.1.1. This means that the multiplication operator $\sigma I : \tilde{\boldsymbol{L}}^{2,s}_\sigma \to \boldsymbol{L}^{2,s}_\varphi$ is an isometric isomorphism. A consequence of this is that the relations (A)–(F) of Section 3.1.1 remain true if $S^{\gamma,\delta}_n$, $L^{\gamma,\delta}_n$, and $\boldsymbol{L}^{2,s}_{\gamma,\delta}$ are substituted by P_n, M_n, and $\tilde{\boldsymbol{L}}^{2,s}_\sigma$, respectively. Hence, we get the following theorem.

Theorem 3.7. *Let the operator* $A = aI + bS : \tilde{\boldsymbol{L}}^{2,s_0}_\sigma \to \tilde{\boldsymbol{L}}^{2,s_0}_\sigma$ *be continuous for some* $s_0 > \frac{1}{2}$. *If the sequence* $\{A_n\} = \{M_n A P_n\}$ *is stable in* $\boldsymbol{L}^2_\sigma$ *and if the solution* u^* *of* $Au = f$ *belongs to* $\tilde{\boldsymbol{L}}^{2,s}_\sigma$ *for some* $s > \frac{1}{2}$, *then*

$$\|u^*_n - u^*\|_{t,\sim} \leqslant \text{const}\, n^{t-s} \|u^*\|_{s,\sim}, \quad 0 \leqslant t \leqslant s,$$

where $u^*_n \in \operatorname{im} P_n$ *is the solution of* (3.16) *and the constant does not depend on* n, t, *and* u^*.

Of course, the condition $u^* \in \tilde{\boldsymbol{L}}^{2,s}_\sigma$ is satisfied for some sufficiently large s only if u^* is smooth in the interior of the integration interval and behaves near the endpoints like the ansatz functions $\tilde{u}_n$. Thus, in order to be more flexible, this fact requires the investigation of similar collocation methods, for example of collocation methods again w.r.t. Chebyshev nodes but with other classical Jacobi weights in the functions $\tilde{u}_n$ (see, for example [15,18]).

3.2.3. Computational aspects

A suitable implementation of the collocation method (3.16) will enable us to solve the resulting system of linear equations with a fast algorithm that requires only $O(n^2)$ operations and $O(n)$ storage due to the special structure of the system matrix. At first we write u_n in (3.16) in the form

$$u_n(t) = \varphi(t) \sum_{k=1}^{n} \xi_{kn} \ell_{kn}(t) \quad \text{with } \ell_{kn}(t) = \frac{T_n(t)}{(t - t^\sigma_{kn}) T'_n(t^\sigma_{kn})}. \tag{3.17}$$

Since, for $t \neq t^\sigma_{kn}$, $k = 1, \ldots, n$,

$$\begin{aligned} (S\ell_{kn})(t) &= \frac{1}{T'_n(t^\sigma_{kn})} \frac{1}{\pi \mathrm{i}} \int_{-1}^{1} \frac{\varphi(s) T_n(s)}{(s-t)(s-t^\sigma_{kn})}\, \mathrm{d}s \\ &= \frac{1}{T'_n(t^\sigma_{kn})} \frac{1}{t^\sigma_{kn} - t} \frac{1}{\pi \mathrm{i}} \int_{-1}^{1} \left(\frac{1}{s - t^\sigma_{kn}} - \frac{1}{s-t} \right) \varphi(s) T_n(s)\, \mathrm{d}s \\ &= \frac{1}{T'_n(t^\sigma_{kn})} \frac{\rho_n(t^\sigma_{kn}) - \rho_n(t)}{t^\sigma_{kn} - t}, \end{aligned}$$

where $\rho_n = S\varphi T_n = \frac{1}{2}S\varphi(U_n - U_{n-2}) = \frac{\mathrm{i}}{2}(T_{n+1} - T_{n-1})$, the collocation method (3.16) can be written in the form

$$\sum_{k=1}^{n} \alpha_{jk} \frac{\xi_{kn}}{T_n'(t_{kn}^{\sigma})} = f_j, \quad j = 1, \ldots, n, \tag{3.18}$$

where $f_j = f(t_{jn}^{\varphi})$ and

$$\alpha_{jk} = \frac{b(t_{jn}^{\varphi})\rho_n(t_{kn}^{\sigma}) - a(t_{jn}^{\varphi})\varphi(t_{jn}^{\varphi})T_n(t_{jn}^{\varphi}) - b(t_{jn}^{\varphi})\rho_n(t_{jn}^{\varphi})}{t_{kn}^{\sigma} - t_{jn}^{\varphi}}.$$

Here we assume that n is an even number to guarantee $t_{kn}^{\sigma} \neq t_{jn}^{\varphi}$. We see that the coefficients α_{jk} can be written in the form

$$\alpha_{jk} = \frac{\varepsilon_j \alpha_k - \beta_j}{\gamma_k - \delta_j}.$$

Thus, the system matrix has a Löwner structure, which gives us the possibility to apply the idea of UV-reduction presented in [10] to establish an algorithm of $\mathrm{O}(n^2)$-complexity to solve the system (3.18). The details one can found in [19, Section 9].

3.2.4. *Systems of CSIEs*

We consider a system of CSIEs of the form

$$\sum_{k=1}^{\ell} (a_{jk}I + b_{jk}S)u_k = f_j, \quad j = 1, \ldots, \ell.$$

Let $\underline{a} = [a_{jk}]_{j,k=1}^{\ell}$, $\underline{b} = [b_{jk}]_{j,k=1}^{\ell}$, and denote by $\underline{S}$ the diagonal operator $[\delta_{jk}S]_{j,k=1}^{\ell}$ in $(\boldsymbol{L}_{\sigma}^2)^{\ell}$. Analogously, $\underline{I}$, $\underline{P}_n$, and $\underline{M}_n$ are defined. Then, the respective operator sequence of the collocation method can be written in the form $\{\underline{M}_n(\underline{a}\,\underline{I} + \underline{b}\,\underline{S})\underline{P}_n\}$. Now from Theorem 3.6 one can easily obtain the following result.

Theorem 3.8. *Let $a_{jk}, b_{jk} \in \boldsymbol{PC}[-1,1]$. If $a_{jk}(\pm 1) = b_{jk}(\pm 1) = 0$ for all $j < k$ (or for all $j > k$), then the sequence $\{\underline{M}_n(\underline{a}\,\underline{I} + \underline{b}\,\underline{S})\underline{P}_n\}$ is stable in $(\boldsymbol{L}_{\sigma}^2)^{\ell}$ if and only if the operators $\underline{a}\,\underline{I} \pm \underline{b}\,\underline{S}$ are invertible in $(\boldsymbol{L}_{\sigma}^2)^{\ell}$ and if $|a_{kk}(\pm 1)| > |b_{kk}(\pm 1)|$ for all $k = 1, \ldots, \ell$.*

Another result one can find in [19, Section 7].

3.2.5. *Application to nonlinear CSIEs*

In a series of applications nonlinear CSIEs of the form

$$F(t, u(t)) + \frac{1}{\pi}\int_{-1}^{1} \frac{u(s)}{s - t}\,\mathrm{d}s + d = 0, \quad x \in (-1, 1), \tag{3.19}$$

occur, where $F : [-1,1] \times \mathbb{R} \to \mathbb{R}$ is given and $u : [-1,1] \to \mathbb{R}$ satisfying $u(-1) = u(1) = 0$ as well as the real number d are sought. Thus, if we are interested in a linearization of Eq. (3.19) (for example, in order to apply a Newton iteration method), then we have to solve more general equations than (1.1). A possible description is the following. Let $\boldsymbol{V}$ be a finite-dimensional subspace of $\boldsymbol{L}^2_\sigma$ and look for a solution $(u,v)\in \boldsymbol{L}^2_\sigma\times\boldsymbol{V}$ of the equation

$$(aI+bS)u+v=f, \tag{3.20}$$

where $a,b\in \boldsymbol{PC}[-1,1]$ and $f\in\boldsymbol{L}^2_\sigma$ are given functions. We try to apply the collocation method (3.16) to this situation. Obviously, $\boldsymbol{X}:=\boldsymbol{L}^2\times\boldsymbol{V}$ equipped with the norm $\|(u,v)\|_X:=\sqrt{\|u\|^2_{\boldsymbol{L}^2_\sigma}+\|v\|^2_{\boldsymbol{L}^2_\sigma}}$ is a Banach space. Let $\{\tilde v_0,\dots,\tilde v_{m-1}\}$ be a basis in $\boldsymbol{V}$ and define the isomorphism $J_m:\boldsymbol{X}\to\boldsymbol{L}^2_\sigma$ by

$$J_m\left(u,\sum_{k=0}^{m-1}\alpha_k\tilde v_k\right):=\sum_{j=0}^{\infty}\langle u,\tilde u_j\rangle_\sigma\tilde u_{j+m}+\sum_{k=0}^{m-1}\alpha_k\tilde u_k=V^m u+\sum_{k=0}^{m-1}\alpha_k\tilde u_k,$$

where V denotes the shift operator $V\tilde u_n=\tilde u_{n+1}$, $n=0,1,2,\dots$. Then the inverse operator $J^{-1}:\boldsymbol{L}^2_\sigma\to\boldsymbol{X}$ is given by

$$J_m^{-1}f=\left((V^*)^m f,\sum_{k=0}^{m-1}\langle f,\tilde u_k\rangle_\sigma\tilde v_k\right).$$

Thus, setting $w=J_m(u,v)$ Eq. (3.20) is equivalent to

$$(aI+bS)(V^*)^m w+Kw=f,\quad w\in\boldsymbol{L}^2_\sigma,$$

where the operator $K:\boldsymbol{L}^2_\sigma\to\boldsymbol{L}^2_\sigma$ is defined by $Kw:=\sum_{K=0}^{m-1}\langle w,\tilde u_k\rangle_\sigma\tilde v_k$. The respective collocation method for Eq. (3.20) reads as

$$M_n(aI+bS)P_nu_n+M_nv_n=M_nf,\quad u_n\in\operatorname{im}P_{n-m},\ v_n\in\boldsymbol{V},$$

or equivalently as

$$B_nw_n:=M_n[(aI+bS)(V^*)^m+K]P_n=M_nf,\quad w_n\in\operatorname{im}P_n.$$

It turns out that $V=\varphi I-\mathrm{i}\psi S$, $\psi(t)=t$. Since $B_n=M_n(aI+bS)P_n(M_nV^*P_n)^m+M_nKP_n$ and $V^*=\varphi I+\mathrm{i}\psi S$ one concludes that the sequence $\{B_n\}$ belongs to the algebra $\mathscr{A}$ considered in Section 3.2.1 and can apply Theorem 3.6 if the basis functions $\tilde v_k$ of $\boldsymbol{V}$ are sufficiently regular.

Theorem 3.9 (Junghanns and Müller [16, Theorem 2.7]). *Let $a,b\in\boldsymbol{PC}[-1,1]$, $m\in\{1,2,\dots\}$, and assume that the functions $\tilde v_k$, $k=0,\dots,m-1$, are locally Riemann integrable and satisfy*

$$|\tilde v_k(t)|\leqslant \mathrm{const}\,(1-t^2)^{-\varepsilon},\quad t\in(-1,1),\quad \textit{for some } \varepsilon\in(0,\tfrac14).$$

Then the sequence $\{B_n\}$ is stable in $\boldsymbol{L}^2_\sigma$ if and only if the operators $(aI+bS)(V^*)^m+K$ and $(aI-bS)V^m$ are invertible in $\mathscr{L}(\boldsymbol{L}^2_\sigma)$.

Let us investigate the collocation method

$$F(t^\varphi_{jn},u_n(t^\varphi_{jn}))+\frac{1}{\pi}\int_{-1}^{1}\frac{u_n(s)}{s-t^\varphi_{jn}}\,\mathrm{d}s+d_n=0,\quad j=1,\dots,n,\ (u_n,d_n)\in\boldsymbol{X}_n. \tag{3.21}$$

$\boldsymbol{X}_n:=\operatorname{im}P_{n-1}\times\mathbb{C}$. To solve this collocation equation approximately we try to determine a sequence $\{(u_n^{(m)},d_n^{(m)})\}_{m=0}^\infty\subset\boldsymbol{X}_n$ with $u_n^{(m+1)}=u_n^{(m)}+\Delta u_n^{(m)}$ by a modified Newton method

$$a_0(t^\varphi_{jn})\Delta u_n^{(m)}(t^\varphi_{jn})+\frac{1}{\pi}\int_{-1}^{1}\frac{u_n^{(m+1)}(s)}{s-t^\varphi_{jn}}\,\mathrm{d}s+d_n^{m+1}=-F(t^\varphi_{jn},u_n^{(m)}(t^\varphi_{jn})), \tag{3.22}$$

$j=1,\dots,n$, where $a_0(t)=F_u(t,u_n^{(0)}(t))$.

Theorem 3.10 (Junghanns and Müller [16, Theorem 3.9]). *Let* (u^*, d^*) *be a solution of* (3.19). *Assume that* $a(t) := a_0(t)$ *and* $b(t) \equiv \mathrm{i}$ *satisfy the conditions of Theorem* 3.9 *for* $m = 1$ *and* $\tilde{v}_0(t) \equiv 1$, *that the function* $F_u : [-1, 1] \times \mathbb{R} \to \mathbb{R}$ *is continuous and satisfies the Hölder condition*

$$|F_u(t, u^1) - F_u(t, u^2)| \leqslant \text{const}\, |u^1 - u^2|, \quad t \in [-1, 1],\ u^1, u^2 \in \mathbb{R}$$

and that $\varphi^{-1}u^*$ *belongs to* $\boldsymbol{C}^{\gamma,\delta}_{1/2,1/2}$ *for some* $\gamma > \frac{1}{2}$ (*cf. Section* 3.1.3). *Then, for all sufficiently large n, there exists an element* $(u_n^{(0)}, d_n^{(0)}) \in \boldsymbol{X}_n$ *such that Eq.* (3.22) *possesses a unique solution* $(\Delta u_n^{(m)}, d_n^{m+1)})$ *for all* $m = 1, 2, \ldots$. *The sequence* $\{(u_n^{(m)}, d_n^{(m)})\}$ *converges for* $m \to \infty$ *in the norm of* $\boldsymbol{X} := \boldsymbol{L}^2_\sigma \times \mathbb{C}$ *to a solution* (u_n^*, d_n^*) *of Eq.* (3.21) *and, for* $0 < \varepsilon < \gamma - \frac{1}{2}$,

$$\|(u_n^*, d_n^*) - (u^*, d^*)\|_X \leqslant \text{const}\, n^{\varepsilon - \gamma} \log^\delta(n + 1),$$

where the constant does not depend on n.

References

[1] B.A. Amosov, On the approximate solution of elliptic pseudodifferential equations on smooth curves, Z. Anal. Anwendungen 9 (1990) 545–563 (in Russian).

[2] D. Berthold, W. Hoppe, B. Silbermann, A fast algorithm for solving the generalized airfoil equation, J. Comput. Appl. Math. 43 (1992) 185–219.

[3] D. Berthold, B. Silbermann, Corrected collocation methods for periodic pseudodifferential equations, Numer. Math. 70 (1995) 397–425.

[4] D. Berthold, B. Silbermann, The fast solution of periodic pseudodifferential equations, Appl. Anal. 63 (1996) 3–23.

[5] M.R. Capobianco, G. Criscuolo, P. Junghanns, A fast algorithm for Prandtl's integro-differential equation, J. Comput. Appl. Math. 77 (1997) 103–128.

[6] M.R. Capobianco, G. Criscuolo, P. Junghanns, U. Luther, Uniform convergence of the collocation method for Prandtl's integro-differential equation, Proceedings of the David Elliott Conference, Hobart, 1997.

[7] M.R. Capobianco, P. Junghanns, U. Luther, G. Mastroianni, Weighted uniform convergence of the quadrature method for Cauchy singular integral equations, in: A. Bottcher, I. Gohberg (Eds.), Singular Integral Operators and Related Topics, Birkhäuser, Basel, 1996, pp. 153–181.

[8] I. Gohberg, N. Krupnik, One-Dimensional Linear Singular Integral Equations, Birkhäuser, Basel, 1992.

[9] R. Hagen, S. Roch, B. Silbermann, Spectral Theory of Approximation Methods for Convolution Equations, Birkhäuser, Basel, 1994.

[10] G. Heinig, K. Rost, Algebraic Methods for Toeplitz-like Matrices and Operators, Birkhäuser, Basel, 1984.

[11] P. Junghanns, Product integration for the generalized airfoil equation, in: E. Schock (Ed.), Beiträge zur Angewandten Analysis und Informatik, Shaker Verlag, 1994, pp. 171–188.

[12] P. Junghanns, U. Luther, Cauchy singular integral equations in spaces of continuous functions and methods for their numerical solution, J. Comput. Appl. Math. 77 (1997) 201–237.

[13] P. Junghanns, U. Luther, Uniform convergence of a fast algorithm for Cauchy singular integral equations, Linear Algebra Appl. 275–276 (1998) 327–347.

[14] P. Junghanns, U. Luther, Uniform convergence of the quadrature method for Cauchy singular integral equations with weakly singular perturbation kernels, Proceedings of the Third International Conference on Funct. Anal. Appr. Theory, Maratea, 1996, Vol. II, pp. 551–566.

[15] P. Junghanns, G. Mastroianni, On the stability of collocation methods for Cauchy singular integral equations on an interval, to appear in: Operator Theory, Advances and Applications (Proceedings of the 11th TMP, Chemnitz, 1999), Birkhaüser, Basel.

[16] P. Junghanns, K. Müller, A collocation method for nonlinear Cauchy singular integral equations, J. Comput. Appl. Math. 115 (2000) 283–300.

[17] P. Junghanns, S. Roch, B. Silbermann, Stability of collocation methods for systems of Cauchy singular integral equations on an interval, in preparation.

[18] P. Junghanns, U. Weber, Banach algebra techniques for Cauchy singular integral equations on an interval, in: J.I. Frankel, C.A. Brebbia, M.A.H. Aliabadi (Eds.), Boundary Element Technology XII, Computational Mechanics Publications, Southampton, Boston, 1997, pp. 419–428.
[19] P. Junghanns, U. Weber, Local theory of projection methods for Cauchy singular integral equations on an interval, in: M. Golberg (Ed.), Boundary Integral Methods: Numerical and Mathematical Aspects, Computational Mechanics Publications, Series: Computational Engineering, Vol. 1, Boston, Southampton, 1998, pp. 217–256.
[20] G. Monegato, S. Prössdorf, Uniform convergence estimates for a collocation and a discrete collocation method for the generalized airfoil equation, in: A.G. Agarval (Ed.), Contributions to Numerical Mathematics, World Scientific, Singapore, 1993, pp. 285–299.
[21] S. Prössdorf, B. Silbermann, Numerical Analysis for Integral and Related Operator Equations, Birkhäuser, Basel, 1991.
[22] S. Roch, Lokale Theorie des Reduktionsverfahrens für singuläre Integraloperatoren mit Carlemanschen Verschiebungen, Ph.D. Thesis, TU Karl-Marx-Stadt (now Chemnitz), 1987.
[23] S. Roch, B. Silbermann, Asymptotic Moore-Penrose invertibility of singular integral operators, Integral Equations Operator Theory 26 (1996) 81–101.
[24] J. Saranen, G. Vainikko, Fast solvers of integral and pseudodifferential equations on closed curves, Math. Comp. 67 (1998) 1473–1491.

ELSEVIER Journal of Computational and Applied Mathematics 125 (2000) 423–437

JOURNAL OF COMPUTATIONAL AND APPLIED MATHEMATICS

www.elsevier.nl/locate/cam

Numerical methods for integral equations of Mellin type

J. Elschner[a,*], I.G. Graham[b]

[a] *Weierstraß -Institut für Angewandte Analysis und Stochastik, Mohrenstraße 39, 10117 Berlin, Germany*
[b] *Department of Mathematical Sciences, University of Bath, Bath BA2 7AY, UK*

Received 27 August 1999

Abstract

We present a survey of numerical methods (based on piecewise polynomial approximation) for integral equations of Mellin type, including examples arising in boundary integral methods for partial differential equations on polygonal domains. © 2000 Elsevier Science B.V. All rights reserved.

1. Introduction

In the last 30 years or so a great deal of interest has focused on the numerical analysis of boundary integral equations arising from PDEs on nonsmooth domains (see [49] for one of the pioneering papers in this field). Here the chief difficulties are not only the loss of smoothness of the solution near nonsmooth boundary points, but also (and more crucially) the singularity induced in the integral operator itself. The development of a proper understanding of these singularities has a huge practical motivation due to the large range of applications — particularly in engineering — and even the geometrically simple case of a polygonal domain still contains open problems of considerable mathematical subtlety. This survey concentrates on the numerical analysis of a class of equations which arises generically in such problems, namely the equations of *Mellin type*. The simplest case of such an equation contains the operator

$$Kv(s) = \int_0^1 \kappa\left(\frac{s}{\sigma}\right) v(\sigma)\frac{d\sigma}{\sigma}, \quad s \in [0,1], \tag{1.1}$$

where the *kernel* κ is a given function on $\mathbb{R}^+:=[0,\infty)$. Often κ is a smooth function on $(0,\infty)$ satisfying certain asymptotic estimates at 0 and ∞, in which case $\kappa(s/\sigma)\sigma^{-1}$ is smooth at $s=\sigma>0$ but blows up with $O(\sigma^{-1})$ when $s=\sigma\to 0$ (i.e., the operator (1.1) has a *fixed singularity* at the

* Corresponding author.
E-mail address: elschner@wias-berlin.de (J. Elschner).

PII: S0377-0427(00)00483-0

origin). Note that the upper limit of integration in (1.1) is to some extent arbitrary, since the operator $K_\varepsilon v(s) := \int_0^\varepsilon \kappa(s/\sigma)v(\sigma)\,d\sigma/\sigma$, $s \in [0,\varepsilon]$ can easily be reduced to (1.1) via the transformation $\sigma \to \varepsilon\sigma$.

The operator K (or more generally K_ε, for $\varepsilon > 0$) can be considered as a localised version of the operator: $\mathscr{K}v(s) := \int_0^\infty \kappa(s/\sigma)\,d\sigma/\sigma$, $s \in \mathbb{R}^+$, which is normally treated using the *Mellin transform*: $\tilde{v}(z) = \int_0^\infty s^{z-1}v(s)\,ds$, for $z \in \mathbb{C}$. The convolution theorem then states that (for suitably well-behaved κ and v) we have $\widetilde{\mathscr{K}v} = \tilde{\kappa}\tilde{v}$, and from this it is easily shown that $\|K\|_2 \leqslant \|\mathscr{K}\|_2 = \sup_{\mathrm{Re}(z)=\frac{1}{2}} |\tilde{\kappa}(z)|$. (Here $\|\cdot\|_2$ denotes the operator norm on the space L_2 of square-integrable functions.) Moreover, K is also bounded on L_∞ and on C (the continuous functions on $[0,1]$), with $\|K\|_\infty = \int_0^\infty |\kappa(s)|\,ds/s < \infty$ (provided this integral exists), in which case

$$\lim_{s\to 0} Kv(s) = \tilde{\kappa}(0)v(0) \quad \text{for } v \in C. \tag{1.2}$$

Using (1.2) the following simple argument ([3]) shows that K is noncompact in C: For each $n \in \mathbb{N} := \{1,2,\ldots\}$, let $v_n : [0,1] \to \mathbb{R}$ denote a continuous function with $v_n(0) = 1 = \|v_n\|_\infty$ and $\operatorname{supp} v_n \subset [0,1/n]$. If K were compact on C then the sequence $\{Kv_n\}$ would contain a convergent subsequence, $\{Kv_{n_j}\}$ in C. However, (1.2) implies that $Kv_{n_j}(0) = \tilde{\kappa}(0)$ for all j. Moreover, for $s > 0$ we can employ the change of variable $x = s/\sigma$ to obtain $|Kv_{n_j}(s)| \leqslant \int_{n_j s}^\infty |\kappa(x)|\,dx/x \to 0$ as $j \to \infty$, demonstrating that $\{Kv_{n_j}\}$ cannot have a continuous limit when $\tilde{\kappa}(0) \neq 0$. In fact the spectrum of K contains all the values of $\tilde{\kappa}(z)$ for $\mathrm{Re}(z) = 0$, and K is not compact on any L_p space either (see Section 3 for a discussion of this).

All the problems which we shall consider in this paper have as their heart the solution of second-kind equations of the form

$$(I - K)u = f \tag{1.3}$$

with K as defined in (1.1). An important role in the theory of these equations is played by the *finite section operator* KT^τ, where T^τ is the truncation operator satisfying $T^\tau v(s) = 0$, for $s < \tau$ and $T^\tau v(s) = v(s)$ for $s \geqslant \tau$. Then, for $\tau \in (0,1]$, we have $KT^\tau v(s) = \int_\tau^1 \kappa(s/\sigma)v(\sigma)\,d\sigma/\sigma$. At various points in this review we will require assumptions on (i) the well-posedness of (1.3) and (ii) the stability of the corresponding finite section operators, i.e.,

$$\text{(i) } \|(I-K)^{-1}\| \leqslant C \quad \text{and} \quad \text{(ii) } \|(I-KT^\tau)^{-1}\| \leqslant C \quad \text{as } \tau \to 0, \tag{1.4}$$

for some norm $\|\cdot\|$. Throughout the paper we let $C, C_1, C_2, \ldots$ denote generic constants in the usual way.

To analyse (1.3), we introduce for $\alpha \in \mathbb{R}$ and $r \in \mathbb{N}$, the space $C^{r,\alpha}$ comprising the completion of the infinitely smooth functions on $(0,1]$ with respect to the norm $\|v\|_{r,\alpha} := \sup_{s\in(0,1],\, l=0,\ldots r} |s^{[l-\alpha]}D^l v(s)|$, where $[\beta] = \beta$ for $\beta \geqslant 0$ and $[\beta] = 0$ for $\beta < 0$. In general the solution u of (1.3) (or perhaps the higher derivatives of u) will have a singularity at $s = 0$, and thus will lie in $C^{r,\alpha}$ with the size of α depending on the zeros of the symbol $1 - \tilde{\kappa}(z)$, for $z \in \mathbb{C}$ (see, for example, [12] or [38, pp. 172–174]).

To approximate (1.3), we introduce piecewise polynomial spaces on $[0,1]$ as follows. For any integer $n \geqslant 1$, introduce a mesh $0 = x_0 < x_1 < \cdots < x_n = 1$. Then for $r > d+1 \geqslant 0$, $S_n^{r,d}$ denotes the functions which reduce to polynomials of degree $r-1$ on each interval $I_i = (x_{i-1}, x_i)$ and have d continuous derivatives globally on $[0,1]$. Thus, for $r > 0$, $S_n^{r,-1}$ denotes the piecewise polynomials of degree $r-1$ which may be discontinuous at each x_i, $i = 1,\ldots,n-1$, whereas $S_n^{r,r-2}$ denotes the smoothest splines on $[0,1]$ (without any end-point conditions). We shall also need the 2π-periodic smoothest splines of degree $r-1$ (and C^{r-2} continuity), which we denote $S^r_{n,p}$. There is a well-worked literature on approximation in these spaces (see, e.g., [18,38,46]).

To deal with the singularity in u, one approach is to consider graded meshes constructed (either analytically or adaptively) to satisfy the inequalities

$$h_i \leqslant C_1(1/n)(i/n)^{q-1} \quad \text{and} \quad x_i \geqslant C_2(i/n)^q, \quad i=1,\ldots,n, \tag{1.5}$$

for some *grading exponent* $q\geqslant 1$, where $h_i = x_i - x_{i-1}$. These inequalities imply that near $x=0$ mesh subintervals are of length $O((1/n)^q)$ whereas near $x=1$ they are of length $O(1/n)$ as $n\to\infty$. We call meshes which satisfy (1.5) "q-*graded* at 0". A standard example of such a mesh is [42] $x_i{:=}(i/n)^q$, $i=0,\ldots,n$, which satisfies (1.5) with $C_1=q$ and $C_2=1$.

To illustrate the properties of such meshes, consider approximating a function $u\in C^{r,\alpha}$ by $S^{r,-1}_n$ (where $r\geqslant 1$), and suppose for convenience that $\alpha\in(0,1]$. Then standard Taylor series estimates show that there exists a function $\phi_n\in S^{r,-1}_n$ such that $\|u-\phi_n\|_{\infty,I_i}\leqslant h_i^r\|D^r u\|_{\infty,I_i}$, provided the norm on the right-hand side is finite. Thus, for $i\geqslant 2$ making use of (1.5), we have $\|u-\phi_n\|_{\infty,I_i}\leqslant Ch_i^r x_{i-1}^{\alpha-r}\|u\|_{r,\alpha}\leqslant C(1/n)^r((i-1)/n)^{q\alpha-r}\|u\|_{r,\alpha}\leqslant C(1/n)^r\|u\|_{r,\alpha}$, where the final inequality requires that the grading exponent q should be sufficiently large, namely $q\geqslant r/\alpha$. On the other hand, for $s\in I_1$, elementary arguments show that $|u(s)-u(0)|\leqslant Cs^\alpha\|u\|_{r,\alpha}\leqslant C(1/n)^r\|u\|_{r,\alpha}$, again provided $q\geqslant r/\alpha$. So, setting $\phi_n\equiv u(0)$ on I_1 we see that $\|u-\phi_n\|_\infty$ is of optimal order $O(1/n)^r$. In some examples the solution u of (1.3) is not continuous but instead has an infinite singularity of order $s^{\alpha-1}$ (as $s\to 0$) for some $\alpha\in(\frac{1}{2},1)$. Then analogous arguments to those given above (but in the L_2 context) [17] show that there exists $\phi_n\in S^{r,-1}_n$ with $\phi_n\equiv 0$ on I_1 such that $\|u-\phi_n\|_2=O(n^{-r})$ provided $q>r/(\alpha-\frac{1}{2})$. Both the L_2 and uniform estimates also extend to the case of approximation by splines of arbitrary smoothness [18].

An alternative way of dealing with a singularity in the solution $u(s)$ of (1.3) at $s=0$ (and a method which we shall consider in more detail below) involves a change of variable $s=\gamma(x)$, where $\gamma:[0,1]\to[0,1]$ is an increasing function, with $\gamma(0)=0$, $\gamma(1)=1$ and $\gamma(x)$ having a zero of an appropriately high order at $x=0$. For example if $u\in C^{r,\alpha}$ where $\alpha\in(0,1]$ and if γ has r continuous derivatives on $[0,1]$ with $(D^j\gamma)(x)=O(x^{q-j})$ for $j=0,\ldots,r$, then it is easily shown that the function $u\circ\gamma$ has r continuous derivatives, provided $q\geqslant r/\alpha$. This function can then be approximated by a piecewise polynomial ϕ_n of desired smoothness with respect to the *uniform mesh* $x_i=i/n$, yielding (after inverse transformation) an optimal order approximation $\phi_n(\gamma^{-1}(s))$ to $u(s)$. If (1.3) has a solution which blows up at $s=0$ (for example the function $u(s)=s^{\alpha-1}$ with $\alpha\in(\frac{1}{2},1)$), then the straight substitution $s=\gamma(x)$ with $\gamma(x)$ given above makes it worse rather than better-behaved. This difficulty can be circumvented by considering instead the function $w(x)=(u\circ\gamma)(x)|\gamma'(x)|$, with γ as above, which arises naturally when u appears inside an integral. Then it is easily shown that $w(s)$ has r continuous derivatives provided $q\geqslant(r+1)/\alpha$ (see, e.g., [21]). Such nonlinear change of variables techniques can be combined with piecewise polynomial approximation schemes or indeed global schemes involving algebraic or trigonometric polynomials. A selection of results can be found for example in [5,16,32,36,37,41].

A third method of obtaining optimal convergence for singular solutions (which we shall not discuss at length here) is to augment the approximating spaces with some of the singular terms occurring in the expansion of the solution (see, e.g., [12,33,34,51]).

However the chief difficulty in solving (1.3) is not the approximation of the singular solution u but rather proving the stability of the chosen numerical method, with the main theoretical barrier being the noncompactness of the operator K. This was emphasised in [10], where it was shown that there exist piecewise polynomial collocation methods which converge optimally when K is compact but which actually diverge for (1.3) when K is given by (1.1). In [10] a way around this barrier was found by considering a modified method (which excluded the counterexample but which was nevertheless very close to a standard collocation method) and proving stability and convergence for it. Subsequently, this modification technique has been applied to a great variety of spline approximation methods for (1.3) (see [38] for an extensive review), and as far as we are aware, it is still the standard way of proving stability and convergence for practical methods for integral equations of Mellin type. In particular, this approach has been successfully extended to collocation and quadrature methods based on global algebraic and trigonometric polynomials. Examples of results which use the modification technique to prove stability (in conjunction with mesh grading) are [9–11,15,17,19,20,24,27,29,39], whereas the same technique is used in conjunction with a nonlinear change of parametrisation in [21–23,25,26,30–32,35,47]. The modification technique for proving stability later found a more practical use as a parameter for accelerating the convergence of multigrid-type algorithms [4,40].

It is important to point out that in the case of classical Galerkin methods for boundary integral equations on corner domains (where a variational formulation of the underlying integral equation is exploited and errors due to quadrature are not taken in to account), the stability analysis is not difficult provided one restricts to the energy norm. The numerical analysis then reduces to finding efficient ways of approximating the singular solution. In this context the literature is older and includes, for example, [51,12]. The papers [6,7] also concern the Galerkin method but analyse errors in the uniform norm and therefore require a more sophisticated stability analysis.

We begin this survey in Section 2 by illustrating the use of the modification technique in the (relatively simple) context of discontinuous piecewise polynomial collocation methods for (1.3). The modified method can be thought of as the discretization of the finite section approximation of K, and then a perturbation argument is the key to proving stability. In Section 3 we explain how this idea can be extended to a unified convergence theory of spline approximation methods for Eq. (1.3). Section 4 is devoted to some examples of second- and first-kind boundary integral equations for elliptic PDEs on corner domains leading to the model Eq. (1.3) (more precisely systems of such equations), with emphasis on Laplace's equation. In Section 5 we give a survey of results on Symm's integral equation and related first-kind equations.

2. Introduction to modification techniques

To illustrate the technique of modification (mentioned in Section 1) in a simple setting, consider Eq. (1.3) and suppose that assumption (1.4) holds in the essential supremum norm. Assume also that κ satisfies the estimates

$$\int_0^\infty s^k |D^k \kappa(s)| \,\mathrm{d}s/s < \infty \quad \text{for all integers } k \geqslant 0. \tag{2.1}$$

To solve (1.3), we consider classical piecewise polynomial collocation methods in $S_n^{r,-1}$. To define the collocation procedure, choose r points $0 \leqslant \xi_1 \leqslant \cdots \leqslant \xi_r \leqslant 1$ in the reference domain $[0,1]$ and map these to each I_i with the formula $x_{ij} = x_{i-1} + \xi_j h_i$, $i = 1,\ldots,n$, $j = 1,\ldots,r$. Defining the interpolatory projection P_n onto $S_n^{r,-1}$ by requiring that $P_n v(x_{ij}) = v(x_{ij})$ for all i,j it follows that P_n converges pointwise to the identity on C and has uniform norm bounded as $n \to \infty$. The classical collocation method for (1.3) seeks an approximate solution $u_n \in S_n^{r,-1}$ such that

$$(I - P_n K)u_n = P_n f. \tag{2.2}$$

To focus on the difficulty in analysing (2.2), recall that if $K : L_\infty \to C$ were compact, then $\|(I-K)-(I-P_nK)\|_\infty = \|(I-P_n)K\|_\infty \to 0$ as $n \to \infty$ (since pointwise convergence is uniform on compact sets). Hence, by the Banach perturbation lemma (applied in L_∞) and the well-posedness assumption (1.4)(i), $(I - P_nK)^{-1}$ exists for n sufficiently large and has uniform norm bounded as $n \to \infty$. In this case a unique collocation solution u_n exists, and $u - u_n = (I - P_nK)^{-1}(u - P_nu)$, from which we obtain the usual error estimate $\|u - u_n\|_\infty \leqslant C\|u - P_nu\|_\infty$.

As mentioned in Section 1, this argument fails for the noncompact operator K in (1.1), an observation which led in [10] to the introduction of the (slightly more general) *modified collocation method.* Here, for simplicity, we shall introduce this technique in the special case where the solution u of (1.3) satisfies $u(0) = 0$, although — as we indicate in Section 3 — the principle can be applied in the general case also. In its simplest form the modification technique involves choosing an integer $i* \geqslant 0$ and seeking $u_n \in S_n^{r,-1}$ satisfying $u_n \equiv 0$ on $[0, x_{i*}]$ and (instead of (2.2)) the modified collocation equations:

$$(I - K)u_n(x_{ij}) = f(x_{ij}), \quad j = 1,\dots r, \quad i = i* + 1,\dots,n.$$

In operator form this can be written

$$u_n \in S_n^{r,-1} : (I - P_nT^{x_{i*}}K)u_n = P_nT^{x_{i*}}f, \tag{2.3}$$

which is clearly equivalent to (1.3) when $i* = 0$.

To analyse (2.3), the first step is to recall the formal identity: $(I - T^{x_{i*}}K)^{-1} = I + T^{x_{i*}}(I - KT^{x_{i*}})^{-1}K$. Using this, together with the assumption (1.4)(ii) and the identity $\|T^\tau\|_\infty = 1$, it follows that, for fixed $i* \geqslant 0$, $\|(I - T^{x_{i*}}K)^{-1}\|_\infty$ is uniformly bounded as $n \to \infty$. Then, attempting to mimick the argument in the compact case, we can show that (2.3) is well-posed provided we show that $\|(I - T^{x_{i*}}K) - (I - P_nT^{x_{i*}}K)\|_\infty = \|(I - P_n)T^{x_{i*}}K\|_\infty$ is sufficiently small. Although this quantity does not approach zero as $n \to \infty$, we shall see in the next lemma that it can be made arbitrarily small independent of n by an appropriate choice of parameter $i*$.

Lemma 2.1. *There exists a constant C independent of n and $i*$ such that* $\|(I - P_n)T^{x_{i*}}K\|_\infty \leqslant C(1/i*)^r$.

Proof. Let $v \in L_\infty$. For $i > i*$, we have, using (1.5),

$$\begin{aligned}\|(I - P_n)Kv\|_{\infty,I_i} &\leqslant Ch_i^r\|D^rKv\|_{\infty,I_i} \leqslant Ch_i^r x_{i-1}^{-r}\|s^r(D^rKv)(s)\|_{\infty,I_i}\\ &\leqslant C(1/i*)^r\|s^r(D^rKv)(s)\|_{\infty,I_i}.\end{aligned} \tag{2.4}$$

Now, by assumption (2.1) the Mellin convolution operator s^rD^rK (with kernel $s^rD^r\kappa$) is bounded on L_∞, and (2.4) proves the lemma. □

From this we can prove the stability of (2.3) using the Banach lemma by taking $i*$ sufficiently large:

Theorem 2.2. *There exists $i* \geqslant 0$ such that for all n sufficiently large, the modified collocation equations* (2.3) *have a unique solution u_n and satisfy the error estimate $\|u - u_n\|_\infty \leqslant C\|u - P_n T^{x_{i*}} u\|_\infty$.*

If $u \in C^{r,\alpha}$ with $0 < \alpha \leqslant 1$ and $u(0) = 0$ then, as described in Section 1, the above error estimate implies convergence with optimal order $O(n^{-r})$ provided $q \geqslant r/\alpha$. Note the philosophy of the argument: Lemma 2.1 shows that there exists a modification parameter $i*$ (fixed with respect to n) which ensures stability. Then Theorem 2.2 shows that the resulting modified method converges optimally provided the mesh is appropriately graded. The choice of any fixed $i*$ for stability does not affect the rate of convergence as $n \to \infty$, although it does affect the asymptotic constant in the error estimate.

3. Further results for second-kind equations

We now give a survey of results on piecewise polynomial collocation methods and their iterated and discrete versions for Eq. (1.3). Using graded meshes and modified spline spaces as described in the previous section, it is possible to obtain stability (provided $(I - K)$ is well-posed) and the same optimal orders of convergence as in the case of second-kind equations with smooth kernels. We present here general convergence results in the space $L_p = L_p(0,1)$, $1 \leqslant p \leqslant \infty$, for which we need the following assumptions:

(A1) For all $k \geqslant 0$, $\int_0^\infty s^{1/p+k} |D^k \kappa(s)| \, ds/s < \infty$.
(A2) The symbol $1 - \tilde{\kappa}(z)$ does not vanish on $\mathrm{Re}(z) = 1/p$, and the winding number of this function with respect to the origin is equal to 0.
(A3) For some $1 \geqslant \alpha > -1/p$, $u \in C^{k,\alpha}$ for all k.

Note that (A1) (with $k = 0$) ensures that K is bounded on L_p and $\tilde{\kappa}(z)$ is a continuous function on $\mathrm{Re}(z) = 1/p$ vanishing at infinity. It turns out that (A2) is then equivalent to each of the conditions (i) and (ii) in (1.4) for the L_p norm. This follows from known results on Wiener–Hopf integral equations (see [38]). The assumption (A3) holds if the right-hand side f of (1.3) is (infinitely) smooth on $[0,1]$ and $(1 - \tilde{\kappa}(z))^{-1}$ is analytic in the strip $-\alpha \leqslant \mathrm{Re}(z) \leqslant 1/p$; see [20] for precise regularity results. We first consider the modified collocation method (2.3) again and extend Theorem 2.2 to the L_p case [19].

Theorem 3.1. *Assume the mesh $\{x_i\}$ is q-graded at 0, and suppose $i*$ is sufficiently large. Then the collocation method* (2.3) *is stable in L_p. It converges in L_p with optimal order $O(n^{-r})$ provided $q > r/(\alpha + 1/p)$ (although when $\alpha \in [0,1]$, the additional assumption $u(0) = 0$ is required for convergence).*

The proof is analogous to that of Theorem 2.2. Note that the crucial estimate (2.4) (for the L_p norm) follows from the boundedness of the operators $s^r D^r K$ (ensured by (A1)) and the standard *local approximation property* of $S_n^{r,-1}$, i.e., $\|(I - P_n)v\|_{p,I_i} \leqslant C h_i^r \|D^r v\|_{p,I_i}$, with C independent of i, n and v (see, e.g., [46]). To obtain consistency of the method in the case $\alpha \in [0,1], u(0) \neq 0$, more general modifications of the spline spaces instead of the simple cut-off by zero on $[0, x_{i*}]$ should be used; see [10] for a version including the piecewise constants on the first $i*$ subintervals and [20] for a method based on splines from $S_n^{r,0}$ which reduce to a (global) constant on $[0, x_{i*}]$. However, in general the stability of these methods cannot be obtained from (1.4)(ii) by small perturbation. To get around this problem, either an additional condition on the norm of K should be imposed [10] or another approach based on Wiener–Hopf factorization can be employed [20].

As mentioned in the introduction, in several important classical second kind boundary integral equations on corner domains (which have localisation of form (1.3) — see Section 4), the operator to be approximated turns out to be strongly elliptic and the Galerkin method is stable without modification in the energy norm. If one wants to prove convergence in other norms (e.g., the uniform norm) more delicate analyses are needed (e.g., [6,7,9]). More generally, for problem (1.3) under the assumptions (A1)–(A3) above, we must again consider modifications in order to prove stability even for the Galerkin method (see [18,20]). Indeed the unmodified method is in general unstable for operators satisfying only (A2) [20].

The collocation method is of practical interest because its implementation requires less numerical integration than the Galerkin method. However, even the collocation method generally requires quadrature for its implementation, and this should be included in an error anaylsis. Thus we now discuss a fully discrete version of the collocation method (2.3), which also turns out to be closely related to the classical Nyström method. To define this method, introduce an r point interpolatory quadrature rule on $[0,1]$: $\int_0^1 v \cong \sum_{j=1}^r \omega_j v(\xi_j)$ with weights ω_j and points $0 \leqslant \xi_1 < \cdots < \xi_r \leqslant 1$. Let R be the *order* of this rule so that $R \geqslant r$ and $R = 2r$ if and only if ξ_j are the r Gauss–Legendre points on $[0,1]$. Define x_{ij} as in Section 2 and set $Q = \{(i,j)\colon i = i*+1,\dots,n;\ j = 1,\dots,r\}$. Then the (modified) *composite quadrature rule* obtained by shifting the above rule on $[0,1]$ to each I_i, and summing over $i > i*$, is $\int_0^1 v \cong \sum_Q \omega_j v(x_{ij}) h_i$. The integral operator K in (1.3) will be approximated by

$$K_n v(s) = \sum_Q \omega_j \kappa(s/x_{ij}) v(x_{ij}) h_i / x_{ij}. \tag{3.1}$$

The (modified) *discrete collocation method* for (1.3) seeks an approximate solution $u_n \in S_n^{r,-1}$ satisfying $u_n \equiv 0$ on $[0, x_{i*}]$ such that

$$(I - P_n T^{x_{i*}} K_n) u_n = P_n T^{x_{i*}} f, \tag{3.2}$$

where P_n is the interpolatory projection defined in Section 2. The *Nyström* (or *discrete iterated collocation*) solution $u_n^*(s)$ to (1.3) is then defined by $u_n^* = f + K_n u_n$, and it satisfies

$$(I - K_n) u_n^* = f; \tag{3.3}$$

note that $P_n T^{x_{i*}} u_n^* = u_n$. By collocation at $s = x_{ij}$, $(i,j) \in Q$, (3.3) is reduced to the linear system (3.2) for $u_n^*(x_{ij}) = u_n(x_{ij})$, the solution of which in turn gives $u_n^*(s)$ for all $s \in [0,1]$. The following result extends Theorem 3.1 to the discrete collocation method (3.2) and establishes *superconvergence* for the Nyström method (3.3).

Theorem 3.2. *Under the assumptions of the preceding theorem, the method* (3.2) *is stable and optimally convergent in* L_p. *Moreover, if the grading exponent satisfies the* (*possibly stronger*) *requirement* $q > R/(\alpha + 1/p)$, *then the Nyström solution converges with the error bound* $\|u - u_n^*\|_p = O(n^{-R})$ *as* $n \to \infty$.

For details of the proof of Theorem 3.2, we refer to [19,27]. To give a brief overview of the proof, we remark that the stability of (3.2) can be obtained from that of the collocation method by small perturbation in the operator norm, as described in [19]. It is also possible to approach (3.3) directly in the case $p=\infty$. In [27] it is shown that the operator K_n defined in (3.1) is uniformly bounded on C. This allows a more straightforward approach to stability by regarding $I-K_n$ as a small perturbation of the finite section operator $I-KT^{x_{i*}}$. The error bound for the Nyström solution follows from the estimate $\|u-u_n^*\|_p \leqslant C\|(K-K_n)u\|_p$, where the last term is of order $O(n^{-R})$, provided that $u\in C^{R,\alpha}$, for $0\leqslant\alpha\leqslant 1$, $u(0)=0$, and the grading exponent satisfies $q\geqslant R/\alpha$.

For the model problem (1.3) it is simple to extend all the above methods to the case when $u(0)\neq 0$. Using (1.3) together with (1.2), it follows that $(1-\tilde{\kappa}(0))u(0)=f(0)$. Then it is easy to see that the function $v:=u-u(0)$ satisfies $v(0)=0$ and can be computed by solving (1.3) with the modified right-hand side $f(s)-f(0)(1-K1(s))/(1-\tilde{\kappa}(0))$ (where 1 is the unit function on $[0,1]$). In more general situations (such as the second-kind boundary integral equations described in Section 4), Eq. (1.3) appears only as a localised model problem in a coupled system and in this context it is not possible to compute $u(0)$ explicitly. Nevertheless, stable and consistent methods can be constructed by considering appropriate extended systems [27].

All the results mentioned in this section can be generalised to systems of equations of the form (1.3). In particular, the stability of the methods can be again obtained from the stability of the finite section operators by small perturbation. However, for matrix operators, condition (1.4)(ii) is no longer equivalent to the well-posedness of $(I-K)$ and requires the invertibility of an additional Mellin convolution operator; see [38] for a discussion of this in the case of Wiener–Hopf operators. Fortunately, there is an important special case where (1.4)(ii) is always satisfied in the L_2 norm, namely the case of a strongly elliptic (matrix) symbol, i.e., $\mathrm{Re}(I-\tilde{\kappa}(z))$ is uniformly positive definite for $\mathrm{Re}(z)=\frac{1}{2}$. Together with (A1) and (A3) (for $p=2$), this implies stability and optimal convergence for the modified collocation and quadrature methods, whereas Galerkin's method is of course stable with $i*=0$. We indicate an important application of this technique in Section 5.

Finally we want to emphasise that the simple perturbation argument presented in Lemma 2.1 is restricted to the case of continuous symbols. The stability analysis of more general classes of convolution operators (containing singular integral operators of Cauchy type for example) requires more sophisticated methods combining Mellin and local Banach algebra techniques; see, e.g., [39,38,14,29].

4. Boundary integral equations on corner domains

Boundary value problems for linear elliptic PDE's can be reduced to boundary integral equations through the use of a fundamental solution. For Laplace's equation in 2D this is the function $G(\boldsymbol{x},\boldsymbol{y})=(2\pi)^{-1}\log|\boldsymbol{x}-\boldsymbol{y}|^{-1}$. If U satisfies Laplace's equation in a bounded polygonal domain Ω with boundary Γ then the Cauchy data $u:=U|_\Gamma$ and $v:=\partial_n U|_\Gamma$ satisfy *Green's identity*

$$\mathscr{V}v(\boldsymbol{x})-\mathscr{W}u(\boldsymbol{x})=-(\tfrac{1}{2})u(\boldsymbol{x}),\quad \boldsymbol{x}\in\Gamma,$$

for all smooth points $\boldsymbol{x}$ of Γ, where $\mathscr{V}v(\boldsymbol{x})=\int_\Gamma G(\boldsymbol{x},\boldsymbol{y})v(\boldsymbol{y})\,\mathrm{d}\Gamma(\boldsymbol{y})$ is the single layer potential, $\mathscr{W}u(\boldsymbol{x})=\int_\Gamma \partial_{n(\boldsymbol{y})}G(\boldsymbol{x},\boldsymbol{y})u(\boldsymbol{y})\,\mathrm{d}\Gamma(\boldsymbol{y})$ is the double layer potential, and ∂_n denotes differentiation in the outward normal direction from Ω. This identity can be extended to all $\boldsymbol{x}\in\Gamma$ by taking appropriate limits. An analogous relation holds for exterior problems. For the Neumann problem, with v given, we have to solve the second-kind equation

$$u(\boldsymbol{x})-2\mathscr{W}u(\boldsymbol{x})=g(\boldsymbol{x}):=-2\mathscr{V}u(\boldsymbol{x}) \tag{4.1}$$

for the Dirichlet data u. For the Dirichlet problem with u given, we have to solve the first-kind equation

$$2\mathscr{V}v(\boldsymbol{x}) = g(\boldsymbol{x}) := -u(\boldsymbol{x}) + 2\mathscr{W}u(\boldsymbol{x}), \tag{4.2}$$

for the Neumann data, and for mixed Dirichlet–Neumann conditions a first–second kind system arises. Analogous equations arise from the classical indirect boundary integral method [2]. A rigorous justification of the underlying potential theory in nonsmooth domains can be found in [12]. The method is of course applicable to much more general PDEs (e.g., [50]).

To see how the model problem (1.3) arises from these applications, consider the case that Γ is (infinitely) smooth with the exception of a corner, without loss of generality situated at the origin $\mathbf{0}$. We further assume that Γ in the neighbourhood of $\mathbf{0}$ consists of two straight lines intersecting with an interior angle $(1-\chi)\pi$, $0<|\chi|<1$. Consider a parametrisation $\gamma(s) : [-\pi,\pi] \to \Gamma$, $|\gamma'(s)| > 0$ for $s \in [\pi,\pi]$, which near $s=0$ may be given by

$$\gamma(s) = \begin{cases} (-\cos\chi\pi, \sin\chi\pi)|s|, & s \in [-\varepsilon, 0], \\ (1,0)|s|, & s \in [0,\varepsilon]. \end{cases} \tag{4.3}$$

Considering first of all the relatively straightforward case (4.1), let ψ be a smooth function on Γ with $\psi(\boldsymbol{x}) \equiv 1$ when $|\boldsymbol{x}| \leqslant \varepsilon/2$ and $\psi(\boldsymbol{x}) \equiv 0$ when $|\boldsymbol{x}| > \varepsilon$ and observe that $2\mathscr{W} - \psi 2\mathscr{W}\psi$ is an operator with smooth kernel. The behaviour of (4.1) is thus dominated by the localised operator, $I - \psi 2\mathscr{W}\psi$. A short calculation shows that

$$(\psi 2\mathscr{W}\psi)u(\gamma(s)) = \begin{cases} \displaystyle\int_0^{\varepsilon} \kappa\left(\frac{s}{\sigma}\right) u(\gamma(\sigma))\frac{\mathrm{d}\sigma}{\sigma}, & s \in [-\varepsilon, 0], \\ \displaystyle -\int_{-\varepsilon}^{0} \kappa\left(\frac{s}{\sigma}\right) u(\gamma(\sigma))\frac{\mathrm{d}\sigma}{\sigma}, & s \in [0,\varepsilon], \end{cases}$$

where

$$\kappa(s) := \frac{\sin\chi\pi}{\pi}\left\{\frac{s}{1-2s\cos\chi\pi + s^2}\right\}.$$

Thus $\psi 2\mathscr{W}\psi$ corresponds to a matrix of operators of form (1.1) and the analysis of (1.3) is the key to understanding (4.1). The above argument can be extended to the case of many corners in an obvious way.

Now let us turn to the first-kind Eq. (4.2). The connection to (1.3) here is much less obvious. With the parametrisation $\gamma : [-\pi,\pi] \to \Gamma$ introduced above, we can write

$$(2\mathscr{V}u)(\gamma(s)) = \frac{1}{\pi}\int_{-\pi}^{\pi} \log\frac{1}{|\gamma(s)-\gamma(\sigma)|} w(\sigma)\,\mathrm{d}\sigma =: Vw(s), \tag{4.4}$$

where $w(\sigma) = u(\gamma(\sigma))|\gamma'(\sigma)|$. (Note that here we take the Jacobian into the unknown. As indicated in Section 1, this is useful when nonlinear parametrisations are used to treat corner singularities.) In the theory of V a special role is played by the operator

$$Aw(s) := \frac{1}{\pi}\int_{-\pi}^{\pi} \log\frac{1}{|2\sin(s-\sigma)/2|} w(\sigma)\,\mathrm{d}\sigma + Jw, \tag{4.5}$$

where $Jw = (1/2\pi)\int_{-\pi}^{\pi} w(\sigma)\,\mathrm{d}\sigma$. The first term in the expression for A is simply the operator V restricted to the unit circle $\gamma(s) = (\cos s, \sin s)$. The additional compact perturbation J is added to make A invertible with the result that A is an isometry from H^k onto H^{k+1} for all k (where H^k denotes the usual 2π—periodic Sobolev space of order k). It is a special feature of V that it can in some sense be conveniently regularised by the operator A^{-1}. More precisely, we can write

$$A^{-1}Vw = A^{-1}(A + (V - A))v =: (I + M)v, \tag{4.6}$$

where $M = A^{-1}(V - A)$ and

$$(V - A)w(s) = \frac{1}{\pi}\int_\Gamma \log\frac{|2\sin(s-\sigma)/2|}{|\gamma(s) - \gamma(\sigma)|} w(\sigma)\,\mathrm{d}\sigma - Jw.$$

When Γ is smooth (e.g., C^∞), the kernel of the first term in $V - A$ has a removable singularity and it can be shown [53] that the operator $V - A$ maps L_2 to H^k for all $k \geqslant 0$ and hence that M is compact from L_2 to H^k for all k. Thus in the smooth case the first-kind Eq. (4.2) is equivalent to the nonstandard second-kind equation

$$(I + M)w = f := A^{-1}g. \tag{4.7}$$

When Γ is polygonal the regularization (4.6) can still be carried out, but the resulting operator M is no longer compact. In fact, local to each corner of Γ, M turns out to be composed of Mellin convolution operators of the form (1.1). To see this we need some more details about the operator A. We have the well-known relations (see e.g., [38]) $DA = H$ and $A^{-1} = -HD + J$, where H is the 2π-periodic Hilbert transform $Hv(s) = -(2\pi)^{-1}\int_{-\pi}^{\pi}\cot((s-\sigma)/2)v(\sigma)\,\mathrm{d}\sigma$ (with the integral to be interpreted in the Cauchy principle value sense) and D is the 2π-periodic differentiation operator. Hence, the essential behaviour of M near each corner can be found by studying $HD(V - A)$. To compute this, we observe that

$$DVw(s) = -\frac{1}{\pi}\int_{-\pi}^{\pi}\frac{(\gamma(s)-\gamma(\sigma))\cdot\gamma'(s)}{|\gamma(s)-\gamma(\sigma)|^2}w(\sigma)\,\mathrm{d}\sigma.$$

For w locally supported near $\mathbf{0}$, i.e., supp $w \subset [-\varepsilon, \varepsilon]$, we have the representation

$$DVw(s) = \begin{cases} -\dfrac{1}{\pi}\displaystyle\int_{-\varepsilon}^{0}\frac{w(\sigma)}{s-\sigma}\,\mathrm{d}\sigma + \int_0^{\varepsilon}\kappa_1\left(\frac{s}{\sigma}\right)w(\sigma)\frac{\mathrm{d}\sigma}{\sigma}, & s \in [-\varepsilon, 0] \\ \displaystyle\int_{-\varepsilon}^{0}\kappa_1\left(\frac{s}{\sigma}\right)w(\sigma)\frac{\mathrm{d}\sigma}{\sigma} - \frac{1}{\pi}\int_0^{\varepsilon}\frac{w(\sigma)}{s-\sigma}\,\mathrm{d}\sigma, & s \in [0, \varepsilon] \end{cases}$$

where

$$\kappa_1(s) = \frac{1}{\pi}\left\{\frac{\cos\chi\pi - s}{1 - 2s\cos\chi\pi + s^2}\right\}.$$

This calculation, which shows that $D(V - A) = DV - H$ can be represented as a matrix of operators of the form (1.1), was first given in [53] and shows that $A^{-1}(V - A)$ is represented (local to each corner) as a product of H with operators of the form (1.1). From this a numerical analysis of collocation methods followed [28,52]. However, this analysis was somewhat restricted, mainly because $M = A^{-1}(V - A) = -HD(V - A)$ (modulo compact operators) and although the operator $D(V - A)$ was well-understood (as above) the important product $HD(V - A)$ was not. In [21] this product was computed using the symbolic calculus for Mellin operators. This is possible since (local to the corner $\mathbf{0}$) we can write (modulo a compact operator)

$$Hv(s) = \int_{-\varepsilon}^{\varepsilon}\frac{1}{s-\sigma}v(\sigma)\,\mathrm{d}\sigma = \int_{-\varepsilon}^{\varepsilon}(s/\sigma - 1)^{-1}v(\sigma)\frac{\mathrm{d}\sigma}{\sigma}, \quad s \in [-\varepsilon, \varepsilon]$$

which can be also treated using the Mellin transform. In fact in [21] more general results than this were obtained. Following the "parametrisation" method for handling singularities outlined in Section 1, [21] considered parametrisations of Γ which varied more slowly than arc length near each corner. An example is to parametrise Γ near $\mathbf{0}$ by replacing $|s|$ with $|s|^q$ in (4.3). The above calculation first of V in (4.4) (which now depends on q) and of $A^{-1}(V-A)$ can again be performed and yields again a representation of M near each corner involving operators of the form (1.1).

In this section we have shown that model problem (1.3) arises in both standard and nonstandard ways from localisations of boundary integral equations on nonsmooth domains. (Here we have restricted to the Laplace equation but similar local problems arise, for example, from the Helmholtz equation [11] and in linear elasticity [24].) For the classical second-kind boundary integral equations (such as (4.1)) on polygonal domains it is possible to give a complete error analysis of (modified) methods, using the knowledge of numerical methods for the local model problem (1.3) outlined in Sections 2, 3 — see, for example [27]. However, for first kind equations such as (4.2) which are connected to the model problem (1.3) in less standard way, the numerical analysis is more complicated. In the final section we give a brief survey of this area with pointers to the literature where the reader can find more details.

5. Results for first-kind equations

We first discuss the numerical solution of Symm's integral equation (4.2) on polygonal domains by high-order spline collocation methods. To approximate the singularities of solutions at the corner points, the first idea that comes to one's mind is to attack this equation directly by using splines on graded meshes as in the case of the double-layer potential equation (4.1). This approach was taken in [13] where stability and optimal convergence rates for piecewise linear break-point collocation were proved with respect to a weighted Sobolev norm. So far these results have not been generalized to higher-order splines.

On the other hand, if Γ is smooth then the operator in (4.2) is a classical periodic pseudodifferential operator, and thus the full force of the general convergence theory developed in [1,44,45] for collocation methods with smooth splines (mostly) on uniform grids becomes available; see also the review in [48] for these and related methods and the detailed presentation in [38]. Although the piecewise constant mid-point collocation method was shown to converge for quite general meshes in [8], this analysis is restricted to smooth boundaries and there is still no general convergence analysis for (4.2) for general boundaries and general piecewise polynomial approximation schemes.

This situation essentially motivated the approach in [21] where the use of a nonlinear parametrisation (or mesh grading transformation) of the boundary curve together with a uniform mesh has allowed a first stability and convergence analysis of high-order collocation methods in the presence of corners.

To illustrate this type of result, we retain the notation of the preceding section and parametrise the boundary Γ with one corner at $\mathbf{0}$ by $\gamma : [-\pi,\pi] \to \Gamma$ such that $\gamma(0)=\mathbf{0}$, and, near $s=0$,

$$\gamma(s) = \begin{cases} (-\cos\chi\pi,\ \sin\chi\pi)|s|^q, & s\in[-\varepsilon,0], \\ (1,0)|s|^q, & s\in[0,\varepsilon]. \end{cases} \tag{5.1}$$

Here the grading exponent q is an integer $\geqslant 1$. The Eq. (4.2) transforms to

$$Vw(s) = g(s), \quad s \in [-\pi, \pi], \tag{5.2}$$

where V and w are defined as in (4.4) (but using the nonlinear parametrisation (5.1)), and $g(s) := g(\gamma(s))$. By appropriate choice of q, the solution w of (5.2) can be made smooth local to the corner (provided g is smooth), and hence w can be optimally approximated using splines from $S^r_{n,p}$ (the 2π-periodic smoothest splines of degree $r-1$ on the uniform mesh $x_i = ih$, $i = 0, \ldots, n$, with meshsize $h = 2\pi/n$). To discretise (5.2), introduce the interpolant $Q_n v \in S^r_{n,p}$ by requiring
(a) when r is odd $Q_n v(t_i) = v(t_i)$, $i = 1, \ldots, n$,
(b) when r is even $Q_n v(x_i) = v(x_i)$, $i = 0, \ldots, n-1$,
where t_i are the mid-points of subintervals. Then the collocation method for (5.2) seeks $w_n \in S^r_{n,p}$ such that

$$Q_n V w_n = Q_n g. \tag{5.3}$$

The approach to the analysis of (5.3) is analogous to that used in Section 4 where (4.2) is transformed to the nonstandard second kind Eq. (4.7). In fact (5.3) can be rewritten as a nonstandard projection method for (4.7) as follows. For any $v \in H^0$, let $P_n v \in S^r_{n,p}$ solve the collocation equations $Q_n A(P_n v) = Q_n A v$ for the circle operator A defined in (4.5). It is well-known (see [38, pp. 492–493] and the references listed at the beginning of this section) that this prescription defines a (uniformly) bounded projection operator $P_n : H^0 \to S^r_{n,p}$. It is then straightforward to see that (5.3) is equivalent to

$$(I + P_n M) w_n = P_n f, \quad \text{with } M = A^{-1}(V - A), \quad f = A^{-1} g. \tag{5.4}$$

To overcome the difficulty in the stability analysis of (5.3), or equivalently (5.4), one may introduce an analogous cut-off procedure in the vicinity of the corner as in the case of the model second-kind Eq. (1.3). To describe the modification, introduce the truncation $T^\tau v$ as $T^\tau v(s) = 0$, for $|s| < \tau$, and $T^\tau v(s) = v(s)$ for $\tau < |s| < \pi$. Then, for any fixed $i* \geqslant 0$, consider the method

$$Q_n(A + (V - A) T^{i*h}) w_n = Q_n g, \tag{5.5}$$

which coincides with (5.3) when $i* = 0$. By mimicking the derivation of (5.4) from (5.3), it is easily seen that (5.5) is equivalent to

$$(I + P_n M T^{i*h}) w_n = P_n f. \tag{5.6}$$

Applying the technique outlined in Sections 2 and 3 to the projection method (5.6) and employing the (nontrivial) Mellin analysis of the operator M discussed in the previous section, one then can prove the following convergence result for the modified collocation method [21].

Theorem 5.1. *Suppose the grading exponent satisfies $q > (r + \frac{1}{2})(1 + |\chi|)$, where $(1 - \chi)\pi$ is the interior angle at the corner. Then there exists $i*$ such that (5.5) has a unique solution for all n sufficiently large and is optimally convergent in the L_2 norm, i.e., $\|w - w_n\|_2 = O(n^{-r})$ as $n \to \infty$.*

A crucial prerequisite for this result is the strong ellipticity of the second-kind operator $I + M$, i.e., $\mathrm{Re}(I + M)$ is positive definite in $H^0 = L_2$, modulo compact operators. Together with a uniqueness result for the transformed integral equation (5.2), this implies the analogue of (1.4) in this setting, i.e., the well-posedness of $I + M$ and the stability of the finite section operators $I + MT^{i*h}$

(as $h \to 0$) in H^0. The final step in the stability proof for (5.6) is again a perturbation argument similar to that of Lemma 2.1, which however, requires a thorough study of the Mellin convolution kernel of the operator M localised to the corner (see [21], with improvements given in [23]). The optimal error estimate then follows from standard spline approximation results since, as it was also shown in [21], the solution of (5.2) satisfies $w \in H^r$ and has appropriate decay as $s \to 0$ provided the grading exponent q is sufficiently large.

The above stability and convergence results may be extended to various related parametrisation methods and to other first-kind equations on polygonal boundaries. In [23] it was shown that Theorem 5.1 remains true when the collocation integrals are approximated using singularity subtraction and a suitable composite quadrature rule. A fully discrete trigonometric collocation method is given in [26]. This method is based on the trapezoidal rule and is easier to implement than the quadrature–collocation scheme of [23]. More general results on discrete qualocation methods can be found in [31]. Parametrisation methods based on global algebraic polynomials have recently been applied to Symm's equation [35] and to the generalized airfoil equation for an airfoil with a flap [36,47]. A convergence analysis of the trigonometric collocation method applied to mixed boundary value problems on corner domains is presented in [25].

In conclusion we remark that the numerical analysis of these 2D corner problems is still not as satisfactory as in the case of smooth boundaries where even fully discrete high-order methods of almost linear computational complexity are known. However, fast solution methods for classical first-kind integral equations on open arcs have recently been obtained applying the cosine transform and discrete trigonometric collocation (see [43] and the references therein). The development of analogous methods for more general problems on corner domains remains a challenge for the future. Moreover, a stability theory of such methods which avoids the use of any modification technique appears to be another intriguing open problem. At present this is even not known in the relatively straightforward model case of global (algebraic or trigonometric) polynomial collocation methods applied to second-kind Mellin convolution equations.

References

[1] D.N. Arnold, W.L. Wendland, The convergence of spline collocation for strongly elliptic equations on curves, Numer. Math. 47 (1985) 317–341.

[2] K.E. Atkinson, The Numerical Solution of Integral Equations of the Second Kind, Cambridge University Press, Cambridge, 1997.

[3] K.E. Atkinson, F. De Hoog, The numerical solution of Laplace's equation on a wedge, IMA J. Numer. Anal. 14 (1984) 19–41.

[4] K.E. Atkinson, I.G. Graham, Iterative solution of linear systems arising from the boundary integral method, SIAM J. Sci. Statist. Comput. 13 (1992) 694–722.

[5] K. Bühring, A quadrature method for the singular integral equation on curves with corner points, Math. Nachr. 167 (1994) 43–81.

[6] G.A. Chandler, Galerkin's method for boundary integral equations on polygonal domains, J. Austral. Math. Soc. Ser. B 26 (1984) 1–13.

[7] G.A. Chandler, Superconvergent approximations to the solution of a boundary integral equation, SIAM J. Numer. Anal. 23 (1986) 1214–1229.

[8] G.A. Chandler, Optimal order convergence of midpoint collocation for first kind equations, Preprint, University of Queensland, 1991.

[9] G.A. Chandler, I.G. Graham, Uniform convergence of Galerkin solutions to non-compact integral operator equations, IMA J. Numer. Anal. 7 (1987) 327–334.
[10] G.A. Chandler, I.G. Graham, Product integration – collocation methods for noncompact integral operator equations, Math. Comp. 50 (1988) 125–138.
[11] K. Chen, S. Amini, Numerical analysis of boundary integral solution of the Helmholtz equation in domains with nonsmooth boundaries, IMA J. Numer. Anal. 13 (1993) 43–66.
[12] M. Costabel, E.P. Stephan, Boundary integral equations for mixed boundary value problems in polygonal domains and Galerkin approximation, in: W. Fiszden, K. Wilmanski, (Eds.), Banach Centre Publications, Vol. 15, PWN-Polish Scientific Publishers, Warsaw, 1985, pp. 175–251.
[13] M. Costabel, E.P. Stephan, On the convergence of collocation methods for boundary integral equations on polygons, Math. Comp. 49 (1987) 467–478.
[14] V.D. Didenko, S. Roch, B. Silbermann, Approximation methods for singular integral equations with conjugation on curves with corners, SIAM J. Numer. Anal. 32 (1995) 1910–1939.
[15] V.D. Didenko, B. Silbermann, An approximate method on non-equidistant partitions for double layer potential equation, Appl. Numer. Math. 26 (1998) 41–48.
[16] D. Elliott, S. Prössdorf, An algorithm for the approximate solution of integral equations of Mellin type, Numer. Math. 70 (1995) 427–452.
[17] J. Elschner, On spline approximation for a class of integral equations I: Galerkin and collocation methods with piecewise polynomials, Math. Methods Appl. Sci. 10 (1985) 543–559.
[18] J. Elschner, On spline approximation for a class of integral equations. II: Galerkin's method with smooth splines, Math. Nachr. 140 (1989) 273–283.
[19] J. Elschner, On spline collocation for convolution equations, Integral Equations Operator Theory 12 (1989) 486–510.
[20] J. Elschner, On spline approximation for a class of non-compact integral equations, Math. Nachr. 146 (1990) 271–321.
[21] J. Elschner, I.G. Graham, An optimal order collocation method for first kind boundary integral equations on polygons, Numer. Math. 70 (1995) 1–31.
[22] J. Elschner, I.G. Graham, Parametrization Methods for First Kind Integral Equations on Non-smooth Boundaries, Lecture Notes in Pure and Applied Mathematics, Vol. 167, Marcel Dekker, New York, 1995, pp. 81–99.
[23] J. Elschner, I.G. Graham, Quadrature methods for Symm's integral equation on polygons, IMA J. Numer. Anal. 17 (1997) 643–664.
[24] J. Elschner, O. Hansen, A collocation method for the solution of the first boundary value problem of elasticity in a polygonal domain in R^2, J. Integral Equations Appl. 11 (1999) 141–196.
[25] J. Elschner, Y. Jeon, I.H. Sloan, E.P. Stephan, The collocation method for mixed boundary value problems on domains with curved polygonal boundaries, Numer. Math. 76 (1997) 547–571.
[26] J. Elschner, E.P. Stephan, A discrete collocation method for Symm's integral equation on curves with corners, J. Comput. Appl. Math. 75 (1996) 131–146.
[27] I.G. Graham, G.A. Chandler, High order methods for linear functionals of solutions of second kind integral equations, SIAM J. Numer. Anal. 25 (1988) 1118–1137.
[28] I.G. Graham, Y. Yan, Piecewise constant collocation for first kind boundary integral equations, J. Austr. Math. Soc. Ser. B 33 (1991) 39–64.
[29] R. Hagen, S. Roch, B. Silbermann, Spectral Theory of Approximation Methods for Convolution Equations, Birkhäuser, Basel, 1995.
[30] Y. Jeon, A Nyström method for boundary integral equations in domains with a piecewise smooth boundary, J. Integral Equations Appl. 5 (1993) 221–242.
[31] Y. Jeon, I.H. Sloan, E.P. Stephan, J. Elschner, Discrete qualocation methods for logarithmic kernel integral equations on a piecewise smooth boundary, Adv. Comput. Math. 7 (1997) 547–571.
[32] R. Kress, A Nyström method for boundary integral equations in domains with corners, Numer. Math. 58 (1990) 145–161.
[33] P. Laubin, High order convergence for collocation of second kind boundary integral equations on polygons, Numer. Math. 79 (1998) 107–140.
[34] P. Laubin, M. Baiwir, Spline collocation for a boundary integral equation on polygons with cuts, SIAM J. Numer. Anal. 35 (1998) 1452–1472.

[35] G. Monegato, L. Scuderi, Global polynomial approximation for Symm's equation on polygons, Numer. Math., to appear.

[36] G. Monegato, I.H. Sloan, Numerical solution of the generalized airfoil equation for an airfoil with a flap, SIAM J. Numer. Anal. 34 (1997) 2288–2305.

[37] S. Prössdorf, A. Rathsfeld, Quadrature methods for strongly elliptic Cauchy singular integral equations on an interval, in: H. Dym, et al. (Eds.), Operator Theory: Advances and Applications, Vol. 41, Birkhäuser, Basel, 1989, pp. 435–471.

[38] S. Prössdorf, B. Silbermann, Numerical Analysis for Integral and Related Operator Equations, Akademie-Verlag, Berlin, 1991.

[39] A. Rathsfeld, Eine Quadraturformelmethode für Mellin–Operatoren nullter Ordnung, Math. Nachr. 137 (1988) 321–354.

[40] A. Rathsfeld, Iterative solution of linear systems arising from the Nyström method for the double-layer potential equation over curves with corners, Math. Methods Appl. Sci. 16 (1993) 443–455.

[41] A. Rathsfeld, Error estimates and extrapolation for the numerical solution of Mellin convolution equations, IMA J. Numer. Anal. 16 (1996) 217–255.

[42] J.R. Rice, On the degree of convergence of nonlinear spline approximation, in: I.J. Schoenberg (Ed.), Approximation with Special Emphasis on Spline Functions, Academic Press, New York, 1969, pp. 349–365.

[43] J. Saranen, G. Vainikko, Fast collocation solvers for integral equations on open arcs, J. Integral Equations Appl. 11 (1999) 57–102.

[44] J. Saranen, W.L. Wendland, On the asymptotic convergence of collocation methods with spline functions of even degree, Math. Comp. 45 (1985) 91–108.

[45] G. Schmidt, On spline collocation methods for boundary integral equations in the plane, Math. Methods Appl. Sci. 7 (1985) 74–89.

[46] L.L. Schumaker, Spline Functions. Basic Theory, Wiley, New York, 1981.

[47] L. Scuderi, A collocation method for the generalized airfoil equation for an airfoil with a flap, SIAM J. Numer. Anal. 35 (1998) 1725–1739.

[48] I.H. Sloan, Error analysis for boundary integral methods, in: A. Iserles (Ed.), Acta Numerica, Vol. 1, Cambridge University Press, Cambridge, 1992, pp. 287–339.

[49] W.L. Wendland, Die Behandlung von Randwertaufgaben in $\mathbb{R}^3$ mit Hilfe von Einfach- und Doppelschichtpotentialen, Numer. Math. 11 (1968) 380–404.

[50] W.L. Wendland, On some mathematical aspects of boundary element methods for elliptic problems, in: J.R. Whiteman (Ed.), MAFELAP V, Academic Press, New York, 1985, pp. 193–227.

[51] W.L. Wendland, E.P. Stephan, G.C. Hsiao, On the integral equation method for the plane mixed boundary value problem of the Laplacian, Math. Methods Appl. Sci. 1 (1979) 265–321.

[52] Y. Yan, The collocation method for first kind boundary integral equations on polygonal domains, Math. Comp. 54 (1990) 139–154.

[53] Y. Yan, I.H. Sloan, On integral equations of the first kind with logarithmic kernels, J. Integral Equations Appl. 1 (1988) 549–579.

ELSEVIER
Journal of Computational and Applied Mathematics 125 (2000) 439–460

JOURNAL OF COMPUTATIONAL AND APPLIED MATHEMATICS

www.elsevier.nl/locate/cam

Quadrature methods for 2D and 3D problems ☆

A. Rathsfeld

Weierstraß-Institut für Angewandte Analysis und Stochastik, Mohrenstr. 39, D-10117 Berlin, Germany

Received 22 June 1999

Abstract

In this paper we give an overview on well-known stability and convergence results for simple quadrature methods based on low-order composite quadrature rules and applied to the numerical solution of integral equations over smooth manifolds. First, we explain the methods for the case of second-kind equations. Then we discuss what is known for the analysis of pseudodifferential equations. We explain why these simple methods are not recommended for integral equations over domains with dimension higher than one. Finally, for the solution of a two-dimensional singular integral equation, we prove a new result on a quadrature method based on product rules. © 2000 Elsevier Science B.V. All rights reserved.

1. Introduction

A major task in numerical analysis is to provide methods for the solution of integral equations. For instance, the popular boundary element method consists in transforming a boundary value problem for a partial differential equation into an equivalent boundary integral equation and in solving this boundary equation numerically. Usually, collocation or Galerkin schemes are applied for the discretization of integral equations. If no analytic formulas for the integrals appearing in the discretized matrix operators are known, then, in a further discretization step, the integrals are to be replaced by quadrature formulas. Therefore, methods like Galerkin's, collocation or qualocation are called semi-discretization schemes. To get efficient numerical methods, the question arises how to choose optimal quadrature rules. This essential question is discussed in a lot of papers in the engineering literature, and mathematicians have analyzed and systematized these quadrature algorithms (cf., e.g., [2,10,19–22,26,28,40,41,50,53,56,59]).

However, right from the start (cf., e.g., [39]) fully discrete schemes have been proposed. Applying these so-called quadrature methods, the integrals in the original integral equation are directly replaced by a quadrature rule. The entries of the resulting linear system can be expressed as linear combinations of kernel function values with quadrature weights as coefficients. The advantage of quadrature

☆ Supported by Deutsche Forschungsgemeinschaft under grant numbers PR 336/5-1 and Pr 336/5-2.

PII: S0377-0427(00)00484-2

methods is that they require less time for writing codes and a little bit less time for computation. On the other hand, as a rule of thumb, the approximation errors of quadrature methods are a little bit larger than those of Galerkin or collocation schemes. Especially, the errors measured in negative Sobolev norms may be essentially larger than those for Galerkin methods. However, there are cases when quadrature methods can compete with the accuracy of other schemes. Quadrature methods can be recommended for univariate integral equations of the second kind with smooth kernels. For univariate equations of the first kind and nonsmooth kernels, quadrature methods often require modifications, and their analysis is much more involved. Note that first kind equations with smooth kernels are ill-posed, and the methods of their regularization will not be discussed in this paper. In case of higher-dimensional equations, the simple quadrature methods can be recommended only for second-kind equations with smooth kernels. For the general case, more complicated quadrature methods like methods based on product integration are needed. The latter, however, are very close to Galerkin or collocation methods with quadrature approximated entries in the stiffness matrix. Note that, in general, there is no big difference between a quadrature method and a collocation scheme combined with an efficient quadrature algorithm. Only the "singular" integrals in the main diagonal of the stiffness matrix and the "almost singular" integrals corresponding to the neighbor elements are treated differently. Unfortunately, this small difference is essential for the convergence analysis and the error estimates.

Similarly to the semi-discretized schemes, the quadrature methods can be divided into h-methods, p-methods, and h–p-methods according to the underlying quadrature rule. If the last is exact for high-order polynomials, i.e., a variant of a Gauß rule, then the quadrature algorithm is called a p-method. These p-versions of quadrature are known to be useful for second-kind equations, and they have been studied very extensively for Cauchy singular integral equations over the interval (cf. the results and references in [17,18,44]). Quite recently they have been applied to different one-dimensional operator equations as well (cf. [33,37] and see also [58] for a comparable approach). If the underlying quadrature rule is a low-order composite rule, i.e., if the domain of integration is subdivided into small domains of step size less or equal to h and if a low-order rule like the trapezoidal rule or Simpson's rule is applied to each subdomain, then we call the quadrature method an h-method. h-methods for second- and first-kind equations have been well analyzed (cf., e.g., [3,4,15,16,18,23,44] and the references in these publications). Clearly, due to the fixed polynomial accuracy, these h-versions of quadrature methods are designed for problems with finite degree of smoothness. Finally, a combination of the composite technique with quadrature rules over the subdomains of variable orders (cf. [55]) is called an h–p method. Note that p- and h–p-methods seem to be very promising even for equations with a finite degree of smoothness. The analysis of these methods for general equations, however, seems to be a challenging problem.

In this paper we give an overview on more or less well-known results for the h-version of quadrature methods. In Section 2 we shall introduce the notion of simple quadrature methods and that of quadrature methods with product integration. We shall formulate some convergence results for second-kind equations with smooth kernel functions and smooth solutions. In Section 3 we shall apply simple quadrature methods to pseudodifferential equations, i.e., to first-kind integral equations over smooth curves and over the torus. Note that, for these methods, a sort of "Fourier analysis" is required to derive stability and convergence. If the integral operator is defined over nonsmooth boundary curves, then the "Fourier analysis" of the approximation methods is much more involved, and we refer, e.g., to [16,44] for more details. In Section 4 we show how the concept of mesh gradings for higher-dimensional quadrature methods leads naturally to fully discretized collocation schemes. We explain why simple quadrature methods may not converge in case of second-kind boundary integral equations over curves and surfaces with corners and edges. Finally, we explain that, from the view point of complexity, simple quadrature methods over graded meshes are not optimal for the approximation of these higher-dimensional integral equations. In Section 5 we consider a quadrature method based on product integration for the numerical solution of two-dimensional strongly singular integral equations.

2. Quadrature methods and Fredholm integral equations of the second kind

A lot of boundary value problems over domains with smooth boundary can be converted into a Fredholm integral equation of the second kind (cf., e.g., [34]) the numerical theory for which is well known (cf., e.g., [3,4,23]). Let us begin with the simplest one-dimensional case. Suppose we have to solve the equation

$$x(t) + \int_0^1 k(t,\tau)x(\tau)\,\mathrm{d}\tau = y(t), \quad 0 \leqslant t \leqslant 1, \tag{2.1}$$

where y and k are given smooth functions and x is to be determined. Replacing the integral by the rectangle rule, we obtain the Nyström method for (2.1).

$$\tilde{x}(t) + \sum_{l=0}^{n-1} k\left(t, \frac{l+1/2}{n}\right)\tilde{x}\left(\frac{l+1/2}{n}\right)\frac{1}{n} = y(t), \quad 0 \leqslant t \leqslant 1. \tag{2.2}$$

The solution of this continuous equation over the interval $[0,1]$ consists of two steps. First, one has to solve the quadratic linear system for the values $\tilde{x}((j+\frac{1}{2})/n)$, $j=0,\dots,n-1$

$$\tilde{x}\left(\frac{j+1/2}{n}\right) + \sum_{l=0}^{n-1} k\left(\frac{j+1/2}{n}, \frac{l+1/2}{n}\right)\tilde{x}\left(\frac{l}{n}\right)\frac{1}{n} = y\left(\frac{j}{n}\right), \quad j=0,1,\dots,n-1. \tag{2.3}$$

Then, knowing the values $\tilde{x}(j/n)$, $\tilde{x}$ is to be computed via Nyström's interpolation

$$\tilde{x}(t) = y(t) - \sum_{l=0}^{n-1} k\left(t, \frac{l+1/2}{n}\right)\tilde{x}\left(\frac{l+1/2}{n}\right)\frac{1}{n}, \quad 0 \leqslant t \leqslant 1. \tag{2.4}$$

Using, e.g., the theory of collectively compact operators, one can prove that (2.2) is stable, i.e., that (2.3) has a unique solution for n large enough and that the spectral norm of the inverse matrix is uniformly bounded. The approximate solution $\tilde{x}$ converges to x with the same order as the quadrature rule in (2.2) approximates the integral in (2.1).

Next, we generalize this method. Suppose Γ is a compact manifold which is embedded in a Euclidean space and which is either closed or open. One should think of closed smooth curves or two-dimensional closed surfaces (i.e., boundary surfaces of open domains) or pieces of these two. Over the manifold we consider the integral equation

$$a(t)x(t) + \int_\Gamma k(t,\tau)x(\tau)\,\mathrm{d}_\Gamma\tau = y(t), \quad t \in \Gamma \tag{2.5}$$

including the kernel k and the coefficient function a. For first-kind equations, a is zero. If a is a bounded nonvanishing function, then we can divide the equation by a. Thus we may suppose that a is a constant. A good example for the kernel k is the two-dimensional double-layer kernel which corresponds to three-dimensional boundary value problems for Laplace's equation and which is defined by the formula $k(t,\tau):=(t-\tau)\cdot \nu_\tau/(4\pi\|t-\tau\|^3)$, where $\|t-\tau\|$ is the Euclidean distance from t to τ and ν_τ stands for the unit normal to the manifold Γ taken at the point τ. Note that the integral operator corresponding to this double-layer kernel is a pseudodifferential operator of order minus one (cf., e.g., [9,25]). The double-layer equation including this integral operator is an equation of the second kind with $a=0.5$. In order to discretize (2.5) we introduce a partition $\Gamma=\bigcup_{k=1}^K \Gamma_k$ of Γ into small submanifolds Γ_k of diameter less than a prescribed small positive number h. Fixing a small integer L and choosing quadrature knots $t_{k,l}\in\Gamma_l$, $l=1,\dots,L$, and nonnegative quadrature weights $\omega_{k,l}$, $l=1,\dots,L$, for a quadrature over Γ_k, we arrive at the composite quadrature rule

$$\int_\Gamma f(\tau)\,\mathrm{d}_\Gamma\tau=\sum_{k=1}^K\int_{\Gamma_k} f(\tau)\,\mathrm{d}_{\Gamma_k}\tau \sim \sum_{k=1}^K\sum_{l=1}^L f(t_{k,l})\omega_{k,l}. \tag{2.6}$$

Note that, for finer and finer approximations, K tends to infinity, the maximum h of the diameters $\operatorname{diam}\Gamma_k$, $k=1,\dots,K$, tends to zero but L is supposed to be fixed. By m_Q we denote the order of convergence defined by

$$\left|\int_\Gamma f(\tau)\,\mathrm{d}_\Gamma\tau-\sum_{k=1}^K\sum_{l=1}^L f(t_{k,l})\omega_{k,l}\right|\leqslant Ch^{m_Q}. \tag{2.7}$$

For example, the partition could be a triangulation of a two-dimensional polyhedron and the quadrature rule the mid-point rule $\int_{\Gamma_k} f\sim f(t_{k,1})\omega_{K,1}$ with $t_{k,1}$ the centroid of triangle Γ_k and $\omega_{k,1}:=\int_{\Gamma_k}1$ or the three-point rule using the mid-points of the sides of the triangle as quadrature knots and the weights $\omega_{k,l}=\int_{\Gamma_k}1/3$. Note that the mid-point rule is exact for linear functions whereas the three-point rule is exact for quadratic functions over the subtriangles of the triangulation which leads to an order of convergence of $m_Q=2$ and 3, respectively. For polygons the subdomains are intervals, and one could take the trapezoidal rule and Simpson's rule, which are exact for linear and cubic polynomials, respectively. In other words $m_Q=2$ and 4, respectively. However, for periodic functions over the interval, the order m_Q of the trapezoidal rule is even ∞. In case of curved polygons or polyhedra, we can introduce parametrization mappings $\gamma:\Omega\to\Gamma$ to reduce the integral $\int_{\Gamma_k} f$ to the integral $\int_{\Omega_k} f\circ\gamma\cdot|\gamma'|$ over a subdomain $\Omega_k=\gamma^{-1}(\Gamma_k)$ of Ω which is a subtriangle or subinterval. Applying the just mentioned rules to the transformed integral, we end up with a rule of the form (2.6).

Now, we replace the integration in (2.5) by quadrature (2.6) and arrive at the corresponding simple quadrature method (cf. (2.3))

$$a\tilde{x}_h(t_{k',l'})+\sum_{k=1}^K\sum_{l=1}^L k(t_{k',l'},t_{k,l})\tilde{x}_h(t_{k,l})\omega_{k,l}=y(t_{k',l'}),\quad k'=1,\dots,K,\ l'=1,\dots,L. \tag{2.8}$$

If the constant coefficient a is not zero and if the linear system (2.8) is solved, then we even can define the Nyström interpolant (cf. (2.4))

$$\tilde{x}_h(t):=\frac{1}{a}\left\{y(t)-\sum_{k=1}^K\sum_{l=1}^L k\,(t,t_{k,l})\,\tilde{x}_h\,(t_{k,l})\,\omega_{k,l}\right\},\quad t\in\Gamma. \tag{2.9}$$

Theorem 1. *Suppose that the compact manifold Γ is m_Q+1 times continuously differentiable and the right-hand side y is m_Q times continuously differentiable. Furthermore, suppose that the kernel*

k is m_Q times continuously differentiable with respect to each of its variable such that even the mixed derivatives $\partial_t^\alpha \partial_\tau^\beta k(t,\tau)$ with order α and β less than or equal to m_Q are bounded. Finally, assume that the constant a is not zero and that, for $y \equiv 0$, the integral equation (2.5) has only the trivial solution $x \equiv 0$. Then the linear system of the quadrature method (2.8) is uniquely solvable for any right-hand side y at least if the step size of discretization h is sufficiently small. The approximate solution $\tilde{x}_h$ converges uniformly to the exact solution x and

$$\sup_{t \in \Gamma} |\tilde{x}_h(t) - x(t)| \leqslant C h^{m_Q} \tag{2.10}$$

with a constant C independent of the discretization parameters h and K.

Note that in case of quasi-uniform partitions, i.e., in case that there exists a constant $c > 1$ with $c^{-1} h \leqslant \operatorname{irad} \Gamma_k \leqslant \operatorname{diam} \Gamma_k \leqslant ch$,

$$\operatorname{diam} \Gamma_k := \sup\{|t - \tau|:\ t, \tau \in \Gamma_k\},$$

$$\operatorname{irad} \Gamma_k := \sup\{\epsilon:\ \exists \tau \in \Gamma_k \text{ s.t. } |t - \tau| \leqslant \epsilon \Rightarrow t \in \Gamma_k\},$$

then the number of degrees of freedom is of order $O(h^{-d})$ with $d = 1$ and 2 for boundary curves and two-dimensional surfaces, respectively.

Theorem 2. *Suppose that k is the kernel of a classical pseudodifferential operator of negative order $-m$. Furthermore, suppose that $m_Q \geqslant m > 0$, that the compact manifold Γ is $m+1$ times continuously differentiable, and the right-hand side y is m times continuously differentiable. Finally, assume that a is a nonzero constant and that, for $y \equiv 0$, the integral equation (2.5) has only the trivial solution $x \equiv 0$. Then the linear system of the quadrature method (2.8) is uniquely solvable for any right-hand side y at least if the step size of discretization h is sufficiently small. The approximate solution $\tilde{x}_h$ converges uniformly to the exact solution x and*

$$\sup_{t \in \Gamma} |\tilde{x}_h(t) - x(t)| \leqslant C \log h^{-1} h^m \tag{2.11}$$

with a constant C independent of the discretization parameters h and K.

In particular, the quadrature method applied to the double-layer equation over a two-dimensional boundary manifold converges with order $O(h \log h^{-1})$. To prove the results of the last two theorems, one first shows stability of the discretized operators on the right-hand side of (2.8). This can be done, for instance, by the principle of collective compactness. Once stability is shown, the convergence order is derived from the order of convergence of the quadrature. For details we refer, e.g., to [3,4,18,23]. The reason for the restrictive order of convergence in Theorem 2 is the singular behavior of the kernel which can be characterized by the so called Calderón–Zygmund estimate

$$|\partial_t^\alpha \partial_\tau^\beta k(t,\tau)| < C\ |t - \tau|^{-d+m-|\alpha|-|\beta|}, \tag{2.12}$$

valid for all derivatives of order α and β such that $-d + m - |\alpha| - |\beta| < 0$. Here $-m$ is the order of the pseudodifferential operator and d the dimension of the underlying manifold Γ. The order in Theorem 2 can be improved if a slightly modified quadrature method is considered. This modification is called singularity subtraction or regularization (cf., e.g., [18]). To introduce this method we write (2.5) as

$$b(t)x(t) + \int_{\Gamma} k(t,\tau)[x(\tau) - x(t)]\,\mathrm{d}_{\Gamma}\tau = y(t), \quad t \in \Gamma,$$

$$b(t) := a + \int_{\Gamma} k(t,\tau)\,\mathrm{d}_{\Gamma}\tau. \tag{2.13}$$

Thus, we assume that we are able to compute the function b explicitly. For example, for the double-layer equation over smooth surfaces, constant functions are known to be eigenfunctions of the integral operator corresponding to the eigenvalue one-half, and (2.13) takes the form

$$x(t) + \frac{1}{4\pi}\int_{\Gamma} \frac{n_\tau \cdot (t-\tau)}{\|\tau - t\|^3}[x(\tau) - x(t)]\,\mathrm{d}_{\Gamma}\tau = y(t), \quad t \in \Gamma. \tag{2.14}$$

If we replace the integration in (2.13) by quadrature (2.6), we obtain the following quadrature method and the following Nyström interpolation step:

$$b(t_{k',l'})\tilde{x}_h(t_{k',l'}) + \sum_{k=1}^{K}\sum_{l=1}^{L} k(t_{k',l'}, t_{k,l})[\tilde{x}_h(t_{k,l}) - \tilde{x}_h(t_{k',l'})]\omega_{k,l} = y(t_{k',l'}),$$

$$k' = 1,\ldots,K, \; l' = 1,\ldots,L. \tag{2.15}$$

$$\tilde{x}_h(t) := \frac{y(t) - \sum_{k=1}^{K}\sum_{l=1}^{L} k(t,t_{k,l})\tilde{x}_h(t_{k,l})\omega_{k,l}}{b(t) - \sum_{k=1}^{K}\sum_{l=1}^{L} k(t,t_{k,l})\omega_{k,l}}, \quad t \in \Gamma. \tag{2.16}$$

Theorem 3. *Suppose that k is the kernel of a classical pseudodifferential operator of negative order $-m$. Furthermore, suppose that $m_Q \geqslant m+1 > 0$, that the compact manifold Γ is $m+2$ times continuously differentiable, and that the right-hand side y is $m+1$ times continuously differentiable. Finally, assume that a is a nonzero constant and that, for $y \equiv 0$, the integral equation (2.5) has only the trivial solution $x \equiv 0$. Then the linear system of the quadrature method (2.15) is uniquely solvable for any right-hand side y and the denominator in (2.16) does not vanish at least if the step size of discretization h is sufficiently small. The approximate solution $\tilde{x}_h$ converges uniformly to the exact solution x and*

$$\sup_{t\in\Gamma} |\tilde{x}_h(t) - x(t)| \leqslant C \log h^{-1} h^{m+1} \tag{2.17}$$

with a constant C independent of the discretization parameters h and K.

Another way to improve quadrature methods for nonsmooth kernels is to apply quadrature rules of product type (cf., e.g., [3,18,30]). Indeed, in many applications the kernel function $k(t,\tau)$ is singular but it admits a factorization

$$k(t,\tau) = k_{\mathrm{sm}}(t,\tau)k_{\mathrm{si}}(t,\tau), \tag{2.18}$$

where the first factor k_{sm} has at least a finite degree of smoothness and where the singularity of k is contained in k_{si}. Moreover, we suppose that the singular kernel k_{si} is simpler such that the integral of k_{si} can be computed by analytic formulae. Or we suppose that $k_{\mathrm{si}}(t,\tau)$ is analytic with respect to τ for $\tau \neq t$ such that the integral of k_{si} can be computed by higher-order Gauß rules and other techniques (cf., e.g., [28,53,55,56]). Note that an additional additive perturbation by a smooth kernel function can be treated easily. For the sake of simplicity, however, we drop this additional term.

One example for a factorization of the form (2.18) is the representation of one-dimensional potential kernels for the Helmholtz equation. In particular, the single-layer kernel $k_{\boldsymbol{k}}$ corresponding to the equation with wave number $\boldsymbol{k}$ and transformed to the 2π periodic interval (cf. [30]) takes the form

$$k_{\boldsymbol{k}}(t,\tau)=M_1(t,\tau)\log\left|4\sin^2\frac{t-\tau}{2}\right|+M_2(t,\tau),$$

$$M_1(t,\tau):=-\frac{1}{2\pi}J_0(\boldsymbol{k}|\gamma(t)-\gamma(\tau)|),$$

$$M_2(t,\tau):=\frac{\mathrm{i}}{2}H_0^{(1)}(\boldsymbol{k}|\gamma(t)-\gamma(\tau)|)-H_1(t,\tau)\log\left|4\sin^2\frac{t-\tau}{2}\right|, \tag{2.19}$$

where $\gamma:[0,2\pi]\to\Gamma$ is the parametrization of the boundary curve, J_0 is the Bessel function of order zero, and $H_0^{(1)}$ is the Hankel function of order one. The factors M_1 and M_2 in (2.19) are analytic (resp. smooth) functions if the parametrization γ is analytic (resp. smooth). Another example for a factorization is the representation $k(t,\tau)=k_0(t,\tau)|t-\tau|^{-\alpha}$ for a typical boundary integral kernel over a smooth boundary curve $\tilde{\Gamma}$, where k_0 is an analytic function and where $\alpha>0$ is a certain degree of singularity. If $\gamma:\Gamma=[0,2\pi]\to\tilde{\Gamma}$ denotes the parametrization of the boundary manifold and if γ_0 is the parametrization of the unit circle, then we get a factorization of the form (2.18) for the kernel transformed to the 2π-periodic interval setting

$$k_{\mathrm{sm}}(\gamma(t),\gamma(\tau)):=k_0(\gamma(t),\gamma(\tau))\frac{|\gamma(t)-\gamma(\tau)|^{\alpha}}{|\gamma_0(t)-\gamma_0(\tau)|^{\alpha}}|\gamma'(\tau)|,$$

$$k_{\mathrm{si}}(\gamma(t),\gamma(\tau)):=|\gamma_0(t)-\gamma_0(\tau)|^{-\alpha}. \tag{2.20}$$

Unfortunately, such a factorization does not work for the higher-dimensional case. In the higher-dimensional case, the structure of singularity is more involved and depends strongly on the geometry. Thus factorization (2.18) is to be defined by $k_{\mathrm{sm}}(t,\tau)=k_0(t,\tau)$ and $k_{\mathrm{si}}(t,\tau)=|t-\tau|^{-\alpha}$ (cf., Section 5 for more details). Then, in the case of curved boundaries, there are no analytic formulas available for the integration of k_{si}. However, if the boundary manifold is piecewise analytic, then the integral of k_{si} can be computed by tensor products of Gaussian quadratures. For general boundaries of finite degree of smoothness, the parametrization γ can be replaced by a piecewise polynomial interpolant $\tilde{\gamma}$ which is polynomial at least over each subdomain Γ_k of the corresponding partition of the quadrature method. After this substitution the integral over the kernel $k_{\mathrm{si}}(\tilde{\gamma}(t),\tilde{\gamma}(\tau))=|\tilde{\gamma}(t)-\tilde{\gamma}(\tau)|^{-\alpha}$ can again be computed by tensor products of Gaussian quadratures (for more details in some special case cf., e.g., [14]).

Now, we choose points $\tau_{k,l}\in\Gamma_k$ and interpolating polynomials $\varphi_{k,l}$ over Γ_k such that $\varphi_{k,l}(\tau_{k,l'})=\delta_{l,l'}$. Polynomial means here polynomial with respect to a given parametrization of the boundary manifold. We consider the quadrature rule

$$\int_\Gamma k(t,\tau)x(\tau)\,\mathrm{d}\tau=\int_\Gamma k_{\mathrm{si}}(t,\tau)[k_{\mathrm{sm}}(t,\tau)x(\tau)]\,\mathrm{d}\tau$$

$$\sim \sum_{k=1}^{K} \int_{\Gamma_k} k_{\mathrm{si}}(t,\tau) \sum_{l=1}^{L} [k_{\mathrm{sm}}(t,\tau_{k,l}) x(\tau_{k,l})] \varphi_{k,l}(\tau)\, \mathrm{d}\tau$$

$$= \sum_{k=1}^{K} \sum_{l=1}^{L} k_{\mathrm{sm}}(t,\tau_{k,l}) x(\tau_{k,l}) \omega_{k,l}^{p}, \quad \omega_{k,l}^{p} := \int_{\Gamma_k} k_{\mathrm{si}}(t,\tau) \varphi_{k,l}(\tau)\, \mathrm{d}\tau. \tag{2.21}$$

In order to simplify the assumptions, we assume that the manifold Γ is a curve or a surface given by a single parametrization $\gamma : \Omega \to \Gamma$ and that the preimages $\Omega_k := \gamma^{-1}(\Gamma_k)$ of the subdomains Γ_k are intervals or triangles. Moreover, we suppose that the L parameter points $\sigma_{k,l}$ corresponding to the quadrature knots $\tau_{k,l} = \gamma(\sigma_{k,l}) \in \Gamma_k$ are defined as the affine images of fixed points σ_l, $l = 1, \ldots, L$ in the standard interval $[0,1]$ (resp., in the standard triangle $\{(s_1,s_2)\colon 0 \leqslant s_2 \leqslant s_1 \leqslant 1\}$). Likewise, the polynomials $\varphi_{k,l}$ are supposed to be the pull backs of interpolatory polynomials φ_l defined over the standard interval or triangle. If this basis spans a space containing all polynomials of degree less than m_p, than the convergence order of the quadrature rule is $m_Q = m_p$. Applying the corresponding product rule to (2.5), we arrive at the quadrature method

$$a\tilde{x}_h(\tau_{k',l'}) + \sum_{k=1}^{K} \sum_{l=1}^{L} k_{\mathrm{sm}}(\tau_{k',l'},\tau_{k,l}) \tilde{x}_h(\tau_{k,l}) \omega_{k,l}^{p} = y(\tau_{k',l'}), \quad k' = 1, \ldots, K,\ l' = 1, \ldots, L. \tag{2.22}$$

Let us note that, for the special choice $k_{\mathrm{sm}} \equiv 1$, method (2.16) coincides with the piecewise polynomial collocation method, where the trial space is the span of the $\{\varphi_{k,l},\ k = 1, \ldots, K,\ l = 1, \ldots, L\}$. In other words, the quadrature method with product rule is already a compromise between quadrature and collocation method.

Theorem 4. *Suppose that the kernel* k *admits a factorization* (2.18), *where* k_{sm} *is* m_p *times continuously differentiable with respect to both variables such that even the mixed derivatives* $\partial_t^{\alpha} \partial_{\tau}^{\beta} k(t,\tau)$ *with order* α *and* β *less than or equal to* m_p *are bounded. For* $k_{\mathrm{si}}(t,\tau)$, *we suppose the same degree of differentiability for* $t \neq \tau$ *and, for* $t \to \tau$ *and the same orders of differentiation, estimates* (2.12) *where* $m > 0$. *Furthermore, suppose that the compact manifold* Γ *is* $m_p + 1$ *times continuously differentiable, and that the exact solution* x *and the right-hand side* y *are* m_p *times continuously differentiable. Finally, assume that* a *is a nonzero constant and that, for* $y \equiv 0$, *the integral equation* (2.5) *has only the trivial solution* $x \equiv 0$. *Then the linear system of the quadrature method* (2.22) *is uniquely solvable for any right-hand side* y *at least if the step size of discretization* h *is sufficiently small. The approximate solution* $\tilde{x}_h$ *converges uniformly to the exact solution* x *and*

$$\sup_{t \in \{\tau_{k,l}: k = 1,\ldots,K,\, l = 1,\ldots,L\}} |\tilde{x}_h(t) - x(t)| \leqslant C h^{m_p} \tag{2.23}$$

with a constant C *independent of the discretization parameters* h *and* K.

3. Quadrature methods for pseudodifferential equations over smooth boundaries

Boundary integral operators over smooth boundaries belong to the class of classical pseudodifferential operators (cf., e.g., [9,25]). If the order of such an operator is nonnegative, then the kernels of the integral operators are strongly singular or even hypersingular. The convergence of simple quadrature methods applied to such functions is not guaranteed. In fact, in many situations the straightforward quadrature methods do not converge. We present here convergent variants of quadrature methods, only. All these methods rely on a singularity subtraction step (cf., e.g., (3.7), (3.8), and [6,49] for operators of order minus one) though, at first glance, this may not be visible. Let us start with the simplest case, i.e., with a Cauchy singular integral equation over the unit circle $\mathbb{T}$

$$Ax(t) := a(t)x(t) + b(t)\frac{1}{\pi i}\int_{\mathbb{T}} \frac{x(\tau)}{\tau - t}\,d\tau + \int_{\mathbb{T}} k(t,\tau)x(\tau)\,d\tau = y(t), \quad t \in \mathbb{T}. \tag{3.1}$$

Here a, b, k, and y are given functions and x is to be determined. Using the ideas developed for second-kind integral equations, it is not hard to reduce the problem for arbitrary kernel functions k to the case $k \equiv 0$. Moreover, for simplicity, we suppose a and b to be continuous. We choose an even positive integer n, set $t_k := e^{i2\pi k/n}$, and consider the following quadrature rules:

$$\int_{\mathbb{T}} f(\tau)\,d\tau = \int_0^{2\pi} f(e^{is})ie^{is}\,ds \sim \sum_{l=0}^{n-1} f(t_l)t_l\frac{2\pi i}{n}, \tag{3.2}$$

$$\int_{\mathbb{T}} f(\tau)\,d\tau \sim \sum_{\substack{l=0,\dots,n-1 \\ l \equiv k+1 \bmod 2}} f(t_l)t_l\frac{4\pi i}{n}. \tag{3.3}$$

Note that rule (3.3) has doubled step size in comparison with (3.2). However, it is appropriate to functions f having a singularity at t_k and will lead to optimal quadrature methods. Thus, we consider (3.1) for $t = t_k$, $k = 0, \dots, n-1$, replace the integral by rule (3.3) to obtain the quadrature method

$$a(t_k)\tilde{x}(t_k) + b(t_k)\frac{1}{\pi i}\sum_{\substack{l=0,\dots,n-1 \\ l \equiv k+1 \bmod 2}} \frac{\tilde{x}(t_l)}{t_l - t_k}t_l\frac{4\pi i}{n} = y(t_k), \quad k = 0, \dots, n-1. \tag{3.4}$$

We call this quadrature method stable if, at least for sufficiently large n, Eq. (3.4) are uniquely solvable for any right-hand side and if the Euclidean matrix norms of the matrices of the linear systems in (3.4) and the norms of their inverses are uniformly bounded with respect to n. The method is called convergent if the trigonometric interpolation

$$L_n\tilde{x}(t) := \sum_{k=1}^{n-1} \tilde{x}(t_k)\frac{1}{n}\sum_{l=-n/2}^{n/2-1} \frac{t^l}{t_k^l} \tag{3.5}$$

tends in the L^2 norm to the exact solution x of (3.1) for all continuous right-hand sides y. Note that stability is an important condition for solving the linear system of equations. Moreover, it is necessary for the method to be convergent. We get (cf. [44] and compare the analogous results in [5,32]).

Theorem 5. *If the singular integral operator A is invertible, then the quadrature method* (3.4) *is stable and convergent. For a right-hand side which is m times differentiable such that the mth derivative is square integrable, the L^2 error $\|L_n\tilde{x} - x\|$ is less than a constant times n^{-m}.*

Proof. We assume a and b to be constant. The general case can be treated by well-known localization techniques (cf., e.g., [44]). Set $e_k(t) := t^k$, denote the span of the e_k, $k = -n/2, \dots, n/2-1$ by T_n, and recall that L_n stands for the interpolation projection of (3.5). Now it is well known that e_k is an eigenfunction of A corresponding to the eigenvalue $a + b\,\mathrm{sign}(k + \frac{1}{2})$. Hence, T_n is an invariant subspace for A. The collocation solution $x_n \in T_n$ is defined by $Ax_n(t_k)=y(t_k)$, $k=0,\dots,n-1$, i.e., by $L_nAx_n=L_ny$. Consequently, we get $Ax_n=L_ny$ and the collocation solution $x_n=A^{-1}L_ny$ converges to the exact solution $x=A^{-1}y$. Thus in order to prove our theorem, it is sufficient to show the equivalence of method (3.4) and the collocation method.

The solution $\tilde{x}$ of (3.4) is a discrete function over $\{t_k,\ k=0,\dots,n-1\}$. We identify $\tilde{x}$ with the linear interpolation $L_n\tilde{x}$. Then our proof is finished if we show

$$Ax_n(t_k)=ax_n(t_k)+b\frac{1}{\pi i}\sum_{\substack{l=0,\dots,n-1\\ l\equiv k+1 \bmod 2}}\frac{x_n(t_l)}{t_l-t_k}t_l\frac{4\pi i}{n}.$$

We have to prove that, for $x_n\in T_n$,

$$\frac{1}{\pi i}\int_{\mathbb{T}}\frac{x_n(\tau)}{\tau-t_k}\mathrm{d}\tau=\frac{1}{\pi i}\sum_{\substack{l=0,\dots,n-1\\ l\equiv k+1 \bmod 2}}\frac{x_n(t_l)}{t_l-t_k}t_l\frac{4\pi i}{n}. \tag{3.6}$$

We arrive at

$$\frac{1}{\pi i}\int_{\mathbb{T}}\frac{x_n(\tau)}{\tau-t_k}\mathrm{d}\tau=\frac{1}{\pi i}\int_{\mathbb{T}}\frac{x_n(\tau)-x_n(t_k)}{\tau-t_k}\mathrm{d}\tau+x_n(t_k)\frac{1}{\pi i}\int_{\mathbb{T}}\frac{1}{\tau-t_k}\mathrm{d}\tau$$

$$=\frac{1}{\pi i}\sum_{\substack{l=0,\dots,n-1\\ l\equiv k+1 \bmod 2}}\frac{x_n(t_l)-x_n(t_k)}{t_l-t_k}t_l\frac{4\pi i}{n}+x_n(t_k), \tag{3.7}$$

where we have used that $e_0\equiv 1$ is an eigenfunction corresponding to the eigenvalue 1, that $\{x_n(t)-x_n(t_k)\}/\{t-t_k\}$ is in the span$\{e_k,\ k=-n/2,\dots,n/2-2\}$ and that (3.3) is exact on span$\{e_k,\ k=-n/2,\dots,n/2-2\}$. Note that the exactness of (3.3) is a simple consequence of the formula for the geometric series. Now (3.6) follows from (3.7) by a straightforward computation. The convergence order can be derived by standard methods (cf., e.g., [44]). □

Theorem 5 can be generalized to nonuniform partitions (cf. [8,38,54]). An analogous result holds for the one-dimensional hypersingular equation ([27], cf. also [7,11,29]). However, the singularity subtraction step (3.7) is to be replaced by the following regularization of the finite part integral:

$$\frac{1}{2\pi}\int_{\mathbb{T}}\frac{\tilde{x}(\tau)}{|\tau-t_k|^2}|\mathrm{d}\tau|=\frac{1}{2\pi}\int_{\mathbb{T}}\frac{\tilde{x}(\tau)-\tilde{x}(t_k)-\tilde{x}'(t_k)(\tau-t_k)}{|\tau-t_k|^2}|\mathrm{d}\tau|$$

$$+\tilde{x}(t_k)\frac{1}{2\pi}\int_{\mathbb{T}}\frac{1}{|\tau-t_k|^2}|\mathrm{d}\tau|+\tilde{x}'(t_k)\frac{1}{2\pi}\int_{\mathbb{T}}\frac{\tau-t_k}{|\tau-t_k|^2}|\mathrm{d}\tau|. \tag{3.8}$$

Applying (3.3) to the first integral on the right-hand side, computing the others and performing some easy calculations, we arrive at the quadrature approximation

$$\frac{1}{2\pi}\int_{\mathbb{T}}\frac{\tilde{x}(\tau)}{|\tau-t_k|^2}|\mathrm{d}\tau|\sim\frac{n}{8}\tilde{x}(t_k)+\sum_{\substack{l=0,\dots,n-1\\ l\equiv k+1 \bmod 2}}\frac{\tilde{x}(t_l)}{|t_l-t_k|^2}t_l\frac{2}{n}, \tag{3.9}$$

which, again, is exact for $\tilde{x}=x_n\in T_n$. Note that a regularization like in (3.8) is necessary in order to obtain a convergent quadrature method.

Now, let us consider the generalized single-layer equation

$$Ax(t):=\int_{\mathbb{T}} k(t,\tau)x(\tau)|d\tau| = y(t), \quad t\in\mathbb{T}, \tag{3.10}$$

$$k(t,\tau):=a(t)k_0(t,\tau)+b(t)k_1(t,\tau)+k_2(t,\tau),$$

$$k_0(t,\tau):=-\frac{1}{\pi}\log|t-\tau|, \quad k_1(t,\tau):=k_1(t/\tau),$$

$$k_1(e^{i2\pi u}):=\begin{cases} i[u-0.5] & \text{if } 0<u<1,\\ 0 & \text{if } u=0,\end{cases} \tag{3.11}$$

where a, b and k_2 are smooth functions. Operator A is a pseudodifferential operator with principal symbol $\sigma_A(t,\xi)=[a(t)+b(t)\,\mathrm{sign}(\xi)]|\xi|^{-1}$. Replacing integration by quadrature (3.2), we arrive at the quadrature method

$$\sum_{l=0}^{n-1} k(t_k,t_l)\tilde{x}(t_l)\frac{2\pi}{n} = y(t_k), \quad k=0,\dots,n-1. \tag{3.12}$$

Note that $k_0(t,t):=\lim_{\tau\to t} k_0(t,\tau)=\infty$. Thus, in the last formula, we need to fix an artificial finite value for $k_0(t,t)=k_0(1,1)$. Due to the factor $2\pi/n$ this value is of no importance for the consistency of the quadrature. However, the choice of this value is essential for the stability and the order of convergence. We take $k_0(1,1)=-\log n/\pi$ which corresponds to the quadrature method modified by singularity subtraction [6,49]. For quadrature methods applied to the general pseudodifferential equation (3.10) of order -1 and analogous methods applied to other pseudodifferential equations of negative order, the method of the proof to Theorem 5 fails. The theory of collectively compact operators is helpful to treat the compact perturbations $\int k_2(t,\tau)x(\tau)|d\tau|$. The stability of the main part of the equation, however, requires new techniques. Of course, stability is to be understood not in terms of the Euclidean matrix norm but in terms of a more general operator norm induced by the norms of the Sobolev spaces in which A and its inverse are bounded. The first method of proof is the so-called localization principle. The second is the Fourier analysis or circulant technique. For example, the stability of the discretized weakly singular operator defined by the left-hand side of (3.12) can be reduced by localization to the stability of the corresponding matrices with frozen functions a, b, and k_2. The matrix with constant a, b, and k_2 is a circulant and takes the form

$$V_n:=\left(-a\log|1-t_{k-l}|\frac{2}{n}+bk_1(t_{k-l})\frac{2\pi}{n}+k_2\frac{2\pi}{n}\right)_{k,l=0}^{n-1}. \tag{3.13}$$

In general, a matrix $(a_{k,l})_{k,l=0}^{n-1}$ is called a circulant if $a_{k,l}=a_{k-l}$ and $a_{k-l}=a_{k-l\pm n}$. The eigenvalues $\{\lambda_l,\ l=-n/2,-n/2+1,\dots,n/2-1\}$ of the circulant $(a_{k-l})_{k,l=0}^{n-1}$ are connected with the entries by

$$\lambda_k=\sum_{l=0}^{n-1} e^{i2\pi lk/n}a_l. \tag{3.14}$$

Using (3.14), writing $|1-t_l|=2|\sin(\pi l/n)|$, and substituting $\sin(\pi x)$ by $\pi x\prod_{j=1}^{\infty}(1-x^2/j^2)$, it is not hard to verify that the matrix V_n has the eigenvalues

$$\lambda_l^n=\begin{cases}[s(t_l)]/n \text{ if } l=-\frac{n}{2},\dots,-1,\ 1,\dots,\frac{n}{2}-1,\\ 2\pi k_2 \qquad \text{if } l=0,\end{cases} \tag{3.15}$$

where the numerical symbol function $\boldsymbol{s}$ is defined by $\boldsymbol{s}(t){:=}a\boldsymbol{f}(t)+b\boldsymbol{g}(t)$,

$$\boldsymbol{f}(t){:=}2\log(2\pi)-2\sum_{l\in\mathbb{Z},\ l\neq 0}\log|l|\,t^l,\qquad \boldsymbol{g}(\mathrm{e}^{\mathrm{i}2\pi u}){:=}\pi\cot(\pi u).$$

Note that function $\boldsymbol{f}$ is smooth except at $t=1$. Multiplying $\boldsymbol{f}(t)$ by $(t-1)^2$ we get an absolutely convergent series. By the way, $\boldsymbol{f}$ is the symbol function of the Toeplitz matrix $(-2\log|k-l|)_{k,l=-\infty}^{\infty}$ which is the quadrature discretization of step size one for the logarithmic equation over the real axis. Comparing eigenvalues (3.15) with the eigenvalues $\mu_0=2\pi k_2$ and $\mu_l=(a+\mathrm{sign}(l+\frac{1}{2}))|l|^{-1}$, $l=\pm 1,\pm 2,\ldots$ corresponding to operator A, we observe the consistency property $\lambda_l^n/\mu_l\to 1$ for any fixed l and for $n\to\infty$. The stability is equivalent to the existence of a constant $c>1$ such that $c^{-1}|\mu_l|\leqslant|\lambda_l^n|\leqslant c|\mu_l|$ holds for sufficiently large n and $l=-n/2,\ldots,n/2-1$. We arrive at the following typical theorem.

Theorem 6. *If the pseudodifferential operator A of order -1 is invertible, if $k_0(1,1)$ is chosen to be $-\log n/\pi$, and if $a(t)+\lambda b(t)\neq 0$ for all $t\in\mathbb{T}$ and $-1\leqslant\lambda\leqslant 1$, then the quadrature method* (3.4) *is stable and convergent. For a right-hand side which is four times continuously differentiable or at least contained in the Sobolev space $H^4(\mathbb{T})$, we get the estimate (cf.* (3.5))

$$\sup_{t\in\mathbb{T}}|L_n\tilde{x}-x|\leqslant C\,||L_n\tilde{x}-x||_{H^1(\mathbb{T})}\leqslant Cn^{-2}\,||y||_{H^4(\mathbb{T})}. \tag{3.16}$$

The assumption $a(t)+\lambda b(t)\neq 0,\ -1\leqslant\lambda\leqslant 1$ means that operator A is strongly elliptic. Note that the interval $[-1,1]$ in this condition originates from $[-1,1]=\{\boldsymbol{g}(t)/\boldsymbol{f}(t)\colon t\in\mathbb{T}\}$. The convergence order two in estimate (3.16) can be derived from the symbol function $\boldsymbol{s}$, too. Namely, if there exists a constant $\alpha>0$ such that $|x|\boldsymbol{f}(x)=1+\mathrm{O}(|x|^\alpha)$ and $x\boldsymbol{g}(x)=1+\mathrm{O}(|x|^\alpha)$ for $x\to\pm 0$, then $|\lambda_l^n-\mu_l|/|\mu_l|\leqslant\mathrm{O}(n^{-\alpha})$ and the order of convergence is α. In fact this constant exists, and is equal to two. As it is well known (cf. [6,49]), the convergence order is even three in the case that the coefficient b vanishes identically. To improve the order of convergence, one can use, for instance, an end-point correction for the rectangle rule (cf. [1]). More details and different modifications to improve convergence can be found in [6,18,31,35,36,44,49,52,57].

In general, for the stability of the quadrature method applied to a first-kind integral operator, the invertibility of the operator is not sufficient. Often strong ellipticity turns out to be the necessary and sufficient stability condition. For one-dimensional pseudodifferential operators of order less than -1, the quadrature method can also be considered as a Galerkin method with Dirac-δ ansatz functions (cf. [51]). In this case standard techniques for the Galerkin approximation of strongly elliptic operators can be applied.

Finally, let us remark that there is not much known for quadrature methods applied to pseudodifferential equations over the boundaries of higher-dimensional domains. The only paper in this direction we know about is due to Saad Abdel-Fattah [48]. To report this result, we consider the two-dimensional pseudodifferential operator of order zero over the torus $\mathbb{T}^2:=\{(t_1,t_2)\in\mathbb{R}^2\colon 0\leqslant t_i<1,\ i=1,2\}$

$$Ax(t):=a(t)x(t)+\int_{\mathbb{T}^2}k(t,\tau)x(\tau)\mathrm{d}\tau=y(t),\quad t\in\mathbb{T}^2,$$

$$k(t,\tau):=k_0(t,t-\tau)+k_2(t,\tau), \tag{3.17}$$

where the coefficient a and the kernel function k_2 are supposed to be smooth and one-periodic functions and where the singular kernel $k_0(t,\sigma)$ satisfies

$$k_0(t,\varrho\sigma)=\text{sign}(\varrho)k_0(t,\sigma)\varrho^{-2}, \quad \varrho\neq 0 \tag{3.18}$$

and is smooth and one-periodic with respect to $t\in\mathbb{T}^2$ and smooth with respect to σ for $|\sigma|=1$. Applying the tensor product trapezoidal rule to (3.17), we arrive at the quadrature method

$$a\left(\frac{k_1}{n},\frac{k_2}{n}\right)x\left(\frac{k_1}{n},\frac{k_2}{n}\right)+\sum_{l_1,l_2=0}^{n-1}k\left(\left(\frac{k_1}{n},\frac{k_2}{n}\right),\left(\frac{l_1}{n},\frac{l_2}{n}\right)\right)x\left(\frac{l_1}{n},\frac{l_2}{n}\right)\frac{1}{n^2}$$

$$=y\left(\frac{k_1}{n},\frac{k_2}{n}\right), \quad k_1,k_2=0,\ldots,n-1. \tag{3.19}$$

Here the singular value $k(t,t)$ is set to zero. Using localization techniques and two-dimensional Fourier analysis, Saad proved the following theorem.

Theorem 7. *If the singular integral operator A is invertible and satisfies condition* (3.18), *then method* (3.19) *is stable and convergent in the same sense as method* (3.4) *in Theorem* 5.

The convergence order for smooth right-hand sides is one. Note that this result is not important as a result for the artificial torus but it is important as a local analysis of the quadrature method over regular tensor product grids. Of course, for a full understanding of quadrature methods a lot of further local cases have to be studied, and these cases seem to be much more involved.

4. Negative results for quadrature methods applied to higher-dimensional equations

Now, let us have a look at quadrature methods for the solution of general integral equations over two-dimensional manifolds. If the kernel function and the manifold are smooth, then we have nice results for second-kind equations like in Theorem 1 and a lot of problems with first-kind equations which are severely ill-posed and not discussed here. In many important applications, however, the kernel $k(t,\tau)$ is singular in the sense of (2.12). In this case, even if the quadrature method is stable, the convergence of the quadratures and, consequently, that of the approximate solutions is very poor (cf., e.g., Theorem 7) if the method is not, in fact, diverging. We only mention here the lack of convergence for the simplest quadrature method applied to the double-layer equation over polyhedra. Without loss of generalization, we choose the simplest example and consider the equation

$$2[1-d_c(t)]x(t)+\frac{1}{2\pi}\int_\Gamma \frac{\nu_\tau\cdot(t-\tau)}{|t-\tau|^3}x(\tau)\,\mathrm{d}_\tau\Gamma=y(t), \quad t\in\Gamma \tag{4.1}$$

over the boundary Γ of $C=\{(x_1,x_2,x_3)\in\mathbb{R}^3: 0\leqslant x_i\leqslant 1,\ i=1,2,3\}$. Here ν_τ is the unit normal to Γ at τ and $d_C(t)$ is the normalized solid angle of C at the boundary points, i.e., $d_C(t)=\frac{1}{8}$ for vertex points, $d_C(t)=\frac{1}{4}$ for edge points, and $d_C(t)=\frac{1}{2}$ else. Note that the double layer kernel is strongly singular at the edge and vertex points and (2.12) holds with $m=0$. For simplicity, we choose $\Gamma=\bigcup_{k=1}^K\Gamma_k$ to be the partition of Γ into $K=6n^2$ uniform squares of side length $h=1/n$, and we suppose that rule (2.6) in method (2.8) is the mid-point rule. Then the error $\sup_{k,l}|\tilde{x}_h(t_{k,l})-x(t_{k,l})|$ need not tend to zero even if the right-hand side y is smooth. This follows from the fact that the quadrature error does not turn to zero uniformly. Indeed, choose $t_{k',l'}=(0.5h,0,0.5h)$ and x to be one over $\{(x_1,x_2,0)\in\mathbb{R}^3\colon 0\leqslant x_i\leqslant 1,\ i=1,2\}$ and zero over the rest of Γ and consider the quadrature error

$$\int_\Gamma k(t_{k',l'},\tau)x(\tau)\mathrm{d}_\Gamma\tau - \sum_{k=1}^{K}\sum_{l=1}^{1} k(t_{k',l'},t_{k,l})x(t_{k,l})\omega_{k,l}$$

$$=\frac{1}{2\pi}\int_0^1\int_0^1 \frac{(0,0,1)\cdot((0.5h,0,0.5h)-(x_1,x_2,0))}{|(0.5h,0,0.5h)-(x_1,x_2,0)|^3}\,\mathrm{d}x_1\,\mathrm{d}x_2$$

$$-\frac{1}{2\pi}\sum_{k_1,k_2=1}^{n}\frac{(0,0,1)\cdot((0.5h,0,0.5h)-([k_1-0.5]h,[k_2-0.5]h,0))}{|(0.5h,0,0.5h)-([k_1-0.5]h,[k_2-0.5]h,0)|^3}h^2$$

$$=\frac{1}{2\pi}\int_0^n\int_0^n \frac{0.5}{|(0.5,0,0.5)-(x_1,x_2,0)|^3}\,\mathrm{d}x_1\,\mathrm{d}x_2$$

$$-\frac{1}{2\pi}\sum_{k_1,k_2=1}^{n}\frac{0.5}{|(0.5,0,0.5)-([k_1-0.5],[k_2-0.5],0)|^3}.$$

Obviously, this tends to

$$\frac{1}{2\pi}\int_0^\infty\int_0^\infty \frac{0.5}{|(0.5,0,0.5)-(x_1,x_2,0)|^3}\,\mathrm{d}x_1\,\mathrm{d}x_2$$

$$-\frac{1}{2\pi}\sum_{k_1,k_2=1}^{\infty}\frac{0.5}{|(0.5,0,0.5)-([k_1-0.5],[k_2-0.5],0)|^3},$$

i.e., to the quadrature error over an unbounded conical boundary manifold. This quadrature error with step size $h=1$ is different from zero. Now, the convergence properties of the quadrature method correspond to those of the quadrature rule, and method (2.8) does not converge with respect to the supremum norm. Similar homogeneity arguments apply to quadrature methods including special graded meshes and double-layer equations over general piecewise smooth boundary manifold. To get a converging quadrature method, it is sufficient to choose the version with singularity subtraction (2.15) (cf. [47]). Analogous arguments can be used also for disproving the convergence in the case of strongly singular integral equations.

Now turn again to general quadrature methods over two-dimensional manifolds. To improve a low order of convergence, one has to adapt the quadrature to the singular behavior of the kernel function $\tau\mapsto k(t,\tau)$. We shall discuss two methods, mesh gradings in this section and product rules in Section 5. The first and simplest way of adaption is to use a mesh grading towards the singularity point t of the kernel. In other words, the quadrature rule employed for the numerical method should not be a fixed rule but it should depend on the source point t. Such an improved method seeks approximate values $\tilde{x}(t)$ for the unknown solution x over the points t of a fixed grid G. For each point $t\in G$, we have to approximate the integral $\int k(t,\tau)x(\tau)\,\mathrm{d}\tau$ in (2.5) by a quadrature rule over a refined grid G_t the points of which accumulate around t. Hence function values $x(\tau)$ at the quadrature knots $\tau\in G_t$ of this refined grid are required, and these can be obtained by interpolating the fixed set of approximate values $\{\tilde{x}(t),\ t\in G\}$. If the interpolant is $\tilde{x}_I$, i.e., if the values $\tilde{x}(\tau)$, $\tau\in G_t$ are approximated by $\tilde{x}_I(\tau)$, $\tau\in G_t$, then the quadrature approximation to $\int k(t,\tau)x(\tau)\,\mathrm{d}\tau$ is a discretization of $\int k(t,\tau)\tilde{x}_I(\tau)\,d\tau$. In other words, the resulting scheme in its simplest form is rather not a quadrature method but rather a fully discretized collocation method.

An exception, where a refined mesh can lead to an improved quadrature method in the sense of (2.15), is the case of second-kind integral equations over nonsmooth but piecewise smooth surfaces. Let the piecewise smooth surface $\Gamma \in \mathbb{R}^3$ take the form $\bigcup_{m=1}^{m_\Gamma} \Gamma^m$, where all the patches Γ^m are smooth. Then, for instance, the kernel $k(t,\tau)$ of the double-layer equation satisfies (2.12) with $m=0$ for points t,τ from different patches Γ^m and $\Gamma^{m'}$. For points of the same smooth boundary patch Γ^m, estimate (2.12) holds with $m=1$. Moreover, if the patches Γ^m are planar, then the double-layer kernel vanishes. Hence, one can choose a fixed mesh G graded towards the boundaries of the patches Γ^m, and, for each $t \in G$, the grids G_t can be chosen to be G. The mesh grading means that the diameter of the partition domain has to be small when the domain is close to the edge, i.e., to the boundary of the smooth patches Γ^m. Unfortunately, a partition with subdomains small only in the direction toward the edge and larger in the direction parallel to the edge is not sufficient. The resulting quadrature methods take the form (2.15). The number of subdomains and the corresponding number of degrees of freedom corresponding to such gradings is usually in the order $[h^{-1}]^\alpha$ where h is the maximal mesh size and where $\alpha > 2$ depends on the smoothness of the solution or, equivalently, on the geometry of Γ. Thus, substantially more degrees of freedom are necessary than the $[h^{-1}]^2$ for methods over uniform grids. The corresponding quadrature methods are analyzed in [45,47].

To evaluate this quadrature method over graded meshes we turn to the complexity. Let us suppose that β is the order of complexity for Nyström's methods over regular grids, i.e., suppose that the number of necessary arithmetic operations to compute an approximate solution with a supremum norm error less than a prescribed $\varepsilon > 0$ is less than $O([\varepsilon^{-1}]^\beta)$. Here the β depends on the singularities of the exact solution to the double-layer equation. It turns out that, using an appropriately graded mesh, the order of complexity of Nyström's method can be reduced to $\beta/2$. In contrast to this higher-dimensional result for quadrature methods, the complexity order of the univariate quadrature method and that of higher-dimensional discretized collocation or Galerkin methods can be reduced to an arbitrarily small number if only a quadrature rule (resp. a trial space) of sufficiently high order is used and if the mesh is appropriately graded. In particular, in case of the two-dimensional collocation method, the graded meshes can be chosen to include subdomains which are of small size in direction to the closest edge and which have a larger size in the perpendicular direction. Hence, the number of degrees of freedom can be estimated by $[h^{-1}]^2$ and, at least asymtotically, the order of complexity can be reduced to an arbitrarily small number. Consequently, even in the case of second-kind equations, the fully discretized collocation or Galerkin methods are more efficient than the simple quadrature methods (2.8) and (2.15).

5. Product quadrature for two-dimensional singular equations

Suppose Γ is a smooth two-dimensional manifold. Over Γ we consider the integral equation $Ax = y$ from (2.5) with A an operator invertible in the space $L^2(\Gamma)$. We suppose that the kernel k admits a factorization $k(t,\tau) = k_{\mathrm{sm}}(t,\tau)k_{\mathrm{si}}(t,\tau)$ with the factor k_{sm} of finite degree of smoothness and with the singularity factor k_{si}, which satisfies (2.12) with $m=0$ and $d=2$. We assume that, in contrast to the integration of k, the integration of k_{si} is easy to perform. Using the factorization, we can consider the product quadrature rule (2.21) of order m_p and the corresponding quadrature method (2.22) from Section 2.

Let us discuss one important example. Operator A could be a classical pseudodifferential operator of order zero. Clearly, the corresponding equation is of the form (3.17) with $\mathbb{T}^2$ replaced by Γ. To enable an explicit factorization, we consider singular kernels k_0 (cf. (3.17)) of the form

$$k_0(t,\tau)=k_{00}(t,\tau)\frac{p(t-\tau)}{|t-\tau|^{\alpha}}, \tag{5.1}$$

where α is an integer greater or equal to two and where p is a homogeneous polynomial of degree $\alpha-2$. Using (5.1), we define the factorization $k_0(t,\tau)=k_{\mathrm{sm}}(t,\tau)k_{\mathrm{si}}(t,\tau)$ by

$$k_{\mathrm{sm}}(\gamma(s),\gamma(s')):=k_{00}(\gamma(s),\gamma(s'))\cdot|\gamma'(s')|,$$

$$k_{\mathrm{si}}(\gamma(s),\gamma(s')):=\frac{p(\gamma(s)-\gamma(s'))}{|\gamma'(s')|\cdot|\gamma(s)-\gamma(s')|^{\alpha}}, \tag{5.2}$$

where $\gamma:\Omega\to\Gamma$ is the parametrization of Γ and where $|\gamma'(s)|$ with $s=(s_1,s_2)$ stands for the Jacobian determinant $|\partial_{s_1}\gamma(s)\times\partial_{s_2}\gamma(s)|$ of the parametrization. To simplify the formulas, for the case that there is no global parametrization, we suppose that Ω is the disjoint union of the parameter domains corresponding to local parametrization patches.

For our example, we now consider a quadrature partition $\Gamma=\bigcup_{k=1}^{K}\Gamma_k$ which corresponds to a triangulation of the parameter domain Ω. We may suppose that the parametrization γ is analytic over each panel Γ_k since otherwise we can replace γ by a piecewise polynomial parametrization which is polynomial over the parametrization domain $\gamma^{-1}(\Gamma_k)$ (for an estimate of such an replacement cf., e.g., [14]). Note that the integrand $s'\mapsto k_{\mathrm{si}}(\gamma(s),\gamma(s'))|\gamma'(s')|$ is analytic over all triangular subdomains $\gamma^{-1}(\Gamma_k)$ with a possible singularity at $s'=s$. The degree of smoothness of k_{sm} is determined by the degree of smoothness of k_{00} and of γ.

Let us turn back to the general case. To simplify the notation, we suppose from now on, that the knots $\tau_{k,l}$ are located in the interior of the triangular panels Γ_k. Moreover, we shall call the triangulation $\Gamma=\bigcup_{k=1}^{K}\Gamma_k$ locally quasi-uniform if

(i) There is an $\varepsilon>0$ such that the interior angles of the triangles $\gamma^{-1}(\Gamma_k)$ are all bounded between ε and $\pi-\varepsilon$.

(ii) There exists constants $c>0$ and $\beta\geqslant 1$ such that the quadrature step size $h:=\max\{\operatorname{diam}\Gamma_k\colon k=1,\dots,K\}$ satisfies the estimate $ch^{\beta}\leqslant\min\{\operatorname{diam}\Gamma_k\colon k=1,\dots,K\}$.

(iii) There is a constant $C>0$ such that, for any two nonneighbor subdomains Γ_k and $\Gamma_{k'}$, we have $\operatorname{diam}\Gamma_k\leqslant C\operatorname{dist}(\Gamma_k,\Gamma_{k'})$.

As before, we call method (2.22) stable, if (2.22) is uniquely solvable for any right-hand side at least for sufficiently small h and if the norm of the matrix

$$[(a\delta_{(k,l),(k',l')}+k_{\mathrm{sm}}(\tau_{k,l},\tau_{k',l'})\omega^{p}_{k',l'})_{(k,l),(k',l')}]^{-1}$$

inverse to the matrix of system (2.22) is uniformly bounded for all locally quasi-uniform partitions with sufficiently small step size h. The norm of the matrix is the one induced by the L^2 space. Since

$$\left\|\sum_{k=1}^{K}\sum_{l=1}^{L}\xi_{k,l}\varphi_{k,l}\right\|_{L^2(\Gamma)}\sim\sqrt{\sum_{k=1}^{K}\sum_{l=1}^{L}\varrho_k^2|\xi_{k,l}|^2},\quad \varrho_k:=\sqrt{\int_{\Gamma_k}1\,\mathrm{d}_{\Gamma}t} \tag{5.3}$$

holds for any sequence of numbers $\xi_{k,l}$, the norm of the matrix is the Euclidean matrix norm of

$$[aI+(\varrho_k k_{\mathrm{sm}}(\tau_{k,l},\tau_{k',l'})\omega^{p}_{k',l'}\varrho_{k'}^{-1})_{(k,l),(k',l')}]^{-1}.$$

As mentioned in Section 2, the quadrature method based on product integration is a perturbation of the collocation method where the trial functions are functions spanned by $\varphi_{k,l}$. More precisely, the collocation method seeks an approximate solution $\tilde{x}$ for the exact solution x of (2.5) in the span of the functions $\varphi_{k,l}$ such that $A\tilde{x}(\tau_{k,l}) = y(\tau_{k,l})$ holds for any point $\tau_{k,l}$. The coefficients of $\tilde{x}$ with respect to the basis functions $\varphi_{k,l}$ are to be determined from a system of linear equations including the so-called stiffness matrix $(A\varphi_{k',l'}(\tau_{k,l}))_{(k,l),(k',l')}$. Analogously to the quadrature method, the collocation is called stable if the stiffness matrix is invertible at least for small step size h and if the Euclidean matrix norm of the inverse matrices $(\varrho_k A\varphi_{k',l'}(\tau_{k,l})\varrho_{k'}^{-1})^{-1}_{(k,l),(k',l')}$ are uniformly bounded. The stability analysis of these collocation methods for two-dimensional manifolds is a difficult task. It seems, there exist only very few results for special cases (cf. [24,42] and, for similar operator equations, cf. [3,12,13,43,46,60]). On the other hand, many engineers use collocation methods successfully without observing any stability problem. If stability is true, then the derivation of the usual convergence results for the collocation is not difficult.

Theorem 8. *We suppose that the partition* $\Gamma = \bigcup_{k=1}^{K} \Gamma_k$ is *locally quasi-uniform. Furthermore, we suppose that the parametrization* γ *is analytic over each subdomain* Γ_k *and* $m_p + 1$ *times continuously differentiable. Recall that* $m_p \geqslant 2$ *is the order of approximation of the interpolation* $f \mapsto \sum_{k,l} f(\tau_{k,l})\varphi_{k,l}$ *and the order of the product rule* (2.21). *We assume that the kernel of* (2.5) *admits a factorization* $k(t,\tau) = k_{\rm sm}(t,\tau)k_{\rm si}(t,\tau)$ *such that the factor* $k_{\rm sm}$ *is* m_p *times continuously differentiable and that* $k_{\rm si}$ *satisfies* (2.12) *with* $m = 0$ *and* $d = 2$. *For the exact solution* x *of* (2.5), *we suppose the existence of square integrable derivatives up to order* m_p. *Finally, we suppose that the integral operator on the right-hand side of* (2.5) *is invertible and that the collocation method based on the trial basis functions* $\varphi_{k,l}$ *and the collocation points* $\tau_{k,l}$ *is stable. Then the quadrature method* (2.22) *based on product quadrature is stable, too. Moreover, we get the error estimate*

$$\|\tilde{x}_h - x\|_{L^2(\Gamma)} \leqslant Ch^{m_p} \log h^{-1}, \quad \tilde{x}_h(t) := \sum_{l=1}^{L} \tilde{x}_h(\tau_{k,l})\varphi_{k,l}(t) \quad if\ t \in \Gamma_k. \tag{5.4}$$

Proof. We have to show two things. First, to obtain stability, we have to prove that the matrix of the quadrature method is a small perturbation of the collocation matrix with respect to the norm. Second, to show the error estimate, we have to derive consistency, i.e., we have to consider the difference of the quadrature discretized operator applied to the exact solution minus the operator applied to the exact solution and to prove that the result can be estimated by the right-hand side of the estimate in (5.4).

For the difference of the matrix entries corresponding to the quadrature and collocation matrices, we get

$$\begin{aligned} d_{(k,l),(k',l')} &:= k_{\rm sm}(\tau_{k,l}, \tau_{k',l'})\omega^p_{k',l'} - \int_{\Gamma_{k'}} k(\tau_{k,l}, t)\varphi_{k',l'}(t)\,\mathrm{d}_\Gamma t \\ &= \int_{\Gamma_{k'}} [k_{\rm sm}(\tau_{k,l}, t_{k',l'}) - k_{\rm sm}(\tau_{k,l}, t)]k_{\rm si}(\tau_{k,l}, t)\varphi_{k',l'}(t)\,\mathrm{d}_\Gamma t. \end{aligned} \tag{5.5}$$

In view of the local uniformness of the mesh, we conclude, for $t \in \Gamma_k$ and $t' \in \Gamma_{k'}$ with disjoint Γ_k and $\Gamma_{k'}$, that (cf. condition (iii) of the local uniformness)

$$\operatorname{dist}(\Gamma_k, \Gamma_{k'}) \leqslant |t - t'| \leqslant \operatorname{dist}(\Gamma_k, \Gamma_{k'}) + \operatorname{diam}\Gamma_k + \operatorname{diam}\Gamma_{k'}$$

$$\leqslant (1+2C)\mathrm{dist}(\Gamma_k, \Gamma_{k'}),$$

$$|t-t'|^{-2} \sim |\tau_{k,l} - \tau_{k',l'}|^{-2}. \tag{5.6}$$

Similarly, for neighbors Γ_k and $\Gamma_{k'}$, the definition of the points $\tau_{k,l} = \gamma(\sigma_{k,l})$ as affine images of interior points σ_k in the standard triangle, implies

$$|t-\tau_{k',l'}|^{-2} \sim |\tau_{k,l} - \tau_{k',l'}|^{-2} \tag{5.7}$$

for any $t \in \Gamma_k$. Using this, the estimate $|\varphi_{k,l'}(t)| \leqslant C\,\mathrm{diam}(\Gamma_k)^{-1}|t-\tau_{k,l}|$ valid for $l \neq l'$, condition (2.12), representation (5.5), and the differentiability of kernel k_{sm}, we arrive at

$$|d_{(k,l),(k',l')}| \leqslant C \begin{cases} h\,|\tau_{k,l} - \tau_{k',l'}|^{-2}\varrho_{k'}^2 & \text{if } k \neq k', \\ h & \text{if } k = k'. \end{cases} \tag{5.8}$$

We estimate the norm of the corresponding matrix by Schur's lemma to get

$$n := \|(\varrho_k d_{(k,l),(k',l')} \varrho_{k'}^{-1})_{(k,l),(k',l')}\|$$

$$\leqslant \sup_{k,l} \left\{ \sum_{k',l'} |d_{(k,l),(k',l')}| \right\} \sup_{k',l'} \left\{ \sum_{k,l} \varrho_k^2 |d_{(k,l),(k',l')}| \varrho_{k'}^{-2} \right\}. \tag{5.9}$$

Now, inequality (5.8) together with (5.6),(5.7) and property (iii) of local uniformness of the quadrature partition lead to

$$\sum_{k',l'} |d_{(k,l),(k',l')}| \leqslant Ch + Ch \sum_{k',l'} |\tau_{k',l'} - \tau_{k,l}|^{-2} \varrho_{k'}^2$$

$$\leqslant Ch + Ch \int_{\Gamma \setminus \Gamma_k} |t - \tau_{k,l}|^{-2} \,\mathrm{d}_\Gamma t \leqslant Ch \log h^{-1},$$

$$\sum_{k,l} \varrho_k^2 |d_{(k,l),(k',l')}| \varrho_{k'}^{-2} \leqslant Ch + Ch \sum_{k,l} |\tau_{k',l'} - \tau_{k,l}|^{-2} \varrho_k^2 \leqslant Ch \log h^{-1}.$$

Hence, the difference of the quadrature discretized operator minus the collocation discretized operator has a norm n less than $Ch \log h^{-1}$.

Next, we turn to the estimation of the difference of the quadrature discretized operator minus the full operator applied to the exact solution. Thus, we have to estimate the norm of $\sum_{k,l} d_{k,l}\varphi_{k,l}$ with

$$d_{k,l} := \sum_{k',l'} k_{\mathrm{sm}}(\tau_{k,l}, \tau_{k',l'}) x(\tau_{k',l'}) \omega^p_{k',l'} - \int_\Gamma k(\tau_{k,l}, t) x(t) \,\mathrm{d}_\Gamma t$$

$$= \int_\Gamma k_{\mathrm{si}}(\tau_{k,l}, t) \{ L[k_{\mathrm{sm}}(\tau_{k,l}, t) x(t)] - [k_{\mathrm{sm}}(\tau_{k,l}, t) x(t)] \} \,\mathrm{d}_\Gamma t,$$

where L stands for the interpolatory projection, i.e., $L[k_{\mathrm{sm}}(\tau_{k,l}, t)x(t)] := \sum_{k',l'} k_{\mathrm{sm}}(\tau_{k,l}, \tau_{k',l'}) x(\tau_{k',l'})$ $\varphi_{k',l'}(t)$. We split $d_{k,l} = d^1_{k,l} + d^2_{k,l}$ with

$$d^1_{k,l} := \int_{\Gamma_k} k_{\mathrm{si}}(\tau_{k,l}, t) \{ L[k_{\mathrm{sm}}(\tau_{k,l}, t) x(t)] - [k_{\mathrm{sm}}(\tau_{k,l}, t) x(t)] \} \,\mathrm{d}_\Gamma t,$$

$$d^2_{k,l} := d_{k,l} - d^1_{k,l}$$

and estimate the norms of $\sum_{k,l} d^1_{k,l}\varphi_{k,l}$ and $\sum_{k,l} d^2_{k,l}\varphi_{k,l}$ separately. Using the approximation property of the interpolation as well as the smoothness assumptions for k_{sm} and x, we arrive at

$$|d^2_{k,l}| \leqslant C \sum_{k'\neq k,l'} |k_{\mathrm{si}}(\tau_{k,l},\tau_{k',l'})|\varrho_{k'}\sqrt{\int_{\Gamma_{k'}} |L[k_{\mathrm{sm}}(\tau_{k,l},t)x(t)] - [k_{\mathrm{sm}}(\tau_{k,l},t)x(t)]|^2\,\mathrm{d}_\Gamma t}$$

$$\leqslant C \sum_{k'\neq k,l'} |k_{\mathrm{si}}(\tau_{k,l},\tau_{k',l'})|\varrho_{k'}h^{m_p}\sqrt{\sum_{n=0}^{m_p}\int_{\Gamma_{k'}} |\nabla^n x(t)|^2\,\mathrm{d}_\Gamma t}.$$

This expression can be looked at as the result of multiplying the vector $(\sqrt{\sum_n \int_{\Gamma_{k'}} |\nabla^n x(t)|^2\,\mathrm{d}t})_{(k',l')}$ by a matrix. Hence, in view of (2.12) and (5.3), the norm of $\sum_{k,l} d^2_{k,l}\varphi_{k,l}$ is less than

$$Ch^{m_p}\|(\varrho_k|\tau_{k,l}-\tau_{k',l'}|^{-2}\varrho_{k'})_{(k,l),(k',l')}\|\sqrt{\sum_{n=0}^{m_p}\int_\Gamma |\nabla^n x(t)|^2\,\mathrm{d}_\Gamma t}. \tag{5.10}$$

Analogously to the estimate $h\log h^{-1}$ for (5.9), we get the estimate $C\log h^{-1}$ for the matrix norm in (5.10). Finally, the norm of $\sum_{k,l} d^2_{k,l}\varphi_{k,l}$ is less than the expression $Ch^{m_p}\log h^{-1}$ on the right-hand side of the estimate in (5.4).

Let us turn to $\sum_{k,l} d^1_{k,l}\varphi_{k,l}$. Over an arbitrary smooth and bounded two-dimensional manifold $\tilde{\Gamma}$, the functions of the Sobolev space H^2 are known to be Lipschitz, and we get that, for a fixed constant $C>0$, for any $\tilde{\tau}\in\tilde{\Gamma}$, and for any function $\tilde{f}$ on $\tilde{\Gamma}$,

$$\int_{\tilde{\Gamma}} |\tilde{\tau}-\tilde{t}|^{-2}|\tilde{f}(\tilde{\tau})-\tilde{f}(\tilde{t})|\,\mathrm{d}_{\tilde{\Gamma}}\tilde{t} \leqslant C\sqrt{\int_{\tilde{\Gamma}} |\nabla\tilde{f}(\tilde{t})|^2\,\mathrm{d}_{\tilde{\Gamma}}\tilde{t}} + C\sqrt{\int_{\tilde{\Gamma}} |\nabla^2\tilde{f}(\tilde{t})|^2\,\mathrm{d}_{\tilde{\Gamma}}\tilde{t}}.$$

Choosing $\tilde{\Gamma}:=\{\tilde{t}=t/\mathrm{diam}\,\Gamma_k\colon t\in\Gamma_k\}$, substituting the variable of integration $\tilde{t}$ by $t/\mathrm{diam}\,\Gamma_k$, and setting $f(t)=\tilde{f}(\tilde{t})$ and $f(\tau)=\tilde{f}(\tilde{\tau})$, we arrive at

$$\int_{\Gamma_k} |\tau-t|^{-2}|f(\tau)-f(t)|\,\mathrm{d}_{\Gamma_k}t \leqslant C\sqrt{\int_{\Gamma_k} |\nabla f(t)|^2\,\mathrm{d}_{\Gamma_k}t}$$

$$+\,C\,\mathrm{diam}\,\Gamma_k\sqrt{\int_{\Gamma_k} |\nabla^2 f(t)|^2\,\mathrm{d}_{\Gamma_k}t}, \quad \tau\in\Gamma_k.$$

Using this and the approximation property for projection L, we obtain that

$$|d^1_{k,l}| \leqslant C\left|\int_{\Gamma_k} |\tau_{k,l}-t|^{-2}|L[k_{\mathrm{sm}}(\tau_{k,l},t)x(t)] - [k_{\mathrm{sm}}(\tau_{k,l},t)x(t)]|\,\mathrm{d}_\Gamma t\right|$$

$$\leqslant C\sqrt{\int_{\Gamma_k} |\nabla\{L[k_{\mathrm{sm}}(\tau_{k,l},t)x(t)] - [k_{\mathrm{sm}}(\tau_{k,l},t)x(t)]\}|^2\,\mathrm{d}_\Gamma t}$$

$$+\,Ch\sqrt{\int_{\Gamma_k} |\nabla^2\{L[k_{\mathrm{sm}}(\tau_{k,l},t)x(t)] - [k_{\mathrm{sm}}(\tau_{k,l},t)x(t)]\}|^2\,\mathrm{d}_\Gamma t}$$

$$\leqslant Ch^{m_p-1}\sum_{n=0}^{m_p}\sqrt{\int_{\Gamma_k} |\nabla^n x(t)|^2\,\mathrm{d}_\Gamma t}.$$

Hence, in view of $\varrho_k \leqslant Ch$ and (5.3), we get that the norm of $\sum_{k,l} d^1_{k,l}\varphi_{k,l}$ is less than Ch^{m_p}, and the consistency order of (5.4) is shown. □

Note that, in view of the last proof we can relax the assumptions of Theorem 8. The global differentiability of γ and k_{sm} can be replaced by differentiability over each subdomain Γ_k together with the global boundedness of these local derivatives. This weaker assumption holds true when a parametrization is replaced by its piecewise polynomial interpolation. Furthermore, Theorem 8 remains true if the solution x has a weak singularity at a finite number of points. In this case, the mesh should be graded toward these points such that the larger values for the $\sqrt{\int_{\Gamma_k} |\nabla^{m_p} x|^2}$ in the estimates for the interpolation error $x - \sum x(\tau_{k,l})\varphi_{k,l}$ are compensated by the factors $[\operatorname{diam} \Gamma_k]^{m_p}$ which are smaller than h^{m_p}.

Further, we remark that the logarithm in the error estimate (5.4) can be dropped if the integral operator with the kernel function $|k(t,\tau)|$ is bounded in L^2. This last assumption holds true, e.g., for operators of double-layer type defined over non-smooth domains. Finally, a generalization of Theorem 8 to operators of order minus one and to piecewise linear collocation over regular grids has been treated in [14]. In that paper even a fast quadrature algorithm for a wavelet approach has been derived.

References

[1] B.K. Alpert, Hybrid Gauss-trapezoidal quadrature rules, SIAM J. Sci. Statist. Comput. 20 (1999) 1551–1584.

[2] B.A. Amosov, On the approximate solution of elliptic pseudodifferential equations on smooth curves, Z. Anal. Angew. 9 (1990) 545–563.

[3] K.E. Atkinson, The Numerical Solution of Integral Equations of the Second Kind, Cambridge University Press, Cambridge, 1997.

[4] C.T.H. Baker, The Numerical Treatment of Integral Equations, Oxford University Press, London, 1977.

[5] S.M. Belotserkovskij, I.K. Lifanov, Numerical Methods in Singular Integral Equations and their Application to Aerodynamics, Elasticity Theory and Electrodynamics, Nauka, Moscow, 1985 (in Russian).

[6] B. Bialecki, Y. Yan, A rectangular quadrature method for logarithmically singular integral equations of the first kind, J. Integral Equations Appl. 4 (1992) 337–369.

[7] K. Bühring, A quadrature method for a hypersingular integral equation on an interval, J. Integral Equations Appl. 7 (1995) 263–301.

[8] G.A. Chandler, C.B. Schneider, B. Weilbächer, A quadrature method for a Cauchy singular integral equation with weakly singular solution, Z. Angew. Math. Mech. 75 (Suppl. 2) (1995) 619–620.

[9] G. Chen, J. Zhou, Boundary Element Methods, Academic Press, London, 1992.

[10] D. Chien, Numerical evaluation of surface integrals in three dimensions, Math. Comp. 64 (1995) 727–743.

[11] D. Chien, K. Atkinson, A discrete Galerkin method for a hypersingular boundary integral equation, IMA J. Numer. Anal. 17 (1997) 463–478.

[12] M. Costabel, W. McLean, Spline collocation for strongly elliptic equations on the torus, Numer. Math. 62 (1992) 511–538.

[13] W. Dahmen, S. Prößdorf, R. Schneider, Wavelet approximation for pseudodifferential equations I: Stability and convergence, Math. Z. 215 (1994) 583–620.

[14] S. Ehrich, A. Rathsfeld, Piecewise linear wavelet collocation on triangular grids, Approximation of the boundary manifold and quadrature, WIAS preprint, 434, Berlin, 1998.

[15] J. Elschner, On spline approximation for a class of non-compact integral equations, Math. Nachr. 146 (1990) 271–321.

[16] J. Elschner, I.G. Graham, Numerical methods for integral equations of Mellin type, this issue, J. Comp. Appl. Meth. 250 (2000) 423–437.

[17] A. Gerasoulis, Nyström's iterative variant methods for the solution of Cauchy singular integral equations, SIAM J. Numer. Anal. 26 (1989) 430–441.
[18] M.A. Golberg, C.S. Chen, Discrete Projection Methods, Computational Mechanics Publications, Southampton, 1997.
[19] I. Graham, W. Hackbusch, S. Sauter, Discrete boundary element methods on general meshes in 3D, Bath mathematics preprints 97/05, 1997.
[20] I. Graham, W. Hackbusch, S. Sauter, Hybrid Galerkin boundary elements on degenerate meshes, Proceedings of Minisymposium on Mathematical Aspects of Boundary Elements Methods in honour of V.G. Maz'ya, Paris, 1998.
[21] I. Graham, W. Hackbusch, S. Sauter, Hybrid Galerkin boundary elements: Theory and Implementation, Ber. Math. Sem. Kiel 98-6, 1998.
[22] J.-L. Guermond, Numerical quadratures for layer potentials over curved domains in $\mathbb{R}^3$, SIAM J. Numer. Anal. 29 (1992) 1347–1369.
[23] W. Hackbusch, Integral Equations. Theory and Numerical Treatment, Birkhäuser, Basel, 1995.
[24] W. Hoppe, Stabilität von Splineapproximationsverfahren für singuläre Integralgleichungen auf kompakten, glatten Mannigfaltigkeiten ohne Rand, Dissertationsschrift, TU Chemnitz, 1994.
[25] L. Hörmander, The Analysis of Linear Partial Differential Equations. III Pseudo-Differential Operators, Springer, Berlin, 1985.
[26] C.G.L. Johnson, L.R. Scott, An analysis of quadrature errors in second-kind boundary integral methods, SIAM J. Numer. Anal. 26 (1989) 1356–1382.
[27] R. Kieser, B. Kleemann, A. Rathsfeld, On a full discretization scheme for a hypersingular boundary integral equation over smooth curves, Z. Anal. Angew. 4 (1992) 385–396.
[28] R. Kieser, C. Schwab, W.L. Wendland, Numerical evaluation of singular and finite-part integrals on curved surfaces using symbolic manipulation, Computing 49 (1992) 279–301.
[29] R. Kress, On the numerical solution of a hypersingular integral equation in scattering theory, J. Comput. Appl. Math. 61 (1995) 345–360.
[30] R. Kress, I.H. Sloan, On the numerical solution of a logarithmic integral equation of the first kind, Numer. Math. 66 (1993) 199–214.
[31] R. Kussmaul, Ein numerisches Verfahren zur Lösung des Neumannschen Außenraumproblems für die Helmholtzsche Schwingungsgleichung, Computing 4 (1969) 246–273.
[32] I.K. Lifanov, Singular Integral Equations and Discrete Vortices, Utrecht, VSP ix, 1996.
[33] G. Mastroianni, G. Monegato, Nyström interpolants based on the zeros of Legendre polynomials for a non-compact integral operator equation, IMA J. Numer. Anal. 14 (1993) 81–95.
[34] V.G. Mazya, Boundary Integral Equations, Encyclopaedia of Math. Sciences, Vol. 27, Analysis IV, Springer, Berlin, 1991.
[35] W. McLean, Fully-discrete collocation methods for an integral equation of the first kind, J. Integral Equations Appl. 6 (1994) 537–571.
[36] W. McLean, I.H. Sloan, A fully discrete and symmetric boundary element method, IMA J. Numer. Anal. 14 (1994) 311–345.
[37] G. Monegato, L. Scuderi, High order methods for weakly singular integral equation with non-smooth input functions, Math. Comp. 67 (1998) 1493–1515.
[38] H.N. Mülthei, C.B. Schneider, Discrete collocation for a first kind Cauchy singular integral equation with weakly singular solution, J. Integral Equations Appl. 9 (1997) 341–371.
[39] E.J. Nyström, Über die praktische Auflösung von linearen Integralgleichungen mit Anwendungen auf Randwertaufgaben der Potentialtheorie, Comment. Phys.-Math. 4 (1928) 1–25.
[40] F. Penzel, Error estimates for discretized Galerkin method for a boundary integral equation in two dimensions, Numer. Methods Partial Differential Equations 8 (1992) 405–421.
[41] F. Penzel, Error estimates for discretized Galerkin and collocation boundary element methods for time harmonic Dirichlet screen problems in R^3, Numerical Analysis and Mathematical modelling, Banach Center Publications 29 (1994) 115–134.
[42] S. Prößdorf, R. Schneider, A spline collocation method for multidimensional strongly elliptic pseudodifferential operators of order zero, Integral Equations Operator Theory 14 (1991) 399–435.
[43] S. Prößdorf, R. Schneider, Spline approximation methods for multidimensional periodic pseudodifferential equations, Integral Equations Operator Theory 15 (1992) 287–626.

[44] S. Prößdorf, B. Silbermann, Numerical Analysis of Integral and Related Operator Equations, Akademie Verlag, Berlin, 1991.

[45] A. Rathsfeld, On quadrature methods for the double layer potential equation over the boundary of a polyhedron, Numer. Math. 66 (1993) 67–95.

[46] A. Rathsfeld, Piecewise polynomial collocation for the double layer potential equation over polyhedral boundaries, in: M. Dauge, M. Costabel, S. Nicaise (Eds.), Boundary Value Problems and Integral Equations in Non-Smooth Domains, Lecture Notes in Applied Mathematics, Marcel Dekker, Basel, 1994, pp. 219–253.

[47] A. Rathsfeld, Nyström's method and iterative solvers for the solution of the double layer potential equation over polyhedral boundaries, SIAM J. Numer. Anal. 32 (1995) 924–951.

[48] I. Saad Abdel-Fattah, Stability analysis of quadrature methods for two-dimensional singular equations, Ph.D. Thesis, University of El-Mansoura, 1996.

[49] J. Saranen, On the effect of numerical quadratures in solving boundary integral equations, Notes on Numerical Fluid Mechanics 21 (1988) 196–209.

[50] J. Saranen, The modified quadrature method for logarithmic-kernel integral equations on closed curves, J. Integral Equations Appl. 3 (1991) 575–600.

[51] J. Saranen, L. Schroderus, Quadrature methods for strongly elliptic equations of negative order on smooth closed curves, SIAM J. Numer. Anal. 30 (1993) 1769–1795.

[52] J. Saranen, I.H. Sloan, Quadrature methods for logarithmic-kernel integral equations on closed curves, IMA J. Numer. Anal. 12 (1992) 167–187.

[53] S. Sauter, Cubature techniques for 3-D Galerkin BEM, in: W. Hackbusch, G. Wittum (Eds.), Boundary Elements: Implementation and Analysis of Advanced Algorithms, Notes on Numerical Fluid Mechanics, Vol. 54, Vieweg Verlag, Braunschweig, Wiesbaden, 1996.

[54] C.B. Schneider, Inversion formulas for the discrete Hilbert transform on the unit circle, SIAM J. Numer. Anal. 35 (1998) 71–77.

[55] C. Schwab, Variable order composite quadrature of singular and nearly singular integrals, Computing 53 (1994) 173–194.

[56] C. Schwab, W.L. Wendland, On numerical cubatures of singular surface integrals in boundary element methods, Numer. Math. 62 (1992) 343–369.

[57] I.H. Sloan, B.J. Burn, An unconventional quadrature method for logarithmic kernel integral equations on closed curves, J. Integral Equations Appl. 4 (1992) 117–151.

[58] F. Stenger, Numerical Methods Based on Sinc and Analytic Functions, Springer Series in Computational Mathematics, Springer, New York, 1993.

[59] G. Vainikko, Periodic integral and pseudodifferential equations, Helsinki Univ. of Technology, Inst. of Math., Research Report C13, Helsinki, 1996.

[60] W. Wendland, Die Behandlung von Randwertaufgaben im $\mathbb{R}^3$ mit Hilfe von Einfach- und Doppelschichtpotentialen, Numer. Math. 11 (1968) 308–404.

JOURNAL OF
COMPUTATIONAL AND
APPLIED MATHEMATICS

ELSEVIER Journal of Computational and Applied Mathematics 125 (2000) 461–478

www.elsevier.nl/locate/cam

Qualocation

Ian H. Sloan

School of Mathematics, University of New South Wales, Sydney 2052, Australia

Received 1 October 1999

Abstract

The qualocation method for boundary integral equations on smooth curves is reviewed, with an emphasis on recent developments, including 'second-generation' qualocation rules and 'tolerant qualocation'. Using smoothest splines and uniform meshes, this version of the qualocation method can achieve the same convergence order as the Galerkin method, with no additional smoothness requirement on the solution, for a wide range of boundary integral operators. Included are singular integral equations with nonconstant coefficients. © 2000 Elsevier Science B.V. All rights reserved.

1. Introduction

Qualocation was introduced in the late 1980s [5,12,13,30,36] as a compromise between the Galerkin and collocation methods: roughly speaking, it aimed, in the context of spline approximation methods for boundary integral equations on curves, to achieve the benefits of the Galerkin method at a cost comparable to the collocation method.

At the one extreme, the Galerkin methods (reviewed, for example, in [25,31,41]) have a robust and elegant stability analysis, and beautiful convergence and superconvergence properties; yet they are costly to implement, and indeed are usually impossible to implement without further approximation. At the other extreme the collocation method (see [3,4,28,31]) is simpler to implement, but has a more delicate stability analysis, and does not in general (as we shall see) have the same superconvergence properties. The qualocation method (the name means 'quadrature modified collocation method') aims to achieve the best properties of both.

In structure, the qualocation method is similar to the Galerkin method, but with the exact inner products in the Galerkin method replaced by carefully tailored quadrature rules. For earlier reviews of qualocation, see [31,32].

To allow us to define the qualocation method more precisely, some definitions are needed. We suppose that our boundary integral equation (of which examples are given later) lives on a smooth curve Γ which is the boundary of a simply connected bounded domain $\Omega \subseteq \mathbb{R}^2$. After suitable

PII: S0377-0427(00)00485-4

parametrisation of the curve Γ, the boundary integral equation can be written as

$$(Au)(x) = f(x), \quad x \in [0,1], \tag{1.1}$$

where A is an integral operator on a space of (real- or complex-valued) 1-periodic functions, f is a given function, and u is the unknown function.

Spline methods begin with the introduction of a partition of the (periodic) interval $[0,1]$:

$$0 = x_0 < x_1 < \cdots < x_{n-1} < 1. \tag{1.2}$$

Using the periodic labelling convention, $x_{k+n} = x_k$ for all k, we define $h_k := x_{k+1} - x_k$ as the length of the kth interval and $h := \max_k h_k$ as the maximum length of an interval. Unless stated otherwise, we shall assume that the partition is uniform, i.e. $x_k = k/n$ for $k = 0, \ldots, n-1$, and $h = h_k = 1/n$. (In particular, irregular meshes are not allowed, an important limitation on qualocation methods.)

Now let S_h denote the space of smoothest splines of order $r \geqslant 1$ on the partition $\{x_k\}$. That is to say, $v_h \in S_h$ is a polynomial of degree $\leqslant r-1$ on each sub-interval $[x_k, x_{k+1}]$, and $v_h \in C^{r-2}$. Thus S_h is a space of piecewise constant functions if $r = 1$, of continuous piecewise-linear functions if $r = 2$, and of cubic splines if $r = 4$. (In principle splines of lesser smoothness, such as continuous quadratics, can be used, but the obstacles in the way of the analysis are considerable, see [22], and qualocation methods for such spaces have so far not been developed.)

The Galerkin method for Eq. (1.1) and the space S_h is: find $u_h^{\mathrm{G}} \in S_h$ such that

$$(Au_h^{\mathrm{G}}, \chi_h) = (f, \chi_h) \quad \forall \chi_h \in S_h, \tag{1.3}$$

where $(\cdot,\cdot)$ denotes the L_2 inner product

$$(f,g) := \int_0^1 f(x)\overline{g(x)}\,\mathrm{d}x.$$

In contrast, the qualocation method is: find $u_h^{\mathrm{Q}} \in S_h$ such that

$$\langle Au_h^{\mathrm{Q}}, \chi_h \rangle = \langle f, \chi_h \rangle \quad \forall \chi_h \in S_h, \tag{1.4}$$

where

$$\langle f,g \rangle := \sum_{k=0}^{n-1} h_k \sum_{j=1}^{J} w_j (f\bar{g})(x_k + \xi_j h_k) \tag{1.5}$$

and the parameters ξ_j and w_j for $j = 1, \ldots, J$ are fixed numbers satisfying

$$0 \leqslant \xi_1 < \cdots < \xi_J < 1, \quad w_j > 0 \quad \text{for } j = 1, \ldots, J. \tag{1.6}$$

The expression on the right-hand side of (1.5) can be thought of as the composite quadrature rule that results from applying to the product $(f\bar{g})$ a scaled version of the J-point quadrature formula

$$qF := \sum_{j=1}^{J} w_j F(\xi_j) \approx \int_0^1 F(x)\,\mathrm{d}x \tag{1.7}$$

on each sub-interval $[x_k, x_{k+1}]$. The real story of qualocation begins with the observation that we have great freedom in the choice of the rule q. If we wish we can design new rules that depend on A and S_h, rather than rely on any of the usual choices such as Simpson's rule, Gauss rule, etc. We shall return to the question of the choice of q shortly.

For the practical implementation of either the Galerkin or qualocation methods we first need to select a basis $\{\phi_1,\dots,\phi_n\}\subseteq S_h$. In the case $r=2$, for example, one may take the familiar 'hat functions', or for the case $r=4$ the cubic B-splines. Then (taking the Galerkin case first) the approximation may be written as

$$u_h^{\mathrm{G}}=\sum_{i=1}^{n} a_i\phi_i,$$

where (from (1.3)) the coefficients satisfy the $n\times n$ linear system

$$\sum_{i=1}^{n}(A\phi_i,\phi_k)a_i=(f,\phi_k),\quad k=1,\dots,n.$$

The $n\times n$ matrix with elements $(A\phi_i,\phi_k)$ is invariably dense, expensive to compute with the spline bases used here, and generally impossible to compute exactly, since each element is a double integral (one, the 'inner' integral, comes from the operation of A on ϕ_i, the other, the 'outer' integral, is the inner-product integral).

The qualocation matrix, instead, has elements $\langle A\phi_i,\phi_k\rangle$, in which the inner integral still has to be evaluated, but the outer integral is now replaced by the quadrature rule (1.5) with Jn points, meaning that $A\phi_i$ need be evaluated only at these points.

It is in the choice of the 'qualocation rule' q that the distinctive character of the qualocation method lies. In this review we shall be mainly concerned with recent developments, especially the 'second generation' qualocation rules of [37], see Section 4, which allow the method to be applied much more widely, and the 'tolerant' variant of the qualocation method, which allows the full results of the Galerkin method to be recovered. However, in this introductory section we indicate the possibilities in the simplest way, by describing instead the oldest qualocation method (introduced in [30]), applied to one of the simplest boundary integral equations.

Consider, then, the particular case of the boundary integral equation

$$-\frac{1}{\pi}\int_\Gamma \log|X-Y|U(Y)\,\mathrm{d}l(Y)=F(X),\quad X\in\Gamma, \tag{1.8}$$

where $\mathrm{d}l(Y)$ is the element of arc length, and $|X-Y|$ the Euclidean distance between points X and Y on Γ. This equation (discussed in detail in [31]) arises, for example, if the solution of the Dirichlet problem for the Laplacian, i.e., $\Delta\Phi=0$ in Ω, with $\Phi=F$ on Γ, is expressed as a 'single-layer potential'

$$\Phi(X_0)=-\frac{1}{\pi}\int_\Gamma \log|X_0-Y|U(Y)\,\mathrm{d}l(Y),\quad X_0\in\Omega, \tag{1.9}$$

where U is an unknown 'charge-density'. We shall therefore refer to (1.8) as the 'single-layer equation for the Laplacian', or just 'the single-layer equation'.

To allow us to put (1.8) into the required form (1.1), the first step is to parametrise the smooth curve Γ by a 1-periodic function $v\in C^\infty(\mathbb{R})$, $v:[0,1]\to\Gamma$, with $|v'(x)|\neq 0$, in which case (1.8) becomes

$$-2\int_0^1 \log|v(x)-v(y)|u(y)\,\mathrm{d}y=f(x), \tag{1.10}$$

where

$$f(x) = F(v(x)), \qquad u(x) = \frac{1}{2\pi}U(v(x))|v'(x)|, \quad x \in \mathbb{R}.$$

This is of form (1.1) with $A = V$, where

$$(Vu)(x) := -2\int_0^1 \log|v(x) - v(y)|u(y)\,\mathrm{d}y, \quad x \in \mathbb{R}. \tag{1.11}$$

Suppose, now, that Eq. (1.10) is approximated by the Galerkin method described above, with $r=2$ so that u_h^{G} is a continuous piecewise-linear function on the uniform mesh $\{x_k\}$. What is so wonderful about the Galerkin method? To understand this we need to appreciate that the error $u - u_h^{\mathrm{G}}$ can be measured in many different ways. (In the following we quote well-known results, see for example [15].) At the simplest level, the Galerkin equation has a unique solution for h sufficiently small, and the error in the L_2 norm has the optimal order, namely

$$\|u - u_h^{\mathrm{G}}\|_0 \leqslant Ch^2\|u\|_2.$$

(Here C is a generic constant, $\|v\|_0$ denotes the L_2 norm $\|v\|_0 = (v, v)^{1/2}$, and $\|\cdot\|_2$ is the norm in the Sobolev space H^2, of L_2 functions whose second derivatives are in L_2; we are therefore assuming implicitly that u is smooth enough to belong to H^2.) This unsurprising result in the L_2 norm is easily achieved by many other approximation methods, including the collocation method, see below.

Suppose, however, that we are interested not in u itself, but rather in $\Phi(X_0)$ at some point $X_0 \in \Omega$, where $\Phi(X_0)$ is the potential given by (1.9), or equivalently given in terms of u by

$$\Phi(X_0) = -\frac{1}{\pi}\int_0^1 \log|X_0 - v(y)|u(y)\,\mathrm{d}y.$$

If we compute the Galerkin approximation to $\Phi(X_0)$ by

$$\Phi_h^{\mathrm{G}}(X_0) := -\frac{1}{\pi}\int_0^1 \log|X_0 - v(y)|u_h^{\mathrm{G}}(y)\,\mathrm{d}y \tag{1.12}$$

with the integral evaluated exactly, then it follows from results in [15] that the error $\Phi(X_0) - \Phi_h^{\mathrm{G}}(X_0)$ is of the remarkably high order h^5, that is,

$$|\Phi(X_0) - \Phi_h^{\mathrm{G}}(X_0)| \leqslant Ch^5\|u\|_2, \tag{1.13}$$

where C is a constant which depends on X_0 but not on h.

More generally, the h^5 order of convergence of the Galerkin method seen in (1.13) is experienced whenever we compute not u itself but rather an inner product

$$(u, w) := \int_0^1 u(x)\overline{w(x)}\,\mathrm{d}x$$

for w a suitably smooth function: the error, if w has three square-integrable derivatives, is from [15]

$$|(u, w) - (u_h^{\mathrm{G}}, w)| = |(u - u_h^{\mathrm{G}}, w)| \leqslant Ch^5\|u\|_2\|w\|_3. \tag{1.14}$$

This hidden convergence of the Galerkin method, which can be seen only if u is post-processed in some such way as above, is conveniently expressed through the use of 'negative norms', which may be defined for $s > 0$ by

$$\|v\|_{-s} := \sup_{0 \neq w \in H^s} \frac{|(v, w)|}{\|w\|_s}. \tag{1.15}$$

We may now write, equivalent to (1.14),

$$\|u - u_h^G\|_{-3} \leqslant Ch^5 \|u\|_2. \tag{1.16}$$

This is the form in which such results are stated in [15]. (Here and elsewhere in the paper to make the results as simple as possible we shall generally state only the strongest results available. Analogous results are available for $u \in H^t$ with $t < 2$, or in norms $\|u - u_h^G\|_s$ with $s > -3$, at the expense of corresponding reductions in the exponent of h.)

For most practical methods the $O(h^5)$ error of the Galerkin method in appropriate negative norms is hard to achieve. One that does achieve this order is the oldest qualocation method, from [30], which uses a 2-point rule (i.e., $J = 2$), with one quadrature point at the end-point and the other at the mid-point, and to that extent (when we think of applying this rule to sub-intervals put together) is like Simpson's rule. But the weights are not the Simpson weights: for the case $r = 2$ and the single-layer equation (1.10) the rule is

$$qF = \tfrac{3}{7}F(0) + \tfrac{4}{7}F(\tfrac{1}{2}).$$

This '$\frac{3}{7}, \frac{4}{7}$' rule, as we shall call it, is tailor-made to ensure that the qualocation method for the single-layer potential for the Laplacian yields

$$\|u - u_h^Q\|_{-3} \leqslant Ch^5 \|u\|_4, \tag{1.17}$$

which has the same high order of convergence as the Galerkin method.

Admittedly, the high order of convergence in (1.17) comes at a certain cost: we see that u is now required to have four derivatives in L_2 instead of just 2. The 'extra smoothness' requirement in the estimate (1.17) is a characteristic defect of the original qualocation method, and is a serious limitation whenever the exact solution u is not smooth. Later in this review we shall see how the defect can be overcome, by changing to the 'tolerant' version of the qualocation method, see Section 6, and employing 'second generation' qualocation rules, (see Section 3).

For the present, though, we content ourselves by seeing that the high order of convergence really is achievable in practice with the $\frac{3}{7}, \frac{4}{7}$ qualocation rule. In this simple example (taken from [5]) the curve Γ is taken to be the circle of radius $\frac{1}{2}$ centred at the origin and the boundary function F is the restriction to Γ of a harmonic function

$$v(y) = \tfrac{1}{2}(\cos 2\pi y, \sin 2\pi y), \quad F(X) = F(X_1, X_2) = X_1$$

and the quantity computed is $\Phi(X_0)$, where $X_0 = (0.1, 0.2)$. In Table 1 we show, for various n, the error $|\Phi(X_0) - \Phi_h^Q(X_0)|$ (with $\Phi_h^Q(X_0)$ defined in analogy with (1.12), with u_h^G replaced by u_h^Q); and

Table 1
Errors and apparent orders of convergence ρ, single-layer equation for circle

n	Qualocation ($\frac{3}{7}, \frac{4}{7}$ rule)	ρ	Collocation	ρ
8	1.31 (−5)		5.06 (−4)	
16	3.71 (−7)	5.14	5.98 (−5)	3.08
32	1.12 (−8)	5.04	7.37 (−6)	3.02
64	3.50 (−10)	5.00	9.18 (−7)	3.00

we also show the apparent order of convergence ρ, computed as $\log_2$ of the ratio of errors for two successive values of n. Very clearly, the error from the qualocation method is small and converging rapidly, and the apparent order of convergence is approaching 5 in completely convincing way.

In the same table we show also the errors achieved with the (breakpoint) collocation method. In this method we again take an approximation $u_h^C \in S_h$, where S_h is the space of continuous piecewise linear functions on the uniform partition, but this time with the approximation being determined by collocation at the breakpoints; i.e. by

$$u_h^C \in S_h, \quad (Vu_h^C)(x_k) = f(x_k), \quad k = 0, \ldots, n-1. \tag{1.18}$$

Very clearly, the collocation error in Table 1 is *not* of order h^5: rather, it is of order h^3. This is consistent with a result in [3] (or see [31]), where it is proved that the (best) result available for the breakpoint collocation method with piecewise linear splines is

$$\|u - u_h^C\|_{-1} \leqslant Ch^3 \|u\|_2.$$

This is a good moment to observe that the collocation method (1.18) is in fact a special case of the qualocation method: it is easily seen that the qualocation equation (1.4) is equivalent (mathematically, but not necessarily numerically) to the collocation equation (1.18), if we take q to be the 1-point (i.e., $J = 1$) rule $qF = F(0)$. In this article we are mainly interested in qualocation methods that have better properties than the collocation method, thus we will from now on require $J \geqslant 2$.

Before finishing this introductory section we make one minor extension to the Galerkin and qualocation methods as described above: in both methods it is sometimes desirable, or even essential, to use not one but two spaces of splines: one in which to seek the solution, the other in which to 'test' it. Thus, just as we have defined a space S_h of smoothest splines of order $r \geqslant 1$, now we fix also $r' \geqslant 1$ and define a space S_h' of smoothest splines of order r'. Then the Petrov–Galerkin scheme studied, for example, by Saranen [26] is, instead of (1.3): find $u_h^G \in S_h$ such that

$$(Au_h^G, \chi_h) = (f, \chi_h) \quad \forall \chi_h \in S_h' \tag{1.19}$$

and the qualocation scheme, instead of (1.4) is: find $u_h^Q \in S_h$ such that

$$\langle Au_h^Q, \chi_h \rangle = \langle f, \chi_h \rangle \quad \forall \chi_h \in S_h'. \tag{1.20}$$

These are the forms we shall assume from now on.

The remainder of the paper is arranged as follows. In the next section we consider some of the boundary integral operators that can occur in practice, so that we can understand the kinds of problems that the qualocation methods need to solve. Then in Section 3, we introduce the 'second-generation' qualocation schemes. Briefly, these have the same form as before, and so are defined by (1.20), (1.5) and (1.7), but now have qualocation rules that are much more robust than the $\frac{3}{7}, \frac{4}{7}$ rule above. These second-generation rules are designed to handle a much wider range of boundary integral operators, including operators with nonconstant coefficients. In Section 4, we explain in briefest outline the theory behind the second-generation schemes, and then in somewhat more detail point to tables of such rules, and explain how further second-generation qualocation rules can be computed. A numerical illustration, for singular integral equations with nonconstant coefficients, is given in Section 5.

In Section 6, we introduce an easy yet important modification of the qualocation method, called (for reasons that will soon become clear) 'tolerant qualocation'. The tolerant version of qualocation differs only in that the inner products on the right-hand side of (1.20) are integrated exactly (or at any rate to high accuracy), which is computationally easy to do since all the real work is associated with computing the left-hand side matrix, and perhaps also with solving the linear system. Yet we shall see that this small change has the effect (when used in conjunction with second-generation rules) of overcoming completely the principal defect of the qualocation method, namely the additional smoothness requirements on the exact solution (when compared to the Galerkin method) in error bounds such as (1.17). *Tolerant qualocation, used in conjunction with second-generation rules, is the version of the qualocation method we would recommend in practice.*

Finally, in Section 7 we point to extensions to more difficult problems, such as the treatment of corners, splines other than smoothest splines, irregular meshes, fully discrete methods, and (above all in importance) extension to three-dimensional problems. In each such case the available results are fragmentary, and almost all of the interesting questions are open.

2. Boundary integral operators we may meet

The purpose of this section is not to give a systematic exposition of boundary integral operators, but rather to explain the principal properties of some important boundary integral operators on curves, because these influence very much the design and selection of qualocation methods (by which we mean the selection of r and r', as well as the qualocation rule q). For further discussion of boundary integral operators see [14,24,28,41], and Section 3 of Lamp et al. [17].

Starting with the single-layer operator (1.11), an essential aspect of this operator for a smooth curve is revealed if we take the particular case of the circle, and ask: what does the operator do to the Fourier mode $\psi_k(x) := e^{2\pi ikx}$, for $k \neq 0$? The surprisingly simple answer (see, for example, [19] or [31] Proposition 1), is $V\psi_k = |k|^{-1}\psi_k$, so that ψ_k is an eigenfunction with eigenvalue $|k|^{-1}$. In the language of pseudo-differential operators, or ψdo's (for which we refer to [1,20,29], and see also below) V is a ψdo of order -1 with (constant) 'symbol' $|\xi|^{-1}$. (The order is -1 because this is the degree of the positive homogeneous function $|\xi|^{-1}$.) Operators with nonconstant symbol are also important; one such is the operator B defined by $(Bu)(x) = a(x)(Vu)(x)$, where a is a given C^∞ 1-periodic function. Now the symbol of B (again based on the action of the operator on ψ_k for $k \neq 0$) is $a(x)|\xi|^{-1}$.

In general, if an operator L operating on 1-periodic functions has the result

$$(L\psi_k)(x) = \sigma(x;k)\psi_k, \quad k \neq 0,$$

when operating on $\psi_k(x) = e^{2\pi ikx}$, where $\sigma : \mathbb{R} \times \mathbb{R} \to \mathbb{C}$ is a C^∞ function of the first variable, and with respect to the second variable has the property

$$\sigma(x;\lambda\xi) = \lambda^\beta \sigma(x;\xi) \quad \text{for } \lambda \in (0,1) \quad \text{and} \quad \xi \in \mathbb{R}\setminus\{0\},$$

then L is a ψdo of order β and symbol $\sigma(x;\xi)$; see [29].

It is useful to think of ψdo's of order -1 as 'once-smoothing' operators. Another operator of order -1 is the operator J defined for $k \neq 0$ by $J\psi_k = k^{-1}\psi_k$, so that (apart from a constant factor) J is the operator of anti-differentiation. In this case the symbol is ξ^{-1}. Unlike the previous examples, in this case the symbol is 'odd', i.e., $\sigma(x,-\xi) = -\sigma(x,\xi)$ for $\xi \neq 0$, whereas the symbol in the previous example was 'even', i.e., $\sigma(x,-\xi) = \sigma(x,\xi)$.

The simplest example of an operator of order 0 is the identity operator, for which the symbol is of course 1, which is even. Another ψdo of order 0 is the Hilbert transform S, given by

$$(SU)(X) = \frac{1}{\pi \mathrm{i}} \int_\Gamma \frac{1}{Y-X} U(Y)\,\mathrm{d}Y,$$

where Γ is now a curve in the complex plane $\mathbb{C}$, and the integral is to be understood in the principal-value sense. The symbol for the Hilbert transform is sign ξ for $\xi \neq 0$ (see [19]) which is odd.

Pseudo-differential operators of order $+1$ are also important. One such is the operation of (tangential) differentiation, for which the symbol is $2\pi \mathrm{i}\xi$ for $\xi \neq 0$, which is odd. Another is the 'hypersingular' operator, for which the symbol is $|\xi|$, which is of course even.

So far the operators we have mentioned have had symbols which are either even or odd, and most have also been constant. However, many pseudo-differential operators of importance lack at least one of these properties. For example, the important 'singular integral operators', of the form $aI + bS$, are neither even nor odd if both a and b are nonzero, and if either a or b is nonconstant then the symbol is nonconstant: for the symbol is $a(x) + b(x)\,\mathrm{sign}\,\xi$. We shall consider singular integral operators in the next section.

3. Second-generation qualocation

The earliest qualocation rules, see [5], had limited applicability, in that they assumed the operator A in the boundary integral equation (1.1) to be (in the language explained in the preceding section) a pseudo-differential operator, or ψdo, with a principal symbol (i.e. the symbol of the highest order part of the operator) either even or odd, and constant. Moreover, any perturbation of the principal part of the operator was assumed to be an integral operator with an unlimited smoothing capability.

While some problems are of this type (a notable example being the single-layer equation for the Laplacian for a smooth curve Γ), most are not.

The 'second-generation' qualocation rules are designed to handle much more general problems, and for that reason are much more robust. In this section we explain the class of problems to be tackled, and specify more precisely the convergence results that are to be sought for second-generation qualocation methods. Our principal reference here is the recent paper [39].

As a first illustration of the broader problems to be tackled, consider the 'singular integral equation'

$$A(X)U(X) + \frac{B(X)}{\mathrm{i}\pi} \int_\Gamma \frac{U(Y)}{Y-X}\,\mathrm{d}Y + C(X) \int_\Gamma K(X,Y)U(Y)\,|\mathrm{d}Y| = F(X), \quad X \in \Gamma, \tag{3.1}$$

where Γ is a smooth curve in the complex plane, A, B, C are smooth complex-valued functions, and K is a given weakly singular kernel, which for the sake of definiteness we take to be

$$K(x,y) := \log|x-y| + K'(x,y) \tag{3.2}$$

with K' a C^∞ function of both variables. With Γ parametrised by the C^∞ function v and with $a(x) := A(v(\mathrm{e}^{2\pi \mathrm{i}x}))$ and $b(x) := B(v(\mathrm{e}^{2\pi \mathrm{i}x}))$, the principal symbol in this case is $a(x) + b(x)\,\mathrm{sign}\,\xi$ (see Section 2), which is of order 0, but in general neither even nor odd.

This example is more complicated than the examples considered in [5] in three ways: the symbol is neither even nor odd (unless A or B is absent); the symbol is not constant, since it depends on x; and the principal part of the operator (i.e. the part corresponding to the principal symbol) is accompanied, through (3.2), by a ψdo of lower order (namely, in this case, order -1).

To be able to apply the qualocation method we need to assume (as with the corresponding Galerkin method) that (3.1) is either 'strongly elliptic' or 'oddly elliptic', as defined below.

First, though, we state the problem in greater generality, by assuming, following [39], that the operator A in (1.1) has the form, for some $\beta \in \mathbb{Z}$,

$$Au := b_+(x)L_+^\beta u + b_-(x)L_-^\beta u + Ku, \tag{3.3}$$

where $b_\pm$ are 1-periodic complex-valued C^∞ functions, and L_+^β and L_-^β are constant-coefficient ψdo's of order β and (constant) symbol $|\xi|^\beta$ and $|\xi|^\beta \operatorname{sign} \xi$, respectively. (That is, $L_+^\beta \psi_k = |k|^\beta \psi_k$ and $L_-^\beta = |k|^\beta \operatorname{sign} k \psi_k$ for $k \neq 0$, with $\psi_k(x) = e^{2\pi i k x}$.) Concrete characterisations of the operators $L_\pm^\beta$ for $\beta = -1, 0$ and 1 are given in Section 2. Moreover, we assume that K can include any combination of ψdo's of lower integer order:

$$K := \sum_{i=1}^{\infty} (a_{i,+}(x)L_+^{\beta-i} + a_{i,-}(x)L_-^{\beta-i}) + K', \tag{3.4}$$

with $a_{i,\pm} \in C^\infty$ and with only a finite number of the $a_{i,\pm}$ allowed to be nonzero, and K' an integral operator with a kernel which is a C^∞ function of both variables.

We also assume, using the definitions in [39], that the operator in (3.3) is either 'strongly elliptic' or 'oddly elliptic': the operator is said to be (uniformly) strongly elliptic if there exists a C^∞ complex-valued function Θ such that

$$\inf_{x \in \mathbb{R}} \min\{\operatorname{Re}(\Theta(x)(b_+(x) + b_-(x))),\ \operatorname{Re}(\Theta(x)(b_+(x) - b_-(x)))\} > 0.$$

If $b_\pm$ are real, this is equivalent to the assumption

$$b_+(x) + \lambda b_-(x) \neq 0 \quad \text{for all } \lambda \in [-1, 1] \quad \text{and} \quad x \in \mathbb{R}.$$

It is (uniformly) oddly elliptic if the same property holds with b_+ and b_- interchanged. Roughly, the operator A is strongly elliptic if the first term of (3.3) dominates, it is oddly elliptic if the second term dominates.

Assumption 3.1. If A is strongly elliptic then the spline orders r and r' are chosen either both even or both odd. If A is oddly elliptic then r and r' are chosen of opposite parity.

This assumption is natural even for the Galerkin method. (If A is *both* strongly elliptic and oddly elliptic, see [37], then either choice can be made.)

Even if r and r' satisfy Assumption 3.1, it is not necessarily the case that an arbitrary qualocation rule q gives a stable method. The issue of the stability of qualocation rules is studied in [37], in the context of constant-coefficient operators, by means of Fourier analysis.

In the old-style qualocation methods, applied to constant-coefficient operators which are either even or odd, stability is always guaranteed for rules with a number of points $J \geqslant 2$ (see Theorem 3 of Chandler and Sloan [5]). With the more complicated problems discussed here this is no longer true. Rather, for each candidate qualocation rule q stability is something that has to be determined computationally, by testing the strict positivity of a certain real-valued function. It is reported in

[37] that of all the rules tested there approximately half were stable. Only stable rules are recorded in that paper.

In [37] it is shown that if the method is stable, then the qualocation solution u_h^{Q} of the corresponding constant-coefficient operators (either strongly or oddly elliptic, as appropriate), always satisfies the basic error estimate

$$\|u - u_h^{\mathrm{Q}}\|_\beta \leqslant Ch^{r-\beta}\|u\|_r \tag{3.5}$$

if $u \in H^r$. This observation will allow us to avoid in this survey a formal definition of stability: for the present discussion it is enough to say that a particular qualocation method is *stable* if it satisfies the error estimate (3.5) in the corresponding constant-coefficient case.

The basic error estimate (3.5) is also achieved by the appropriate collocation methods, even for the variable-coefficient case, see [3,4,31]. For the qualocation method we are seeking, for the full variable-coefficient case, not only estimates of form (3.5), but also, as explained in Section 1, higher orders of convergence, through a judicious choice of the qualocation rule q. That leads us to define an *additional order of smoothness* associated with each qualocation method: we say that the additional order of smoothness is b, with $0 \leqslant b \leqslant r'$, if the best possible estimate for the error in the qualocation solution for a constant-coefficient operator of form (3.3) with $K = 0$ is

$$\|u - u_k^{\mathrm{Q}}\|_{\beta-b} \leqslant Ch^{r-\beta+b}\|u\|_{r+b}. \tag{3.6}$$

(Here we see again the additional smoothness requirement on u when $b > 0$. We repeat that this can be eliminated completely, provided we are using second-generation qualocation rules, by changing to the 'tolerant' version of the qualocation method as defined in Section 6.)

The highest value of b, namely $b = r'$, the order of the test space, yields the full $\mathrm{O}(h^{r-\beta+r'})$ order of convergence of the (Petrov–)Galerkin method with trial and test spaces of orders r and r', respectively.

4. Second-generation qualocation — the rules and a theorem

In this section we specify precisely the properties required of the qualocation rules if they are to yield estimates of form (3.6) when applied to equations $Au = f$ with A of the full generality given in (3.3) and (3.4), and we explain where to find tables of such rules, or how to compute new ones. We finish by stating precisely a convergence theorem (from [39]) that applies to such rules.

The first requirement, explored in [37] for the case in which the coefficients in (3.3) are constant and K is an integral operator with smooth kernel, is that the rule q must integrate exactly certain 1-periodic functions G_α, defined for $\alpha > 1$ by the absolutely convergent series

$$G_\alpha(x) = \sum_{p=1}^{\infty} \frac{1}{p^\alpha} \cos 2\pi px.$$

(Compared to the usual definition, we have dropped an unnecessary factor of 2 from the definition of G_α.) Note two obvious properties, that the exact integral over $[0,1]$ is 0, and that G_α is symmetric about $x = \frac{1}{2}$. For methods of computing G_α and further properties see [21].

The precise necessary and sufficient condition established in [37] for b to be the additional order of smoothness in (3.6) is

$$qG_{r-\beta+\ell} := \sum_{j=1}^{J} w_j G_{r-\beta+\ell}(\xi_j) = 0 \quad \text{for } \ell = 0, 1, \ldots, b-1. \tag{4.1}$$

The requirement on q that is manifested by (4.1) emerges from the detailed analysis by Fourier series arguments in [37]. Space does not permit us to detail the arguments here, but analogous arguments, in the context of simpler operators that are either even or odd, are considered in the earlier review [32].

Condition (4.1), while still needed when the coefficients in (3.3) are no longer constant, is then no longer enough: it is found in [39] that now q must also have a classical degree of precision of $r-\beta+b-1$ — that is, the rule q must now also integrate exactly all polynomials of degree $\leqslant r-\beta+b-1$.

There is a significant overlap between these two conditions, as becomes clear from the observation that when α is even G_α is just a constant times the Bernoulli polynomial B_α of (even) degree α (see [5,21]).

In practice, it is convenient to require the quadrature rule q to be *symmetric*, by which we mean that if $\xi \in (0,1)$ is a quadrature point then so is $1-\xi$, and the weight associated with $1-\xi$ is the same as that associated with ξ. (If q is symmetric and $\xi_1 = 0$ then it is really the associated quadrature rule

$$\tilde{q}F := \frac{1}{2} w_1 (F(0) + F(1)) + \sum_{j=2}^{J} w_j F(\xi_j) \tag{4.2}$$

which is symmetric in the ordinary sense. Because the mesh is uniform, the rules $\tilde{q}$ and q have exactly the same effects.)

Because a symmetric quadrature rule automatically integrates to zero an odd-degree polynomial which is symmetric about the mid-point $\frac{1}{2}$, and because (4.1) already holds, we see that the requirement that the rule q be of precision $r-\beta+b-1$ is satisfied if conditions (4.1) are supplemented by

$$qB_{2m} = 0, \quad 0 \leqslant 2m < r-\beta. \tag{4.3}$$

Tables listing qualocation quadrature rules satisfying the constant-coefficient condition (4.1), and which are also *stable* in the sense that (3.5) holds for appropriate constant-coefficient operators, are published in [37]. Separate tables are given there (numbered 1 to 6) for the strongly elliptic and oddly elliptic cases, and for operators of orders $\beta = -1, 0$ and $+1$. *Some* of the rules in these tables also satisfy (4.3), and are therefore available for use in second-generation qualocation rules. Those are the rules labelled there $G_{J,b,\alpha}$ and $L_{J,b,\alpha}$ (with $\alpha = r-\beta$), with capital letters G and L rather than lower case. (The rules labelled $G_{J,b,\alpha}$ are 'Gauss-like' in that end-points are not included. Those labelled $L_{J,b,\alpha}$ are 'Lobatto-like' when expressed in the equivalent form (4.2).) The quadrature points and weights for the rules $G_{J,b,\alpha}$ and $L_{J,b,\alpha}$ are listed in Tables A and B of Sloan and Wendland [37].

The convergence properties of the qualocation method with the second-generation qualocation rules described above, applied to variable-coefficient operators of form (3.3), are summed up in the following theorem.

Theorem 4.1 (Sloan and Wendland [39]). *Let A be given by* (3.3) *and* (3.4), *and be either strongly elliptic or oddly elliptic. Moreover, let r,r' satisfy Assumption* 3.1, *with $r>\beta+1$ and $r'\geqslant 2$, and assume that the qualocation rule characterised by r,r' and q is stable, in the sense that* (3.5) *holds in the corresponding constant-coefficient case with $K=0$. Then the qualocation solution $u_h^{\mathrm{Q}}\in S_h$ exists and is unique for h sufficiently small, and satisfies* (3.5). *Moreover, if q satisfies* (4.1) *and* (4.3) *then u_h^{Q} also satisfies*

$$\|u-u_h^{\mathrm{Q}}\|_{\beta-b}\leqslant Ch^{r-\beta+b}\|u\|_{r+b}. \tag{4.4}$$

Space allows us only the merest sketch of the proof of the theorem. Essentially there are two parts, of which the first and most delicate is to establish that the basic convergence estimate (3.5) extends to the variable-coefficient case. This proof, which in essence extends stability to the variable-coefficient case, follows the model of that of Arnold and Wendland [4] for the collocation case. It makes essential use of Korn's trick, and of the fact that the commutator of a smooth function with a pseudo-differential operator is a pseudo-differential operator of lower order. However, there is one major difference from the collocation case. This we now attempt to explain.

First, we must write the qualocation method in a different way, by introducing the 'qualocation projection', which is the projection R_h onto the space S_h' defined by

$$R_hg\in S_h',\quad \langle R_hg,\chi_h\rangle=\langle g,\chi_h\rangle\quad \forall\chi_h\in S_h'.$$

(Since this equation is just the qualocation method for the case of the identity operator with $r=r'$, the existence and uniqueness of R_hg for $J\geqslant 2$ is ensured by Theorem 3 in [5].) The qualocation approximation (1.20) to $Au=f$ can now be written as

$$u_h^{\mathrm{Q}}\in S_h,\quad R_hAu_h^{\mathrm{Q}}=R_hf.$$

The following 'commutator property', from [38], plays a crucial role in the proof of Theorem 4.1:

$$\|R_h(aL_\pm^\beta v_h)-aR_h(L_\pm^\beta v_h)\|_0\leqslant c(a)h\|v_h\|_\beta \tag{4.5}$$

for $v_h\in S_h$ and a a smooth function. That result is essentially a spline property, in that it does *not* hold if the space S_h' is replaced by a space of trigonometric polynomials. This fact notwithstanding, the proof in [38] is by way of a highly technical Fourier series argument. (For a related commutator property, obtained by a different argument, see [11].)

The significance of the commutator result is that it allows a smooth non-constant function to be, in effect, 'moved through' R_h, at the cost of only a small error. The commutator property is not needed in the collocation argument in [4], for the simple reason that the commutator in that case vanishes at each collocation point x_k: for in that case we have, for all continuous functions F, $(R_hF)(x_k)=F(x_k)$, and hence

$$(R_h(aF))(x_k)=a(x_k)F(x_k)=(a(R_hF))(x_k),\quad k=0,\ldots,n-1.$$

The second part of the proof is to prove (4.4), knowing already the estimate (3.5). This is done by a duality argument, using both the Aubin–Nitsche trick as in the corresponding proof for the Galerkin method, see [15], and a technical quadrature estimate, (see [39], Theorem 4.1) for the difference $\langle aL_\pm^\beta w_h,v_h\rangle-(aL_\pm^\beta w_h,v_h)$, for a smooth and $w_h\in S_h$, $v_h\in S_h'$. For further details we refer to Sloan and Wendland [39].

5. Numerical illustration

In this numerical example, taken from [39], Eq. (3.1) is solved with $C(X)=0$, for the case in which Γ is the unit circle. The circle is parametrised by $X=e^{2\pi i x}$, so that with

$$u(x)=U(X), \quad f(x)=F(X), \quad a(x)=A(X) \quad \text{and} \quad b(x)=B(X),$$

the equation becomes

$$a(x)u(x)+2b(x)\int_0^1 \frac{u(y)}{e^{2\pi i y}-e^{2\pi i x}}e^{2\pi i y}\,dy = f(x), \quad x\in\mathbb{R}.$$

For the coefficients $a(x), b(x)$ we choose

$$a(x)=3+2\sin 2\pi x, \quad b(x)=1, \quad x\in\mathbb{R},$$

making the singular integral operator strongly elliptic, and choose f so that the exact solution is

$$u(x)=\cos 2\pi x + i|\sin 2\pi x|.$$

For the approximation method we choose $r=r'=2$, so satisfying Assumption 3.1, and as qualocation rule we select, from Table 2 of [37], the 3-point rule $G_{3,2,2}$, which has the points and weights

$$\xi_1=0.1057\ 4635\ 7567, \quad w_1=0.2680\ 6328\ 1387,$$

$$\xi_2=0.5, \quad w_2=1-2w_1,$$

$$\xi_3=1-\xi_1, \quad w_3=w_1.$$

Table 2 shows the computed errors for this problem in the H^0, H^{-1} and H^{-2} norms, for $n=16, 32$ and 64, together with the apparent order of convergence ρ. (The norms were estimated by evaluating quadrature approximations to the Fourier coefficients of the error, and inserting them into the Fourier series definition of the norm, see [31]). From the results it is clear that the errors are satisfyingly small, and moreover that the orders of convergence (remembering that $\beta=0$ in this case) are as predicted by Theorem 4.1, for the case $u\in H^4$, namely

$$\|u-u_h^Q\|_0=O(h^2), \quad \|u-u_h^Q\|_{-1}=O(h^3), \quad \|u-u_h^Q\|_{-2}=O(h^4).$$

(In fact in this example u is not in H^4, nor even in H^2; we do not understand why the full $O(h^4)$ order is nevertheless maintained.)

Table 2
Errors in H^s norms and apparent orders of convergence ρ

n	$s=0$	ρ	$s=-1$	ρ	$s=-2$	ρ
16	0.584 (−2)		0.365 (−3)		0.434 (−4)	
32	0.144 (−2)	2.02	0.440 (−4)	3.05	0.249 (−5)	4.12
64	0.360 (−3)	2.00	0.546 (−5)	3.01	0.151 (−6)	4.04

6. The tolerant version of qualocation

In this section we describe a simple modification, introduced in the qualocation context in [40], that has the surprising effect of overcoming the principal defect in the qualocation method, namely the high smoothness requirement on the exact solution in Theorem 4.1.

Tolerant qualocation differs in only one way from the standard method described above, namely that the inner product on the right-hand side of (1.20) is now evaluated exactly, rather than by using the qualocation rule q. Thus the method is: find $u_h^{\mathrm{T}} \in S_h$ such that

$$\langle Au_h^{\mathrm{T}}, \chi_h \rangle = (f, \chi_h) \quad \forall \chi_h \in S_h'. \tag{6.6}$$

The benefit of evaluating the right-hand side exactly was first pointed out (in the context of a fully discrete qualocation method) by Saranen and Sloan [27]. Exact evaluation of the right-hand side was exploited also in [23].

For the purposes of the present review, where we are emphasizing second-generation qualocation rules, it is preferable to focus not on [40], but rather on the more recent paper [35], which considered the full range of operators allowed in [39] (see Section 3) and came to the following nice conclusions: first, that the additional smoothness requirement on u seen in (4.4) is entirely eliminated; and second, that *the qualocation rules needed to achieve a given order of convergence are exactly the same second-generation rules already considered in Section* 4. (The conclusions of Tran and Sloan [40] were more complicated, with the tolerant method imposing extra requirements on the rule q. That is because only simple operators were considered there, and the qualocation rules in [40] were not the robust second-generation rules considered in [35].)

The conclusions of Sloan and Tran [35] are summed up in this companion to Theorem 4.1.

Theorem 6.1 (Sloan and Tran [35]). *Let A, r and r' be as in Theorem* 4.1. *Then the tolerant qualocation solution $u_h^{\mathrm{T}} \in S_h$ exists and is unique for h sufficiently small, and satisfies*

$$\|u - u_h^{\mathrm{T}}\|_\beta \leqslant Ch^{r-\beta}\|u\|_r. \tag{6.7}$$

Moreover, if q satisfies (4.1) *and* (4.3) *then u_h^{T} also satisfies the estimate*

$$\|u - u_h^{\mathrm{T}}\|_{\beta-b} \leqslant Ch^{r-\beta+b}\|u\|_r. \tag{6.8}$$

The result in (6.8), in contrast to that in (4.4), is optimal for the given norms on the left and right. If b has its maximum value r' then the result in (6.8) is *exactly* that achieved by the Petrov–Galerkin method with the same trial and test spaces.

To prove Theorem 6.1, we note first that the existence and uniqueness of u_h^{T} for small h is almost obvious, given that it is only the right-hand side of the qualocation equation that has been changed. For a similar reason, it turns out after some argument that we are able to derive the basic stability result (6.7) from the corresponding result for the standard qualocation method in Theorem 4.1. The main result (6.8) rests again on a duality argument, and uses many of the ideas of Sloan and Wendland [39], including the delicate quadrature estimate mentioned at the end of Section 4. Of course there are some differences, the most important being that the tolerant qualocation equation (6.6), unlike (1.20), is not 'consistent': if $u \in S_h$ it does not follow that $u_h^{\mathrm{T}} = u$.

Why is it that the change to the right-hand side of the qualocation equation leads to such an improvement? This is hard to answer, because of the highly technical nature of the arguments.

Perhaps the best that can be said is that the standard qualocation method, because it involves the application of a quadrature approximation to f (or more precisely, to $f\chi_h$), inevitably imposes a smoothness requirement on u, given that $f = Au$: at the very least we need Au continuous. The tolerant version makes no such demands in u, and it turns out that no higher smoothness requirement arises in the proof.

In the implementation of tolerant qualocation it is of course usually impossible to integrate the right-hand side of (6.6) exactly. However, since most of the computational work is associated with the left-hand side (either in setting up the matrix, or in obtaining the solution of the linear system), it is not difficult to compute the right-hand side to any desired accuracy.

The following numerical example, extended from [40], is for the single-layer equation (1.8), for the case in which Γ is an ellipse centred at the origin with major axis of length 4 along the x-axis, and minor axis of length 2, parametrised by $X = (2\cos 2\pi x, \sin 2\pi x)$, $x \in [0,1]$, and with $F(X) = f(x) = \sqrt{x(1-x)}$, so that the solution $u(x)$ is not smooth at $x = 0$. The quantity computed is $\Phi(X_0)$, see (1.9), with $X_0 = (1.0, 0.3)$.

The operator in this example is strongly elliptic and of order -1. We choose $r' = r = 2$ (the piecewise-linear case), and select from Table 1 of Sloan and Wendland [37] a rule q with the maximum additional order of convergence $b = r' = 2$. From the possible rules $G_{4,2,3}$ and $L_{4,2,3}$ we choose the latter, which is given explicitly by

$$\xi_1 = 0, \quad w_1 = 0.0798\ 0806\ 8152, \tag{6.9}$$

$$\xi_2 = 0.1557\ 3169\ 6555, \quad w_2 = 0.2673\ 8508\ 6074, \tag{6.10}$$

$$\xi_3 = 0.5, \quad w_3 = 1 - w_1 - 2w_2, \tag{6.11}$$

$$\xi_4 = 1 - \xi_2, \quad w_4 = w_2. \tag{6.12}$$

As noted above, because $f(x) = \sqrt{x(1-x)}$ is not smooth, neither is u: in fact (see [40]) $u \in H^{-\varepsilon}$ for $\varepsilon > 0$ but $u \notin H^0$. Because of this limited smoothness of u the best convergence order for the tolerant qualocation method (from (6.8), suitably modified), is

$$\|u - u_h^{\mathrm{T}}\|_{-3} \leqslant Ch^{3-\varepsilon}\|u\|_{-\varepsilon}.$$

In contrast, the best result available from traditional qualocation (see (3.5)) is

$$\|u - u_h^{\mathrm{Q}}\|_{-1} \leqslant Ch^{1-\varepsilon}\|u\|_{-\varepsilon}.$$

In Table 3 we show the (estimated) errors for firstly the tolerant method with the qualocation rule $L_{4,2,3}$; secondly the standard qualocation method with the same rule $L_{4,2,3}$; and thirdly the standard qualocation method with the $\frac{3}{7}, \frac{4}{7}$ rule as described in Section 1. (Note that if u is smooth all rules are predicted to give errors of order $\mathrm{O}(h^5)$; this is confirmed numerically for the first and third cases in [40].)

From the results in Table 3 it appears that the order of convergence of the tolerant method is better than the predicted order $\mathrm{O}(h^{3-\varepsilon})$, and the order of the standard qualocation rules better than the predicted order $\mathrm{O}(h^{1-\varepsilon})$. Still, the story is clear, that the tolerant version of qualocation does give improved accuracy when the solution is not smooth. This supports our recommendation, that the tolerant version of the qualocation method, used in conjunction with second generation qualocation rules, is the one to be preferred in practice. Additionally, a comparison of the two results for the

Table 3
Estimated errors and apparent convergence order ρ for the potential $\Phi(X_0)$, for a nonsmooth solution u

n	Tolerant qualocation ($L_{4,2,3}$)	ρ	Standard qualocation ($L_{4,2,3}$)	ρ	Standard qualocation ($\frac{3}{7},\frac{4}{7}$)	ρ
32	0.659 (−5)		0.471 (−5)		0.278 (−4)	
64	0.371 (−7)	7.47	0.606 (−6)	2.96	0.115 (−4)	1.28
128	0.307 (−8)	3.59	0.219 (−6)	1.47	0.398 (−5)	1.53
256	0.289 (−9)	3.41	0.716 (−7)	1.61	0.129 (−5)	1.62

standard qualocation method shows that improving the quality of the underlying rule q can improve the accuracy, even if not the asymptotic order of convergence.

7. Extensions and open questions

The principal remaining limitation of the qualocation method, in its tolerant second-generation form as presented in Section 6, is that it requires a uniform mesh. Can the method be extended to a wider class of meshes, or even to arbitrary meshes? Possibly, but a different approach to stability would be needed, one not based on Fourier series. Is it conceivable that the same second-generation qualocation rules q as in Section 4 would still be valid? (For rules that include the endpoint, form (4.2) should then be understood.) Perhaps, but even experimental evidence is lacking. For the older style qualocation rules a numerical experiment in [5] with an irregular mesh provides a counterexample, but it is clear that the second-generation qualocation rules have much better local behaviour.

Can the methods be extended to curves with corners? One application of the qualocation method (for the case of the single-layer equation on a polygon) has been studied in [6]. There the mesh was in effect 'graded' near the corner by introducing a carefully tailored transformation that spreads out both sides of the corner, and then using a uniform mesh with respect to the new variable. The arguments used there extend to second-generation qualocation methods, but for the tolerant version a separate analysis would be needed.

Are there qualocation methods for splines other than smoothest splines? Most promising, perhaps, are the C^0 splines. For these the stability properties of discrete orthogonal projection (or equivalently, of the qualocation approximation for the case of the identity operator) was studied in [8] (with no requirement of mesh uniformity), and commutator properties analogous to (4.5) were studied in [10].

A preliminary study of qualocation with C^0 splines on quasi-uniform meshes, applied to strongly elliptic singular integral equations with piecewise-smooth coefficients, has recently been completed [9].

Fully discrete versions of qualocation, not discussed here, were studied by Sloan and Burn [34], Saranen and Sloan [27], McLean and Sloan [23] and Jeon et al. [16]. The paper [23] employed a direct two-dimensional quadrature scheme, allowing the discrete approximation to preserve the self-adjointness property of the operator A in a case such as the single-layer operator.

Another approach to stability, based on the stability of the Galerkin method, has been used in the study of qualocation-like methods for the single-layer equation on Lipschitz curves, see [2,33]. In this work, however, higher-order convergence was not explored.

Finally, can qualocation methods be developed for problems in which Γ is a surface rather than a curve? This is undoubtedly the biggest challenge, but the way forward is not clear, since uniform meshes and Fourier methods are no longer applicable. The only 3D studies known to the writer in which the methods might be thought of as true qualocation methods are two papers [7,18] which extend the Galerkin-based arguments of Sloan and Atkinson [33] to the single-layer equation for a plate, using piecewise-constant elements. Again higher-order convergence was not established.

In all the directions mentioned in this section it seems fair to say that there are many more questions than answers.

Acknowledgements

The support of the Australian Research Council is gratefully acknowledged, as is the hospitality of the University of Bath during the writing of the paper.

References

[1] M.S. Agranovich, Spectral properties of elliptic pseudodifferential operators on a closed curve, Funct. Anal. Appl. 13 (1979) 279–281.

[2] M. Ainsworth, R.D. Grigorieff, I.H. Sloan, Semi-discrete Galerkin approximation of the single layer equation by general splines, Numer. Math. 79 (1998) 157–174.

[3] D.N. Arnold, W.L. Wendland, On the asymptotic convergence of collocation methods, Math. Comp. 41 (1983) 349–381.

[4] D.N. Arnold, W.L. Wendland, The convergence of spline collocation for strongly elliptic equations on curves, Numer. Math. 47 (1985) 317–341.

[5] G.A. Chandler, I.H. Sloan, Spline qualocation methods for boundary integral equations, Numer. Math. 58 (1990) 537–567.

[6] J. Elschner, S. Prössdorf, I.H. Sloan, The qualocation method for Symm's integral equation on a polygon, Math. Nachr. 177 (1996) 81–108.

[7] R.D. Grigorieff, I.H. Sloan, Galerkin approximation with quadrature for the screen problem in R^3, J. Integral Equations Appl. 9 (1997) 293–319.

[8] R.D. Grigorieff, I.H. Sloan, Stability of discrete orthogonal projections for continuous splines on L_p spaces, Bull. Austral. Math. Soc. 58 (1998) 307–332.

[9] R.D. Grigorieff, I.H. Sloan, On qualocation and collocation methods for singular integral equations with piecewise continuous coefficients, using continuous splines on quasi-uniform meshes, In: J. Elschner (Ed.), Siegfried Prössdorf Memorial Volume. Series Operator Theory: Advances and Applications, Birkhäuser Verlag, to appear.

[10] R.D. Grigorieff, I.H. Sloan, J. Brandts, Superapproximation and commutator properties of discrete orthogonal projections for continuous splines, submitted for publication.

[11] R. Hagen, S. Roch, B. Silbermann, Spectral Theory of Approximation Methods for Convolution Equations, Birkhäuser, Basel, 1995.

[12] R. Hagen, B. Silbermann, On the stability of the qualocation method, Seminar Analysis, Operator Equations and Numerical Analysis 1987/1988, Karl-Weierstrass-Institut für Mathematik, Berlin, pp. 43–52.

[13] R. Hagen, B. Silbermann, On the convergence of the qualocation method, preprint 207, Technische Universität Chemnitz, 1991.

[14] G.C. Hsiao, W.L. Wendland, A finite element method for some integral equations of the first kind, J. Math. Anal. Appl. 58 (1977) 449–481.

[15] G.C. Hsiao, W.L. Wendland, The Aubin–Nitsche lemma for integral equations, J. Integral Equations 3 (1981) 299–315.

[16] Y. Jeon, I.H. Sloan, E.P. Stephan, J. Elschner, Discrete qualocation methods for logarithmic-kernel integral equations on a piecewise smooth boundary, Adv. Comput. Math. 7 (1997) 547–571.

[17] U. Lamp, K.-T. Schleicher, W.L. Wendland, The fast Fourier transform and the numerical solution of one-dimensional boundary integral equations, Numer. Math. 47 (1985) 15–38.

[18] D. Mauersberger, I.H. Sloan, A simplified approach to the semi-discrete Galerkin method for the single-layer equation for a plate, in: M. Bennet, A.M. Sändig, W.L. Wendland (Eds.), Mathematical Aspects of Boundary Element Methods, Chapman and Hall, 1999, pp. 178–190.

[19] W. McLean, A spectral Galerkin method for a boundary integral equation, Math. Comp. 47 (1986) 597–607.

[20] W. McLean, Local and global descriptions of periodic pseudodifferential operators, Math. Nachr. 150 (1991) 151–161.

[21] W. McLean, Numerical evaluation of some trigonometric series, Math. Comp. 63 (1994) 271–275.

[22] W. McLean, S. Prössdorf, Boundary element collocation methods using splines with multiple knots, Numer. Math. 74 (1996) 419–451.

[23] W. McLean, I.H. Sloan, A fully discrete and symmetric boundary integral method, IMA J. Numer. Anal. 14 (1994) 311–345.

[24] S. Prössdorf, S.G. Mikhlin, Singular Integral Operators, Akademie-Verlag, Berlin, 1986.

[25] S. Prössdorf, B. Silbermann, Numerical Analysis for Integral and Related Operator Equations, Akademie Verlag, Berlin, 1991.

[26] J. Saranen, Local error estimates for some Petrov–Galerkin methods applied to strongly elliptic equations on curves, Math. Comp. 48 (1987) 485–502.

[27] J. Saranen, I.H. Sloan, Quadrature methods for logarithmic kernel integral equations on closed curves, IMA J. Numer. Anal. 12 (1992) 162–187.

[28] J. Saranen, W.L. Wendland, On the asymptotic convergence of collocation methods with spline functions of even degree, Math. Comp. 45 (1985) 91–108.

[29] J. Saranen, W.L. Wendland, The Fourier series representation of pseudo-differential operators on closed curves, Complex Variables Theory Appl. 8 (1987) 55–64.

[30] I.H. Sloan, A quadrature-based approach to improving the collocation method, Numer. Math. 54 (1988) 41–56.

[31] I.H. Sloan, Error analysis of boundary integral methods, Acta Numer. 1 (1991) 287–339.

[32] I.H. Sloan, Boundary element methods, in: M. Ainsworth, J. Levesley, W.A. Light, M. Marletta (Eds.), Theory and Numerics of Ordinary and Partial Differential Equations, Clarendon Press, Oxford, 1995, pp. 143–180.

[33] I.H. Sloan, K.E. Atkinson, Semi-discrete Galerkin methods for the single layer equation on Lipschitz curves, J. Integral Equations Appl. 9 (1997) 279–292.

[34] I.H. Sloan, B.J. Burn, An unconventional quadrature method for logarithmic-kernel integral equations on closed curves, J. Integral Equations Appl. 4 (1992) 117–151.

[35] I.H. Sloan, T. Tran, The tolerant qualocation method for variable-coefficient elliptic equations on curves, submitted for publication.

[36] I.H. Sloan, W.L. Wendland, A quadrature-based approach to improving the collocation method for splines of even degree, Z. Anal. Ihre Anwendung. 8 (1989) 361–376.

[37] I.H. Sloan, W.L. Wendland, Qualocation methods for elliptic boundary integral equations, Numer. Math. 79 (1998) 451–483.

[38] I.H. Sloan, W.L. Wendland, Commutator properties for periodic splines, J. Approx. Theory 97 (1999) 254–281.

[39] I.H. Sloan, W.L. Wendland, Spline qualocation methods for variable-coefficient elliptic equations on curves, Numer. Math. 83 (1999) 497–533.

[40] T. Tran, I.H. Sloan, Tolerant qualocation — a qualocation method for boundary integral equations with reduced regularity, J. Integral Equations Appl. 10 (1998) 85–115.

[41] W.L. Wendland, Boundary element methods for elliptic problems, in: A.H. Schatz, V. Thomée, W.L. Wendland (Eds.), Mathematical Theory of Finite and Boundary Element Methods, Birkhäuser, Basel, 1990, pp. 219–276.

ELSEVIER Journal of Computational and Applied Mathematics 125 (2000) 479–501

JOURNAL OF COMPUTATIONAL AND APPLIED MATHEMATICS

www.elsevier.nl/locate/cam

A sparse $\mathscr{H}$-matrix arithmetic: general complexity estimates

W. Hackbusch*, B.N. Khoromskij

Max-Planck-Institut, Mathematik in den Naturwissenschaften, Inselstr. 22-26, D-04103 Leipzig, Germany

Received 30 June 1999

Abstract

In a preceding paper (Hackbusch, Computing 62 (1999) 89–108), a class of matrices ($\mathscr{H}$-matrices) has been introduced which are data-sparse and allow an approximate matrix arithmetic of almost linear complexity. Several types of $\mathscr{H}$-matrices have been analysed in Hackbusch (Computing 62 (1999) 89–108) and Hackbusch and Khoromskij (Preprint MPI, No. 22, Leipzig, 1999; Computing 64 (2000) 21–47) which are able to approximate integral (nonlocal) operators in FEM and BEM applications in the case of quasi-uniform unstructured meshes. In the present paper, the general construction of $\mathscr{H}$-matrices on rectangular and triangular meshes is proposed and analysed. First, the reliability of $\mathscr{H}$-matrices in BEM is discussed. Then, we prove the optimal complexity of storage and matrix–vector multiplication in the case of rather arbitrary admissibility parameters $\eta < 1$ and for finite elements up to the order 1 defined on quasi-uniform rectangular/triangular meshes in $\mathbb{R}^d$, $d=1,2,3$. The almost linear complexity of the matrix addition, multiplication and inversion of $\mathscr{H}$-matrices is also verified.

MSC: 65F05; 65F30; 65F50

Keywords: Hierarchical matrices; Data-sparse approximations; Formatted matrix operations; Fast solvers; BEM; FEM

1. Introduction

A class of hierarchical ($\mathscr{H}$) matrices has been recently introduced in [5]. They are shown to provide an efficient tool for a data-sparse approximation to large and fully populated stiffness matrices arising in BEM and FEM applications. In fact, the storage and matrix–vector multiplication complexity of the rank-k $\mathscr{H}$-matrices associated with quasi-uniform grids are estimated by $O(kn\log n)$, where n is the problem size, see [5,6]. Moreover, these matrices also allow the arithmetic of optimal complexity. In particular, the "formatted" matrix–matrix addition, product as well as the inversion for a class of $\mathscr{H}$-matrices were proven to have almost linear complexity $O(n\log^q n)$ with moderate $q\geqslant 0$. In this way the approach may be applied for the data-sparse approximation and fast

* Corresponding author.

0377-0427/00/$ - see front matter
PII: S0377-0427(00)00486-6

solution of the linear integral/pseudodifferential equations which arise in the FE/BE methods for elliptic problems.

First, we discuss the principal ingredients of the $\mathscr{H}$-matrix techniques. We then show the existence of optimal order approximations by $\mathscr{H}$-matrices for a class of integral operators in FEM/BEM applications. We prove the almost linear complexity of various $\mathscr{H}$-matrix operations. In particular, we study the complexity of hierarchical matrices in the following cases:

(i) arbitrary constant $\eta < 1$ in the admissibility criterion;
(ii) quasi-uniform quadrangular/triangular meshes in $\mathbb{R}^d$, $d = 1, 2, 3$;
(iii) piecewise constant/linear/bilinear elements.

Our results for the storage and matrix–vector multiplication expenses are given with asymptotically sharp constants which depend explicitly upon the spatial dimension d, the parameter η and the problem size. We prove the linear-logarithmic complexity of the formatted *addition, multiplication and inverse* of $\mathscr{H}$-matrices.

We also stress that our constructions apply to unstructured quasi-uniform meshes as well, using the techniques from [6]. The extension to the case of graded meshes was discussed in [7]. A class of $\mathscr{H}^2$-matrices having the linear complexity $O(n)$ was developed in [9]. The systematic approach to build optimal order degenerate approximations (wire-basket expansions of the order $O(\log^{d-1} n)$) for a class of kernels in FEM and BEM applications has been considered in [8].

2. Introduction to $\mathscr{H}$-matrices

2.1. A motivation for data-sparse approximations in BEM

In this section, we discuss simple examples illustrating the principal ideas of $\mathscr{H}$-matrix approximations in BEM. The nonlocal operators to be approximated arise in both FEM and BEM applications. FE/FD approximations of elliptic PDEs result in sparse stiffness matrices. In such applications, we are interested in the data-sparse approximation of the inverse to discrete elliptic operators or to the Schur-complement matrices with respect to a certain subset of degrees of freedom. In both cases, we actually deal with a discretisation of an integral (pseudodifferential) operator with implicitly given Schwartz kernel. Below, we consider three examples of integral operators

$$(A_\gamma u)(x) = \int_\Sigma s_\gamma(x, y) u(y)\,dy, \quad x \in \Sigma := [0, 1] \tag{1}$$

with $\gamma = 1, 2, 3$, where

$$s_1(x, y) := \log(1 + (x - y)^2), \quad s_2(x, y) := \log(x + y), \quad s_3(x, y) := \log|x - y|. \tag{2}$$

The FE Galerkin discretisation of (1) with piecewise constant basis functions defined for the uniform grid (partitioning)

$$X_i = [(i - 1)h,\ ih], \quad h := n^{-1},\ 1 \leqslant i \leqslant n$$

leads to the full stiffness matrix

$$M = (m_{ij})_{i,j \in I}, \quad m_{ij} := \int_{X_i \times X_j} s_\gamma(x, y)\,dx\,dy,$$

where $I=\{1,\dots,n\}$ is the corresponding index set of the Galerkin ansatz functions $\{\varphi_i\}_{i\in I}$. Assume a hierarchical p-level structure of the grid by imposing $n=2^p$. The $\mathscr{H}$-matrix approximation to M will provide a matrix $M_{\mathscr{H}}$ such that the error $M-M_{\mathscr{H}}$ is of the same order $\varepsilon=h^{\delta}$, $\delta>0$, as for the Galerkin error related to M. However, both the storage and the matrix–vector multiplication costs for $M_{\mathscr{H}}$ will amount to $O(n\log^q n)$ instead of $O(n^2)$, with a moderate $q\geqslant 0$ discussed below.

For the first example in (2) the desired approximation $M_{\mathscr{H}}$ can be obtained exploiting the global smoothness of the kernel in the product domain $\Sigma\times\Sigma$. Due to classical approximation theory there exists a simple approximation of $s_1(x,y)$ by a short sum $\tilde{s}_1:=\sum_{\beta=1}^{k} a_\beta(x)c_\beta(y)$ of separable functions (e.g., by the Taylor expansion or by the ortho–projection onto polynomials, see also Section 3) such that

$$\left|s_1(x,y)-\sum_{\beta=1}^{k} a_\beta(x)c_\beta(y)\right| \lesssim \varepsilon \tag{3}$$

with $k=O(\log\varepsilon^{-1})$. The corresponding stiffness matrix

$$M_{\mathscr{H}}:=(\tilde{m}_{ij})_{i,j\in I},\quad \tilde{m}_{ij}:=\int_{X_i\times X_j}\tilde{s}_1(x,y)\,\mathrm{d}x\,\mathrm{d}y,$$

provides the required approximation of M on the one hand, and also it has the data-sparse structure (indeed, it is a matrix of rank k) of the complexity $O(kn)$, on the other hand. Therefore, the global smoothness of s_1 allows a data-sparse approximation of M by an $n\times n$ low-rank matrix.

The singular kernels in the second and third examples allow instead of a global only blockwise degenerate approximations. In this way, the above construction is applied locally in a hierarchical manner and it is based on an *admissible partitioning* of the product index set $I\times I$. Such an admissible partitioning is described below using hierarchical cluster trees of I and $I\times I$.

2.2. The cluster trees of I and $I\times I$

Starting with the full index set $I_1^0:=I$ of level 0, we then split it into two equal subsets I_1^1 and I_2^1 and then apply this procedure to each part successively such that at level p, we reach the one-element sets $I_1^p=\{1\},\dots,I_n^p=\{n\}$. In general, at level ℓ, we have the set of tree vertices (clusters)

$$I_j^\ell:=\{(j-1)2^{p-\ell}+1,\dots,j2^{p-\ell}\}\quad\text{for } 0\leqslant\ell\leqslant p,\ 1\leqslant j\leqslant 2^\ell.$$

In the following, the vertices are called the *clusters*. Each cluster $\tau=I_j^\ell$ has exactly two sons, $I_{j'}^{\ell+1}$ and $I_{j''}^{\ell+1}$ with $j'=2j+1$ and $j''=2j+2$, obtained by halving the parent vertex. The set of all clusters I_j^ℓ together with the tree structure is called the *cluster tree* $T(I)$. In this example, $T(I)$ is a binary tree of depth p. I is the root of $T(I)$ and the sets I_i^p, $i=1,\dots,n$, are the leaves of $T(I)$ (one-element vertices). Introducing the isomorphism between the index set I and the interval decomposition $\{X_i\}_{i\in I}$ by $i\leftrightarrow J_i$, one can define diameters and the distance between two clusters τ and σ just measuring the Euclidean diameter $\mathrm{diam}(X(\tau))$ and the distance $\mathrm{dist}(X(\tau),X(\sigma))$, where $X(\tau):=\bigcup\{X_\alpha:\ \alpha\in\tau\}$.

Having in hands the cluster tree $T_1:=T(I)$, we then construct the corresponding hierarchical tree $T_2:=T(I\times I)$ on the product index-set $I\times I$ and with the same number p of levels. In our particular case, we have the following set of vertices:

$$\boldsymbol{I}_{ij}^\ell:=I_i^\ell\times I_j^\ell\quad\text{for } 0\leqslant\ell\leqslant p,\ 1\leqslant i,j\leqslant 2^\ell.$$

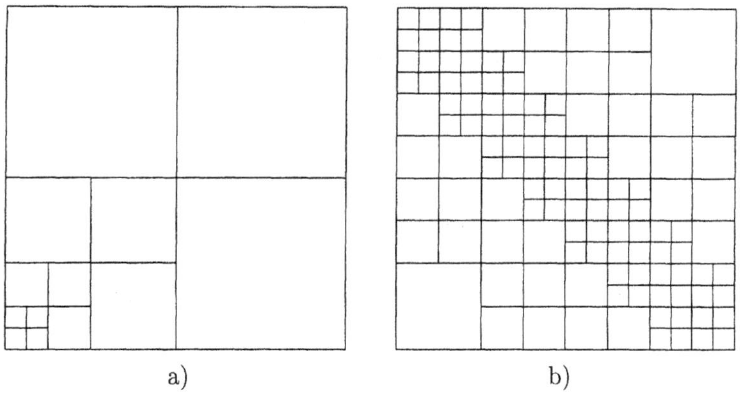

Fig. 1. Block-structure for $\mathcal{M}_{\mathcal{N}}$ (a) and $\mathcal{M}_{\mathcal{D}}$ (b) formats.

The set of sons $S_2(t)$ of $t = \boldsymbol{I}^{\ell}_{ij} \in T_2$ is given by $S_2(t) := \{\tau \times \sigma : \tau \in S_1(I^{\ell}_i),\ \sigma \in S_1(I^{\ell}_j)\}$, where $S_1(f)$ is the set of sons belonging the parent cluster $f \in T_1$. This construction inherits the hierarchical structure of $T(I)$ and provides the recursive data access of optimal complexity. The tree T_2 contains a variety of large and small blocks. The block decomposition described later on will use only blocks contained in T_2. Note that the general construction of hierarchical trees $T_1 = T(I)$ and $T_2 = T(I \times I)$ for an arbitrary index set I is introduced in [5,6]. Here we concentrate only on the particular examples which, however, illustrate the main features of the general framework.

The hierarchical format of an $\mathcal{H}$-matrix is based on a particular partitioning P_2 of $I \times I$ satisfying certain admissibility conditions. The latter will guarantee the optimal approximation.

2.3. Admissible block partitionings P_2 and $\mathcal{H}$-matrices

A partitioning $P_2 \subset T_2$ is a set of disjoint blocks $b \in T_2$ such that the union of all blocks from P_2 yields $I \times I$. The partitioning P_2 is usually built by a recursive construction involving implicitly an admissibility condition. The latter incorporates characteristics of the singularity locations of the kernel function $s(x, y)$, $x, y \in \Sigma$, and provides the balance between the size of matrix blocks and their distance from the singularity points.

For a globally smooth kernel as the *first example* s_1 in (2), we need no admissibility restriction; therefore the biggest block $I \times I$ is already admissible resulting in the simplest partitioning $P_2 = \{I \times I\}$. As we have seen above, this block will be filled by a *rank-k* matrix.

In the *second example* (kernel s_2), we use the following admissibility condition: a block $\tau \times \sigma$ with $\tau, \sigma \in T_1$ belongs to P_2 if

$$\min\{\mathrm{diam}(\tau), \mathrm{diam}(\sigma)\} \leqslant 2\eta \max(\mathrm{dist}(\tau, 0), \mathrm{dist}(\sigma, 0)), \tag{4}$$

where $\eta \leqslant 1$ is a given threshold parameter responsible for the approximation. Let, e.g., $\eta = \frac{1}{2}$. The block $I \times I$ is not admissible and must be decomposed into its four sons (see Fig. 1a). Three of them already satisfy (4), and only one must be refined further on. Finally, we obtain the block partitioning

$$P_2 = \{\boldsymbol{I}^{\ell}_{ij} \in T_2 \colon 0 < \ell < p,\ \max\{i, j\} = 2\} \cup \{\boldsymbol{I}^{p}_{11}\}.$$

The block-matrix corresponding to $b \in P_2$ is denoted by $M^b := (m_{\alpha\beta})_{(\alpha,\beta)\in b}$. The level number ℓ of a block b is written as level(b).

In the case of $s_3(x, y)$, the admissibility condition is more restrictive because we have the singularity of the kernel in each diagonal point $x = y$ of the product domain $\Sigma \times \Sigma$. Now $\tau \times \sigma$ belongs to P_2 if

$$\min\{\mathrm{diam}(\tau), \mathrm{diam}(\sigma)\} \leqslant 2\eta \,\mathrm{dist}(\tau, \sigma), \quad \eta < 1. \tag{5}$$

For the choice $\eta = \frac{1}{2}$, we obtain a block partitioning $P_2 := \bigcup_{\ell=2}^{p} P_2^\ell$, where $P_2^2 = \{I_{14}^2\} \cup \{I_{41}^2\}$ and

$$P_2^\ell = \{I_{ij}^\ell \in T_2\colon\ |i - j| \geqslant 1 \text{ and } I_{ij}^\ell \cap P_2^{\ell'} = \emptyset,\ \ell' < \ell\} \quad \text{for } \ell = 3, \ldots, p.$$

So far, we have given an explicit definition of the partitioning P_2. In the following, we describe a recursive definition [1] which leads to the same partitioning.

Now, we consider families of three different matrix formats: $\mathcal{R}$, $\mathcal{N}$, and $\mathcal{D}$ which correspond to P_2-partitionings in the above-mentioned examples. Here "$\mathcal{D}$" is the abbreviation for the case with diagonal singularities. $\mathcal{R}$-matrices are matrices of rank $\leqslant k$. The value of k is thought to be much less than the problem (or block) size, in particular, the choice $k = \mathrm{O}(\log n)$ is sufficient for the optimal order approximation. The $\mathcal{R}$-matrices can be represented in the form

$$\sum_{i=1}^{k} [a_i, c_i] \quad \text{where } [a_i, c_i] := a_i * c_i^H,$$

with column vectors a_i and row vectors c_i^H. We abbreviate by $n_\ell = 2^{p-\ell}$ the problem size on the level ℓ. The set of real $\mathcal{R}$-matrices of the size n_ℓ is denoted by $\mathcal{M}_{\mathcal{R}} \subset \mathbb{R}^{n_\ell \times n_\ell}$. This class gives the trivial example of $\mathcal{H}$-matrices of the rank k.

The class $\mathcal{M}_{\mathcal{N}} \subset \mathbb{R}^{n_\ell \times n_\ell}$, $\ell = p, \ldots, 1$, of $\mathcal{N}$-matrices serves for the approximation of the operators with the kernel $s_2(x, y)$ having only one singularity point $x = y = 0$ in $\Sigma \times \Sigma$. For $\ell = p$, $\mathcal{N}$-matrices are simple 1×1 matrices. Then we define the $\mathcal{N}$-format recursively for the levels $\ell = p - 1, \ldots, 1$. An $n_\ell \times n_\ell$ matrix M has the $\mathcal{N}$-format if

$$M = \begin{bmatrix} M_{11} & M_{12} \\ M_{21} & M_{22} \end{bmatrix} \quad \text{with } \frac{n_\ell}{2} \times \frac{n_\ell}{2}\text{-blocks } M_{ij}, \quad i, j = 1, 2, \tag{6}$$

where $M_{11}, M_{12}, M_{22} \in \mathcal{M}_{\mathcal{R}}$ and $M_{21} \in \mathcal{M}_{\mathcal{N}}$. Similarly, we define the transposed format: M is an $\mathcal{N}^*$-matrix if M^{T} has the $\mathcal{N}$-format. This format may be applied in the case of one singular point of $s(x, y)$ at $x = y = 1$. The sets of $\mathcal{N}$- and $\mathcal{N}^*$-matrices are denoted by $\mathcal{M}_{\mathcal{N}}$ and $\mathcal{M}_{\mathcal{N}^*}$, respectively.

Finally, the class $\mathcal{M}_{\mathcal{D}}$ of $\mathcal{H}$-matrices of the $\mathcal{D}$-format is defined by the following recursion. Let $M \in \mathbb{R}^{n_\ell \times n_\ell}$ with $\ell = p, \ldots, 1$. For $\ell = p$, $\mathcal{M}_{\mathcal{D}}$ contains all 1×1 matrices. For $\ell = p - 1, \ldots, 1$, an $n_\ell \times n_\ell$-matrix M belongs to $\mathcal{M}_{\mathcal{D}}$ if

$$M = \begin{bmatrix} M_{11} & M_{12} \\ M_{21} & M_{22} \end{bmatrix} \quad \text{with } M_{11}, M_{22} \in \mathcal{M}_{\mathcal{D}},\ M_{12} \in \mathcal{M}_{\mathcal{N}},\ M_{21} \in \mathcal{M}_{\mathcal{N}^*}, \tag{7}$$

where all block-matrices M_{ij} are of the size $n_\ell/2 \times n_\ell/2$. In the case of $p = 4$, the resulting block structure of an 16×16 matrix is given in Fig. 1. The partitionings defined above correspond to the

[1] The explicit and recursive definitions are possible for the model problems discussed. In general, there is an algorithm for computing the minimal admissible partitioning (see [3]).

choice $\eta = \frac{1}{2}$ in the related admissibility conditions (4) and (5). This provides the approximation order $O(\eta^m)$ with the appropriate choice $m = O(\log n)$, see Section 3. Note that if the partitioning P_2 is given a priori, then, we obtain the following explicit definition of $\mathscr{H}$-matrices.

Definition 1. Let a block partitioning P_2 of $I \times I$ and $k < n = 2^p$ be given. The set of real $\mathscr{H}$-matrices induced by P_2 and k is

$$\mathcal{M}_{\mathscr{H},k}(I \times I, P_2) := \{M \in \mathbb{R}^{I\times I} \colon \forall b \in P_2, \text{ there holds rank } (M^b) \leqslant k\}. \tag{8}$$

Note that $\mathscr{H}$-matrices with block-dependent rank (e.g., $k(b) := a_1 \text{level}(b) + a_2$) can also be considered, cf. [9]). In [9], a special hierarchical construction of bases $\{a_i\}, \{c_i\}$ for the block-matrices M^b leads to an $O(n)$ complexity of both the memory and the matrix–vector multiplication.

3. Reliability of $\mathscr{H}$-matrix approximations in BEM

The $\mathscr{H}$-matrices provide sparse discretisations of integral operators. In this section, we show that the hierarchical matrices are also dense enough, i.e., they lead to the same asymptotically optimal approximations as the exact FE/BE Galerkin schemes. We consider the typical BEM applications, where integral operators of the form

$$(Au)(x) = \int_\Sigma s(x,y)u(y)\,\mathrm{d}y, \quad x \in \Sigma$$

occur with s being the fundamental solution (singularity function) associated with the p.d.e. under consideration or with s replaced by a suitable directional derivatives Ds of s. Here Σ is either a bounded $(d-1)$-dimensional manifold (surface) or a bounded domain in $\mathbb{R}^d$, $d = 2, 3$. The $\mathscr{H}$-matrix techniques exploit the block-wise approximation of s by a degenerate kernel based on the smoothness properties of the singularity function s (cf. [4, Definition 3.3.3]). This holds for s as well as for $\partial s(x,y)/\partial n(x)$ or $\partial s(x,y)/\partial n(y)$ (double-layer kernel and its adjoint; cf. [4, (8.1.31a,b)]) even if the normal vector n is nonsmooth (because of the nonsmoothness of the surface Σ). More precisely, we assume that the singularity function s satisfies [2]

$$|\partial_x^\alpha \partial_y^\beta s(x,y)| \leqslant c(|\alpha|,|\beta|)|x-y|^{-|\alpha|-|\beta|}g(x,y) \quad \text{for all } |\alpha|,|\beta| \leqslant m \tag{9}$$

and for all $x, y \in \mathbb{R}^d$, $x \neq y$, where α, β are multi-indices with $|\alpha| = \alpha_1 + \cdots + \alpha_d$. We consider two particular choices of the (singular) function $g \geqslant 0$ defined also on $\Gamma \times \Gamma$. The first case $g(x,y) = |s(x,y)|$ is discussed in [6]. The second choice to be discussed is $g(x,y) = |x-y|^{1-d-2r}$. Here $2r \in \mathbb{R}$ is the order of the integral operator $A \colon H^r(\Gamma) \to H^{-r}(\Gamma)$. Similar smoothness prerequisites are usually required in the wavelet or multi-resolution techniques (cf. [1,13]). We shall give a simple example how the above assumption on the kernel implies the local expansions of the form

$$s_{\tau,\sigma} = \sum_{j=1}^{k} a_j(x)c_j(y), \quad (x,y) \in \tau \times \sigma \tag{10}$$

[2] In the case $g(x,y) = |s(x,y)|$, estimate (9) is a bit simplified. It covers most of the situations, e.g., the case of the singularity function $(1/4\pi)|x-y|^{-1}$ for $d = 3$. As soon as logarithmic terms appear (as for $d = 2$; $s(x,y) = \log(x-y)/2\pi$), one has to modify (9). A simple modification is also required for the single-layer potential on polyhedrons.

for each cluster $\tau \times \sigma \in P_2$, where k is the order of expansion. Then, we prove the consistency error estimate. We refer to [2] on the familiar multipole expansions of the form (10) applied in the case of the Laplace equation.

By Definition 1, $\mathscr{H}$-matrices are composed locally (blockwise) of rank-k matrices. These low rank matrices can be constructed by means of separable representations (10). In turn, the latter can be obtained, for example, by polynomial approximation with the Taylor expansion[3] of $s(x, y)$. Alternatively, the local L^2-projection onto the set of polynomials as well as the multipole-type expansions (the latter are only available for special kernels like $(1/4\pi)|x-y|^{-1}$ for $d=3$) may be also applied.

Let x, y vary in the respective sets $X(\tau)$ and $X(\sigma)$ corresponding to the admissible clusters $\tau, \sigma \in T_1$ (cf. Section 2.2) and assume, without loss of generality, that $\operatorname{diam}(X(\sigma)) \leqslant \operatorname{diam}(X(\tau))$. The optimal centre of expansion is the Chebyshev centre[4] y_* of $X(\sigma)$, since then $\|y-y_*\| \leqslant \frac{1}{2}\operatorname{diam}(X(\sigma))$ for all $y \in X(\sigma)$. The Taylor expansion reads $s(x,y)=\tilde{s}(x,y)+R$ with the polynomial

$$\tilde{s}(x,y)=\sum_{|\nu|=0}^{m-1}\frac{1}{\nu!}(y_*-y)^{\nu}\frac{\partial^{\nu}s(x,y_*)}{\partial y^{\nu}} \tag{11}$$

and the remainder R, which can be estimated by

$$|R|=|s(x,y)-\tilde{s}(x,y)| \leqslant \frac{1}{m!}\|y_*-y\|^m \max_{\zeta\in X(\sigma),\ |\gamma|=m}\left|\frac{\partial^{\gamma}s(x,\zeta)}{\partial\zeta^{\gamma}}\right|. \tag{12}$$

Below, we recall the familiar approximation results based on the Taylor expansions (see, e.g., [6] for the proof).

Lemma 2. *Assume that* (9) *is valid and that the admissibility condition* (5) *holds with* η *satisfying* $c(0,1)\eta<1$. *Then for* $m \geqslant 1$, *the remainder* (12) *satisfies the estimate*

$$|s(x,y)-\tilde{s}(x,y)| \leqslant \frac{c(0,m)}{m!}\eta^m \max_{y\in X(\sigma)}|g(x,y)|, \quad x\in X(\tau),\quad y\in X(\sigma). \tag{13}$$

Let $A_{\mathscr{H}}$ be the integral operator with s replaced by $\tilde{s}(x,y)$ for $(x,y)\in X(\tau)\times X(\sigma)$ provided that $\tau\times\sigma\in P_2$ is an admissible block and no leaf. Construct the Galerkin system matrix from $A_{\mathscr{H}}$ instead of A. The perturbation of the matrix induced by $A_{\mathscr{H}}-A$ yields a perturbed discrete solution of the initial variational equation

$$\langle(\lambda I+A)u,v\rangle=\langle f,v\rangle \quad \forall v\in W:=H^r(\Sigma),\ \ r\leqslant 1,$$

where $\lambda\in\mathbb{R}$ is a given parameter. The effect of this perturbation in the panel clustering methods is studied in several papers (cf. [10–12]). Here we give the consistency error estimate for the $\mathscr{H}$-matrix approximation. Define the integral operator $\hat{A}$ with the kernel

$$\hat{s}(x,y):=\begin{cases}\max_{y\in\sigma}|g(x,y)| & \text{for } (x,y)\in X(\tau)\times X(\sigma),\ \tau\times\sigma\in P_2,\\ 0 & \text{otherwise.}\end{cases} \tag{14}$$

[3] This does not require that the practical implementation has to use the Taylor expansion. If the singular-value decomposition technique from [5] is applied, the estimates are at least as good as the particular ones for the Taylor expansion.

[4] Given a set X, the Chebyshev sphere is the minimal one containing X. Its centre is called the *Chebyshev centre*.

For the given ansatz space $W_h \subset W$ of piecewise constant/linear FEs, consider the perturbed Galerkin equation for $u_{\mathscr{H}} \in W_h$,

$$\langle (\lambda I + A_{\mathscr{H}})u_{\mathscr{H}}, v\rangle = \langle f, v\rangle \quad \forall v \in W_h.$$

In the following we use a bound on the discrete operator norm $||\hat{A}||_{W_h \to W_h'}$ appearing in

$$|\langle \hat{A}u, v\rangle| \leqslant ||\hat{A}||_{W_h \to W_h'} ||u||_W ||v||_W, \quad \forall u,\ v \in W_h. \tag{15}$$

Lemma 3. *Assume that* (9) *is valid. Suppose that the operator* $\lambda I + A \in \mathscr{L}(W, W')$ *is* W*-elliptic. Then there holds*

$$||u - u_{\mathscr{H}}||_W \leqslant c\left\{ \inf_{v_h \in V_h} ||u - v_h||_W + \frac{c(0,m)}{m!}\eta^m ||\hat{A}||_{W_h \to W_h'} ||u||_W \right\}.$$

The norm of $\hat{A}$ *is estimated by*

$$||\hat{A}||_{W_h \to W_h'} \lesssim \begin{cases} ||A|| & \text{if } g = s(x,y) \wedge s(x,y) \geqslant 0, \\ \delta(d,r) h^{\min\{0,r\}} & \text{if } g = |x-y|^{1-d-2r}, \end{cases} \tag{16}$$

where (with $\varepsilon = 1 - d - 2r$*)*

$$\delta(d,r) := \left(\sum_{l=0}^{p} 2^{2(l-p)\varepsilon} \right)^{1/2} = \begin{cases} O(1), & \varepsilon > 0, \\ O(p), & \varepsilon = 0, \\ O(h^{\varepsilon}), & \varepsilon < 0. \end{cases}$$

Proof. The continuity and strong ellipticity of A imply

$$||u - u_{\mathscr{H}}||_W \lesssim \inf_{v \in W_h} ||u - v||_W + \sup_{u,v \in W_h} \frac{|\langle (A - A_{\mathscr{H}})u, v\rangle|}{||u||_W ||v||_W} ||u_{\mathscr{H}}||_W$$

(cf. first Strang Lemma). On the other hand, under assumption (9), Lemma 2 yields

$$|\langle (A - A_{\mathscr{H}})u, v\rangle| \lesssim \frac{c(0,m)}{m!}\eta^m ||\hat{A}||_{W_h \to W_h'} ||u||_W ||v||_W, \quad u, v \in W_h.$$

Indeed,

$$\begin{aligned} |\langle (A - A_{\mathscr{H}})u, v\rangle| &\lesssim \frac{c(0,m)}{m!}\eta^m \sum_{\tau \times \sigma \in P_2} \int_{X(\tau)\times X(\sigma)} |\hat{s}(x,y)u(y)v(x)|\,\mathrm{d}x\,\mathrm{d}y \\ &\lesssim \frac{c(0,m)}{m!}\eta^m ||\hat{A}||_{W_h \to W_h'} ||u||_W ||v||_W. \end{aligned} \tag{17}$$

Now, assuming that $(c(0,m)/m!)\eta^m ||\hat{A}||_{W_h \to W_h'}$ is sufficiently small, estimate (16) and $\eta < 1$ imply the strong ellipticity of the discrete Galerkin operator yielding the stability $||u_{\mathscr{H}}||_W \leqslant c||u||_W$. Note that in the case $g = s(x,y)$, the first assertion in (16) follows from the bound $|||u|||_W \leqslant ||u||_W$ for all $u \in W_h$. In the case $g = |x-y|^{1-d-2r}$ and $r \geqslant 0$, bound (16) follows from the direct estimate based on the essential properties of the admissible partitioning P_2: $\mathrm{diam}(\tau) = O(2^{-\ell})$, $\tau \in P_2^{\ell}$ and $\#P_2^{\ell} = O(2^{d\ell})$. In the case $r < 0$, we first obtain an estimate with the constant $\delta(d,r)$ in the L^2-norm. Then, applying the inverse inequality $||v||_{L^2(\Gamma)} \lesssim h^r ||v||_{H^r(\Gamma)}$, $v \in W_h$ completes our proof. □

The block Rk-approximation in the Galerkin method may be computed as the block entry $\mathscr{A}_{\mathscr{H}}^{\tau\times\sigma}$ of the stiffness matrix $\mathscr{A}_{\mathscr{H}}:=\{\langle A_{\mathscr{H}}\varphi_i,\varphi_j\rangle\}_{i,j=1}^{N}$ associated with each cluster $\tau\times\sigma$ on the level ℓ may be presented as a rank-k matrix

$$\mathscr{A}_{\mathscr{H}}^{\tau\times\sigma}=\sum_{|\nu|=0}^{m-1} a_\nu * b_\nu^T \quad \text{where } k:=\binom{d_\Sigma+m-1}{m-1}=\mathrm{O}((m-1)^{d_\Sigma})$$

is the number of terms and

$$a_\nu=\left\{\int_{X(\tau)}(y-y_*)^\nu\varphi_i(y)\,\mathrm{d}y\right\}_{i=1}^{N_\tau},\quad b_\nu=\left\{\int_{X(\sigma)}\frac{\partial^\nu s(x,y_*)}{\partial y^\nu}\varphi_j(x)\,\mathrm{d}x\right\}_{j=1}^{N_\sigma}.$$

Here $N_\tau=\#\tau=\mathrm{O}(2^{d_\Sigma(p-\ell)})$ (resp. $N_\sigma=\#\sigma=\mathrm{O}(2^{d_\Sigma(p-\ell)})$) is the cardinality of τ (resp. σ). Note that in BEM applications, we have $d_\Sigma=d-1$, while for volume integral calculations there holds $d_\Sigma=d$.

4. $\mathscr{H}$-matrices on tensor-product meshes

4.1. Partitioning of tensor-product index set]Proof

In $\Omega=(0,1)^d$ with $d=1,2,3$, we consider the regular grid

$$I=\{\boldsymbol{i}=(i_1,\ldots,i_d):\ 1\leqslant i_k\leqslant N,\ k=1,\ldots,d\},\quad N=2^p. \tag{18}$$

We define the norms $|\boldsymbol{i}|_\infty=\max_{1\leqslant n\leqslant d}|i_n|$ and $|\boldsymbol{i}|_1=\sum_{n=1}^d|i_n|$. Each index $\boldsymbol{i}\in I$ is identified with the (collocation) point $\xi_{i_1\ldots i_d}=((i_1-\frac{1}{2})h,\ldots,(i_d-\frac{1}{2})h)\in\mathbb{R}^d$, where $h:=1/N$ and the value $\xi_{\boldsymbol{i}}=\xi_{i_1\ldots i_d}$ is the midpoint of the support $X_{\boldsymbol{i}}$ of the basis function $\varphi_{\boldsymbol{i}}$ in the FE or BE method considered (cf. (19) below).

The cluster tree $T_1=T(I)$ of I uses a division of the underlying cubes into 2^d subcubes. The blocks

$$t_{\boldsymbol{j}}^\ell=\{\boldsymbol{i}:\ 2^{p-\ell}j_1+1\leqslant i_1\leqslant 2^{p-\ell}(j_1+1),\ldots,\ 2^{p-\ell}j_d+1\leqslant i_d\leqslant 2^{p-\ell}(j_d+1)\}$$

for $\boldsymbol{j}\in\{0,\ldots,2^\ell-1\}^d$ belong to level ℓ. $S_1(t_{\boldsymbol{j}'}^{\ell-1}):=\{t_{\boldsymbol{j}}^\ell:\ 0\leqslant 2j_k'-i_k\leqslant 1\ (1\leqslant k\leqslant d)\}$ defines the set of sons of the cluster $t_{\boldsymbol{j}'}^{\ell-1}$. Hence, the tree T_1 consisting of all blocks at all levels $\ell\in\{0,\ldots,p\}$ is a binary, quad- or octree for $d=1,2,3$, respectively. The number of clusters on level ℓ equals $\mathrm{O}(2^{d\ell})$.

Each index $\boldsymbol{i}\in I$ is associated with the d-dimensional cube[5]

$$X_{\boldsymbol{i}}:=\{(x_1,\ldots,x_d):(i_1-1)h\leqslant x_1\leqslant i_1h,\ldots,(i_d-1)h\leqslant x_d\leqslant i_dh\}, \tag{19}$$

which may be considered as the support of the piecewise constant function for the index $\boldsymbol{i}$. Using the Euclidean norm, we obtain the diameter $\mathrm{diam}(t)=\sqrt{d}2^{p-\ell}h=\sqrt{d}/2^\ell$ for blocks of level ℓ. Let τ,σ be two blocks of level ℓ characterised by $\boldsymbol{j}$ and $\boldsymbol{j}'$, i.e., $\tau=t_{\boldsymbol{j}}^\ell$, $\sigma=t_{\boldsymbol{j}'}^\ell$. Then

$$\mathrm{dist}(\tau,\sigma)=2^{-\ell}\sqrt{\delta(j_1-j_1')^2+\cdots+\delta(j_d-j_d')^2} \tag{20}$$

[5] The grid can also be associated with a regular *triangulation* and, e.g., the supports $X_{\boldsymbol{i}}$ of piecewise linear functions, see Section 5. The asymptotic complexities turn out to be the same as for the present choice.

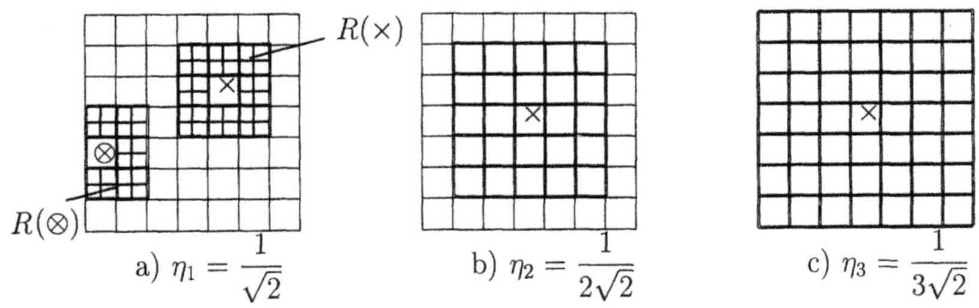

Fig. 2. Unacceptable region for the given clusters "×", "⊗" depending on the threshold constant η.

with $\delta(\xi):=\max\{0,|\xi|-1\}$. Let the block-cluster tree $T_2 = T(I\times I)$ be defined in accordance with the cluster tree $T_1 = T(I)$ (see [5] for more details). An important property is stated in

Remark 4. Let $\tau\times\sigma\in T(I\times I)$. Then $\tau,\sigma\in T(I)$ belong to the same level $\ell\in\{0,\dots,p\}$.

In view of this remark, for $\ell\in\{0,\dots,p\}$, we denote by T_2^ℓ the set of clusters $\tau\times\sigma\in T_2$ such that blocks τ,σ belong to level ℓ. In particular, $T_2^0=\{I\times I\}$ is the root of T_2 and $T_2^p=\{\{(x,y)\}:x,y\in I\}$ is the set of leaves. The set of clusters $t\in T(I)$ from level ℓ is called T_1^ℓ. In the following we consider the choice

$$\eta=\eta_\mu=\frac{\sqrt{d}}{2\mu},\qquad \mu\in\mathbb{N} \tag{21}$$

of η. Note that increasing μ yields arbitrarily small values of η.

Using $\min\{\mathrm{diam}(t_1),\mathrm{diam}(t_2)\}=\sqrt{d}/2^\ell$ and $\mathrm{dist}(t_1,t_2)$ from (20), we observe that $t\in T(I\times I)$ is admissible for the choice (21) if the squares $X_1=X(\tau)$, $X_2=X(\sigma)$, $\tau,\sigma\in T(I)$ have a relative position as indicated in Figs. 2a–c corresponding to $\mu=1,2$ and 3, respectively, with $d=2$. The square $X(\tau)$ corresponding to τ is the crossed square, while $X(\sigma)$ must be outside the bold area. In the case of $d=2$ and $\eta=1/\sqrt{2}$, i.e., for $\mu=1$, the admissible T_2-partitioning P_2 was described in details in [6]. Note that the general Definition 5 of $\mathcal{M}_p^\square(p,\eta)$-formats given below generalises the particular examples for $d=1,2,3$ from [5,6].

4.2. Basic definitions

In this section, we introduce the general formats for matrices operating in the vector space $\mathbb{K}^I$ for the cell-centred tensor product grid $I=I_h^d$ in $\Omega=(0,1)^d$ with the mesh-size $h=2^{-p}$, $\#I=2^{dp}$ and $d=1,2,3$. The natural notation of indices from $I=I_h^d$ is by multiindices $\boldsymbol{i}=(i_1,\dots,i_d)$ with $1\leqslant i_n\leqslant N=2^p$.

As in the particular cases in [5,6], we can describe the partitioning by a number of formats $\mathcal{M}_q^{\boldsymbol{j}}=\mathcal{M}_q^{\boldsymbol{j}}(p,\eta)$, where $q\in\{0,\dots,p\}$, η is parametrised by (21) and the multiindex $\boldsymbol{j}=(j_1,\dots,j_d)$ with $|\boldsymbol{j}|_\infty\leqslant\mu$ indicates a translation in the following sense. Let $b=\tau\times\tau'\in T_2^\ell$ be a block of level $\ell=p-q$. If $\tau=\tau'$, we have a diagonal block corresponding to the vanishing shift, i.e., $\boldsymbol{j}=\boldsymbol{0}=(0,\dots,0)$. For these blocks we shall introduce the top format $\mathcal{M}_q^{\boldsymbol{0}}=\mathcal{M}_q^{\boldsymbol{0}}(p,\eta)$. In general, let $\tau=(i_{01},\dots,i_{0d})+\{(i_1,\dots,i_d)\text{: }1\leqslant i_n\leqslant 2^\ell\}$ and $\tau'=(i_{01}+j_12^\ell,\dots,i_{0d}+j_d2^\ell)+\{(i_1,\dots,i_d):1\leqslant i_n\leqslant 2^\ell\}$ be two clusters (cubes of length 2^ℓ). Then their relation is given by the translation in direction

$\vec{j}=(j_1,\dots,j_d)^{\mathrm{T}}$. We write $\tau'=\mathscr{T}^{j}_{\ell}\tau$, where $\mathscr{T}^{j}_{\ell}$ is the translation operator with respect to the vector $h_\ell\cdot\vec{j}$, $|\boldsymbol{j}|_\infty\leqslant 2^\ell-1$ (due to (21), we actually have the bound $|\boldsymbol{j}|_\infty\leqslant\mu$ for non-admissible clusters), where $h_\ell=2^{-\ell}$.

Let $\tau\in T_1$ be a cluster from level ℓ. The corresponding set of sons, $S_1(\tau)=\{\sigma_i\}_{i\in I_d}$ is associated with the set of multiindices I_d, where

$$I_d=\{\boldsymbol{k}\in\mathbb{N}^d : |\boldsymbol{k}|_\infty=1\wedge|\boldsymbol{k}|_1=d\},\quad \dim I_d=2^d,$$

as depicted in Fig. 3a for $d=3$, and Fig. 3b for $d=2$. Equivalently, $S_1(\tau)=\{a,b,c,d,e,f,g,h\}$. This multiindex block numbering indicates the location of sons with respect to the centre of gravity of the parent cluster: $\mathrm{cent}(\sigma_i)=\mathrm{cent}(\tau)+\frac{1}{2}h_{\ell+1}\boldsymbol{i}$ for $\sigma_i\in S_1(\tau)$. For example, there holds $\sigma_{i_a}=a$ and $\sigma_{i_b}=b$ with the vector notations $\boldsymbol{i}_a=(-1,1,1)$, $\boldsymbol{i}_b=(1,1,1)$. The block-matrix with columns from a and rows from b is denoted by $A_{i_a i_b}\in\mathbb{K}^{a\times b}$. The examples of two-dimensional vectors are drawn in Fig. 3b, where, e.g., $\boldsymbol{i}_1=(-1,1),\boldsymbol{i}_2=(1,1)$.

For block-clusters $\sigma\times\sigma'\in T_2^\ell$ from level $\ell=p-q$, where $\sigma'=\mathscr{T}^{j}_{\ell}\sigma$, $|\boldsymbol{j}|_\infty\leqslant\mu$, we define recursively for $q=0,\dots,p$ the formats $\mathscr{M}^{j}_{q}=\mathscr{M}^{j}_{q}(p,\eta)$ of $\mathscr{H}$-matrices from $\mathbb{K}^{\sigma\times\sigma'}$ starting from $q=0$ and ending with $q=p$. In this way, a family of auxiliary formats $\mathscr{M}^{j}_{q}$, with $|\boldsymbol{j}|_\infty\neq 0$ is involved, e.g., "next neighbours" ($|\boldsymbol{j}|_\infty=1$), "2-layer neighbours" ($|\boldsymbol{j}|_\infty=2$) and so on. In Definition 5 below these formats contain the same construction at the next level ("self-reference") and other formats as depicted in the graph generalising the corresponding picture from [6]:

top format	$\boldsymbol{j}=0$		self-reference $=2^d$
	$\downarrow$	$\searrow$	
next neighbours:	$\lvert\boldsymbol{j}\rvert_\infty=1$	…	self-reference $\leqslant 2^{d-1}$
	$\downarrow$	$\searrow$	
second-layer neighbours:	$\lvert\boldsymbol{j}\rvert_\infty=2$	…	self-reference $\leqslant 2^{d-2}$
	$\downarrow$	$\searrow$	
"large-distance" formats:	$\lvert\boldsymbol{j}\rvert_\infty>2$	…	self-reference $=0$ ($\lvert\boldsymbol{j}\rvert_\infty>\frac{\mu}{2}+1$)
	$\downarrow$		
leaves	Rk		self-reference $=0$.

We underline that the matrix format $\mathscr{M}^{j}_{q}$ does not depend on the particular choice of the cluster σ but it is only determined by the translation operator $\mathscr{T}^{j}_{p-q}$. Roughly speaking, each *format* under consideration actually specifies (in general, recursively) the location and size of *Rk*-blocks in the matrix array from $\mathbb{K}^{I\times I}$ corresponding to the given admissible partitioning P_2 of $I\times I$. The partitioning P_2 itself is generated implicitly by Definition 5 below. Here the basic parameters $p\in\mathbb{N}$ and $\mu\in\mathbb{N}$ are both fixed, so, we may skip them in the notation $\mathscr{M}^{j}_{q}$ without ambiguity.

We recall that $\mathscr{M}^{R}_{q}$ is a set of *Rk*-matrices of the size $2^{dq}\times 2^{dq}$, $q=0,1,\dots,p$. Now, we define our format in the following range of parameters: $q=0,1,\dots,p$ and $|\boldsymbol{j}|_\infty\leqslant 2^\ell-1$, where $\ell=p-q$.

Definition 5. (a) For $q=0,\dots,p$ and for all $|\boldsymbol{j}|_\infty\geqslant\mu+1$, define the format $\mathscr{M}^{j}_{q}$ by $\mathscr{M}^{j}_{q}=\mathscr{M}^{R}_{q}$.

(b) For $q=0$, define $\mathscr{M}^{j}_{0}$ as the set of 1×1 matrices for all $|\boldsymbol{j}|_\infty\leqslant\mu$.

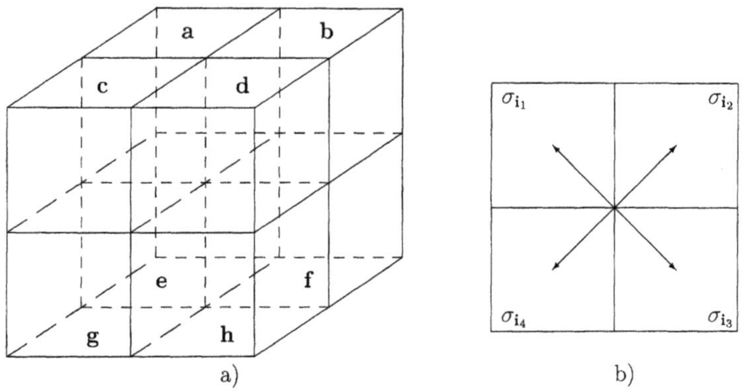

Fig. 3. (a) Multiindex labelling of sons of the 3D cluster, where $a=(-1,1,1)$, $b=(1,1,1)$, $c=(-1,-1,1)$, $e=(-1,1,-1)$, $d=(1,-1,1)$, $f=(1,1,-1)$, $g=(-1,-1,-1)$, $h=(1,-1,-1)$. (b) The ordering by local translations for 2D cell.

(c) Consider the case $q=1,\dots,p$ and $1 \leqslant |\boldsymbol{j}|_\infty \leqslant \mu$. To describe the recursion step, assume that for each $q \leqslant q_0$ with some $q_0 \geqslant 0$, the format $\mathscr{M}_q^{\boldsymbol{j}}$ is already defined for all $1 \leqslant |\boldsymbol{j}|_\infty \leqslant \mu$. In the following we define the format $\mathscr{M}_q^{\boldsymbol{j}}$ for $q=q_0+1$. Consider indices $\boldsymbol{j}$ with $1 \leqslant |\boldsymbol{j}|_\infty \leqslant \mu$ and blocks $\sigma \times \sigma' \in T_2$ of level $l=p-q$ such that $\sigma' = \mathscr{T}_\ell^{\boldsymbol{j}}\sigma$.[6]

For the matrices from $\mathbb{K}^{\sigma\times\sigma'}$, we say that $A_{\sigma,\sigma'} = \{A_{\boldsymbol{i}\boldsymbol{i}'}\}_{\boldsymbol{i}\in I_d, \boldsymbol{i}'\in I_d'}$ belongs to $\mathscr{M}_q^{\boldsymbol{j}}$, if $A_{\boldsymbol{i}\boldsymbol{i}'} \in \mathscr{M}_{q-1}^{\boldsymbol{i}'-\boldsymbol{i}+2\boldsymbol{j}}$, where, due to the admissibility condition there holds $\mathscr{M}_{q-1}^{\boldsymbol{i}'-\boldsymbol{i}+2\boldsymbol{j}} \in \mathscr{M}_{q-1}^R$ for all indices from the range $|\boldsymbol{i}'-\boldsymbol{i}+2\boldsymbol{j}|_\infty \geqslant \mu+1$.

(d) Finally, for $\boldsymbol{j}=\boldsymbol{0}$, define the top formats $\mathscr{M}_q^{\boldsymbol{0}}$ for $q=1,\dots,p$. Let $\sigma \in T_1$ be from level $\ell = p-q$. Then we say that $A_{\sigma,\sigma} = \{A_{\boldsymbol{i}\boldsymbol{i}'}\}_{\boldsymbol{i}\in I_d, \boldsymbol{i}'\in I_d}$ belongs to $\mathscr{M}_q^{\boldsymbol{0}}$ if there holds $A_{\boldsymbol{i}\boldsymbol{i}} \in \mathscr{M}_{q-1}^{\boldsymbol{0}}$ and $A_{\boldsymbol{i}\boldsymbol{i}'} \in \mathscr{M}_{q-1}^{\boldsymbol{i}'-\boldsymbol{i}}$ for $\boldsymbol{i}' \neq \boldsymbol{i}$, where the auxiliary formats are already defined in item (c).

Note that the format $\mathscr{M}_p^{\boldsymbol{0}}$ introduced by Definition 5 reproduces (with different abbreviations) the particular constructions from [5,6] given for $d=1,2,3$ and for $\mu=1$.

4.3. Complexity estimates

In the following, we discuss the storage requirements $\mathscr{N}_{\mathrm{st}}$ and the cost $\mathscr{N}_{\mathrm{MV}}$ of the matrix–vector multiplication for the general $\mathscr{M}_p^{\boldsymbol{0}}(p,\eta)$ formats. The corresponding results for the particular cases $\mathscr{M}_p^{\boldsymbol{0}}(p,\sqrt{d}/2)$ were presented in [5,6].

Note that the maximal level number p is $\leqslant O(|\log h|)$. In the following, we call a pair of one addition and one multiplication a *coupled operation*.

Theorem 6. *Let $d \in \{1,2,3\}$, $A \in \mathscr{M}_{p,k}^{\boldsymbol{0}}(p,\eta)$ and $\eta = \eta_\mu := \sqrt{d}/2\mu$, $\mu \in \mathbb{N}$. Then the matrix–vector multiplication complexity is bounded by*

$$\mathscr{N}_{\mathrm{MV}} \leqslant (2^d-1)(\sqrt{d}\eta^{-1}+1)^d\, pkn \tag{22}$$

[6] As above, we use the local numbering of sons $S(\sigma) = \{\sigma_{\boldsymbol{i}}\}_{\boldsymbol{i}\in I_d}$ and $S(\sigma') = \{\sigma'_{\boldsymbol{i}'}\}_{\boldsymbol{i}'\in I_d'}$, where $I_d' = \mathscr{T}_\ell^{\boldsymbol{j}} I_d$.

coupled operations. There holds

$$\mathcal{N}_{\text{st}} \leqslant (2^d - 1)(\sqrt{d}\eta^{-1} + 1)^d \, pkn \tag{23}$$

for the storage requirements. Both estimates are asymptotically sharp.

Proof. Recall that the matrix–vector multiplication with matrices from $\mathcal{M}_p^R$ costs $2kn$ multiplications and kn additions. For each $\tau \in T_1^\ell$, we introduce the set of nonadmissible clusters $R(\tau)$ by

$$R(\tau) := \{\tau' \in P_1^\ell,\ \tau' \neq \tau:\ \operatorname{diam}(\tau') > 2\eta_\mu \operatorname{dist}(\tau', \tau)\}.$$

For any son $\sigma \in S(\tau)$, the number $Q_\sigma := \#\{b \in P_2:\ b = \tau \times \sigma\}$ of Rk-blocks in the block-matrix row of A and associated with a cluster position σ is majorised by the corresponding one for the case of purely "interior" cluster τ.[7] Particularly, Q_σ equals the number of sons $\sigma_{\tau'} \in S(\tau')$ from the set of clusters $\tau' \in T_1^\ell$, which are neighboured to τ and satisfy the admissibility condition with σ,

$$Q_\sigma := \#\{\sigma_{\tau'}:\ \sigma_{\tau'} \in S(\tau'),\ \tau' \in R(\tau),\ \sigma \times \sigma_{\tau'} \text{ satisfies (5)}\}.$$

In the case of purely "interior" clusters, the direct calculation shows that this number is equal to $Q_\tau = (2^d - 1)(2\mu + 1)^d$. Now the multiplication complexity of all Rk-blocks from the given level ℓ amounts to $2^\ell Q_\tau$ multiplications of Rk-blocks with a vector of the dimension $n2^{-\ell}$. Moreover, we have the summation of intermediate results located in the block columns which costs $(k-1)Q_\tau$ additions of full n-dimensional vectors. This exactly results in the constant 2 for counting the *coupled* operations. To prove the sharpness of this bound, we note that the number of "nearly boundary" clusters[8] on each level ℓ is estimated by $O(2^{(d-1)\ell})$. Thus, the complexity count for the corresponding matrix blocks is dominated by the value $O(\sum_{\ell=0}^{p} 2^{-\ell} kn) = O(kn)$ which shows that (22) is asymptotically sharp. The bound (23) is proven along the same line. □

Remark 7. It is clear by the construction that using linear/bilinear elements disturbs the parameter η only slightly. In fact, the perturbed parameter is estimated by $\eta_{\text{new}} = \eta + ch < 1$ for small enough h. Then all the previous constructions remain verbatim with the corresponding modifications.

In view of above remark, we need also the construction based on the truncated tree. For a level number $p_0 \in \{0, \dots, p\}$, we call the T_2-partitioning P_2^* a p_0-truncation of P_2 if it is obtained from the smaller tree $T_2^* \subset T_2$ by deleting all vertices belonging to levels $\ell > p - p_0$ and inserting the sons of size 1×1 (leaves) for all nonadmissible blocks of the initial tree T_2 at level $\ell = p - p_0$, i.e., $\tau \in T_2^{p-p_0}$ has the sons $S(\tau) = \{\{i\}:\ i \in \tau\}$. By assumption, all nonadmissible blocks of level $\ell = p - p_0$ are full submatrices. Clearly, a treatment of these blocks costs $2^{dp_0}(2\mu + 1)^d n$ operations. This yields the following estimates (24) and (25) for the p_0-truncated partitioning: the matrix–vector product costs

$$\mathcal{N}_{\text{MV}} \leqslant (2^d - 1)(\sqrt{d}\eta^{-1} + 1)^d (p - p_0) kn + 2^{dp_0 - 1}(\sqrt{d}\eta^{-1} + 1)^d n \tag{24}$$

[7] The purely "interior" cluster $\sigma \in T_1$ from level ℓ is defined to satisfy $\operatorname{dist}(\sigma, \partial\Omega) \geqslant \mu 2^{-\ell}$, see an example with the cluster "×" in Fig. 2.

[8] The "nearly boundary" cluster $\sigma \in T_1$ from level ℓ is defined to satisfy $\operatorname{dist}(\sigma, \partial\Omega) \leqslant (\mu - 1)2^{-\ell}$, see an example in Fig. 2.

coupled operations; for the storage needs there holds

$$\mathcal{N}_{st} \leqslant (2^d - 1)(\sqrt{d}\eta^{-1} + 1)^d (p - p_0)kn + 2^{dp_0}(\sqrt{d}\eta^{-1} + 1)^d n. \tag{25}$$

Remark 8. Bounds (24) and (25) allow the optimal choice $p_0 = O(\log k)$ of the parameter, which provides a balance between both summands on the right-hand sides. On the other hand, along the line of Section 3.6 in [6] and taking into account Theorem 6, we conclude that $\eta = O(1) < 1$ and $k = \log^{\beta} n$ with some $\beta = \beta(d)$ would be the optimal choice retaining the approximation order $O(h^{\alpha})$, $\alpha > 0$, of the exact Galerkin scheme.

5. $\mathcal{H}$-matrices on triangular meshes

5.1. Translation operators on the index set $I_{\triangle}$

The computational domain Ω is assumed to be composed of a finite number M of macrotriangles $\Omega_1, \dots, \Omega_M$. For the ease of presentation, we restrict our considerations to $\Omega = \Omega_1$, i.e., $M = 1$. We consider the index set $I = I_{\triangle}$ associated with the supports of piecewise constant elements. The index structure for the hierarchical triangulation is defined in accordance with Fig. 4. Fig. 4c illustrates the non admissible clusters with respect to $\tau_1 \in T(I)$ taken as a crossed triangle. Here all admissible clusters τ_2 must be outside the bold area restricted by Γ_{μ} and composed of μ cluster layers, where η is parametrised by $\eta = \eta_{\mu} = 2/3\mu$ with $\mu = 1, 2, 3, \dots$.

The cluster tree $T_1 = T(I)$ is defined by a subdivision of each triangle into four equal parts. The admissible partitionings from the block cluster tree $T(I \times I)$ are determined by (4) with the constant $\eta = \eta_{\mu}$, see also Figs. 4b and c.

We identify the sons of a cluster $\sigma \in T_1^{\ell}$ in accordance with their relative locations, which will be described by the proper translation/reflection operators. In this way, we introduce the oriented clusters from $T_1 = \Lambda \cup \Upsilon$: the subset Λ contains clusters with the "standard" orientation, see Fig. 4a, while Υ contains the set of reflected clusters with respect to the centre of gravity (e.g., σ_4 in Fig. 4a). Accordingly, we write $\Lambda^{\ell} = \Lambda \cap T_1^{\ell}$ and $\Upsilon^{\ell} = \Upsilon \cap T_1^{\ell}$. We also distinct the orientationally dependent and orientationally invariant transforms. The latter include simple translations π to be specified later on. The orientationally dependent (converting) maps include the identity operators E_{Λ} and E_{Υ} in the classes Λ and Υ, respectively, as well as reflection operators $\mathcal{S}_m : \Lambda \to \Upsilon$, $\mathcal{S}_m^{T} : \Upsilon \to \Lambda$, $m = 1, 2, 3$

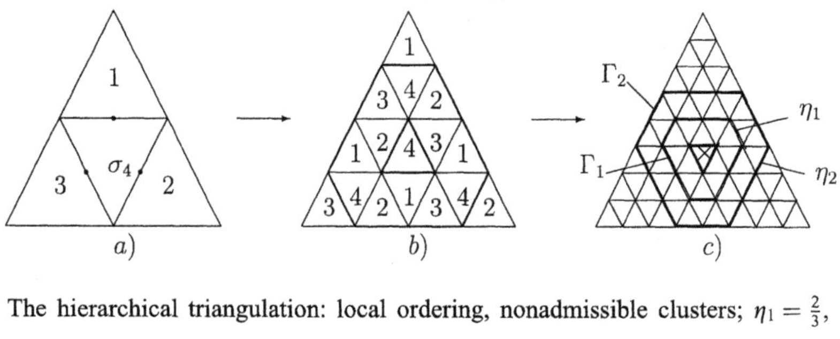

Fig. 4. The hierarchical triangulation: local ordering, nonadmissible clusters; $\eta_1 = \frac{2}{3}$, $\eta_2 = \frac{1}{3}$.

defined below. We shall also distinguish the mapping classes $\mathcal{T}_\Lambda$ and $\mathcal{T}_\Upsilon$ containing the maps from $\Lambda^\ell \to T_1^\ell$ and $\Upsilon^\ell \to T_1^\ell$, respectively.

Assume that the target cluster σ belongs to Λ. The son $\sigma_4 \in S(\sigma)$ (see Fig. 4a) belongs to $\Upsilon^{\ell+1}$ and it corresponds to the trivial translation operator E_Υ, while σ_i, $i = 1,2,3$ belong to $\Lambda^{\ell+1}$. Let ξ_i, $i = 1,\dots,4$, be the centres of gravity for the corresponding clusters σ_i providing $\xi_4 = \mathrm{cent}(\sigma)$. Introduce the vectors $\boldsymbol{j}_{nm} = \xi_m - \xi_n$ and reflection transforms $\mathcal{S}_m$ and $\mathcal{S}_m^{\mathrm{T}}$ with $n,m = 1,2,3$. $\mathcal{S}_m$ maps the cluster σ_m into its symmetric image σ_4 with respect to the centre of common edge. Similarly, the transposed (inverse) mapping $\mathcal{S}_m^{\mathrm{T}}: \sigma_4 \to \sigma_m$ may be introduced. The general *translation* $\pi = \pi^{\boldsymbol{j}_\alpha}$ is defined as a shift by the vector $h_\ell \boldsymbol{j}_\alpha$. Here, $\alpha \in \mathbb{N}_0^3$ such that $\boldsymbol{j}_\alpha := \alpha_1 \boldsymbol{j}_{13} + \alpha_2 \boldsymbol{j}_{21} + \alpha_3 \boldsymbol{j}_{32}$, where $\boldsymbol{j}_{13} + \boldsymbol{j}_{21} + \boldsymbol{j}_{32} = 0$. The general *transforms* $\mathcal{T}_1 \in \mathcal{T}_\Lambda$, $\mathcal{T}_2 \in \mathcal{T}_\Upsilon$ now take the form

$$\mathcal{T}_1 := \pi^{\boldsymbol{j}_\alpha}(\mathcal{S}_m)^\beta, \quad \mathcal{T}_2 := \pi^{\boldsymbol{j}_\alpha}(\mathcal{S}_m^{\mathrm{T}})^\beta, \quad \beta \in \{0,1\}, \ m \in \{1,2,3\}. \tag{26}$$

We call $\mathcal{T} \in \boldsymbol{S}_\mu^\ell$ if $|\mathcal{T}| \leqslant \mu$, where the "norm" is defined by $|\mathcal{T}| = \max\{|\beta|, |\alpha|_\infty\}$. This value measures the translation distance (shift) between σ and $\sigma' = \mathcal{T}\sigma$. Note that the transposed transform is defined by (say, for $\mathcal{T} \in \mathcal{T}_\Lambda$)

$$\mathcal{T}^{\mathrm{T}} := (\mathcal{S}_m^{\mathrm{T}})^\beta \pi^{-\boldsymbol{j}_\alpha}$$

yielding $\mathcal{T}\mathcal{T}^{\mathrm{T}} = E_\Upsilon$, $\mathcal{T}^{\mathrm{T}}\mathcal{T} = E_\Lambda$.

With the given $\mu \geqslant 1$, the nonadmissible area for the underlying cluster σ is then defined by

$$R(\sigma) := \{\mathcal{T}\sigma : 1 \leqslant |\mathcal{T}| \leqslant \mu\}.$$

For example, let $\sigma = \sigma_4 \in \Lambda^\ell$ be the smallest triangle located in the centre of the reference triangle drawn in Fig. 4b and choose $\eta_1 = \frac{2}{3}$. Then, nonadmissible clusters within the bold area $R(\sigma_4)$ are associated with the set of transforms $\{\pi^{\pm \boldsymbol{j}_1}, \pi^{\pm \boldsymbol{j}_2}, \pi^{\pm \boldsymbol{j}_3}, \mathcal{S}_1, \mathcal{S}_2, \mathcal{S}_3, \pi^{\boldsymbol{j}_3}\mathcal{S}_1, \pi^{\boldsymbol{j}_1}\mathcal{S}_2, \pi^{\boldsymbol{j}_2}\mathcal{S}_3\} \in \boldsymbol{S}_1^\ell \setminus E_\Lambda$ corresponding to $\mu = 1$. Let $\sigma, \sigma' \in T_1^\ell$ with $\sigma' = \mathcal{T}\sigma$, where $\mathcal{T} \in \boldsymbol{S}_\mu^\ell$. For the matrix block $\sigma \times \sigma' \in \mathbb{K}^{\sigma \times \sigma'}$, we construct the family of formats $\mathcal{M}_{p-\ell}^{\mathcal{T}}(p,\mu) = \mathcal{M}_{p-\ell}^{\mathcal{T}}$, where the case $|\mathcal{T}| = 0$, i.e., $\mathcal{T} \in \{E_\Lambda, E_\Upsilon\} \in \boldsymbol{S}_0^\ell$, corresponds to the top format $\mathcal{M}_{p-\ell}^{\triangle}(p,\mu) = \mathcal{M}_{p-\ell}^{\triangle}$ if $\mathcal{T} = E_\Lambda$ and $\mathcal{M}_{p-\ell}^{\nabla}(p,\mu) = \mathcal{M}_{p-\ell}^{\nabla}$ if $\mathcal{T} = E_\Upsilon$.

To have a constructive definition, we need the recursive representation of $\mathcal{M}_{p-\ell}^{\mathcal{T}}$ in terms of matrices with smaller subindex $p - \ell - 1$. To that end, with each $\sigma' = \mathcal{T}_\ell \sigma$, we associate a 4×4 matrix of transforms on level $\ell + 1$ generated from $\mathcal{T}_\ell$ by a lifting mapping

$$\mathcal{L}^\ell : \mathcal{T}_\ell \to \{\mathcal{T}_{\ell+1}^{jj'}\}_{j,j'=1}^4, \quad \mathcal{T}_{\ell+1}^{jj'} \in \boldsymbol{S}_\nu^{\ell+1}, \quad 0 \leqslant \nu \leqslant 2|\mathcal{T}_\ell| + 1,$$

where $\mathcal{T}_{\ell+1}^{jj'} : \sigma_j \to \sigma'_{j'}$, $\sigma_j \in S(\sigma)$, $\sigma'_{j'} \in S(\sigma')$. All the transforms $\mathcal{T}_{\ell+1}^{jj'}$ belong to class (26), where the specific parameters α, β and m are uniquely determined by the corresponding characteristics of $\mathcal{T}_\ell$ and by the choice of j and j'. In particular, according to Fig. 4a, the matrix-valued operator $\mathcal{L}^l(E_\Lambda) := \{\mathcal{T}^{jj'}\}$ has the form

$$\mathcal{L}^\ell(E_\Lambda) := \begin{array}{|c|c|c|c|} E_\Lambda & \pi^{\boldsymbol{j}_2} & \pi^{-\boldsymbol{j}_1} & \mathcal{S}_1 \\ \hline \pi^{-\boldsymbol{j}_2} & E_\Lambda & \pi^{\boldsymbol{j}_3} & \mathcal{S}_2 \\ \hline \pi^{\boldsymbol{j}_1} & \pi^{-\boldsymbol{j}_3} & E_\Lambda & \mathcal{S}_3 \\ \hline \mathcal{S}_1^{\mathrm{T}} & \mathcal{S}_2^{\mathrm{T}} & \mathcal{S}_3^{\mathrm{T}} & E_\Upsilon \end{array}, \quad \mathcal{L}^\ell(E_\Upsilon) := \begin{array}{|c|c|c|c|} E_\Upsilon & \pi^{-\boldsymbol{j}_2} & \pi^{\boldsymbol{j}_1} & \mathcal{S}_1^{\mathrm{T}} \\ \hline \pi^{\boldsymbol{j}_2} & E_\Upsilon & \pi^{-\boldsymbol{j}_3} & \mathcal{S}_2^{\mathrm{T}} \\ \hline \pi^{-\boldsymbol{j}_1} & \pi^{\boldsymbol{j}_3} & E_\Upsilon & \mathcal{S}_3^{\mathrm{T}} \\ \hline \mathcal{S}_1 & \mathcal{S}_2 & \mathcal{S}_3 & E_\Lambda \end{array}, \tag{27}$$

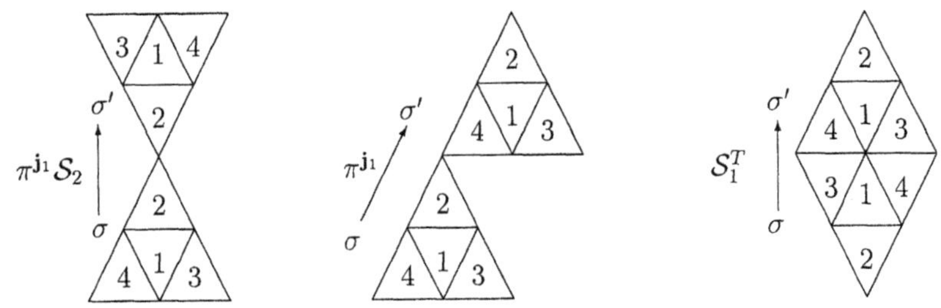

Fig. 5. Coupling of clusters corresponding to typical translations for $\mu = 1$.

where $\mathscr{T}^{jj'} \in \boldsymbol{S}_1^{\ell+1}$, $j, j' = 1, \ldots, 4$. Having defined the lifting mapping $\mathscr{L}^\ell$, we are looking for the recursive representation of the matrix structure (format) of the block $b = \sigma \times \mathscr{T}\sigma$ for $l = 0, \ldots, p-1$ and $\sigma \in T_1^\ell$,

$$A_{\sigma,\sigma'} = \{A_{jj'}\}_{j,j'=1,\ldots,4} \in \mathscr{M}_{p-\ell}^{\mathscr{T}_\ell} \quad \text{if } A_{jj'} \in \mathscr{M}_{p-\ell-1}^{\mathscr{T}_{\ell+1}^{jj'}}.$$

While all the far-distance formats with $|\mathscr{T}_{\ell+1}^{jj'}| > \mu$ are supposed to have a $2^{p-\ell-1} \times 2^{p-\ell-1}$ *Rk*-matrix structure, the blocks corresponding to nonadmissible area $|\mathscr{T}_{\ell+1}^{jj'}| \leqslant \mu$ are to be defined in the next recurrence steps.

For example, let us consider the recursive block structure of a particular format $\mathscr{M}_p^{\triangle}(p, \frac{2}{3})$, where the initial index set belongs to the class Λ. We choose $\eta_1 = 2/3$ and build the matrix-valued lifting transforms $\mathscr{L}(\mathscr{T}_\ell)$, $\mathscr{T}_\ell \in \mathscr{S}_1^\ell$ for typical neighbouring translations with $|\mathscr{T}_\ell| = 1$. Here and in the following, R denotes the class of translations with $|\mathscr{T}| \geqslant \mu + 1$ resulting in the *Rk*-matrix blocks on the corresponding level. The following schemes illustrate typical lifting transforms in the case of two clusters with one common vertex, see Fig. 5 (left and middle):

$$\mathscr{L}^\ell(\pi^{j_1}\mathscr{S}_2) := \begin{array}{c|c|c|c|} \pi^{j_1}\mathscr{S}_2 & R & R & R \\ \hline R & R & R & R \\ \hline R & R & R & R \\ \hline R & R & R & R \end{array}, \quad \mathscr{L}^\ell(\pi^{j_1}) := \begin{array}{c|c|c|c|} R & R & \pi^{j_1} & R \\ \hline R & R & R & R \\ \hline R & R & R & R \\ \hline R & R & R & R \end{array}. \tag{28}$$

The translation of sons for two adjacent clusters with one common edge has the following block (recursive) structure, see Fig. 5(right):

$$\mathscr{L}^\ell(\mathscr{S}_1^{\mathrm{T}}) := \begin{array}{c|c|c|c|} R & R & R & R \\ \hline R & \pi^{-j_3}\mathscr{S}_1^{\mathrm{T}} & \mathscr{S}_1^{\mathrm{T}} & \pi^{j_3} \\ \hline R & \mathscr{S}_1^{\mathrm{T}} & \pi^{j_3}\mathscr{S}_1^{\mathrm{T}} & \pi^{j_2} \\ \hline R & \pi^{j_1} & \pi^{-j_2} & \pi^{j_1}\mathscr{S}_2 \end{array}. \tag{29}$$

Using the nondiagonal lifting transforms defined by (28) and (29), we can describe the recursion for the identity transforms, see (27), which then generates the top formats, see the diagram in Fig. 6.

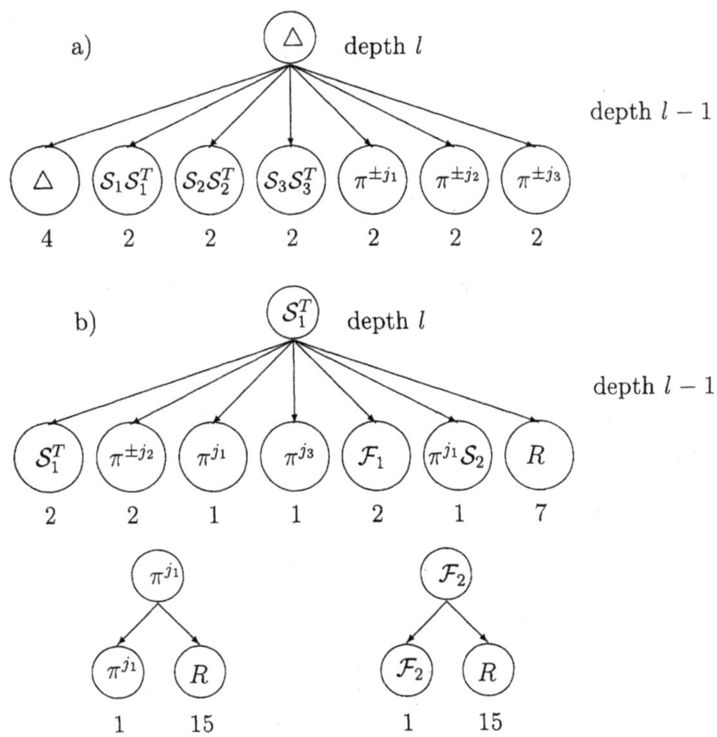

Fig. 6. The subtrees of the diagonal and typical auxiliary formats on $I_\triangle$, where $\mathscr{F}_1 = \pi^{\pm j_3}\mathscr{S}_1^{\mathrm{T}}$, $\mathscr{F}_2 = \pi^{j_1}\mathscr{S}_2$.

5.2. General definition and complexity of $\mathscr{M}_p^\triangle$-formats

Corresponding to the case $\Omega \in \Lambda$, we introduce the general $\mathscr{M}_p^\triangle$-format, where the level number $p \in \mathbb{N}$ is a fixed parameter. If the target domain $\Omega \in \Upsilon$, the format $\mathscr{M}_p^\nabla$ may be defined along the same line. Recall $\mathscr{M}_q^R$ as a set of Rk-matrices of the size $4^q \times 4^q$, $q = 0, 1, \ldots, p$.

Definition 9. (a) For $q = 0, \ldots, p$ define $\mathscr{M}_q^{\mathscr{T}} = \mathscr{M}_q^R$ for all $|\mathscr{T}| \geqslant \mu + 1$.

(b) For $q = 0$, define $\mathscr{M}_0^\triangle(p, \eta)$ and $\mathscr{M}_0^{\mathscr{T}}(p, \eta)$ as the sets of 1×1 matrices for all $|\mathscr{T}| \leqslant \mu$.

(c) Consider the case $q = 1, \ldots, p$ and $1 \leqslant |\mathscr{T}| \leqslant \mu$. Assume that for each $q \leqslant q_0$ the format $\mathscr{M}_q^{\mathscr{T}}$ is already defined for all $\mathscr{T} : |\mathscr{T}| \leqslant \mu$, and define the formats $\mathscr{M}_q^{\mathscr{T}}$ for $q = q_0 + 1$. Consider translation $\mathscr{T} \in \boldsymbol{S}_\mu^\ell$ with $|\mathscr{T}| = \mu$, $\mu - 1, \ldots, 1$ and blocks $\sigma, \sigma' \in T_1^\ell$ of level $\ell = p - q$ such that $\sigma' = \mathscr{T}\sigma$. For the matrices from $\mathbb{R}^{\sigma \times \sigma'}$, we say that $A = \{A_{jj'}\}_{\sigma_j \in S(\sigma), \sigma'_{j'} \in S(\mathscr{T}\sigma)}$ belongs to $\mathscr{M}_q^{\mathscr{T}}$ if

$$A_{jj'} \in \mathscr{M}_{q-1}^{\mathscr{T}_{l+1}^{jj'}} \quad \text{for } |\mathscr{T}_{l+1}^{jj'}| \leqslant \mu$$

and (due to item (a))

$$A_{jj'} \in \mathscr{M}_{q-1}^R \quad \text{for } |\mathscr{T}_{l+1}^{jj'}| \geqslant \mu + 1,$$

where $\mathscr{T}_{\ell+1}^{jj'} := (\mathscr{L}^\ell(\mathscr{T}))_{jj'}$.

(d) Finally, define the top format $\mathcal{M}_q^{\triangle}$ for $q=1,\dots,p$. Let $\sigma\in\Lambda^\ell$ be from level $\ell=p-q$ and set $\mathcal{T}=E_\Lambda\in \boldsymbol{S}_0^\ell$. Then we say that $A=\{A_{jj'}\}_{\sigma_j,\sigma_{j'}\in S(\sigma)}$ belongs to $\mathcal{M}_q^{\triangle}$ if there holds $A_{jj}\in\mathcal{M}_{q-1}^{\triangle}$ and $A_{jj'}\in\mathcal{M}_{q-1}^{\mathcal{L}^\ell(E_\Lambda)_{jj'}}$ for $j\neq j'$. The same construction is applied for $\sigma\in\Upsilon^\ell$, $\mathcal{T}=E_\Upsilon$.

The following statement gives sharp complexity bounds for the above-defined family of formats. Here we use the generalised construction based on the p_0-truncated partitioning as in Section 4.

Theorem 10. *Let $A\in\mathcal{M}_p^{\triangle}(p,\eta)$ with $\eta_\mu=2/3\mu$, $\mu=1,2,\dots$. With given $p_0\in\{0,\dots,p-1\}$, suppose that $\mathcal{M}_p^{\triangle}$ corresponds to the p_0-truncated partitioning P_2^*. Then the complexity of the matrix–vector multiplication is bounded by*

$$\mathcal{N}_{\mathrm{MV}}^{\triangle}\leqslant 6(6\mu^2+6\mu+1)(p-p_0)kn+2^{2p_0-1}(6\mu^2+6\mu+1)kn$$

coupled operations. Moreover,

$$\mathcal{N}_{\mathrm{st}}^{\triangle}\leqslant 3(6\mu^2+6\mu+1)(p-p_0)kn+4^{p_0}(6\mu^2+6\mu+1)kn.$$

The constants in both relations are asymptotically sharp.

Proof. The proof is similar to those from Theorem 6. In fact, let $\sigma\in S(\tau)$ be an arbitrary son for each "purely interior" cluster $\tau\in T_1^\ell$. Then, the number of sons $\sigma'\in S(\tau')$ from the set of neighbouring to τ clusters $\tau'\in T_1^\ell$, i.e., $\tau'\in R(\tau)$, and satisfying with σ the admissibility condition (4) on level $\ell+1$, is equal to $Q_\sigma=(2^2-1)((1+3\mu)^2-3\mu^2)$. Then the assertions follow. □

Remark 11. Combining Definition 9 with the corresponding results from Section 4, we obtain formats of the optimal complexity for the right triangular prism elements in 3D. Further extensions of construction from above to the 3D case are based on breaking the tetrahedron into 8 or 27 parts.

Remark 12. Due to larger nonadmissible area in the construction of the $\mathcal{M}_p^{\triangle}(p,\eta)$-format, see Theorem 10, the corresponding constants in $\mathcal{N}_{\mathrm{st}}^{\triangle}$ and $\mathcal{N}_{\mathrm{MV}}^{\triangle}$ appear to be bigger than in the case of $\mathcal{M}_p^{\square}$-formats.

When using grid (18) for finite difference or finite element discretisations of the second-order PDEs, we obtain a five-, seven-, or nine-point formula as discretisation matrix for $d=2$ (similar for $d=3$). The next lemma implies that such a matrix can be represented *exactly* as an $\mathcal{H}$-matrix, see [6] for the proof in the case of $\mathcal{M}_p^{\square}$-format.

Lemma 13. *The FE stiffness matrix A_h is in the set*[9] *$\mathcal{M}_{\mathcal{H},k}(I\times I,P_2)$ for any $k\geqslant 1$.*

As a consequence, the approximate inverse of $A_h\in\mathcal{M}_p^{\square}$ as well as of $A_h\in\mathcal{M}_p^{\triangle}$ can be computed with the complexity $O(p^2k^2n)$, where $n=\#I$, see Section 6.3.

[9] If I is as in (18) with fixed p, $\mathcal{M}_{\mathcal{H},k}(I\times I,P_2)$ equals $\mathcal{M}_p^{\square}(p,\frac{1}{\sqrt{2}})$. However, this lemma holds for rather general $\mathcal{H}$-partitionings.

6. Matrix addition, multiplication and inversion

6.1. Matrix addition

In this Section, we study the complexity of matrix addition, multiplication and inverse-to-matrix operations for the principal case $\mu=1$ and with $d=2$. As in [5], one can introduce the approximate *addition* $+_\Box$, *multiplication* $*_\Box$, and *inversion* to the matrices from $\mathcal{M}_p^\Box = \mathcal{M}_p^{\mathbf{0}}(p, \frac{1}{\sqrt{2}})$. The complexity analysis of formatted addition $+_\Box$ is rather simple (it operates with the same types of formats in a blockwise sense) and yields $\mathcal{N}_{\Box+\Box}(p) = \mathrm{O}(pn)$, where $n = 2^{dp}$. Indeed, let us denote by symbols $\bigcirc$ and $\times$ each set of formats $\mathcal{M}_p^{\boldsymbol{j}}$ where $|\boldsymbol{j}|_1 = 1$ and $|\boldsymbol{j}|_1 = 2 \wedge |\boldsymbol{j}|_\infty = 1$, respectively. Then the recursion

$$\mathcal{N}_{\Box+\Box}(p) = 4\mathcal{N}_{\Box+\Box}(p-1) + 8\mathcal{N}_{\circ+\circ}(p-1) + 4\mathcal{N}_{\times+\times}(p-1) \tag{30}$$

follows from (28), see Fig. 6a. In turn, the recursive Definition 5 easily implies

$$\mathcal{N}_{\circ+\circ}(p) = 2\mathcal{N}_{\circ+\circ}(p-1) + 2\mathcal{N}_{\times+\times}(p-1) + 12\mathcal{N}_{R1+R1}(p-1),$$

$$\mathcal{N}_{\times+\times}(p) = \mathcal{N}_{\times+\times}(p-1) + 15\mathcal{N}_{R1+R1}(p-1).$$

The latter two relations lead to the bounds

$$\mathcal{N}_{\times+\times}(p) = \mathrm{O}(n), \quad \mathcal{N}_{\circ+\circ}(p) = \mathrm{O}(n). \tag{31}$$

Substitution of (31) into (30) implies the desired complexity estimate for $\mathcal{N}_{\Box+\Box}(p)$, taking into account $\mathcal{N}_{R+R}(p) = 21n + \mathrm{O}(1)$, see [5].

6.2. Complexity of matrix multiplication

The proof of $\mathcal{N}_{\Box*\Box}(p) = \mathrm{O}(p^2k^2n)$ is more lengthy, since various combinations of factors occur. First, we introduce the formatted matrix–matrix multiplication procedure. The recursive definition of formatted multiplication of two matrices A and B from $\mathcal{M}_p^\Box$ is similar to Definition 5 above. For the precise description, we use the following notations and remark. We call $\boldsymbol{j}_1 \prec \boldsymbol{j}_2$ if either $|\boldsymbol{j}_1|_\infty < |\boldsymbol{j}_2|_\infty$ or $|\boldsymbol{j}_1|_\infty = |\boldsymbol{j}_2|_\infty \wedge |\boldsymbol{j}_1|_1 < |\boldsymbol{j}_2|_1$ and define $\boldsymbol{j}_1 \approx \boldsymbol{j}_2$, otherwise.

Remark 14. Any *Rk*-matrix of the size $2^{dq} \times 2^{dq}$ may be exactly converted to each of the formats $\mathcal{M}_q^{\boldsymbol{j}}$, $|\boldsymbol{j}|_\infty \leqslant 1$, so we have the imbedding $\mathcal{M}_q^R \hookrightarrow \mathcal{M}_q^{\boldsymbol{j}}$. We also assume that either $\mathcal{M}_q^{\boldsymbol{j}_1} \hookrightarrow \mathcal{M}_q^{\boldsymbol{j}_2}$ if $\boldsymbol{j}_2 \prec \boldsymbol{j}_1 \vee \boldsymbol{j}_1 \approx \boldsymbol{j}_2$ or (if the above imbedding is not the case) $\mathcal{M}_q^{\boldsymbol{j}_1}$ may be approximately converted to the format $\mathcal{M}_q^{\boldsymbol{j}_2}$ with almost linear cost, where $\mathcal{M}_q^{\boldsymbol{j}} \hookrightarrow \mathcal{M}_q^R$ for $|\boldsymbol{j}|_\infty > 1$. This assumption is based on the properties of the particular format $\mathcal{M}_p^\Box$ under consideration.

Definition 15 (Recursion step). Assume that for some $q < p$ and for each $A \in \mathcal{M}_q^{\boldsymbol{j}_1}$, $B \in \mathcal{M}_q^{\boldsymbol{j}_2}$ the $\mathcal{M}_q^{\boldsymbol{j}_3}$-formatted product $C = A *_{\mathscr{H}} B \in \mathcal{M}_q^{\boldsymbol{j}_3}$ is already defined for $|\boldsymbol{j}_m| \geqslant 0$, $m = 1,2,3$.

Then, for each matrix $A \in \mathcal{M}_{q+1}^{\boldsymbol{j}_1}$ and $B \in \mathcal{M}_{q+1}^{\boldsymbol{j}_2}$ with the recursive block structure $A = \{A_{ij}\}_{i,j \in I_d}$, $B = \{B_{ij}\}_{i,j \in I_d}$, we define $C = A *_{\mathscr{H}} B := \{C_{\boldsymbol{km}}\}_{\boldsymbol{k},\boldsymbol{m} \in I_d}$ with $C_{\boldsymbol{km}} \in \mathcal{M}_{q+1}^{\boldsymbol{k}-\boldsymbol{m}+2\boldsymbol{j}_3}$ by

$$C_{\boldsymbol{km}} = \sum_{\boldsymbol{i} \in I_d} A_{\boldsymbol{ki}} *_{\mathscr{H}} B_{\boldsymbol{im}}, \quad A_{\boldsymbol{ki}} \in \mathcal{M}_q^{\boldsymbol{k}-\boldsymbol{i}+2\boldsymbol{j}_1}, \quad B_{\boldsymbol{im}} \in \mathcal{M}_q^{\boldsymbol{i}-\boldsymbol{m}+2\boldsymbol{j}_2}.$$

Here the formatted addition $+_{\mathscr{H}}$ is understood as the operation within the format $\mathscr{M}_q^{k-m+2j_3}$ in view of Remark 14. In particular, if $j_1 = j_2 = j_3 = 0$, we obtain the multiplication procedure for the top format.

In view of Definition 15 and taking into account the particular structure of $\mathscr{M}_p^{\square}$-format, the complexity estimate $\mathscr{N}_{\square * \square}(p)$ on the level p is reduced recursively to the following operation counts: $\mathscr{N}_{\square * \square}(p-1), \mathscr{N}_{\square * \circ}(p-1), \mathscr{N}_{\square * \times}(p-1)$, $\mathscr{N}_{\circ * \times}(p-1)$, $\mathscr{N}_{\circ * \circ}(p-1)$ and $\mathscr{N}_{\times * \times}(p-1)$. The latter may be further reduced to the already known estimates for $\mathscr{N}_{R*R}(p-2)$ and $\mathscr{N}_{R+R}(p-2)$, see the proof of Lemma 16.

Lemma 16. *The following complexity bounds hold*:

$$\mathscr{N}_{\square + \square}(p) = \mathrm{O}(pkn), \quad \mathscr{N}_{\square * \square}(p) = \mathrm{O}(p^2k^2n) + \mathrm{O}(k^3n). \tag{32}$$

Proof. The first assertion is proved in Section 6.1. The bound for $\square * \square$ is based on the recurrence

$$\begin{aligned}\mathscr{N}_{\square * \square}(p) = {} & 4\mathscr{N}_{\square * \square}(p-1) + 16\mathscr{N}_{\square * \circ}(p-1) + 8\mathscr{N}_{\square * +}(p-1) \\ & + 16\mathscr{N}_{\circ * \circ}(p-1) + 16\mathscr{N}_{\circ * +}(p-1) + 4\mathscr{N}_{+ * +}(p-1) + \sum_{\alpha,\beta \in \aleph} \mathscr{N}_{\alpha+\beta}(p-1),\end{aligned} \tag{33}$$

where $\aleph := \{\square, \circ, \times, R\}$. To proceed with, we then estimate the remaining terms on the right-hand side above. In this way we use the relations

$$\begin{aligned}\mathscr{N}_{\square * \circ}(p) = {} & 2\mathscr{N}_{\square * \circ}(p-1) + 2\mathscr{N}_{\square * +}(p-1) \\ & + 12\mathscr{N}_{\square * R}(p-1) + 4\mathscr{N}_{\circ * \circ}(p-1) + 24\mathscr{N}_{\circ * R}(p-1) \\ & + 6\mathscr{N}_{\circ * +}(p-1) + 2\mathscr{N}_{+ * +}(p-1) + 12\mathscr{N}_{+ * R}(p-1) + \sum_{\alpha,\beta \in \aleph} \mathscr{N}_{\alpha+\beta}(p-1),\end{aligned}$$

$$\begin{aligned}\mathscr{N}_{\square * +}(p) = {} & \mathscr{N}_{\square * +}(p-1) + 2\mathscr{N}_{\circ * +}(p-1) + \mathscr{N}_{+ * +}(p-1) \\ & +15(\mathscr{N}_{\square * R}(p-1) + 2\mathscr{N}_{\circ * R}(p-1) + \mathscr{N}_{+ * R}(p-1)) + \sum_{\alpha,\beta \in \aleph} \mathscr{N}_{\alpha+\beta}(p-1),\end{aligned}$$

$$\mathscr{N}_{\circ * \circ}(p) = 4\mathscr{N}_{\circ * R}(p-1) + 12\mathscr{N}_{+ * R}(p-1) + 40\mathscr{N}_{R * R}(p-1) + \sum_{\alpha,\beta \in \aleph} \mathscr{N}_{\alpha+\beta}(p-1),$$

$$\mathscr{N}_{\circ * +}(p) = 4\mathscr{N}_{\circ * R}(p-1) + 8\mathscr{N}_{+ * R}(p-1) + 52\mathscr{N}_{R * R}(p-1),$$

$$\mathscr{N}_{+ * +}(p) = 7\mathscr{N}_{+ * R}(p-1) + 57\mathscr{N}_{R * R}(p-1) + \sum_{\alpha,\beta \in \aleph} \mathscr{N}_{\alpha+\beta}(p-1),$$

$$\mathscr{N}_{+ * R}(p) = 4\mathscr{N}_{+ * R}(p-1) + 60\mathscr{N}_{R * R}(p-1) + \sum_{\alpha,\beta \in \aleph} \mathscr{N}_{\alpha+\beta}(p-1).$$

Note that $\mathscr{N}_{\alpha+\beta}(p) = \mathrm{O}(n)$ for $\alpha, \beta \in \{\circ, \times, R\}$, while $\mathscr{N}_{\square+\alpha}(p) = \mathrm{O}(pn)$ for $\alpha \in \{\circ, \times, R\}$. Substituting these results into above recurrences and taking into account $\mathscr{N}_{R*R}(p) = 3n - 1$, see [5], we obtain

$$\mathscr{N}_{\alpha * \beta}(p) = \mathrm{O}(n), \quad \mathscr{N}_{\square * \alpha}(p) = \mathrm{O}(pn), \quad \alpha, \beta \in \{\circ, \times, R\}.$$

Finally, Eq. (33) results in the recursion $\mathcal{N}_{\square*\square}(p) = 4\mathcal{N}_{\square*\square}(p-1) + O(pn)$, which yields the desired assertion. In fact, the term $O(k^3 n)$ results from the cost of eigenvalue problem solvers (or the singular-value decomposition) within the implementation of Rk-matrix arithmetic, see [5]. □

6.3. Matrix inversion

The recursive inversion is based on blockwise transformations and the Schur-complement calculations involving the addition and multiplication addressed above, see [5] for more details. While in [5] the $\mathcal{H}$-matrix was treated as a 2×2 block matrix, now the refinement format has a 4×4 block pattern. This does not change the complexity order $\mathcal{N}_{\text{Inversion}}(p) = O(p^2 n)$ obtained there with $k = O(1)$.

As an alternative, here we discuss in more details the nonrecursive construction of the inverse of an $\mathcal{H}$-matrix based on the iterative correction and formatted matrix–matrix multiplication. We propose to apply the nonlinear iterations for computation of A^{-1}. The proper initial guess X_0 may be obtained by the recursive Schur-complement algorithm from [5]. Assume that A is invertible. Let us solve the nonlinear operator equation in the corresponding normed space $\boldsymbol{Y} := \mathbb{R}^{n \times n}$ of square matrices

$$F(X) := X^{-1} - A = 0, \quad X \in \boldsymbol{Y}$$

by the Newton's method, which results in the iterations

$$X_{i+1} = X_i(2I - A \cdot X_i), \quad X_0 \text{ given}, \quad i = 1, 2, \ldots . \tag{34}$$

For this scheme, which is well known from the literature, we give a simple direct convergence analysis.

Lemma 17. *Let $A \in \boldsymbol{Y}$ be invertible and assume that the initial guess in* (34) *satisfies*

$$||A|| \, ||X_0 - A^{-1}|| = q < 1. \tag{35}$$

Then iteration (34) *converges quadratically,*

$$||X_{i+1} - A^{-1}|| \leqslant c\, q^{2^i}, \quad i = 1, 2, \ldots . \tag{36}$$

Suppose that A and X_0 are both the s.p.d. matrices and X_0 satisfies $0 < X_0 < A^{-1}$. Then the iteration (34) *yields $X_i = X_i^{\mathrm{T}} > 0$ for all $i = 1, 2, \ldots$.*

Proof. Denote $X_i = A^{-1} - \delta_i$. By definition

$$X_{i+1} = 2(A^{-1} - \delta_i) - (A^{-1} - \delta_i) A (A^{-1} - \delta_i) = A^{-1} - \delta_i A \delta_i,$$

which implies

$$\delta_{i+1} = \delta_i \, A \, \delta_i, \quad i = 1, 2, \ldots . \tag{37}$$

Therefore, the first assertion follows:

$$||\delta_{i+1}|| \leqslant ||A||^{(1+2+2^2+2^3+\cdots+2^{i-1})} \, ||\delta_0||^{2^i} \leqslant c\, q^{2^i}.$$

In the case of s.p.d. matrices, we have $A^{-1} > \delta_0 > 0$ by assumption. Furthermore, assume by induction, that $A^{-1} > \delta_i > 0$. Then X_{i+1} is symmetric and (37) yields $\delta_{i+1} = A^{-1} - X_{i+1} > 0$. Moreover, the inequality

$$A^{1/2}\delta_{i+1}A^{1/2} = (A^{1/2}\delta_i A^{1/2})^2 < I$$

implies $\delta_{i+1} < A^{-1}$ yielding $X_{i+1} > 0$. This proves the induction step. □

Due to the quadratic convergence of the scheme proposed, we need only $\log\log\varepsilon^{-1}$ iterative steps which results in the $O(k^2 p^2 n \log\log n)$ complexity of the iterative correction algorithm.

A specific truncation error analysis of the $*_\square$-multiplication and of the inversion will not be considered in this paper. However, the background to create efficient calculus of $\mathcal{H}$-matrices is based on the observation that for many practically important problem classes the product or the sum of pseudodifferential operators A and B as well as the inverse operator A^{-1} have the integral representations which ensure the existence of the proper $\mathcal{H}$-matrix approximations to $A+B$, $B*A$ and A^{-1} themselves. Having in hands the linear complexity multiplication/inversion algorithms, one may use then two basic strategies for fast solution of the operator equation $Au = f$:

(a) Direct method based on the $\mathcal{H}$-matrix approximation to the operator A^{-1} by the recursive Schur-complement scheme. Here the approximation of A^{-1} must be sufficiently good.
(b) Computation of a rather rough inverse $B \approx A^{-1}$ and correction by few steps of $u^{i+1} = u^i - B *_{\mathcal{H}} (Au^i - f)$.

Both approaches provide almost linear complexity algorithms for solving a wide class of integral/pseudodifferential equations.

To complete the discussion, we note that all the $\mathcal{H}$-matrix formats considered may be extended to the case of quasi-uniform unstructured meshes. A possible construction is based on the *fictitious* uniform tensor-product or triangular grids discussed in the previous section, see [6]. We do not claim that such a construction is optimal, but it leads to a straightforward proof of the almost linear complexity bounds. The $\mathcal{H}$-matrices on graded meshes have been analysed in [7]. Numerical experiments mainly confirm the approximation and complexity results for the $\mathcal{H}$-matrix techniques applied to the boundary integral operators in 3D as well as for the data-sparse approximation to inverse of the discrete Laplacian. These results will be reported in a forthcoming paper.

References

[1] W. Dahmen, S. Prössdorf, R. Schneider, Wavelet approximation methods for pseudodifferential equations II: Matrix compression and fast solution, Adv. Comput. Math. 1 (1993) 259–335.

[2] L. Greengard, V. Rokhlin, A new version of the fast multipole method for the Laplace equation in three dimensions, Acta Numer. 6 (1997) 229–269.

[3] W. Hackbusch, The panel clustering algorithm, in: J.R. Whiteman (Ed.), MAFELAP 1990, Academic Press, London, 1990, pp. 339–348.

[4] W. Hackbusch, Integral Equations. Theory and Numerical Treatment, Birkhäuser, Basel, 1995. ISNM 128.

[5] W. Hackbusch, A sparse matrix arithmetic based on $\mathcal{H}$-matrices. Part I: introduction to $\mathcal{H}$-matrices, Computing 62 (1999) 89–108.

[6] W. Hackbusch, B.N. Khoromskij, A sparse $\mathcal{H}$-matrix arithmetic. Part II. Application to multi-dimensional problems, preprint MPI, No. 22, Leipzig, 1999, Computing 64 (2000) 21–47.

[7] W. Hackbusch, B.N. Khoromskij, $\mathscr{H}$-matrix approximation on graded meshes, preprint MPI, No. 54, Leipzig, 1999, in: J.R. Whiteman (Ed.), Proceedings of MAFELAP 1999, to appear.
[8] W. Hackbusch, B.N. Khoromskij, Towards $\mathscr{H}$-matrix approximation of linear complexity, in: Siegfried Prößdorf Memorial Volume Proceedings of the 11th TMP Conference, Birkhäuser-Verlag, Basel, 2000, to appear.
[9] W. Hackbusch, B.N. Khoromskij, S. Sauter, On $\mathscr{H}^2$-matrices, In: Lectures on Applied Mathematics, H.-J. Bungartz, R. Hoppe, C. Zenger (Eds.), Springer Verlag, Berlin, 2000, 9–30.
[10] W. Hackbusch, Z.P. Nowak, On the fast matrix multiplication in the boundary element method by panel clustering, Numer. Math. 54 (1989) 463–491.
[11] W. Hackbusch, S.A. Sauter, On the efficient use of the Galerkin method to solve Fredholm integral equations, Appl. Math. 38 (1993) 301–322.
[12] S.A. Sauter, Über die effiziente Verwendung des Galerkin-Verfahrens zur Lösung Fredholmscher Integralgleichungen, Ph.D. Thesis, Universität, Kiel, 1992.
[13] T. von Petersdorff, C. Schwab, Wavelet approximations for first kind boundary integral equations, Numer. Math. 74 (1996) 479–516.

ELSEVIER Journal of Computational and Applied Mathematics 125 (2000) 503–519

JOURNAL OF COMPUTATIONAL AND APPLIED MATHEMATICS

www.elsevier.nl/locate/cam

Multilevel methods for the h-, p-, and hp-versions of the boundary element method ☆

Ernst P. Stephan *

Institut für Angewandte Mathematik, Universität Hannover, Welfengarten 1, D-30167 Hannover, Germany

Received 9 September 1999; received in revised form 15 February 2000

Abstract

In this paper we give an overview on the definition of finite element spaces for the h-, p-, and hp-version of the BEM along with preconditioners of additive Schwarz type. We consider screen problems (with a hypersingular or a weakly singular integral equation of first kind on an open surface Γ) as model problems. For the hypersingular integral equation and the h-version with piecewise bilinear functions on a coarse and a fine grid we analyze a preconditioner by iterative substructuring based on a non-overlapping decomposition of Γ. We prove that the condition number of the preconditioned linear system behaves polylogarithmically in H/h. Here H is the size of the subdomains and h is the size of the elements. For the hp-version and the hypersingular integral equation we comment in detail on an additive Schwarz preconditioner which uses piecewise polynomials of high degree on the fine grid and yields also a polylogarithmically growing condition number. For the weakly singular integral equation, where no continuity of test and trial functions across the element boundaries has to been enforced, the method works for nonuniform degree distributions as well. Numerical results supporting our theory are reported. © 2000 Elsevier Science B.V. All rights reserved.

MSC: 65N55; 65N38

Keywords: h-p version of the boundary element method; Schwarz methods; Preconditioning

1. Introduction

The use of piecewise polynomials of high degree guarantees high accuracy of Galerkin solutions for elliptic boundary value problems even with singularities [8]. This holds both for the finite element method (FEM) [9] as well as for the boundary element method (BEM) [41], i.e., for Galerkin schemes to solve corresponding integral equations. The convergence analysis of the

☆ Supported by the German Research Foundation (DFG) under grant no. Ste 573/4-1.

* Tel.: +49-511-7623360.

PII: S0377-0427(00)00487-8

hp-version of the BEM for integral equations on polygons is analyzed in [1,11,20,21,23] within the framework of Mellin convolution operators. For three-dimensional problems, i.e., integral equations on polyhedral/open surfaces see [19,26,27,29]. The solution of the weakly singular integral equation of the first kind with the single-layer potential belongs to the countably normed spaces $B^1_\beta(\Gamma)$ when Γ is a polyhedron. The solution of the hypersingular integral equation with the operator of the normal derivative of the double-layer potential belongs to $B^2_\beta(\Gamma)$ when the given data are piecewise analytic on Γ. The Dirichlet problem for the Laplace operator in a polyhedral domain, which is converted into the above weakly singular integral equation, is analyzed in [29], whereas the corresponding Neumann problem leads to the hypersingular integral equation considered in [30]. We show in both cases that the solution of the boundary integral equation can be approximated exponentially fast by appropriately chosen piecewise polynomials on a geometric mesh which is refined towards the edges and corners of the polyhedral surface. The trial functions can be chosen as tensor products of Legendre polynomials and their antiderivatives, respectively. For further reference compare the survey article [41].

The development of efficient adaptive refinement strategies for BEM to solve 3D problems is of high practical importance. Residual error estimators for the *h*-version have been studied in [5,6] extending to BEM the Eriksson/Johnson approach for FEM. Another strategy to define error indicators uses hierarchical multilevel decompositions of the trial spaces (for curves see [4,46], and for weakly singular integral equations on surfaces see [32]). The framework of adaptive multilevel decompositions seems also to be suitable for the construction of *p*- and *hp*-adaptive methods (for numerical experiments of the BEM see [28]). For a complete theoretical study the corresponding multilevel decompositions need to be analyzed which do not only localize the subspaces containing the trial functions with high degrees but these subspaces must be further decomposed. To the authors' knowledge this is still an open problem. Nevertheless, a sequence of preliminary work has been developed recently, examining the preconditioners for domain decomposition techniques belonging to *h*-, *p*-, and the *hp*-versions BEM. A lot of work has been done for preconditioning techniques for the pure *h*-version [22,42,47]; for the *p*-version see [13,14,16,48] and for the *hp*-version see [24,25]. Here in Section 2 we will report on [24]. So far there seems to be no theoretical results available for domain decomposition methods for the *hp*-version of the BEM for 3D problems on nonuniform meshes with anisotropic elements. First results for two-level decompositions with respect to the polynomial degree are in [28] which can be used for adaptive refinements. The above-mentioned references deal with symmetric positive definite problems. Domain decompositions for nonsymmetric or indefinite systems for the BEM on curves (*h*- and *p*-version) are investigated in [44,45]. These techniques can also be applied to 3D problems [15] and be used for adaptive steering of indefinite boundary element problems [28]. There is a rapidly growing literature on the above topic for the FEM. For brevity the given references address only the BEM (only some papers for the FEM are cited). For the *h*-version BEM there are further preconditioning techniques which however do not use subspace decompositions of the boundary element space (see [39] and the references therein).

The paper is organized as follows. In Section 2 we present the additive Schwarz method for the *hp*-version of the Galerkin boundary element method applied to first kind integral equations on surfaces. In Section 3 for the *h*-version we prove the polylogarithmic growth of the condition number for the preconditioned system of the hypersingular integral equation. In Section 4 we give some numerical experiments showing the influence of preconditioning for various *p*-versions.

2. Schwarz methods for boundary integral equations of first kind

Additive Schwarz methods for the h- and p-versions of the BEM applied to weakly singular and hypersingular integral equations of first kind in $\mathbb{R}^2$ are studied in [14,25,47,48]. For the p-version [14,48] it could be shown that the condition number of the additive Schwarz operator grows at most like $O(\log^2 p)$ where p denotes the polynomial degree (for the hp-version see below this section). For the h-version we could show in [47] that the condition number of the additive Schwarz operator grows at most like $h^{-\varepsilon}$ for $\varepsilon > 0$ arbitrarily small where h denotes the mesh size. The corresponding result for methods with hierarchical basis functions in $\mathbb{R}^2$ is derived in [46] with growth $O(|\log h|)$. For a summary of the results compare the survey article [42]. The multilevel method from [47] could be generalized to hypersingular integral operators on surfaces in [15]. The results show bounded condition number of the preconditioned system for closed surfaces and an upper bound $O(|\log^{1/2} h|)$ for the condition number in case of open surfaces. Additive Schwarz decompositions with two levels and independent coarse grid and fine grids were analyzed in [22] for hypersingular integral operators on surfaces. As described below the corresponding condition numbers of additive Schwarz operators grow at most like $O(|\log^2 H/h|)$ where H denotes the size of the subdomains.

Domain decompositions and additive Schwarz methods for the p-version of the BEM in $\mathbb{R}^3$ are discussed in [13,16–18]. For nonoverlapping decompositions and weakly singular integral operators it could be shown in [18] that the condition number of the corresponding additive Schwarz operator grows at most like $O(\log^2 Hp/h)$. In [16] special, nonhierarchical basis functions have been used to define decompositions for hypersingular operators where the condition number of the additive Schwarz operator grows at most polylogarithmically in p. Overlapping decompositions are analyzed for 2D problems in [49] and for 3D in [17].

As a model problem we consider the weak form of the hypersingular integral equation

$$\langle Du, v\rangle_{L^2(\Gamma)} = \langle f, v\rangle_{L^2(\Gamma)} \quad \text{for all } v \in \tilde{H}^{1/2}(\Gamma) \tag{1}$$

on a plane rectangular surface piece $\Gamma \subset \mathbb{R}^3$ where $f \in H^{-1/2}(\Gamma)$ is a given function. Here D is the hypersingular integral operator

$$Du(x) = \frac{1}{4\pi}\frac{\partial}{\partial n_x}\int_\Gamma u(y)\frac{\partial}{\partial n_y}\frac{1}{|x-y|}\,\mathrm{d}S_y, \quad x \in \Gamma$$

which is a continuous and positive-definite mapping from $\tilde{H}^{1/2}(\Gamma)$ onto $H^{-1/2}(\Gamma)$, cf. [40]. Hence, there holds the equivalence of norms

$$\langle Dv, v\rangle_{L^2(\Gamma)} \simeq ||v||^2_{\tilde{H}^{1/2}(\Gamma)} \quad \text{for all } v \in \tilde{H}^{1/2}(\Gamma).$$

The Sobolev spaces $\tilde{H}^{1/2}(\Gamma)$ and $H^{-1/2}(\Gamma)$ are defined in the next section. The solution u of (1) is the jump across Γ of the solution of a Neumann problem for the Laplacian in $\mathbb{R}^3\backslash\bar{\Gamma}$, cf. [40]. The extension of our results to hypersingular integral equations on closed, polyhedral surfaces [37] and to more practical problems like exterior traction problems in linear elasticity is essentially straightforward and for ease of presentation we concentrate on the generic model problem for the Laplacian.

The Galerkin scheme for (1) reads as follows. Given a finite-dimensional subspace $\Psi \subset \tilde{H}^{1/2}(\Gamma)$ with $\dim \Psi = N$ find $u_N \in \Psi$ such that

$$\langle Du_N, v\rangle_{L^2(\Gamma)} = \langle f, v\rangle_{L^2(\Gamma)} \quad \text{for all } v \in \Psi. \tag{2}$$

The solution u of (1) behaves singularly at the edges and corners of Γ, cf. [37,43]. Due to these singularities the standard h- and p-versions of the Galerkin method converge at a rather low rate. On the other hand, when appropriately combining mesh refinements and polynomial-degree distributions in a nonuniform fashion, even an exponential rate of convergence is achievable, cf. [19,29].

The approach from [24] described here is a first step towards preconditioning methods for the general *hp*-version of the boundary element method in three-dimensions. We consider nonuniform meshes as well as nonuniform degree distributions. However, we require that the elements are shape regular, i.e., they are not too distorted, and locally quasi-uniform. Moreover, we assume that the polynomial degrees vary not too much within elements, i.e., the ratio of maximum and minimum polynomial degrees is bounded on individual elements. Since the polynomial degrees on neighboring elements are coupled by the continuity of the basis functions this boundedness of the ratio then holds also on patches of adjacent elements. Therefore, we call this nonuniform p version locally uniform.

In any case, the stiffness matrices for the *hp*-version in (2) are ill-conditioned and a preconditioner is necessary for an efficient solution. The method of choice for solving positive definite linear systems is the conjugate gradient method. Let A denote the stiffness matrix of the linear system with spectral condition number κ. Then a bound on the decrease of the energy norm of the error, after k steps, is given by

$$2\left(\frac{\sqrt{\kappa}-1}{\sqrt{\kappa}+1}\right)^k \quad \text{where } \kappa = \frac{\lambda_{\max}(A)}{\lambda_{\min}(A)}.$$

The goal, now, is to investigate a preconditioner for A which yields good bounds for κ. Provided the stiffness matrix is given in a suitable basis with appropriate numbering of basis functions, we present a preconditioner which amounts to a block-Jacobi step where some of the blocks may overlap with others. Each block of this method corresponds to a discretization of the integral operator for a given subspace of the full approximation space Ψ. Therefore, the preconditioner is related to a decomposition of Ψ into subspaces. The main structure of this decomposition is given by a three-level method. There are two levels for the piecewise polynomials of lowest degree corresponding to a coarse and a fine mesh. The fine level subspace then is further decomposed into a couple of subspaces associated with the wire basket and individual elements of the coarse mesh. Finally, the third level is given by the full space of piecewise polynomials of high degrees and is also further decomposed. The subspaces of the latter decomposition belong to the wire basket and individual elements of the fine mesh.

The three-level decomposition of our method is analogous to that proposed by Guo and Cao [10] for the finite element method in three-dimensions. However, we have to deal with a hypersingular integral operator on surfaces that means we have to consider trace spaces of H^1. We take polynomials with minimal L^2-norm as given in [36] and use the L^2-bilinear form on the wire basket.

Let us note that [24] extends the results in [22] where the pure h version of the boundary element method on quasi-uniform meshes is considered. Indeed, the first two levels of our preconditioner are identical with the subspace decomposition of the h version in [22] where, however, in each case the original bilinear form is used as preconditioner. In our previous paper [25] we also deal with preconditioners for the *hp*-version of the boundary element method where, in particular, geometrically graded meshes and nonuniform polynomial degrees are considered. However, that paper only deals with two-dimensinal problems on polygonal domains, i.e., with integral operators on curves, and the subspace decompositions and technical tools are not as sophisticated as for three-dimensional problems.

The ansatz space Ψ consists of piecewise polynomials of varying degree on a locally quasi-uniform mesh. This mesh is defined by two levels.

One level is the *coarse mesh* which is given by a regular family of triangles or quadrilaterals, $\bar{\Gamma} = \bigcup_{j=1}^{J} \bar{\Gamma}_j$. The elements Γ_j of this level are called *subdomains* and its nodal points are referred to as *vertices* whereas the lines between the vertices are the *edges*. The subdomains must be shape regular, i.e., they are not too distorted. The coarse mesh can be nonuniform and the diameter of Γ_j is denoted by H_j.

The second level of the mesh is the *fine mesh*. It is given by partitioning each subdomain into a number of quasi uniform quadrilaterals Γ_{ji} (the *elements*) which are shape regular and of diameter h_j on Γ_j. The nodal points of the fine mesh are called *nodes* and the lines between the nodes are the *sides*.

Having defined the mesh on Γ the h–p approximation space is completely determined by defining basis functions locally on the reference element $\Gamma_{\text{ref}}:=(-1,1)^2$ and by specifying polynomial degrees on the elements of the mesh.

On the reference element we use the vector of polynomial degrees

$$\mathscr{P} = (p_{\gamma_1}, \ldots, p_{\gamma_4}, p_{I_1}, p_{I_2}),$$

where p_{γ_j} and p_{I_j} are the degrees associated with the sides and the interior of Γ_{ref} (x_1- and x_2-direction), respectively, in a certain order. The maximum polynomial degree on Γ_{ref} is denoted by $p_{\max}$. The elements Γ_{ji} are associated with degree vectors $\mathscr{P}_{ji}$, and we assume that the ratio of the maximum and minimum polynomial degrees on individual elements is bounded uniformly on the fine mesh.

Now let us define the basis functions. As in the standard p version we make a distinction between nodal, side, and internal shape functions. Let φ_0^p denote the pth degree polynomial on $I:=(-1,1)$ with $\varphi_0^p(-1)=0$ and $\varphi_0^p(1)=1$ which minimizes the $L^2(I)$-norm over the space of all pth degree polynomials (subject to the same boundary conditions).

- One of the four nodal shape functions, the one for node $V_1 = (-1,-1)$, is given by

$$\varphi_0^{p_{\gamma_1}}(-x_1)\varphi_0^{p_{\gamma_4}}(-x_2).$$

- One of the sets of side shape functions, for the edge $E_1 = \{(x_1,x_2);\ x_2 = -1\}$, is given by

$$\psi^{p_{\gamma_1}}(x_1)\varphi_0^{p_{I_2}}(-x_2)$$

  where $\psi^{p_{\gamma_1}}$ is a polynomial of degree p_{γ_1} such that $\psi^{p_{\gamma_1}}(-1) = \psi^{p_{\gamma_1}}(1) = 0$.
- The interior shape functions are polynomials of degree p_{I_1} in x_1 and of degree p_{I_2} in x_2 which vanish at the sides of Γ_{ref}.

Using the shape functions defined above we introduce a polynomial space on Γ_{ref} by

$$\Psi_{\mathscr{P}}(\Gamma_{\text{ref}}) = \Psi_{\mathscr{P}}^{[N]}(\Gamma_{\text{ref}}) + \bigcup_{l=1}^{4} \Psi_{\mathscr{P}}^{[\gamma_l]}(\Gamma_{\text{ref}}) + \Psi_{\mathscr{P}}^{[I]}(\Gamma_{\text{ref}}).$$

Here, $\Psi_{\mathscr{P}}^{[N]}(\Gamma_{\text{ref}})$ is the space of nodal shape functions on the reference element Γ_{ref}, $\Psi_{\mathscr{P}}^{[\gamma_l]}(\Gamma_{\text{ref}})$ is the space of side shape functions on the side γ_l of Γ_{ref}, and $\Psi_{\mathscr{P}}^{[I]}(\Gamma_{\text{ref}})$ denotes the space of interior shape functions on Γ_{ref}.

The full h–p approximation space on Γ is now defined by taking affine transformations onto the elements Γ_{ji} of the polynomial space $\Psi_{\mathscr{P}}(\Gamma_{\text{ref}})$. Using the notation previously introduced we have on each element the representation

$$\Psi_{\mathscr{P}_{ji}}(\Gamma_{ji}) = \Psi_{\mathscr{P}_{ji}}^{[N]}(\Gamma_{ji}) + \bigcup_{l=1}^{4} \Psi_{\mathscr{P}_{ji}}^{[\gamma_l]}(\Gamma_{ji}) + \Psi_{\mathscr{P}_{ji}}^{[I]}(\Gamma_{ji})$$

and the full space

$$\Psi(\Gamma) = \{\psi;\ \psi|_{\Gamma_{ji}} \in \Psi_{\mathscr{P}_{ji}}(\Gamma_{ji})\} \cap \tilde{H}^{1/2}(\Gamma).$$

We decompose the approximation space by

$$\Psi(\Gamma) = \Psi_H(\Gamma) + \Psi_{\mathscr{W}}(\Gamma) + \bigcup_{j} \Psi_{\Gamma_j} + \Psi_W(\Gamma) + \bigcup_{j,i} \Psi_{\mathscr{P}_{ji}}^{[I]}(\Gamma_{ji}). \tag{3}$$

Here, $\Psi_H(\Gamma) = \Pi_h \Psi_H^*(\Gamma)$ where $\Psi_H^*(\Gamma)$ is the space of piecewise linear/bilinear functions on the coarse mesh and

$$\Pi_h : \Psi(\Gamma) \to \Psi_h(\Gamma)$$

is the interpolation operator onto the space of piecewise bilinear functions $\Psi_h(\Gamma)$ on the fine mesh. Further, $\Psi_{\mathscr{W}}(\Gamma)$ is the space of piecewise bilinear functions on the fine mesh which are zero at the nodes which are not on the wire basket $\mathscr{W}$. The piecewise bilinear functions on the fine mesh which are nonzero only on the subdomain Γ_j span the space Ψ_{Γ_j}. The remaining spaces, $\Psi_W(\Gamma)$ and $\Psi_{\mathscr{P}_{ji}}^{[I]}(\Gamma_{ji})$, represent a decomposition of the space of high-degree polynomials. $\Psi_W(\Gamma)$ is spanned by all the side and nodal functions and $\Psi_{\mathscr{P}_{ji}}^{[I]}(\Gamma_{ji})$ comprises all polynomials of the specified degrees on Γ_{ji} that vanish on $\Gamma \backslash \Gamma_{ji}$.

Note that the above decomposition amounts to a three-level method. The first two levels, $\Psi_H(\Gamma)$ and $\Psi_{\mathscr{W}}(\Gamma) + \bigcup_j \Psi_{\Gamma_j}$, represent a two-level decomposition of the piecewise polynomials of lowest degree whereas the third level, $\Psi_W(\Gamma) + \bigcup_{j,i} \Psi_{\mathscr{P}_{ji}}^{[I]}(\Gamma_{ji})$, contains all piecewise polynomials of higher degrees.

For ease of presentation we use, instead of (3), also the notation

$$\Psi(\Gamma) = H_1 + \cdots + H_k,$$

for the three-level decomposition of $\Psi(\Gamma)$ where the number of subspaces k equals to three plus the number of subdomains Γ_j plus the number of elements Γ_{ji} (if the polynomial degrees are large enough such that all the subspaces are nonempty).

The additive Schwarz method consists in solving, by an iterative method, the equation

$$Pu_N := (P_1 + P_2 + \cdots + P_k)u_N = f_N, \tag{4}$$

where the projections $P_j : \Psi(\Gamma) \to H_j$, $j=1,\ldots,k$, are defined for any $v \in \Psi(\Gamma)$ by

$$a_j(P_j v, \varphi) = \langle Dv, \varphi \rangle_{L^2(\Gamma)} \quad \text{for any } \varphi \in H_j.$$

Here, a_j, $j=1,\ldots,k$, are given bilinear forms. On all but the wire basket spaces $\Psi_{\mathscr{W}}(\Gamma)$ and $\Psi_W(\Gamma)$ we use the original bilinear form given by the integral operator, i.e.,

$$a_j(v,w) := \langle Dv, w \rangle_{L^2(\Gamma)} \quad \text{for } v,w \text{ both in } \Psi_H(\Gamma), \Psi_{\Gamma_j} \text{ or } \Psi^{[I]}_{\mathscr{P}_{ji}}(\Gamma_{ji}).$$

On the wire basket spaces $\Psi_{\mathscr{W}}(\Gamma)$ and $\Psi_W(\Gamma)$ we use the L^2-bilinear form over the wire baskets, i.e.,

$$a_j(v,w) := \langle v, w \rangle_{L^2(\mathscr{W})} \quad \text{for } v,w \in \Psi_{\mathscr{W}}(\Gamma) \tag{5}$$

and

$$a_j(v,w) := \langle v, w \rangle_{L^2(W)} \quad \text{for } v,w \in \Psi_W(\Gamma). \tag{6}$$

The right-hand side of (4), $f_N = \sum_{j=1}^k P_j u_N$, can be computed without knowing the solution u_N of (2) by

$$a_j(P_j u_N, \varphi) = \langle f, \varphi \rangle_{L^2(\Gamma)} \quad \text{for any } \varphi \in H_j,\ j=1,\ldots,k.$$

Eq. (4) is the preconditioned linear system and an estimate of its condition number is given by the next theorem.

Theorem 1. *There exist positive constants c_1, c_2 which are independent of H_j, h_j, and p_j such that for all $v \in \Psi$ there holds*

$$c_1 \min_j \left(1 + \log \frac{H_j}{h_j} p_j\right)^{-2} \langle Dv, v \rangle_{L^2(\Gamma)} \leqslant \langle DPv, v \rangle_{L^2(\Gamma)} \leqslant c_2 \langle Dv, v \rangle_{L^2(\Gamma)}.$$

P is the additive Schwarz operator defined by the decomposition of the ansatz space Ψ and by the given bilinear forms.

The proof of Theorem 1 needs a far amount of technical details (see [24]). For simplicity we present in the next section the proof for the pure h-version, i.e., $p_j = 1$, from [22]. Nevertheless, this simpler case should still suffice to highlighten the various building blocks of our analysis.

A prototype of a weakly singular integral equation is

$$\langle Vu, v \rangle = \langle f, v \rangle \quad \forall v \in \tilde{H}^{-1/2}(\Gamma), \tag{7}$$

with the single-layer potential operator

$$Vu(x) := \frac{1}{4\pi} \int_\Gamma \frac{u(y)}{|x-y|} \, \mathrm{d}s_y.$$

This pseudodifferential operator has order -1, and the corresponding energy space is the dual space $\tilde{H}^{-1/2}(\Gamma)$ of $H^{1/2}(\Gamma)$ where the latter is an interpolation space between $L^2(\Gamma)$ $H^1(\Gamma)$. Hence to obtain bounds for the condition number of the additive Schwarz operator for weakly singular operators inequalities of the form

$$c_1 \sum_{i=1}^N \|v_i\|^2_{\tilde{H}^{-1/2}(\Gamma)} \leqslant \|v\|^2_{\tilde{H}^{-1/2}(\Gamma)} \leqslant c_2 \sum_{i=1}^N \|v_i\|^2_{\tilde{H}^{-1/2}(\Gamma)}$$

are of central importance. The function v belongs to the BEM trialspace which is now a subspace of $\tilde{H}^{-1/2}(\Gamma)$ and the representation $v=\sum_{i=1}^{N} v_i$ belongs to an appropriate subspace splitting. In contrary to the hypersingular operator the conformity in case of the weakly singular operator requires no continuity of the piecewise polynomial trial functions and the components of v can, e.g., be defined by restrictions.

Let us consider the Galerkin scheme for (7) and concentrate on the p-version of BEM for which we introduce a nonoverlapping method which is almost optimal. To define the additive Schwarz preconditioners for our model problem let $\bar{\Gamma}_h=\bigcup_{j=1}^{J}\bar{\Gamma}_j$ be a given mesh of J rectangles which define implicitly the subspace $\Psi\subset\tilde{H}^{-1/2}(\Gamma)$ of piecewise polynomials on Γ_h by specifying the polynomial degrees. For the decomposition of Ψ we choose a coarse mesh $\bar{\Gamma}_H=\bigcup_{j=1}^{n}\bar{G}_j$ of size $H\geqslant h$, assuming that Γ_H is compatible with the boundary element mesh Γ_h. We decompose

$$\Psi = H_0 \oplus H_1 \oplus \cdots \oplus H_n, \tag{8}$$

where H_0 is the space of piecewise constant functions on the coarse mesh Γ_H and

$$H_j:=\{v_{|G_j}:\ v\in\Psi\backslash H_0,\ \langle v,1\rangle_{L^2(G_j)}=0\},\quad j=1,\ldots,n.$$

Theorem 2 (Heuer [15,18]). *There exists a constant $c>0$ independent of h, H, and p such that for the condition number of the additive Schwarz operator P implicitly defined by the decomposition* (8) *there holds*

$$\kappa(P)\leqslant c\left(1+\log\left(\frac{H}{h}(p+1)\right)\right)^2.$$

Remark 1. Above the same degree p is used everywhere, for simplicity, but here the method works for nonuniform degree distributions as well. The above Theorem can be directly applied to the h-version (using piecewise constant trial functions), cf. [32]. Since the boundary element functions for the weakly singular integral equations need not to be continuous across the inner element boundaries the proof of Theorem 2 only consists of a detailed analysis of the Sobolev norms involved. No special care has to be taken of the basis functions. In contrast, Theorem 1 covers the hypersingular integral operator and there the trial space Ψ is a subspace of $\tilde{H}^{1/2}(\Gamma)$ and therefore continuity of the boundary element functions across the element boundaries is required.

In the two-dimensional situation, when dealing with integral equations on curves, additive Schwarz methods for weakly singular operators directly correspond to additive Schwarz methods for hypersingular operators and viceversa. This is due to the existence of simple isomorphisms between $\tilde{H}^{1/2}(\Gamma)$ and

$$\tilde{H}_0^{-1/2}(\Gamma):=\left\{\psi\in\tilde{H}^{-1/2}(\Gamma):\ \int_\Gamma \psi\,\mathrm{d}s=0\right\},$$

which are the energy spaces of operators of orders 1 and -1, respectively.

The extensions of standard differentiation and integration, which preserve polynomials, onto $\tilde{H}^{1/2}(\Gamma)$ and $\tilde{H}_0^{-1/2}(\Gamma)$, respectively can be taken. By these mappings, any subspace decomposition of an ansatz space for hypersingular operators gives a related subspace decomposition of the ansatz space of differentiated functions for weakly singular operators, and vice versa. Both decompositions then provide the same spectral properties of the corresponding additive Schwarz methods.

Such an easy isomorphism which preserves polynomials on surfaces in $\mathbb{R}^3$ is not known. For example $(-\triangle)^{1/2}$ and its inverse would be candidates but they are only pseudo-differential operators which in general do not map polynomials onto polynomials. Therefore, on surfaces we use different tools to analyze Schwarz preconditioners for operators of order one and of order -1.

Let us mention some other approaches for preconditioning linear systems arising from the h-version of BEM. Im [34,35] norm equivalences are proved for finite element multilevel splittings both in $H^{1/2}(\Gamma)$ and $H^{-1/2}(\Gamma)$ which yield estimates for multilevel additive Schwarz preconditioners applied to BEM [32].

Further, we mention the method by Steinbach [39] who uses operators of opposite orders to construct preconditioners. This method is especially worth being considered when one deals with systems where all the needed operators occur. Then there is no extra work to construct the needed stiffness matrices. In the framework of domain decomposition this approach has also been proposed by Xu and Zhang, see [51]. Here, the explicit representation of the inverse of the Steklov–Poincaré operator by a weakly singular operator, which is well-known in the boundary element literature, see, e.g., [38], is used to precondition the Steklov–Poincaré operator which is hypersingular. Finally, we mention that multiplicative Schwarz methods for the BEM are studied in [12,31].

3. Proof of Theorem 1 for the *h*-version

We return to the Galerkin scheme (2) and analyze for piecewise linear elements ($p=1$) its additive Schwarz preconditioner belonging to (3). But now we use on all subspaces the energy bilinear form $a_j(\cdot,\cdot)=\langle D\cdot,\cdot\rangle$. For simplicity we restrict our considerations to uniform rectangular meshes Γ_h. The decomposition of Γ is given by a uniform rectangular mesh Γ_H which is assumed to be compatible with Γ_h, i.e., the nodes of Γ_H are also nodes of Γ_h. Then the decomposition (3) becomes

$$S_h^1(\Gamma)=S_H^1(\Gamma)\cup S_{h,H}^1(\Gamma)\cup\bigcup_{j=1}^{J}S_h^1(\Gamma_j). \tag{9}$$

The spaces $S_H^1(\Gamma)$ and $S_h^1(\Gamma)$ consist of the usual continuous piecewise bilinear functions on the meshes Γ_H and Γ_h of size H and h on Γ. $S_{h,H}^1(\Gamma)$ is the so-called wire basket space which is spanned by the piecewise bilinear hat functions of $S_h^1(\Gamma)$ which are concentrated at the nodes lying on the element boundaries of the mesh of size H. The spaces $S_h^1(\Gamma_j)$ are spanned by the piecewise bilinear hat functions concentrated at the nodes interior to the restricted meshes $\Gamma_h|_{\Gamma_j}=\Gamma_{j,h}$, $j=1,\dots,J$.

We are now in the position to state and prove Theorem 1 which says that the preconditioner implicitly defined by the decomposition (9) is almost optimal for general mesh sizes h and H and that it is optimal if one fixes the ratio H/h and the polynomial degree.

The following two lemmas present abstract bounds for the minimum and maximum eigenvalues of the additive Schwarz operator corresponding to the h-version, i.e., we choose

$$\Psi(\Gamma)=S_h^1(\Gamma)=H_1+\cdots+H_k$$

according to (9) and we take the energy bilinear form $a(u,v)=\langle Du,v\rangle$. The proofs can be found, e.g., in [33,50].

Lemma 1. *If there exists a constant* C_1 *such that for any* $\varphi \in S_h^1(\Gamma)$ *there exist* $\varphi_j \in H_j$, $j=1,\dots,k$, *satisfying* $\varphi = \sum_{j=1}^k \varphi_j$ *and*

$$\sum_{j=1}^{k} a(\varphi_j, \varphi_j) \leqslant C_1^{-1} a(\varphi, \varphi),$$

then

$$\lambda_{\min}(P) \geqslant C_1.$$

Lemma 2. *If there exists a constant* C_2 *such that for any* $\varphi \in S_h^1(\Gamma)$ *and* $\varphi_j \in H_j$, $j=1,\dots,k$, *satisfying* $\varphi = \sum_{j=1}^k \varphi_j$ *and*

$$a(\varphi, \varphi) \leqslant C_2 \sum_{j=1}^{k} a(\varphi_j, \varphi_j)$$

then

$$\lambda_{\max}(P) \leqslant C_2.$$

Let us define the Sobolev spaces that are in use. The space $H^1(\Gamma)$ is endowed with the usual norm

$$\|\cdot\|_{H^1(\Gamma)}^2 = c\|\cdot\|_{L^2(\Gamma)}^2 + \|\partial_{x_1}\cdot\|_{L^2(\Gamma)}^2 + \|\partial_{x_2}\cdot\|_{L^2(\Gamma)}^2$$

where ∂_{x_1} and ∂_{x_2} denote the partial derivatives with respect to the Cartesian coordinates x_1 and x_2 on Γ. For fixed domains Γ or Ω the constant $c=1$ is taken. However, if we consider subdomains of diameter H the constant $c = 1/H^2$ is taken. This is to ensure appropriate scaling properties of the norms. The space $H_0^1(\Gamma)$ is the completion of $C_0^\infty(\Gamma)$ with respect to the norm $\|\cdot\|_{H^1(\Gamma)}$. For nonintegral s we use the K-method of the interpolation theory as described in [2]. For two normed spaces A_0 and A_1 the interpolation space $A_s = [A_0, A_1]_s$ $(0<s<1)$ is equipped with the norm

$$\|a\|_{[A_0,A_1]_s} := \left(\int_0^\infty \left(t^{-s} \inf_{a=a_0+a_1}(\|a_0\|_{A_0} + t\|a_1\|_{A_1})\right)^2 \frac{\mathrm{d}t}{t}\right)^{1/2}.$$

For $0<s<1$ we define

$$H^s(\Gamma) = [L^2(\Gamma), H^1(\Gamma)]_s, \qquad \tilde{H}^s(\Gamma) = [L^2(\Gamma), H_0^1(\Gamma)]_s.$$

The spaces $H^{-s}(\Gamma)$ and $\tilde{H}^{-s}(\Gamma)$ are the dual spaces with respect to the L^2-inner product

$$H^{-s}(\Gamma) = (\tilde{H}^s(\Gamma))', \qquad \tilde{H}^{-s}(\Gamma) = (H^s(\Gamma))'.$$

The spaces $H^s(\Gamma)$ and $\tilde{H}^s(\Gamma)$ for $|s|>1$ can be defined analogously by interpolating between $H^{m-1}(\Gamma)$ and $H^m(\Gamma)$ or $H_0^m(\Gamma)$ for the smallest integer $m > |s|$. The following lemma is used for estimating the largest eigenvalue of the additive Schwarz operator.

Lemma 3 (Heuer [13, Lemma 4]). *Let* $\{\Gamma_j;\ j=1,\dots,J\}$ *be a finite covering of* Γ *by subdomains* Γ_j *with Lipschitz boundary,*

$$\bar{\Gamma} = \bigcup_{j=1}^{J} \bar{\Gamma}_j,$$

with a covering constant J_c, *i.e., we can color* $\{\Gamma_j;\ j=1,\dots,J\}$ *using at most* J_c *colors in such a way that subdomains of the same color are disjoint. Let* $\varphi=\sum_{j=1}^{J}\varphi_j\in\tilde{H}^s(\Gamma)$ *for real* s *with* $\varphi_j\in\tilde{H}^s(\Gamma_j)$, $j=1,\dots,J$. *Then there holds*

$$\|\varphi\|^2_{\tilde{H}^s(\Gamma)}\leqslant J_c\sum_{j=1}^{J}\|\varphi_j\|^2_{\tilde{H}^s(\Gamma_j)}.$$

It is crucial for substructuring techniques to split global norms into norms over subdomains. This is straightforward for Sobolev norms of integral order. For norms of nonintegral order which typically appear in the boundary element method for first kind integral equations this is not trivial. The following lemma is used in the proof of the main theorem.

Lemma 4 (Heuer [16, Lemma 3.3]). *Let* $s>0$ *and* $\varphi\in\tilde{H}^s(\Gamma)$ *with* $\varphi_j{:=}\varphi|_{\Gamma_j}\in\tilde{H}^s(\Gamma_j)$, $j=1,\dots,J$. *There exist constants* $C_1,C_2>0$ *which are independent of* φ *and* J *such that*

$$C_1\sum_{j=1}^{J}\|\varphi_j\|^2_{H^s(\Gamma_j)}\leqslant\|\varphi\|^2_{\tilde{H}^s(\Gamma)}\leqslant C_2\sum_{j=1}^{J}\|\varphi_j\|^2_{\tilde{H}^s(\Gamma_j)}.$$

To bound the maximum eigenvalue of P we take for a given $\phi\in S_h^1(\Gamma)$ an arbitrary representation

$$\phi=\phi_H+\phi_{h,H}+\sum_{j=1}^{J}\phi_{j,h}$$

according to the decomposition (9) By the triangle inequality and by applying a colouring argument (Lemma 3) to the third component $\sum_{j=1}^{J}\phi_{j,h}$ we obtain

$$\|\phi\|^2_{\tilde{H}^{1/2}(\Gamma)}\leqslant C\left(\|\phi_H\|^2_{\tilde{H}^{1/2}(\Gamma)}+\|\phi_{h,H}\|^2_{\tilde{H}^{1/2}(\Gamma)}+\sum_{j=1}^{J}\|\phi_{j,h}\|^2_{\tilde{H}^{1/2}(\Gamma_j)}\right).$$

Thus, due to Lemma 3, we proved the boundedness of the maximum eigenvalue,

$$\lambda_{\max}(P)\leqslant C.$$

In order to derive a lower bound for the minimum eigenvalue of P we apply Lemma 1 to a specific representation for an arbitrary function $\phi\in S_h^1(\Gamma)$. We choose

$$\phi_H{:=}(Q_H\mathscr{E}\phi)|_\Gamma\in S_H^1(\Gamma),$$

where Q_H is the L^2-projector onto $S_H^1(\Omega)$. Here, $\mathscr{E}$ is the discretely harmonic extension operator from Γ onto $\Omega{:=}(-H,H)\times\Gamma$ (cf. (11)). By the trace theorem, the stability of Q_H in $H^1(\Omega)$, and the extension theorem for discretely harmonic functions we obtain

$$\|\phi_H\|^2_{\tilde{H}^{1/2}(\Gamma)}\leqslant C|Q_H\mathscr{E}\phi|^2_{H^1(\Omega)}\leqslant C|\mathscr{E}\phi|^2_{H^1(\Omega)}\leqslant C\|\phi\|^2_{\tilde{H}^{1/2}(\Gamma)}. \tag{10}$$

Here, we made use of the fact that piecewise trilinear functions in Ω with respect to the mesh size H are discretely harmonic.

Let us use the notation

$$w_h{:=}\phi-\phi_H.$$

To define the component $\phi_{h,H}$ of ϕ which belongs to $S_{h,H}^1(\Gamma)$ we need some more notations.

By W we denote the wire basket of the mesh Γ_H of size H on Γ, i.e. the union of the edges of the elements of Γ_H. We neglect the edges which are on the boundary of Γ since we need zero boundary conditions for subspaces of $\tilde{H}^{1/2}(\Gamma)$. The nodes of the mesh Γ_h of size h which belong to the wire basket W are denoted by W_{nodes}. The nodes of Γ_h which do not belong to the boundary of Γ are denoted by Γ_{nodes}. Of course, these sets depend on the mesh sizes H and h. Now we define the component $\phi_{h,H}$ by the following relations:

$$\phi_{h,H} \in S_h^1(\Gamma), \qquad \phi_{h,H}(x) := \begin{cases} w_h(x) & \text{for all } x \in W_{\text{nodes}}, \\ 0 & \text{for all } x \in \Gamma_{\text{nodes}} \setminus W_{\text{nodes}}. \end{cases}$$

Obviously, the function $\phi_{h,H}$ belongs to $S_{h,H}^1(\Gamma)$. Let $\mathscr{E}\phi_{h,H}$ denote the discretely harmonic extension of $\phi_{h,H}$ onto $\Omega := (-H,H) \times \Gamma$. More precisely we embed the mesh Γ_h in the three-dimensional mesh Ω_h of cubes of size h which is defined on $\Omega = (-H,H) \times \Gamma$. We identify $\Gamma = \Omega|_{z=0}$ and use the notations $\Omega_1 := \Omega|_{z<0}$ and $\Omega_2 := \Omega|_{z>0}$. Then we define

$$\mathscr{E}\phi_{h,H} \in S_h^1(\Omega), \qquad \mathscr{E}\phi_{h,H}|_\Gamma = \phi_{h,H},$$

$$\int_\Omega \nabla \mathscr{E}\phi_{h,H} \nabla \varphi \, \mathrm{d}(x,y,z) = 0 \quad \text{for all } \varphi \in S_h^1(\Omega_i), \ i = 1,2.$$

Using the trace theorem and [7, Lemma 4.7] we deduce

$$\|\phi_{h,H}\|^2_{\tilde{H}^{1/2}(\Gamma)} \leqslant C|\mathscr{E}\phi_{h,H}|^2_{H^1(\Omega)} \leqslant C\|\phi_{h,H}\|^2_{L^2(W)}. \tag{11}$$

Since $\phi_{h,H} = w_h$ on the wire basket W we obtain by again using the discretely harmonic extension operator $\mathscr{E}$ and by [7, Lemma 4.3]

$$\|\phi_{h,H}\|^2_{L^2(W_j)} \leqslant C\left(1 + \log\frac{H}{h}\right)\|\mathscr{E}w_h\|^2_{H^1(\Omega_{i,j})}, \quad i = 1,2, \ j = 1,\dots,J. \tag{12}$$

Here $\bar{\Omega}_i = \bigcup_{j=1}^J \bar{\Omega}_{i,j}$ is a covering of Ω_i, $i = 1,2$, which is compatible with the decomposition of Γ into subdomains Γ_j, $j = 1,\dots,J$, and $W_j := \partial\Gamma_j$. By the approximation property of the projection operator Q_H, and by using the identity $\mathscr{E}((Q_H\mathscr{E}\phi)|_\Gamma) = Q_H\mathscr{E}\phi$, there holds

$$\|\mathscr{E}w_h\|^2_{L^2(\Omega_{i,j})} = \|\mathscr{E}\phi - Q_H\mathscr{E}\phi\|^2_{L^2(\Omega_{i,j})} \leqslant CH^2|\mathscr{E}\phi|^2_{H^1(\Omega_{i,j})},$$

and by [3, (3.11)] we obtain for discretely harmonic functions φ

$$|\varphi|^2_{H^1(\Omega_i)} \leqslant C\|\varphi\|^2_{\tilde{H}^{1/2}(\partial\Omega_i)}.$$

Therefore, together with (10), we obtain

$$\begin{aligned}\sum_{j=1}^J \|\mathscr{E}w_h\|^2_{H^1(\Omega_{i,j})} &= \sum_{j=1}^J (H^{-2}\|\mathscr{E}w_h\|^2_{L^2(\Omega_{i,j})} + |\mathscr{E}w_h|^2_{H^1(\Omega_{i,j})}) \\ &\leqslant C(|\mathscr{E}\phi|^2_{H^1(\Omega_i)} + |\mathscr{E}w_h|^2_{H^1(\Omega_i)}) \\ &\leqslant C(\|w_h\|^2_{\tilde{H}^{1/2}(\Gamma)} + \|\phi\|^2_{\tilde{H}^{1/2}(\Gamma)}) \leqslant C\|\phi\|^2_{\tilde{H}^{1/2}(\Gamma)}.\end{aligned} \tag{13}$$

Combining (11)–(13) we obtain

$$\|\phi_{h,H}\|^2_{\tilde{H}^{1/2}(\Gamma)} \leqslant C\left(1 + \log\frac{H}{h}\right)\|\phi\|^2_{\tilde{H}^{1/2}(\Gamma)}. \tag{14}$$

In the last step we define the components of ϕ belonging to the spaces $S_h^1(\Gamma_j)$, $j=1,\dots,J$, by

$$\phi_{j,h} := \begin{cases} w_h - \phi_{h,H} & \text{on } \Gamma_j, \\ 0 & \text{elsewhere.} \end{cases}$$

Since $w_h = \phi_{h,H}$ on the boundaries of the subdomains the functions $\phi_{j,h}$ are continuous on Γ and therefore belong to $S_h^1(\Gamma)$ and the corresponding subspace $S_h^1(\Gamma_j)$ as well. Thus we have

$$\phi_h := \sum_{j=1}^{J} \phi_{j,h} \in S_h^1(\Gamma) \subset \tilde{H}^{1/2}(\Gamma)$$

and

$$\phi_h|_{\Gamma_j} = \phi_{j,h} \in S_h^1(\Gamma_j) \subset \tilde{H}^{1/2}(\Gamma_j), \quad j=1,\dots,J.$$

By Lemma 4 in [24] there holds

$$||\phi_{j,h}||_{\tilde{H}^{1/2}(\Gamma_j)} \leqslant C\left(1+\log\frac{H}{h}\right)||w_h||_{\tilde{H}^{1/2}(\Gamma_j)}.$$

Therefore, we obtain by Lemma 4 and (10)

$$\begin{aligned}\sum_{j=1}^{J} ||\phi_{j,h}||^2_{\tilde{H}^{1/2}(\Gamma_j)} &\leqslant C\left(1+\log\frac{H}{h}\right)^2 ||w_h||^2_{\tilde{H}^{1/2}(\Gamma)} \\ &\leqslant C\left(1+\log\frac{H}{h}\right)^2 ||\phi||^2_{\tilde{H}^{1/2}(\Gamma)}. \end{aligned} \tag{15}$$

Since

$$\begin{aligned}\phi_H + \phi_{h,H} + \sum_{j=1}^{J} \phi_{j,h} &= \phi_H + \phi_{h,H} + w_h - \phi_{h,H} \\ &= \phi_H + \phi - \phi_H = \phi\end{aligned}$$

we defined a representation of ϕ and using (10), (14) and (15) we proved that

$$\begin{aligned}&||\phi_H||^2_{\tilde{H}^{1/2}(\Gamma)} + ||\phi_{h,H}||^2_{\tilde{H}^{1/2}(\Gamma)} + \sum_{j=1}^{J} ||\phi_{j,h}||^2_{\tilde{H}^{1/2}(\Gamma)} \\ &\quad \leqslant C||\phi||^2_{\tilde{H}^{1/2}(\Gamma)} + C\left(1+\log\frac{H}{h}\right)^2 ||\phi||^2_{\tilde{H}^{1/2}(\Gamma)} \leqslant C\left(1+\log\frac{H}{h}\right)^2 ||\phi||^2_{\tilde{H}^{1/2}(\Gamma)}.\end{aligned}$$

Therefore, due to Lemma 1,

$$\lambda_{\min}(P) \geqslant C\left(1+\log\frac{H}{h}\right)^{-2}$$

and

$$\kappa(P) = \lambda_{\max}(P)/\lambda_{\min}(P) \leqslant C\left(1+\log\frac{H}{h}\right)^2.$$

4. Numerical results

To demonstrate the efficiency of our preconditioning method and to underline the theoretical estimates we present some experimental results for the extremum eigenvalues and the condition numbers of the preconditioned systems belonging to the Galerkin p-version for the hypersingular integral equation (1).

We emphasize that the boundary element method produces stiffness matrices which are in general fully occupied which means that even functions with disjoint supports are coupled via the integral operator. Therefore, when performing a domain decomposition to create a preconditioner, one not only decouples adjacent subdomains but also neglects the coupling of functions in subdomains which are not adjacent. The latter coupling is not present in the finite element method. Therefore, in the boundary element method, the theoretical bounds for the extremum eigenvalues are most often just asymptotically obtained and are not as obvious as in the finite element method from the experimental results.

Due to Theorem 1 we expect for the uniform and locally uniform methods bounded maximum eigenvalues and minimum eigenvalues which behave like $(1+\log p_{\max}H/h)^{-2}$.

For our model problem we choose the domain $\Gamma=(-1/2,1/2)^2\times\{0\}$ and take a uniform mesh of squares with length h. For the concrete choice of the trial spaces see [24]. The polylogarithmic behavior in p of the condition number is checked with Fig. 1. Here we consider the mesh $h=1/3$ and $H/h=1$. Both cases, uniform and locally uniform p-version, as well as the results for the nonuniform p-version are shown. In the uniform case $p=7$ corresponds to $N=400$ number of unknowns. In the locally uniform case we have only $N=202$ for $p=7$ and in the nonuniform example $p=7$ means $N=82$. However, we observe that all curves are quite close which means that the condition number essentially depends on the maximum polynomial degree. The efficiency of the preconditioner seems

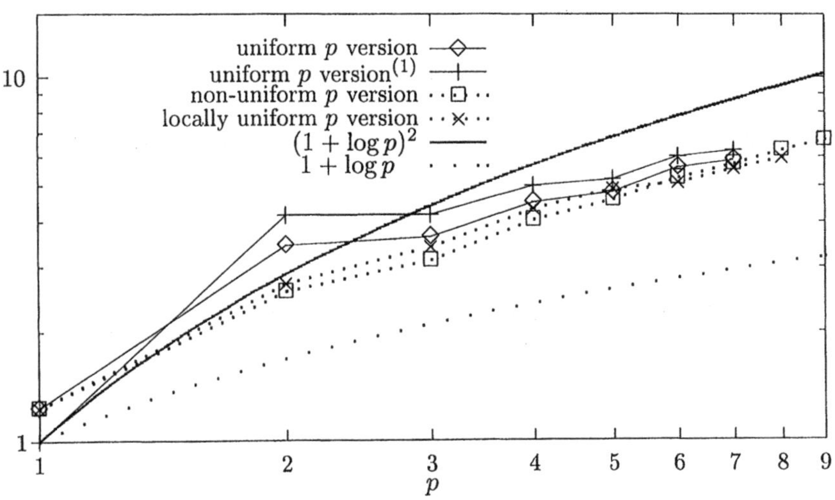

Fig. 1. Condition numbers for the preconditioned uniform and nonuniform/locally uniform p-version ($1/h=3$, $H/h=1$). The results marked by (1) (only the uniform p-version) are obtained by using the original bilinear form instead of the L^2-bilinear form in the definition of the preconditioner.

to be independent of the actual distribution of the polynomial degrees, which is restricted in our theory. Moreover, in all cases the theoretical bound $(1+\log p)^2$ is numerically fulfilled. Further, let us note that sometimes it is natural, e.g., when the full stiffness matrix is available, to use the original bilinear form instead of the L^2-bilinear form on the wire baskets, cf. (5) and (6). This replacement yields a different preconditioner whose implementation does not require additional inner products. Although this method is not covered by our theory the results in Fig. 1 for this preconditioner (indicated by (1)) show the same asymptotic behavior as the theoretically justified method.

References

[1] I. Babuška, B. Guo, E.P. Stephan, The *hp*-version for boundary element Galerkin methods on polygons, Math. Methods Appl. Sci. 12 (1990) 413–427.

[2] J. Bergh, J. Löfström, in: Interpolation Spaces, Grundlehren der mathematischen Wissenschaften, Vol. 223, Springer, Berlin, 1976.

[3] J.H. Bramble, J.E. Pasciak, A.H. Schatz, The construction of preconditioners for elliptic problems by substructuring IV, Math. Comp. 53 (1989) 1–24.

[4] T. Cao, Adaptive and additive multilevel methods for boundary integral equations, Ph.D. Thesis, School of Mathematics, University of New South Wales, Sydney, Australia, 1995.

[5] C. Carstensen, E.P. Stephan, A posteriori error estimates for boundary element methods, Math. Comp. 64 (1995) 483–500.

[6] C. Carstensen, E.P. Stephan, Adaptive boundary element methods for some first kind integral equations, SIAM J. Numer. Anal. 33 (1996) 2166–2183.

[7] M. Dryja, B. Smith, O.B. Widlund, Schwarz analysis of iterative substructuring algorithms for elliptic problems in three dimensions, SIAM J. Numer. Anal. 31 (1994) 1662–1694.

[8] B. Guo, The *hp*-version of the finite element method for solving boundary value problems in polyhedral domains, in: M. Costabel, M. Dauge, S. Nicaise (Eds.), Boundary Value Problems and Integral Equations in Non-smooth Domains (Luminy, 1993), Lecture Notes in Pure and Applied Mathematics, Vol. 167, Dekker, New York, 1995. pp. 101–120.

[9] B. Guo, I. Babuška, The *hp*-version of the finite element method, Part 1: The basic approximation results, Comput. Mech. 1 (1986) 21–41.

[10] B. Guo, W. Cao, An additive Schwarz method for the $h-p$ version of the finite element method in three dimensions, SIAM J. Numer. Anal. 35 (1998) 632–654.

[11] B. Guo, N. Heuer, E.P. Stephan, The *hp*-version of the boundary element method for transmission problems with piecewise analytic data, SIAM J. Numer. Anal. 33 (1996) 789–808.

[12] M. Hahne, E.P. Stephan, Schwarz iterations for the efficient solution of screen problems with boundary elements, Computing 56 (1996) 61–85.

[13] N. Heuer, Additive Schwarz methods for weakly singular integral equations in $\mathbb{R}^3$ – the *p*-version, in: W. Hackbusch, G. Wittum (Eds.), Boundary Elements: Implementation and Analysis of Advanced Algorithms, Vieweg-Verlag, Braunschweig, Wiesbaden, 1996, pp. 126–135.

[14] N. Heuer, Efficient algorithms for the *p*-version of the boundary element method, J. Integral Equations Appl. 8 (1996) 337–361.

[15] N. Heuer, Preconditioners for the *p*-version of the boundary element Galerkin method in $\mathbb{R}^3$, Habilitation Thesis, Institut für Angewandte Mathematik, Universität Hannover, Germany, 1998.

[16] N. Heuer, An iterative substructuring method for the *p*-version of the boundary element method for hypersingular integral equations in three dimensions, Numer. Math. 79 (1998) 371–396.

[17] N. Heuer, Additive Schwarz methods for indefinite hypersingular integral equations in $\mathbb{R}^3$ – the *p*-version, Appl. Anal. 72 (1999) 411–437.

[18] N. Heuer, Additive Schwarz method for the p-version of the boundary element method for the single layer potential operator on a plane screen, Numer. Math., to appear.

[19] N. Heuer, M. Maischak, E.P. Stephan, Exponential convergence of the *hp*-version for the boundary element method on open surfaces, Numer. Math. 83 (1999) 641–666.

[20] N. Heuer, E.P. Stephan, The *hp*-version of the boundary element method on polygons, J. Integral Equations Appl. 8 (1996) 173–212.

[21] N. Heuer, E.P. Stephan, Boundary Integral operators in countably normed spaces, Math. Nachr. 191 (1998) 123–151.

[22] N. Heuer, E.P. Stephan, Iterative substructuring for hypersingular integral equations in $\mathbb{R}^3$, SIAM J. Sci. Comput. 20 (1999) 739–749.

[23] N. Heuer, E.P. Stephan, The Poincaré-Steklov operator within countably normed spaces, in: M. Bonnet, A.-M. Sändig, W.L. Wendland (Eds.), Mathematical Aspects of Boundary Element Methods, Research Notes in Mathematics, Vol. 414, Chapman & Hall, London, 1999,, pp. 152–164.

[24] N. Heuer, E.P. Stephan, An additive Schwarz method for the *hp*-version of the boundary element method for hypersingular integral equations in $\mathbb{R}^3$, IMA J. Numer. Analys., to appear.

[25] N. Heuer, E.P. Stephan, T. Tran, Multilevel additive Schwarz method for the *hp*-version of the Galerkin boundary element method, Math. Comp. 67 (1998) 501–518.

[26] H. Holm, M. Maischak, E.P. Stephan, The *hp*-version of the boundary element method for the Helmholtz screen problems, Computing 57 (1996) 105–134.

[27] H. Holm, M. Maischak, E.P. Stephan, Exponential convergence of the *hp*-version boundary element method for mixed boundary value problems on polyhedrons, Math. Methods Appl. Sci. to appear.

[28] M. Maischak, P. Mund, E.P. Stephan, Adaptive multilevel BEM for acoustic scattering, Comput. Methods Appl. Mech. Eng. 150 (1997) 351–367.

[29] M. Maischak, E.P. Stephan, The *hp*-version of the boundary element method in $\mathbb{R}^3$. The basic approximation results, Math. Meth. Appl. Sci. 20 (1997) 461–476.

[30] M. Maischak, E.P. Stephan, The *hp*-version of the boundary element method in $\mathbb{R}^3$, Part II: Approximation in countably normed spaces, to appear.

[31] M. Maischak, E.P. Stephan, T. Tran, Multiplicative Schwarz algorithms for the Galerkin boundary element methods, SIAM J. Numer. Anal., to appear.

[32] P. Mund, E.P. Stephan, J. Weiße, Two level methods for the single layer potential in $\mathbb{R}^3$, Computing 60 (1998) 243–266.

[33] S.V. Nepomnyaschikh, Domain decomposition and Schwarz methods in a subspace for approximate solution of elliptic boundary value problems, Ph.D. Thesis, Computing Center of the Siberian Branch of the USSR, Academy of Sciences, Novosibirsk, USSR, 1986.

[34] P. Oswald, Multilevel Finite Element Approximation, Theory and Applications, Teubner Skripten zur Numerik, Teubner, Stuttgart, 1994.

[35] P. Oswald, Multilevel norms for $H^{-1/2}$, Computing 61 (1998) 235–255.

[36] L.F. Pavarino, O.B. Widlund, Iterative substructuring methods for spectral elements: problems in three dimensions based on numerical quadrature. Approximation theory and applications, Comput. Math. Appl. 33 (1–2) (1997) 193–209.

[37] T. von Petersdorff, E.P. Stephan, Regularity of mixed boundary value problems in $\mathbb{R}^3$ and boundary element methods on graded meshes, Math. Methods Appl. Sci. 12 (1990) 229–249.

[38] A.H. Schatz, V. Thomée, W.L. Wendland, Mathematical Theory of Finite and Boundary Element Methods, Birkhäuser, Basel, 1990.

[39] O. Steinbach, Fast solvers for the symmetric boundary element method, in: W. Hackbusch, G. Wittum (Eds.), Boundary Elements: Implementation and Analysis of Advanced Algorithms, Notes on Numerical Fluid Mechanics, Vieweg-Verlag, Braunschweig, Wiesbaden, 1996, pp. 232–242.

[40] E.P. Stephan, Boundary integral equations for screen problems in $\mathbb{R}^3$, Integral Equations Operator Theory 10 (1987) 257–263.

[41] E.P. Stephan, The *hp* boundary element method for solving 2- and 3-dimensional problems, Comput. Methods Appl. Mech. Eng. 133 (1996) 183–208.

[42] E.P. Stephan, Additive Schwarz methods for integral equations of the first kind, in: J.R. Whiteman (Ed.), The Mathematics of Finite Elements and Applications, Wiley, Chicester, 1997, pp. 123–144.

[43] E.P. Stephan, T. von Petersdorff, Decomposition in edge and corner singularities for the solution of the Dirichlet problem of the Laplacian in a polyhedron, Math. Nachr. 149 (1990) 71–104.

[44] E.P. Stephan, T. Tran, Domain decomposition algorithms for indefinite hypersingular integral equations – the *h*- and *p*-versions, SIAM J. Sci. Comput. 19 (1998) 1139–1153.

[45] E.P. Stephan, T. Tran, Domain decomposition algorithms for indefinite weakly singular equations – the h- and p-versions, IMA J. Numer. Analysis 20 (2000) 1–24.
[46] E.P. Stephan, T. Tran, P. Mund, Hierarchical basis preconditioners for first kind integral equations, Appl. Anal. 65 (3–4) (1997) 353–372.
[47] T. Tran, E.P. Stephan, Additive Schwarz method for the h-version boundary element method, Appl. Anal. 60 (1996) 63–84.
[48] T. Tran, E.P. Stephan, Additive Schwarz algorithms for the p-version of the Galerkin boundary element method, Numer. Math. 85 (2000) 433–468.
[49] T. Tran, E.P. Stephan, Two-level additive Schwarz preconditioners for the hp-version of the Galerkin boundary element method, to appear.
[50] O.B. Widlund, Optimal iterative refinement methods, in: T.F. Chan, R. Glowinski, J. Périaux, O.B. Widlund (Eds.), Domain Decomposition Methods for Partial Differential Equations, SIAM, Philadelphia, 1989, pp. 114–125.
[51] J. Xu, S. Zhang, Preconditioning the Poincaré-Steklov operator by using Green's function, Math. Comp. 166 (1997) 125–138.

ELSEVIER

Journal of Computational and Applied Mathematics 125 (2000) 521–537

JOURNAL OF COMPUTATIONAL AND APPLIED MATHEMATICS

www.elsevier.nl/locate/cam

Domain decomposition methods via boundary integral equations

G.C. Hsiao[a], O. Steinbach[b], W.L. Wendland[b,*]

[a]*Department of Mathematical Sciences, University of Delaware, Newark, Delaware 19716, USA*
[b]*Mathematisches Institut A, Universität Stuttgart, Pfaffenwaldring 57, 70569 Stuttgart, Germany*

Received 12 March 1999

Abstract

Domain decomposition methods are designed to deal with coupled or transmission problems for partial differential equations. Since the original boundary value problem is replaced by local problems in substructures, domain decomposition methods are well suited for both parallelization and coupling of different discretization schemes. In general, the coupled problem is reduced to the Schur complement equation on the skeleton of the domain decomposition. Boundary integral equations are used to describe the local Steklov–Poincaré operators which are basic for the local Dirichlet–Neumann maps. Using different representations of the Steklov–Poincaré operators we formulate and analyze various boundary element methods employed in local discretization schemes. We give sufficient conditions for the global stability and derive corresponding a priori error estimates. For the solution of the resulting linear systems we describe appropriate iterative solution strategies using both local and global preconditioning techniques. © 2000 Elsevier Science B.V. All rights reserved.

Keywords: Domain decomposition; Boundary integral equations; Boundary element methods; Preconditioning techniques

1. Introduction

Domain decomposition methods were originally designed to solve boundary value problems in complicated domains. We mention here only the famous alternating Schwarz method [25]. Since modern parallel computers are available, these methods have become very useful in the numerical analysis of partial differential equations, in particular, with respect to the development of efficient algorithms for the numerical solution of complicated problems, see e.g. [35]. Due to the decomposition into substructures, domain decomposition methods are well suited for the coupling of different discretization schemes such as finite and boundary element methods, see e.g. [5,7,11]. In finite element methods,

* Corresponding author.
E-mail addresses: hsiao@math.udel.edu (G.C. Hsiao), steinbach@mathematik.uni-stuttgart.de (O. Steinbach), wendland@mathematik.uni.stuttgart.de (W.L. Wendland).

PII: S0377-0427(00)00488-X

the domain decomposition approach is often applied to construct efficient preconditioners for parallel computations. This is mainly based on a splitting of the global trial space into local ones arising from the domain decomposition. Applying these ideas to boundary integral equations leads to additive Schwarz methods based on a decomposition of the boundary into overlapping or nonoverlapping parts, see e.g. [12,19,21,32].

Here we will concentrate our considerations to geometry-based domain decomposition methods where the original boundary value problem is reduced to local subproblems involving appropriate coupling conditions. When assuming boundary conditions of either Dirichlet or Neumann type on the local subdomain boundaries, the solution of the local subproblems defines local Dirichlet–Neumann or Neumann–Dirichlet maps. Hence, in domain decomposition methods we need to find the complete Cauchy data on the skeleton. This results in a variational formulation to find either the Dirichlet or Neumann data on the skeleton, and the remaining data are determined by the local problems and the coupling conditions. Using boundary integral equations we are able to describe the Dirichlet–Neumann map by the Steklov–Poincaré operator which admits different representations. Analyzing the mapping properties of local boundary integral operators [8,9,34], we get unique solvability of the resulting boundary integral variational problem. Moreover, applying a standard Galerkin scheme, we get stability and quasi-optimal a priori error estimates for the approximate solution. However, boundary integral representations of the Steklov–Poincaré operator involve inverse integral operators. Hence we are not able to compute the corresponding stiffness matrices exactly. Therefore we have to define suitable boundary element approximations and we need to derive related stability and error estimates, see e.g. [13,24,28]. Finally, we will discuss the efficient solution of the resulting linear systems by appropriate iterative methods in parallel. Here we need local and global preconditioning matrices.

2. Domain decomposition methods

As a model problem, we consider the Dirichlet boundary value problem

$$L(x)u(x) = f(x) \quad \text{for } x \in \Omega, \qquad u(x) = g(x) \quad \text{for } x \in \Gamma. \tag{2.1}$$

Here $\Omega \subset \mathbb{R}^n$, $n=2$ or 3 is a bounded domain with Lipschitz boundary $\Gamma = \partial\Omega$ and $L(\cdot)$ is a formally positive elliptic partial differential operator of second order. Applications of (2.1) are, for example, boundary value problems in potential theory and in elastostatics. In domain decomposition methods, we begin with the decomposition of Ω. Let

$$\bar{\Omega} = \bigcup_{i=1}^{p} \bar{\Omega}_i, \tag{2.2}$$

be a subdivision into p nonoverlapping subdomains Ω_i. Note that this decomposition can be done either due to the geometrical form of Ω or due to some properties of the partial differential operator involved in (2.1). In particular, for $x \in \Omega_i$ we assume that $L(x) = L_i$ is a partial differential operator with constant coefficients which can be different in different subdomains. Without loss of generality, we also assume that the local subdomain boundaries $\Gamma_i = \partial\Omega_i$ are strong Lipschitz. We denote by

$\Gamma_{ij}=\Gamma_i\cap\Gamma_j$ for $i,j=1,\dots,p$ local coupling boundaries, and define the skeleton Γ_S of the domain decomposition (2.2) by

$$\Gamma_S=\bigcup_{i=1}^{p}\Gamma_i=\Gamma\cup\bigcup_{i,j=1}^{p}\Gamma_{ij}. \tag{2.3}$$

Defining $u_i(x)=u(x)$ for $x\in\Omega_i$, instead of (2.1) we need to consider local boundary value problems

$$L_i u_i(x)=f(x)\quad\text{for } x\in\Omega_i,\qquad u_i(x)=g(x)\quad\text{for } x\in\Gamma_i\cap\Gamma. \tag{2.4}$$

In addition to the boundary conditions in (2.4), we need also appropriate coupling conditions across all local coupling boundaries Γ_{ij}. More precisely, let $T_i u_i(x)$ denote the conormal derivative of u_i defined for $x\in\Gamma_i$ almost everywhere. Then, the natural coupling conditions, induced by (2.1), are

$$u_i(x)=u_j(x),\qquad (T_i u_i)(x)+(T_j u_j)(x)=0\quad\text{for } x\in\Gamma_{ij}. \tag{2.5}$$

As will be seen, the essence of the domain decomposition methods amounts to reduce the solution of the original boundary value problem (2.1) to the solutions of local boundary value problems (2.4), (2.5). According to (2.5) we may formulate different domain decomposition methods, additive and multiplicative Schwarz methods, leading to different discretization techniques as well. In what follows we will restrict ourselves to the case that the first coupling condition in (2.5), $u_i(x)=u_j(x)$ for $x\in\Gamma_{ij}$ is required to be satisfied pointwise, while the second condition will be required in a week sense only.

We now need some function spaces. We denote by $H^{1/2}(\Gamma_S)$ the trace space of $H^1(\Omega)$ equipped with the norm

$$\|u\|_{H^{1/2}(\Gamma_S)}:=\left\{\sum_{i=1}^{p}\|u_{|\Gamma_i}\|^2_{H^{1/2}(\Gamma_i)}\right\}^{1/2}. \tag{2.6}$$

Let $u\in H^{1/2}(\Gamma_S)$ with $u(x)=g(x)$ for $x\in\Gamma$. Then we define the restrictions $u_i(x)=u(x)$ for $x\in\Gamma_i$ which implies that $u_i(x)=u_j(x)$ for $x\in\Gamma_{ij}$. Now we consider local Dirichlet boundary value problems

$$L_i u_i(x)=f(x)\quad\text{for } x\in\Omega_i,\qquad u_i(x)=u(x)\quad\text{for } x\in\Gamma_i \tag{2.7}$$

and define the corresponding local Dirichlet–Neumann maps

$$\boldsymbol{T}_i u(x):=(T_i u_i)(x)\quad\text{for } x\in\Gamma_i. \tag{2.8}$$

The latter implies that the Neumann coupling condition in (2.5) can be rewritten as

$$\boldsymbol{T}_i u(x)+\boldsymbol{T}_j u(x)=0\quad\text{for } x\in\Gamma_{ij}. \tag{2.9}$$

Let $\tilde g\in H^{1/2}(\Gamma_S)$ be an arbitrary but fixed extension of the given Dirichlet data g satisfying $\tilde g(x)=g(x)$ for $x\in\Gamma$. By defining the test function space

$$W:=\{v\in H^{1/2}(\Gamma_S)\colon v(x)=0\quad\text{for } x\in\Gamma\}, \tag{2.10}$$

we have the variational formulation of (2.9) to find $\tilde u\in W$ such that $u=\tilde u+\tilde g$ and

$$\sum_{i=1}^{p}\int_{\Gamma_i}\boldsymbol{T}_i u(x)\cdot v(x)\,\mathrm{d}s_x=0\quad\text{for all } v\in W. \tag{2.11}$$

In what follows we will describe a boundary integral approach to express the local Dirichlet–Neumann maps (2.8) by using boundary integral operators, see e.g. [15,16]. Based on mapping properties of local boundary integral operators we show unique solvability of (2.11). Note that the local Dirichlet–Neumann maps can be expressed in terms of local domain bilinear forms for which the unique solvability of (2.11) follows directly based on the corresponding result of (2.1). In fact, using domain bilinear forms in some subdomains Ω_i for $i = 1,\ldots,q < p$, leads to a coupled finite and boundary element formulation.

3. Boundary integral operators

We now assume that for each subdomain Ω_i there *exists a corresponding fundamental solution* $U^i(x,y)$, see [23, Section 2.3] for a general discussion. Then the solution of the local subproblems (2.7) is given by the representation formulae

$$u_i(x) = \int_{\Gamma_i} U^i(x,y)(T_i u)(y)\,\mathrm{d}s_y - \int_{\Gamma_i} T_{i,y} U^i(x,y) u(y)\,\mathrm{d}s_y + \int_{\Omega_i} U^i(x,y) f(y)\,\mathrm{d}y \quad \text{for } x \in \Omega_i. \tag{3.1}$$

Now we define the standard boundary integral operators locally for $x \in \Gamma_i$, the single-layer potential operator

$$(V_i t_i)(x) = \int_{\Gamma_i} U^i(x,y) t_i(y)\,\mathrm{d}s_y, \tag{3.2}$$

the double-layer potential operator

$$(K_i u_i)(x) = \int_{\Gamma_i} T_i U^i(x,y) u_i(y)\,\mathrm{d}s_y \tag{3.3}$$

and the adjoint double-layer potential

$$(K_i' t_i)(x) = \int_{\Gamma_i} T_{i,x} U^i(x,y) t_i(y)\,\mathrm{d}s_y \tag{3.4}$$

as well as the hypersingular integral operator

$$(D_i u_i)(x) = -T_{i,x} \int_{\Gamma_i} T_i U^i(x,y) u_i(y)\,\mathrm{d}s_y. \tag{3.5}$$

The mapping properties of all local boundary integral operators defined above are well known, see e.g. [8,9]. In particular, the boundary integral operators are bounded for $|s| \leqslant 1$:

$$\begin{aligned} V_i &: H^{-1/2+s}(\Gamma_i) \to H^{1/2+s}(\Gamma_i), \\ D_i &: H^{1/2+s}(\Gamma_i) \to H^{-1/2+s}(\Gamma_i), \\ K_i &: H^{1/2+s}(\Gamma_i) \to H^{1/2+s}(\Gamma_i), \\ K_i' &: H^{-1/2+s}(\Gamma_i) \to H^{-1/2+s}(\Gamma_i). \end{aligned}$$

Moreover, without loss of generality, we assume that the local single layer potentials V_i are $H^{-1/2}(\Gamma_i)$–elliptic satisfying

$$\langle V_i w_i, w_i \rangle_{L_2(\Gamma_i)} \geqslant c \; \|w_i\|^2_{H^{-1/2}(\Gamma_i)} \quad \text{for all } w_i \in H^{-1/2}(\Gamma_i). \tag{3.6}$$

The local hypersingular integral operators D_i are assumed to be $H^{1/2}(\Gamma_i)$ semi-elliptic,

$$\langle D_i u_i, u_i \rangle_{L_2(\Gamma_i)} \geqslant c \; \|u_i\|^2_{H^{1/2}(\Gamma_i)} \quad \text{for all } u_i \in H^{1/2}(\Gamma_i)_{/\boldsymbol{R}_i}. \tag{3.7}$$

Here, $\boldsymbol{R}_i$ is the solution space of the local homogeneous Neumann boundary value problems defined by $L_i u_i = 0$ in Ω_i and $T_i u_i = 0$ on Γ_i.

In addition to the boundary integral operators defined above we will use the local Newton potentials given by

$$(N_i f)(x) = \int_{\Omega_i} U^i(x,y) f(y)\, \mathrm{d}y \quad \text{for } x \in \Gamma_i. \tag{3.8}$$

Then, the standard boundary integral equation related to the local partial differential equation in (2.7) is

$$(V_i t_i)(x) = (\tfrac{1}{2}I + K_i)u_i(x) - (N_i f)(x) \quad \text{for } x \in \Gamma_i. \tag{3.9}$$

Since the local single-layer potential operators V_i are assumed to be invertible we can describe the local Dirichlet–Neumann map by

$$t_i(x) = (S_i u_i)(x) - V_i^{-1}(N_i f)(x) \quad \text{for } x \in \Gamma_i, \tag{3.10}$$

using the Steklov–Poincaré operator

$$(S_i u_i)(x) = V_i^{-1}(\tfrac{1}{2}I + K_i)u_i(x) \tag{3.11}$$

$$= [D_i + (\tfrac{1}{2}I + K_i')V_i^{-1}(\tfrac{1}{2}I + K_i)]u_i(x) \tag{3.12}$$

$$= (\tfrac{1}{2}I - K_i')^{-1} D_i u_i(x). \tag{3.13}$$

Hence, the Dirichlet–Neumann map (2.8) can be written as

$$\boldsymbol{T}_i u(x) = (S_i u)(x) - V_i^{-1}(N_i f)(x) \quad \text{for } x \in \Gamma_i. \tag{3.14}$$

Inserting (3.14) into the variational problem (2.11) we get the boundary integral variational formulation: to find $\tilde{u} \in W$ such that

$$\sum_{i=1}^{p} \int_{\Gamma_i} (S_i \tilde{u})(x) v(x)\, \mathrm{d}s_x = \sum_{i=1}^{p} \int_{\Gamma_i} [V_i^{-1}(N_i f)(x) - (S_i \tilde{g})(x)] v(x)\, \mathrm{d}s_x \tag{3.15}$$

holds for all $v \in W$.

Theorem 1 (Carstensen et al. [5], Costabel [7], Hsiao et al. [14], Hsiao and Wendland [16]). *The global boundary integral bilinear form*

$$a(v,w) = \sum_{i=1}^{p} \int_{\Gamma_i} (S_i v)(x) w(x)\, \mathrm{d}s_x \tag{3.16}$$

is bounded in $H^{1/2}(\Gamma_S)$ *and* W*-elliptic, i.e.*,

$$a(v,v) \geqslant c_1^S \cdot \|v\|^2_{H^{1/2}(\Gamma_S)} \quad \text{for all } v \in W. \tag{3.17}$$

Proof. From the mapping properties of the local boundary integral operators we get

$$\|S_i u_i\|_{H^{-1/2}(\Gamma_i)} \leqslant c\|u_i\|_{H^{1/2}(\Gamma_i)} \quad \text{for all } u_i \in H^{1/2}(\Gamma_i).$$

Therefore,

$$\begin{aligned} |a(u,v)| &\leqslant \sum_{i=1}^{p} |\langle S_i u_{|\Gamma_i}, v_{|\Gamma_i}\rangle_{L_2(\Gamma_i)}| \leqslant c \sum_{i=1}^{p} \|u_{|\Gamma_i}\|_{H^{1/2}(\Gamma_i)} \|v_{|\Gamma_i}\|_{H^{1/2}(\Gamma_i)} \\ &\leqslant c \left(\sum_{i=1}^{p} \|u_{|\Gamma_i}\|^2_{H^{1/2}(\Gamma_i)}\right)^{1/2} \left(\sum_{i=1}^{p} \|v_{|\Gamma_i}\|^2_{H^{1/2}(\Gamma_i)}\right)^{1/2} \\ &= c\, \|u\|_{H^{1/2}(\Gamma_S)} \|v\|_{H^{1/2}(\Gamma_S)} \quad \text{for all } u,v \in H^{1/2}(\Gamma_S). \end{aligned}$$

For $u \in W$ we have $u(x) = 0$ for $x \in \Gamma$. Since there is at least one subdomain boundary Γ_{i^*} with $\Gamma_{i^*} \cap \Gamma \neq \emptyset$ we conclude $u_{|\Gamma_{i^*}} \in H^{1/2}(\Gamma_{i^*})_{/\mathbf{R}_i}$. We can repeat this argument recursively to get $u_{|\Gamma_i} \in H^{1/2}(\Gamma_i)_{/\mathbf{R}_i}$ for all $i = 1,\ldots,p$. Hence we have, using the symmetric representation (3.12),

$$\langle S_i u_{|\Gamma_i}, u_{\Gamma_i}\rangle_{L_2(\Gamma_i)} \geqslant \langle D_i u_{|\Gamma_i}, u_{|\Gamma_i}\rangle_{L_2(\Gamma_i)} \geqslant c\, \|u_{|\Gamma_i}\|^2_{H^{1/2}(\Gamma_i)}.$$

Summation over $i = 1,\ldots,p$ gives (3.17). □

With Theorem 1, all assumptions of the Lax–Milgram lemma are satisfied, hence there exists a unique solution $\tilde{u} \in W$ satisfying the variational problem (3.15).

4. Boundary element methods

Let

$$W_h := \text{span}\{\varphi_k\}_{k=1}^{M} \subset W, \tag{4.1}$$

be a boundary element trial space with piecewise polynomial basis functions φ_k of polynomial degree μ. A suitable choice is the use of piecewise linear trial functions with $\mu = 1$. For convenience, we define also local restrictions of W_h onto Γ_i, in particular,

$$W_{h,i} = \text{span}\{\varphi_{k,i}\}_{k=1}^{M_i}. \tag{4.2}$$

Obviously, for any $\varphi_{k,i} \in W_{h,i}$ there exists a unique basis function $\varphi_k \in W_h$ with $\varphi_{k,i} = \varphi_{k|\Gamma_i}$. By using the isomorphisms

$$\underline{u}_i \in \mathbb{R}^{M_i} \leftrightarrow u_{h,i} = \sum_{k=1}^{M_i} u_{i,k}\varphi_{k,i} \in W_{h,i}, \qquad \underline{u} \in \mathbb{R}^{M} \leftrightarrow u_h = \sum_{k=1}^{M} u_k \varphi_k \in W_h,$$

there exist connectivity matrices $A_i \in \mathbb{R}^{M_i \times M}$ such that

$$\underline{u}_i = A_i \underline{u}. \tag{4.3}$$

We assume that there holds an approximation property of W_h in W,

$$\inf_{v_h \in W_h} \|v - v_h\|_{H^{1/2}(\Gamma_S)} \leqslant \left(\sum_{i=1}^{p} h_i^{2s-1} \|v\|^2_{H^s(\Gamma_i)} \right)^{1/2} \tag{4.4}$$

for all $v \in W \cap \prod_{i=1}^{p} H^s(\Gamma_i)$ and $s \leqslant \mu + 1$ where h_i is the local mesh size of the underlying boundary element mesh on Γ_i.

The Galerkin variational problem of (3.15) is to find a boundary element approximation $\tilde{u}_h \in W_h$ satisfying

$$\sum_{i=1}^{p} \int_{\Gamma_i} (S_i \tilde{u}_h)(x) v_h(x)\, \mathrm{d}s_x = \sum_{i=1}^{p} \int_{\Gamma_i} [V_i^{-1}(N_i f)(x) - (S_i \tilde{g})(x)] v_h(x)\, \mathrm{d}s_x \tag{4.5}$$

for all test functions $v_h \in W_h$. This is equivalent to a system of linear equations, $S_h \underline{\tilde{u}} = \underline{f}$, with a stiffness matrix S_h defined by

$$S_h[\ell, k] = \sum_{i=1}^{p} \langle S_i \varphi_k, \varphi_\ell \rangle_{L_2(\Gamma_i)} = \sum_{i=1}^{p} S_{h,i}[\ell, k] \quad \text{for } k, \ell = 1, \ldots, M. \tag{4.6}$$

Since the associated bilinear form is W-elliptic, Cea's lemma provides the quasi-optimal error estimate

$$\|\tilde{u} - \tilde{u}_h\|_{H^{1/2}(\Gamma_S)} \leqslant c \inf_{v_h \in W_h} \|\tilde{u} - v_h\|_{H^{1/2}(\Gamma_S)} \tag{4.7}$$

and, hence, convergence due to the approximation property of $W_h \subset W$. In fact, in order to assemble (4.6) we have to compute the local stiffness matrices defined by

$$S_{h,i}[\ell, k] = \langle S_i \varphi_{k|\Gamma_i}, \varphi_{\ell|\Gamma_i} \rangle_{L_2(\Gamma_i)}, \tag{4.8}$$

using the definition of the local Steklov–Poincaré operators S_i. Note that all of these representations include a composition of different boundary integral operators including some inverse operators as well. Hence, the Galerkin scheme (4.5) cannot be realized exactly in general. Instead, we have to introduce some local approximations $\tilde{S}_i$ leading to a computable scheme yielding almost optimal error estimates as in the exact Galerkin scheme. Therefore we may consider an approximated variational problem to find $\hat{u}_h \in W_h$ satisfying

$$\sum_{i=1}^{p} \int_{\Gamma_i} (\tilde{S}_i \hat{u}_h)(x) v_h(x)\, \mathrm{d}s_x = \sum_{i=1}^{p} \int_{\Gamma_i} [V_i^{-1}(N_i f)(x) - (S_i \tilde{g})(x)] v_h(x)\, \mathrm{d}s_x \tag{4.9}$$

for all test functions $v_h \in W_h$.

Theorem 2. *Let*

$$\tilde{a}(u_h, v_h) = \sum_{i=1}^{p} \int_{\Gamma_i} (\tilde{S}_i u_h)(x) v_h(x)\, \mathrm{d}s_x \tag{4.10}$$

be bounded in $H^{1/2}(\Gamma_S)$ *and* W_h*-elliptic, i.e.*

$$\tilde{a}(v_h, v_h) \geqslant \tilde{c} \cdot \|v_h\|^2_{H^{1/2}(\Gamma_S)} \quad \textit{for all } v_h \in W_h.$$

Then there exists a unique solution of the approximate variational problem (4.9) *satisfying the error estimate*

$$\|\tilde{u} - \hat{u}_h\|_{H^{1/2}(\Gamma_S)} \leqslant c \cdot \left\{ \|\tilde{u} - \tilde{u}_h\|_{H^{1/2}(\Gamma_S)} + \sum_{i=1}^{p} \|(S_i - \tilde{S}_i)u\|_{H^{1/2}(\Gamma_S)} \right\}. \tag{4.11}$$

Note that Theorem 2 is a variant of the first Strang lemma for some perturbation of an elliptic bilinear form, see [6, Theorem 4.1.1].

To define suitable local approximations $\tilde{S}_i$ of the Steklov–Poincaré operators S_i, we first define local trial spaces

$$Z_{h,i} = \text{span}\{\psi_{\tilde{k}}^i\}_{\tilde{k}=1}^{N_i} \subset H^{-1/2}(\Gamma_i) \quad \text{for } i = 1, \ldots, p. \tag{4.12}$$

Again we may use piecewise polynomial trial functions of polynomial degree v, for example trial functions with piecewise constant basis functions where $v = 0$. We assume that for each $Z_{h,i}$ there holds an approximation property:

$$\inf_{\tau_{h,i} \in Z_{h,i}} \|w_i - \tau_{h,i}\|_{H^{-1/2}(\Gamma_i)} \leqslant c h_i^{\sigma+1/2} \|w_i\|_{H^{\sigma}(\Gamma_i)} \tag{4.13}$$

for all $w_i \in H^{\sigma}(\Gamma_i)$ with $\sigma \leqslant v + 1$.

4.1. Symmetric approximation

For an arbitrarily given function $u_i \in H^{1/2}(\Gamma_i)$ the application of the Steklov–Poincaré operator can be written, using the symmetric representation (3.12), as

$$(S_i u_i)(x) = (D_i u_i)(x) + (\tfrac{1}{2}I + K_i')w_i(x) \quad \text{for } x \in \Gamma_i,$$

where w_i satisfies the equation

$$\langle V w_i, \tau_i \rangle_{L_2(\Gamma_i)} = \langle (\tfrac{1}{2}I + K_i)u_i, \tau_i \rangle_{L_2(\Gamma_i)} \quad \text{for all } \tau_i \in H^{-1/2}(\Gamma_i). \tag{4.14}$$

This motivates us to define suitable approximations $\tilde{S}_i$ of the local Steklov–Poincaré operators S_i as follows: The Galerkin discretization of (4.14) is to find $w_{h,i} \in Z_{h,i}$ satisfying

$$\langle V w_{h,i}, \tau_{h,i} \rangle_{L_2(\Gamma_i)} = \langle (\tfrac{1}{2}I + K_i)u_i, \tau_{h,i} \rangle_{L_2(\Gamma_i)} \quad \text{for all } \tau_{h,i} \in Z_{h,i}. \tag{4.15}$$

Applying standard arguments we get by Cea's lemma the quasi-optimal error estimate

$$\|w_i - w_{h,i}\|_{H^{-1/2}(\Gamma_i)} \leqslant c_i \cdot \inf_{\tau_{h,i}} \|w_i - \tau_{h,i}\|_{H^{-1/2}(\Gamma_i)}, \tag{4.16}$$

yielding convergence by the approximation property of the trial space $Z_{h,i}$. Now we can define an approximate Steklov–Poincaré operator as

$$(\tilde{S}_i u_i)(x) := (D_i u_i)(x) + (\tfrac{1}{2}I + K_i')w_{h,i}(x) \quad \text{for } x \in \Gamma_i. \tag{4.17}$$

Note that from (4.17) with (4.16) we get

$$\|(S_i - \tilde{S}_i)u_i\|_{H^{-1/2}(\Gamma_i)} \leqslant \|w_i - w_{i,h}\|_{H^{-1/2}(\Gamma_i)}. \tag{4.18}$$

In case of the symmetric approximation (4.17) of the local Steklov–Poincaré operators S_i the following theorem is valid, see also [1,14,24].

Theorem 3. *Let the approximated bilinear form* (4.10) *be defined by the use of the symmetric approximation* (4.17) *of the local Steklov–Poincaré operators* S_i. *Then it follows that the assumptions of Theorem 2 are satisfied, and in particular, there holds the quasi-optimal error estimate*

$$\|\tilde{u}-\hat{u}_h\|_{H^{1/2}(\Gamma_S)} \leqslant c\left\{\inf_{v_h\in W_h}\|\tilde{u}-v_h\|_{H^{1/2}(\Gamma_S)}+\sum_{i=1}^{p}\inf_{\tau_{h,i}\in Z_{h,i}}\|S_i\tilde{u}_i-\tau_{h,i}\|_{H^{-1/2}(\Gamma_i)}\right\}. \tag{4.19}$$

Proof. From (4.15) we conclude the stability estimate

$$\|w_{h,i}\|_{H^{-1/2}(\Gamma_i)} \leqslant c\cdot\|u_i\|_{H^{1/2}(\Gamma_i)}$$

and therefore

$$\begin{aligned}\|\tilde{S}_i u_i\|_{H^{-1/2}(\Gamma_i)} &\leqslant \|D_i u_i\|_{H^{-1/2}(\Gamma_i)} + \|(\tfrac{1}{2}I+K_i')w_{h,i}\|_{H^{-1/2}(\Gamma_i)}\\ &\leqslant c\{\|u_i\|_{H^{1/2}(\Gamma_i)}+\|w_{h,i}\|_{H^{-1/2}(\Gamma_i)}\}\leqslant c\|u_i\|_{H^{1/2}(\Gamma_i)}.\end{aligned}$$

Hence, for $u,v\in W$ we have, with the help of the Cauchy–Schwarz inequality,

$$\begin{aligned}|\tilde{a}(u,v)| &\leqslant \sum_{i=1}^{p}|\langle\tilde{S}_i u,v\rangle_{L_2(\Gamma_i)}| \leqslant \sum_{i=1}^{p}\|\tilde{S}_i u\|_{H^{-1/2}(\Gamma_i)}\|v\|_{H^{1/2}(\Gamma_i)}\\ &\leqslant c\sum_{i=1}^{p}\|u\|_{H^{1/2}(\Gamma_i)}\|v\|_{H^{1/2}(\Gamma_i)}\leqslant c\|u\|_{H^{1/2}(\Gamma_S)}\|v\|_{H^{1/2}(\Gamma_S)}\end{aligned}$$

and therefore the boundedness of $\tilde{a}(\cdot,\cdot)$. Since the local single-layer potentials V_i are $H^{-1/2}(\Gamma_i)$-elliptic, this gives with (4.15)

$$\begin{aligned}\langle\tilde{S}_i v,v\rangle_{L_2(\Gamma_i)} &= \langle D_i v,v\rangle_{L_2(\Gamma_i)} + \langle(\tfrac{1}{2}I+K_i')w_{h,i},v\rangle_{L_2(\Gamma_i)}\\ &= \langle D_i v,v\rangle_{L_2(\Gamma_i)} + \langle V w_{h,i},w_{h,i}\rangle_{L_2(\Gamma_i)} \geqslant \langle D_i v,v\rangle_{L_2(\Gamma_i)}\end{aligned}$$

and therefore

$$\tilde{a}(v,v)\geqslant\sum_{i=1}^{p}\langle D_i v,v\rangle_{L_2(\Gamma_i)}.$$

Hence, the W-ellipticity of $\tilde{a}(\cdot,\cdot)$ follows from the mapping properties of the assembled local hypersingular integral operators D_i. Now we can apply Theorem 2 to get the error estimate (4.11). Finally, (4.19) follows from (4.7), (4.18) and (4.16). □

Note that in the symmetric approximation case the assumptions of Theorem 2 and therefore Theorem 3 hold without any restrictions on the definition of the trial spaces W_h and $Z_{h,i}$, only approximation properties have to be assumed. It turns out, to guarantee an optimal order of convergence, that the polynomial degree of the local trial spaces $Z_{h,i}$ should be chosen one degree less than the polynomial degree of the global trial space W_h. For example, one may use piecewise linear basis functions to define W_h while we can take piecewise constant trial functions for describing $Z_{h,i}$.

According to the symmetric approximation (4.17) of the local Steklov–Poincaré operators S_i we define local stiffness matrices as

$$D_{h,i}[\ell,k] = \langle D_i\varphi_{k,i},\varphi_{\ell,i}\rangle_{L_2(\Gamma_i)}, \quad K_{h,i}[\tilde{\ell},k] = \langle K_i\varphi_{k,i},\psi_{\tilde{\ell},i}\rangle_{L_2(\Gamma_i)},$$
$$V_{h,i}[\tilde{\ell},\tilde{k}] = \langle V_i\psi_{\tilde{k},i},\psi_{\tilde{\ell},i}\rangle_{L_2(\Gamma_i)}, \quad M_{h,i}[\tilde{\ell},k] = \langle \varphi_{k,i},\psi_{\tilde{\ell},i}\rangle_{L_2(\Gamma_i)}$$

for $k,\ell = 1,\ldots,M_i$ and $\tilde{k},\tilde{\ell} = 1,\ldots,N_i$. Then, the Galerkin discretization of the approximate Steklov–Poincaré operator $\tilde{S}_i$ reads as

$$\tilde{S}_{h,i} = D_{h,i} + (\tfrac{1}{2}M_{h,i}^{\top} + K_{h,i}^{\top})V_{h,i}^{-1}(\tfrac{1}{2}M_{h,i} + K_{h,i}) \quad \text{for } i = 1,\ldots,p. \tag{4.20}$$

Hence, the approximated Galerkin formulation (4.9) is equivalent to the system of linear equations given by

$$\tilde{S}_h\underline{u} := \sum_{i=1}^{p} A_i^{\top}\tilde{S}_{h,i}A_i\underline{\hat{u}} = \sum_{i=1}^{p} A_i^{\top}\underline{f}_i =: \underline{f} \tag{4.21}$$

with the connectivity matrices A_i as introduced in (4.3) and with local vectors $\underline{f}_i$ defined by

$$f_{i,k} = \langle V_i^{-1}N_i f - S_i\tilde{g},\varphi_{k,i}\rangle_{L_2(\Gamma_i)} \quad \text{for } k = 1,\ldots,M_i; i = 1,\ldots,p.$$

The stiffness matrix $\tilde{S}_h$ in (4.21) is symmetric and positive definite, hence we can use a standard preconditioned conjugate gradient scheme in parallel to solve (4.21) efficiently. The construction of appropriate preconditioning techniques will be discussed later in Section 5.

Defining

$$D_h = \sum_{i=1}^{p} A_i^{\top}D_{h,i}A_i, \quad K_h = \sum_{i=1}^{p} K_{h,i}A_i,$$
$$V_h = \operatorname{diag}(V_{h,i})_{i=1}^{p}, \quad M_h = \sum_{i=1}^{p} M_{h,i}A_i,$$

the linear system (4.21) can be written as a block system of the form

$$\begin{pmatrix} V_h & -\tfrac{1}{2}M_h - K_h \\ \tfrac{1}{2}M_h^{\top} + K_h^{\top} & D_h \end{pmatrix}\begin{pmatrix} \underline{w} \\ \underline{\hat{u}} \end{pmatrix} = \begin{pmatrix} \underline{0} \\ \underline{f} \end{pmatrix}. \tag{4.22}$$

Note that the stiffness matrix in (4.22) is either block skew–symmetric and positive definite, or by simple manipulations, symmetric but indefinite. Hence, for the iterative solution of (4.22) one may use any appropriate solver such as BiCGStab or GMRES applicable to nonsymmetric or indefinite systems. Instead, following [3,5] one can transform (4.22) into a symmetric and positive-definite system. Let $C_{V,i}$ be local preconditioning matrices for the discrete single-layer potential operators satisfying the spectral equivalence inequalities

$$c_1^{V_i}(C_{V,i}\underline{w}_i,\underline{w}_i) \leqslant (V_{h,i}\underline{w}_i,\underline{w}_i) \leqslant c_2^{V_i}(C_{V,i}\underline{w}_i,\underline{w}_i) \quad \text{for all } \underline{w}_i \in \mathbb{R}^{N_i} \tag{4.23}$$

with positive constants $c_1^{V_i}$ and $c_2^{V_i}$. In addition, we assume $c_1^{V_i} > 1$. This can be accomplished in general by some scaling of the preconditioning matrices $C_{V,i}$. Defining $C_V = \operatorname{diag}(C_{V_i})_{i=1}^{p}$ we then obtain the spectral equivalence inequalities

$$c_1^V(C_V\underline{w},\underline{w})\leqslant(V_h\underline{w},\underline{w})\leqslant c_2^V(C_V\underline{w},\underline{w})\quad\text{for all }\underline{w}\in\mathbb{R}^N \tag{4.24}$$

with $N=\sum_{i=1}^p N_i$ and positive constants

$$1<c_1^V:=\min_{1\leqslant i\leqslant p}c_1^{V_i},\quad c_2^V:=\max_{1\leqslant i\leqslant p}c_2^{V_i}.$$

Due to the assumption $c_1^V>1$, instead of (4.22), we may solve the transformed linear system

$$\begin{pmatrix} V_hC_V^{-1}-I & 0\\ -(\frac{1}{2}M_h^\top+K_h^\top)C_V^{-1} & I\end{pmatrix}\begin{pmatrix} V_h & -\frac{1}{2}M_h-K_h\\ \frac{1}{2}M_h^\top+K_h^\top & D_h\end{pmatrix}\begin{pmatrix}\underline{w}\\ \underline{\hat{u}}\end{pmatrix}$$
$$=\begin{pmatrix} V_hC_V^{-1}-I & 0\\ -(\frac{1}{2}M_h^\top+K_h^\top)C_V^{-1} & I\end{pmatrix}\begin{pmatrix}\underline{0}\\ \underline{f}\end{pmatrix}. \tag{4.25}$$

It turns out, see [3] for details, that the transformed stiffness matrix in (4.25) is now symmetric and positve definite. Hence we can use the preconditioned conjuate gradient scheme to solve (4.25) efficiently.

4.2. Hybrid approximation techniques

Instead of the symmetric approximation (4.17) based on the symmetric representation (3.12) one may use any other boundary element approximation of the local Steklov–Poincaré operators S_i as for example, the local representations (3.11) or (3.13). Following [27,33] we will describe a non-symmetric and a so-called "hybrid" boundary element scheme by discretizing the Steklov–Poincaré operator representation (3.11) (see also [10]).

For an arbitrarily given function $u_i\in H^{1/2}(\Gamma_i)$, the application of the Steklov–Poincaré operator S_i in view of (3.11) reads as

$$(S_iu_i)(x)=w_i(x)\quad\text{for }x\in\Gamma_i,$$

where w_i is, as in the symmetric approximation, the unique solution of the variational problem (4.14). As in (4.15) we can define a corresponding Galerkin solution $w_{h,i}\in Z_{h,i}$. Therefore, an approximate Steklov–Poincaré operator is here given by

$$(\tilde{S}u_i)(x)=w_{h,i}(x)\quad\text{for }x\in\Gamma_i,\ i=1,\dots,p. \tag{4.26}$$

Obviously, the error estimate (4.18) for $\|(S-\tilde{S})u_i\|_{H^{-1/2}(\Gamma_i)}$ remains valid. As in the proof of Theorem 3 we can conclude that the bilinear form $\tilde{a}(\cdot,\cdot)$ defined by the local approximations (4.26) is bounded in $H^{1/2}(\Gamma_S)$.

Theorem 4. *Let H_i be the local mesh size of the trial space W_h while h_i is the local mesh size of $Z_{h,i}$ respectively. Let the inverse inequality in W_h be valid locally,*

$$\|w_{h|\Gamma_i}\|_{H^s(\Gamma_i)}\leqslant c\,H_i^{1/2-s}\|w_{h,i}\|_{H^{1/2}(\Gamma_i)}. \tag{4.27}$$

If $h_i\leqslant c_{0,i}H_i$ is satisfied with positive, sufficiently small constants $c_{0,i}\leqslant 1$, then the bilinear form $\tilde{a}(\cdot,\cdot)$ defined by the approximation (4.26) is W_h-elliptic.

Proof. For $v_h \in W_h$ we have by (3.17), (4.18), (4.16) and the inverse inequality, for $s \leqslant v+1$,

$$\begin{aligned}
c_i^S \|v_h\|^2_{H^{1/2}(\Gamma_S)} &\leqslant \sum_{i=1}^p \langle S_i v_h, v_h\rangle_{L_2(\Gamma_i)} \\
&\leqslant \sum_{i=1}^p \langle \tilde{S}_i v_h, v_h\rangle_{L_2(\Gamma_i)} + \sum_{i=1}^p \langle (S_i - \tilde{S}_i) v_h, v_h\rangle_{L_2(\Gamma_i)} \\
&\leqslant \sum_{i=1}^p \langle \tilde{S}_i v_h, v_h\rangle_{L_2(\Gamma_i)} + \sum_{i=1}^p \|(S_i - \tilde{S}_i) v_h\|_{H^{-1/2}(\Gamma_i)} \|v_h\|_{H^{1/2}(\Gamma_i)} \\
&\leqslant \sum_{i=1}^p \langle \tilde{S}_i v_h, v_h\rangle_{L_2(\Gamma_i)} + \sum_{i=1}^p c_i h_i^{s+1/2} \|v_h\|_{H^{s+1}(\Gamma_i)} \|v_h\|_{H^{1/2}(\Gamma_i)} \\
&\leqslant \sum_{i=1}^p \langle \tilde{S}_i v_h, v_h\rangle_{L_2(\Gamma_i)} + \sum_{i=1}^p \tilde{c}_i (h_i/H_i)_i^{s+1/2} \|v_h\|^2_{H^{1/2}(\Gamma_i)}
\end{aligned}$$

Hence, if $\tilde{c}_i(h_i/H_i)^{s+1/2} \leqslant c_1^S/2$ is satisfied the theorem is proved. □

When using the approximation $\tilde{S}_i$ as defined in (4.26) then the local Galerkin discretization is given by

$$\tilde{S}_{h,i} = M_{h,i}^\top V_{h,i}^{-1}(\tfrac{1}{2}M_{h,i} + K_{h,i}), \tag{4.28}$$

while the global system is given as in (4.21) by

$$\tilde{S}_h \underline{\hat{u}} = \sum_{i=1}^p A_i^\top \tilde{S}_{h,i} A_i \underline{\hat{u}} = \underline{f}. \tag{4.29}$$

The assembled stiffness matrix $\tilde{S}_h$ is still positive definite but, in general, not symmetric. Therefore, we recommend a suitable preconditioned BiCGStab or GMRES algorithm for an efficient solution strategy. Moreover, the local stiffness matrices $\tilde{S}_{h,i}$ as given in (4.28) are nonsymmetric perturbations of an originally symmetric stiffness matrix $S_{h,i}$. To keep the symmetry in the approximation of local Steklov–Poincaré operators, which is important when coupling boundary elements with a symmetric finite element scheme, one can introduce a modified hybrid discretization scheme [10,28]. That is again based on the representation (3.11) but on the formulation of the local Steklov–Poincaré operator S_i as

$$S_i = V_i^{-1}(\tfrac{1}{2}I + K_i)V_i V_i^{-1} = V_i^{-1} F_i V_i^{-1} \tag{4.30}$$

with the self-adjoint and computable operator

$$F_i = (\tfrac{1}{2}I + K_i)V_i. \tag{4.31}$$

As before, we can introduce an appropriate approximation of S_i, now based on representation (4.30). Then, the local Galerkin discretization is given by

$$\tilde{S}_{h,i} = M_{h,i}^\top V_{h,i}^{-1} F_{h,i} V_{h,i}^{-1} M_{h,i}, \tag{4.32}$$

which is now a symmetric and positive-definite matrix provided $F_{h,i}$ can be computed accurately. We remark that the computation of

$$F_{h,i}[\tilde{\ell}, \tilde{k}] = \langle F_i \psi_{\tilde{k},i}, \psi_{\tilde{\ell},i}\rangle_{L_2(\Gamma_i)}$$

Table 1
Errors for the boundary element solution

M	Case i		Case ii		Case iii	
	N	$\|\|u-u_h\|\|_{L^2}$	N	$\|\|u-u_h\|\|_{L^2}$	N	$\|\|u-u_h\|\|_{L^2}$
32	64	2.04 − 2	64	2.19 − 2	38	1.74 − 2
64	128	5.10 − 3	128	5.41 − 3	70	4.37 − 3
128	256	1.28 − 3	256	1.35 − 3	134	1.11 − 3
256	512	3.20 − 4	512	3.36 − 4	262	2.79 − 4
512	1024	8.02 − 5	1024	8.40 − 5	518	7.02 − 5

for $\tilde{k},\tilde{\ell}=1,\ldots,N_i$ requires the evaluation of two boundary integral operators per matrix element. To ensure stability of the hybrid discretization scheme (4.32) we have to assume the stability assumption,

$$c\|v_{h,i}\|_{H^{1/2}(\Gamma_i)} \leqslant \sup_{w_{h,i}\in Z_{h,i}} \frac{|\langle v_{h,i}, w_{h,i}\rangle_{L_2(\Gamma_i)}|}{\|w_{h,i}\|_{H^{-1/2}(\Gamma_i)}} \tag{4.33}$$

for all $v_{h,i} \in W_{h,i}$, see [28] for details. In fact, for a local trial space $W_{h,i}$ we have to define trial spaces $Z_{h,i}$ in such a way that (4.33) is satisfied. Note that for a given $W_{h,i}$, the construction of $Z_{h,i}$ is not unique. We will describe three possible choices of $Z_{h,i}$ for the case that $W_{h,i}$ is spanned by piecewise linear continuous basis functions, see also [26].

i. *Mesh refinement.* As in Theorem 4 we can define $Z_{h,i}$ by using piecewise constant basis functions with respect to a sufficiently refined boundary element mesh compared with the underlying mesh of $W_{h,i}$. In this case we have to assume an inverse inequality, see (4.27). Therefore, this approach is applicable for quasi-uniform boundary element meshes only. For more details, see e.g. [14,28,33].
ii. *Iso-parametric trial functions.* We first consider the case $\tilde{Z}_{h,i}=W_{h,i}$. Then the stability property (4.33) is strongly related to the stability of the corresponding L_2 projection Q_h onto $W_{h,i}$ in $H^{1/2}(\Gamma)$. The latter holds for a rather large class of nonuniform refinements based on adaptive strategies provided that certain local conditions are satisfied. We refer to [27] for a detailed discussion. Now we define $Z_{h,i}$ to be the trial space of piecewise linear but discontinuous basis functions. Obviously, $\tilde{Z}_{h,i}\subset Z_{h,i}$ and the stability condition (4.33) remains valid.
iii. *Nonmatching boundary meshes.* In both cases described above, the definition of $Z_{h,i}$ requires the use of appropriate trial functions satisfying (4.33) which implies a significant growth of the dimension N_i of the trial space $Z_{h,i}$. In view of (4.19), the optimal choice seems to be, to define $Z_{h,i}$ by piecewise constant basis functions where the mesh size of $W_{h,i}$ and $Z_{h,i}$ is almost equal. However, it is not possible to define $Z_{h,i}$ with respect to the same boundary element mesh as $W_{h,i}$, since then the corresponding mass matrix $M_{h,i}$ would become singular. Instead we can define $Z_{h,i}$ with respect to the mesh dual to that of $W_{h,i}$. In this case, (4.33) is satisfied again, also for nonuniform boundary element meshes; for a further discussion see [26,29].

For comparison we consider a simple numerical example. Let Ω be an L-shaped domain with boundary Γ. We solve a mixed boundary value problem in one subdomain by using approximation (4.26). In Table 1 we give the approximation errors for the boundary element solution according to Theorem 2 while in Table 2 we give the errors of the approximations of the Steklov–Poincaré operator. In

Table 2
Errors for the approximation of the Steklov–Poincaré operator

M	Case i		Case ii		Case iii	
	N	$\|(S-\tilde{S})u\|_{L^2}$	N	$\|(S-\tilde{S})u\|_{L^2}$	N	$\|(S-\tilde{S})u\|_{L^2}$
32	64	4.24 − 1	64	3.20 − 1	38	6.41 − 1
64	128	1.84 − 1	128	9.87 − 2	70	3.30 − 1
128	256	8.70 − 2	256	2.74 − 2	134	1.67 − 1
256	512	4.27 − 2	512	7.33 − 3	262	8.42 − 2
512	1024	2.13 − 2	1024	1.99 − 3	518	4.23 − 2

both the tables, M is the number of all boundary nodes while N is the degree of freedom needed for the definition of $\tilde{S}$. Note that with respect to both, computational work as well as accuracy, the approach based on the dual mesh is favourable.

5. Preconditioning techniques

For the iterative solution of linear systems (4.21) or (4.22) resulting from the symmetric approximation or (4.29) in case of the nonsymmetric approximation, we need to use some appropriate preconditioning techniques to reduce the number of iterations needed. In particular, we assume that there are given local preconditioning matrices $C_{V,i}$ satisfying the spectral equivalence inequalities

$$\gamma_1^{V_i}(C_{V,i}\underline{w}_i,\underline{w}_i)\leqslant(V_{h,i}\underline{w}_i,\underline{w}_i)\leqslant\gamma_2^{V_i}(C_{V,i}\underline{w}_i,\underline{w})\quad\text{for all } \underline{w}_i\in\mathbb{R}^{N_i} \tag{5.1}$$

and $i=1,\dots,p$, as well as a global preconditioning matrix C_S satisfying

$$\gamma_1^S(C_S\underline{u},\underline{u})\leqslant(\tilde{S}_h\underline{u},\underline{u})\leqslant\gamma_2^S(C_S\underline{u},\underline{u})\quad\text{for all } \underline{u}\in\mathbb{R}^M. \tag{5.2}$$

5.1. Local preconditioners

To define local preconditioners $C_{V,i}$ for the local single-layer potential operators V_i satisfying (5.1), one can apply different strategies. One approach is based on the use of geometrically similar and rotational symmetric domains which leads to block circulant matrices which can be used as local preconditioners [20]. Here, a proper ordering of the degrees of freedom has to be assumed. A classical approach, as in finite element methods, is the use of multigrid preconditioners for the local single-layer potentials, which are operators of order −1 [2]. Another strategy is the use of multilevel methods such as additive or multiplicative Schwarz methods [19]. However, in both multigrid and multilevel approaches a suitable mesh hierarchy has to be assumed. Here we will describe an approach [18,31] which neither requires a proper ordering of the degrees of freedom nor a given mesh hierarchy. From the mapping properties of the local single-layer potential operators V_i we get the spectral equivalence inequalities

$$c_1^{V_i}\|w_i\|^2_{H^{-1/2}(\Gamma_i)}\leqslant\langle Vw_i,w_i\rangle_{L_2(\Gamma_i)}\leqslant c_2^{V_i}\|w_i\|^2_{H^{-1/2}(\Gamma_i)} \tag{5.3}$$

for all $w_i\in H^{-1/2}(\Gamma_i)$. On the other hand, there hold the spectral equivalence inequalities

$$c_1^{D_i}\|u_i\|^2_{H^{1/2}(\Gamma_i)}\leqslant\langle Du_i,u_i\rangle_{L_2(\Gamma_i)}\leqslant c_2^{D_i}\|u_i\|^2_{H^{1/2}(\Gamma_i)} \tag{5.4}$$

for all $u_i \in H^{1/2}(\Gamma_i)_{/\mathbf{R}_i}$. Hence, it follows that

$$I + D_i : H^{1/2}(\Gamma_i) \to H^{-1/2}(\Gamma_i)$$

is bounded and $H^{1/2}(\Gamma_i)$-elliptic. Therefore, with (5.3), the spectral equivalence inequalities

$$\gamma_1^i \langle (I + D_i)^{-1} w_i, w_i \rangle_{L_2(\Gamma_i)} \leqslant \langle V_i w_i, w_i \rangle_{L_2(\Gamma_i)} \leqslant \gamma_2^i \langle (I + D_i)^{-1} w_i, w_i \rangle_{L_2(\Gamma_i)} \tag{5.5}$$

hold for all $w_i \in H^{-1/2}(\Gamma_i)$. For the preconditioning matrix C_{V_i} defined by

$$C_{V_i}[\tilde{\ell}, \tilde{k}] = \langle (I + D_i)^{-1} \psi_{\tilde{k},i}, \psi_{\tilde{\ell},i} \rangle_{L_2(\Gamma_i)} \tag{5.6}$$

for $\tilde{k}, \tilde{\ell} = 1, \ldots, N_i$, the spectral equivalence inequalities (5.1) then follow from (5.5) with the positive constants $c_1^{V_i} = \gamma_1^i, c_2^{V_i} = \gamma_2^i$. Similar as for the Steklov–Poincaré operators S_i, in general one is not able to compute the matrix elements (5.6) directly. Instead we use an approximation

$$\tilde{C}_{V_i} = \bar{M}_{h,i}^{\top} (\tilde{M}_{h,i} + \tilde{D}_{h,i})^{-1} \bar{M}_{h,i} \tag{5.7}$$

in terms of the local matrices

$$\tilde{D}_{h,i}[\tilde{\ell}, \tilde{k}] = \langle D_i \tilde{\varphi}_{\tilde{k},i}, \tilde{\varphi}_{\tilde{\ell},i} \rangle_{L_2(\Gamma_i)},$$
$$\tilde{M}_{h,i}[\tilde{\ell}, \tilde{k}] = \langle \tilde{\varphi}_{\tilde{k},i}, \tilde{\varphi}_{\tilde{\ell},i} \rangle_{L_2(\Gamma_i)},$$
$$\bar{M}_{h,i}[\tilde{\ell}, \tilde{k}] = \langle \tilde{\varphi}_{\tilde{k},i}, \psi_{\tilde{\ell},i} \rangle_{L_2(\Gamma_i)}$$

where $\tilde{W}_{h,i} := \text{span}\{\tilde{\varphi}_{\tilde{k},i}\}_{\tilde{k}=1}^{N_i} \subset H^{1/2}(\Gamma_i)$ is an appropriate trial space to be used for the discretization of the local hypersingular integral operators D_i. As it was shown in [31], there holds the upper estimate

$$(\tilde{C}_{V_i} \underline{w}_i, \underline{w}_i) \leqslant (C_{V_i} \underline{w}_i, \underline{w}_i) \quad \text{for all } \underline{w}_i \in \mathbb{R}^{N_i}. \tag{5.8}$$

Theorem 5 (Steinbach and Wendland [31]). *Assume the stability condition*

$$c_0 \|u_{h,i}\|_{H^{1/2}(\Gamma)_i} \leqslant \sup_{w_{h,i} \in Z_{h,i}} \frac{|\langle w_{h,i}, u_{h,i} \rangle_{L_2(\Gamma_i)}|}{\|w_{h,i}\|_{H^{-1/2}(\Gamma_i)}} \quad \textit{for all } u_{h,i} \in \tilde{W}_{h,i}. \tag{5.9}$$

Then,

$$\gamma_0 (C_{V_i} \underline{w}_i, \underline{w}_i) \leqslant (\tilde{C}_{V_i} \underline{w}_i, \underline{w}_i) \quad \textit{for all } \underline{w}_i \in \mathbb{R}^{N_i}. \tag{5.10}$$

Note that the stability condition (5.9) is similar to the stability condition (4.33) needed in hybrid discretizations of the Steklov–Poincaré operators locally. Since (5.9) ensures the invertibility of the mass matrix $\bar{M}_{h,i}$, as a consequence we have from (5.7)

$$\tilde{C}_{V_i}^{-1} = \bar{M}_{h,i}^{-1} (\tilde{M}_{h,i} + \tilde{D}_{h,i}) \bar{M}_{h,i}^{-\top}. \tag{5.11}$$

5.2. Parallel preconditioners

To construct a global preconditioning matrix C_S satisfying the spectral equivalence inequalities (5.2) we first note that there hold the spectral equivalence inequalities

$$c_1^{S_i} (S_{h,i} \underline{u}_i, \underline{u}_i) \leqslant (\tilde{S}_{h,i} \underline{u}_i, \underline{u}_i) \leqslant c_2^{S_i} (S_{h,i} \underline{u}_i, \underline{u}_i) \tag{5.12}$$

for all $\underline{u}_i \in \mathbb{R}^{M_i}$. In case of the symmetric approximation given by (4.20), (5.12) follows from Theorem 3, since the assumptions of Theorem 2 are satisfied. When using either the nonsymmetric approximation (4.28) or the hybrid approximation (4.32) we need to assume (4.33) to ensure (5.12). Hence, instead of (5.2) it is sufficient to construct a global preconditioning matrix C_S which is spectrally equivalent to the global bilinear form (3.16) [17]. Moreover, since the local Steklov–Poincaré operators S_i are spectrally equivalent to the local hypersingular integral operators D_i, we need only to find a preconditioning matrix for the modified bilinear form

$$\tilde{a}(u,v) := \sum_{i=1}^{p} \langle D_i u_{|\Gamma_i}, v_{|\Gamma_i} \rangle_{L_2(\Gamma_i)} \quad \text{for } u,v \in W. \tag{5.13}$$

When using the symmetric approximation (4.20), the local Galerkin discretization of the hypersingular integral operators is already computed. Hence, the action of the preconditioner can be defined by the solution $\underline{v}$ of

$$\sum_{i=1}^{p} A_i^\top D_{h,i} A_i \underline{v} = \underline{r} \tag{5.14}$$

by any available efficient method, this defines an optimal preconditioning strategy. For example, we can use a standard multigrid scheme as in [22] for the hypersingular integral operator to solve (5.14) in parallel, see for example [5]. Alternatively, we may solve (5.14) approximately by some suitable iterative scheme using some appropriate preconditioning strategy for the assembled Galerkin matrix D_h. Again we can use multigrid or multilevel preconditioners, or some additive Schwarz methods as described in [4] (for an application of the latter case, see [30]).

References

[1] M. Bonnet, G. Maier, C. Polizzotto, Symmetric Galerkin boundary element methods, Appl. Mech. Rev. 51 (1998) 669–704.

[2] J.H. Bramble, Z. Leyk, J.E. Pasciak, The analysis of multigrid algorithms for pseudodifferential operators of order minus one, Math. Comp. 63 (1994) 461–478.

[3] J.H. Bramble, J.E. Pasciak, A preconditioning technique for indefinite systems resulting from mixed approximations of elliptic systems, Math. Comp. 50 (1988) 1–17.

[4] J.H. Bramble, J.E. Pasciak, A.H. Schatz, The construction of preconditioners for elliptic problems by substructuring I, Math. Comp. 47 (1987) 103–134.

[5] C. Carstensen, M. Kuhn, U. Langer, Fast parallel solvers for symmetric boundary element domain decomposition methods, Numer. Math. 79 (1998) 321–347.

[6] P.G. Ciarlet, The Finite Element Method for Elliptic Problems, North-Holland, Amsterdam, 1978.

[7] M. Costabel, Symmetric methods for the coupling of finite and boundary elements, in: C.A. Brebbia et al. (Eds.), Boundary Elements IX, Springer, Berlin, 1987, pp. 411–420.

[8] M. Costabel, Boundary integral operators on Lipschitz domains: Elementary results, SIAM J. Math. Anal. 19 (1988) 613–626.

[9] M. Costabel, W.L. Wendland, Strong ellipticity of boundary integral operators, Crelle's J. Reine Angew. Math. 372 (1986) 34–63.

[10] C. Fiedler, L. Gaul, The hybrid boundary element method in elastodynamics, Mech. Res. Comm. 23 (1996) 3074–3079.

[11] G.N. Gatica, G.C. Hsiao, Boundary-Field Equation Methods for a Class of Nonlinear Problems, Pitman Research Notes in Mathematics Series, Vol. 331, Longman, London, 1995.

[12] N. Heuer, E.P. Stephan, Iterative substructuring for hypersingular integral equations in $\mathbb{R}^3$, SIAM J. Sci. Comput. 20 (1999) 739–749.
[13] G.C. Hsiao, E. Schnack, W.L. Wendland, A hybrid coupled finite-boundary element method in elasticity, Comput. Methods Appl. Mech. Eng. 173 (1999) 287–316.
[14] G.C. Hsiao, E. Schnack, W.L. Wendland, Hybrid coupled finite-boundary element methods for elliptic systems of second order, Comput. Mechanics Advances, to appear.
[15] G.C. Hsiao, W.L. Wendland, Domain decomposition methods in boundary element methods, in: R. Glowinski et al. (Eds.), Proceedings of the Fourth International Conference on Domain Decomposition Methods, SIAM, Philadelphia, PA, 1990, pp. 41–49.
[16] G.C. Hsiao, W.L. Wendland, Domain decomposition via boundary element methods, in: H. Alder et al. (Eds.), Numerical Methods in Engineering and Applied Sciences I, CIMNE, Barcelona, 1992, pp. 198–207.
[17] B.N. Khoromskij, W.L. Wendland, Spectrally equivalent preconditioners in substructuring techniques, East–West J. Numer. Math. 1 (1992) 1–25.
[18] W. McLean, T. Tran, A preconditioning strategy for boundary element Galerkin methods, Numer. Methods Partial Differential Equations 13 (1997) 283–301.
[19] M. Maischak, E.P. Stephan, T. Tran, Domain decomposition for integral equations of the first kind: numerical results, Appl. Anal. 63 (1997) 111–132.
[20] A. Meyer, S. Rjasanow, An effective direct solution method for certain boundary element equations in 3D, Math. Methods Appl. Sci. 13 (1990) 43–53.
[21] P. Mund, E.P. Stephan, J. Weisse, Two-level methods for the single layer potential in $\mathbb{R}^3$, Computing 60 (1998) 243–266.
[22] T. von Petersdorff, E.P. Stephan, On the convergence of the multigrid method for a hypersingular integral equation of the first kind, Numer. Math. 57 (1990) 379–391.
[23] A. Pomp, The Boundary-Domain Integral Method for Elliptic Systems. With an Application to Shells, Lecture Notes in Mathematics, Vol. 1683, Springer, Berlin, 1998.
[24] G. Schmidt, Boundary element discretization of Poincaré–Steklov operators, Numer. Math. 69 (1994) 83–101.
[25] H.A. Schwarz, Über einige Abbildungsaufgaben, Ges. Math. Abh. 11 (1869) 65–83.
[26] O. Steinbach, Stable boundary element approximations of Steklov–Poincaré operators, in: M. Bonnet et al. (Eds.), Mathematical Aspects of Boundary Element Methods, Research Notes in Mathematics, Vol. 414, Chapman & Hall, London, 1999, pp. 296–305.
[27] O. Steinbach, On the stability of the L_2 projection in fractional Sobolev spaces, Numer. Math., to appear.
[28] O. Steinbach, On a hybrid boundary element method, Numer. Math. 84 (2000) 679–695.
[29] O. Steinbach, On a generalized L_2 projection and some related stability conditions, submitted.
[30] O. Steinbach, W.L. Wendland, Hierarchical boundary element preconditioners in domain decomposition methods, in: P. Bjorstad, M.S. Espedal, D.E. Keyes (Eds.), Domain Decomposition Methods in Sciences and Engineering, Ninth International Conference, Bergen, Norway, Domain Decomposition Press, Bergen, 1998, pp. 497–503.
[31] O. Steinbach, W.L. Wendland, The construction of some efficient preconditioners in the boundary element method, Adv. Comput. Math. 9 (1998) 191–216.
[32] E.P. Stephan, T. Tran, Additive Schwarz method for the h-version boundary element method, Appl. Anal. 60 (1996) 63–84.
[33] W.L. Wendland, On asymptotic error estimates for combined BEM and FEM, in: E. Stein, W.L. Wendland (Eds.), Finite Element and Boundary Element Techniques from Mathematical and Engineering Point of View, CISM Courses and Lectures, Vol. 301, Springer, Berlin, 1988, pp. 273–333.
[34] W.L. Wendland, On boundary integral equations and applications, Tricomi's Ideas and Contemporary Applied Mathematics, Vol. 147, Accd. Naz. dei Lincei. Atti dei Covegni Lincei, Roma, 1998, pp. 49–71.
[35] R. Glowinski et al. (Eds.), Domain Decomposition Methods in Science and Engineering, Vols. 1–12, Proceedings of International Conferences, 1987–1999, SIAM, AMS, Wiley, DD Press.

JOURNAL OF COMPUTATIONAL AND APPLIED MATHEMATICS

ELSEVIER Journal of Computational and Applied Mathematics 125 (2000) 539–540

www.elsevier.nl/locate/cam

Author Index Volume 125 (2000)

PII: S0377-0427(00)00608-7

www.ingramcontent.com/pod-product-compliance
Ingram Content Group UK Ltd.
Pitfield, Milton Keynes, MK11 3LW, UK
UKHW051456290926
15471UKWH00008B/1225

9 780444 506009